ENCYCLOPEDIA OF TIME

Editorial Board

ENCYCLOPEDIA OF TIME

Science, Philosophy, Theology, & Culture

H. James Birx *Editor*

Canisius College | State University of New York at Geneseo | Buffalo Museum of Science

2

Los Angeles • London • New Delhi • Singapore • Washington DC

A SAGE Reference Publication

For information:

SAGE Publications, Inc.
2455 Teller Road
Thousand Oaks, California 91320
E-Mail: order@sagepub.com

SAGE Publications Ltd.
1 Oliver's Yard
55 City Road
London EC1Y 1SP
United Kingdom

SAGE Publications India Pvt. Ltd.
B 1/I 1 Mohan Cooperative Industrial Area
Mathura Road, New Delhi 110 044
India

SAGE Publications Asia-Pacific Pte. Ltd.
33 Pekin Street #02-01
Far East Square
Singapore 048763

Printed in the United States of America.

Library of Congress Cataloging-in-Publication Data

Encyclopedia of time : science, philosophy, theology, and culture / H. James Birx, editor.
p. cm.
Includes bibliographical references and index.
ISBN 978-1-4129-4164-8 (cloth : alk. paper) 1. Time—Encyclopedias. I. Birx, H. James.

BD638.E525 2009
115.03—dc22 2008030694

This book is printed on acid-free paper.

09 10 11 12 13 10 9 8 7 6 5 4 3 2 1

Publisher:	Rolf A. Janke
Assistant to the Publisher:	Michele Thompson
Developmental Editor:	Sanford Robinson
Reference Systems Manager:	Leticia Gutierrez
Production Editor:	Kate Schroeder
Copy Editors:	Kristin Bergstad, Colleen Brennan, Cate Huisman
Typesetter:	C&M Digitals (P) Ltd.
Proofreaders:	Kristin Bergstad, Kevin Gleason, Penelope Sippel
Indexer:	David Luljak
Cover Designer:	Ravi Balasuriya
Marketing Manager:	Amberlyn Erzinger

Contents

Volume 2

List of Entries

Reader's Guide

Biography
Abelard, Peter
Albertus Magnus
Alexander the Great
Alexander, Samuel
Alighieri, Dante
Anaximander
Anaximines
Anselm of Canterbury
Apollodorus of Athens
Aquinas, Saint Thomas
Aristotle
Asimov, Isaac
Attila the Hun
Augustine of Hippo, Saint
Avicenna
Baer, Karl Ernst Ritter von
Bakhtin, Mikhail Mikhailovich
Barth, Karl
Baxter, Stephen
Bede the Venerable, Saint
Bergson, Henri
Berkeley, George
Boethius, Anicius
Bohm, David
Bonaparte, Napoleon
Boscovich, Roger Joseph
Boucher de Perthes, Jacques
Bradbury, Ray
Bruno, Giordano
Caesar, Gaius Julius
Calvin, John
Campanella, Tommaso
Carroll, Lewis
Cartan, Élie Joseph
Chambers, Robert
Charlemagne
Chaucer, Geoffrey
Clarke, Arthur C.
Coleridge, Samuel Taylor
Columbus, Christopher
Comte, Auguste
Copernicus, Nicolaus
Dali, Salvador
Darwin, Charles
Deleuze, Gilles
Derrida, Jacques
Descartes, René
Diderot, Denis
Dilthey, Wilhelm
Donne, John
Dostoevsky, Fyodor M.
Doyle, Arthur Conan
Duns Scotus, John
Durkheim, Emile
Eckhart, Meister
Einstein, Albert
Eliade, Mircea
Eliot, T. S.
Empedocles
Engels, Friedrich
Eriugena, Johannus Scotus
Farber, Marvin
Feuerbach, Ludwig
Fichte, Johann Gottlieb
Flaubert, Gustave
Frege, Gottlob
Galilei, Galileo
Gamow, George
Gehlen, Arnold
Genghis Khan
Gibran, Kahlil
Gödel, Kurt
Goethe, Johann Wolfgang von
Gosse, Philip Henry
Haeckel, Ernst
Harris, Marvin
Harrison, John
Hartshorne, Charles
Hawking, Stephen
Hegel, Georg Wilhelm Friedrich
Heidegger, Martin
Heraclitus
Herder, Johann Gottfried von
Herodotus
Hesiod
Hitler, Adolf
Homer
Hume, David
Husserl, Edmund
Hutton, James
Huxley, Thomas Henry
Jaspers, Karl
Josephus, Flavius
Joyce, James
Kafka, Franz
Kant, Immanuel
Kierkegaard, Søren Aabye
Knezevic´, Bozidar
Kropotkin, Peter A.
Kuhn, Thomas S.
Lamarck, Jean-Baptiste de
Laplace, Marquis Pierre-Simon de
Leibniz, Gottfried Wilhelm von
Lemaître, Georges Édouard
Lenin, Vladimir Ilich
Lucretius
Luther, Martin
Lydgate, John
Lyell, Charles
Lysenko, Trofim D.
Mach, Ernst
Machiavelli, Niccolò
Malthus, Thomas

Biology/Evolution

Culture/History

Geology/Paleontology

Philosophy

Kant, Immanuel
Kierkegaard, Søren Aabye
Knezević, Bozidar
Kropotkin, Peter A.
Leibniz, Gottfried Wilhelm von
Lenin, Vladimir Ilich
Leucippus
Logical Depth
Lucretius
Mach, Ernst
Machiavelli, Niccolò
Maxwell's Demon
Marx, Karl
Materialism
McTaggart, John M. E.
Mellor, David Hugh
Merleau-Ponty, Maurice
Metaphysics
Morality
More, Saint Thomas
Nabokov, Vladimir
Newton, Isaac
Newton and Leibniz
Nicholas of Cusa (Cusanus)
Nietzsche, Friedrich
Nietzsche and Heraclitus
Nothingness
Now, Eternal
Ontology
Ovid
Paley, William
Parmenides of Elea
Plato
Plotinus
Poincaré, Henri
Postmodernism
Presocratic Age
Predestination
Predeterminism
Prigogine, Ilya
Progress
Rahner, Karl
Rawls, John
Regress, Infinite
Ricoeur, Paul
Rousseau, Jean-Jacques
Russell, Bertrand
Santayana, George
Scheler, Max
Schelling, Friedrich W. J. von
Schopenhauer, Arthur
Schopenhauer and Kant
Simmel, Georg
Sloterdijk, Peter
Solipsism
Spencer, Herbert
Spinoza, Baruch de
Structuralism
Teilhard de Chardin, Pierre
Teleology
Thales
Theodicy
Tillich, Paul
Time, Cyclical
Time, Illusion of
Time, Imaginary
Time, Logics of
Time, Nonexistence of
Time, Objective Flux of
Time, Observations of
Time, Operational Definition of
Time, Perspectives of
Time, Phenomenology of
Time, Problems of
Time, Real
Time, Relativity of
Time, Subjective Flow of
Unamuno y Jugo, Miguel de
Values and Time
Virtual Reality
Weber, Max
Whitehead, Alfred North
William of Conches
William of Ockham
Xenophanes
Zeno of Elea
Zoroaster
Zurvan

Physics/Chemistry

Astrolabes
Attosecond and Nanosecond
Aurora Borealis
Big Bang Theory
Big Crunch Theory
Black Holes
Bohm, David
Calendar, Astronomical
Causality
Chemical Reactions
Chemistry
Chronometry
Chronology
Clocks, Atomic
Clocks, Mechanical
Comets
Copernicus, Nicolaus
Cosmogony
Cosmological Arguments
Cosmology, Cyclic
Cosmology, Inflationary
Cryonics
Dating Techniques
Decay, Organic
Decay, Radioactive
Determinism
DNA
Dying and Death
Earth, Age of
Earth, Revolution of
Earth, Rotation of
Eclipses
Einstein, Albert
Entropy
Equinoxes
Eternity
Event, First
Evolution, Chemical
Evolution, Issues in
Finitude
Forces, Four Fundamental
Fossil Fuels
Galilei, Galileo
Gamow, George
Global Warming
Hawking, Stephen
Heat Death, Cosmic
Histories, Alternative
Infinity
Laplace, Marquis Pierre-Simon de
Latitude
Leap Years
Lemaître, Georges Édouard
Life, Origin of
Light, Speed of
Longitude
Mach, Ernst

Psychology/Literature

Religion/Theology

Theories/Concepts

Galaxies, Formation of

See Nebular Hypothesis

Galilei, Galileo (1564–1642)

Italian mathematician, astronomer, and physicist Galileo Galilei is considered to be the founder of the modern scientific method. He pioneered verification by experimentation and critical analysis of phenomena. Galileo was the first person to use a telescope to make and interpret systematic astronomical observations and made many discoveries regarding the solar system. Galileo's observations of the eclipses of Jupiter's moons led to his discovery of a cosmic clock that, in effect, recorded absolute time. The Galilean transformations of space and time variables led to the development of the Newtonian laws of mechanics. His theoretical work in physics laid the groundwork for the future exploration of relativity and the laws of motion. Although Galileo himself did not invent the pendulum-regulated clock, his initial designs inspired others to do so. His preliminary research and design of an escapement mechanism led to the development of the first pendulum-regulated timepiece. Galileo's quest to measure very small quantities of time accurately paved the way for discoveries about sound and light waves that eventually led to modern investigations of quantum physics.

Galileo was born in Pisa, Italy, on February 15, 1564. He was the first of six, possibly seven children born to Vincenzo Galilei, a musician, composer, and wool trader; their mother was Giulia Ammannati of Pesscia. Three days later the famous artist Michelangelo died. Leonardo

Galileo first put forth his observations on the moons of Jupiter in his book Siderius Nuncius. *He studied Jupiter over the course of a month and was able to show the movement of the satellites around Jupiter.*

Source: Library of Congress, Prints & Photographs Division, LC-USZ62-7923.

da Vinci had passed away 45 years prior to Galileo's birth. Nicholas Copernicus had been dead for 21 years. William Shakespeare would be born 2 months later. This was the time of the Renaissance. A new awakening had arrived for philosophy, music, art, the sciences, literature, and discovery.

Early Life

At age 7, Galileo was sent off to a monastery to prepare to study medicine. Galileo enjoyed life at the monastery and soon decided he wanted to become a monk. His father did not agree and, complaining about his son's untreated eye infection, removed Galileo from the monastery. Back at home young Galileo was strongly influenced by his father's experiments on the nonlinear relationship between the tension and pitch of stretched lute strings. Working with his father, Galileo learned how to experiment and gather data. By attaching carefully measured weights to a range of strings of different lengths and thicknesses, the Galileis listened to the tones produced. Each modification altered the frequency of the vibration, producing a different note. The length of the string altered the pitch, or cycles per second of the vibrating string, in much the same way the rate of the swinging of a pendulum was related to the length of its cord. This led to the discovery that the interval between two notes was related to the inverse squares of the length of the string, when the same weight was attached and the same interval observed. For a vibrating string, the frequency or sound heard is inversely proportional to the square root of the string's weight per unit of length, so thicker, heavier strings produce lower notes. This mathematical law contradicted traditional musical theory. Galileo learned from his father that it was foolish to accept anything as truth without examining the evidence in support of it. He was taught in the tradition of Plato's student Aristotle that theory must follow facts. It has been argued that Galileo's devotion to the Catholic Church inspired him to seek evidence about the world in order to protect the church from disseminating misinformation.

Contributions to Medicine

In 1582, at age 18, Galileo began his study of medicine at the University of Pisa. Years later, Galileo wrote about the state of medical education by describing an anatomical dissection. The topic was the origin of the nervous system. According to Aristotle, the heart was the source of the nerves, but the anatomist clearly demonstrated the brain was the true source of the nerves. The Aristotelian philosopher replied that the evidence before his eyes would clearly indicate the brain as the origin of the nerves and he would believe it to be so, if it were not for the words of Aristotle!

While attending services at Pisa's cathedral, Galileo noticed a swinging lamp and, using his resting heart rate as a timepiece, he observed that each swing of the lamp appeared to take the same number of pulses in his wrist and therefore approximately the same amount of time, regardless of the length of the swing. One could imagine him reminiscing about weights hanging from lute strings at home. His observations and discussions of this isochronicity of the pendulum led a friend of his, Santorre Santorio, a physician in Venice, to design a small pendulum that could be used to calibrate the human heartbeat rate. This device, called the *pulsilogium*, could be used to get an objective measurement of the heart rate of a medical patient. By observing changes in the pulse rate, physicians could now obtain data on the vitality of their patients and the efficacy of their treatments. In addition, Galileo was the first to invent a device to measure changes in temperature. With Santorio, Galileo worked to improve the science of experimental medicine, including the study of human metabolism. Although Galileo chose not to complete his study of medicine, his research led to major contributions to the scientific study of human anatomy and physiology.

Physical and Mathematical Investigations

Leaving the study of the medical arts behind, Galileo embarked on a lifelong journey in mathematics. Galileo's exploration of the geometry of Euclid of Alexandria and the physics of Archimedes of Syracuse and of Aristarchus of Samos inspired his creative genius. Stimulated by exploration of the foundations of astronomy from Hipparchus of Samos and Claudius Ptolemaeus, Galileo's mind was preparing to wrangle with ideas of cosmic proportions. Here Galileo cultivated a philosophy of scientific realism and a belief that there were explanations for natural phenomena that are revealed through observation and reason.

Inspired by the story of Archimedes and the golden crown of suspect purity, Galileo experimented with floating objects and developed a precise balance scale he called the *hydrostatic balance,* which could compare the weight of an object to an equal volume of water. He described this invention in a book he titled *The Sensitive Balance* (*La Bilancetta*) in 1586. He would later use measured volumes of water to calibrate elapsed time precisely in his acceleration experiments.

In 1587, Galileo was asked to cast a horoscope for Francesco de Medici, the Grand Duke of Tuscany. Galileo had disdain for astrology, which he considered to be heretical, but felt it was unwise to thwart the wishes of this generous benefactor. Drawing up a chart for the 46-year-old Grand Duke, Galileo announced that it indicated a long and fruitful life. Within a week the Duke was dead. In January of 2007, an analysis of the Duke's remains revealed an extremely elevated level of arsenic that was indicative of deliberate poisoning.

By 1593, Galileo had moved to the University of Padua where he taught mathematics and military engineering. It was here that he met and fell in love with Maria Gamba. While never married, they did have three children. His two daughters entered the convent and his son Vincenzo was by his side when he died. Galileo worked on many inventions, including a device that used horses to raise water from aquifers. In 1604, a "New Star," as it was called then, appeared in the constellation Sagittarius. Here was evidence for all to see that the heavens were not fixed and permanent, as Aristotle had decreed. If the great philosopher Aristotle could be wrong about this fundamental quality of the heavens, what else could he be in error about? We now know that what Galileo had witnessed was not a new (Nova) star, but rather a very old star in its last stages of stellar evolution.

Financial success came from Galileo's invention of a military compass that could be used to calculate the ideal firing angle for cannons as well as the gunpowder charge and projectile weight for maximum effect and accuracy. A civilian compass model that could be used for land surveying followed. This invention is considered by many to be the world's first pocket calculator. His text describing the use of the compass, titled *Operations of the Geometric and Military Compass (Le Operazioni del Compasso Geometrico e Militare)* was published in 1606. This invention also introduced Galileo to the unpleasant world of patent infringement, as he eventually had to prove that others had copied his work.

Galileo's lecture series on the location, shape, and dimensions of Dante's Inferno helped to earn him a 3-year appointment at the University of Pisa to teach mathematics. It was there, according to legend, that Galileo demonstrated that falling bodies of varying sizes and weights fell the 54 meters from the top of the Leaning Tower of Pisa at the same rate. By disproving one of Aristotle's alleged laws of nature, Galileo showed that blind acceptance of doctrine must yield to scientific experimentation. Galileo was adept at thought experiments. He reasoned that if two weights, one heavier than the other, were supposed to fall at different speeds, then why did hailstones of a wide range of weights fall together? He also pondered the question of whether, if a lighter and heavier weight were tied together, their rate of descent would change. Galileo asked whether, if the two were tied together, would the lesser one subtract velocity from the fall or would the two weights added together increase the rate of fall? If the lighter one subtracted velocity from the heavier one, then they should fall more slowly. Or, if the two together now weighed more than the original, would they fall all the faster? His conflicting results led inevitably to the conclusion that they must fall at the same rate regardless of their weight. Galileo used these kinds of thought experiments to help others to visualize the fundamental elements of motion.

When the objection was raised that a feather did indeed fall much more slowly than a cannonball, Galileo realized that air resistance accounted for the difference and that if this variable could be controlled, the two objects would fall at the same rate.

Galileo experimented with resistance in different media such as water and oil. He was not able to perform the experiment in a vacuum, the production of which was then technically unattainable. In July of 1971, 365 years later, Apollo 15 astronaut David Scott dropped a falcon feather weighing 0.03 kilograms and a geological hammer weighing 1.32 kilograms from a height of 1.6 meters on the moon. With no air resistance, they landed simultaneously, and Commander Scott duly noted that Galileo had been right.

Galileo sought an accurate way to measure the change in the speed of a falling object over time. Galileo, a talented musician and composer, had a well-developed internal rhythm and would be able count off beats in his head quite accurately to measure seconds. The tools available to measure time in those days were extremely limited in accuracy and reliability. Sundials and hourglasses would be useless to measure the brief intervals that Galileo sought to investigate. Mechanical clocks had appeared in Western Europe around 1330. A few years later a clock was built by Giovanni de'Dondi in Padua, Italy, that displayed the position of the sun, moon, planets, and the timing of eclipses. It was beautiful to behold, but its accuracy was limited. The timepieces available to Galileo could measure hours with reasonable accuracy; however, the quantities of time involved in calibrating acceleration would require accuracy not just to the minute, but to the "second minute," which we now know as the "second." It appeared to be virtually impossible to calculate events occurring in fractions of seconds.

Galileo needed a way to quantify time objectively. He tackled this daunting problem from two directions. First he devised a way to slow the falling object's rate of speed. By building a diagonal ramp he, in effect, diluted gravity. Now he could study acceleration at a more leisurely pace by rolling a ball down the highly polished inclined plane. Using his military compass, he could carefully determine the angle of descent of the plane. The second part of the problem involved measuring very small intervals of time. His familiarity with the lute would most likely lead him to include frets or slightly raised ridges on some of his inclined planes. By doing so, he could hear the ball striking the frets as it descended. By spacing the frets in a way that the time interval between each sound was identical, Galileo would have a measurable unit of distance to indicate acceleration. He had previously measured the isochronicity of a pendulum's swing using his own resting pulse rate; but from his observations with the pulsilogium, he knew that the pulse rate was too variable and therefore not a reliable enough clock to gauge acceleration. Added to this was another element arguing against using his own pulse as a clock. As his experiments began to reveal the laws of motion, his excitement would no doubt raise his pulse rate, rendering the measurement useless.

Another of his inventions, the hydrostatic balance scale, would provide the inspiration for a quantifiable unit of acceleration. Galileo designed a water clock to measure velocity indirectly. By starting the ball down the ramp and beginning the release of water simultaneously, and stopping the flow of water when the ball reached the end of the ramp, he could weigh the amount of water released in a given time. In this fashion Galileo was able to make an accurate comparison of the amount of time a rolling object spent in each portion of its descent down the ramp. He found that the distance the ball rolled down the plane was proportional to the square of the elapsed time. Galileo observed that balls of different weights increased their speed at the same rate. The development of modern science is based on the idea of mathematically measurable sequences. One such measure, a unit of acceleration, is known as a "Galileo." Galileo's struggle with the accurate measurement of time laid the foundation for those who followed. His calculations would be used in 1687 by Sir Isaac Newton to formalize the laws of motion in *Principia Mathematica*.

Another observation of nature that stimulated Galileo's inquisitive mind was the apparent difference between the speed of light and the speed of sound. Since lightning precedes thunder and the cannon's flash precedes the boom, Galileo knew that here was another mystery that could be solved mathematically. He attempted to design an experiment using lanterns spaced miles apart but could report only that light traveled so much faster than sound that light speed could not be measured with the instruments available at the time. Today, scientists can measure time to the attosecond, which is one quintillionth of a second!

Optics and Astronomy

While visiting Venice in 1609, Galileo first heard of the spyglass that a Dutch spectacle maker had invented. Galileo realized that a device that made distant objects appear closer had obvious military and potential financial value. His prior experience with the military compass and artillery inspired him to improve and capitalize on this invention. By experimenting with various combinations of concave and convex lenses he was able to improve on the original design by increasing its magnifying ability and righting the image. Without the proper combination and spacing of lenses, images appeared upside-down. Viewing an upside-down ship with the sea above and the sky below was disconcerting and detracted from the general usefulness of the spyglass. Galileo presented his improved 10-power telescope to the senate of Venice and demonstrated how ships at sea could be identified as friend or foe hours before a look-out without such a device could make such an identification. He was proclaimed a genius and given a generous salary increase and lifetime tenure. This granting of tenure, along with the appearance shortly thereafter of a large influx of cheap spyglasses from Northern Europe, angered some and may have contributed to problems he would face later in life.

Galileo continued to improve his telescope. With higher magnifying power and improved lens shaping and polishing techniques, he began his observations of the heavens. His preliminary investigations of the moon immediately revealed that it was uneven, rough, and full of cavities, craters, and prominences. It was not the smooth, polished, perfect heavenly body everyone believed it to be. By carefully observing the shadows cast by lunar mountains, he was able to make estimates of their altitudes. To those who insisted that the moon was covered with a smooth transparent crystal, he replied that they should grant him the equal courtesy of constructing with that same crystal mountains, valleys, and craters.

Galileo's explorations of the night sky brought new discoveries every clear evening. The Milky Way, which was considered to be a pale vapor of light, revealed itself to be an uncountable number of stars vast distances away. The Seven Sisters, a star cluster also known as the Pleiades, in the constellation Taurus the Bull, when magnified became hundreds of stars. The Great Sword of Orion, when closely examined displayed a marvelous cloud embedded with tiny newborn stars. Galileo's examination of Ursa Major, the Great Bear or Big Dipper or Plough, revealed an amazing double star that he observed during a failed attempt to measure parallax and therefore demonstrate Earth's revolution around the Sun.

In January 1610 Galileo turned his new and improved telescope on Jupiter. With modifications to the lenses, Galileo was now able to magnify the apparent diameter of an image 30 times. He noted a star to the left of Jupiter and two on the right, all in a straight line. One could only imagine Galileo's amazement when on the next evening he saw that all three of the stars were now on the left and still in a straight line. Weeks of observations and further improvements that widened the field of view of his telescope revealed a fourth star that circled around Jupiter. Galileo realized that these were moons that orbited Jupiter just as our moon orbited Earth. Here was evidence that not everything revolved around the earth. Seeing Jupiter's moons revolve around Jupiter also discredited the idea that if the earth revolved around the sun it would leave its moon behind. Additional calibration of the orbits of Jupiter's moons inspired Galileo to consider that their regular orbital periods could serve as a cosmic clock for ships at sea. This could aid in the accurate timekeeping that was essential for the determination of longitude. Galileo continued his telescopic observations and developed a table of the eclipses of Jupiter's moons to be used by ship's captains as a cosmic timepiece to determine longitude while out of sight of land. He even developed a telescopic device one could wear like a hat to observe Jupiter while keeping the hands free to pilot the ship. Apparently it was difficult to use and was abandoned. Galileo did notice a slight abnormality in the timing of eclipses of Jupiter's moons. It was not until 66 years later, in 1676, that the Danish astronomer Oleaus Romer was able to calculate that the 10-minute systematic error of Jupiter's observed synodic period was because light does not travel instantaneously. Years later it would be understood as a function of the varying distance between Earth and Jupiter. This parallax effect is a manifestation of our changing viewpoint as we revolve around the sun. Galileo named Jupiter's moons the "Medician Moons" in honor of his benefactors. Simon Mayr created the names we use today—Io, Europa, Ganemede, and Callisto—in 1614.

Galileo observed the planet Mars and saw that its apparent diameter increased when it was closer and diminished significantly when it was farther away from Earth. This was additional evidence that Mars revolved around the sun. When Mars and Earth were on the same side of the sun, Mars appeared twice as big as it did when Mars was on the far side of its orbit around the sun. This could not be clearly seen with the unaided eye, but was readily apparent when viewed through the telescope. In his honor, there is a crater on Mars called Galilaei as well as an asteroid named Galilea.

Another important observation first made by Galileo was that the planet Venus showed phases like the moon. When it was at its greatest distance from Earth, and more directly illuminated by the sun, it appeared as a small sphere. As its orbit took it around the sun and closer to Earth, it appeared as a progressively larger waning crescent. These observations lent credence to the Copernican idea that the sun was at the center of the solar system. Countless hours of watching the planets convinced Galileo of the validity of the Copernican heliocentric view. Seeing sunspots parading across the solar disc clearly demonstrated that our star, the sun, was not the perfect heavenly object described by the ancients.

Galileo was fascinated and perplexed by his observations of Saturn. The limited resolving ability of his telescope rendered a tiny, blurry image of Saturn and its ring system that was not a sphere like Jupiter but looked rather elongated or shaped like an American football. Could this be more evidence against the Aristotelian belief in perfectly spherical heavenly bodies? Galileo assumed that he was seeing three separate bodies. It is interesting to note that Galileo's observations of the planet Saturn led to an ironic cryptic message. In a coded letter, Galileo wrote that the last planet was triune or three-mooned. In the year 2005, the Hubble space telescope revealed that Pluto, which was considered at the time to be the last planet in our solar system, had three moons, which were named Charon, Hydra, and Nix. A contemporary analysis of Galileo's notes indicates that he probably was the first person to observe the planet Neptune, in the year 1612.

Galileo described his astronomical observations made using a telescope in a small book he called *The Starry Messenger (Siderius Nuncius)* in March of 1610. It received a great deal of attention and generated much heated discussion. In it, Galileo explained his view of scientific realism. He was certain that there were explanations of natural phenomena that would be revealed through observation and reasoning. There were those who felt that some of Galileo's findings contradicted the teachings of the Catholic Church. Objections were raised that certain passages in the Bible appeared to indicate that the sun went around the earth and that to believe otherwise was heresy. Galileo responded that while the Bible could never be wrong, it was not meant to be taken literally and that mistakes of interpretation could lead to confusion. This led others to complain that only the clergy could interpret the Bible and that Galileo had to be stopped. With the Protestant Reformation of Martin Luther and John Calvin and The Thirty Years War to contend with, the Catholic Church had little tolerance for dissension within the ranks. Galileo had met Giordano Bruno, who in 1600 was found guilty of heresy and burned to death. He knew others who had suffered at the hands of the Inquisition and was aware that he had better tread lightly. An investigation of the charges against Galileo found him innocent of heresy, but he was cautioned not to teach the Copernican system as a proven fact.

In 1618, three comets were visible over Europe. A Jesuit mathematician, Father Horiatio (Orazio) Grassi, who wrote using the pseudonym Lotario Sarsi, argued that the highly elliptical orbit of the comets argued against Copernicanism, which postulated circular orbits. Galileo replied in an essay titled *The Assayer (Il Saggiatore)*, published in 1623. Here Galileo established norms and rules for the investigation of nature. It is considered to be one of the great works of scientific literature. In it, Galileo describes how the grand book of the universe is written in the language of mathematics.

Geocentrism, Heliocentrism, and Conflict With Established Authority

Galileo's attraction to the sea and things nautical led him to a contemplation of the causes of the tides. He believed that here he would develop the strongest evidence for Earth's motion around its axis and around the sun. As it turned out, he was wrong in discounting the moon's influence on the tides, which is much greater than the influence of the sun. Galileo sought permission from Pope Urban VIII to write a book about the motions of the solar system. The pope agreed, provided that the book gave a balanced view of the two conflicting theories of geocentrism and heliocentrism. The pope also requested that Galileo mention the pope's personal views that the heavenly bodies may move in ways that man cannot comprehend. Instead of writing his findings in the form of a scientific report, Galileo choose to present his ideas as a conversation among three individuals. *His "Dialogue Concerning the Two Chief World Systems: Ptolemaic and Copernican" (Dialogo sopra i due massimi sistemi del mundo, tolemaico e copernicano)* was published in 1632. The December 2006 issue of *Discover* magazine listed Galileo's *Dialogue* as the fourth greatest scientific book of all time. This dialogue featured a character named Salviati, a proponent of Galileo's ideas. Salviati's heliocentric or sun-centered views were presented as witty, intelligent, and well informed. Sagredo, the bystander, who served as the mediator, was usually persuaded by Sagredo. Simplicio, the proponent of geocentrism or an Earth-centered solar system, was portrayed as somewhat slow-witted and easily befuddled. As an expert on Aristotelian thought, Simplicio represented those who ignored evidence and preferred to cling to dogma rather than explore new ideas. The dialogue on the two great systems of the world presented Copernican theory as the logical and intelligent man's preference. It was considered by many to be a literary and philosophical masterpiece. It was considered by Pope Urban VIII to be a grievous insult to have his views presented by the dim-witted character Simplicio, the simpleton.

Word reached Rome that Galileo was teaching Copernicanism; and worse, it was suggested that he had modeled the fool in his dialogue after the pope. Some said that *The Dialogue* made a mockery of the pope's intellectual authority and undermined his temporal power. Galileo had, in effect, challenged the Catholic Church's authority in the interpretation of scientific knowledge. He was ordered to appear before the Holy Office of the Inquisition to face charges of heresy. Aware of the potential for torture and death, Galileo confessed that he had been wrong to say that the earth moved around the sun. Galileo was found guilty of suspicion of heresy and forced to recant his heliocentric beliefs. He was sentenced to life imprisonment, later commuted to house arrest, and forbidden to discuss his views with anyone. Publication of anything he had written or would write in the future was forbidden. It is interesting to note that Copernicus had never been accused of heresy and that his book, *De Revolutionibus,* was not banned but rather withdrawn for corrections. In his 2001 book, *Galileo's Mistake,* author Wade Roland argues that the church's main problem with Galileo was not his belief in the Copernican system but rather in Galileo's belief in a mechanistic, materialistic philosophy that seeing is believing.

While confined to his home, Galileo continued to investigate other areas of science. He applied mathematics to a variety of problems. He explored geometry and went from the study of lengths, areas, and volumes to the contemplation of motion, mass, and time. For his last, and some consider his greatest, literary masterpiece he returned to the literary device of three gentlemen discussing a wide variety of issues and arguing their points of view. Here, Galileo developed the fundamentals of relativity. He went into great detail regarding the nature of matter. Galileo was able to have his notes smuggled out of Italy and published in 1638 by a Dutch publisher named Luis Elzevir as *Discourses About Two New Sciences* (*Discorsi e dimostrazioni matematiche intorno a due nuove scienze*). This work proved to be the foundation for the modern science of physics.

Galileo's last astronomical discovery was the lunar librations. Galileo discovered that the moon's equator is inclined to its orbital plane. This causes a slight wobble in the moon's axis and allows us to see a bit of the far side of the moon periodically. A lunar crater 15 kilometers in diameter is named in honor of Galileo. It is located just west of the one named for Copernicus.

Galileo and Modernity

At the end of his life, Galileo, who had seen farther than any man before him, became completely blind. He passed away on January 8, 1642, with his son and students by his side. Isaac Newton was born 11 months later. Newton referred to Galileo when he said that the reason he had seen farther was because he had stood on the shoulders of a giant. Stephen Hawking, the Lucasian Professor of Mathematics at Cambridge University and author of *A Brief History of Time,* describes Galileo as the single individual most responsible for the birth of modern science. He notes that Galileo was one of the first to argue that man can understand how the world works by observing the real world.

It was not until 99 years after Galileo's death, in the year 1741, that Pope Benedict XIV lifted the ban on Galileo's scientific works. In 1979 Pope John Paul II asked the Pontifical Academy of Sciences to conduct an in-depth study of the Galileo case. In 1992, the church formally and publicly cleared Galileo of any wrongdoing, 350 years after his death. Pope John Paul II expressed regret for how the Galileo affair had been handled. In his summary of the conclusions he noted that Galileo showed himself to be more perceptive of the criteria for scriptural interpretation than the theologians who opposed him. The pontiff paraphrased Saint Augustine's words that truth can never contradict truth, and that where the Holy Scriptures appear to contradict the natural world, it is the error of interpretation that must be resolved.

Edward J. Mahoney

See also Aristotle; Bruno, Giordano; Clocks, Mechanical; Copernicus, Nicolaus; Einstein, Albert; Hawking, Stephen; Kuhn, Thomas S.; Newton, Isaac; Nicholas of Cusa (Cusanus); Telescopes; Time, Measurements of

Further Readings

Bixby, W. (1964). *The universe of Galileo and Newton.* New York: American Heritage.

Drake, S. (1978). *Galileo at work: His scientific biography.* Chicago: University of Chicago Press.

Frova, A., & Marenzana, M. (2006). *Thus spoke Galileo.* Oxford, UK: Oxford University Press.

Hilliam, R. (2005). *Galileo Galilei, father of modern science.* New York: Rosen.

Reston, J., Jr. (1994). *Galileo, a life.* New York: HarperCollins.

Rowland, W. (2001). *Galileo's mistake.* New York: Arcade.

Gamow, George (1904–1968)

George Gamow, Russian-born physicist, was noted for his contributions to interdisciplinary understanding and for his synthesis of modern physics with both a cosmological and evolutionary framework. Taking a comprehensive view of modern physics, Gamow presented the evolution of the universe and human life as a chance product of chemical interaction/reaction within the spatiotemporal parameters of the universe. From the initial theory of the big bang that resulted in our expanding universe, the conceptual framework of time, space, and distance poses unique problems for both traditional physics and scientific epistemology. Gamow explained the impact of these problems and the theoretical basis for our current understanding of the universe and its implications for life on this planet.

The conceptual framework of time and humankind's perception of the natural world became the basis for continual theoretical advances. Gamow articulated this open-ended perspective by illustrating the historical progress, both philosophical and scientific, made in mathematical understanding of the physics that govern the universe. Contrary to cultural perceptions of the temporal and static nature of the universe, Gamow depicted the spatiotemporal nature of the cosmos as an expanding and temporally changing universe, filled with innumerable planets, stars, and galaxies. This uniform expansion of the universe was suggested as being approximately 2 to 3 billion years ago, with comparable ages of the oldest celestial bodies. The age of our solar system, specifically our planet and the sun, has deep implications for life. Although our sun and planet were estimated to be relatively young, around 3 to 4 billon years and 2 billion years, respectively,

Gamow's calculations put the lifespan of our sun at around 50 billion years. Fueled by nuclear reactions within the bending of time and space, the birth and death of star(s) becomes a tethered line for life on this planet and possible life on other worlds. Though Gamow speculated on the probability of life elsewhere in the universe (including the immensity of distance between planets), the chemical sequence and the emergence of life from inorganic matter became a probability and a particular point of scientific wonder.

The spatiotemporal nature of the universe is paradoxical. Concepts of infinity within finitude are deeply rooted within the human psyche. Although new developments in mathematics, physics, and chemistry continue to inform our ever-growing understanding of the universe, these dual concepts of time seem to preclude any definitive and comprehensive theory of both life and the physics of the universe by which life itself is governed. Gamow's substantial contributions to cosmology, even in light of recent advances, allow us to appreciate more fully both the finitude of human existence and the need for further understanding of the complex relationships that obtain within the universe.

David Alexander Lukaszek

See also Big Bang Theory; Hawking, Stephen; Lemaître, Georges Edouard; Time, Emergence of; Universe, Evolving; Universe, Origin of

Further Readings

Gamow, G. (1954). *One two three . . . Infinity.* New York: Viking Press.

Gamow, G. (1961). *The creation of the universe.* New York: Viking Press.

Gamow, G. (1972). *Cosmology, fusion & matter: George Gamow memorial volume.* Boulder: Colorado Associated University Press.

Gehlen, Arnold (1904–1976)

Arnold Gehlen is known as a cofounder of philosophical anthropology and was one of Germany's leading postwar sociologists; he was also a significant time diagnostician. His understanding of man as an organically "deficient being" paved the way for a theory of institutions that is not only substantial but also empirically adaptable. His anthropological views served asa foundation for his contemporary analyses of Western industrial societies, which were farsighted and, as a result of his conspicuous conservatism, also controversial at the same time.

Gehlen received his PhD in 1927 after studying under the philosopher and biologist Hans Driesch; 3 years later he qualified for a tenured professorship under, among others, Hans Freyer, after having written his habilitation. During the Third Reich, Gehlen had a shining career and quickly received professorships in Leipzig, Königsberg, and Vienna. After serving in the army's administrative council (1941–1942), he was sent to the front and was severely wounded. In 1947, after his denazification, he received his first teaching position as a university professor in Speyer, then later in Aachen, where he taught sociology from 1962 until his retirement.

Like most of his colleagues from Leipzig, Gehlen joined the Nazi Party in 1933 and became a member of the National Socialist German University Lecturers League. He demonstrated his approval of the National Socialistic regime in several ways; for example, in his inauguration lecture in Leipzig 1935. His reference to the "highest systems of leadership" in his first anthropological study (1940) can also be understood as opportunistic. Such declarations of loyalty were never part of Gehlen's scientific thinking, however. His understanding of man as a deficient being (*Mängelwesen*) did not refer to any racial differences. In his opinion, from a biological point of view the "Aryan" is as inadequate as every other human race. Consequently, Gehlen's anthropology was contrary to official Nazi ideology from the beginning; and he himself was always considered an "uncertain type" by Nazi leadership.

Gehlen's ideological home was not totalitarian National Socialism. It was the world of conservative thinking and order. So it is possible to find him referring specifically to the tradition of political thought where the state was the center of focus—as in the work of Hobbes or Hegel. But Gehlen's philosophy was also influenced by many other ideas that were quite diverse: His first phase was shaped partly by existentialist ideas, which are dealt with in his 1931 dissertation. His intensive study of German idealism, especially Fichte, influenced his theory of the freedom of the will and characterized, to a certain extent, the second phase of his work. His anthropological phase began specifically in the mid-1930s. During that period of time Gehlen was one of the first German philosophers to discover American pragmatism, particularly citing the works of John Dewey and George Herbert Mead. The philosophy of life (Schopenhauer, Nietzsche, Bergson) had a major influence on Gehlen's research; he also dealt with Driesch's concept of neovitalism in his dissertation. In the end, the idea of combining philosophical and anthropological studies may have been enhanced by the transdisciplinary atmosphere at the University of Leipzig, which included, among others, the sociologists Hans Freyer and Helmut Schelsky, the psychiatrist Hans Bürger-Prinz, and the philosopher Gotthard Günther.

Works and Ideas

Gehlen's main anthropological work, *Man, His Nature and Place in the World,* was first published in 1940 and has been reprinted several times. In his study of late-modern civilization (1956) Gehlen worked on the deficits of his concept of action, which had essentially been instrumentally conceived, and presented a differentiated institutional theory. His highly acclaimed writings *Man in the Age of Technology* (1957) and his 1969 work on ethics were particularly relevant from the perspective of critical time diagnostics.

The main concern of philosophical anthropology consists of discussing questions that deal directly with the way humans see themselves. According to Gehlen, it is in this context that it is of primary importance to cast off idealistic ballast and to overcome the dualism of the body and the soul. Gehlen begins with a concept of action: Like Nietzsche, he sees in man the "still undetermined animal," a being whose life is at risk. Man cannot feel secure in his environment because he lacks protective instincts. As a result of his deficient biological makeup there is no natural environment for him. Everything and everyone can be his enemy. So man has no other alternative but to create his own relationship to the world around him and to himself through his actions. Man's nature is civilization: Man not only has a life, he needs to lead a life—and thus compensates for his deficient being.

Yet, following Gehlen, man is not only at the mercy of his environment; he is also dangerous. Vague but driven by physical desires, he is latently subject to the danger of mutating, becoming an enemy of his own kind or even of himself. So he doesn't just have to lead his own life, he also has to be led, in particular by institutions. Institutions are the substitute for missing instincts; they give man something to hold onto by requiring him to act in a certain way. Their "unquestioning manner" of guiding human behavior in certain directions relieves the individual of the duty to constantly make decisions. For instance, a letter must be answered—at least in times of postal communication; it requires contemplation, at best concerning the content; but this is also simplified by use of socially standardized forms of address, endings, gratitude, and more. In Gehlen's opinion, institutions are therefore different from mere organizations. They don't serve just one particular purpose and then lose their significance when the job is done. They are created by mutual social interrelations within a community that they also symbolize. Consequently, institutions possess an inherent worth that commits the individual and motivates him to act.

In his writings on time diagnostics, Gehlen especially points out the social processes that develop a force capable of destroying institutions: They are closely connected to a technology that pervades all areas of existence and leads to a degradation of human senses and an intellectualization of life. Everything seems possible, everything can be tried out—in this sense, "anything goes" exemplifies the typical contemporary attitude, in Gehlen's opinion; it is of an experimental nature but also nonbinding and formal, supported by a subjectivism that constantly calls for self-identity: Everyone can imagine everything. The real world becomes virtual in the process of a continuous "psychologization." In order to evoke a response, the cause has to be catchy and easy to remember, never subtle, and preferably shrill. Today's world, as Gehlen said back then, is trivial and busy—unproductiveness at high speed, a racing standstill.

So it is easy to assume that modern society has come to an end. Things could still be recycled and combined, but man has essentially reached the end of history. Man in modern times has come to accept this inevitability from which he can no longer be lured—"*rien ne va plus.*"

Gehlen's thesis has been extensively criticized, one objection to his anthropology of the deficient being is that it is too pessimistic. The deficits of man are exaggerated in comparison to his rational abilities, which enable him to be highly flexible and adaptable. Furthermore, by profoundly emphasizing the differences between man and animal, the variations of human existence are completely neglected. Specifically from an ethical point of view, it is of utmost importance to differentiate between the human and the inhuman, but Gehlen's theory leaves no room for such thoughts. Similarly, a great deal of criticism targets the lack of criteria used in his institutional theory. Gehlen's approach is concerned only with the stability and justification of the status quo. From today's perspective, a large part of this criticism seems justified; basically it is necessary to criticize his philosophical as well as his sociological ideas for excluding substantial normative principles without omitting extensive critical evaluations of society. Nevertheless, people often do not realize that his action theory stresses man's openness and productivity in his relationship to the world and to himself. In that sense, it is possible to imagine that his institutional theory can be critically reconstructed within the framework of a normative intention.

Gehlen's primary achievement is that he developed a theoretical view of society that does not simply deal with individual rational actions or the structures of social systems. It also points out the links between the individual and the system that serve as the basis for his time diagnostic analyses. By using these analyses, it is possible to discover the grayness of an apparently modern conformity amidst the colorful world of new postmodernism.

Oliver W. Lembcke

See also Anthropology; Bergson, Henri; Fichte, Johann Gottlieb; Hegel, Georg Wilhelm Friedrich; History, End of; Nietzsche, Friedrich; Schopenhauer, Arthur

Further Readings

Berger, P. L., & Kellner, H. (1965). Arnold Gehlen and the theory of institutions. *Social Research, 32*(1), 110–115.

Gehlen, A. (1978ff). *Gesamtausgabe* [Works]. (K.-S. Rehberg, Ed.). Frankfurt a.M.: Klostermann.

Gehlen, A. (1980). *Man in the age of technology* (P. Lipscomb, Trans.). New York: Columbia University Press. (Original work published 1957 as *Die Seele im technischen Zeitalter.* Hamburg: Rowohl)

Gehlen, A. (1988). *Man, his nature and place in the world* (C. McMillan & K. Pillemer, Trans.). New York: Columbia University Press (Original work published 1940 as *Der Mensch: Seine Natur und Stellung in der Welt.* Berlin: Junker and Dünnhaupt)

Thies, C. (2000). *Arnold Gehlen zur Einführung* [Introduction to Arnold Gehlen]. Hamburg: Junius.

Weiss, D. M. (Ed.). (2002). *Interpreting man.* Aurora, CO: Davies.

Genesis, Book of

Genesis is the first book of the Hebrew Bible and Christian Old Testament. The basic premise of this book is to recount the stories of the Hebrew patriarchs and their covenant relationship with God. The Book of Genesis connects the God of the patriarchs with the beginning of time as the creator of the universe.

The book is a book of beginnings, as its Hebrew title suggests—*bĕrē'šît* meaning "In the beginning." The name "Genesis" comes from the transliteration of the Greek for the first word of the book, which means "origins." Both titles appropriately describe the contents of the Book of Genesis. The book tells of the creation of the universe, the dawn of humanity, the origin of sin, and the beginning of the Hebrew people.

The book centers around the promise God gave to Abram (renamed Abraham, Gen 17:5), which is to make him the father of a multitude of people who would become a nation (12:1–3) living in a specific land (13:14–15). This promise divides the book into two sections. The first part (chapters 1–11) explains how the world came to be, why it is the way it is at the time of Abraham, and that God was involved the entire time. Time passes quickly in these chapters through genealogies that connect the stories of creation, the introduction of sin, the flood account, and God's promise to Abram. This section contains two stories of God creating the world, each having a different sequence of events. In the first account (1:1–2:3) God made light and separated it from darkness. Next he made the heavens, sea, and land, then the plants, then the sun, moon, and stars, and then he created the birds, fish, and land animals. Finally, God created humans, both male and female. This event completed Creation. In the second Creation account (2:4–25), God made the heavens and Earth and then he created a man named Adam. Next, God created Eden, an idyllic garden full of plants, where he placed Adam. Afterward, God created animals and birds and finally he made a woman whom Adam named Eve.

While Eve lived in the garden, a serpent had a conversation with her and deceived her. Adam supported her and silently consented. They yielded to temptation by eating forbidden fruit from the tree of the knowledge of good and evil and thereby introduced sin into the world. Sin quickly spread throughout Creation and intensified, as demonstrated in fratricide among Adam and Eve's two sons, Cain and Abel. During 10 long generations sin increased to such an intolerable point that God sent a flood to destroy the world with the exception of a righteous man, Noah, and his family. Noah built an ark that saved him from the destruction of the flood. After the flood, one of Noah's sons sins, an indication that the flood did not rid the world of this problem; thus God would have to find another solution. This dilemma leads to the promise to Abram, the next attempt at the problem of sin. Genealogies following the flood story provide an explanation for all the different people groups, nations, and languages in the world at the time of Abram and set the stage for Abram's appearance.

Genesis 12–50 is the second part of the book. Time in this section slows as the author concentrates on the stories of four generations of patriarchs—Abraham and his descendants—and their struggles to fulfill God's promise. Abraham, the original recipient of the promise, had various difficulties in producing children and staying in the promised land. His son Isaac had similar problems but managed to have twin sons who were in competition for their father's inheritance and blessing. Jacob, the victorious twin, had his own problems establishing a place for himself. However, God reiterated the promise to Jacob and changed his name to Israel. Jacob married two sisters and their two servants and produced 12 sons and a daughter. These sons became the fathers of and namesakes of the Twelve Tribes of Israel and they represent the birth of the nation.

The Book of Genesis ends with a series of stories about Joseph, one of Jacob's sons, and his adventures in Egypt. By the end of the book, not only was Joseph in Egypt but his father, all his brothers, and their families and flocks as well. This ending sets the stage for the Book of Exodus and God's deliverance of his people from Egypt to the promised land.

Thus the Book of Genesis describes God as the creator of the universe who created everything including the first humans. Because humanity's sin brought about the fall of Creation, Genesis tells that God took an active role in humanity's salvation and in world events. In doing so, the book covers a time from the beginning of the universe to the second millennium BCE.

Stories from the Book of Genesis, especially the Creation stories, have been used as common themes in literature throughout history from John Milton's *Paradise Lost* to Gary Larson's *Far Side* cartoons to the *Star Trek* movies.

Terry W. Eddinger

See also Adam, Creation of; Bible and Time; Christianity; Cosmological Arguments; Creationism; God as Creator; Milton, John; Moses; Noah; Sin, Original; Time, Sacred

Further Readings

Towner, W. S. (2001). *Genesis* (Westminster Bible Companion Series). Louisville, KY: Westminster John Knox.

Wenham, G. J. (1987). Genesis 1–15. *Word biblical commentary, Vol. 1*. Waco, TX: Word Books.

Wenham, G. J. (1994). Genesis 16–50. *Word biblical commentary, Vol. 2*. Waco, TX: Word Books.

GENGHIS KHAN (C. 1162–1227)

The Mongolian emperor Genghis Khan's audacity and ingenuity fueled the vast expansion of the Mongol Empire during the 12th and 13th centuries, securing him a place of note in world history. His empire eventually extended over most of Asia, including portions of Russia.

Genghis Khan (born Temujin) remains among the most influential people of all time, having revolutionized the conduct of war and establishment of laws in addition to instilling fear and awe in those he led and those he encountered on the battlefield. As for military convention, Genghis Khan implemented tactics that helped the Mongol forces attain one of the largest empires in history. Dependent on well-trained cavalry, Genghis Khan's forces used hit-and-run tactics to disrupt and slowly carve away at enemy forces, beginning with enemy commanders. Genghis Khan's strategies made it possible to attack and disperse larger forces, reflecting the daring and brilliance of the Mongol leader so many came to fear. Arguably, it is these cunning tactics, along with his methodology of concealing his army's size and whereabouts, that ultimately won Genghis Khan land and reputation.

Though Genghis Khan's military tactics instilled fear in his enemies, it is his compilation of laws (*Yassa*) that reflects his understanding of people and adds considerable depth to his legacy. Genghis Khan incorporated longstanding traditions and his own decrees to formulate one canon of laws. This combining of the traditional and the contemporary ensured legitimacy for the Yassa and, consequently, for Genghis Khan's reign.

Chinggis, or Genghis, Khan is the title bestowed on Temujin on his attaining the Mongol throne. The year of Chinggis Khan's birth and the exact location of his tomb remain unknown, though several dates and locations have been postulated. Further, a complete copy of Chinggis Khan's Yassa has yet to be found, making it difficult to ascertain the extent of his transformation of traditions to generate his laws. What is known is that he created one of the greatest empires in history, employing both harsh and ingenious methods to do so. Chinggis Khan's death in 1227 is documented as to the day and his final actions as ruler, yet even with documentation from a number of texts, questions remain.

Portrait of warrior-ruler Genghis Khan (1167–1227), founder of the Mongolian Empire, 1200. One of history's more charismatic and dynamic leaders, Genghis Khan during his lifetime conquered more territory than any other conqueror, and his successors established the largest contiguous empire in history.

Source: Time & Life Pictures/Getty Images.

Neil Patrick O'Donnell

See also Attila the Hun; Nevsky, Saint Alexander

Further Readings

Grousset, R. (1966). *Conqueror of the world.* New York: Orion Press.

May, T. (2007). Genghis Khan: Secrets of success. *Military History, 24*(5), 42–49.

Morgan, D. O. (1986). The great Yasa of Chingiz Khan and Mongol law in the Ilkhanate. *Bulletin of the School of Oriental and African Studies, University of London, 49*(1), 163–176.

Sinopoli, C. M. (1994). The archeology of empires. *Annual Review of Anthropology, 23,* 159–180.

Vernadsky, G. (1938, December). The scope and contents of Chingis Khan's Yasa. *Harvard Journal of Asiatic Studies, 3*(3/4), 337–360.

Geological Column

The geological column is a composite diagram that shows in a single column the vertical or chronologic arrangement of the subdivisions of geologic time, or the sequence of rock units of a given region. Geologic time includes the part of the earth's history that is represented by and recorded in the successions of rocks, or the time extending from the formation of the earth as a separate planetary body to the beginning of written history. Earth scientists use a common language to talk about geologic time. That common language is standard, and it is ruled by the Geologic Timescale (GTS). The modern GTS consists of two different scales: the relative timescale, which is made of chronostratigraphic units, and the chronometrical or absolute timescale, which consists of geochronologic units. Due to the complexity and duration of geologic history, both scales are divided into hierarchical levels that are used by historical geology to analyze the history of our planet and of life on Earth.

Principles and Development

The standard geological column represents an ideal succession containing rocks from all ages, the earliest rocks on Earth at the bottom of the column and the youngest ones at the top. The construction of the geological column is based on the underlying principles of stratigraphy, first proposed by Nicolaus Steno around 1669. According to his *principle of superposition,* the oldest stratigraphic units are located at the bottom of the column and the youngest at the top, with dips adjusted to the horizontal. The resulting geological column indicates the relations between the stratigraphic units and the subdivisions of geologic time, and their relative positions to each other. The principle of superposition is the basis for establishing the relative ages of all strata and the fossils that they contain.

The geological column was developed largely during the early 19th century, and its origin probably begins with the story of the first geological map of England, published by William Smith in 1815. Smith was the first to realize that fossils were arranged in order and regularly in strata, always in the same order from the bottom to the top of a section, each stratum being characterized by particular types of fossils. These observations led him to propose the *principle of faunal succession.* Furthermore, the relative order of the formations was proved to be the same even in distant locations of Great Britain. The application of these two simple principles (superposition and faunal succession) led to the construction of the first geological column. In addition, the geological column was based on the uniformitarian principles (the present is the key to the past, i.e., processes operating in the past were constrained by the same laws of physics that operate today) first proposed by James Hutton in the mid-18th century and further developed by Charles Lyell.

The standard geological column aims to establish a classification system to organize systematically the rocks of the earth's crust into formal units corresponding to intervals of geologic time. Such units must be of global extent to allow correlation. Among the formal units for stratigraphic classification, chronostratigraphic units—units based on the time of formation of the rock bodies—offer the greatest potential for worldwide application because they are based on their time of formation, and are therefore the most accepted units to mark positions in the stratigraphic column. Other units such as lithostratigraphic, biostratigraphic, and unconformity-bounded units are all of limited areal extent and thus unsatisfactory for global synthesis. The biostratigraphic units are nevertheless unique in the sense that the fossils they contain show evolutionary changes through geologic time that are not repeated in the stratigraphic record. Due to the irreversibility of evolutionary change, biostratigraphic units are indicative of geologic age. However, owing to the imperfection or incompleteness of the fossil record, and the dependence of the fossil-producing organisms on biogeography and depositional facies, the boundaries of the biostratigraphical units commonly lie at different stratigraphic horizons and, similar to unconformity-bounded units, they may be diachronous and represent all or parts of one or several chronostratigraphic units. Magnetostratigraphic polarity units approach synchronous horizons because their boundaries record the rapid reversals of the earth's magnetic field. Although magnetostratigraphic polarity units may be useful guides for chronostratigraphic position and have a potentially worldwide extent, they have relatively little individuality because one reversal looks like another, and they can usually be identified only by supporting age evidence. Therefore, magnetostratigraphic polarity units require extrinsic data such as biostratigraphic data or stable isotope analyses for their recognition and dating. All these stratigraphic units are based on one property each, and they will not necessarily coincide with those based on another.

Chronostratigraphic Scale

For convenience, a chronostratigraphic scale has been created to divide the rock record into chronostratigraphic units, which are relative time units. Chronostratigraphic units are divisions of rock bodies based on geologic time. They are studied in Chronostratigraphy, the branch of Stratigraphy that deals with the relative time relations and ages of rock bodies. The purpose of the chronostratigraphic classification is to organize systematically the rocks of the earth's crust into chronostratigraphic units corresponding to intervals of geologic time. These intervals of geologic time are called geochronologic units, and they actually measure time in years before the present. The geochronologic scale helps to calibrate the chronostratigraphic scale to linear time. Chronostratigraphy aims to provide a basis for time correlation and to create a reference system to record events of geologic history; in order to achieve these goals, the scale is standardized by the International Commission on Stratigraphy (ICS).

Chronostratigraphic units are tangible stratigraphic units because they encompass all the rocks, layer upon layer, formed within a certain time span of the earth's history regardless of their compositions or properties. By definition, they are worldwide in extent, and their boundaries, which are called chronostratigraphic horizons or chronohorizons, are synchronous, everywhere the same age. Whereas other kinds of stratigraphic units are identified on the basis of observable physical features, chronostratigraphic units are distinguished and established on the basis of their time of formation as interpreted from these observable properties. Several hierarchical levels may be distinguished among chronostratigraphic units, namely eonothem, erathem, system, series, and stage (from the most to the least comprehensive levels). Their rank and relative magnitude are a function of the time interval represented by their rocks. The oldest eonothem is the Archean, and it is followed by the Proterozoic, and by the most recent Phanerozoic. The names of the different erathems are related to the ideas of evolution, representing major changes of the development of life. The oldest erathem of the Phanerozoic is thus called Paleozoic, which means "ancient life," and it is followed by the Mesozoic, meaning "middle life," and by the Cenozoic, which means "recent life." The names of most formal stratigraphic units more commonly consist of an appropriate geographic name (usually, the geographical regions where they were first found and studied) combined with an appropriate term indicating the kind and rank of the unit. Position within a chronostratigraphic unit is expressed by adjectives indicative of position, such as basal, lower, middle, upper, and so on. Stages can be subdivided into substages or grouped into superstages. A stage is defined by its boundary stratotypes, that is, sections that contain a designated point in a stratigraphic sequence of almost continuous deposition, chosen for its correlation potential. The lower and upper boundary stratotypes represent specific moments in geologic time, and the geologic time between them is the time span of a stage (generally between 2 and 10 million years). Special attention is paid to the selection of the lower boundaries of chronostratigraphic units, since the upper boundary of a given chronostratigraphic unit corresponds to the lower boundary of the succeeding unit. Therefore, each chronostratigraphic unit is defined in the rock record by a boundary stratotype that is formally known as a Global Stratotype Section and Point (GSSP), which provides an unequivocal definition of the chronostratigraphic units in the stratigraphic record. If possible, boundary stratotypes must be identified in marine, fossiliferous, and continuous sections that are well exposed and easily accessible, and they should contain synchronous marker horizons that allow long-distance correlation. An example of a geologically synchronous boundary stratotype is the Cretaceous/Tertiary boundary stratotype, whose GSSP is located at the El Kef section in Tunisia. This boundary stratotype contains multiple markers, including evidence of the mass extinction of marine and terrestrial groups such as calcareous plankton or dinosaurs, the restructuring of other faunal groups such as benthic foraminifera, the deposition of a rusty-red layer with an anomalous concentration of iridium, microtektites, shocked quartz grains, and Ni-rich spinels, isotope anomalies (negative shift in C-13), and so forth. These palaeontological, mineralogical, geochemical markers are related to a global event, the impact of an asteroid on Earth, and they allow worldwide correlation of the Cretaceous/ Tertiary boundary. Boundary stratotypes are important because, apart from defining stages, they also define series (whose time spans range from 13 to 35 million years) and systems (normally from 30 to 80 million years each).

A chronozone is a chronostratigraphic unit of unspecified rank, and it includes all rocks formed everywhere during the time span of some designated stratigraphic unit or geologic feature. Although chronozones are formal chronostratigraphic units, they are not part of the hierarchical chronostratigraphic classification.

Geochronologic Scale

The time during which a chronostratigraphic unit was formed corresponds to a geochronologic unit, which corresponds to a unit of time, a subdivision of geologic time. The geologic timescale is based on geochronologic units, and it is the time-intangible-equivalent to the physical geological column, which is based on chronostratigraphic units. Time cannot be found in a rock body, but we can assign a certain age to rock bodies through the analysis of tangible features. Therefore, the geological column can also be divided into geochronologic units. The geologic timescale is usually presented in the form of a chart showing the names of the various stratigraphic units, including chronostratigraphic units and geochronologic units. Unlike the chronostratigraphic scale, which is based on relative time units, the geochronologic or chronometric scale measures time in years before the present. The geologic timescale results from joining the chronostratigraphic and the geochronologic scales.

As in the chronostratigraphic units, several hierarchical levels can be distinguished among geochronologic units, namely eon, era, period, epoch, and age, with eons being the most comprehensive levels and ages the smallest levels. Eras are divided into periods, the periods are further divided into epochs, and the latter into ages. These units are equivalent to chronostratigraphic units: for example, an age represents the time during which a stage was formed and it takes the same name as the corresponding stage, and an epoch is the geochronologic equivalent of a series. Eras and eons take the same name as their corresponding erathems and eonothems. Position within a geochronologic unit is expressed by adjectives indicative of time, such as early, late, latest, and so on. The time span during which a chronozone was deposited corresponds to a chron. Although the *International Stratigraphic Guide* states that "a chronozone includes all rocks formed everywhere during the time span of some designated stratigraphic unit or geologic feature," most chronozones and their corresponding chrons are derived from previously established biozones or biostratigraphical units (bodies of stratified rocks that are characterized by their fossil content).

Standard Global Chronostratigraphic (Geochronologic) Scale

All units of the standard chronostratigraphic and geochronologic hierarchies are theoretically worldwide in extent. They provide a standard scale of reference, what is known as the Standard Global Chronostratigraphic (Geochronologic) Scale, that aims to date all the rocks everywhere and to relate all rocks everywhere to the earth's geologic history. The standard geological column and its equivalent geologic age system have been built up by superposition of local columns from many different localities. A local geologic column is called a "geologic section," and it consists of any sequence of rock units found in a given region either at the surface or below it. Although the geologic column is not found complete at any place on Earth and the representative sediments common to all the major divisions cannot be found all together in a single section, the relative order of the formations still remains the same; in addition, such relative order also fits the geologic column.

Geochronometry

The quantitative (numerical) measurement of geologic time is dealt with by geochronometry, a branch of geochronology. The improvement of radiometric techniques since 1917 has allowed scientists to determine the absolute ages of rocks and to work out the duration of the intervals of geologic time, which had been previously established by means of fossils. More recently, enhanced methods of extracting linear time from the rock record have enabled high-precision age assignments. Apart from high-resolution radiometric dating, some of these calibration methods include the use of geochemical variations, Milankovitch climate cycles, and magnetic reversals. Radioactivity allowed scientists to date chronostratigraphical units, contributing to the development of the modern GTS. Technologic advances in measuring magnetic properties of rocks, together with the intense drilling of oceanic sediments and their biostratigraphical calibration by means of microfossils, have led to an improved chronology (magnetobiochronology). Moreover, the rapid development of cyclostratigraphy during the past decades has led to the construction of an astronomical timescale for dating events in the geologic record, based mainly on the relation between sedimentological, geochemical, or palaeontological cycles and variations in the earth's orbital parameters.

Recent Developments and Future Directions

The construction of the geological column has been under way for the last 2 centuries, and it has overcome several obstacles such as the precision and accuracy of correlation and dating tools, the limits of the stratigraphic database, or problems of nomenclature. In March 2005 the current available stratigraphic and geochronologic information was compiled by the project "Geologic Time Scale 2004" and published by a team designated by the International Commission on Stratigraphy. The results of this project summarized the history and status of boundary definitions of all geologic stages, compiled integrated stratigraphy (biologic, chemical, sea-level, magnetic, etc.) for each period, and assembled a numerical age scale from an array of astronomical tuning and radiometric ages. A combination of zones, polarity chrons, stages, and ages was carried out in order to calculate the best possible timescale. Earth scientists thus keep concentrating their efforts on the construction and improvement of the GTS. Research on Ocean Drilling Program (ODP) cores, and on Integrated Ocean Drilling Program (IODP) cores in the near future, will improve the calibration of various biostratigraphical scales, and together with the development of new tuning strategies will probably extend the astronomical timescale downward. Even so, there are still some stratigraphic and geochronologic issues to be resolved in the next updated version of the Geologic Time Scale, the GTS2008, which is expected to include a consensus on all stage boundary stratotypes, which is one of the main challenges for the future.

Laia Alegret

See also Chronostratigraphy; Dating Techniques; Geologic Timescale; Geology; Hutton, James; Lyell, Charles; Smith, William; Steno, Nicolaus; Time, Planetary

Further Readings

Gradstein, F. M., Ogg, J. G., & Smith, A. G. (2004). *A geologic time scale 2004.* Cambridge, UK: Cambridge University Press.

Hedberg, H. D. (Ed.). (1976). *A guide to stratigraphic classification, terminology and procedures.* New York: Wiley.

Holmes, A. (1960). A raised geological time-scale. *Transactions of the Geological Society of Edinburgh, 17,* 183–216.

NASCSN (North American Commission on Stratigraphic Nomenclature). (1983). North American stratigraphic code. *American Association of Petroleum Geologists Bulletin, 67,* 841–875.

Geologic Timescale

The geologic timescale is the framework for deciphering the history of planet Earth. It is used by geologists and other scientists to describe the timing and relationships between events that have occurred during the history of the earth.

Nomenclature

The history of the earth is broken up into a hierarchical set of divisions for describing geologic time. In increasingly smaller units of time, the generally accepted divisions are eon, era, period, epoch, and age. The Phanerozoic eon represents the time during which the majority of macroscopic organisms, algal, fungal, plant, and animal, lived. When first proposed as a division of geologic time, the beginning of the Phanerozoic, approximately 542 million years ago (mya), was thought to coincide with the beginning of life. In reality, this eon coincides with the appearance of animals that evolved external skeletons, like shells, and the somewhat later animals that formed internal skeletons, such as the bony elements of vertebrates. The time before the Phanerozoic is usually referred to as the Precambrian. The Phanerozoic consists of three major divisions: the Cenozoic, the Mesozoic, and the Paleozoic eras. The *zoic* part of the word comes from the root *zoo,* which means animal. *Cen* means recent, *meso* means middle, and *paleo* means ancient. These divisions reflect major changes in the composition of ancient faunas, each era being recognized by its domination by a particular group of animals. The Cenozoic has sometimes been called the age of mammals, the Mesozoic the age of dinosaurs, and the Paleozoic the age of fishes. This is an overly simplified view; it has some value for the newcomer but can be a bit misleading. For instance, other groups of animals lived during the Mesozoic. In addition to the dinosaurs, animals such as mammals, turtles, crocodiles, frogs, and countless varieties of insects also lived on land. In addition, there were many kinds of plants living in the past that no longer live today. Ancient floras went through great changes too, and not always at the same times that the animal groups changed.

Few discussions in geology can occur without reference to geologic time, which is often discussed in two forms: (1) Relative time (chronostratic), subdivisions of the earth's geology in a specific order based upon relative age relationships; these subdivisions are given names, most of which can be recognized globally, usually on the basis of fossils. (2) Absolute time (chronometric), numerical ages in millions of years or some other measurement. These are most commonly obtained via radiometric dating methods performed on appropriate rock types.

History

The first people who needed to understand the geological relationships of different rock units were miners. Mining had been of commercial interest since at least the days of the Romans, but it wasn't until the 1500s and 1600s that these efforts produced an interest in local rock relationships. By noting the relationships of different rock units, Nicolaus Steno in 1669 described two basic geologic principles. The first stated that sedimentary rocks are laid down in a horizontal manner, and the second stated that younger rock units were deposited on top of older rock units. To envision this latter principle, think of the layers of paint on a wall. The oldest layer was put on first and is at the bottom, while the newest layer is at the top. An additional concept was introduced by James Hutton in 1795, and later emphasized by Charles Lyell in the early 1800s. This was the idea that natural geologic processes were uniform in frequency and magnitude throughout time, an idea known as the principle of uniformitarianism. Steno's principles allowed workers in the 1600s and early 1700s to begin to recognize rock successions. However, because rocks were locally described by the color, texture, or even smell, comparisons between rock sequences of different areas were often not possible. Fossils provided the opportunity for workers to correlate geographically distinct areas. This contribution was possible because fossils are found over wide regions of the earth's crust.

For the next major contribution to the geologic timescale we turn to William Smith, a surveyor, canal builder, and amateur geologist in England. In 1815 Smith produced a geologic map of

England in which he successfully demonstrated the validity of the principle of faunal succession. This principle simply stated that fossils are found in rocks in a very definite order. This principle led others who followed to use fossils to define increments within a relative timescale.

Arthur Holmes (1890–1965) was the first to combine radiometric ages with geologic formations in order to create a geologic timescale. His book, *The Age of the Earth,* written when he was only 22, had a major impact on those interested in geochronology. For his pioneering scale, Holmes carefully plotted four radiometric dates, one in the Eocene and three in the Paleozoic, from radiogenic helium and lead in uranium minerals, against estimates of the accumulated maximum thickness of Phanerozoic sediments. If we ignore sizable error margins, the base of the Cambrian interpolates at 600 mya, curiously close to modern estimates. The new approach was a major improvement over a previous "hourglass" method that tried to estimate maximum thickness of strata per period to determine their relative duration, but had no way of estimating rates of sedimentation independently. In 1960, Holmes compiled a revised version of the age-versus-thickness scale. Compared with the initial 1913 scale, the projected durations of the Jurassic and Permian are more or less doubled, the Triassic and Carboniferous are extended about 50%, and the Cambrian gains 20 million years at the expense of the Ordovician.

W. B. Harland and E. H. Francis as part of a Phanerozoic timescale symposium coordinated a systematic, numbered radiometric database with critical evaluations. Items in The Phanerozoic Time-Scale: A Symposium, were listed in the order as received by the editors. Supplements of items were assembled by the Geological Society's Phanerozoic Time-Scale Sub-Committee from publications omitted from the previous volume or published between 1964 and 1968, and items relating specifically to the Pleistocene were provided primarily by N. J. Shackleton. The compilation of these additional items with critical evaluations was included in *The Phanerozoic Time-Scale: A Supplement* published in 1971 by Harland and Francis. In 1978, R. L. Armstrong published a reevaluation and continuation of The Phanerozoic Time-Scale database. This publication did not include abstracting and critical commentary. These catalogs of items and of Armstrong's continuation of items were denoted "PTS" and "A," respectively, in later publications.

In 1976, the Subcommission on Geochronology recommended an intercalibrated set of decay constants and isotopic abundances for the U-Th-Pb, Rb-Sr, and K-Ar systems with the uranium decay constants by Jaffey et al. in 1971 as the mainstay for the standard set. This new set of decay constants necessitated systematic upward or downward revisions of previous radiometric ages by 1%–2%.

In *A Geological Time Scale,* Harland et al. standardized the Mesozoic–Paleozoic portion of the previous PTS-A series to the new decay constants and included a few additional ages. Simultaneously, in 1982 G. S. Odin supervised a major compilation and critical review of 251 radiometric dating studies as Part II of *Numerical Dating in Stratigraphy.* This "NDS" compilation also reevaluated many of the dates included in the previous "PTS–A" series. A volume of papers on The Chronology of the Geological Record from a 1982 symposium included reassessments of the combined PTS–NDS database with additional data for different time intervals. After applying rigorous selection criteria to the PTS–A and NDS databases and incorporating many additional studies (mainly between 1981 and 1988) in a statistical evaluation, Harland and coworkers presented *A Geological Time Scale 1989.*

The statistical method of timescale building employed by *GTS82* and refined by *GTS89* derived from the marriage of the chronogram concept with the chron concept, both of which represented an original path to a more reproducible and objective scale. Having created a high-temperature radiometric age data set, the chronogram method was applied that minimizes the misfit of stratigraphically inconsistent radiometric age dates around trial boundary ages to arrive at an estimated age of stage boundaries. From the error functions, a set of age/stage plots was created (Appendix 4 in *GTS89*) that depicts the best age estimates for Paleozoic, Mesozoic, and Cenozoic stage boundaries. Because of wide errors, particularly in Paleozoic and Mesozoic dates, *GTS89* plotted the chronogram ages for stage boundaries against the same stages with relative duration scaled proportionally to their component chrons. For convenience, chrons were equated with biostratigraphic zones. The chron concept in *GTS89* implied equal duration of zones in prominent biozonal schemes, such as a conodont scheme for the Devonian.

The Bureau de Récherches Géologiques et Minières and the Société Géologique de France published a stratigraphic scale and timescale compiled by Odin and Odin. Of more than 90 Phanerozoic stage boundaries, 20 lacked adequate radiometric constraints, the majority of which were in the Paleozoic.

The International Stratigraphic Chart is an important document for stratigraphic nomenclature (including Precambrian), and included a summary of age estimates for stratigraphic boundaries.

During the 1990s, a series of developments in integrated stratigraphy and isotopic methodology enabled relative and linear geochronology at unprecedented high resolution. Magnetostratigraphy provided correlation of biostratigraphic datums to marine magnetic anomalies for the Late Jurassic through Cenozoic. Argon–argon dating of sanidine crystals and new techniques of uranium–lead dating of individual zircon crystals yielded ages for sediment-hosted volcanic ashes with analytical precessions less than 1%. Comparison of volcanic-derived ages to those obtained from glauconite grains yielded systematically younger ages, thereby removing a former method of obtaining direct ages on stratigraphic levels. Pelagic sediments record features from the regular climate oscillations produced by changes in the earth's orbit, and recognition of these "Milankovich" cycles allowed precise tuning of the associated stratigraphy to astronomical constants.

Aspects of the *GTS89* compilation began a trend in which different portions of the geologic timescale were calibrated by different methods. The Paleozoic and early Mesozoic portions continued to be dominated by refinements of integrating biostratigraphy with radiometric tie points, whereas the Late Mesozoic and Cenozoic also utilized oceanic magnetic anomaly patterns and astronomical tuning. A listing of the radiometric dates and discussion of specific methods employed in building *GTS2004* can be found in Gradstein, Ogg, and Smith's *A Geologic Time Scale 2004*.

Calibration

Because the timescale is the main tool of the geological trade, insight on its construction, strengths, and limitations greatly enhances its function and its utility. According to Gradstein, all scientists should understand how the evolving timescales are constructed and calibrated, rather than merely using the numbers in them.

The calibration to linear time of the succession of events recorded in the rock record has three components: (1) The international stratigraphic divisions and their correlation in the global rock record, (2) the means of measuring linear time or elapsed durations from the rock record, and (3) the methods of effectively joining the two scales.

For convenience in international communication, the rock record of Earth's history is subdivided into a chronostratigraphic scale of standardized global stratigraphic units, such as "Paleogene," "Eocene," "*Morozovella velascoensis* planktic foraminifera zone," or "polarity Chron C24r." Unlike the continuous ticking clock of the chronometric scale (measured in years before the present), the chronostratigraphic scale is based on relative time units in which global reference points at boundary stratotypes define the limits of the main formalized units, such as Neogene. The chronostratigraphic scale is an agreed convention, whereas its calibration to linear time is a matter for discovery or estimation. By contrast, Precambrian stratigraphy is formally classified chronometrically; that is, the base of each Precambrian eon, era, and period is assigned a numerical age.

Continual improvement in data coverage, methodology, and standardization of chronostratigraphic units implies that no geologic timescale can be final. *A Geologic Time Scale 2004* (*GTS2004*) provides an overview of the status of the geological timescale and is the successor to *GTS1989*.

Since 1989, there have been several mayor developments. Stratigraphic standardization through the work of the International Commission on Stratigraphy (ICS) has greatly refined the international chronostratigraphic scale. In some cases, traditional European-based geological stages have been replaced with new subdivisions that allow global correlation. New or enhanced methods of extracting linear time from the rock record have enabled high-precision age assignments. An abundance of high-resolution radiometric dates has been generated and has led to improved age assignments of key geologic stage boundaries. Global geochemical variations, Milankovitch climate cycles, and magnetic reversals have become important calibration tools. Statistical techniques of extrapolating ages and associated uncertainties to stratigraphic events have evolved to meet the challenge of more accurate age dates and more precise zonal assignments. Fossil event databases with multiple stratigraphic sections through the globe can be integrated into composite standards.

The compilation of *GTS2004* has involved a large number of specialists, including contributions by past and present chairs of different subcommissions of ICS, geochemists working with radiometric and stable isotopes, stratigraphers using diverse tools from traditional fossils to astronomical cycles to database programming, and geomathematicians. The set of chronostratigraphic units (stages, eras) and their computed ages, which constitute the main framework for *A Geologic Time Scale 2004*, are summarized in the chart available online from the ICS.

Eustoquio Molina

See also Chronostratigraphy; Darwin, Charles; Dating Techniques; Earth, Age of; Geological Column; Geology; Neogene; Paleogene; Synchronicity, Geological; Time, Measurements of

Further Readings

Berry, W. (1987). *Growth of a prehistoric time scale: Based on organic evolution* (Rev. ed.). Palo Alto, CA: Blackwell Scientific Publications.

Gradstein, F. (2004). Introduction. In F. Gradstein, J. Ogg, & A. Smith (Eds.), *A geologic time scale 2004* (pp. 3–19). Cambridge, UK: Cambridge University Press.

Gradstein, F., Ogg, J., & Smith, A. (Eds.). *A geologic time scale 2004.* New York: Cambridge University Press.

Harland, W. B., Armstrong, R. L., Cox, A. V., et al. (1990). *A geologic time scale 1989.* New York: Cambridge University Press.

International Commission on Stratigraphy. (2004). *International stratigraphic chart.* Available from http://www.stratigraphy.org/chus.pdf

Remane, J. (2000). *International stratigraphic chart, with explanatory note.* Sponsored by International Commission on Stratigraphy (ICS), International Union of Geological Sciences (IUGS), and UNESCO. 31st International Geological Congress, Rio de Janeiro.

Geology

Geology is the scientific study of planet Earth and its history, through 4,600 million years to the present. This natural science is traditionally divided into two branches: physical geology and historical geology. Physical geology focuses on physical structure, materials, and geological processes of the earth. Historical geology examines the origin of our planet and life, and all the climatic, geographic, oceanographic, and biological events that have taken place across geological time. This dual division is rather arbitrary, therefore both points of view (physical and historical) are found currently integrated within the framework of plate tectonics, the current paradigm of geological science.

Physical Geology

Physical geology includes such disciplines as: geophysics (applies principles of physics to the study of the earth); geochemistry (the study of the chemical characteristics of minerals and rocks); mineralogy and petrology (the study of the origin, properties, structure, and classification of minerals and rocks, respectively); hydrogeology (the study of the origin, occurrence, and movement of water masses); structural geology (the study of the deformational history of rocks and regions and of the forces responsible); geomorphology (the study of the origin and modification of landforms); volcanology (the study of volcanoes and magma formation processes); sedimentology (the study of sedimentary rocks and the processes by which they were formed); and engineering geology (the study of the interactions of the earth's crust with human-made structures such as tunnels and mines).

Some areas of specialization for professional geologists related to physical geology include exploration and extraction of natural resources (mineral deposits, coal, oil, etc.), prediction and evaluation of geological hazards (landslides, earthquakes, volcanic eruptions, or meteoritic impacts), evaluation of the stability of construction sites, the search for supplies of clean water, and analysis of environmental problems such as soil and coastal erosion.

Historical Geology

A consubstantial part of geology is the study of how Earth's materials and continents, surface environments, processes, and organisms have changed over geological time. Processes, fossils, and geological events are recorded in rocks. Thus the main objective of historical geology is the analysis of the geological record in order to reconstitute and understand the earth's history. Historical geology is based on paleontology (the study of life in the past from the fossil record, including evolutionary relationships, and its applications in environmental reconstructions and in the relative dating of rocks); stratigraphy (the study of stratified rocks in terms of mode of origin, original succession, relative dating, and geologic history), and paleogeography (the reconstruction of the ancient geography of the earth's surface). Other important disciplines closely related to historical geology are: paleoclimatology (the application of geological science to determine past climatic conditions) and paleoceanography (the reconstruction of the history of the oceans, with regard to circulation, chemistry, or patterns of sedimentation).

When geologists come to interpret the earth's history, they rely on two complementary types of dating of rocks: relative dating and absolute dating. Relative dating places historical events in their correct temporal order, and absolute dating provides a numerical age for a rock and establishes how many years ago a geological event took place.

Relative Dating

The relative ages of rocks and events in geologic sequences can be established by interpreting the fossil record and utilizing several basic principles in stratigraphy. The four most important are: the principle of original horizontality (sedimentary rocks are formed in essentially horizontal beds named strata; nonhorizontal strata have been disturbed after lithification), the principle of superposition (in any undisturbed stratigraphic sequence, older strata are buried beneath younger strata), the principle of intersection (when a fault or igneous intrusion cuts across a formation of sedimentary rocks, the fault or the intrusion is younger than these strata), and the principle of inclusions (the inclusion of a rocky body in a sequence of strata is older than the sedimentary rocks that contain it). Since the 19th century, the application of these rules to establish the relative ages of rocks contributed to developing the standard geological column in geological sites undisturbed or minimally disturbed, such as the Zumaya stratigraphic section in Spain.

Geologists soon understood that to develop a global geological timescale, a comparison of rocks of similar age located in different regions or continents was required. This process is known in stratigraphy as "correlation." Correlation involves matching up rock layers of similar age that are in different regions. When conditions of exposure are good, the litho-correlation across short distances can be done by applying the principle of lateral continuity (sediments are deposited forming strata over a large area in a continuous sheet). When correlation involves a long distance (even

Overview to coastal Zumaya stratigraphic section, a classic European locality for the study of the Cretaceous and Paleogene periods, due to its richness in fossils, continuity of sedimentation, and good exposure of rocks.

Source: Photo by Asier Hilario Orus. Used with permission.

between two different continents), geologists depend on the fossil remains of ancient organisms. In this case the bio-correlation is carried out using the principle of faunal succession (in a stratigraphic sequence, fossil species succeed one another in a definite and determinable order, so any time period can be recognized by its fossil content).

Fossil species appear and disappear throughout the fossil record as a consequence of the evolution (speciation and extinction) of species. As each fossil species lived during a specific interval of geological time, its presence (or sometimes absence) may be used to provide a relative age for the strata in which it is found. Each fossil species also lived at the same time in a more or less extensive geographical region. For this reason, paleontologists can use fossils to establish a bio-correlation of strata among different localities, where fossils of one species or of a species assemblage were present.

Correlation based on fossils is the focus of biostratigraphy, a discipline that deals with the distribution of fossils in the stratigraphic record, and the organization of strata into correlatable units (biozones) on the basis of the fossils they contain. Most species lived for tens of millions of years before they became extinct or evolved into new species. Nevertheless, some species lived for only hundreds of thousands of years or a few million years. If they were limited to a short period of time, had a wide geographic distribution, and are abundant in the fossil record, these species are considered "index fossils." Index fossils provide a precise means for estimating the relative age of sedimentary rock, and for correlating biozones. For example, many species of trilobites, ammonites, or foraminifera are ideal for biostratigraphy in marine series, and micro-mammals, grains of pollen, and spores are good index fossils for continental deposits.

Absolute Dating

The discovery of radioactivity and the development of the mass spectrometer in the 20th century permitted many radioactive elements to be used as geologic clocks. These techniques are based on the natural decay of radioactive elements (unstable isotopes) that cause the radioactive parent elements to decay to stable daughter elements.

Through the radioactive decay of isotopes in rock, exotic daughter elements are introduced over time. By measuring the concentration of the stable end products of decay, coupled with knowledge of the half-life and initial concentration of the decaying elements, the age of a rock can be calculated. This technique is known as radiometric dating, a powerful tool for reconstructing the earth's history.

Potassium-40, for example, decays into argon-40 after a half-life of 1.25 billion years, so that after 1.25 billion years half of the potassium-40 in a rock will have become argon-40. This means that if a rock sample contained equal amounts of potassium-40 and argon-40, it would be 1.25 billion years old. Other isotopes used in radiometric dating are: uranium-235, uranium-238, thorium-232, and rubidium-87.

Relative Versus Absolute Dating

Relative dating provides a relative timescale formed by bio- and chronostratigraphical units. Absolute dating provides an absolute timescale composed of geochronological units. But how are both scales correlated and integrated in the geological timescale if fossils occur in sedimentary rocks, and most minerals that contain radioactive isotopes are in igneous rocks?

The simplest method to correlate both scales is the radiometric dating of volcanic ash or of oceanic basaltic layers that are interbedded between sedimentary rocks. The age of the volcanic rock is younger than the underlying sedimentary rocks and older than the overlying sedimentary rocks. If these sedimentary rocks contain fossils, they are very relevant since they can be correlated with the biozones defined in this area. Thus, rocks whose ages have been determined by absolute dating can be incorporated into a succession of strata determined by relative dating. Then geologists can use correlations to infer the ages of rocks and fossils that cannot be directly dated. In fact, the combined absolute/relative timescale is always being revised in order to produce an even more precise picture of Earth's history.

A Brief History of the Earth

From the modern methods of radiometric age dating, we know that the earth is around 4.6 billion years old. Historical geologists divide all this geological time into three major divisions (eons):

Archaean, Proterozoic, and Phanerozoic, although a fourth pregeologic eon is considered: Hadean. All of them are made up of eras that usually ended with profound changes in the disposition of the earth's continents and oceans, and are characterized by the emergence of new forms of life or by the disappearance of ancient ones.

Hadean Eon (4550–3900 Million Years Ago)

The Hadean eon is the geologic time extending from the birth of the solar system and the earth's formation 4,600–4,550 million years ago, to the formation of the oldest rocks 3,900–3,800 million years ago (mya). The Hadean is the first eon in the earth's History, but very little geological record was preserved because the earth's surface was molten.

The solar system's planets, including the earth, were formed by accretion from a cloud of gas and dust known as solar nebula. Grains in orbit around the central primitive sun (protoplanetary disk) began to join, collecting in bodies called planetesimals. In a few million years, several large planets grew through low-velocity collisions between nearby planetesimals. The planets and satellites formed in the inner solar system—Earth, Mars, and Venus—were composed mainly of materials with high melting points, such as silicates and metals (iron and nickel).

Since the Hadean lacks any official status, this eon has been arbitrarily subdivided in three informal periods, taking into account the primitive geological history of the moon: Cryptic (4,550–4,500 mya), Ryderian (4,500–4,100 mya), and Nectarian (4,100–3,900 mya). The first period includes the accretion time of Earth from the solar disk, the big whack event, and the formation of the moon. The big whack is a hypothetical event that occurred roughly 4,530 mya in which a Mars-sized body, usually called Theia, impacted the proto-Earth at an oblique angle. The giant impact destroyed Theia, ejecting into space most of its mass together with a significant portion of the earth's silicate mantle. A ring of debris began to orbit near the earth's equator, coalescing into the moon about 100 years after the huge impact. The Ryderian era includes the progressive cooling of the earth (and the moon) and the process of differentiation of the earth's core, mantle, and protocrust, when the earth acquired a primitive inner structure.

A great number of asteroidal and cometary objects remained among the newborn planets, starting a well-known period called the Late Heavy Bombardment (LHB) during the Nectarian era. The LHB happened about 4,100–3,800 mya, resulting in a large number of impact craters on the earth as well as on the rest of the planets and satellites in the solar system. This cataclysm period created an enormous amount of heat, completely melting the earth and allowing its materials to separate definitively into three main layers: iron core, silicate mantle, and thin outer crust. This intensive bombardment probably destroyed the primitive earth's protocrust, retaining only individual zircon crystals that were redeposited in most modern sediments. The oldest known zircons were radiometrically dated to about 4,400 million years and are found in the Acasta Gneiss in western Canada. The study of some of these zircons suggests that there was liquid water at that period, indicating that primitive atmosphere and oceans must have existed then. The rock vapor might condense around the young Earth, resulting in a dense atmosphere of carbon dioxide, water, methane, nitrogen, and hydrogen.

Archean Eon (3,900–2,500 Million Years Ago)

The Archean eon is the geologic time that extends from the formation of the oldest rocks 3,900–3,800 mya to 2,500 mya. It is formally subdivided into four eras: Eoarchaean (3,900–3,600 mya), Paleoarchean (3,600–3,200 mya), Mesoarchean (3,200–2,800 ma) and Neoarchean (2,800–2,500 mya). A significant event occurred at the beginning of the Archaean eon: the origin of life, in a scenario with an atmosphere composed mostly of carbon dioxide.

The oldest known rocks on Earth are found in the Issua Greenstone Belt (southwestern Greenland) and include well-preserved volcanic, metamorphic, and sedimentary rocks dated at 3,800–3,700 million years. Archean volcanic activity was probably considerably more intensive than it is today, and plate tectonics were surely very active because the inner earth was much hotter at that time. A greater rate of recycling of crustal material should have occurred then, preventing the formation of continents. For this reason, only some Archean rocks survive, including metamorphized igneous rocks, such as granites, peridotites, and unusual ultramafic mantle-derived volcanic rocks called komatiites. Archean rocks also include stromatolites (the oldest fossil traces of prokaryotic organisms), and hard metamorphized deepwater sediments.

Only when the mantle cooled and convection slowed down could the tectonic activity slow down. Small protocontinents (*cratons*) were the norm during the Archean eon. Several Archean cratons have been identified, including a hypothetical first continent now called Vaalbara. According to radiometric dating, the Vaalbara continent existed at least 3,300 mya during the Paleoarchean era. This continent collected the two only known Eoarchean cratons: Kaapvaal craton (South Africa) and Pilbara craton (Western Australia). Other continents were probably formed 3,000 mya in the Mesoarchean era, grouping small cratons of present western Australia, eastern India, eastern Antarctica, and southeastern Africa. It is now called the Ur continent.

During the Neoarchean era, 2,700 mya, the earliest known supercontinent, called Kenorland, was born. It formed as result of a series of events of accretion, comprising several cratons: Laurentia (including Canada and Greenland), Baltica (including Scandinavia and the Baltic area), Yilgarn (Western Australia), and Kalahari (Botswana, South Africa, and Namibia).

Proterozoic Eon (2,500–542 Million Years Ago)

The Proterozoic eon began 2,500 mya, and ended with the disappearance of the complex Ediacaran biota 542 mya. The geological record of the Proterozoic is much better known than that of the Archean, since Proterozoic rocks are less metamorphized and are more abundant. Many Proterozoic sedimentary rocks were deposited in extensive shallow epicontinental seas, and their study suggests that Proterozoic plate tectonics were both massive and rapid.

The Proterozoic eon has been formally subdivided into three eras: Paleoproterozoic (2,500–1,600 mya), Mesoproterozoic (1,600–1,000 mya), and Neoproterozoic (1,000–542 mya). Both the start and the end of the Proterozoic were marked by widespread glaciation: the Huronian and Varangian glaciations, respectively.

Paleoproterozoic Era (2,500–1,600 Million Years Ago)

The Huronian glaciation began 2,400 mya and lasted until 2,100 mya. It was one of the most severe ice ages, and it is possible that the earth's surface was entirely covered by ice. It is perhaps related to the decrease of atmospheric carbon dioxide, consumed and captured by photosynthetic microorganisms, and the virtual disappearance of the greenhouse gas methane due to chemical oxidation.

The neoarchaic Kenorland supercontinent began to break up at the start of the Siderian period (2,500–2,300 mya) when two Baltic fragments, the Kola and Karelia cratons (today in Russia), began to drift apart. During the Rhyacian (2,300–2,050 mya), several continents were created out of Kenorland when it broke up: Arctica (including today's Canada and Siberia), Atlantica (including today's eastern South America and western Africa), and Baltica (northern Europe). Cyanobacteria developed in the Siderian and Rhyacian seas, producing a great quantity of photosynthetic oxygen that resulted in a large increase of this gas in the atmosphere. The combination of oxygen with the iron dissolved in the oceans formed a distinctive Archaic–Paleoproterozoic type of rock called banded iron formations (BIFs) that are composed of iron oxides like magnetite and hematite.

The atmosphere became oxygen rich during the Orosirian period (2,050–1,800 mya) with a progressive decrease in iron dissolved in the ocean. The formation of BIFs ceased, but the deposition of the so-called red beds began. These rocks are tinged with hematite, indicating an increase in atmospheric oxygen. The progressive accumulation of oxygen in the atmosphere might have caused the extinction of numerous groups of anaerobic bacteria, the only type of life that existed up to that time. This event has sometimes been called the oxygen catastrophe. From the point of view of plate tectonics, collisions between cratons were common during this period, forming multiple orogens worldwide that were the prelude to the formation of the supercontinent Columbia.

The last period of the Paleoproterozoic, the Statherian (1,800–1,600 mya), was characterized by its tectonic stability. At the beginning of the Statherian, the supercontinent Columbia was formed and lasted from approximately 1,800 to 1,500 mya. It consisted of almost all of Earth's continents: Laurentia (proto-North America), Nena (Arctica, Baltica, and eastern Antarctica), Atlantica (Amazonia and western Africa), Ur (Australia, India, and northern Antarctica), and possibly northern China and Kalahari (southern Africa). At the beginning of the Statherian a new type of cell appeared, eukaryotic cells, that experienced great evolutionary exits in later periods of mass extinction.

Mesoproterozoic Era (1,600–1,000 Million Years Ago)

The Mesoproterozoic era began with the breakup of Columbia and ended with the formation of a new supercontinent: Rodinia. By about the Mesoproterozoic era, 80% of the earth's continental crust had been formed. Columbia began to fragment at the start of the Calymnian period (1,600–1,400 mya), forming several continents that drifted apart, including Laurentia, Nena, Australia-Antarctica, Amazonia, Sahara (northern Africa), Congo (west-central Africa), Kalahari (southern Africa), and Sao Francisco (southeastern South America).

Rodinia began forming in the Ectasian period (1,400–1,200 mya) and finished assembling during the Stenian period (1200–1000 mya). The Laurentian continent was the core of Rodinia, and was sandwiched between two large blocks: East Gondwana (the original continent of Ur) and West Gondwana (the original continent of Atlantica). Moreover, Baltica and Siberia were nearby, to the southeast and northeast of Laurentia, respectively. Rodinia was surrounded at that time by a great superocean called Mirovia, in which the first cells with sexual reproduction appeared.

Neoproterozoic Era (1,000–542 Million Years Ago)

The Neoproterozoic era was a time of complex continental drift that followed the breakup of Rodinia. The breakup of this supercontinent started in the Tonian period (1,000–850 mya) and continued through the Cryogenian period (850–630 mya). In the first phase, 800 mya, Rodinia split into two large segments. The lands of East Gondwana and Congo (the Proto-Gondwana continent) moved north while rotating counterclockwise, beginning the formation of the Panthalassic superocean. The great block composed of Laurentia, Siberia, Baltica, South China, and West Gondwana (the Proto-Laurasia continent) drifted southward, rotating clockwise. The Panafrican Ocean formed between these macrocontinents.

Recent studies indicate that at least seven independent continents must have existed around 750 mya: Laurentia-Baltica-West Gondwana, North China, Siberia, Australia-East Antarctica, South China, Malani (India, Iran, Arabia, and Madagascar), and Congo-Kalahari. During the Ediacaran period (630–542 mya), most of the earth's large continents came together again in the southern hemisphere. They formed the Pannotia supercontinent, which was short lived since it lasted only about 60 million years.

The Cryogenian period is characterized by the massive Varanginian glaciations that are represented by worldwide tillite deposits, suggesting that the earth suffered the most severe ice age of its history with glaciers extending as far as the equator. It is believed that all the planetary oceans were deeply frozen, a phenomenon known as Snowball Earth. Finally, the last Neoproterozoic period, the Ediacaran, is unusual because it contains strange soft bodied fossils known as Ediacaran biota. The severe Neoproterozoic glaciations caused profound changes in oxygen levels and ocean chemistry, which could explain why life developed intensively during the Ediacaran and later during the Cambrian.

Phanerozoic Eon (542 Million Years Ago to the Present)

The Phanerozoic eon left a rich fossil record, starting with the Cambrian explosion about 540 mya. It is formally subdivided into three eras: Paleozoic (542–251 mya), Mesozoic (251–65 mya), and Cenozoic (65 mya to the present).

Earth seems to have gone through alternating icehouse (with ice caps) and greenhouse (without ice caps) phases during the Phanerozoic, perhaps partly controlled by how the continents and oceans were distributed. At least three icehouse phases are known: Late Ordovician–Silurian (about 460–416 mya), Late Carboniferous–Permian (about 318–251 mya), and the Oligocene–Neogene period (34 mya to the present). Throughout the Phanerozoic new continents and oceans appear and disappear, assembling and separating until reaching their present locations. This is a complex history, and the reader is invited to visit the Paleomap project's Web page where several full-color paleogeographic maps are available that show the changing distribution of lands and seas, as well as the various continental and oceanic plates that developed in the earth's recent history.

Paleozoic Era (542–251 Million Years Ago)

The Paleozoic era covers the geological time from the first occurrence of an abundant fossil record (Cambrian Explosion) to the greatest mass extinction event in the earth's history: the Permian–Triassic boundary event. The Cambrian Explosion was the greatest evolutionary radiation in the earth's history, bringing forth nearly all the major groups or phyla of animals, including the trilobites. The Paleozoic began with the breakup of the Pannotia supercontinent and ended with the formation the last known supercontinent, Pangea.

Pannotia started to break up at the beginning of the Cambrian period (542–488 mya) and formed four continents: Laurentia (North America), Baltica (Northern Europe), Siberia, and Gondwana. The three first drifted toward the north and Gondwana drifted toward the south, with most of the land staying in equatorial latitudes. The Panthalassic superocean covered most the northern hemisphere, and two new minor oceans began to form: Proto-Tethys and Iapetus. Proto-Tethys formed between the proto-Laurasian continents (Laurentia, Baltica, and Siberia) and Gondwana. Iapetus formed between Baltica and Laurentia. The Cambrian oceans seem to have been broad and shallow, causing a climate significantly warmer than that of the preceding periods dominated by ice ages.

At the start of the Ordovician period (488–443 mya), Gondwana began to drift toward the South Pole. The Ordovician climate was very warm, and Laurentia, Baltica, and Gondwana were widely covered by warm shallow seas, allowing the development of shelled organisms and the deposition of great amounts of biogenic limestone. While the Panthalassic superocean still covered most of the northern hemisphere, another ocean was born, the Paleo-Tethys, which was taking over territory at the expense of ancient Proto-Tethys. The new Rheic Ocean was the result of the formation of the Avalonia microcontinent (composed mainly of Newfoundland and England) that had broken off from Gondwana and drifted toward Baltica. That ocean was gaining territory at the expense of the old Iapetus Ocean. By the end of the Ordovician, Gondwana was approaching the South Pole and was largely glaciated. The extensive glaciation of Gondwana and the subsequent fall of the sea level may have triggered the Late Ordovician mass extinction event between 447–444 mya, in which the 85% of species died off.

The icecaps of the Silurian period (443–416 mya) were less extensive than those of the Late Ordovician, and the melting of the Silurian icecaps and glaciers contributed to a rise in the sea level. During this period, continents drifted near the equator and the earth's climate entered a long greenhouse phase with warm shallow seas covering much of the equatorial landmasses. Coral reefs expanded, and land plants began to colonize the barren continents. Baltica and Laurentia started to collide, completely closing off the Iapetus Ocean and forming the Euramerica macrocontinent. Such collisions folded the sediments deposited in the previous Iapetus basin, forming the Caledonian orogen (Appalachian Mountains, the Anti-Atlas in Morocco, and the Caledonian Mountains in Great Britain and Scandinavia). At that time, the Rheic and Paleo-Tethys oceans occupied the area between Euramerica and Gondwana, and between Siberia and Gondwana, respectively. During the Silurian a new ocean was formed, the Ural Ocean, located between Siberia and Baltica.

The Devonian period (416–359 mya) was a time of great tectonic activity as the continents drew closer together. The Ural Ocean disappeared toward the end of the Devonian when Siberia collided with the Baltica coast of Euramerica, forming the Ural Mountains and the Laurussia macrocontinent, also called the Old Red Continent. Sea levels were high worldwide during the Devonian, with broad areas of continents widely submerged under shallow seas where tropical reef organisms were plentiful. The deep waters of the giant Panthalassic superocean covered most of the earth, Paleo-Tethys continued spreading, and other minor oceans the Rheic Ocean began to close off. The Devonian ended with a major mass extinction event that affected up to 82% of all species, probably caused by several meteorite impacts like those forming the 120-kilometer–diameter Woodleigh crater (Australia).

The well-known Pangea supercontinent began to form in the Carboniferous period (359–299 mya). Laurussia and Gondwana collided during this period, definitively closing the Rheic Ocean. The collision occurred along the eastern coast of North America (where the Alleghenian orogen was formed) and the northwestern coast of South America, Africa, and South Europe (where the Hercinian orogen emerged). Finally, North China collided with Siberia at the end of the Carboniferous, closing off another old ocean, the Proto-Tethys. Most the continents were grouped together toward the end of the Carboniferous, although South China was still separated from Laurussia by the Paleo-Tethys Ocean. The Carboniferous was a period of active mountain-building. There was also a drop in south polar temperatures, glaciating the southern portion of Gondwana. Nevertheless, the tropical latitudes were warm and humid, allowing for the development of extensive forests and swamps.

Except for the South China continent, all of Earth's major land masses were grouped together as Pangea in the Permian period (299–251 mya); Pangea reached from the equator toward both poles. This affected ocean currents in the Panthalassic superocean and the Paleo-Tethys, which was located between the Asian part of Laurussia and Gondwana. At the beginning of the Permian, a rift started to open from the north of Gondwana to form a new ocean: the Tethys Ocean. The Permian ended with the most extensive mass extinction event in geological history: the Permian–Triassic extinction event, affecting more than 90% of species. The Late Permian glaciations, Siberian Traps volcanism, severe drop in sea levels, and even several large meteorite impacts, among other causes, have been proposed to explain the massive extinction.

Mesozoic Era (251–65 Million Years Ago)

The Mesozoic era spans geological time from the Permian–Triassic boundary event 251 mya to the Cretaceous–Tertiary boundary event 65 mya. It was an era of intensive tectonic, climatic, and evolutionary activity.

During the Triassic period (251–199 mya), almost all of the earth's landmasses had collected into a single supercontinent, Pangea, that was surrounded by the Panthalassic superocean. The recently formed Tethys Ocean continued opening at the expense of the Paleo-Tethys. Since Pangea's large size limited the moderating effect of the oceans, the Triassic climate became very hot and dry, forming typical red bed sandstones and gypsum. This kind of terrestrial climate was suitable for reptiles. The Triassic ended with another major mass extinction episode that affected mainly marine environments. The cause of this extinction is uncertain, but global cooling or even meteorite impacts have been proposed. These extinctions allowed the dinosaurs to expand, initiating the Age of Dinosaurs that spanned the Jurassic (199–145 mya) and the Cretaceous (145–65 mya).

In the Early Jurassic, the oceanic crust of the Paleo-Tethys was completely subducted. This episode resulted in the collision of South China with Laurussia, initiating the Cimmerian orogeny that created mountain ranges as high as today's Himalayas. Pangea was shaped like a "C" at that time, curling around the Tethys Ocean with the concave area occupied by the Tethys Ocean. Both Pangea and Tethys were surrounded by the huge Panthalassic superocean. As in the Triassic, the Jurassic climate was warm; there is no evidence of glaciation since no continent was near either pole. During this period, Pangea began to break up into two major landmasses: Laurasia (North America and Eurasia) and Gondwana (South America, Africa, Australia, Antarctica, and India), which began the opening of the Central Atlantic Ocean. Toward the end of the Jurassic, the Panthalassic superocean converted into the current Pacific Ocean.

Pangea was definitively broken up during the Cretaceous into today's continents, although the continents had positions substantially different from today's. The progressive drift of the Laurasian and Gondwanan landmasses opened the western Tethys Ocean (today's Mediterranean Sea) and separated North America and South America, forming a continuous ocean current around the equator. The North Atlantic Ocean opened, separating Iberia from Newfoundland and England-Scandinavia from Greenland. The separation of South America and Africa formed the South Atlantic Ocean. Finally, the drift of India northward formed the Indian Ocean, which was gaining territory at the expense of eastern Tethys. This active rifting during the opening of the Atlantic Ocean raised mountain ranges (including the giant American Cordillera: the Rocky Mountains, the Sierra Madre, and the Andes Mountains) around the entire coastline of old Pangea. This intensive tectonic activity raised the sea level, forming broad shallow seas over North America (the Western Interior Seaway) and Europe. The Cretaceous climate was very warm, devoid of ice at the poles. The Cretaceous ended with the most recent major mass extinction event: the Cretaceous–Tertiary event 65 mya, affecting more than 75% of species. A meteorite impact that formed the approximately 180-kilometer-diameter Chicxulub crater (Yucatan, Mexico) appears to be its cause.

Cenozoic Era (65 Million Years Ago to the Present)

The Cenozoic era covers geological time from the Cretaceous–Tertiary mass extinction event to the present. During the Cenozoic, mammals evolved from a few small insectivores that had survived the Cretaceous–Tertiary extinction. The continents moved to their current positions, causing remarkable climatic and oceanographic changes.

During the Paleogene period (65–23 mya), the Laurasian and Gondwanan plates formed in the Cretaceous continued to split apart, with North America, Eurasia, South America, Africa, India, and Antarctica-Australia pulling away from each other. India continued its migration toward central Eurasia, and Africa also headed north toward western Eurasia, slowly closing the Tethys Ocean during the Paleocene epoch (65–56 mya). The rifting and splitting apart of Eurasia, Greenland, and North America increased hydrothermal and volcanic activity in the North Atlantic. This tectonic episode combined with the movement of the Indian plate northward, restricted the Tethys oceanic current, and helped to trigger the Paleocene–Eocene Thermal Maximum event 56 mya. During the Eocene epoch (56–34 mya), Antarctica and Australia began to split, leaving Antarctica isolated in its current location at the South Pole. Moreover, the Indian microcontinent collided with central Eurasia, folding the Himalayas upward and closing off the eastern Tethys Ocean. The western Tethys was converting into the present Mediterranean Sea, which was being progressively narrowed by the drift of Africa (including Arabia) toward Europe. The Alps started to rise in Europe as the African continent continued to push north into the Eurasian plate. All these continental movements culminated in the Oligocene epoch (34–23 mya). At that time Antarctica was isolated definitively, forming the circumantarctic ocean current and allowing a permanent ice cap to develop. A global cooling occurred during the Late Eocene and the Oligocene, causing a gradual major extinction episode.

The Neogene period covers the past 23 million years, during which modern birds and mammals, including humans, evolved. Continents continued to drift toward their current positions. During the Miocene epoch (23–5.3 mya), global mountain building continued to take place, raising the western American cordilleras, the Alps, and the Himalayas. Arabia, which became part of the Africa plate, collided with Eurasia, forming the Caucasus Mountains, separating the Indian Ocean and the Mediterranean Sea and closing off the remnants of the old Tethys. The rise of mountains in the western Mediterranean (the Betic Cordillera in southern Spain and the Rif Mountain in northern Morocco), combined with the global drop in sea level due to formation of Antarctica ice cap, caused what is known as the Messinian Salinity Crisis in the Mediterranean Sea approximately 6 mya. The Mediterranean Sea has dried up several times due to repeated closings of the old Straits of Gibraltar, thus forming enormous evaporative deposits throughout the Mediterranean. When the present Strait of Gibraltar eventually opened, the Atlantic would have poured a vast volume of water into the dry Mediterranean basin in a gigantic waterfall much more than 1,000 meters high and far more powerful than Niagara Falls. During the Pliocene epoch (5.3–1.8 mya), South America and North America joined, creating the Isthmus of Panama. This tectonic episode had major consequences for global temperatures: warm equatorial ocean currents were cut off and the climate became cooler and drier, resulting in the formation of the Arctic ice cap 2 mya. Both the Antarctic and the Arctic became much colder. During the Pleistocene epoch (1.8 million to 11,500 years ago), the modern continents were essentially at their present positions, initiating repeated glacial cycles. Four major glacial episodes have been identified (usually called Günz, Mindel, Riss, and Würm in the Alps; Nebraskan, Kansan, Illinoian, and Wisconsin in North America; Weichsel or Vistula, Saale, Elster, and Menapian in Northern Europe; and Devensian, Wolstonian, Anglian, and Beestonian in Britain), separated by interglacial episodes. The Holocene is the geological epoch that spans the last 11,500 years of the earth's history (from 9,500 BCE to the present), starting with the retreat of the Pleistocene glaciers. It was preceded by the Younger Dryas cold period, the final part of the Pleistocene, and it is characterized by global warming. The optimum climate of the Holocene has favored the flourishing of human civilizations.

José Antonio Arz

See also Chronostratigraphy; Decay, Radioactive; Earth, Age of; Fossil Record; Geological Column; Geologic Timescale; Hutton, James; Lyell, Charles; Paleontology; Plate Tectonics; Steno, Nicolaus; Stratigraphy; Wegener, Alfred

Further Readings

Grotzinger, J., Jordan T., Press F., & Siever R. (2006). *Understanding Earth* (5th ed.). New York: Freeman.

Poort, J. M., & Carlson, R. J. (2004). *Historical geology: Interpretations and applications* (6th ed.). New York: Prentice Hall.

Stanley, S. M. (2004). *Earth system history* (2nd ed.). New York: Freeman.

Wicander, R., & Monroe J. S. (2003). *Historical geology: Evolution of Earth and life through time* (4th ed.). London: Brooks/Cole.

Web Sites

Paleomap Project: http://scotese.com/earth.htm

GERONTOLOGY

Although research into human aging has been conducted sporadically for many years, it is only in the past several decades that it has gained enough momentum in the social sciences to become an established program of formal academic study—under the title "gerontology." Gerontology is the study of the processes of aging. The term comes from the Greek *γεροντοσ*, an old man and *λογοσ*, word, science, or study. A closely related term is *geriatrics*, the medical care of the elderly, coined by Ignatius Leo Nascher in 1909 from the Greek *γεροντοσ* and *γστροσ*, to cure.

Historical Perspective

The care of aging and injured individuals has a long record in the (pre)history of humankind. Some of the earliest evidence for such care comes from Neanderthal remains in Europe and the Middle East: for example the La Chappelle-aux-Saints fossils in France and those from Shanidar Cave in Iraq. In the case of the individual ("the Old Man") from La Chappelle-aux-Saints (the type specimen of *Homo neanderthalensis*), the bones indicate an individual who suffered from arthritis of the jaws, spine, and legs. It is unlikely that such an individual would have been capable of food procurement in a Paleolithic society of the type supposed for Neanderthals. The only way such an individual could survive to a relatively advanced age was through the care of others.

Ralph Solecki excavated the remains of an individual at Shanidar Cave in Iraq who appears to have survived the amputation of his right arm. He had been injured and possibly blinded in the left eye. He also showed a healed injury to the right parietal bone. Although he survived these injuries, he was eventually killed by a slab of falling limestone, while he stood upright. It is not unreasonable, therefore, to assume that there was some sort of care for such older (by Neanderthal standards) individuals who most likely would have been unable to forage for themselves.

In a brief history of gerontology, Joseph Freeman delineates nine periods in the "scientific" study of old age in the past 5,000 years; that is, since the dawn of recorded history. These periods include the Archaic period, which extended from the emergence of writing to the development of early civilizations. His second period lasted from the efflorescence of Mesopotamian, Biblical, and Egyptian cultures to the advent of Minoan and Greek civilizations and includes descriptions of the aged and codes of behavior toward them. The third period, the Greco-Roman, included Hippocrates and Aristotle among the Greeks and Galen and Cicero in Rome. During this period these individuals began to assemble a body of literature pertaining to the treatment of ailments that attend aging. The general belief was that "innate heat" from the heart began to diminish over time, and that this "cooling" led to aging.

During Freeman's Judeo-Arabic period, Moses Maimonides, Arnoldus de Villa Nova, and Avicenna described the differences between young and old and recommended regimens for older people, including less frequent blood-letting—only once per year for septuagenarians.

The period of European Emergence brought with it the first book specifically devoted to geriatrics: Gabrielle Zerbi's *Gerontocomia,* published in 1489. Roger Bacon argued reasonably in *De retardanis senectutis accedentibus, et de sensibus conservandis,* printed in Oxford in 1590 (300 years after Bacon's death) that if people were as zealous in their efforts to conserve health as to restore it, they would lead longer lives free of disease. Bacon recommended the use of magnifying glasses for older people with poor vision shortly before the appearance, in Italy, of vision-correcting eyeglasses.

A Renaissance publication in 1534, *The Castel of Healthe,* by Sir Thomas Elyot, advised the elderly to follow a prudent diet consisting of a number of small meals per day rather than a few large ones. Sir John Floyer wrote in *Medicina Gerocomica* that old age is the result of an imbalance of bodily humors, with a preponderance of cold and dry.

During the seventh period, Benjamin Franklin published translations of earlier works on senescence and expressed interest in the variety of ways with which to increase health and stave off old age. Shortly after Franklin's death, Christoph Wilhelm Hufeland published *The Art of Prolonging the Life of Man* in which he prescribed a *Makrobiotik* approach.

In 1804, Sir John Sinclair translated classical works, reviewed statistics, and summarized much of the previous work on aging. Later, Sir Anthony Carlisle published *Essay on the Disorders of Old Age* and advised young people to take care of themselves early in life if they wished to secure longevity. Carl Canstatt wrote on general theories of aging in 1839 and hypothesized that the death of cells led to irreplaceable tissue death. In Paris, Jean-Martin Charcot lectured that the diseases that accompany aging have a latency period, which, it appears, some such as cancers do. The primary concern still was to look upon senescence as pathology, rather than as a normal continuance of the lifelong aging process, but by the 1890s C. A. Stephens opened a laboratory in Maine for the study of old age and began a magazine called *Long Life.*

Although quite a bit of data on biological and environmental factors in longevity were generated during the 1940s, advances in gerontology have accelerated since 1950 because of advances in medicine, emphasizing physical and mental health in later years. Demographic and economic studies have been concerned with the effects changes in the age structure would have on the social and financial sectors. For example, what changes would occur in taxation structures should the mandatory retirement age change from 65 to 70 or to 60 years?

Recent Theories and Research

A major impetus to aging studies was disengagement theory introduced by Elaine Cumming and W. E. Henry in *Growing Old: The Process of Disengagement,* based upon the Kansas City Study of Adult Life. Previously, much of the literature on aging was material gathered from research on other topics, or it was concerned with cataloging characteristics of the aged. Cumming and Henry proposed that elderly people around the world perform mutual disengagement from their societies. Disengagement was supposed to be an inevitable developmental event in the life cycle, although its start and pattern might vary from culture to culture. This was thought to lead to (or at least correlate with) passivity in later life.

However, David L. Gutmann's study of the highland Druze demonstrated that disengagement is not inevitable, and passivity does not necessarily accompany it. Gutmann felt that in traditional folk societies with strong religious orientations, older people's passivity may be a central and necessary component of their *new* engagement with social roles and traditions associated with their new status. What is taken for passivity in modern societies is a move toward religious engagement—mastery of the supernatural realm—in traditional societies. According to Gutmann, what Cumming and Henry found in American society is not universal, but is an "artifact of secular society" that rejects a normative order dependent upon older persons' traditional, moral roles after passing the "parental and productive life periods." Gutmann argues that this shift to Passive and Magical Mastery does not necessarily lead to disengagement and death, but to social rebirth through the religious role that turns destructive passivity in older persons into vehicles of social power.

Activity theory is an alternative to disengagement based upon the observation that some people not only do not *choose* to disengage, but also that they *do not* disengage. Activity theory, or "reengagement," like disengagement theory, is concerned with the relationship between levels of activity and psychological health in the last trimester of life. Activity theory proposes that there may be a natural tendency for older people to associate with other persons and to be active in community affairs. Contemporary retirement practices block this natural inclination and may lead to poor adjustment. By allowing continued engagement in community activities, or by prompting reengagement in society, older persons may maintain (or regain) psychological health. Robert Havighurst noted that for the three dimensions of aging studied by the Kansas City group, activity, satisfaction, and personality, it is personality that is the key factor in patterns of aging and life satisfaction. He concluded that both activity and disengagement theories are unsatisfactory, as they do not deal with personality differences within the older population.

Continuity theorists, such as Robert Atchley, feel that a simple relationship between activity levels and psychological health is insufficient to explain social aging. Continuity theorists propose that one's levels of engagement and psychological well-being are the results of lifelong patterns of activity or inactivity and psychological health or ill health. Linda George argues that continuity theory has value not only for historical interest, but because it accounts for individual differences—something that previous theories did not. George tested the impact of personality and social status variables upon levels of activity and psychological well-being in 380 white males and females aged 50 to 75 years to determine if one may predict continuity across life stages. She found that, contrary to disengagement and activity theories, there was only a weak correlation between activity levels and psychological well-being; that is to say, different variables predict the two phenomena. Personality factors were better predictors of psychological well-being than were social status factors, and social status factors better predicted activity.

Thus, one may conclude that neither disengagement nor activity and psychological well-being are outgrowths of the aging process. Rather, they are the products of long-term personality or social processes that are effective on an individual basis. To put it more simply, some older people have high or low activity levels because they always have had high or low activity levels; the same may be said for psychological well-being.

After the Second World War, modernization theorists and age stratification theorists began to view societies developmentally with regard to urbanization and industrialization. People lived longer in more developed cultures, which resulted in more older people surviving. This resulted in competition for jobs, forced retirement, and/or disengagement, as technological advances caused the need for relocation and the loss of their status as older people became more common and were no longer something "special." One possible result of this is a collapse of traditional family structures, according to Donald Cowgill and Lowell Holmes.

Matilda White Riley's theory of age stratification from the 1970s looked at the relationship between social structure and age across the life course. Those individuals who are born in the same age cohort (roughly similar to age sets in many traditional societies) will have many experiences in common. These experiences will be different for each generation even when the experiences have been generated by the same events, for example when a population has experienced a war or natural disaster: Children are impacted differently than are adults; older adults are impacted differently than are younger adults. This theory allows us to view the different ways in which age cohorts may react to life events, and how such life events affect the structure of the society.

By the 1990s aging research had shifted to family gerontology, according to Katherine R. Allen, Rosemary Blieszner, and Karen A. Roberto. Allen and her colleagues examined 908 articles and 30 books on family gerontology published in the 1990s and discovered a shift in the field of aging individuals with regard to family social relations. By far the largest focus of the articles was on caregiving (32.6%); the next largest category was social support and social networks at only 13.7%, followed by parent–adult child relations at 10.1%, and marital status transitions at 9.5%. Further, they found that the students of later life are developing an appreciation for pluralism and resilience strengthened by the incorporation of feminist and life course approaches. Likewise, there has been greater sophistication in the use of longitudinal data, especially as the focus has shifted from exclusively studying the aging individual to the individual within the familial matrix. This, they say, forces a chronological variable into the research, especially as there is no agreed-upon chronological definition of middle age (e.g., becoming a parent or grandparent) and major life events no longer can be predicted even by gender). Indeed, the family within which one ages may be a chosen family rather than a biological one, through adoption or choosing a favorite niece or nephew to become analogous to a daughter or son. The feminist approach began to focus on female intergenerational dyads (mother–daughter ties); daughters are three times more likely to give care to aging parents than are sons, even if the sons live closer. Ironically, daughters-in-law are more likely to be the caregivers than are sons, as demonstrated by Kathleen Lynch's and Eithne McLaughlin's observations on caring labor and love labor in modern Ireland.

For Allen and colleagues, the life course perspective is the major theoretical advance in gerontology during the 1990s. Life course studies focusing on historical and social processes and their impacts upon individuals answer two important considerations: how individuals change over time and how their changes are linked to other members of their families. Longitudinal approaches increasingly are employing narrative methods to discover subjective meanings that individuals use to construct their lives. Through the narrative interplay of historical events and self-perception we can create our "selves" as we need them to be.

Michael J. Simonton

See also Dying and Death; Longevity; Malthus, Thomas

Further Readings

Aldrich, C. K. (1964). Personality factors and mortality in the relocation of the aged. *The Gerontologist, 4(2)*, 92–93.

Aldrich, C. K.,& Mendkoff, E. (1963). Relocation of the aged and disabled: A mortality study. *Journal of the American Geriatric Society, 11*, 185–194.

Allen, K. R., Blieszner, R., & Roberto, K. A. (2000). Families in the middle and later years: A review and critique of research in the 1990s. *Journal of Marriage and the Family, 62*, 911–926.

Botwinick, J. (1978). *Aging and behavior: A comprehensive integration of research findings*. New York: Springer.

Bromley, D. B. (1974). *The psychology of human ageing*. Harmondsworth, UK: Penguin.

Cumming, E., & Henry, W. E. (1961). *Growing old: The process of disengagement*. New York: Basic Books.

Freeman, J. T. (1979). *Aging: Its history and literature*. New York: Human Sciences Press.

George, L. K. (1978). The impact of personality and social status factors upon levels of activity and psychological well-being. *Journal of Gerontology, 33*(6), 84.0–84.7.

Gormley, M. (1964, Winter). The care of the aged in Ireland. *Administration, 12*, 297–324.

Gutmann, D. L. (1974). Alternatives to disengagement: The old men of the highland Druze. In R. A. LeVine (Ed.), *Culture and personality: Contemporary readings* (pp. 232–245). Chicago: Aldine.

Havighurst, R. J. (1968). Personality and patterns of aging. *The Gerontologist, 8*(2, II), 20–23.

Loustaunau, M. (2006). *Aging in society*. San Diego, CA: National Social Science Press.

Lynch, K., & McLaughlin, E. (1995). Caring labour and love labour. In P. Clancy, S. Drudy, K. Lynch, & L. O'Dowd (Eds.), *Irish society: Sociological perspectives* (pp. 250–292). Dublin: Institute of Public Administration.

Streib, G. F. (1968). Old age in Ireland: Demographic and sociological aspects. *The Gerontologist, 8*(4), 227–235.

Streib, G. F. (1970). Farmers and urbanites: Attitudes toward intergenerational relations in Ireland. *Rural Sociology, 35*(1), 26–39.

Gestation Period

The gestation period is the length of time, in viviparous (live-bearing) animals) needed for a fetus to develop inside the body of its mother. Gestation begins with conception (the implantation of the embryo in the uterus) and ends with labor and birth.

The length of the gestational period is species-specific, correlating somewhat with the body size for any given species, with larger species generally having a more prolonged gestational period. The gestational period is therefore quite short for very small mammals (21 days for mice), is longer for medium-sized animals (approximately 60 days for cats, dogs, and raccoons), and is quite lengthy for such large animals as rhinoceros (16 months) or whales (18 months). The shortest gestation period is that of the opossum (12 days), and the longest known gestation period for any animal is that of the elephant (18–22 months). For humans, the gestation period is normally 40 weeks (280 days), measured from the first day of the mother's last menstrual period.

Neonate Type

Although the length of the gestation period is generally correlated to an animal's body size, it is also determined by the state of development at birth for offspring of that species. Among mammals, there are three different groups that can be defined by the level of gestational development of their offspring at birth, and these groupings have proven (through fossil evidence) to be remarkably consistent throughout evolutionary history. Marsupials, which have the shortest gestation periods of all animals, are considered "superaltricial," with offspring that are very underdeveloped at birth and usually crawl into a pouch of some kind to continue their development before being able to survive detached from their mother's body. A second group of mammals includes insectivores, carnivores, and rodents. They have somewhat longer gestational periods and give birth to "altricial" offspring, which are generally poorly developed and typically must develop in a nest or burrow for some period after birth so that they can continue to grow and learn how to move around independently. Altricial offspring are still underdeveloped at birth, being usually hairless and with their eyes and ears sealed off by membranes. Altricial animals bear multiple offspring in litters and tend to have shorter gestation periods, reflecting the fact that the mother has limited space in her womb for completing development prior to birth. In contrast, a third group of mammals (primates, whales, dolphins, and hoofed mammals) has very long gestational periods and usually gives birth to only one or two well-developed "precocial" offspring that are relatively active soon after birth; they are born with fully developed vision and hearing, and are protected by fur and hair. This type of gestation is most common in animals that live out in the open, exposed to the elements, and are unable to conceal their young in burrows or caves.

Each main mammal group is therefore characterized by the type of neonate (altricial, superaltricial, or precocial) that its members produce, and by the length of gestational period typical of that group. Insectivores, tree shrews, carnivores, and many rodents have short gestational periods and bear altricial offspring; hoofed mammals, hyraxes, elephants, cetaceans (whales), pinnipeds (walruses, seals, and sea lions), primates, and hystricomorph rodents (Old World porcupines) bear precocial offspring after a fairly long gestational period. In fact, phylogenetic groupings have more impact on the length of the gestational period than animal body size does; the gestation period for an animal that bears precocial offspring will typically be three times longer than that for an animal of the same body size that bears altricial offspring.

Gestational Age

The gestational age of a fetus is a measure of how far along it is in the gestational period. Since the exact time of conception is not always known, the beginning of gestation is usually dated from a particular point in the reproductive cycle of any given species, or is estimated based on the size of the developing fetus. In humans, for example, gestational age is projected by a measure of time elapsed since the first day of the mother's last menstrual period, or is calculated with an ultrasound examination (usually administered between the 8th and 18th weeks of pregnancy), which uses the size of the developing fetus as the most accurate measure for dating conception and projecting when full development and birth will occur.

The gestational period is divided into three major phases: the preimplantation phase (from fertilization to implantation in the mother's womb), an embryonic phase (from implantation to the formation of recognizable organs), and a fetal phase (from organ formation to birth). These phases are called trimesters. In humans, each phase lasts about 3 months; the first trimester lasts from weeks 1 through 12, the second from weeks 13 through 27, and the third from weeks 28 to 40. In some mammals, the gestation period is extended to include an additional phase known as delayed implantation, or embryonic diapause, that precedes the embryonic phase. In these instances, the fertilized egg develops into a blastocyst (forming a hollow ball of cells), but then it stops developing and remains free in the womb rather than implanting in the wall of the womb (the usual next step in gestational development for most mammals). Delayed implantation is a strategy used by animals to delay the birth of a fetus until such external variables as food or weather are favorable for the survival of offspring, or in some cases as a way of holding potential offspring in reserve to immediately replace older offspring that do not survive long after birth. Other factors that influence the length of the gestational period in minor ways include gender (males take a few days longer than females), the number of young (single young take a few days longer than multiples), and heredity through the maternal line. In general, though, there is very little variation in the length of the gestational period within the same species, that is, 4% or less variability.

Helen Theresa Salmon

See also Birthrates, Human; Clocks, Biological; Duration; Evolution, Organic; Fertility Cycle; Gestation Period; Life Cycle; Maturation

Further Readings

Assali, N. S. (1968). *Biology of gestation: Vols. 1 and 2.* Burlington, MA: Academic Press.

Davies, J. (1960). *Survey of research in gestation and the developmental sciences.* Baltimore, MD: Williams & Wilkins.

Gilbert, S. F. (2006). *Developmental biology.* Sunderland, MA: Sinauer Associates.

Grzimek, B., Schlager, N., & Olendorf, D. (2003). Life history and reproduction. In D. G. Kleiman, V. Geist, & M. C. McDade (Eds.), *Grzimek's animal life encyclopedia* (Vol. 12, pp. 95–96). Farmington Hills, MI: Thomson Gale.

Gibran, Kahlil (1883–1931)

The writings of Kahlil Gibran, the popular 20th-century poet, philosopher, and artist, have been admired by millions and translated into over a dozen languages. By the time of his death, he had established a reputation in both the English and Arabic-speaking worlds. He never won critical or scholarly acclaim, but his style prompted the coining of a literary term, *Gibranism*, to describe the poised and poetic quality of his prose, which expressed an intense need for self-reflection and self-fulfillment. Throughout his life, he was primarily religiously nondenominational, seeking wisdom and truth from within instead. His writings reflect a preoccupation with mysteries and with deep spiritual meaning; he describes and reflects on his encounters with, and understandings of, godliness and addresses perennial themes such as Time.

Gibran was born in Bsherri (Bechari), Lebanon, in 1883. When he was 12, his family moved to the United States and settled in Boston, Massachusetts. After 2 years, Gibran went to Paris to study art under Auguste Rodin, at the famous Ecole des Beaux Arts. In Paris, Gibran also began to write plays and short stories in the Arabic language.

In 1904, he moved back to the United States and spent time living in both Boston and New York. While working in New York, he wrote his most popular literary works, *The Prophet* (1923), and *Jesus the Son of Man: His Words and His*

Deeds as Told and Recorded by Those Who Knew Him (1928). *The Prophet* included a series of poems on topics such as love, marriage, and children, and included a passage titled *On Time,* in which Gibran says time, like love, is undivided and spaceless.

Gibran's ability to paint and draw enabled him to illustrate many of his own books. Critics often cited the influences biblical literature had on Gibran's writings, and his lyrical passages are commonly read at baptisms, funerals, and weddings throughout the Western world.

Some of his other works include *Nymphs in the Valley* (1948), *Tears and Laughter* (1949), *The Madman: His Parables and Prose Poems* (1918), *Spirits Rebellious* (1946), *The Wanderer: His Parables and His Sayings* (1932), and *The Garden of the Prophet* (1933). His relationship with Mary Haskell, a teacher in Boston, is captured in *Beloved Prophet: The Love Letters of Kahlil Gibran and Mary Haskell, and her Private Journal* (1972).

In 1999, the Arab American Institute founded an annual *Kahlil Gibran Spirit of Humanity Award* that recognizes individuals, organizations, corporations, and communities whose actions demonstrate leadership in supporting and promoting coexistence, diversity, cultural interaction, inclusion, and democratic and humanitarian values across racial, ethnic, and religious groups.

Debra Lucas

See also Bible and Time; Humanism; Poetry; Qur'an

Further Readings

Waterfield, R. (1998). *Prophet: The life and times of Kahlil Gibran.* New York: St. Martin's.

Ginkgo Trees

Ginkgo biloba is the last surviving species of Ginkgophyta, a phylum of long-lived, vascular plants that evolved during the Permian period some 250 million years ago. Ginkgo trees, once thought extinct, are sometimes called living fossils due to the strong resemblance between the modern form and ancient remains preserved in the geologic record. The popularity of ornamental ginkgo trees in Asia has led to their widespread cultivation and dissemination.

The ginkgo tree is one of the best-known examples of a living fossil. This ancient tree represents a primitive family of trees common 160 million years ago in China. The ginkgo thrives in well-drained, loamy soil and is particularly resistant to modern pollutants, wind, pests, and disease.

Source: Michael Pettigrew/iStockphoto.

Ginkgos are medium-large trees that can live for hundreds of years, attaining a height of 30 meters and a trunk diameter of 2 meters. A few exceptional specimens may exceed 1,000 years in age and proportions far in excess of these. They comprise a unique division of the gymnosperms, an important seed-bearing plant group that includes cycads, conifers, and ferns. Ginkgos are easily recognized by their leaves, which look like a two-lobed fan (hence *biloba*).

Ginkgos are dioecious plants, meaning that male and female reproductive organs are borne on separate trees. Male trees form a small cone that produces motile sperm. On the female tree, fertilized seeds mature into a tan, fleshy fruit 25–40 millimeters long. These fruits are rich in butyric acid, which when crushed can release an odor likened to rancid butter. These sorts of naturally occurring chemicals in the tree are thought to have evolved for better resistance to insects and bacteria, or even to help ward off herbivores.

Ginkgophyta trees arose during the Permian, approximately 250 million years ago, and to the present day have left numerous fossils of their distinctive leaves, stems, seeds, and wood. After their emergence the group continued to spread and diversify, surviving the relatively arid Triassic to form part of the great forests of the Jurassic when the genus *Ginkgo* itself first evolved—approximately 170 million years ago. Since the end of the Cretaceous, the ginkgo has suffered from a steady constriction of its range, abundance, and diversity. The precise reasons are unknown, but today the ginkgo is all but extinct in the wild; a few small native populations remain in the Zhejiang province of China, scattered in elevated broadleaved forests.

In some ways, the ginkgo has been reclaiming its former territory. In China the trees have long been cultivated in the gardens of Buddhist temples (and occasionally as a food source). From there they spread to Japan, and in the 17th century to Europe and other Western countries. In modern cities all over the world, they have become a shade tree of choice. The ginkgo is well suited to the urban landscape, as it requires minimal maintenance. It also has a high tolerance for pollution and resistance against disease.

Ginkgo seeds are used in traditional Asian medicine, and the cooked seeds can even be eaten. More recently, ginkgo leaf extract has been packaged to treat a variety of ailments. This utilizes the plant's naturally occurring concentration of flavonoids, a group of antioxidants found in certain fruits and vegetables. Although scientific research has not yet supported the health benefits attributed to ginkgo, some promising lines of pharmaceutical research are investigating its effects on dementia and Alzheimer's disease.

Ginkgos are true icons of long-term survival; their modern form is quite similar to that of fossils 250 million years old. The tree has survived trampling hordes of herbivorous dinosaurs and major environmental catastrophes. Its extremely limited native populations, however, testify to enormous ecological pressure—the ginkgo has been struggling to survive in the present era. In the wild their relatively inefficient reproductive system, slow regeneration, and lack of diversity have exacerbated the damage from habitat loss. *Ginkgo biloba* faces a very high risk of extinction in the wild; it is currently classified as an endangered species. After successfully adapting to nature's challenges over hundreds of millions of years, let us hope that ginkgo remains a *living* fossil rather than a literal one.

Mark James Thompson

See also Coelacanths; Dinosaurs; Evolution, Organic; Extinction and Evolution; Fossils, Living

Further Readings

Hori, T., Ridge, R. W., Tulecke, W., Del Tredici, P., Trémouillaux-Guiller, J., & Tobe, H. (Eds.). (1997). *Ginkgo biloba—A global treasure.* Tokyo: Springer-Verlag.

Sun, W. (1998). *Ginkgo biloba.* In *2006 IUCN red list of threatened species.* Cambridge, UK: IUCN.

Tidwell, W. D. (1998). *Common fossil plants of western North America.* Washington, DC, and London: Smithsonian Institution Press.

Glaciers

Glaciers have been an important feature of the earth's landscape during the present Cenozoic era, which spans about 65 to 70 million years. Cooling near the end of the era began in the Pliocene epoch, and mountain glaciation in the western United States occurred somewhere between 1 and 2 million years ago. The Cenozoic era is divided into the Tertiary and Quaternary periods, with the most recent glaciation occurring during the Quaternary period, which covers approximately one million years before the present (BP). This period is divided into the Pleistocene and Holocene or Recent epochs. During the Pleistocene, four major glacial advances and three interglacial stages occurred. The Holocene or Recent epoch spans a period of approximately 11,000 years BP. During this time, the fourth interglacial stage or a temperature increase has been occurring, including melting ice and a rise in sea level. This may last for tens of thousands of years, and may be followed by another future period of cooling and the growth of glaciers.

Prior to this present age of Pleistocene glaciers, there were at least five other known glacier periods, dating back to two occurrences of glaciation in Precambrian times. Toward the end of the Permian period glaciers appeared again, and they were rather widespread. During the early and late portions of the Cretaceous period glaciers appeared once more. The occurrence of glaciers is evidently a cyclical phenomenon, but the precise cause of their appearance is not known.

During the Pleistocene glaciers were widespread, covering both polar regions and extending outward a considerable distance, covering approximately 30% of the earth's total land area. In the present Holocene epoch, the earth is witnessing a global warming as well as a global dimming: Both of these appear to be antagonized by increased human interference. All glaciers are receding, and have been for thousands of years, but at seemingly a much faster rate now than when they were flowing forth. Consequently, the sea level rose at an average rate of about 6 inches per 100 years during the Holocene. A noticeable increase has been noted during the last several hundred years, however, and this is expected to continue.

In the 19th century, Louis Agassiz, like James Hutton and other scientists in the century before, asserted that the large boulders in the valleys were not placed there by Noah's flood, but were once plucked by erosion out of the advancing ice, sometimes a mile or more deep, and then deposited by a receding ice flow. Agassiz's views were initially met with skepticism and hostility by most of the audience at a meeting of the Swiss Society of Natural Sciences; widespread acceptance came only gradually, later on.

Glaciers are commonly defined as either continental or mountain types. Continental glaciers begin in an area with polar temperatures and extremely large accumulations of snow. There were three areas in the northern hemisphere where this most strikingly occurred during the Pleistocene; in Europe over the northern Baltic Sea and much of Norway, in North America over Hudson Bay and north central Canada, and over northern Siberia. Not to be overlooked are the general expanse of the Arctic icecap including Greenland and other parts of the northern hemisphere, as well as the massive Antarctic continent in the southern hemisphere. These continental glaciers appear to be located on or very near shields consisting of Precambrian igneous and metamorphic rock. With regard to mountain glaciers, the Alps, Andes, Rockies, and the vast Himalayan chain are all fine examples of mountainous glaciated terrains. Also, snow and ice occur on isolated mountain peaks in Hawaii, Japan, and Kenya, to mention a few.

Continental Glaciers

Continental glaciers begin with a decrease in temperature and an accumulation of snow, over a long period, tens of thousands of years. The snow is compacted by its own weight. Eventually, the snow at the deepest level begins to form ice. Glacial or moving ice is subjected to immense pressure and is transformed molecularly into an extremely hard metamorphic rock. Subsequent snows provide added weight and pressure, and the deepest ice begins to move outward by the force of gravity into lobes that move more rapidly along lines of least resistance. The lobes are guided by original terrain features such as mountains, hills, and plains, and the relative resistance of the various types of earth materials and rock structures. The magnitude of accumulation at the glacier's source is believed to be more than 10,000 feet in some cases, and the outward movement is estimated at more than a thousand miles, perhaps at speeds of 30 miles per year or even more.

Continental glaciers sculpt the land by erosion, carving out a variety of landforms. The moving ice deposits the carried sedimentary materials as depositional features upon melting. Some of the eroded features that can be seen today are due to abrasion by rocks carried along by the ice, leaving behind striations, grooves, and polished surfaces. Also left behind are gouged-out valleys that were deeper and wider areas of softer rock, forming depressions such as the Great Lakes, *cuestas* or edges of land surfaces formed by resistant rock strata such as that of the Niagara Cuesta including Niagara Falls, and smaller hills known as *roche moutonée* formed by resistant rock. The materials are deposited in recessional and terminal moraines, or in ridge-like terrains that are accumulated during stand-still periods. As the ice recedes in a sometimes erratic manner, it leaves behind landforms such as ground moraines consisting of glacial till, also known as tillite, along with such features as kames and kettles, drumlins and eskers, resulting in a hummocky topography. Braided streams also result from retreating glaciers, as well as "misfit" valleys where the valley width is much narrower or wider than it should be for the width of the stream currently flowing through it. For example, the upper Mississippi, the Des Moines, and some of the other streams in the midwestern United States are considered misfit streams. In general, the landscape where continental glaciation has taken place is quite depressed compared to unglaciated areas, and exceedingly complex as compared to stream, eolian, or coastal landscapes. This is, in part, due to a yet-to-be-integrated drainage pattern and the periglacial landscape of outwash materials.

Mountain Glaciers

Mountain glaciers begin with a cold temperature and an accumulation of precipitation at the highest elevations. Over a long period of time, as with continental glaciers, snow is compacted, eventually forming ice. The glacial ice begins to move downward by gravity along lines of least resistance, forming tongues, usually in former stream valleys. In many cases these tongues of ice find their way to sea level, and when the ice recedes the fjords remain behind as remnants.

Mountain glaciers sculpt valleys by eroding various types and hardnesses of rock. Then the melting ice leaves the sculpted materials in depositional features. Eroded features include *cirques* or bowl-shaped indentations near the peaks, or matterhorns where the glaciers begin, the knife-like ridges between the valleys called *arêtes*, and low places in the ridges called *cols*. Valleys that were originally V-shaped due to stream action become U-shaped due to ice action, and some have depressions called *tarns*. Retreating glaciers deposit terminal and recessional moraines during stillstands, as well as lateral moraines along their edges as they retreat and medial moraines where two lateral moraines have merged. Glacial till carried from the eroded areas is dropped by the melting ice to form braided streams and outwash plains. In general, mountainous terrain that has been glaciated appears much more concise, complex, and abrupt when compared to unglaciated mountains. For example, the Matterhorn and its descending glaciers in the Swiss Alps are much more sharply unique in appearance as compared to Mount LeConte and other unglaciated peaks and their descending streams in the Great Smoky Mountains of the United States.

Time, distance, and speed play their part in sculpting the earth. Glaciers, like wind, streams, waves, and currents, flow. How fast does the ice flow? How far does the ice flow? How long does the ice take to flow? Winds move much faster than ice and travel the farthest, across land and sea, while streams are restricted to the land, and waves and currents are restricted to the sea. Glaciers move slowly across the land and out into the sea. If they didn't, there would be no icebergs today!

Glacial Movement and Climate Change

Glaciers, at their maximum extent during the Pleistocene, were widespread, covering more than an estimated 45 million square kilometers. Glacial ice moves much like a stream, in that the fastest motion is found in the center of the lobe or tongue. That is, the slippage and flowage is considerably slower closer to the rock formations where plucking and erosion by the ice adds friction. This is tempered by variations in rock hardness and degree of slope. Consequently, there is a huge variation in the mean rate of ice advance. Some of the intermittent advances of mountain glaciers with short duration vary between 2.4 and 32.3 meters per day, with advances between 606 and 4,850 meters per year. For long-term advances of mountain glaciers, the mean rate varies between 80 and 128 meters per year over a period of 100 years.

Continental glacier movement is much more difficult if not impossible to measure, as continental glaciers either no longer exist or are in a recessional mode. It is known that four major ice advances and accompanying recessions occurred during the Pleistocene, with their nomenclature varying depending on location and language. But these major advances included minor advances and recessions. For example, seven substages of the Wisconsin glacial stage in the midwestern United States have been observed. In the western hemisphere, numerous lobes descended from the Laurentide of Canada across the plains of the midwestern United States, forming a complex landscape. From radiocarbon dating it is known that ice receded from north-central Iowa less than 3,000 years before the present, and that near the bottom of the Des Moines lobe the oldest drift is older than 40,000 years before present. This is presumed to be during the Wisconsin, which began about 70,000 years before present. The maximum glacial extent during the Wisconsin occurred about 14,000 years before present, and a considerable amount of glacial activity occurring before and after, including loess deposition and the development of the present day paleosols, all occurred during the interglacial stages. It can be assumed that similar circumstances occurred in Europe and Siberia.

Climate fluctuations during the Pleistocene and Holocene, coupled with periglacial effects, are felt around the world, especially eustasy—worldwide change in sea level. The outwash from glaciers is the most obvious periglacial activity, along with the less obvious permafrost. Also, glacial rebound is evident where the deepest ice has melted away, such as Hudson Bay. Then too, there is the migration of vegetation poleward as temperature and moisture changed. Humankind soon followed the retreating glaciers.

Today, scientists worldwide are monitoring climate change, including rising sea levels, something that has been happening since the late 17th century. Ancient shorelines around the world show that when glaciers were near or at their minimum extent, sea level was about 400 feet higher than it is today. If the sea level continues to rise, then it may reach that stage again; how soon no one knows for sure, but it will continue to rise if global warming and global dimming continue. On the other hand, when and if the present interglacial stage ends, the sea level may fall. During the Illinoisan glacial stage when ice was at its maximum, the sea level is estimated to have been 600 feet below the present level, a condition that is difficult for many people even to imagine.

Although glaciers worldwide are receding at the present time, within the next 90,000 years or so they will again expand across large portions of the earth's landmasses, especially those located in the higher latitudes and altitudes.

Richard A. Stephenson

See also Erosion; Geologic Timescale; Global Warming; Ice Ages; Nuclear Winter

Further Readings

Flint, R. F. (1957). *Glacial and Pleistocene geology.* New York: Wiley.

Martini, I. P., et al. (2001). *Principles of glacial geomorphology and geology.* Upper Saddle River, NJ: Prentice Hall.

Weiner, J. (1986). *Planet Earth.* Toronto, ON, Canada: Bantam Books.

Wright, H. E., Jr., & Frey, D. G. (Eds.). (1965). *The quaternary history of the United States.* Princeton, NJ: Princeton University Press.

Globalization

Globalization refers to the close relationship and interdependence of economic, social, political, cultural, and ecological connections of places in our world. It refers to the intensification of interconnectedness on a global scale. With increased means of communication and travel from 20th-century innovations in technology there is no doubt that our world is shrinking. Never before has it been so easy for people, things, capital, and ideas to move around. Further, globalization has reordered the experience of time and space. Because economic and social processes have shrunk, distance and time are less of a constraint on the organization of human activity. Technological and economic advances have eliminated barriers of space by time and have reorganized time to overcome barriers of space. For example, the same media events are covered all over the world, making it possible for people in, say, the United States and Japan to have access to the same information. Also, this has allowed extremely remote areas of the world to interact with cosmopolitan cities and perhaps with places on the other side of the globe. In other words, time is shortening and space is shrinking. The pace of life is increasing, and the amount of time needed to do things is becoming much shorter. Something that is happening on the other side of the globe can deeply impact people and places vast distances away. Increasingly, economic and social life is becoming standardized around the world.

Globalization in a Historical Context

Forms of globalization were present as early as the 17th century with the rise of the Dutch East India Company, which is said to be the first multinational corporation because it was the first company to share the risk and create joint ownership by offering shares. During the 19th century, liberalization resulted in a rapid growth of international trade and investment between Europe, its colonies, and the United States. This "First Era of Globalization" began to crumble at the onset of World War I and collapsed entirely during the 1920s and 1930s. Arguments and critiques arose against imperialism and the exploitation of developing nations by developed nations.

International trade has grown tremendously since the 1950s; according to estimates, trade has increased from $320 billion to $6.8 trillion. As a result, people from around the world have access to a greater selection of products.

Much of what we are witnessing in the world today concerning globalization started in the 1970s with what is termed a "crisis of overaccumulation" based on the Henry Ford system of mass production. The Ford model, which is based on the mass production of standardized parts, became so successful that Western nations began to overproduce. This resulted in mass layoffs of employees in addition to reducing the demand for products as markets became saturated. There were not enough consumers for all of the goods produced, making corporate profits and government revenues dwindle. Attempts to solve the problem by printing extra money only created inflation. In the midst of this crisis, a new system emerged that depended upon flexibility in labor markets, products, and consumption. During this time new sectors of production emerged and niche markets as well as a greater increase of commercial, technological, and organizational improvements. Labor markets were widened with outsourcing and subcontracting strategies along with the hiring of large numbers of part-time, temporary, and seasonal workers. Markets began to cater to fashion niche markets, recreational activities, and lifestyles, making the pace of economic and social life speed up (popular culture, fashion, sports and leisure, video and children's games). Also, in an attempt to increase profits, many multinational corporations are outsourcing their work. Outsourcing is defined as the delegation of non core activities that the organization was once responsible for but has now hired an outside entity to do. The most common outsourcing jobs include call centers, e-mail services, and payroll. The trend grew during the 1980s as a major way for corporations to save money. Much of the outsourced jobs have gone to China, India, Pakistan, and Vietnam. Advances in technology have reduced the cost of trade tremendously. Satellite communications deployed since the 1970s have increased communications and at a lower cost. Air freight rates have been reduced drastically along with the cost of sea and rail transport. The resulting economic growth also has brought about economic, political, and social disruption. Consequently, the World Trade Organization (WTO) was created to mediate disputes with regard to trading and to create a platform of trading along with GATT (General Agreement on Tariffs and Trade), which was formed to reduce barriers to trade after World War II. In recent times, globalization can be seen in the agreements on trade among countries, such as the one created by NAFTA.

North American Free Trade Agreement

The North American Free Trade Agreement (NAFTA) was created on January 1, 1994, as a free trade agreement between Canada, the United States, and Mexico. The plan was to phase out the majority of tariffs between the countries to encourage trade. Several economists have analyzed the effects of NAFTA and have reported that it has been beneficial because poverty rates have lowered and real income has increased. Over the years trade has increased dramatically between the three countries. Between 1993 and 2004, total trade with U.S. partners increased by 129.3% (Canada 110.1% and Mexico 100.9%).

Social Aspects of Globalization

Interestingly, social life is equally and utterly impacted by these alterations of time and space. Throughout the course of human history, people have interacted with one another mostly face to face. Daily human life is filled with interactions during localized activities. Today, however, more often than not, a person's physical presence is not required. Many tasks can be handled remotely, something made possible by transportation and communication systems. In that sense, social life is altered by the absence of locality, and it fosters relations between people who are too far apart for face-to-face contact. Place becomes less important as the physical setting for social activity, and modern localities are created over long distances. Although place and locale are still significant in daily life, social connections become less dependent on face-to-face interactions, making the local and global intertwined as never before. And, as people continue to live their "local" lives they inevitably are connected globally. Distant events impact local happenings and local developments can have global repercussions that stretch beyond national boundaries.

Culture and Globalization

In the anthropological sense, *culture* refers to a group of individuals—a nation, an ethnicity, a tribe—who share a system of beliefs that define their world and who interpret meaning based on their shared set of beliefs. Traditionally, this definition has been tied to a specific territory. We used terms like "American culture" or "Latin culture" loosely; they brought up images of a specific geographic region or particular country. These images often have certain attributes associated with that particular region. Conversely, in the present context of globalization it is unreasonable to equate a specific culture with a particular geographic area because locales are woven intricately into one another. In other words, the boundaries are blurred. For example, Western cultural objects like clothing, food, and music are now present in nearly all parts of the world. Anthropologists have defined the weakening of the ties between culture and place as "deterritorialization," meaning cultural processes that we once thought of as taking place in a specific region have transcended these specific territorial bounds. This is not to say that cultures are free floating, but there is an uprooting and a reinserting of culture going on due to the flexibility and movement in our current world. Anthropologists have defined cultures as sharing a certain space or "community." A community setting can signify anything from a village to a nation-state. Creating a group of people within a bounded area assumes that this community shares a strong bond within this space. In addition, this space is used to reinforce a worldview and a set of rules to live by, thus creating a certain cultural coherence. In the past, we have understood the "anthropological other" by comparing community with community. Migration is one event that has historically challenged social spaces, such as along the U.S.–Mexico border. Since the 1960s, however, transnational capitalism and migration have strongly influenced remote societies and have changed ideas of culture. For example, in remote villages like Aguililla, Mexico, almost all families had a family member abroad; the local economy depended heavily on the influx of dollars, and many farms were being sustained by family remittances. As large-scale industrial agriculture takes over the Mexican landscape, small-scale subsistence farmers are no longer able to sustain themselves using their traditional practices. The cultivation of maize has been present in Mexico as a primary means of subsistence for 6,000 years. The fact is that people in rural villages can no longer rely solely on local resources to meet their needs. In a globalized world where the local intersects with the global, markets have forced small communities to change and people to migrate. Patterns of transnational migration are prevalent between Mexico and the United States. Most Mexican migrant workers come to work menial jobs in the service sector as dishwashers, janitors, hotel workers, and gardeners. The majority stay in the United States briefly, and those who stay longer remain deeply connected to their families. Their main goal is to finance local dreams by having access to outside funds. Further, the

U.S. economy has become dependent on these workers who occupy society's economic "bottom rung," creating social spaces and communities that transcend geopolitical borders, and finding that their most important kin and friends live hundreds of miles away from them.

Adverse Effects of Globalization

Although trade has increased and indexes for poverty seem to have been lowered according to economists' standards, there are dilemmas with this particular style of free trade. Some would argue that free trade agreements such as NAFTA have damaged and further marginalized large segments of the world's population. It has been demonstrated that although globalization does help to grow economies, it does not do so evenly. The meager share of global income gained by the poorest people in the world has dropped from 2.3% to 1.4% over the past 10 years. While it is true that transportation and communication have made it easier and quicker for people to get around, it is not true for all people in all places. Some people have the social and economic resources to take advantage of these advancements; others do not. Some people have little or no access to electricity, telephone, and Internet service or the opportunity to buy an airplane ticket. Chances are these people will never have this opportunity in their lifetime. Not everybody on the planet is feeling the benefits. There are several places in Latin America and Africa that are literally not even on the map of telecommunications, world trade, and finance. Even so, they can feel the pressure of the market. Such places have few or no circuits connecting them to the outside world other than the communication and transportation that passes over or through them. In this rapid-paced and changing world they are the ones who are forgotten.

Maquiladoras

Maquiladoras are foreign-owned assembly plants in Mexico, primarily located along the border between the United States and Mexico. Maquiladoras first appeared in Mexico in the 1960s; by 1990 there were 2,000 operating with 500,000 workers. These factories produce mainly electronics, clothing, plastics, furniture, appliances, and auto parts. Between 80% and 90% of the goods produced in Mexico are shipped to consumers in the United States. Most people who move from rural settings in search of better opportunities and work in maquiladoras find deplorable living conditions. And, while some company owners have worked to improve the working conditions of their employees, numerous factories misuse their workers. In some instances, reports tell of maquiladora owners subjecting employees to more than 75-hour workweeks without proper breaks. Some 80% of the employees are women because they are fast at repetitive tasks and they get paid a lower wage than men. Minimum wage is $3.40 per day (vs. $6.55 per hour in the U.S.). There have been numerous sexual harassment incidents or the firing of women if they become pregnant. Also, Mexican men resent that women are being hired over men for jobs, leaving women subject to retaliation and violence. In the city of Juarez, alone, more than 200 women who worked for the factories have been murdered. Further, Juarez and other cities lack infrastructure like electricity, sewage, and safe drinking water to sustain the influx of migrants to the cities with factories. For example, since NAFTA the employment rate in Juarez has risen 54%, yet Juarez has no waste treatment facility to treat sewage from the 1.3 million people who live there. Other related problems are that factories set up in these border towns are not held to strict guidelines for environmental protection, which leads to contamination of the air and water supplies. And workers are exposed to dangerous chemicals without any protective wear.

Anti-Globalization or Global Justice Movements

There are many worldwide who do not support the current political and economic system of neoliberal globalization. People involved in global justice movements advocate anarchism, socialism, or social democracy and are against the ruling elites who wish to expand and control world markets for their own gain. These movements call for reforms in the current capitalist model where multinational corporations control 70% of trade in the world market, and for moving toward greater social welfare reforms. Global justice seekers place greater emphasis on equal rights, environmental awareness, and cultural diversity. Alternative ways of growing economies are also present in the movement, where small-scale producers band together to create alternative economies for the goods and services produced. One example is Kallari, an organization of Indigenous Kichwa artisans and farmers in Ecuador who produce coffee, chocolate, handcrafts, and jewelry and sell locally and globally. The Kichwa face strong pressure from the world market through encroachments and a shrinking land base due to outside multinational corporations that drill for oil and mine for gold. Since the Kichwa lacked infrastructure and the support of local and national governments, they needed to organize their own economy. With the assistance of the Jatun Sacha Foundation and its director, Judy Logback, they have been able to enter the world market and to be paid a fair and living wage. In addition, the have opened their own café where they sell their products in Ecuador. Further, they ship goods internationally to niche markets that support environmentally conscious and fairly traded goods. The project has been active for nearly 10 years and has met with great success, with nearly 25 Kichwa communities participating.

Conclusion

For some, globalization is about free trade between countries and the growing of economies that include new job opportunities, while for others it is another way for the rich to prosper at the expense of the poor with little regard for social welfare, human rights, or the environment. Further, globalization has altered space and time for social intercourse and the meaning of community. It remains to be seen how we will redefine our world and our social relationships along with our political economy based upon our values and where we are "located" on the globe.

Luci Latina Fernandes

See also Economics; Evolution, Cultural; Evolution, Social; Materialism; Postmodernism

Further Readings

Fernandes, L. L. (2004). The Kallari Handcraft Cooperative of Ecuador. *Grassroots Economic Organizing: The Newsletter of Democratic Workplaces and Globalization From Below.*

Giddens, A. (1990). *The consequences of modernity.* Stanford, CA: Stanford University Press.

Giddens, A. (1991). *Modernity and self identity.* Cambridge, UK: Polity Press.

Harvey, D. (1989). *The condition of postmodernity.* Oxford, UK: Blackwell.

Inda, J. X., & Rosaldo, R. (Eds.). (2002). *Anthropology of globalization.* Malden, MA: Blackwell.

Rouse, R. (2002). The social space of postmodernism. In *The anthropology of globalization.* Malden, MA: Blackwell.

Global Warming

Even though the global temperature has been changing for millions of years, this is the first century in which we are seeing global changes occur quickly relative to the time that human and animal life activity has been documented. Human-induced changes to global temperature are now evident, whereas the changes in the past were seen as natural phenomena. For example, in the preindustrial era the concentration of CO_2 measured 280 parts per million; by 2005, that had increased to 382 parts per million.

Despite mounting empirical evidence of global temperature change over the past few centuries, it is only relatively recently that scientists have taken note of global temperature changes and publicized them. The 1970s was the first decade in which the general public began to take note of environmental issues. In the United States this occurred through the efforts of environmentalists and scientists who pressured government to institute sweeping changes to U.S. environmental policy. Subsequently, more than 15 national acts of environmental legislation have focused on a range of issues from clean water to endangered species. These new forms of environmental legislation did not, however, take into account increases in fossil fuel emissions that occur with increased global dependence on manufacturing, energy generation, and automobile use. The U.S.

Environmental Protection Agency, created in 1970, did not provide the U.S. auto industry with fuel emission or miles-per-gallon standards, which eventually resulted in millions of tons of carbon entering the atmosphere every year. By not recognizing this serious emission problem, policymakers did not acknowledge changes in potential carbon loads to the atmosphere. Scientists now estimate that global warming will continue, despite our actions to slow the changes. There is concern that we will quickly near what is said to be the ultimate carbon threshold of 480 parts per million in our planet's atmosphere.

The science behind global climate change is complicated. It takes time to establish patterns and understanding of the problem. The U.S. Congress heard first testimony on global warming in 1973; following that, an international campaign was launched to explain the future dangers if such a problem were to be ignored. Scientists predicted global catastrophes such as the melting of the polar ice caps, coastal flooding, and widespread migration of affected populations. We have finally begun to acknowledge the connection between fossil fuel emissions and increased global temperature. However, there remains ongoing debate over whether the problem exists as fact or is a theory.

Increased media coverage of global warming has been due in part to former Vice President Al Gore's 3-decade-long research and investigation into climate change, which culminated in the Academy Award-winning documentary, *An Inconvenient Truth,* a national theatrical film release; a book of the same title; and an international lecture series. Public awareness of this issue has increased, and individual and community efforts to help reduce the impact of global climate change are more widespread. It is not yet clear to what extent social behavior is following public sentiment that now supports societal action to curb the impacts of global climate change. Longitudinal, empirical research will be needed to determine these outcomes.

It has been projected that even if we decrease our carbon output now, it would have very little impact on current and future climate change trends. Notwithstanding, action is still encouraged in order not to escalate the problem. It has taken decades of data collection to determine if a global climate change pattern even exists. Even given this evidence, some still question the validity of the global climate change claims and data. Those who doubt the conclusions argue that longitudinal evidence has not shown that global warming is occurring, let alone creating serious environmental harm. This position, opponents claim, is being fueled by those who are closely allied with the fossil fuel industry and other manufacturers who stand to lose money if their primary source of business is replaced with less environmentally damaging alternatives. Proponents of alternative fuels have been encouraging these companies to invest in alternative, environmentally friendly technology, but with limited success to date.

Much of this debate is centered on time. On the one hand, advocates who support the need to recognize global warming argue that time is of the essence. They advocate for change now, given the likelihood that a planetary overload of carbon will reach dangerous levels in the next 20 years. In 2007, U.S. automobile fuel efficiency standards were 22 miles per gallon (mpg), and the U.S. government requires only a slight increase to 23.5 mpg by 2010. In November 2007, the U.S. Court of Appeals demanded that the Bush administration change the standards and answer some basic questions about why not all vehicles have to meet the current standard; for example, minivans and sport utility vehicles are allowed to be less fuel efficient than cars. The gathering of world leaders for the Kyoto Protocol of 1997 and in Bali, Indonesia, in 2007 indicated an urgent need to establish limits on carbon output to at least slow the effects of global climate change.

On the other hand, opponents argue that more time is needed to establish data trends that support the reality and consequences of global warming. U.S. energy policy does not support the global guidelines set forth in Kyoto, nor are current U.S. policymakers, along with China, willing to support any kind of mandatory carbon and fossil fuel emission reductions. However, there are movements toward tax incentives for voluntary reduction actions.

The full consequences of global warming will emerge only with time. Recognition of the importance of research in this area is reflected in the awarding of the 2007 Nobel Peace Prize to Al Gore and the scientists working with the United Nations International Panel on Climate Change. If momentum toward curbing global warming continues, perhaps the potential disastrous effects on human populations and the environment will be avoided. Yet, even with acknowledgment of the problem, significant changes to the national energy infrastructure have been slow in coming.

Scientists often rely on time through longitudinal studies to develop data trends over which people come to conclusions and take action. In response to the question of how quickly we need to begin to change our lifestyles and behaviors to make a difference in carbon output, some leading scientists claim that we have only 10 remaining years to reverse the effects of global climate change. Many others are calling for an 80% reduction in carbon emissions by 2050. The success of such reforms requires not only the time but also the will to implement far-reaching policy changes such as fuel economy standards, manufacturing caps on carbon, and clean energy options.

Erin E. Robinson-Caskie

See also Ecology; Extinction and Evolution; Glaciers; Ice Ages; Nuclear Winter; Values and Time

Further Readings

Godrej, D. (2001). *The no-nonsense guide to climate change.* Oxford, UK: Verso.

Gore, A. (2006). *An inconvenient truth: The planetary emergency of global warming and what we can do about it.* New York: Rodale Press.

McKibben, B. (2007). *Fight global warming now.* New York: Holt Paperbacks.

Michaels, P. (2005). *Shattered consensus: The true state of global warming.* Lanham, MD: Rowman & Littlefield.

God, Sensorium of

Sensorium of God is a phrase Isaac Newton (1643–1727) used in his discussion to describe his notion of space and time. Newton saw space and time intricately associated with God, thus he used science and religion to explain how the universe functioned.

Newton saw space and time as being absolute and independent of any physical object. He believed space extended without limit in any direction and was immovable. He saw time as infinite both in the past and the future. He also noted that making measurements in absolute space would require using an object within absolute space, thus an absolutely stationary object. This is not possible given the law of gravity, according to which every object, including stellar objects, is influenced by universal gravitational forces. Therefore, no stationary object exists so no such measurement can be made. Instead, objects must be measured relative to another object, both of which are moving.

The sensorium of God was Newton's way of answering an old question in stellar physics: How do particles move? The assumptions of this theory are (1) all objects have substance and therefore have a place in space and (2) these objects are subject to time. Physicists of Newton's time could not imagine another possibility. Newton believed stellar and planetary movement could be determined and predicted using mathematics and mechanics based on Euclidean geometry and his theory of gravity. Building on the work of Galileo, Copernicus, and others, he developed a method of plotting an object in motion. His method uses a coordinate system to plot an object on the three spatial axes and one temporal axis. Thus, one can know the exact location of an object at a particular instant in time. He believed space was geometrically ordered and that objects moved in a straight line unless otherwise influenced by gravity or centripetal force. Thus, Newton's sensorium of God is connected to his famous three laws of motion and his law of universal gravity.

Newton was raised in the Protestant tradition, died a Roman Catholic, but held to a Unitarian belief system that was neither traditional nor orthodox. In his quest to unite knowledge and religious belief, Newton, being somewhat of a theologian as well as a scientist, connected this concept of absolute space with God. He saw God as not restricted to the physical universe; that is, God is both omnipresent (not limited to space) and eternal (beyond the bounds of time). Space and time serve as his sensorium, where God intimately and thoroughly perceives and understands objects and events. God is always present everywhere and is a part of every event. For Newton, God constitutes time and space, being equally present in all space and time.

Newton's work on space and time was revolutionary, with his book *Philosophiae Naturalis Principia Mathematica* considered the most important work ever written in the field of science. Although his attempt to unite belief and science has been largely downplayed, his theories continue to be foundational to the physical sciences.

Terry W. Eddinger

See also Design, Intelligent; Einstein and Newton; God and Time; Newton, Isaac; Newton and Leibniz; Space, Absolute; Time, Absolute

Further Readings

Barbour, J. B. (1989). *Absolute or relative motion: The deep structure of general relativity: Vol. 1. The discovery of dynamics.* Cambridge, UK: Cambridge University Press.

Hodgson, P. E. (2005). *Theology and modern physics* (Ashgate Science and Religion Series). Burlington, VT: Ashgate.

Newton, I. (1999). *The principia* (I. B. Cohen & A. Whitman, Trans.). Berkeley: University of California Press. (Original work published 1687)

God and Time

At the core of most religions is the concept that God, or the Transcendent, may be personal or impersonal. God is the creating power without being created. God is timeless and eternal. With the act of Creation, time came to exist. God is not only the first cause outside of time, but also the basis of time. Especially in Christianity, God is regarded as the starting point of being and time, as well as the end of both. Statements from the ancient world on these ideas can be found in Plato's dialogues (e.g., *The State* or *Timaeus*) and in Aristotle's *Physics* and *Metaphysics.* Classical positions are given in the teaching of Saint Augustine, chiefly based on the Platonic and Neoplatonic traditions, and of Saint Thomas Aquinas, based on Aristotelian philosophy. Rationalistic perspectives are represented by such thinkers as Giordano Bruno, Baruch de Spinoza, and Gottfried Wilhelm von Leibniz. They considered monistic thoughts, as well as pantheistic views, or the question of evil in the world. More recently, Pierre Teilhard de Chardin and Alfred North Whitehead present again considerations on God and time.

Plato and Aristotle

With his idea (form) of the Good, Plato (428–348 BCE) did not directly mention God. But the universal good, as characterized in *The State,* has qualities similar to those of God in the Christian tradition: eternal, indivisible, beyond space and time. In the Platonic dialogue *Timaeus,* God is spoken of as the *dêmiourgós* (demiurge), who built the world out of basic material. But this God is neither the Christian God, nor a creator out of nothing. He has only the function of giving order and reason to the qualities of the cosmos.

Aristotle (384–322 BCE), student-scholar of Plato, speaks in the *Physics* and in the *Metaphysics* of the (first) Unmoved Mover. This first principle, which gives motion to everything in the cosmos, does not itself move. The first principle is loved by everything, so all things in the cosmos move toward it as an aim (*télos*) in circles. By this nonmoving quality, the Unmoved Mover is beyond time, because moving can be measured only within time. The first principle is always the same; it is the "thinking of thinking" (*nóêsis noêseôs*), unchangeable like God. Though it is not the Christian God, the first principle is called "God" (*theós*) by Aristotle.

Augustine and Aquinas

Saint Augustine's (354–430) thoughts on time are embedded in his interpretation of Genesis as he presented them in book X, and especially in book XI, of his *Confessions.* According to Augustine, time (*tempus*) is linked with the Divine Creation (*creatio*). "Time" has no meaning before the Creation. Time did not exist before God created the universe. However, one could say that time was within God before the Creation. So it makes sense to speak of "time" only when there are processes of becoming and vanishing. Time gives natural processes both a basis and a framework in which they can progress.

Augustine is very honest when he says that he knows quite well what time is, as long as he is not asked to describe it. But when he is asked what time is, he does not know how to answer this question: "What is 'time' then? If nobody asks me, I know it; but if I were desirous to explain it to one that should ask me, plainly I know not" (*Quid est ergo tempus? Si nemo ex me quaerat, scio; si quaerenti explicare velim, nescio*). Therefore, one can say that time is very difficult to explain, but at first easy to understand. This sounds a bit paradoxical, but Augustine wants to reach a deeper philosophical understanding of time. So, he goes on with his own investigation.

Time is located in the mind (*animus*): past, present, and future (the three kinds of time) are all represented in mind as dimensions of mind. This also sounds a bit paradoxical, because past, present, and future are basically at the same time in the mind. Though a human only lives in the present, both the past and the future can be represented within the human mind. The past is represented in memory. In a further step, Augustine says that time is measured by the soul: "'Tis in thee, O my mind, that I measure my times" (*In te, anime meus, tempora mea metior*). Therefore, time is an extension of the human mind. Time is a representation in the mind and not an objective physical motion. We can see that Augustine's concept of time is very subjective, because time can be represented and measured only in the subject's mind. In this case, his position is different from the more objective point of view of the Platonic and Neoplatonic traditions.

Augustine points out that Christian eschatology, the doctrine of last things, teaches us that a human cannot fulfill time on his or her own. God will fulfill time when God returns in Christ at the end of our time. So, the perfection of everything can be done only by God, and it is nothing that happens in time; it is, in fact, the perfection of time out of time. Therefore, Augustine wrote that, in time and in the world, we can only use things, and God is the only one who can be enjoyed. In the last books of *The City of God,* especially in book XXII, Augustine shows us that Christian eschatology differs from political eschatology; the latter tries to create an earthly paradise with only political considerations, often linked with political ideologies and violence.

The love of God for a human and the love of a human for God keeps a human, during time, in relationship with God. If one is within the love of God, one can do whatever one wants to do, as it cannot be against the Almighty: "Love, and do what thou wilt."

God, the Sacred Trinity, is independent of time. God is over, behind, or next to time, because God is the eternal, pure being (*esse*) and has never been created. God is and always was, is and always will be; God is timeless without any changes. The universe, created by God, is within time. But the created universe is contingent, it is not a necessary being like God. God, existing out of time, is the only necessary being as a basic ground of contingent being in the universe and in our world.

With his early considerations on time in the 4th and 5th centuries, Augustine is the starting point of Christian theories of time. He influenced later thinkers deeply, especially thinkers during the scholastic period, such as Saint Thomas Aquinas (1224/1225–1274). But Augustine also influenced thoughts about "time" beyond Christianity: modern thinkers like Edmund Husserl (1859–1938), with his phenomenology of time; Henri Bergson (1859–1941), in whose thinking the durative aspect of time (*durée,* duration) is central; and Martin Heidegger (1889–1976), with his main work *Being and Time* (1927), refer directly or indirectly to Augustine's thoughts about time.

Time (*tempus*) for Saint Thomas Aquinas, as for Saint Augustine, is a phenomenon that has to do with the Divine Creation (*creatio*). God creates, as he is the first cause (*prima causa*). All the beings in the created world are creatures in time. Outside the world and the cosmos, time has no meaning and no use. Time appears together with Creation. All natural developments happen in time. Without time, humans could not recognize developments. Also, human processes, even social and moral processes, need time as a framework. As Saint Thomas pointed out in his early commentary to the sentences of Saint Peter Lombardus, time is also necessary for those processes in the mind, because it serves as a measurement for what we have in the mind.

In his main work, *Summa Theologiae,* Saint Thomas investigated time as an important phenomenon. Here, Aquinas differentiates time (*tempus*), which was for him well known as one of the 10 Aristotelian categories, from real eternity (*aeternitas*) and from eternal time (*aevum*). Aevum, "eternal time," is a kind of "infinite time" for Saint Thomas, as an infinite sum of time, which is not a real eternity. It stands in the middle between time and eternity: "And by this way can be measured eternal time, which is the middle between eternity and time" (*Et ideo huiusmodi mensurantur aevo, quod est medium inter aeternitatem et tempus*). Time can be a continuum, or be separated into parts of time. Time can be imagined or real. This distinction is different from Augustine's view, which taught that all three kinds of time (past, present, and future) are only representations in the mind and therefore dimensions of it. But similar to Augustine, Aquinas says that there are three kinds of time: the first time is the beginning, the second time is the following time, and the last time is the end of time ("time has a beginning and an end" (*tempus autem habet principium et finem*), and "between two instants, there is a middle time" (*inter quaelibet duo instantia sit tempus medium*). But again, in contrast to Augustine, Aquinas admits that, according to Aristotle, time is a kind of motion. Time can be measured as the number of changes (e.g., dark to bright, night to day) or of motions ("movement, whose measurement is time" (*motus, cuius mensura est tempus*) and "because time is the number of motions" (*quia tempus est numerus motus*). Time serves also to represent a metaphysical, not only a physical, development or motion: as angels come to their nature at once, humans come to theirs eventually, because they have to discover the nature of a human as a creature of God.

Also for moral actions, time is an important indicator. Humankind needs sufficient time in order to act morally (*Tempus autem est una de circumstantiis quae requiruntur ad actus virtutum*). Therefore, nobody can be forced to immoral action by reducing his or her time to act or to react properly.

Within time, there are temporal things that are contingent, not necessary. These temporal things, which are all material things, come into existence and vanish after a certain amount of time. They cannot exist eternally in time. However, in his mainly philosophical work *De aeternitate mundi* (c. 1271), Saint Thomas shows that the universe itself is eternal, because the substance, the "basic material" (*materia prima*), is not made and is eternal. This does not mean that our world in its present concrete existence is eternal, but the world and the universe are eternal in their "basic material" (*materia prima*). With this argument, Saint Thomas wanted to increase the philosophical value of the pagan philosopher Aristotle to Christian theology. According to Aquinas, species are not eternal; they will, like other temporal beings or material things, appear and vanish in time. Without *materia prima,* the universe would have been a serious vacuum, which is, according to Saint Thomas, impossible.

Furthermore, humans cannot end time. According to Christian eschatology, it is only God who is able to fulfill time, when Christ returns and separates the good from the evil. So, God is the creator and the dominator of time. "To Him belongs time and the ages" (*Eius sunt tempora et aeternitas*), as it is written in the Office of the Easter Vigil of the Roman Catholic Church. God is outside of time and he is the timeless, eternal being (*esse*). God is the creator of everything and has never been created. He is and always was, is and will always be, without any changes. God, the Sacred Trinity, is the origin of time and the basic ground of being. Similar to Augustine, it is not sensible for Aquinas to speak of time before Creation.

Saint Thomas Aquinas deeply influenced Christian doctrine. His considerations on the property of time and other related subjects were philosophically and theologically investigated in many books and treatises. With his thoughts, Saint Thomas stands in the tradition of the *philosophia perennis,* the Everlasting or Permanent Philosophy, which has its starting point in the philosophy of Plato and Aristotle.

In the early 20th century, *neo-Thomism* emerged: neo-Thomistic philosophers and theologians kept up Aquinas's thoughts in a very comprehensive way: the French thinker Jacques Maritain (1882–1973), the Jesuit Erich Przywara (1889–1972), and the German philosophers Max Müller (1906–1994) and Josef Pieper (1904–1997), among others, represent Thomistic thought in our present time.

Bruno, Spinoza, and Leibniz

All three thinkers—Bruno, Spinoza, and Leibniz—have a rationalistic approach to the question of God. Giordano Bruno (1548–1600) was a monk and philosopher who presented a daring cosmology. Some of his thoughts were adopted by two other philosophers, Spinoza and Leibniz. Bruno and Spinoza are, in general, pantheists and monists. Leibniz is a monotheistic thinker but not a monist; however, he has a rationalistic method in common with Bruno and Spinoza.

Bruno presumed that the world and God are, in the end, both together, because reality has its origin in the eternal imagination. For Bruno, the universe is infinite and there are an infinite number of beings in the cosmos. The universe is infinite because God as the creator of the cosmos is almighty and infinite. A finite cosmos would be too sharp a contrast to honor an almighty and infinite God. So, both God and the universe are infinite, but within time. In Bruno's opinion, there exist an infinite number of eternal worlds. Also, according to Bruno, it is possible to measure time accurately in our world. Before Bruno, most of the thinkers followed Aristotle, who was convinced that only in the sphere of the fixed stars could there be a correct measurement of time. Time can be measured by motion; Bruno developed a pendulum system that could measure time accurately by the motion of the pendulum. Furthermore, he was against the Ptolemaic geocentric model of the universe. Because of his heretical opinions, Giordano Bruno was condemned to death by the Roman Catholic Church in 1600.

Influenced by Bruno and Descartes (1596–1650), the philosopher Baruch (or Benedict) de Spinoza (1632–1677) was convinced that the substance of nature and of God is the same, as is pointed out in the sentence *deus sive natura* ("God or Nature"). He is therefore an exponent of monism and pantheism. God is natural and living, and cannot be separated from matter. He is within every being. Some presumed that Spinoza was an atheist, because all materialistic things are finite in contrast to God, who is immaterial and eternal. However, the Spinozistic concept of nature is not only materialistic, but also something much more complex. For Spinoza, nature has a divine quality even though it also has material qualities. God is the necessary, eternal, and infinite substance of nature or the universe. It follows that there is only *one* substance in the universe (monism), because God and nature are of the same substance. For this reason, God exists necessarily and, furthermore, he was forced to create the universe because of this monistic union of God and nature. The main structures of the universe are causality and necessity. Beyond space and time, there is only the real existence of beings in the universe. The world exists as an infinite sequence of natural processes throughout time. God exists as the eternal substance, but within endless time.

Gottfried Wilhelm von Leibniz (1646–1716) was a scientist, mathematician, and philosopher. He argued that there is a preestablished harmony in the world that is based on the infinite number of monads that are the nonmaterial elements of reality. The first monad is God, who has created the world as the best of all possible worlds. The evil in the world is thus reduced to a minimum, because God has compared our universe to every other possible universe, and he has decided this world to be the best possible world. The metaphysical, physical, and moral evils that exist have their origin in the imperfect and finite cosmos. God is the eternal reason and necessary being of the finite world. Space and time are not absolute, but relative as parts of the world. God and the universe are not of the same substance. In contrast to Bruno and Spinoza, Leibniz was neither a monist nor a pantheist. Leibniz earned his great reputation posthumously, perhaps most notably in Bertrand Russell's (1872–1970) writings.

Teilhard de Chardin and Whitehead

The French Jesuit priest, paleontologist, and philosopher Pierre Teilhard de Chardin (1881–1951) attempted to reconcile Christianity, especially the Roman Catholic position, with the results of his scientific research into evolution. God is the alpha and the omega point of the evolving universe, so God is outside of time and its dynamic framework. For Teilhard, the omega point is the last stage of the evolving noosphere, at the end of human time. From this point of view, God is not only a spectator of the evolution of the universe, but also its aim and goal. Teilhard, who was present at the discovery of

Peking man in 1923–1927, gave these arguments in his controversial major work, *The Phenomenon of Man*, which was denied publication by the Roman Holy Office during Teilhard's lifetime. Later, Pope John XXIII rehabilitated him, and his work became an influential contribution to the church's position on evolution.

With his dynamic philosophy presented in his book *Process and Reality: An Essay in Cosmology* (1929), the mathematician and later philosopher Alfred North Whitehead (1861–1947) defended theism and represented the starting point for process theology; for Whitehead, the whole universe is in fluent change. Therefore, God as the creator of the universe is also in a fluent process and is not able to predict each process in the universe precisely. God is also understood within the framework of time, and God makes possible the change, growth, and flux of reality. Whitehead's argument defends theism, but God is not almighty. Therefore, Whitehead's position is different from the Judaic and Christian positions. Nevertheless, theologians such as Charles Hartshorne, John B. Cobb, Jr., and David Ray Griffin followed Whitehead and contributed to process theology.

The positions of Teilhard and Whitehead make it clear that, in the 20th century, the distance between God as the timeless being and the universe as a creation within time became less and less, because the Creator and the cosmos are constantly interrelated in an ongoing process of eternal creativity.

Hans Otto Seitschek

See also Aristotle; Aquinas, Saint Thomas; Augustine of Hippo, Saint; Bruno, Giordano; Christianity; Eschatology; Husserl, Edmund; Leibniz, Gottfried Wilhelm von; Newton, Isaac; Plato; Spinoza, Baruch de; Teilhard de Chardin, Pierre; Time, Sacred; Whitehead, Alfred North

Further Readings

Barnes, J. (Ed.). (1995). *The Cambridge companion to Aristotle*. Cambridge, UK: Cambridge University Press.

DeWeese, G. J. (2004). *God and the nature of time*. Aldershot, UK: Ashgate.

Ganssle, G. E., & Woodruff, D. L. (Eds.). (2001). *God and time: Essays on the divine nature*. Oxford, UK: Oxford University Press.

Garrett, D. (Ed.). (1995). *The Cambridge companion to Spinoza*. Cambridge, UK: Cambridge University Press.

Gatti, H. (Ed.). (2003). *Giordano Bruno: Philosopher of the Renaissance*. Aldershot, UK: Ashgate.

Jolley, N. (Ed.). (1994). *The Cambridge companion to Leibniz*. Cambridge, UK: Cambridge University Press.

Kenny, A. J. P. (2002). *Aquinas on being*. Oxford, UK: Oxford University Press.

Kraut, R. (Ed.). (1992). *The Cambridge companion to Plato*. Cambridge, UK: Cambridge University Press.

Mercier, A. (1996). *God, world and time*. New York: Verlag Peter Lang.

Padgett, A. G. (1992). *God, eternity and the nature of time*. New York: St. Martin's.

Poe, H. L., & Mattson, J. S. (Eds.). (2005). *What God knows: Time, eternity, and divine knowledge*. Waco, TX: Baylor University Press.

Stump, E., & Kretzmann, N. (Eds.). (2001). *The Cambridge companion to Augustine*. Cambridge, UK: Cambridge University Press.

Teilhard de Chardin, P. (1959). *The phenomenon of man* (B. Wall, Trans.). New York: Harper & Row.

Whitehead, A. N. (1979). *Process and reality: An essay in cosmology*. New York: The Free Press.

Wissink, J. B. M. (Ed.). (1990). *The eternity of the world in the thought of Thomas Aquinas and his contemporaries*. Leiden, The Netherlands: Brill Academic.

God as Creator

The idea of God as the Creator is perhaps the most radical theory of time ever conceived, one that imposes a definite beginning to time, along with an explanation of why this should have happened. Before God there was no time, according to this theory, because God created time when he created all things. Theories of God as Creator often, though not invariably, are accompanied by a complementary theory of God overseeing the end of time as well.

Theories of God as Creator are not the common property of humanity, but are the product of specific trends of thought involving monotheism, or the idea of a single god. In cultures that do not subscribe to monotheist notions of God, there is much less evidence of theories of God as Creator or, if such theories can be found, they are much less significant than in monotheist cultures. Creation accounts are also a relatively late development in the evolution of religious thought.

Asian Traditions

When looking at India, for example, amid all the conflicting interpretations one could read into the Vedas, there is little to support the idea of God as a creator from nothing. Among the many speculations is the admission that maybe even the gods do not know, because they emerged after the universe derived its form. The closest parallel to the word *creation* is the Sanskrit word *Srishti,* or "projection." So when it is said that God created things out of nothing, it is meant that the universe is a projection of God, with the extra understanding that the universe creates itself and falls back into itself, in endless cycles, for all time. Other understandings have this projection as a never-ending process of God realizing himself in the universe.

It is also true that no god in the Hindu pantheon was credited with Creation. Brahma is given such an age that even talk of creating the universe seems a paltry exercise. Hinduism divides time into cycles, called *kalpas.* One kalpa consists of 1,000 cycles of 4,320,000 years, which are made up of 12,000 divine years, each of which lasts 360 solar years. And for Brahma one kalpa equals one day.

Another trend in Indian philosophy known as the Kalavada derives its name from *kala,* which originally meant "right moment" but came to mean "time" itself. *Kala* was used in that sense in the Sanskrit writings, where it took on the mantle of being a fundamental principle of the universe and that existed before all other things. It may be that Kali, one of the avatars of Shiva, is also derived from kala. Kali ("the Black One") is, like time, merciless.

In China there is the tale of Pangu (P'an-Ku in the old spelling) who is the child of *yin* and *yang* and who fashioned the cosmos out of the primeval chaos. But he is not a deity to be worshiped, and neither is he credited with actually creating the universe. Then there is the celestial spirit Tai Sui, who presides over the structure of the year. He is, in fact, president of the Celestial Ministry of Time, a prestigious and highly feared office. Tai Sui was venerated as the deity able to influence human destiny. Astrologers were kept busy analyzing dates for auspicious signs that a new project would meet with Tai Sui's approval. The Yuan dynasty (1280–1368) began the practice of sacrificing to Tai Sui before any momentous project was undertaken.

Western Traditions

Among the Greeks, as well as in the great Eastern civilizations, the universe had always existed and creation stories, where they existed at all, told more of fashioning order out of primeval chaos than of creation ordered from nothing. In the *Timaeus,* Plato portrays God as that which imposes order on a preexistent matter. Finding the universe in a state of "inharmonious and disorderly motion," God "reduced it to order from disorder, as he judged that order was in every way better." The same account is given in Book One of Ovid's *Metamorphoses,* and it has been argued that a similar approach can be found in the Hebrew Bible, in, for instance, Psalm 74:12–17 or 89:9–13.

Aristotle was more specific in denying any primal creation role to the gods. Rather, God was the "prime mover" who "moves" the world. As time is eternal, so change is also eternal, which, in turn requires an eternal overseer of that change. This is the prime mover. It is important not to confuse the prime mover with any notion of a personal God as understood in the Christian tradition. The prime mover is nonmaterial and as such occupies no space. It is not distinct from nature, but is the animating principle within nature. Aristotle's God, being eternal, is the sum total of thoughts and animating principles that transcend temporality.

Even more naturalistic was the account of the universe in Lucretius's masterpiece *De rerum natura* (*On the Nature of the Universe*), which was completed in about 60 BCE. Nothing, Lucretius insisted, not even divine power, can create something out of nothing. It was not the principle of divinity that was being challenged but that of supernaturalism. Elsewhere, Lucretius spoke of a limitless universe with no center and composed of indestructible matter.

Nowhere is the difference between monotheist and other cosmologies more apparent than in the idea that the cosmos is the product of a God that operates in history. It is the monotheistic traditions of Judaism, Christianity, and Islam that have the most elaborate accounts of God as Creator. While various interpretations of biblical creation narratives exist, the most influential has been that of *Creatio ex nihilo,* or "creation out of nothing." The Hebrew Bible begins with the assertion: "In the beginning God created the heaven and the earth." There are, however, two accounts of creation in the first books of Genesis. The earlier account can be found in Genesis 2:4–25 and is attributed to an unnamed source called the Yahwist, or simply J, who wrote in the interests of Judah, probably in the eighth or seventh centuries BCE. The J account differs significantly from the later account, in Genesis 1:1–2:4, which is attributed to P, or the Priest, and probably does not predate the Babylonian Captivity of the 6th century BCE.

The J account gives lip service to the creation of the earth and the heavens and gives greater emphasis to the use made of existing materials, such as bringing rain to fructify the earth and creating man out of the dust of the earth. J begins with dry earth from which water later emerges. The later account reverses this and begins with water as the default condition, out of which land emerges. Among other items, this is evidence of P's closer debt to Mesopotamian mythology, which derived from living in a river lowland, unlike the arid land of Palestine, where J wrote. It is also significant that the earlier J Creation account seems contradictory with man being created first, out of the dust (*adam* from *adamah*) whereas the earlier P account has man being created last, as the pinnacle of creation.

These differences in detail and order are symptomatic of a larger difference in underlying attitude. J's account is more organic and closer to the mythological accounts then prevalent. P's account, for instance, makes sure to credit God with the creation of light, departing from many mythological accounts where light is, along with its sister condition darkness, a primordial element of chaos. P's later account is more abstract and rarefied, and serves to strengthen the case of ecclesiastical authority, whose charge it is to interpret these great themes to the unlettered and stand in as their mediators before God.

None of these distinctions played a major role in Christian creation theology before the 19th century. *Creatio ex nihilo* was decreed orthodox theology at the Council of Nicaea in 325 CE and Saint Augustine developed P's account, accentuating the utter majesty and sufficiency of God in overseeing this creation. Augustine also attacked Greek ideas of time having no beginning and there having been no creation. Those who claimed the world was without beginning and consequently not created by God, he wrote, "are strangely deceived, and rave in the incurable madness of impiety." This remained the standard interpretation for more than one and a half thousand years.

The Qur'an is, if anything, even more insistent on the *Creatio ex nihilo* thesis than the Bible. There is little description or explication beyond a bald assertion of Allah's total, entirely sufficient, and generous-hearted creation. Allah is praised as he who "created the heavens and the earth, and made darkness and light." Moreover, he created it with truth, to serve as a sign for the believers and was not fatigued in any way by this creation. These claims are often made in the context of rebuking those who would doubt or warning those who would disbelieve. The Qur'an is no less anthropocentric than the Bible, with several passages allowing for humanity as the highest point of creation by virtue of most resembling Allah the Merciful. As the actual and inviolate words from Allah, this Qur'anic account is, formally at least, the only acceptable Muslim account of the creation of the world.

Despite all the differences among the Greeks, they agreed that *Creatio ex nihilo* made no sense and that creation meant ordering what was already there, rather than starting from scratch. But the triumph of the *Creatio ex nihilo* doctrine changed all that. It also has several important theological ramifications. First, it serves to heighten the focus on God's awesome power to perform such a feat, and our debt to him. Second, it reinforces the distance between the Creator and his creatures, and that God/Allah did not need to create the universe so as to fulfill Himself, or somehow to bring things to a final resolution, because God/Allah was himself already finally resolved.

Third, it extends his power to being above and beyond time.

Perhaps the most specific rendering of God as Creator is the notorious date given by Archbishop James Ussher in 1617. Working from the genealogies of the Hebrew Bible, Ussher dated the creation of the world to October 23, 4004 BCE, with Adam being created 5 days later. This contradicted the Jewish scholars, who decided that October 7, 3761 BCE, was the more likely date. These attempts to discern a specific date for God's act of creation are the ultimate consequence of assuming the biblical narratives to be literally true accounts of what has happened.

Problems with the science and theology of God as Creator idea reappeared with the great advances in science being made in the 17th century. At the very time Archbishop Ussher was devising his chronology based on the biblical genealogies, Galileo, Giordano Bruno, and Isaac Newton were constructing an altogether new account of the universe. This drastic expansion of the universe, and its orientation away from being Earth-centered to sun-centered, had important consequences for ideas about God.

The most imaginative response at the time was deism, which retained for God the role as Creator but attributed everything that has happened since then to the working of nature. Few of the ideas popularized by the deists were original to them. Instead, they reworked ideas articulated by Lucretius, Cicero, and others, in the light of the new understanding of the universe as envisaged by Newton. Problems with deism soon became apparent, in particular, why its vision of God as Creator was any more acceptable than the conventional accounts it sought to replace. There was also the problem that the god of deism was useless for any normal religious function. Another response was pantheism, which subsumed God in Nature, divinizing the latter and secularizing the former. Pantheism was more corrosive to ideas of God as Creator than deism, although it suffered some of the same intellectual objections.

Some scientists in the 19th century, particularly those of a religious persuasion, argued that, as well as the mystery of creation, there was still so much that was unknown and could be attributed only to God's will. But as successive unknown areas have been opened up and provided with naturalistic explanations, the field of God's sphere of influence has been reduced. This was recognized as a problem and by the end of the 19th century was called the "God of the Gaps."

Among nonscientists another response was to relinquish all thoughts of demonstrating the rational and logical veracity of any God-talk, whether in his capacity as Creator or in any other role. Belief in God, in whatever capacity, was a matter of faith alone, in which reason played no part. Variations on this theme were articulated, at the radical end, by the Danish philosopher Søren Kierkegaard, and in more conservative language, by the Oxford movement and Ultramontane Catholicism. A secular variant was arrived at toward the end of the 19th century in early pragmatism, which argued that disputes over the existence of God and his role as Creator were largely irrelevant and incapable of final resolution. What mattered was the benefit such beliefs conferred on the believer.

The most recent response within the religious world to the advances in scientific understanding of the universe and its origins has also been the most conservative. Fundamentalism is the insistence that what is decreed in the scripture of one's religion is literally true in all respects. Thus, God really is the creator of the universe as outlined in the Bible (or the Qur'an), and whatever inconsistencies in the scriptural accounts are either illusory or the result of an unwillingness to believe the truth with the required degree of faith.

Most people, however, have reconciled themselves to the collapse of traditional accounts of God as Creator. By the beginning of the 21st century the general scientific consensus was that the universe is between 13 and 16 billion years old, and came into existence as a result of the big bang. And that the earth is about 4.5 billion years old and is one of countless planets in an expanding universe of unimaginable proportions. Those who have remained religious believers have been content to see the creation accounts as allegories with poetic rather than normative power. There have also been attempts to equate the big bang with accounts of God the Creator, but without success. As several scientific observers have commented, such a move tends to work only if God is reduced to a series of mathematical equations. Those who do not subscribe to any formal religious system simply take the scientific account as the best available account of how things began.

Bill Cooke

See also Aquinas, Saint Thomas; Aristotle; Augustine of Hippo, Saint; Bible and Time; Big Bang Theory; Christianity; Creationism; Genesis, Book of; God and Time; Gosse, Philip Henry; Lucretius; Plato; Qur'an; Time, Sacred

Further Readings

Augustine. (1878). *The city of God* (M. Dods, Trans.). Edinburgh, UK: T & T Clark.

Edis, T. (2002). *The ghost in the universe.* Amherst, NY: Prometheus.

Gribbin, J. (2003). *Science: A history.* London: Penguin.

Plato. (1974). *Timaeus and Critias* (D. Lee, Trans.). London: Penguin.

Watson, P. (2005). *Ideas: A history of thought and invention from fire to Freud.* New York: HarperCollins.

Whitrow, G. J. (1988). *Time in history.* Oxford, UK: Oxford University Press.

Gödel, Kurt (1906–1978)

Kurt Gödel was a mathematical logician who is best known for his incompleteness theorem. He also developed a theory of time travel based on Einstein's theory of relativity.

Gödel was born on April 28, 1906, in Brünn, Austria (now known as Brno, Czech Republic), and was baptized as a Lutheran. He began studying physics at the University of Vienna in 1924, but in 1926 he switched to mathematics, in which he excelled. Eventually he settled into the field of mathematical logic, which originated in the work of Gottlob Frege and was more fully developed by David Hilbert, Bertrand Russell, and Alfred North Whitehead.

In 1926, Gödel began participating in the Vienna Circle, a group of mathematicians and philosophers headed by Moritz Schlick. The group's members devoted themselves to propagating logical positivism, the philosophy that all that can be known about nature or reality must be deduced from immediate sensory experience. In spite of his participation, Gödel, like Albert Einstein, would become a lifelong opponent of positivism, arguing that intuition has a proper role to play in science and mathematics. Both he and Einstein rejected the Kantian notion that one can know only the appearances of things and not the things themselves. Because of Gödel's preference for the idealistic philosophies of Plato and Husserl, the philosophical establishment, dominated by the philosophy of Wittgenstein, either ignored or scorned much of his philosophical work.

In 1930, Gödel obtained his Ph.D. In 1931, Gödel published a response to David Hilbert's formalist attempt to develop a system of first principles (or axioms) from which one could apply rules of syntax to derive all the theorems of a mathematical domain. He undermined Hilbert's program by showing that no set of formalist axioms can ever fully capture the complete set of mathematical truths. There will always be some truths about integers, grasped intuitively, that cannot be proved true or false by any fixed set of axioms. He also showed that a system of axioms for arithmetic could not prove its own consistency. His incompleteness theorem (also known as Gödel's proof) ranks with Einstein's theory of relativity and Heisenberg's uncertainty principle as one of the three most revolutionary scientific findings in the 20th century. The recursive functions that he developed as part of this work were later used by Alan Turing and others in the development of the computer.

From 1933 to 1938, he alternated teaching stints between the University of Vienna and the Institute for Advanced Study (IAS) in Princeton, New Jersey. In 1938, he married Adele Porkert, a divorced nightclub performer. They had no children. After being declared fit for German military service, he and his wife emigrated to America in the winter of 1939–1940 by way of Siberia, Japan, and San Francisco, arriving in March at Princeton where he became a temporary member of the IAS. This status was renewed annually until he became a permanent member in 1946. In 1948, he became a citizen of the United States.

Gödel and Einstein became close friends in 1942 and remained so until Einstein's death in 1955. They joined each other in daily half-hour walks to and from the institute, during which they discussed politics, philosophy, and physics. During this time, Gödel's work focused on proving that Georg Cantor's continuum hypothesis (which dealt with the number of points on a line) was consistent with set theory and therefore could not be disproved.

In 1949, Gödel wrote an essay on the connection between the theory of relativity and idealistic philosophy, which was included in a volume in honor of Einstein's 70th birthday. Gödel developed solutions to the field equations of general relativity that resulted in a possible world (called the Gödel universe) whose spacetime structure is warped or curved so extremely as to form a closed, rotating loop. If a spaceship traveled fast enough along one of the continuous time-like paths in this structure, it could travel to the past or to the future. He then concluded that, if one could travel to the past, then time does not exist as an objective reality. Like Parmenides and Kant, he challenged the intuitive understanding of time as a linear ordering of events in which the past no longer exists, the future is yet to exist, and only the present truly exists. In essence, he showed that if relativity theory is true, then time understood as a succession of never-ending "nows" cannot exist as an objective reality.

Einstein was impressed with the results but doubted that physical data would support the existence of such a universe. In fact, lack of evidence for the rotation of the universe suggests that time travel is not possible in the actual universe (unless scientists can figure out how to create wormholes, which are shortcuts between two points in spacetime). Since its publication, the essay has received some attention from those interested in the topic of time travel, but its conclusion that time is an illusion has been largely ignored. In 1992, Stephen Hawking proposed the "chronology protection conjecture" in order to refute Gödel's argument and other theories of time travel. Hawking postulated that the laws of physics rule out the physical possibility of macroscopic bodies carrying information to the past.

In 1951, Gödel was co-recipient with Julian Schwinger of the first Einstein Award and delivered the Gibbs Lecture for the American Mathematical Society. After that, he became increasingly reclusive and isolated. In 1953, he was elected to the National Academy of Sciences and was promoted to professor at IAS. He published his last paper in 1958. During the 1960s, he developed an ontological argument for the existence of God, building on the arguments of his favorite philosopher, Leibniz. He never published it, perhaps for fear of damage to his reputation if his sympathies toward theism became known. He also continued his quest to develop axioms that would settle the continuum hypothesis. In 1975, he was awarded the National Medal of Science.

After Einstein's death, the economist Oskar Morgenstern became Gödel's friend and caretaker. The logician Hao Wang also became a close associate in the closing years of Gödel's life, and after his death Wang published recollections of conversations with Gödel.

Throughout adulthood, Gödel was plagued by depression, hypochondria, anorexia, and paranoia, and his symptoms worsened as he grew older. He retired from the IAS on July 1, 1976. In 1977, Morgenstern's death and his own wife's deteriorating health exacerbated Gödel's mental and physical disorders. He died of self-starvation on January 14, 1978, weighing only 65 pounds, and was buried in Princeton Cemetery. His wife died in 1981.

Gregory L. Linton

See also Einstein, Albert; Hawking, Stephen; Idealism; Intuition; Kant, Immanuel; Leibniz, Gottfried Wilhelm von; Relativity, General Theory of; Russell, Bertrand; Space and Time; Spacetime, Curvature of; Time, Illusion of; Time, Nonexistence of; Time, Relativity of; Time Machine; Time Travel; Time Warps; Whitehead, Alfred North; Worlds, Possible; Wormholes

Further Readings

Dawson, J. (1997). *Logical dilemmas: The life and work of Kurt Gödel.* Wellesley, MA: A. K. Peters.

Gödel, K. (1949). A remark about the relationship between theory of relativity and Kantian philosophy. In P. A. Schilpp (Ed.), *Albert Einstein: Philosopher-scientist* (pp. 557–562). La Salle, IL: Open Court.

Nagel, E., & Newman, J. R. (2001). *Gödel's proof* (D. Hofstadter, Ed.; Rev. ed.). New York: New York University Press.

Yourgrau, P. (2005). *A world without time: The forgotten legacy of Gödel and Einstein.* New York: Basic Books.

Goethe, Johann Wolfgang von (1749–1832)

Johann Wolfgang von Goethe was born on August 28, 1749, in Frankfurt am Main, and died on March 22, 1832, in Weimar. After studying law in Leipzig, he worked as a lawyer in Frankfurt and then underwent further practical training at the Court of Appeal (Reichskammergericht) of Wetzlar. Influenced by the Sturm und Drang movement, he wrote dramas and poems such as *Prometheus, Götz von Berlichingen,* and *The Sorrows of Young Werther* (*Die Leiden des Jungen Werthers*). A rough draft of what later became *Faust* goes back to his time in Frankfurt; in this drama, the topic of time has an important role.

In 1775, Karl August, the duke of Weimar, asked Goethe to join his court (*Weimarer Hof*) and work with him as a state representative. In 1786, Goethe made his first visit to Italy, which lasted 2 years and had a significant influence on his life. Of particular importance for him was his becoming familiar with classical works of art and the scientific research he made on the journey. A meeting with Friedrich Schiller in 1794 encouraged Goethe to increase his literary productivity again. He wrote *Xenien, Wilhelm Meister's Apprenticeship* (*Wilhelm Meisters Lehrjahre*), began writing his autobiographical *Poetry and Truth* (*Dichtung und Wahrheit*), and continued working on *Faust.* However, international political events such as the French Revolution and in 1806 the fall of the German Reich deeply influenced Goethe's thinking, as did numerous innovations in science and technology and their consequences for the life of the people. He became one of the central figures of the Weimar classical period (*Weimarer Klassik*).

Faust and the Theme of Time

Since Goethe worked on *Faust* throughout his life, this drama reflects the whole continuum of his research and interests, and also reveals the changes his thinking had undergone. An important aspect of the drama is the question concerning the relationship among human beings, time, and eternity. In the *Prolog im Himmel (Faust I),* Goethe's concept of eternity is conventional and depends on the Christian worldview. This concept changes with the last scene, titled *Bergschluchten (Faust II),* into something transcendental. The connection between the periods past, present, and future plays a particularly important role. Goethe assembles the scenes into an order that is remote from ordinary human perceptions and experiences of time. The meeting and union of Faust and Helena encompasses a period of more than 3,000 years, whereas the story time of *Faust I* comprises less than one year. The union of antiquity and modernity is not only the reason for the break in the structure of time. Timelessness also arises because Helena symbolizes eternal beauty while Faust stands for eternal striving.

Another type of time is the lifetime of Faust. It limits the period of impact concerning the bet he made with Mephistopheles. Only as long as Faust is alive can Mephistopheles try to satisfy his desires with earthly pleasures and thereby give him a feeling of complete contentment. Consequently, in Goethe's *Faust,* three important spheres of time are united: eternity; the union of past, present, and future; and Faust's own lifetime.

Faust's desire to experience the eternal is in sharp contrast to his limited existence as a human being. His aim "to get to know everything" fails. He realizes that the studies of science, the Bible, and magic cannot help him in his endeavors, which is the reason why he abandons them and turns to everyday life. As he grasps that the act and not the word is the beginning of realization, Faust becomes aware of the importance of the present moment.

In his pact with Mephistopheles, Faust is interested in the quality of a certain moment he wishes to experience. The problem is that the two partners in the contract have different opinions about time. Mephistopheles affirms this worldliness and inevitable death, so eternity itself is hostile to him. According to Faust, however, time that passes is hostile, because he sees eternity as a cycle of life, growth, death, and rebirth.

In the pact, Faust demands total fulfillment in one moment. Only then is he ready to give his soul to the devil. Mephistopheles tries to win the bet by offering Faust various types of sensual pleasures, but he underestimates Faust's spiritual nature. It is very convenient for Mephistopheles that Faust seems to reject eternity in favor of the moment; this is because God represents the eternal, while Mephistopheles stands for the single moment. In the beginning, Mephistopheles confronts Faust with the pleasant aspects of life in Auerbach's cellar, trying to cheer him up and ease his thoughtful mind.

Concerning individual pleasure, time is a succession of moments that are not related to one another. Faust's striving in the company of Mephistopheles represents the timeless element in human beings. For Faust, momentary pleasures are not sufficient for complete fulfillment. This seems analogous to the beauty of Helena, which is also timeless because beauty is hidden in the law of nature. Therefore, the union with Helena is important for Faust. Faust's situation begins to change when he recognizes past events as important. He develops a new attitude toward sensual experiences, because in his union with Helena he realized how fulfilling a moment can be. His failure to combine eternal beauty, represented by Helena, with present time leads to the need for significant actions. Act IV of the second part of *Faust* shows Faust's development, his willingness to do things for their own sake. At this point, the dangers and consequences of his activities become clear. The need for power and property (land reclamation) brings with it destruction (of landscapes and buildings) and death (Philemon and Baucis).

The process of acceleration is manifested in Faust as well as Mephistopheles. Faust curses slowness; he is marked by precipitated striving, which is cultivated by Mephistopheles. He forges plans and carries them out quickly (money economy at the emperor's court, draining of the coast to gain land). As a consequence, others are damaged and he destroys himself. Lemurs dig Faust's grave under Mephistopheles' supervision. Faust, as an old, blind man, interprets the noises as a continuation of his project to drain the marshland. In this moment, in a vision of a future, free society, Faust asks, with the words, "Stay, thou art so beautiful!" for the moment to stay. He dies, and the devils come to take his soul. But his immortal soul could be saved.

The polarity of moment and eternity corresponds to Goethe's development from youth to old age. The importance of the moment is a quality associated with the Sturm und Drang movement; classical antiquity, however, offers timelessness and permanence.

For the phenomenon of acceleration, Goethe formed the expression *velozifearisch*. It consists of *velocitas* (hurry) and Luzifer (devil). For Goethe, the acceleration of time is the origin of modernity. Modern civilization, which is characterized by rapidity and acceleration, also affects the mental states, thinking, and acting of modern human beings. Faust values progress and so he becomes restless and unscrupulous; omissions, mistakes, and violence result.

Goethe's Historical Interests

In the field of historical research, which was important to Goethe all his life, the polarity of permanence and fast change plays a significant role. In the early decades of his life, Goethe's historical interest can be found in his dramas *Götz von Berlichingen* and *Egmont*.

During his stays in Italy, Goethe realized nature and art were permanent and reliable because history was present in the antiquities. Back in Weimar, he was caught up in the events of world history and his belief in stability began to fade. He regarded the French Revolution as chaotic and as responsible for accelerating the historical process. Goethe wished to understand events and acted accordingly, but felt helpless. He realized that his own desires could not affect the progress of history, and recognized the arbitrariness and unpredictability of events, which reminded him of natural changes. Looking for stability, he discovered his roots and began his autobiographical work, *Poetry and Truth* (*Dichtung und Wahrheit*). Nonetheless, Goethe continued to place a high value on progress.

Botanical Investigations

The ambiguity between Goethe's interests in progress and his own thoughts on development (metamorphosis) come out clearly and are put into perspective. According to Goethe, time plays an important role in nature. One of his fundamental ideas was the organic development from the simple to the absolute. He believed he recognized a special type of original plant (*Urpflanze*) in every plant. To verify this idea, Goethe undertook biological research during his journeys to Italy and at his home, on his idea of an original plant type.

He looked for something that is common in all plants. This led him to make generalizations different from those of Linné (Linnaeus), who tried to find a classification based on the differences between plants. With his generalizations, Goethe progressed from the original plant to the original leaf (*Urorgan Blatt*). In his opinion, all parts of the plant emerged from this leaf. Two laws are responsible for changes in the plant: A law of nature constructs the plant, and a law of the external circumstances modifies it.

Later, Goethe gave up his idea of an original plant type in favor of the theory of types (*Typus*). He subsequently made his theory of the evolution of plants more general so that it included animals. In contrast to the theory of metamorphosis, his type theory represents a constant in natural biodiversity. Nevertheless, both terms (*Leitbegriffe*) in Goethe's scientific studies are connected and lead to biodiversity. The type sets limits that metamorphosis cannot exceed, though it can shine though as an original principle (*Urprinzip*) because of the creation of organisms brought about by metamorphosis.

The concept of duality of moment (type) and eternity (metamorphosis) are implicit or explicit in Goethe's theoretical writings.

Sophie Annerose Naumann

See also Becoming and Being; Eternity; Evolution, Organic; Novels, Time in; Poetry

Further Readings

Goethe, J. W. von. (1987). *Collected works*. Princeton, NJ: Princeton University Press.

Gray, R. (1967). *Goethe: A critical introduction.* Cambridge, UK: Cambridge University Press.

Vincent, D. (1987). *The eternity of being: On the experience of time in Goethe's Faust.* Bonn: Bouvier.

Williams, J. R. (1998). *Life of Goethe: A critical biography.* Malden, MA: Blackwell.

Gospels

The early Christians used the Latin term *evangelium* to refer to the spoken message about Jesus Christ; in Old English, the translation was "god spel" (good tale), which eventually became "Gospel" as a generic label for written narratives of the life, death, and resurrection of Jesus. Four Gospels were accepted into the Christian canon, and around 40 other "apocryphal" Gospels have been identified. The four canonical Gospels were attributed to Matthew, Mark, Luke, and John, and they appear in that order in the Christian canon. Matthew and John were numbered among the 12 disciples of Jesus. Early church tradition associated Mark with Simon Peter, and Luke was an associate of Paul. Scholars generally accept the validity of the early church traditions about the authorship of the Gospels of Mark and Luke, but many are skeptical about the degree to which Matthew and John were responsible for the Gospels associated with their names.

The Gospels bear some similarities with the conventions of ancient biographies, but overall they are unique among ancient writings. As narratives, each Gospel has a beginning, middle, and end, but they do not follow strict chronological order in describing the events of the life of Jesus. Instead, they often group the sayings and deeds of Jesus by theme or topic.

Of the four Gospels, only Matthew and Luke begin their narratives about Jesus by describing his birth. They also include genealogies that describe Jesus' roots in the history of Israel, going back as far as Abraham in the case of Matthew and as far as Adam in the case of Luke. Only Luke describes an event from Jesus' childhood, namely, his visit to the Temple when he was 12 years old.

The first event recorded by Mark and John is the ministry of John the Baptist and his baptism of Jesus. All four Gospels emphasize John's role as a prophet who was preparing God's people for the coming of the Messiah and his kingdom. Immediately after Jesus' baptism, Matthew, Mark, and Luke describe Jesus' temptation by Satan during his 40 days in the wilderness. This initial, decisive confrontation with the forces of evil serves as the overture for his public ministry, during which Jesus repeatedly frees people from the domination of evil by means of healings and exorcisms.

The four Gospels differ in their description of the next events of Jesus' life. Matthew and Mark describe the beginning of his preaching in Galilee and the call of his 12 disciples. Luke also refers to

Jesus' return to Galilee, but then he recounts his sermon in the synagogue in Nazareth. This story serves to introduce important themes developed more fully in the Gospel of Luke and the Acts of the Apostles. After Jesus' baptism by John, the Gospel of John describes Jesus' call of four disciples and the miracle of changing water into wine at the marriage celebration in Cana. After these stories of the beginning of Jesus' ministry, Matthew, Mark, and Luke follow each other closely in terms of content and order. Because of their similarities, they are referred to as the "Synoptic" Gospels, a term that means "to see together." John is significantly different from the other three in both order and content.

All four Gospels describe an event in Jesus' life that serves as the turning point in the narrative. For the Synoptics, this event is Peter's confession of Jesus as Messiah followed by Jesus' first prediction of his suffering, death, and resurrection. That event is followed by the Transfiguration during which the heavenly voice declares that Jesus is God's Son. From then on, each narrative moves inexorably toward Jesus' arrest, trial, and death in Jerusalem. For the Gospel of John, the turning point is associated with Jesus' raising of Lazarus from the dead, an event that crystallized the intention of the Jewish leaders to plot Jesus' death.

All four Gospels contain a lengthy account of the last week of Jesus' life, which is called "the Passion Narrative." This narrative may have existed as a written document before its use by Mark, the first person to write a Gospel. The Passion Narrative begins with Jesus' triumphal entry into Jerusalem on Sunday, continues with various discourses and events that culminate in the Last Supper on Thursday night, and ends with his crucifixion on Friday. All the Gospels conclude with narratives of Jesus' appearances to his disciples after his resurrection on Sunday morning.

The Gospel writers were motivated to write narratives about Jesus because of their conviction that Jesus brought about the beginning of the end of history. Many Jews at that time believed that history would be divided into two parts. This present age, which is dominated by sin and death, would be replaced by the age to come. The event that would usher in the age to come was the Day of the Lord, when the Messiah would establish God's kingly rule on Earth. The resurrection of the dead was associated with the Day of the Lord.

Saint Matthew writing his Gospel under the Inspiration of Christ. From a miniature in a manuscript of the ninth century, in the Burgundian Library, Brussels (drawn by Count H. de Vielcastel).

Christians believed that, through his death and resurrection, Jesus conquered the powers of sin and death and therefore inaugurated the age to come. Because the age to come has broken into the present, the Gospel writers describe the kingdom of God as both present and future. Each Gospel attempted to show how the turning of the ages occurred through Jesus.

The majority view among scholars since the early 1900s is that Mark wrote the first Gospel about 40 years after Jesus' death and that Matthew and Luke used Mark as the framework for their Gospels, which were written 10 to 20 years later. Matthew and Luke supplemented Mark with a written source they had in common, which is called "Q" (from the German *Quelle,* which means "source"). Each included his own distinctive material, which is referred to as "M" and

"L," respectively. A minority of scholars have defended the centuries-old belief that Matthew was the first Gospel. Some have argued against the existence of Q, and some have posited a more complex interaction over time among all the Gospel writers.

Most scholars date the Gospel of John near the end of the first century. The narrative claims to be based on the eyewitness testimony of the Beloved Disciple, who has often been identified with John the son of Zebedee. Various scholars have proposed that the book was revised more than once by different editors. Most would concede that an editor (who was different from the original author or authors) compiled the final version that appears in the Christian canon.

Gregory L. Linton

See also Bible and Time; Christianity; Eschatology; God and Time; Time, Sacred

Further Readings

Bockmuehl, M., & Hagner, D. A. (Eds.). (2005). *The written Gospel.* Cambridge, UK: Cambridge University Press.

Hengel, M. (2000). *The four Gospels and the one gospel of Jesus Christ: An investigation of the collection and origin of the canonical Gospels* (J. Bowden, Trans.). Harrisburg, PA: Trinity Press International.

Stanton, G. (2002). *The Gospels and Jesus* (2nd ed.). Oxford, UK: Oxford University Press.

Gosse, Philip Henry (1810–1888)

In the middle of the 19th century there emerged a serious concern about time. Ongoing discoveries in geology, paleontology, and archaeology were challenging the then-held age of our Earth, fixity of species, and recent appearance of the human animal on this planet. Rocks, fossils, and artifacts were suggesting a new worldview, in terms of pervasive change throughout time, far different from the traditional static philosophy of nature. There was a glaring discrepancy between the scientific perspective of naturalists and the biblical framework of religionists. Inevitably, a major conflict developed between facts and beliefs. The empirical evidence for evolution contradicted the story of Genesis: Is the earth millions of years old, or was it divinely created only a few thousand years ago? A resolution seemed to be impossible, especially for those fundamentalists who held to a strict and literal interpretation of the Holy Scriptures.

Philip Henry Gosse made a bold and unusual attempt to reconcile the creationism of revealed religion with the evolutionism of the earth sciences. He was a fundamentalist creationist who belonged to the Plymouth Brethren sect in Victorian England. As such, his acceptance of the Mosaic cosmogony included a rigid adherence to both the sudden creation of Adam and the later Noachian deluge, each of these two events caused by the personal God of Christianity. However, Gosse was also an avid naturalist who could not easily dismiss the overwhelming and compelling factual evidence for organic evolution. Nevertheless, he rejected outright the dynamic concepts and disturbing implications (for him) of the evolutionary perspective. Therefore, his own personal beliefs required that he somehow resolve the contradiction between evolutionist science and fundamentalist religion. Gosse presented his incredible resolution in two books, *Life* (1857) and *Omphalos: An Attempt to Untie the Geological Knot* (1857).

Gosse's bizarre explanation for divine creation gave priority to biblical beliefs and immediate experience, rather than to science and reason. It is an interpretation ultimately based on an essential distinction between two different modes of existence: preternatural or ideal time in the perfect and infinite mind of a personal God prior to the act of creation, and the later natural or real time of objects and events in this material world of ongoing change and evolutionary development. In short, there is a crucial difference between prochronic time and diachronic time. Moreover, Gosse claimed that the course of everything inorganic and organic in nature is a circle.

Gosse firmly held that both the geological column of rock strata and the paleontological record of fossil evidence were suddenly created, all at once, along with the earth's existing plants and animals (including our own species) through the divine act of a personal God. In fact, Gosse further argued that this planet suddenly came into existence as an ongoing world already containing a sequence of fossil remains in a series of rock layers. Consequently, rocks and fossils only suggest both an extensive natural past for the earth itself and a vast evolutionary history for all previous life forms on this planet. Briefly, due to this instant creation with its built-in geological column and fossil record, the earth only appears to be as old as the geopaleontological evidence suggests to the evolutionary naturalists; the stratigraphic layers with their fossil remains were suddenly formed merely to suggest an ancient planet and the process of organic evolution.

According to Gosse, the appearance of our Earth together with its life forms occurred at a specific point, which was the violent irruption of natural time, at the beginning of cosmic reality from ideal time. Gosse thought that the empirical evidence of science is deceptive, and the untold eons of planetary time are simply an illusion. Consequently, he believed that his unique interpretation of creation would allow him to accept the evidence for evolution without accepting the process of evolution.

In a fantastic thought experiment, Gosse imagined himself experiencing plants and animals living on a pristine Earth just after the instant of creation. All of these organisms would appear to him to have had a planetary history but, he argued, in reality, this was not the case. He even claimed that Adam had been suddenly created with a navel so that it would appear as if he had had a natural birth! Furthermore, the observed rock strata and fossil record had been deliberately placed in the earth by God so that our planet would appear to have had a vast evolutionary history. It was a divine scheme to test the biblical faith of fundamentalist believers (or perhaps even to deceive them). For Gosse, the evolutionary naturalists are merely discovering the divinely imprinted evidence for geological history and biological evolution, although each had never occurred; neither geological nor biological changes had taken place over the assumed sweeping eons of planetary time.

Not surprisingly, Gosse's explanation for the creation of our Earth did not convince scientists or philosophers or theologians. Naturalists remained evolutionists, philosophers debated his ideas, and theologians grappled with the startling implications of the earth sciences. Gosse himself was devastated because his serious effort to reconcile science and religion was rejected by both evolutionists and fundamentalists. Ironically, since he believed in an instant creation of everything, his own theistic cosmogony is not compatible with either the biblical 6-day account of Genesis in the Holy Bible or the vast perspectives of modern process theologians.

Gosse's worldview is supported by neither science nor reason. Any fixed date for the sudden moment of divine creation would be arbitrary; in fact, one could argue that all reality (including natural time and our species) is still in the mind of God, with the origin of this material universe yet to take place. Invoking the law of parsimony, one would be wise to accept the temporal frameworks of cosmic evolution and earth history over Gosse's improbable worldview. However, the creation/evolution controversy over time continues, despite the ongoing advances in the special sciences. Gosse's desperate attempt at reconciliation remains an interesting relic of an outmoded religious-oriented cosmogony. A progressive understanding of and appreciation for time requires that one remain open to the continuing discoveries in science and the new perspectives in philosophy.

H. James Birx

See also Cosmogony; Creationism; Experiments, Thought; God and Time; God as Creator; Religions and Time

Further Readings

Gosse, E. (2004). *Father and son.* New York: Scribner.

Gosse, P. H. (1998). *Omphalos: An attempt to untie the geological knot.* Woodbridge, CT: Ox Bow Press.

Thwaite, A. (2002). *Glimpses of the wonderful: The life of Philip Henry Gosse, 1810–1888.* London: Faber & Faber.

Grand Canyon

The Grand Canyon, located in Arizona in the United States, is usually regarded as one of the world's seven natural wonders. The canyon, carved by the Colorado River and home to many ecosystems and artifacts, still has scientists wondering exactly how and when it was formed. Each year, nearly 5 million people visit the Grand Canyon.

The canyon is expansive, stretching 277 miles from Lake Powell at the Utah–Arizona border down into Lake Mead, near Las Vegas. The canyon is 6,000 feet at its deepest point and 15 miles across at its widest point.

The wildlife and plant matter found within the Grand Canyon are diverse, attributed in large part to its size, depth, and variation. While the Grand Canyon is very large and wide at some locations, its range of height and size vary along its course and its weather is diverse. Weather varies as much as 5.5 degrees Fahrenheit with each 1,000 foot change in elevation. At certain points, then, the temperature may vary an average of 33 degrees from the upper rim to the river walls. These weather variations allow for the presence of five of the seven life zones and three of the four desert types in North America. The five life zones represented are the Lower Sonoran, Upper Sonoran, Transition, Canadian, and Hudsonian.

The origin of the Grand Canyon is still something of a mystery. Scientists continue to study and hypothesize about its age and origin, but cannot come to a consensus. Despite agreement that the layers of granite at the deepest portions of the gorge are 1.7 billion years old, it is widely accepted that the canyon itself is nowhere near as old.

The Grand Canyon in Arizona. Layers of the Grand Canyon were formed over a period of about 2 million years.

Source: The National Park Service.

Modern geologists are puzzled by the seemingly faster erosion in some portions of the canyon than others. This may be explained by the most widely accepted theory of the canyon's formation—the upper sedimentary layers were the base for several small shallow lakes that existed for hundreds of thousands of years. Over time, the rock rose through tectonic forces to create the Colorado plateau. All the while, the Colorado River cut down through the rock, but was thwarted by the harder portions of igneous rock, which now are some of the narrowest portions of the canyon. This whole process was said to have begun about 70 million years ago. Some researchers argue, however, that the landscape of the canyon is far too immature and must be newer than 5 million years old.

Yet another theory suggests that two separate river systems, not lakes, existed and merged to form what is now the Colorado River. They speculate that the westward river absorbed the eastward river in a process called headward erosion. This theory suggests that the headward erosion process would have begun about 15 million years ago.

Some of the most recent research suggests that the gorge may have been carved within the past million years. Scientists suggest that it was only recently that we have understood just how quickly erosion can take place, drastically affecting the timeline of the creation of the canyon.

Although the age of the Grand Canyon remains in dispute, it is home to some of the oldest, most diverse natural elements in the world.

Amy L. Strauss

See also Erosion; Geological Column; Geologic Timescale; Geology; Ice Ages; Old Faithful

Further Readings

Anderson, M. (Ed.). (2005). A gathering of Grand Canyon historians: Ideas, arguments and first-person accounts: *Proceedings of the inaugural Grand Canyon History Symposium, January 2002*. Grand Canyon, AZ: Grand Canyon Association.

Fletcher, C. (1989). *The man who walked through time: The story of the first trip afoot through the Grand Canyon*. New York: Knopf.

Ranney, W. (2005). *Carving Grand Canyon: Evidence, theories, and mystery.* Grand Canyon, AZ: Grand Canyon Association.

Young, R., & Spamer, E. (Eds.). (2003). *Colorado River: Origin and evolution.* Grand Canyon, AZ: Grand Canyon Association.

Gravity

See Relativity, General Theory of

Greece, Ancient

See Presocratic Age

Grim Reaper

In Western cultures, the Grim Reaper is the most commonly known personification of death. Its current form as a macabre and sinister image has evolved over time, beginning as a feminine figure in the ancient world and transforming into the masculine form in ancient Greek literature. These personifications and images are preserved in art, literature, and film. In the future, the form assumed by this figure will naturally reflect the philosophies, ideologies, and technologies of the culture of the time, and the image of the death guide will reflect the notions of death held by tomorrow's generations. The personified death image may be portrayed as an assistant or travel guide, or it may be a more diabolical instigator, a big brother, or a computer-generated entity that perpetrates the dance of death.

Before the 19th century, the Grim Reaper was more commonly referred to as an Angel of Death. This icon of death was a winged angel that waited to escort the living to their place in the afterlife. The image of this angel was much less menacing and gruesome than today's Grim Reaper, a hooded and cloaked skeleton. The skeleton as a symbol resonates deeply within the human psyche, signifying the importance and the inevitability of death. The Grim Reaper carries a scythe or a sickle, an agricultural instrument that for centuries has been used to cut down and harvest crops. He uses his instruments to harvest or gather the living and take them to their death. Using various disguises such as illness, sudden accidents, and catastrophes, he randomly strikes and unquestioningly takes the lives of young and old, male and female, rich and poor, without discrimination.

The Grim Reaper's modern appearance originates in Greek literature and art, which associates the god Cronus with the passing of time. The god Cronus harvested crops with a sickle, which he later used to castrate his father, Uranus. Cronus later went on to swallow all of his children, except Zeus, out of fear that they might in turn retaliate against him in the same way he had retaliated against his father. Cronus is often seen carrying this scythe, a tool similar to the sickle, and is also depicted with an hourglass in his role as Father Time.

In oral tradition, some similarities between Cronus and the Grim Reaper developed, and that is why we often see the Grim Reaper brandishing the scythe. Sometimes, the Grim Reaper is also seen carrying an hourglass, an image intended to remind us that time never stops and that we are mortal. With each grain of sand that falls, we are moving closer to the end of our time on Earth.

Debra Lucas

See also Cronus; Dying and Death; Father Time; Longevity; Satan and Time; Youth, Fountain of

Further Readings

Lonetto, R. (1982). Personifications of death and death anxiety. *Journal of Personality Assessment, 46*(4), 404–407.

Williamson, J., & Schneidman, E. (1995). *Death: Current perspectives.* Mountain View, CA: Mayfield.

Guth, Alan

See Cosmology, Inflationary

Haeckel, Ernst (1834–1919)

Greatly influenced by reading Charles Darwin's *On the Origin of Species,* Ernst Haeckel accepted the fact of evolution and became known as the "Darwin of Germany" for his own extensive researches and copious publications. He had studied medicine, become interested in botany and zoology (especially marine organisms and comparative embryology), and eagerly extended the consequences of evolution to philosophy and theology. His focus on time change included a serious consideration of the origin and history of our species within a cosmic framework.

Unlike the cautious Darwin, the bold Haeckel did not hesitate to consider both the philosophical implications and theological ramifications of evolution for understanding this universe in general and appreciating life on Earth in particular. He argued for the essential unity of this dynamic cosmos. His worldview acknowledged the origin of life from matter in the remote past, and then, over eons of time, the enormous diversity of plants and animals that appeared throughout organic history.

Haeckel did not know the true age of this universe or of the earth. He was unaware of the vast durations of time represented by the geologic column of our planet, with its fossil record. Even so, his mind always remained open to those new facts, concepts, and perspectives that were being contributed by the evolutionary sciences. Likewise, Haeckel never hesitated to criticize severely those individuals (particularly religionists) who refused to accept the fact of evolution and its devastating consequences for all earth-bound and human-centered interpretations of reality. For him, cosmic immensity and the probability of life forms and intelligent beings existing elsewhere on other worlds was sufficient reason for doubting that our own species occupies a special place in evolving nature.

Before Darwin had done so, Haeckel wrote that the human animal had emerged from an apelike form. He speculated that Asia was the birthplace of the first hominids, hypothesizing that a now-vanished land mass (Lemuria) had been the geographical location where the evolution of humanlike hominoids from apelike hominoids had occurred. He even gave the scientific name *Pithecanthropus alalus* (ape-man without speech) to this assumed "missing link" between fossil apes and the first humans.

Furthermore, Haeckel argued that our species and the three great apes (orangutan, gorilla, and chimpanzee) differ merely in degree rather than in kind, due to the relatively recent separation of the earliest hominid form from a common ancestor with the fossil apes in terms of evolution. He also claimed that all human biological characteristics and mental activities had slowly evolved from earlier apelike forms. Consequently, our species is a recent product of, and totally within, material nature.

Ernst Haeckel's most popular work is *The Riddle of the Universe,* in which he presented the basic ideas of his evolutionary worldview. For him, God and the universe are the same entity,

One of the principal supporters of Charles Darwin's theories of evolution in Germany, Ernst Haeckel wrote wide-ranging books addressed to the general public, and his illustrations inspired many artists.

Source: Getty Images. Photograph by Nicola Perscheid, 1905.

with things endlessly becoming and passing away throughout cosmic time. Moreover, as a result of his lifelong studies in comparative morphology, Haeckel himself drew the first tree of life to illustrate the evolutionary relationships among organisms on earth. The fascinating drawings for his scientific publications on evolution and related subjects may still be seen at the Ernst Haeckel House, now a museum in Jena, Germany.

H. James Birx

See also Darwin, Charles; Evolution, Organic; Huxley, Thomas Henry; Spencer, Herbert; Time, Planetary

Further Readings

Birx, H. J. (1984). *Theories of evolution.* Springfield, IL: Charles C Thomas.

DeGrood, D. H. (1965). *Haeckel's theory of the unity of nature.* Boston: Christopher Publishing.

Haeckel, E. (1992). *The riddle of the universe at the close of the nineteenth century.* Amherst, NY: Prometheus. (Original work published 1899)

Hammurabi, Codex of

The Codex of Hammurabi is one of the oldest sets of laws in the world. The over-2-meter-long slabs with their engraved laws were created at the behest of Hammurabi (presumably c. 1792–1750 BCE), a significant ruler of the Babylonian dynasty and king of Sumer and Akkad. The Codex offers an early document of human law that is defined by its reference to the past; in other words, to the divine commandments and common customs that paved the way for the legitimacy of legal regulations.

Similar to its predecessors, among which the Codex of Ur-Nammu, the Codex of Lipit-Ishtar, and the Codex of Eshnunna are the best-known examples, the concept of a "codex" is misleading, because, in a modern sense, a codex is understood to be a generally valid legal system whose enforcement is guaranteed by the state. The ancient Oriental codices, however, are incomplete collections of individual legal cases that usually express the will of the lawmaker to reform outdated laws or traditions. This also applies to the Codex of Hammurabi. But it is different from earlier codices in that it aspires to exert a standardized system of law on the entire realm of the given ruler. This was unprecedented in the ancient Orient, where a generally binding common law in accordance with state regulations was still unknown; justice was usually administered by virtue of traditional customs. However, there are contradictory reports about whether Hammurabi's project was successful or not: Some historians assume that the individual paragraphs were actually taken from everyday legal practice because many trial certificates express the same point of view in their verdict as the Codex. Other historians consider the Codex to have been a more theoretical manifesto that was never directly put into practice because, among other things, many of the reported verdicts refer to Hammurabi's regulations, but none of them mention the inscriptions.

Also, similar to earlier collections of law, the Codex of Hammurabi consists of a prologue, laws, and an epilogue. Both the prologue and the epilogue serve to legitimize the text: for example, at the beginning, by showing the symbols of power that Hammurabi receives from the hands of the sun god Shamash; and, at the end, by referring to the social justice of the legal ordinances. The collection itself consists of 282 paragraphs that, in addition to regulating a number of classical criminal and civil law cases, deal with such seemingly modern areas as the law of tenants and landlords, building regulations, and tort laws. From a sociological point of view, these reforms were designed to solve problems that derived from the socioeconomic developments of the 2 centuries prior to the Codex, during which Babylon grew from a fortified settlement to a metropolis on the Euphrates with a complex social and economic structure.

The Codex's principles of justice, the talion law as well as mirror punishment, are of particular importance to the history and philosophy of law: Talion (Latin *talio,* revenge; Greek *talios,* equal) was seen as a legal figure who sought to achieve a balance between the damage done to the victim and the damage that was to be done to the perpetrator. The best-known formula is "an eye for an eye," which also found its way into the Judeo-Christian Bible. Mirror punishment goes beyond the idea of enacting the same action upon the criminal as the criminal perpetrated upon the victim. It also implies punishing the part of the perpetrator's body used to commit the crime (e.g., chopping off the hand of a thief).

These punishments not only seem draconian today, they also seemed to some extent draconian in Hammurabi's era in comparison to more ancient codices, such as the Codex of Eshnunna. The Codex of Hammurabi even permitted the death penalty for minor offenses like theft or libel. Certainly one of Hammurabi's central motives was to employ methods of deterrence to consolidate his empire, which covered a large section of Mesopotamia. Another reason might have been that imprisonment was not really part of the nomadic tradition, which essentially knew no prisons. Nevertheless, the concept of justice expressed in the principle of talion was adapted again and again. A number of paragraphs from the Codex of Hammurabi can also be found elsewhere, for example, in the Bible, which is how the Codex exerted an influence on the Occident's sense of right or wrong.

Oliver W. Lembcke

See also Bible and Time; Egypt, Ancient; Ethics; Law; Morality; Ur; Values and Time

Further Readings

Mieroop, M. v. d. (2005). *King Hammurabi of Babylon: A biography.* Oxford, UK: Blackwell.

Richardson, M. E. J. (2000). *Hammurabi's laws: Text, translation and glossary.* Sheffield, UK: Academy Press Sheffield.

Harris, Marvin (1927–2001)

Marvin Harris, American anthropologist, was known for his theoretical contributions to the concept of cultural materialism. His theoretical perspective of cultural materialism, though not a unifying theory of culture, sought to explain variations and commonalities among human cultures. As reflected within previous ethnographic research, the development and advancements of various human cultures exhibit many similarities and differences. Amid documenting these cultural characteristics, anthropologists have formulated various theoretical perspectives to account for these cultural aspects. The comparison of cultures, especially in a spatiotemporal framework, has inherited epistemological and teleological problems. These problems are revealed in the paradigm shifts seen within the history of cultural anthropology. Harris understood these philosophical problems and developed a unique approach to, and interpretation of, humankind's cultural material existence.

Harris questioned the theoretical basis for the idea and evaluation of culture. In cultural materialism, he viewed cultural structure and its components in terms of expressed culture (all aspects) within the process of selection. Focusing on infrastructure, structure, and superstructure, the symbiotic relationships provide the basis for cultural stasis, cultural advancements, and diversification. Lacking an ascribed ontology and teleology, this unique perspective encompasses all aspects of time—for example, synchronic and diachronic—within a social and evolutionary framework. Expressed infrastructural differences between etic/emic modes of production and domestic/political economy are believed to be dynamic variants of the superstructure; albeit the stability of the superstructure appears to remain more dependent on the infrastructure. This would be reflected on the rates of change and the appearance of any contradictions seen between infrastructure components and their related superstructure. These structures, when juxtaposed with social hierarchy, provide a sustaining and practical benefit for its members. Although social and economic equality is elusive, social power (which would include empowerment) was stated as one of the selective forces of social dynamics that influence the complex nature of both infrastructure and superstructure.

Human culture is both unique and complex. From sociobiology to postmodernism, the wide range of theoretical constructs was seen by Harris as a set of segmented perspectives that lack a comprehensive view. Though Harris never claimed cultural materialism as a comprehensive accounting of culture, his modified Marxist approach attempted to provide the best explanation for the conditions of human culture. The categories of infrastructure and superstructure are seen to provide the best possible avenues for objective study, although the exact relationship of the parts to the whole is tenuous and unpredictable. The concept of time, in an evolutionary framework, can be depicted as a fluid continuum by which material culture reflects human behavior. In this manner, his theoretical perspective never claimed to be a social manifesto. Harris defended science (with all its limitations) and human dignity and value from the metaphysical implications of contending cultural theories. His theoretical perspective and social awareness had a profound influence on cultural anthropology and the understanding of human nature and culture.

David Alexander Lukaszek

See also Anthropology; Evolution, Cultural; Evolution, Social; Morgan, Lewis Henry; Tylor, Edward Burnett; White, Leslie A.

Further Readings

Harris, M. (1980). *Cultural materialism.* New York: Vintage Books.

Harris, M. (1991). *Cannibals and kings: Origins of culture.* New York: Vintage Books.

Harris, M. (1999). *Theories of culture in postmodern times.* Thousand Oaks, CA: AltaMira.

Harrison, John (1693–1776)

John Harrison, born in Foulby, England, became one of the world's most renowned horologists. He won the Board of Longitude Prize for developing a chronometer that could be used aboard ship to measure longitude to within 0.5 degree at the end of a voyage to the West Indies. About 1720 he had designed a timepiece that included a compensating apparatus by using different metals for correcting errors due to variations in the weather. Scientifically speaking, Harrison invented a timepiece that allowed for temperature changes or distortion. The first chronometer, which weighed 65 pounds, was completed and submitted to the Board in 1735 and was tested aboard ship the following year. The accuracy of the chronometer was outstanding, but like many inventions it had its detractors. He then built three more; the fourth, in 1761, more than met the standard for the prize, as did the first one. But it wasn't until 1773 that he was fully compensated by the Board.

Up to the early 18th century, ships and their cargoes, along with the mariners, were at extreme risk between ports of call because of not knowing their exact location. After all, an hourglass is not exactly the best timepiece for determining time at sea. John Harrison might well be called the "father" of time at sea, as his invention of an accurate timepiece was the final link in being able to determine longitude, something Galileo thought would be the most precise method for determining an east–west position. In 1730, John Hadley in England, and Thomas Godfrey in America, working independently, perfected the sextant, which accurately found latitude by "shooting" the sun at noon. Latitude, which is part of the earth's coordinate system for measuring relative location, had been rather easily observed and measured for centuries by determining the sun's angle above the horizon, early on using a cross-staff or back-staff. On a globe, latitude is shown as east–west lines that measure distance or position north–south, while longitude is shown as north–south lines that measure distance or position east–west. The earth makes a full revolution on its axis from west to east every 24 hours, or 15° of longitude each hour, the equivalent of one modern-day time zone.

Knowing one's exact longitude at sea is important for determining a ship's position with respect to land. Shipwrecks due to position miscalculations were so common that a Board of Longitude was organized in England in 1714 for the purpose of awarding the sum of £20,000 to anyone who could develop a method for accurately measuring longitude. John Harrison's contribution to measuring this accurately was a timepiece, known as a chronometer, that could be used at sea with a fair amount of accuracy. The timepiece was tested on a voyage to Jamaica with his son William Harrison on board, in 1761–1762, and determined longitude to within 18 nautical miles. Later, in 1764, the chronometer was tested again during a voyage to Barbados with his son on board. The timepiece performed brilliantly and well within the standard prescribed by the Board of Longitude. The fifth chronometer was used by Captain James Cook during his journey across the Pacific Ocean in 1776, although Cook was killed by Hawai'ians before the voyage was completed.

Richard A. Stephenson

See also Astrolabes; Hourglass; Latitude; Time, Measurements of; Timepieces

Further Readings

Bowditch, N. (1966). *American practical navigator.* Washington, DC: Government Printing Office. (Original work published 1802)

Christopherson, R. W. (2004). *Elemental geosystems* (4th ed.). Upper Saddle River, NJ: Prentice Hall.

Dunlap, G. D., & Shufeldt, H. H. (1972). *Dutton's navigation and piloting.* Annapolis, MD: Naval Institute Press.

Hartshorne, Charles (1897–2000)

Charles Hartshorne was one of the 20th century's most distinguished philosophers of religion and metaphysicians. He received his Ph.D. at Harvard University, where he was a student of Alfred North Whitehead. Hartshorne taught at the University of Chicago, at Emory University, and at the University of Texas. His ideas about time were influenced by Charles Sanders Peirce, Alfred North Whitehead, and Henry Bergson, as well as by Edmund Husserl and Martin Heidegger.

Hartshorne rests his philosophy on two bases: Its internal coherence and its adequacy in interpreting the facts of experience. Hartshorne believes that the appeal to observation is the sole possible basis of rational knowledge, which includes the purely mathematical. Therefore, the starting point of one's worldview should be one's daily experiences. The world is full of discontinuities, which are measurable as being greater or lesser. One can see them against a background of continuity, such as the continuity of space, time, and color qualities. Since our direct connection to matter is via sensation, the identity of mind and matter is the clue to the nature of things.

Hartshorne agrees with Baruch Spinoza and Whitehead that the primary physical data are within people. However, there is a veil between us and the "external world," and this veil is physiological. Hartshorne himself calls his philosophy *psychicalism.* This means that his philosophy was not derived from physics, but rather from phenomenological observations of sensations as a special class of feelings. The natural sciences tend to abstract from the mind and experience, even though they are (ultimately) derived from experience. But it is not only science that abstracts; even experiences based on the senses are enormous simplifications of the perceived world. Knowledge is not complete until these abstractions are overcome.

What is given to people has two main forms: previous experience by the same person, and other types of events that are not obvious but nonetheless may still be experiences as well, though not experiences by the same person.

The first form occurs in what we usually call memory; the other in what we call perception—the data that always include events in the series of events that have constituted one's own body.

Experiences are directly conditioned by their immediate data, which temporally precede the occurrence of experiences. They must follow and thus may not occur simultaneously with their data. This one-way dependence is a key to time's arrow and to causality: In contrast to classical determinism—which regards the world as a single, tightly interlocked system in which everything causally implies everything else, backward and forward in time—awareness, as in perception or memory, implies the selective and asymmetrical dependence of experience upon the experienced. In memory and perception the experienced events are temporally prior to the experiences.

This directionality of becoming is lost within classical physics. All conditioning involves the temporal priority of the conditions. The conditions are the data, the directly experienced factors. An experience is never identical with its data. No observation could establish an absolute simultaneity. Our categories are positively applicable only in temporal terms. Determinism destroys the structure of time by negating the asymmetry of becoming. The laws of physics that have given an explanation of time's arrow are stated as statistical approximations. Facts of experience give us direct evidence of the reality of contingency.

What the future distinguishes from the present and the past is that future experiences are characterized by their lack of distinct detail: "The future is in principle a rough blueprint, the past a photograph in full color." But the question must be asked as to whether this asymmetry is essential to time or whether it is merely a fact of human psychology. Hartshorne shows that the "not yet" of the future may have a meaning only for a mind not in full enjoyment of all details of what is to come. The future is that in which full consciousness is lacking. We conceive the unity of the different aspects of time—past, present, and future—in the way illustrated by our experiences of memory and anticipation. Without memory and anticipation, "past" and "future" would be meaningless words.

What is the nature of the relation between the past and the present? One explanation is that the cause is past but the effect is present. The cause is the predecessor of the effect. What, then, is the relation of causality? An objective serial order does not require that there be strictly deterministic causal relations. Whitehead, James, Bergson, Immanuel Kant, and others have all answered David Hume's problem in psychological terms. The simplest positive answer furnished by experience is memory. An objective temporal order is explicable if all reality is some form of experience, with each unit endowed with some form of memory. Hartshorne agrees with Whitehead, who has shown how "extension" as well as temporal succession can be described in psychic terms. For Hartshorne there is a world, a temporal–causal process. Things either intrinsically refer to other things, for example, to past events, or they contain no such internal reference to other things. If there is such reference, then it is at least as if the thing perceived or remembered or felt has existed or even still exists. This means that the world is a temporal–spatial network of events. If there is no such reference, then the world has no real connectedness. There would be no world, no real succession of cause–effect at all. Therefore, either everything must be as if idealism were true, or else as if there were no world, no real temporal–causal system. The subject of experience is a wholeness, which is temporally extended through the "specious present" (James), or the quantum of psychic becoming. Spatially, it is one through the voluminous rather than punctiform character of its perspective, or dynamic relationship with other entities.

Joachim Klose

See also Bergson, Henri; Consciousness; Heidegger, Martin; Hume, David; Husserl, Edmund; Idealism; Kant, Immanuel; Memory; Metaphysics; Spinoza, Baruch de; Whitehead, Alfred North

Further Readings

Hartshorne, C. (1932). Contingency and the new era in metaphysics (I). *Journal of Philosophy, 29,* 421–431.

Hartshorne, C. (1937). *Beyond humanism—Essays in the philosophy of nature.* Lincoln: University of Nebraska Press.

Hartshorne, C. (1997). *The zero fallacy and other essays in neoclassical Philosophy.* Chicago: Open Court.

Hawking, Stephen (1942–)

Stephen William Hawking, probably the best-known physicist since Albert Einstein, is Lucasian

Professor of Mathematics at Cambridge University. Hawking's work with the more exotic areas of theoretical physics and his best-selling book *A Brief History of Time* have earned him a place in the public eye as well as in his own chosen world of physics. He is best known for his contributions to black hole theory and scientific cosmology.

British theoretical physicist Stephen Hawking, Cambridge University Professor and Fellow, circa 1985.
Source: Getty Images.

Physics: The Background

Hawking's ideas of time were influenced by the work of scientists and philosophers over two millennia. Aristotle, for example, made many statements about the arrangement of the universe. The whole world was spherical and finite, he postulated in the 4th century BCE. The earth was at the center of the Aristotelian universe, surrounded by one concentric sphere each for the sun, the moon, Mercury, Mars, Venus, Jupiter, and Saturn, and one sphere for the stars. The outermost sphere, containing the stars, was considered superior to the sphere in which the earth resided, and was supposed to be composed of an element called the ether. This concept of interstellar ether would persist in various forms through the early 20th century. The space occupied by the earth, however, was made of the four classical elements of earth, air, fire, and water. Furthermore, on the surface of the earth, the heavier of two bodies that had the same shape would fall faster. Many of Aristotle's postulates were eventually proved false, but his ideas nevertheless became firmly entrenched in science.

Nicolaus Copernicus, over a millennium later in the early 1500s, made the next great development in astronomy. His theory, based on observation, stated that the earth makes not only one complete rotation on its axis per day (as well as executing a slight wobble around this axis) but also one complete revolution around the sun each year, producing the seasons. The sun was placed uncompromisingly at the center of the universe. The earth took its place among the other planets in orbit around the star, with an orbital period of one year. Copernicus retained the celestial spheres of the Ptolemaic and Aristotelian theories, but he associated a greater orbital radius with a longer planetary year, which necessitated a rearrangement of the five known planets into their proper order.

Early in the tumultuous 17th century, an Italian scientist named Galileo Galilei was at the heart of a battle over the Copernican model of the universe. With his innovative new telescope, Galileo discovered Venus's phases, Jupiter's four largest moons, sunspots, and the moon's topographical features. Venus was of particular interest, since its phases proved that the planets revolved around the sun. In addition, Galileo's observations showed that not everything must orbit the sun directly. His discoveries were not limited to the heavens; Galileo conducted experiments that yielded the laws of both falling and projected bodies. His legendary experiment of dropping two objects from the Leaning Tower of Pisa is just one example of Galileo's many attempts to explain the universe. Galileo's work, together with that of Johannes Kepler, led directly to the epitome of classical physics that was Sir Isaac Newton's lifework.

Newton, who was born in the same year that Galileo died, tied together many of the loose ends of classical physics with strings of mathematics. He is the generally accepted founder of modern calculus. Newton sought to explain natural philosophy, as physics and chemistry were collectively known at the time, with his mathematics. His early work was with optics, and led to the invention of the reflecting telescope, which caused fewer color anomalies than Galileo's refracting model. He later turned his attention to gravity, recognizing that the same force kept the earth orbiting the sun and made apples fall to the ground. Newton derived a law that explained almost every behavior of gravity. The inverse square law calculates the strength of gravity dependent upon the distance between two objects. Newton used this equation to calculate the attraction in the solar system and to explain both the tides and the moon's motions to a high degree of accuracy. Though many of his equations are still used today, Newton's work was based upon a model of the universe that would soon be radically changed. The picture of a three-dimensional space full of Aristotle's ether, with an absolute measure of time that never warped, was shattered irretrievably by a Swiss patent clerk named Albert Einstein.

Einstein made the biggest step forward in cosmology since Copernicus rearranged the solar system. He united Newton's and Huygens's theories of light and wrote the special theory of relativity. This second development, which at the time was understood by only a few physicists, secured Einstein a permanent place in the history of physics. The cornerstone of the theory of relativity is what Einstein called the principle of equivalence: there is no difference between constant acceleration and gravity. In the classic example, a person in an elevator that is moving through space with a constant acceleration equal to the gravitational force on Earth feels no different than if stationary in the elevator on Earth. From this postulate, Einstein realized that there is no absolute reference frame; every frame of reference is equally valid. Any of these frames can be considered stationary if it moves at a constant velocity, and anything within one of these frames will obey Newton's laws of physics. The same holds for any person moving at a constant velocity relative to another frame of reference. Not only did Einstein change the definitions of movement and rest, he also rewrote gravity. Out of the relativity theories came the image of spacetime, a four-dimensional medium through which everything moves and that can be pulled or folded like fabric. Ether was expelled from scientific thought. Gravity was no longer a force, but a warpage of spacetime. Any object trying to travel along a straight line would get caught in a massive object's gravity-caused spacetime warp and follow the circular or elliptical path made by the warp—in other words, it would orbit the massive object. In his later work, Einstein tried to unite the weak and strong nuclear forces with the electromagnetic force and gravity. He failed, and many scientists, including Hawking, are still trying to accomplish the same feat 50 years later.

Life

Hawking's interest in cosmology had its roots in his early childhood, long before his mathematical and scientific aptitudes emerged around age 14. He always liked controlling things, and he built a great number of models during his childhood. Hawking enjoyed the theoretical work inherent in designing the models. When he was a young man, his mother watched him walk home one night after the streetlights were turned out. He watched the stars, and she sensed that he always would, despite his dislike of observational astronomy.

Much of Hawking's best work has been done in conjunction with Roger Penrose, the eminent mathematician. When Hawking was one of Dennis Sciama's Ph.D. students, the postgraduate group benefited greatly from attending a series of lectures that Penrose had given on singularities, the unpredictable points of infinite density found at the center of black holes. Hawking immediately began applying Penrose's ideas to the entire universe. Later in Hawking's career, he worked directly with Penrose to investigate the big bang singularity.

Childhood and Adolescence

Hawking was born on January 8, 1942, in Oxford to Isobel and Frank Hawking. His father was a medical doctor who specialized in tropical diseases. When he was 8, the Hawkings moved to the small town of St. Albans. Hawking has since described the place as conformist. His family felt out of place and was considered outlandish by their neighbors.

Isobel was an ex-Communist who still had strong left-wing sympathies that she passed on to her oldest son. She liked to travel while her husband was in Africa, and the family visited many exotic places. Hawking modeled himself after his often-absent father.

He entered St. Alban's, an excellent secondary school, at age 10. He was awkward, lisped, and was poor at anything physical. He developed passions for mathematics and classical music. Hawking gained a group of friends similar to himself, considered rather geeky by their peers. They built a basic computer called LUCE, which they programmed to solve addition problems. In their teens, the group experimented with the metaphysical. After attending a lecture that outlined problems with reports of ESP success, Hawking decided that everything metaphysical was either false or had a scientific, rather than supernatural, explanation. He has retained the same view throughout his life.

When picking classes one year, Hawking had a dispute with his father over a mathematics class. Hawking's father considered mathematics to have few career options. However, Frank lost the argument and Hawking continued his study of math. Hawking applied to University College at Oxford, his father's alma mater. At 17, he made it through the grueling admissions exams and interrogation-style interviews, was accepted, and received a scholarship.

College and Postgraduate Study

The first year of college did not go very well for Hawking. He suffered from depression, had few friends, and was bored with the work. At the time, he has said, no one worked especially hard in Oxford. The academic load was light; for an intelligent person, college was easy. Hawking has said he worked, on average, about an hour per day through his 4 years at Oxford. The second year, he joined the rowing team and became popular almost immediately. Social events suddenly opened to him, and Hawking threw himself into the center of college life with gusto. He admits to drinking his fair share and playing many practical jokes. In this spirit, Hawking finished his studies and realized that he had not prepared properly for his final exams. In England, not all college degrees are created equal. A graduate could finish with a first, second, or lower-class degree. The class depended upon the student's grade on the final exams. Hawking's score was on the borderline between a first- and second-class degree, which necessitated that Hawking explain his plans to the college authorities. He did so, received a first-class degree, and entered Cambridge in October 1962.

Fred Hoyle was the most famous cosmologist in Britain in the early 1960s. Hawking applied for his Ph.D. study with Hoyle as his advisor, but Hoyle turned him down, and Hawking received Dennis Sciama instead. Once over his original disappointment, Hawking realized that Sciama was a better and more available advisor than Hoyle was. However, he was depressed again during his first semester. The work was difficult. Hawking had an insufficient grasp of mathematics, and although he could still stumble along with everyone else, he wasn't doing as well as he was used to doing. During this period, he also began having motor problems. In the morning, Hawking had trouble getting his hands to work to tie his shoes. At one point, he fell down a flight of stairs at school and had temporary amnesia. He took an IQ exam to make sure no permanent damage had been done, but the trouble continued.

As soon as Hawking returned home for the holiday break, Isobel noticed his clumsiness. His father thought, naturally, that Hawking had picked up a foreign disease on a trip. Hawking missed the beginning of the next school term because he was in the hospital for tests. He was told that he had an unusual case, but his diagnosis didn't arrive until after he returned to school: amyotrophic lateral sclerosis (ALS), better known as Lou Gehrig's disease. Hawking's prognosis was 2 years. The disease began developing quickly, forcing Hawking to walk with a cane. He spent a good deal of the next few months in his dorm room, deeply depressed.

Paradoxically, however, Hawking's life was about to turn around. At his parents' New Year's Eve party in 1962, he met a young woman named Jane Wilde. The two fell in love quickly, and were soon engaged. Hawking realized that he needed his Ph.D. to support a family. Frank, Hawking's father, went to see Sciama about shortening the time requirements for Hawking's Ph.D., but Sciama refused. He treated Hawking no differently from any of his other students, which greatly endeared him to Hawking. Hawking's fellow students—George Ellis, Brandon Carter, and Martin Rees—became his best friends and later his colleagues. Soon Hawking looked for a suitable thesis project. He became involved in solving some equations for a student of Hoyle's that related to the expansion of the universe. The subject interested him greatly. When, shortly thereafter, he attended Penrose's lectures on singularity theory and applied it to the big bang model of the universe, the two ideas coalesced into what would become Hawking's thesis project. He graduated in 1965, after producing a manifesto on expanding universes, and married Wilde in July.

Work and Marriage

Hawking received a fellowship in Gonville and Caius College at Cambridge, in the Department of Applied Mathematics and Theoretical Physics (DAMTP). The young couple moved to a two-story cottage on Little St. Mary's Lane.

The Hawkings entertained frequently. Hawking is known to all of his friends to be a very gregarious person and a trickster. During the mid-1960s, he cultivated an incredibly intelligent, cranky savant image. Though Hawking was still relatively unknown in the physics community, this image grew, along with his reputation for asking penetrating and sometimes uncomfortable questions at lectures, academic evidence of his love for mischief.

The mid 1960s were a very important time for Hawking. In 1967, his first child, Robert, was born. The next year, Hawking became a staff member of the Institute of Theoretical Astronomy. By that time, the ALS had restricted him to a wheelchair. His second child, Lucy, was born in November of 1970. Four years later he moved the entire family to California to do some work at Caltech. It was a happy time for the Hawkings, especially when, on their return to England, the Royal Society invited Hawking to be inducted as a Fellow. This accolade meant a great deal to him.

When the Hawkings returned from their trip to California, they moved to a larger, one-story house on West Road. Though it was a little farther away from Hawking's office at the DAMTP, the single level made it easier for him to move around. The ALS had worsened again. From 1975–1976, Hawking's work drew steadily more attention, earning him many awards: the Eddington Medal from the Royal Astronomy Society, the Pius XI Medal from the Pontifical Academy of Science, and the Royal Society's Hughes Medal, among others. During the second half of the decade, the popular publicity Hawking received also rose dramatically. More honorary titles and awards piled up, including the Albert Einstein Award in 1978, esteemed more highly among physicists than the Nobel Prize.

However, things were not so rosy on the home front. Hawking's marriage was undergoing continuous stresses. Jane, a devout Roman Catholic, was aggravated by her husband's efforts to explain everything with science. Hawking argued with Jane more than once about the role and designated abilities of a divine Creator. In addition, Jane wanted to pursue her own career, and felt that she was a mere appurtenance to Hawking. These problems eventually led to their divorce in 1990. In the 1970s, however, that was still far in the future, and in 1979 Hawking's last child, Timothy, was born. The same year, Hawking was made the Lucasian Professor of Mathematics, Newton's chair at the University of Cambridge.

At this point, with a new baby and two private school tuitions to pay for, there was a dearth of money. Without telling anyone, Hawking began working on a popular cosmology book, but the dream wouldn't bear fruit for years. In the meantime, more accolades poured in. In 1981, Hawking was knighted a Commander of the British Empire, and four colleges, including Notre Dame and Princeton University, made him an honorary doctor of science the next year. In 1985, he embarked on a world lecture tour and had his portrait hung in England's National Gallery. However, the same summer his voice was silenced forever. One night in July, he choked in his hotel room. He was diagnosed with pneumonia and Jane was told that she would have to allow the doctors to cut a hole in his trachea to insert a breathing device. Hawking would no longer be able to speak, but he would live. She approved the operation. A short time later, Walt Waltosz sent a computer voice synthesizer called the Equalizer to Hawking. With this program installed on a wheelchair-mounted computer, Hawking can deliver lectures and speak more intelligibly than before the tracheotomy, although at the reduced rate of about 10 words per minute.

In 1990, Hawking and Jane divorced, and he moved in with one of his nurses, Elaine Mason. The reports of abuse that circulated shortly after their marriage have been strongly denied by Hawking, who refused to press charges.

Major Ideas

Hawking has had many ideas that, cumulatively, have changed the faces and directions of cosmology and black hole research. The earliest idea, the one that eventually blossomed into his Ph.D. dissertation, was to apply Penrose's singularity theory to the universe. An offshoot of this concept was the proof, accomplished with Penrose's help, that the big bang was a naked singularity at the beginning of time. At the time, big bang cosmology contended with Hoyle's pet theory of a "steady-state" universe, which did not recognize an expanding and evolving universe from a cosmic point of origin billions of years ago. Hawking no longer agrees with this steady-state hypothesis or his earlier big bang theory. Instead, he now proposes his own "no-boundaries" theorem, which implies a finite universe without a beginning in time before the big bang. This theorem has had an important impact on cosmology, helping to eliminate the steady-state concept altogether.

Hawking also realized, getting into bed one night in 1970, that the event horizon of a black hole does not shrink, but grows with all matter that enters it. His most famous discovery, Hawking radiation, led from this idea. The result of a 1973 argument with a graduate student about entropy also helped Hawking formulate this discovery. The corollary of Hawking's nocturnal epiphany was the surprising fact that black holes, known for dragging everything in their vicinity past the "point of no return," actually emit particles. Hawking also developed the concept of "mini-holes," tiny black holes that can form only in certain conditions and that radiate massive amounts of energy.

More recently, Hawking has taken Richard Feynman's method of finding the most likely path of a particle in quantum physics, known as the "sum-over-histories" or "path integral" approach, and applied it to the entire universe. This groundbreaking concept led to his no-boundaries theorem, developed in the early 1980s, which states that the big bang did not have a spacetime singularity, and neither will a potential big crunch scenario. Hawking is still working at Cambridge, further developing this concept and that of imaginary time.

Publications

Over the almost 5 decades that Hawking has been researching cosmology and black holes, he has published half a dozen important works. Only one of these, *A Brief History of Time,* has found an audience outside theoretical physicists. The first major paper he wrote, "Singularities and the Geometry of Spacetime," was in conjunction with Penrose. It won the duo the Adams Prize in 1966, when Penrose was working for Birkbeck College. This paper detailed their work on the singularity theorems.

Five years later, Hawking was in print again, this time with fellow Sciama student, George Ellis. *The Large Scale Structure of Spacetime* built on this work, giving a description of the general theory of relativity and then explaining the importance of spacetime curvature. Singularities were explained through the structure of spacetime, and proven to be unavoidable in black holes and at the big bang.

General Relativity: An Einstein Centenary Survey, published in 1979 with Werner Israel, was written as a tribute to one of the greatest scientists of the 20th century. It gave an overview of what had been accomplished with Einstein's mathematical brainchild, the general theory of relativity.

Hawking coauthored and coedited *Superspace and Supergravity* with M. Roček. This compilation of essays, published in the United States in 1981, covered the then-current attempts to create the unified field theory, the set of mathematical formulae that would explain every physical phenomenon. This effort has been foiled every time by nonsensical answers produced when the equations for quantum theory and relativity are combined.

The next year Hawking published *The Very Early Universe,* which covered what was then known about the big bang and the moments shortly thereafter.

His most famous publication, *A Brief History of Time,* was published in spring 1988. Hawking's usual publisher, Cambridge University Press, offered him the largest book deal in their history,

10,000 British pounds. Hawking, however, insisted upon more money and instead accepted Bantam's bid of 250,000 American dollars. Manuscript editing began, and problems arose. After learning to be concise so he could communicate more rapidly using the Equalizer, Hawking needed to use many more words to explain his work to laypeople. Eventually, the book was deemed readable. In the meantime, other countries' offers were pouring in: Germany, Japan, China, Russia, and half a dozen others. Bookstores immediately sold out upon *A Brief History of Time*'s release. Five hundred thousand copies had been sold by summer. It remained on the *New York Times'* best-seller list for 53 weeks, and on the *Sunday Times of London*'s for more than 200 weeks. Soon producers were clamoring for film rights, and a documentary, also called *A Brief History of Time*, was released in 1991.

The following sections review Hawking's work and ideas in greater detail.

The Work

Black Holes

Around the time when Hawking began working in the field of theoretical physics, many things that are now well known had yet to be discovered. White dwarf stars—small, brightly burning remnants of stars like the sun—had been observed, but anything denser than these objects was considered impossible. Quantum theory predicted neutron stars, but since most of these are dim, their existence had not been verified. Theoretically, black holes were known to form at around three solar masses, and physicists believed that one of these exotic phenomena would bend spacetime completely around itself.

Hawking's first research project was on black holes, and as soon as he received his Ph.D., he and Penrose began work on finding out more about singularities. They proved mathematically that in our spacetime, certain situations necessitate singularities. At the time, these were considered to be a glitch in Einstein's theory of relativity and even more absurd than black holes. Hawking and Penrose's breakthrough work gave the proof. Penrose's mathematical method was a perfect match for Hawking's understanding and physical applications.

In the 1960s, Hawking helped establish the theory of black holes and brought them out of the realm of science fiction. Penrose had already proved that a black hole could not form without a singularity at its heart, but the entire concept still garnered skepticism until 1973. That year, the new field of X-ray astronomy turned up Cygnus X-1, an X-ray source that has a 95% probability of being a black hole, according to Hawking.

In 1970, Hawking turned his attention to the event horizon of a black hole. The event horizon is defined as the point where all of the trapped light rays that almost escaped hover, also known as the point of no return for incoming matter. Hawking realized that singularity theory could be applied to black holes. An American graduate student, Jacob Bekenstein, published an article claiming that a black hole's entropy, its measure of disorder, was the same as its event horizon. Hawking was horrified; a black hole couldn't have entropy. According to the second law of thermodynamics, the entropy of the universe should increase with time. Hawking, however, thought entropy around a black hole decreases because the black hole swallows disordered matter, leaving nearby space more orderly. When he tried to prove that Bekenstein was wrong, Hawking instead proved the student partially right. This discovery became known as Hawking radiation. The entropy of a black hole is proportional to its surface area, defined by the event horizon. Furthermore, this area can never decrease, but only increase as matter and energy fall inward. These two ideas, Hawking radiation and the increasing event horizon, were major advances in black hole theory.

Hawking radiation is, simply, the radiation that a black hole emits. This radiation adds disorder to the universe at a rate that perfectly maintains the second law of thermodynamics. It can be described by visualizing a particle and its antiparticle, for instance, an electron and a positron. The particle pair comes into existence by borrowing some of the immense stored energy in the black hole and converting it to matter. Such particles are known as virtual pairs, because they are born and annihilate each other so fast that one cannot detect their existence directly. These pairs turn up very close to the horizon, and one particle is pulled away from the other by the hole's gravity.

This pulling channels more of the black hole's energy into the particle that escapes while the other falls into the hole. The escaping particle therefore appears to have been emitted by the hole, and carries part of the hole's mass away with it. Hawking found that in this manner a black hole would, over eons, lose its mass at an accelerated rate, until it became so small that it exploded, spraying radiation in every direction. Hawking was sure his calculations were wrong, but was convinced otherwise by Penrose and Sciama. Upon further investigation, Hawking found that the temperature of a black hole depends inversely upon its mass, and the higher the temperature, the sooner the hole would explode.

History of the Universe

The origin of the universe was not generally considered in the realm of science until the 1950s. The big bang had been predicted through Einstein's equations, and physicist Alexander Friedmann had even calculated the three universes allowed by Einstein's equations. Einstein tried his best to discredit these ideas, preferring the idea of the "cosmic egg," an immense atom-like mass that exploded, releasing the universe's matter. On the whole, however, the subject was merely ignored.

The big bang was, in fact, given its name by its most famous opponent, Fred Hoyle. He proposed the so-called steady-state hypothesis in which the universe expanded very slightly and had no temporal origin or end. The big bang was a label born of sarcasm. However, the name stuck, and Hawking's doctoral thesis proved several problems with the steady-state concept, soon to be discarded entirely. During his work with Penrose from 1965 to 1970, Hawking helped to prove that the big bang had a singularity from which the entire universe expanded, and that the potential big crunch would also have a similar structure.

In 1975, Hawking focused solely on the big bang. During the previous year, he had met his next set of colleagues at Caltech: Kip Thorne and Don Page. Years before, George Gamow, Ralph Alpher, and Robert Herman made predictions for a ubiquitous radiation background left over from the big bang. In 1965, Arno Penzias and Robert Wilson discovered it. The anticipated temperature, about 2.73° Kelvin, and wavelength, microwave range, were confirmed. The background radiation prediction was a success that helped secure the position of the big bang theory. Hawking probed ever farther back, trying to understand the moment of the big bang itself, the point at which time started. Due to the geometry of the big bang, there is no time before the event, and no one can discover what happened before the Planck time, about 10^{-43} seconds. Hawking kept trying, though, and created a more densely populated view of the early universe than previous scientists did. Cosmologists before Hawking, including Hoyle, had described the conditions and time frame that produced the most abundant elements in the universe: hydrogen, helium, and deuterium. These are the elements that stars burn for fuel, creating the heavier elements, like carbon, by nuclear fusion. However, the image of the first few hours and years after the big bang was a picture of radiation and atomic nuclei floating in an evenly distributed, hot morass of expanding space.

Hawking did not like this picture; if everything was evenly distributed, how did galaxies and stars form in the first place? There was some amount of irregularity, he decided, and was vindicated with the discovery of discrepancies in some areas of the background radiation. Hawking drove farther: If there were irregularities, there would be greater gravity in certain areas. These extremely pressurized areas might create unusually tiny black holes that evaporate quickly when compared to their contemporary stellar counterparts. These should still exist, and some might be exploding close enough for us to detect their gamma ray death throes. Unfortunately, most gamma ray bursts detected have been explained using more standard descriptions than Hawking's mini-holes.

Hawking decided to find out how likely the universe is; literally, what the probability was of the universe developing this way. The only way to deduce this was to consider the entire universe to be one body, the same way that an atom or a proton is considered one unit. To accomplish this calculation, Hawking utilized Richard Feynman's quantum sum-over-histories or path integral approach. This involves calculating all of the paths that a particle can take to get from point A to point Z, and each path's probability. Very different paths usually cancel each other out, leaving a few similar, highly probable choices. That is exactly what happened when Hawking used the path integral for the universe. Working together with Jim Hartle, the duo discovered that a finite universe with no boundaries is one of the most likely types. The easiest way to understand the no-boundaries theory is to think of the earth, which is finite yet has no edges. Hawking was also able to eliminate the big bang singularity by introducing the idea of imaginary time, where the singularity can be thought of as the earth's North pole. The earth grows in circumference from this ordinary point, like the universe did from the big bang. The earth also comes back to one point at the South pole, which represents the potential situation of the universe collapsing into a big crunch state. Hawking discovered, furthermore, that time would not reverse in the big crunch. This universe is completely self-contained, and time always moves forward.

Hawking presented his idea at a Vatican scientific conference. He still could not see back to the very beginning, but there was other progress. Quantum theory stated that the density of the original state was not infinite, which would help to explain the irregularities in the early universe; if the density had been infinite, there would have been no room for density variations.

The "chaotic inflation" concept also developed around this time. It states that there is an infinite universe beyond the boundaries of this one, with areas that are expanding and contracting. Some of these areas grow into their own universes. Alan Guth created the original inflation hypothesis, proposing that the universe is very uniform and its curvature nearly flat because it expanded from an extremely small, unstable state to a softball-sized stable state at a high velocity. Hawking vigorously defends both theories.

Hawking's impact on modern cosmology cannot be overstated. He was instrumental in bringing black holes into the light of physical investigation to be considered seriously and, eventually, discovered. He has also given the physics community new tools and insights into the origin and history of the universe. In addition, Hawking has provided a union of important facets of relativity and quantum theory. Outside of his own direct research, he has also brought many of the obscure concepts and near-incomprehensible methods down to the level of the layperson, making the universe a little more accessible to the public.

In fact, social work is one of the areas where Hawking has used his celebrity most. Hawking has lent support and help multiple times to disabled people in battles against various agencies for heightened accessibility. The social and political awareness that Isobel instilled in her son has flowered in his crusades, where he is using his popularity to bring about change.

Ideas and Beliefs

Cosmological

Hawking has never been a person to leave others in doubt about his opinions, especially concerning cosmology. He believes that his no-boundaries theory is the beginning of the complete union of the titans of physics: relativity and quantum theory. Hawking is confident that the answers to all of the questions humans have ever asked about the universe will be found mathematically, and most likely in the near future. He dislikes the anthropic principle's explanation of all phenomena: Simply put, that everything is how humans observe it because if it was different, there would be no humans to observe it. Hawking wants solid, scientific answers to questions. Therefore, he has put his support behind the superstring theory, hoping that the final results of its calculations are as promising as those already known. However, he is reticent to believe in the extra, submicroscopic dimensions that string theories necessitate.

Hawking has also revised some of his earlier conjectures and discussed some exotic topics more closely associated with science fiction than with physics. He now thinks that mini-holes may be less common than he previously believed. He is also considering the possibility that when a black hole gets very small near the end of its life, it may just disappear from its region of the universe, removing its singularity as well. On the subject of determinism, Hawking thinks everything is probably determined, but no one can ever find out if it is. His opinion of time travel is thoroughly scientific. Since the uncertainty of the position and velocity of particles inherent in quantum physics can never be eliminated, quantum fluctuations would most likely destroy the tiny, opening wormhole necessary for time travel. Furthermore, if surrounding particles could travel through the wormhole, they would begin an ever-accelerating loop that would result in radiation too strong for any human to survive passing through. Hawking has called these ideas his "chronology protection conjecture," an answer to the potential time causality problems produced by time travel.

Social

Hawking's social conscientiousness is often found in his work as well as in his outside activism. An explanation of every question should be attempted, he asserts, no matter how daunting or controversial the subject is. No topic should be consigned to metaphysics or religion, in his opinion. The universe should always be under inspection until humanity as a whole understands everything, and this search for information should be conducted freely, without derision on the part of anyone else. Without passing judgment on the subject, Hawking also asserts that human engineering, with great advances in the capacity for knowledge, will occur very soon despite all efforts to stop it. Directed engineering, he states, will overtake evolution. Eventually, in Hawking's opinion, humans will acquire enough wisdom to stop blowing themselves up, avoiding the complete destruction of the human species.

There are many things that Hawking does not believe in, as well. For instance, he avers that no civilization, regardless of where it is or how advanced it is, will ever be able to control the entire universe. In addition, he doesn't believe that any alien civilization has visited Earth.

When discussing religion, Hawking can be extremely ambiguous. He has stated his belief to be that if a Creator does exist, there was a limited number of ways that the universe, initiated by this figure, could have evolved, regardless of that Being's wishes. Hawking strongly advocates the viewpoint that no question should be left to religion. Yet in Hawking's books there are multiple mentions of God, discussing his powers, knowledge, and role in the universe.

Conclusion

Stephen Hawking still works in his little office in the DAMTP at Cambridge. He oversees a few Ph.D. students and runs the relativity group. However, Hawking is largely free to work on his research. He is still pursuing his concepts of imaginary time and the no-boundary universe as the tools to find final cosmological answers.

Hawking's contributions to physics are frequently ranked with the findings of Newton and Einstein. He has discovered many of the main features of black holes, partially united the theories of quantum mechanics and relativity, and proved enough of his work to win widespread recognition of black holes and singularities as real phenomena. Added to these accomplishments, he has managed to popularize a difficult area of science and developed new methods with which to analyze the beginning of the universe. The most productive era of Hawking's work was several decades ago, but he has not retired yet.

Several areas that Hawking has been involved in are among the most widely investigated new sectors of research. Superstring theory has been gaining popularity since the 1980s, and may hold answers to the remaining unanswered cosmological questions. In string theory, the universe contains 11 dimensions—more or fewer depending upon which of the six interconnected theories one considers. What were particles and waves have been reconfigured into one-dimensional loops or strings of pure energy, so tiny that no equipment available can detect their shape. This theory has shown promise, but it is still being constructed and so cannot be used for all situations.

The short gamma ray bursts that Hawking had hoped would prove the existence of mini-holes are still something of a mystery. Most of these radiation events have been identified as colliding neutron stars, but recent reports of slightly longer bursts, in the range of a few seconds to almost 2 minutes long, cannot be explained so easily.

Gravitational waves, which are formed from the fabric of spacetime and, theoretically, ripple outward from colliding black holes, have been studied theoretically for some time. However, there is no concrete evidence to validate their existence. The subject has experienced a recent revival of interest, and the range of error in the calculations is down to about 20%.

Emily Sobel

See also Aristotle; Big Bang Theory; Big Crunch Theory; Black Holes; Copernicus, Nicolaus; Cosmogony; Cosmology, Inflationary; Einstein, Albert; Experiments, Thought; Galilei, Galileo; Newton, Isaac; Quantum Mechanics; Relativity, General Theory of; Singularities; Time Dilation and Length Contraction; Time Warps; Universe, Origin of; Universes, Baby

Further Readings

Cropper, W. H. (2001). Affliction, fame, and fortune. In *Great physicists: The life and times of leading physicists from Galileo to Hawking* (pp. 452–463). New York: Oxford University Press.

Greene, B. (2004). *The fabric of the cosmos.* New York: Random House.

Hawking, S. (1993). *Black holes and baby universes.* New York: Bantam.

Hawking, S. (1996). *A brief history of time.* New York: Bantam.

Hawking, S. (2001). *The universe in a nutshell.* New York: Bantam.

Hawking, S. (2003). *The illustrated history of everything.* Beverly Hills, CA: New Millennium Press.

McEvoy, J. P., & Zarate, O. (1997). *Introducing Stephen Hawking.* New York: Totem.

Overbye, D. (1991). *Lonely hearts of the cosmos: The story of the scientific quest for the secret of the universe.* New York: HarperCollins.

White, M., & Gribbin, J. (1992). *Stephen Hawking: A life in science.* New York: Dutton.

Healing

Healing is the act of repairing or mending, be it tangible or intangible, visible or invisible, over a period of time, whether so lengthy as to be imperceptibly slow or so brief as to appear instantaneous. Time is a key ingredient in any healing, as this process does not in fact occur instantaneously. Rather, in many cases one needs patience to wait for time to pass before the healing reaches its completion or even until the healing had progressed sufficiently to be observed. There is rarely a set time for healing to be completed; rather, one knows when the healing is completed only by the lack of further change taking place.

Healing, whether physical or emotional, is a change; change cannot occur without the passage of time. The passage of time can be measured by the progress seen in the healing—be it a scar forming, bleeding stopping, or a bone knitting itself back together. Although healing can seem like a reversal of time, changing something back to how it once was, in reality it is, of course, always changing to new.

Emotional healing can be harder to perceive, involving as it does the intangible, the unseen, such as the healing of a broken heart or the healing after the death of a loved one. Because the changes in emotional healing are not as easily seen or measured as in the case of physical healing, one may lose track of the progress toward healing at any point on the path the healing has already taken. Such healing may involve rituals that themselves take time, and through these time-based rituals, such as the Catholic anniversary mass or the Jewish kaddish on the anniversary of death, the healing is aided. This time-oriented set of rituals helps with healing that which cannot often be healed through medicine. Time in its sacred aspects is commonly identified with regeneration and renewal.

Sara Marcus

See also Decay, Organic; Dying and Death; Gerontology; Longevity; Medicine, History of

Further Readings

Imber-Black, E. (1991). Rituals and the healing process. In F. Walsh & M. McGoldrick (Eds.), *Living beyond loss: Death in the family* (pp. 207–223). NY: Norton.

Milburn, M. P. (2001). *The future of healing: Exploring the parallels of Eastern and Western medicine.* Freedom, CA: Crossing Press.

Weiss, B. L. (1993). *Through time into healing.* New York: Simon & Schuster.

Heartbeat

The heart is the muscular organ responsible for circulating blood through the body. The average animal heart will beat several billion times throughout life. The human heart, for example, will beat approximately 2.8 billion times over 75 years at 72 beats per minute. The rate at which the human heart beats is controlled by many different factors. The central nervous system controls heart rate with the medulla oblongata. Activation of the parasympathetic nervous system, using the vagus nerve, causes a decrease in heart rate, while activation of the sympathetic nervous system produces an increase in the rate at which the heart beats. The endocrine system controls the heartbeat with hormones, such as epinephrine (adrenaline), which increases heart rate.

Generally, the size of an animal will correlate with how many times its heart will beat in a minute. Smaller animals tend to have a higher resting heart rate, such as the mouse with a heart rate around 500 beats per minute. Larger animals, like the whale or elephant, tend to have a lower resting heart rate, roughly around 20 and 35 beats per minute, respectively. During times of hibernation, the heartbeat in some animals can drop to rates drastically lower than while not hibernating. During summer months, a black bear has a heart rate between 40 and 50 beats per minute. While hibernating, this can slow to as few as 8 beats per minute.

The embryonic heart in humans begins to beat around 3 weeks after conception, at which time it beats at around 75 beats per minute, a rate near the mother's. It then increases linearly to over 170 beats per minute, peaking 7 weeks after conception. The heart rate then decreases to around 145 beats per minute by the 13th week, where it remains until birth. Heart rate remains high throughout childhood, usually not becoming 70 beats per minute until after adolescence.

Young children are often born with heart murmurs. Although these can indicate a defective heart valve, most heart murmurs are from a more benign cause, a patent foramen ovale (PFO). This is an incomplete closure of the wall between the two atria that closes over time as part of normal neonatal development. If a PFO fails to close, as it does in between 20% to 25% of persons, and persists through adulthood, it can increase risk of stroke. This is because minute blood clots in the deoxygenated blood from the peripheral tissues can bypass the lungs, where they are normally filtered out by the microvasculature of the lungs, and pass through the hole between the two atria. These small aggregates can then find their way to the brain and block blood flow. The result can be either a stroke or a transient ischemic attack (TIA), also know as a mini-stroke.

Blood carries oxygen from the lungs to the peripheral tissues and transports nutrients, hormones, and white blood cells. It also aids in waste removal. By beating, the heart is able to pump blood continuously. The rhythmic contractions of the atria and ventricles occur in a synchronized sequence that ensures efficient blood flow. A single heartbeat begins as the result of a spontaneous, rapid depolarization of the pacemaker cells located in the sinoatrial node on the right atrium of the heart. This generates a stimulus for contraction. Pacemaker cells in humans depolarize 70–80 times per minute, resulting in a heart rate of 70–80 beats per minute.

Modern technology has developed artificial pacemakers for patients who suffer from heart problems in which either the natural pacemaker cells in the sinoatrial node do not signal at a high enough rate, or they fail to transmit a stimulus for contraction altogether. The small electronic device is surgically inserted into the chest and connected to the heart with electrodes. The electrodes send electrical impulses to the heart, stimulating each heartbeat. More advanced artificial pacemakers can even control the rate of the electrical impulses, increasing heart rate during physical exertion and decreasing it during times of rest and sleep.

In the late 1960s and early 1970s, around the same time that artificial pacemakers were being successfully implanted and becoming more reliable, the first artificial hearts were being created. Designed by Domingo Liotta, the first artificial heart was successfully implanted in 1969 by surgeon Denton Cooley, although it was used for only a short period of time until a donor heart became available. The Jarvik-7, designed by Robert Jarvik, was used in about 90 patients as a permanent heart replacement before the practice became banned because of the low survival rate of its recipients. Although most patients lived less than a year with the Jarvik-7 as a permanent replacement, it was still used as a temporary device for patients waiting for donor hearts to become available for transplant. In 2004, CardioWest's temporary Total Artificial Heart (TAH-t) was approved by the Food and Drug Administration, the first implantable artificial heart to receive FDA approval. Developed from the Jarvik-7, it is used only to extend life in patients awaiting heart transplant surgery. One patient lived almost one full year with the TAH-t before receiving a donor heart. Over 75% of patients that receive the TAH-t survive through and after the following human donor transplant surgery, most over 5 years.

As the heart ages over time, it can suffer from an array of diseases that ultimately lead to death. Some diseases can be fatal in a very short time; others can take several years until the heart fails. Coronary heart disease is the result of the buildup of plaque in the arteries that supply oxygen to the heart muscle. This occurs over an extended period, sometimes decades. When the plaque then ruptures, the blood forms clots on the plaque and blocks the passage of blood to the heart tissue. Without oxygen it takes only minutes for the heart tissue cells to begin to die and the heartbeat to stop. This is called a myocardial infarction, or heart attack. The heart can also fail as the result of infection by bacteria or a virus. Even poor diet, hypertension, high cholesterol, and structural defects can affect heartbeat and shorten the lifespan of the heart. Although it is possible to survive for prolonged periods of time with some heart diseases, it generally takes only minutes for oxygen-deprived tissues in the heart to die and the heartbeat to stop.

Michael F. Gengo

See also Decay, Organic; Diseases, Degenerative; Dying and Death; Healing; Medicine, History of

Further Readings

Martini, F., & Bartholomew, E. (2007). *Essentials of anatomy and physiology* (4th ed.). San Francisco: Pearson/Benjamin Cummings.

McMillian, B. (2006). *Human body: A visual guide.* Buffalo, NY: Firefly Books.

Tyson, P. (2000). *Secrets of hibernation.* Retrieved July 12, 2008, from http://www.pbs.org/wgbh/nova/satoyama/hibernation.html

Heat Death, Cosmic

The cosmic heat death of the universe is a prediction that, as everything tends to move from order to chaos, the universe will eventually "run down" like an old clock. The second law of thermodynamics states that entropy tends to increase in an isolated system. The arrow of time points to the fact that all heat in the universe will die eventually. This is borne out by the everyday observation that things tend to move from a higher energy state to a lower energy state. High energy states can also be manifested as more highly organized collections of atoms. Any solid will gradually disperse over time. Given enough time, even all the atoms of the universe will break down.

As various physical phenomena have been studied and understood it has become apparent that over time everything moves to this lower-energy, more dispersed state. From the universe's first moments as a single high-energy point of all matter, it is moving toward dispersal as the energy that binds everything runs down. Scientists currently believe we live in an open universe and that everything will continue over time to expand infinitely in all directions. According to current understandings of particle physics and cosmology, in around 10^{100} years the universe will move to a low-energy, steady state of photons and leptons, dispersed through infinite space.

The death will happen in three phases. First the universe will become dark as the stars go out. The death will follow the pattern of increasing entropy as the energy level and matter density of the universe drop to a level at which galaxies and stars cease to form, around 10 to 100 trillion years from now. Some 14 trillion years after that, the last of the long-lived stars (red dwarves) will go out.

The second phase will be the decay of the complex structure of the material universe. At a point some 10^{14} (1,000 trillion) years from now, the visible structure of the universe will begin to degenerate at its lower levels; planets will begin to decay as they leave their orbits as the result of an accumulated weakening of the gravitational force. These changes will be followed approximately 10^{15} years later by degeneration at the highest levels; galaxies will begin to decay as stars leave their orbits. The final level of visible structural disorder will not take place until around 10^{40} years in the future when matter itself begins to break down. All matter is made of elementary particles, chiefly protons, neutrons, and electrons in various combinations. The predicted half-life of protons is 10^{36} years, which means that by 10^{40} years from now virtually no protons will exist.

The last phases of the universe will be visible only at the subatomic level. When more than 10^{100} years have passed, the last remaining macroscopic structures will disappear as the last black holes evaporate into photons and leptons. The universe will be at an absolute minimum temperature and a featureless sea of subatomic particles subject only to random quantum fluctuations.

John Sisson

See also Black Holes; Cosmogony; Cosmology, Inflationary; Entropy; Russell, Bertrand; Time, End of; Universe, End of; Universe, Evolving; Stars, Evolution of

Further Readings

Chow, T. L. (2008). *Gravity, black holes, and the very early universe: An introduction to general relativity and cosmology.* New York: Springer.

Gribbin, J. (2006). *The origins of the future.* New Haven, CT: Yale University Press.

Hegel, Georg Wilhelm Friedrich (1770–1831)

German philosopher Georg Wilhelm Friedrich Hegel (1770–1831). A professor at Heidelberg (1816–1818) and Berlin (1818–1831), he differed with Immanuel Kant in allowing that mankind possessed absolute knowledge. He influenced Karl Marx and the existentialists.

Source: Library of Congress, Prints & Photographs Division, LC-USZ62–130772.

The philosopher Georg Wilhelm Friedrich Hegel was born in Stuttgart, Germany, and educated there and in Tubingen. Along with Schelling and Fichte, he was one of the prime exponents of German Idealism. Over the course of a distinguished academic career, Hegel held positions at several universities including Jena, Heidelberg, and Berlin, and through his teaching and his published works became one of the most influential thinkers in the history of Western philosophy.

Hegel's Analysis of Time

Nature and Spirit

In his first essays, written in his youth, Hegel defines time in an essentially negative way, as a destiny hostile to human beings, or as a finite reality intended to be transcended by reason in the eternal knowledge of Ideas. Directly following the Platonic tradition, time is deprecated in order to give greater importance to eternity. It was only later, during the years 1803–1806, that Hegel developed a positive conception of time in the successive preliminary sketches of his philosophy of nature, which are marked by the major discovery of the *dialectic of time.* Hegel considered that time as such, in its original springing forth, has to be understood as an immemorial element of Nature, which is not yet transformed, neither by language nor by memory, into the time proper to the Spirit, history. Nature is the opposite of Spirit, Absolute Spirit as the other of itself, or the hidden Spirit. As being-other than the Spirit, Nature maintains a double relationship with the latter. It is opposed to it as it is its negative, the *being-other* than the Spirit. But at the same time, it has a hidden relationship of identity with Spirit, since as being-other *than the Spirit,* it is already itself the life of the Spirit that can, by studying Nature, precisely know itself. Philosophy of nature is this knowledge of Nature through Spirit, understood as progressive recognition of Spirit in the being-other.

In the courses given by Hegel at the University of Jena in 1804 and 1805 (the manuscripts of which have been preserved), time is the first moment of the philosophy of Nature, the first form of the exteriorization of Spirit in nature. Time is defined by two concepts, the infinite and the negative. Negativity characterizes the destructive aspect of time, which he later described in detail in his 1817 work, the *Encyclopedia of the Philosophical Sciences.* As for infinity, it leaves open the possibility for a positive determination of time, which is likely to give itself to Spirit. From this double characterization of time flows the dialectic of the three temporal dimensions: present, future, and past. The first moment of time is the present, the Now. The Now manifests itself but does not last; it is immediately suppressed by the future, which comes to take its place. The Now has for its very being no longer being. Once suppressed, the Now becomes the past. The past, in turn, suppresses itself in the sense that it leaves a place for a new Now to come forth. In other words, the future is the negation of the Now, and the past is the negation of the future, and therefore the negation of the negation of the Now. Since double negation is affirmation, the past is the affirmation of a (new) Now. At this point, Hegel distinguishes two forms of time with the help of his logical theory of the two infinites. Either the new Now is a Now without a past, a pure Now without a relationship to the preceding Nows, or time is nothing other than the indefinite repetition of a Now always identical with itself, the bad infinity proper to Nature. In this unending linear movement, there is neither newness nor progress. As Hegel was to say later in Berlin, in nature there is nothing new under the sun. Or, the new Now is a Now of the past, a present that returns to the past, to include it in itself and to enrich itself even more. The image of the circle replaces that of the line. This "real" time is true infinity, which, in a circular movement, goes from the present to the present, through the future and the past. From this real time flows the principle of historicity: the living conservation of the past in the present.

In his 1805–1806 course on the philosophy of nature, Hegel explains the consequences of the primordial role of the past in the dialectic of real time. The past is not only one of the dimensions of time, it is the truth, the goal of time. According to Hegel, the privileged meaning of time is the past, understood not as a moment isolated from time, but as the culmination of "real" time, the concrete present. Against philosophers such as Schelling, who wanted to denigrate if not abolish time, Hegel affirms both the temporality of all Being, in that time is the supreme power imposing itself on all Being, and the rationality of this temporality, in the sense that the true knowledge of beings must consider the latter in light of their historicity. Philosophical knowledge is not the eternal contemplation of the eternal; it is knowledge of being "in its time," according to the temporality that is proper to it. Hegel reverses the traditional relationship of the subordination of time to eternity. The truth of time is not an absolute eternity, since, on the contrary, it is time itself, elevated to its real form, that is the truth of eternity.

Space, Time, and Negativity

The courses on the philosophy of Nature dating to Hegel's Berlin period (1818–1831), which provide commentary on the second part of Hegel's *Encyclopedia* appearing in 1817, do not call into question the philosophical rehabilitation of time undertaken in Jena. Hegel, however, leaves in the background the dialectic of time in order to concentrate on its negativity. According to Hegel, Nature is the Idea in the form of otherness and exteriority. The form of exteriority is divided into the two forms of space and time. The principal difference between space and time is negativity. Space does not allow negativity to deploy itself within it; negativity remains, as it were, paralyzed. The different parts of space coexist next to each other without cancelling each another out. It is not the same for time, the negativity of which incessantly relates to itself and continually suppresses its own moments. In fact, time is the being that, in being, is not, and in not being, is

(*Encyclopedia*, §258). In the *Science of Logic*, Hegel distinguishes pure indeterminate nothingness, which designates the nothingness that is not at all, from determinate nothingness, the non-Being that contains an essential relationship with Being. Negativity is the negation of negation, which ensures the conversion of determinate nothingness into Being. In the light of these logical determinations, it is clear that time is neither pure nothingness nor determinate nothingness, but a type of negativity. Time is the continual passage of Being into nothing—from the present into the past—and from nothing into Being—from the future into the present. It is this double passage that defines the negativity of time. Only the present is; it enjoys in nature an absolute right. But the present contains within it the negativity of time, such that it does not cease to suppress itself and to disappear. Time is a "going-out-of-itself" (*Außersichkommen*), the "negativity going out of itself" (*die außer sich kommende Negativität*). Time projects Being out of itself; it disperses it in a constellation of present, future, and past moments, all exterior to one another. For Hegel, the negativity of time is ecstatic, not so much in Martin Heidegger's sense, but in the *ekstatikon* of Aristotle, who, in his *Physics,* attributes to time the origin of the corruption that is inherent in all natural movement. In Nature, temporal negativity is essentially destructive, the result being pure, indeterminate nothingness, the irreversible disappearance into the past. Hegel uses a play on words in comparing time (Chronos) to the Greek god Kronos (Cronus) who engenders everything and devours his own children.

Becoming

Time not only involves the irreversible disappearance of events into the past. The moment of the disappearance supposes the continual birth of the moments that are destined to disappear. Time includes within itself a certain generation, which Hegel conceives through the category of Becoming. Time is, more exactly, the "intuitioned Becoming." According to its logical definition, Becoming is an alternation of birth and disappearance; it is formed by the unity of these two concepts. But logical Becoming and temporal Becoming could be confused, and the adjective *intuitioned* is there to distinguish them. Intuition traces the boundary between temporal Becoming and logical Becoming, as pure thought. Being, nothing, and Becoming are thoughts, but time is not of thought, and that is why it is intuitioned Becoming, which is experienced in existence. The concept of time is thought, and it is, like all concepts, eternal. Time itself in its existence is intuitioned. That time is a form of Becoming also means that it is not a fixed, permanent framework in which things happen. The only permanent thing in time is the absence of all permanence. Things are not in time, since time itself is in things in the form of an unceasing negativity that devours them from the inside. As Becoming, time is a river that carries everything away with it, including its own banks.

History and Spirit

Hegel clearly distinguishes the domains of Nature and of Spirit. In nature, the past is an indeterminate nothing; events disappear and for the most part leaving no trace. It is not the same for the domain of the Spirit, which is able to give a new opportunity to the past. Time is the "tomb" of the event, but the Spirit preserves the past. That's why history lives only in the Spirit. Most of the time, German idealist philosophers distinguished time from history, but without enquiring into the connection between these two concepts. The originality of Hegel's philosophy is that it allows one to understand the transformation of time in history. It is in the *Phenomenology of Spirit* that we find the clearest development of the transcending of time, that is, the passage of natural, linear time, the indefinite series of Nows, to historical time, which preserves the past within it. The possibility of the passage of time into history rests on the idea that time is the existing concept. This thesis means that time is not opposed to concept, to Spirit, since, on the contrary, it is the Spirit's mode of existence.

How does the passage from time to history take place? It is achieved principally through three operations of consciousness. Transcending the evanescent Now takes place in language, understood as speech and writing. Speech transforms the negativity of the Now into a stable and universal reality, and writing fixes this speech, still in flux, into a permanent sign. Consciousness also preserves the

Now through the work of the interiorization of memory (*Erinnerung*). Interiorization is conceived by Hegel according to the Christological model of the Resurrection. It is what allows the Absolute Spirit to make the present be relived in the past, saving it from oblivion. Through memory, the Spirit converts the indeterminate nothing of the past into a new present, which is the foundation of history. In the ultimate movement of "Absolute Knowledge," Hegel develops a third form of the transcending of time, that of conceptual thought, which overcomes the negativity of natural time in order to extract from it immanent rationality. It is necessary to make an essential distinction between "effective history," which corresponds to the temporal course of events, and "conceived history," which is the retrospective understanding of events in thought by philosophy. The task of conceived history is to organize the chaos of events, giving them the form of the concept, to manifest the movement of the Spirit, which progresses secretly, like a mole, through events.

Hegel affirms that Spirit is time. This identity of time and Spirit is dialectic. On the one hand, Spirit is opposed to the bad infinity of natural time; on the other hand, it transcends the latter in order to unite itself with time in the movement of history. This identity ultimately has the meaning of "*Aufhebung*," of a transcending by which the Spirit makes itself master of time, by means of language, the interiorization of memory, and conceptual thought. The finite beings that inhabit Nature—inorganic things, plants, and animals—are incapable of transcending the negativity of time that they contain in them, which henceforth manifests itself as a hostile power, a source of destruction and death. For finite beings, time will never be anything other than the destructive negativity of nature—a destiny. On the other hand, human beings possess within themselves the absolute negativity of the concept, which designates thought, and more generally, freedom. The concept is the power of time, in the sense that Spirit transcends its negativity, on the one hand, in the very knowledge of time, and on the other hand, by its capacity to transform natural time into historical time. The distinction, first presented in the *Phenomenology of Spirit*, between effective history and conceived history is made more explicit in the *Philosophy of History*, the aim of which is to decipher and translate, in the language of the concept, the palimpsest of represented history, the *historia rerum gestarum*, in which events and their memories are recorded through writing. The Hegelian conception of time also leads, in its final outcome, to thought of the history and of historicity.

Conclusion

When all is said and done, the Hegelian philosophy of Nature brings at least five original answers to the question of time:

1. *The desubjectivation of time.* Time as such is not an interior form of our consciousness, an "internal sense," but a universal determination of Nature present in all its domains.
2. *The mobility of time.* Time is not a fixed and permanent form in which events take place, but the very Becoming of things. From this point of view, things are not in time; it is time itself that is in all things in the form of their intrinsic negativity.
3. *The negativity of time.* Time is the destructive negativity of Nature; it plunges each being into the non-Being of the past. This concept of negativity allows us to take into account several aspects of time, such as the fleetingness of the instant and the irreversibility of the past.
4. *The dialectic of time.* The negativity of time unfolds according to a dialectic of three moments, from which flow two figures of time: on the one hand the indefinite and repetitive time of Nature, and on the other hand, the progressive and historical time of Spirit.
5. *The logicity of time.* The purpose of the desubjectivation of time is to establish the relationship of Spirit and time on a new basis. Time is not, in fact, a power foreign to Spirit, a destiny, but it can be transcended by Spirit in the sense that it can be spoken of, recollected, interiorized, and thought by Spirit. For this reason, the reflection on time begun in the philosophy of Nature is fully realized only in the philosophy of Spirit.

Christophe Bouton

See also Aristotle; Becoming and Being; Dialectics; Eternity; Hegel and Kant; Heidegger, Martin; Idealism; Intuition; Kant, Immanuel; Leibniz, Gottfried Wilhelm von; Metaphysics; Now, Eternal; Ontology

Further Readings

Hegel, G. W. F. (1970). *Philosophy of nature* (M. J. Petry, Ed. & Trans.). New York: Humanities Press. (Original work published 1817)

Hegel, G. W. F. (1977). *Hegel: The essential writings.* London: HarperPerennial.

Hegel, G. W. F. (1977). *Phenomenology of spirit* (A. V. Miller, Trans.). Oxford, UK: Clarendon Press. (Original work published 1807)

Hegel, G. W. F. (1988). *Introduction to the philosophy of history.* Cambridge, MA: Hackett. (Original work published 1837)

Singer, P. (2001). *Hegel: A very short introduction.* New York: Oxford University Press.

Hegel and Kant

Like Immanuel Kant, who sees time as the formal a priori condition of all phenomena in general, G. W. F. Hegel considers time as an absolutely universal determination of nature, which gives it a certain primacy over space. The *Encyclopedia* of Hegel seems to adopt the same presentation as Kant in his *Transcendental Aesthetic:* Time follows space, and the analysis of each of these two moments culminates in an examination of the corresponding sciences. Hegel refers to Kant several times, affirming that time is, like space, a pure form of sensibility or intuition. But behind these apparent affinities hides a systematic critique of the Kantian theory of time developed in the *Critique of Pure Reason.*

In the *Transcendental Aesthetic,* the objective of the metaphysical exposition of time is to analyze its a priori conditions, that which is necessary and universal. And yet, while taking up the main conclusion of this exposition, the very definition of time as a pure form of sensible intuition, Hegel gives it a resolutely different meaning. First of all, he sets out to show the dialectical relationship between space and time, whereas Kant merely juxtaposes them as two forms of our finite human intuition. But above all, Hegel interprets the pure intuition of space and time not through the subject, but through Nature. He thinks that one ought not to make space and time purely subjective forms of human nature, for time is not a condition of becoming; rather, it is becoming itself intuited, the destructive negativity inherent in Nature.

Certainly, Hegel grants Kant the claim that time, like space, is not something real, and is not, as Leibniz thought, an order of things. But the ideality of time does not make it one of the forms of our sensibility. Hegel expresses this Kantian concept in terms of his own thought and replaces the transcendental ideality of time with another form of ideality, designating its negativity, its power to dissolve all reality into the nothingness of the past. According to Hegel, it is therefore useless to want to classify time among subjective or objective beings. As the universal negativity inherent in nature, time might be described as "objective," but as the existence of the Idea in the mode of Being out-of-itself, it is brought back to the mind, and is, in this sense, also "subjective."

In Kant's writing, the purpose of the transcendental exposition is to show the determinations of a concept that constitutes the principles capable of explaining a priori the possibility of certain sciences. Actually, it is rather in the *Analysis of Principles* that this essential aspect of time is described in detail by Kant, who affirms at the beginning of the *Analogies of Experience* that the three modes of time are permanence, succession, and simultaneity. Permanence is the schema of substance by virtue of which I know a priori that substance persists through all changes of phenomena. This permanence is the condition of two other temporal relations: *succession,* which is the schema of causality, and *simultaneity,* which corresponds to the schema of reciprocal action. And yet, based on his understanding of temporality as destructive negativity, Hegel refutes one by one these three fundamental attributes of Kantian time. Simultaneity is thus completely foreign to time, since the latter is defined by its exact opposite, impossible coexistence. The opposite predicate, succession, does not offer an adequate grasp of time, for it masks its specific negativity more than it highlights it. The concept of succession is defective because it represents time as a series of isolated instants split between two domains: the present Now, which is, and the non-present Nows, which are no longer or which are not yet. Being and non-Being are maintained separately by representation, whereas, in truth, the negativity of time that flows from the Now implies an indissoluble unity of Being and nothingness at the heart of each

Now. From this point of view, the future and the past do not constitute other forms of Now but the presence of a negation in Being itself of each Now. By virtue of this negativity, time prohibits all a priori permanence in itself. Of course, certain beings last, but their duration is always of a relative permanence, a deferred death, even for things that we say defy time; for in time nothing persists, nothing remains. According to Hegel, the absolute nonpermanence of temporal things is due to the nonpermanence of time itself, which is ceaselessly transcending itself. Consequently, one should abandon the representation according to which time is a permanent receptacle in which things take place. Kant believes that things change in time, which is, in itself, immutable and fixed. But this, for Hegel, is failing to grasp the very negativity of time. The understanding of time as negativity and becoming therefore implies, in fact, a radical critique of the Kantian theory of time. Time is not that in which things take place, but their very becoming, their own disappearance, as is stated in §258 of the *Encyclopedia.*

In both the transcendental and metaphysical parts, the Kantian exposition of time in the *Transcendental Aesthetic* is therefore criticized by Hegel. Why? Precisely because the transcendental problematic impedes a real understanding of time as negativity, which is at the heart of Hegel's philosophy. The ambiguity of the Kantian conception of time is that it tries to reveal the conditions of the possibility of experience in general and, at the same time, Newtonian physics in particular. This explains why the Kantian understanding of time is predetermined by categories coming from physics, such as causality, substance, and reciprocal action. For Hegel, time is not principally determined by the categories that make possible the physical science of nature, but by its proper dialectical structure, deployed in the present, the future, and the past. Time is the continual, reciprocal passage of Being into nothing—from the present into the past—and from non-Being into Being—from the future into the present.

Christophe Bouton

See also Becoming and Being; Hegel, Georg Wilhelm Friedrich; Idealism; Kant, Immanuel

Further Readings

Hegel, G. W. F. (1970). *Philosophy of nature* (M. J. Petry, Ed. & Trans.). New York: Humanities Press.

Kant, I. (2003). *Critique of pure reason* (N. K. Smith, Trans.). New York/Basingstoke, UK: Palgrave Macmillan. (Original work published 1781 as *Kritik der reinen Vernunft*)

Heidegger, Martin (1889–1976)

Among philosophers of time, Martin Heidegger is one of the most famous. He describes humans as essentially temporal—that is, as "beings in time." Bringing together philosophical currents including phenomenology, philosophy of life, hermeneutics, and ontology, he developed a new philosophy he called Existentialism. His interpretation of the history of philosophy is critical for understanding his works.

Life and Works

Born on September 26, 1889, in Messkirch, in southwestern Germany, Heidegger intended to become a Roman Catholic priest, but after 2 years of theological studies at Freiburg University he switched to mathematics and natural sciences, finally taking a doctorate in philosophy (1913). After completing his postdoctoral thesis, *Die Kategorien und Bedeutungslehre des Duns Scotus* (Doctrine of Categories and Theory of Meaning in Duns Scotus), Heidegger broke with classical Catholic philosophy in 1919 and became an assistant to Edmund Husserl, the founder of phenomenology. In 1923 he was appointed professor at Marburg University and in 1927 Heidegger published *Sein und Zeit* (Being and Time), regarded as one of the most important and at the same time most controversial books in philosophy. He returned to Freiburg one year later as Husserl's successor, where, in 1929, he published three influential books: *Vom Wesen des Grundes* (On the Essence of Ground), *Kant und das Problem der Metaphysik* (Kant and the Problem of Metaphysics), and *Was ist Metaphysik* (What is Metaphysics).

Heidegger's most controversial years were from 1933 to 1945. As a conservative and staunch anti-Communist, he supported some aspects of Hitler's policies and was elected rector of Freiburg University on April 29, 1933. He joined the party shortly afterward—less out of conviction and more to strengthen his position, and although he never embraced Hitler's anti-Semitism, he remained vague about his relationship with the Nazis even after the war. More positively, as rector he prohibited anti-Jewish posters in the university and protected the Jewish professors Hevesy and Thannhauser. He shared a close personal friendship with the Jewish philosopher and political theorist Hannah Arendt in 1925 and again after 1950. His major work *Being and Time* is dedicated to Edmund Husserl, his Jewish predecessor at the University of Freiburg. Heidegger's tenure as rector lasted less than a year, as he was forced to resign after refusing to remove two deans in disfavor with the Nazis. While he never renounced the party, his distance from it was clear in 1944 when he was declared expendable from the university and sent to dig trenches along the Rhine. The relatively few publications from this period include his important essays on Plato's concept of truth (1942–1943). Heidegger's "philosophical turn," which may have begun during the 1930s, seems more evident after the war in his books on Nietzsche and the English publication in 1950 of *Off the Beaten Track* (*Holzwege*). The nature of this turn is contested among scholars, but the theme of philosophical inquiry as a continuous path of understanding is especially appropriate for Heidegger—one that he used repeatedly to describe his own work. He died in 1976 and was buried in Messkirch.

Philosophy of Being and Time

The title of Heidegger's renowned work *Sein und Zeit* (Being and Time) describes the book's key insight about the essential temporality of human existence. In the book, he reveals a remarkably original philosophy that required a whole new vocabulary in order to transcend inadequate understandings of human existence. His lifelong philosophical project attempted to correct a perceived deficiency of Western philosophies from Plato (427–347 BCE) until the rationalistic and scientific worldviews of the 20th century. Many of those philosophies provide great insight into the human condition, especially the classic works of Plato and Aristotle, but according to Heidegger they went astray by inadequately explaining human existence as the precondition for all understanding.

Various English translations preserve his technical terms by keeping them in the original German or by capitalizing them, and often by using hyphens to string together words that together form a unique idea. "Sein" or "Being" is the indefinable concept that begins to convey insight into the human condition when it has been considered carefully—with "Sorge" or philosophically reflective "Care." "Zeit" or "Time" is paired with Being because humans experience reality only within a temporal framework. We experience reality not as a series of present moments, but rather as creatures with memory of the past, awareness of the present, and expectations for the future. Humans come to know Being temporally, in Time, and achieve self-understanding within specific historical circumstances. This insight is conveyed by the term "Dasein" or "There-Being"; illumination, or "Disclosure" of Being's meaning is achieved with reflection upon how humans think about existence in our necessarily limited and temporal way. Humans are thus "beings in the world" and "beings in time."

According to Heidegger, the Being of the human being is fundamentally temporal. This position is the result of an existential analysis of the human being as mortal—Being in relation to death ("Sein zum Tode"). Heidegger's existential analysis of humans' fundamentally temporal nature includes a reflection on this relation to death. Death is the end of each individual's human possibilities ("Sein zum Tode"). Human beings know about their death and are thus able to anticipate the end of their possibilities as their end. Human beings as Beings-in-the-world and Beings-in-time realize that no one else will die their deaths. So the anticipation of death "invites" human beings to live their lives with awareness. If human beings do not take this invitation seriously, they fail to be what they should be. They fail their true nature.

Just as humans do not choose whether to come into existence, the circumstances of particular human lives are also very much given. Heidegger describes this with the term *Geworfenheit* or "Thrown-ness." This means that individuals as Dasein comes to know that they are in Time without having been their own cause.

> Only an entity which, in its Being, is essentially Futural [oriented to the future] so that it is free for its death and can let itself be thrown back upon its Factical "there" by shattering itself against death—that is to say, only an entity which, as Futural, is equiprimordially in the process of having-been, can, by handing down to itself the possibility it has inherited, take over its own Thrown-ness and be in the moment of vision for "its Time." Only authentic temporality which is at the same time finite, makes possible something like Fate—that is to say, authentic historicality. (*Sein und Zeit,* p. 385, translation by Macquarrie & Robinson)

The person as Dasein is thus substantially structured by temporal relationships—in relationship with the past as memory and history, with the future as anticipations that include the inevitability of death, and with the present as a reality shaped by both past and future. The focus on death and mortal limitations is not meant to be morbid, but rather to deepen the understanding and appreciation of human life. Life appears to be diminished in approaching death, but this process reflects the essence of the human being as a presence becoming absence. Awareness of Dasein limitations such as impending death drives home this reality. On the one hand it is frightening, because death implies the end of all potentialities for the person. But on the other hand, the deep awareness of death's ultimate finality brings with it a call to consciousness. Consciousness is not a moral motivation, it is the deep awareness that makes personal freedom possible. This dynamic gets at Heidegger's idea that "Non-Being" makes Dasein ("There-Being") possible. In terms of Time, the reflection of nothingness transforms time and fills us with a wonder for every moment.

Heidegger's Later Philosophy as Philosophy in Time

By the 1930s Heidegger had already begun to revise his ideas in *Sein und Zeit.* The problem is that his philosophical analysis of Time presented it as an eternal and transcendent truth, even as he was attempting to establish the essentially temporal structure of humans as Dasein. This motivated him to struggle with these themes in the philosophical systems of Schelling and Hegel, as well as with the poetry of Friedrich Hölderlin. His goal was to uncover a thoroughly temporal phenomenology, but he was not satisfied with the historical aspects of his thinking and wanted to uncover a more fundamental understanding of the temporal structure of human "Dasein." He used two words that sound similar in German to emphasize the importance of history and time: *Geschichte* means history and *Geschick* means destiny, future, fortune, but also skill. While it is historical, his philosophy is at the same time a skill that makes it possible for the person to accept the future.

Heidegger's investigations into the history of philosophy focused on the question of understanding itself and on the way historical circumstances shape the whole structure of thought. The image he used to describe this situation was a horizon that is formed by the meeting of the land and the sky, that is, between what is given and what is absent. The absence is as important as the existence, or "Dasein," in philosophy as well. Metaphysics is the philosophical enquiry into the nature of Being that began with Plato. From that time until Hegel, philosophers continued to contemplate the nature of Being as divine. The reversal of this connection that came with Nietzsche had great consequences. The natural sciences and technology progressed with great achievements during this time, but they did so according to greatly mistaken assumptions about the nature of existence.

This can be perhaps best shown with an example. Uranium fission creates an object that, in itself, remains insignificant. Only after its possible uses have been considered, such as the creation of energy, can it be meaningfully understood as fuel for "nuclear energy." Nuclear energy, with its both peaceful and wartime uses, serves as a good example because it demonstrates how human understanding is never neutral. A purely technical and practical understanding of objects is an illusion—and an illusion made possible by the modern scientific mindset that had radically alienated careful attention to Dasein. Even while presuming to be neutral in terms of metaphysics and value, it actually assumed an anti-historical metaphysical standpoint. It presented Being as the eternal present—a never-ending "now." Within this worldview, existence is classified according to a functional rationality of technical production. This idea, which Heidegger describes with the term "Ge-stell," explains how such a dangerous technology such as nuclear bombs could be developed by scientists claiming to be neutral. Even so, the chance for a "turn" is possible with new ways of thinking. According to Heidegger later in his career, a new understanding of human existence could reveal itself that is aware of both existential absence and presence, as well as both the holy and the divine.

This later view was an even more radical conception of existence. Being can be grasped intellectually only when it is aware of what is absent. This reality is demonstrated by serious reflection on human history. Yet, as Heidegger reached the end of his career, he became more skeptical about whether humans would "turn" from their objectification of reality.

Heidegger's Significance and Influences

Heidegger has been generally recognized as the most important philosopher to carefully investigate the nature of time and of being. He reached beyond the usual philosophical history from Plato to Nietzsche where the question of human existence had always been assumed as a given. As such, the understanding of existence itself remained shrouded, especially the peculiarly temporal nature of being that is known by its absence. His philosophy is a phenomenology because it considers how Being is revealed in human experience—often in explicit ways, but also in implicit and overlooked ways. Heidegger's is also an existential philosophy, because it considers the structure of all existence from the particular perspective of human existence. He uncovers the nature of human existence as situated in place and time, but also in how humans are capable of contemplating, discovering, and shaping existence. His philosophy is a metaphysics because it is an understanding of existence developed from what is—in other words, from Being itself—and not from a natural science point of view. His philosophy is a hermeneutics (way of interpretation) because it starts from the position that we come to greater understanding of Being by being made aware of our prejudices, and never by observing Being from a supposedly objective perspective. At the same time, by becoming ever more aware of our own prejudices, we can subject them to examination.

Much of Heidegger's philosophy remains in dispute. Analytical philosophers have ridiculed Heidegger's play with language, and linguists have shattered his etymologies. Yet, French existentialists like Albert Camus and Jean-Paul Sartre embraced it and used it to influence a whole generation of educated Europeans and Americans. The autonomous subject was understood to be capable of constructing its own personal existence—an existence potentially independent of reasonable coherence. Hermeneutic philosophers such as Gadamer have further developed many of Heidegger's ideas, especially the notion that preunderstanding (or prejudices understood in a neutral sense) structures more conscious understanding. Heidegger has also profoundly influenced theologians such as Rudolf Bultmann and Karl Rahner. Bultmann developed an existential biblical theology based upon the conviction that the Bible's meaning must be intelligible for modern people and relevant to their existing concerns. Rahner was a former student of Heidegger's who developed a new existential theology that reflects on death, the meaning of freedom, and the nature of God. Without straying from traditional Roman Catholic teaching, Rahner benefited from Heidegger's understanding of God as absolute mystery that, Rahner believed, is revealed as love in Jesus Christ.

Heidegger also influenced ethical thought, especially with his critique of the modern technological way of thinking that has forgotten about Being. Technological progress warrants suspicion because it professes no values and thus its progress is judged according to implicit and hidden principles. The apparent technological successes of the modern period convinced people of its power, and yet blinded people to the dangers that its worldview represented. As environmental and political disasters have demonstrated, technological resources can be used in dangerous ways when guided by wrongheaded principles. Even Heidegger himself was forgetful of this point during the period of his support for the Nazis. Still, Heidegger's philosophical insight remains important, that our understanding of Being requires that we appreciate its relation to Time—to its past, its future; and that this appreciation will continually shape our present.

Nikolaus Knoepffler and Martin O'Malley

See also Becoming and Being; Husserl, Edmund; Jaspers, Karl; Metaphysics; Nietzsche, Friedrich; Ontology; Rahner, Karl

Further Readings

Guignon, C. (Ed.). (2006). *The Cambridge companion to Heidegger.* Cambridge, UK: Cambridge University Press.

Heidegger, M. (1962). *Being and time* (J. Macquarrie & E. Robinson, Trans.). New York: Harper & Row.

Heidegger, M. (1998). *Pathmarks* (W. McNeill, Ed.). New York: Cambridge University Press.

Heraclitus (c. 530–475 BCE)

Heraclitus is considered among the greatest of the Presocratic philosophers. Flux and time play particularly important roles in his thinking. Even though the fragments of his book *On Nature* had an enormous impact upon such diverse philosophers as Plato, G. W. F. Hegel, Friedrich Nietzsche, and Martin Heidegger, not much is known concerning the particulars of his life. However, we do know that he was born in Ephesus, came from an old aristocratic family, and looked unfavorably upon the masses. According to Apollodrus, he was about 40 years old in the 69th Olympiad (504–501 BCE).

Relativity of Time

The most influential aspect of Heraclitus' thinking about time is the concept of the Great Year or the eternal recurrence of everything, an idea that was taken up later by Zeno of Citium (the founder of the Stoa) and Nietzsche. However, within his philosophy, Heraclitus also clarifies other aspects of time. He was clearly aware of the relativity of time. When he explains that the sun is needed for the alteration between night and day to occur, it becomes clear that he was conscious that daytime and nighttime are dependent upon certain conditions. A certain time exists only within a specific framework or paradigm. If the framework changes, then the concept of time within it changes, too. We would not have daytime within a world without the sun. Time is dependent upon a specific perspective, and many distinctions concerning time cannot be drawn from only a cosmic or universal perspective.

Unity of Opposites

Heraclitus criticized Hesiod for not having the best knowledge concerning daytime and nighttime. Only the masses regard Hesiod as a wise man, but truly he was not. According to Heraclitus, daytime and nighttime are one, which Hesiod had failed to realize. From a global perspective, one cannot distinguish

Heraclitus, *painting by Hendrick ter Brugghen. Heraclitus was a Greek philosopher who proposed that everything is in a state of flux and that fire is the principal element of matter.*

daytime and nighttime. One has to be a participating spectator in order to employ the distinction meaningfully. Even though the distinction in question works well from a pragmatic perspective, this does not imply that it is correct. From a universal perspective, the distinction between day and night is not supposed to make any sense, as God is supposed to represent the unity of opposites; that is, God is supposed to be the unity of day and night, as well as summer and winter.

Time as Metaphor

Even though opposites do not exist, Heraclitus himself employs opposites. Concerning time, he clearly holds that there are people who are connected to the night and others who are linked to the day, and he attributes different values to these two types of paradigms. According to him, only the night-roamers are the initiated ones. They have wisdom and they do not belong to the masses. The masses are uninitiated and are connected to the day. Even though, from a global perspective, night and day are one, nighttime and daytime stand for something different. Here, they represent people who are either initiated or uninitiated into wisdom.

Time and Order

Only the initiated know what time really is. Time is a type of orderly motion with limits and periods. Heraclitus also specifies in more detail what he understands as order concerning time, and he explains that it is important that the same order exists on various levels. However, time cannot be reduced to only one aspect of order, as Heraclitus also identifies time with a playing child; that is, time is the kingdom of a playing child. Even though the aspect of order is necessary for games, there is more to the process of playing a game, as there are also the aspects of playfulness, freedom, and chaos. To stress also the important disorderly element represented by time, Heraclitus attributes to this concept his idea of the unity of opposites. Wherever there is order, there has to be chaos. However, that chaos is relevant might only mean that even though there is one certain order in the universe, we cannot securely predict the future. Even though everything is necessary, from our perspective anything can happen, as it is impossible for us to foresee the future.

Time Is Cyclical

According to Heraclitus, the order of time is the cycle. Periods and cycles appear at various levels of existence. There is the world cycle or Great Year, but there is also a human cycle, the cycle of procreation. Human beings are born, grow up, and give birth to other human beings so that the cycle of human life can start again, which happens approximately every 30 years. In this way, a man becomes a father and then a grandfather. However, the most important idea in the philosophical reception of his thought is Heraclitus' world cycle, referred to as the Great Year, or the eternal recurrence of everything. Analogous to human lives, there is a period or a cycle in the progression of world history. The world is supposed to be an ever-living fire that is kindled and extinguished in regular cycles. One cycle represents a Great Year, which has the (surely metaphorical) duration of 10,800 human years. By presenting the Great Year in his philosophy of time, Heraclitus also reveals an option for an immanent type of immortality. The concept of the Great Year is of relevance on various levels. It may be analyzed from a metaphysical, natural philosophical, scientific, ethical, and religious perspective.

Stefan Lorenz Sorgner

See also Eternal Recurrence; Hegel, Georg Wilhelm Friedrich; Heidegger, Martin; Maha-Kala (Great Time); Nietzsche, Friedrich; Plato; Presocratic Age; Time, Cyclical; Universes, Evolving

Further Readings

Kahn, C. H. (2003). *The art and thought of Heraclitus: A new arrangement and translation of the fragments with literary and philosophical commentary.* Cambridge, UK: Cambridge University Press.

Sorgner, S. L. (2001). Heraclitus and curved space. In Universidad Tecnica Particular de Loja (Ed.), *Proceedings of the Metaphysics for the Third*

Millennium Conference (pp. 165–170). Loja, Ecuador: Universidad Tecnica Particular de Loja.

Wheelwright, P. (1999). *Heraclitus*. Oxford, UK: Oxford University Press.

Herder, Johann Gottfried von (1744–1803)

Born in Mohrungen, Prussia, Johann Gottfried von Herder attended the lectures of Immanuel Kant in Königsberg for 2 years, beginning at 18 years of age. Herder was most fascinated with Kant's early scientific reasoning, especially his philosophy of nature.

At the age of 21, Herder became a Protestant pastor. In Königsberg, Herder became friends with the counter-Enlightenment figure Johann Georg Hamann. Both Herder and Hamann rejected the later Kantian critical philosophy, and had already rejected any notion of a "faculty." Both had little interest in "time" as an abstraction. Instead, all consciousness should be thought of as language-based. Though his friend Hamann was averse to the entire tradition of rationalism, and though he himself never accepted Kant's critical philosophy, Herder was a rationalist and an idealist when history was the subject.

Herder traveled throughout the German-speaking lands and conversed with many of the greatest minds of his time. He was interested in all things natural, human, and divine. On December 18, 1803, Herder passed away in Weimar.

The German Historical Sense and Herder

Johann Gottfried Herder occupies a permanent, important place in the history of human thought concerning time, for it was he (with his contemporary Hamann) who introduced a "historical sense" to the German nation as a whole. Of course, before Herder there were authors with a strong historical sense. But all these influences and predecessors were non-Germans, especially the figures from the French Enlightenment. And, of course, before the Enlightenment, the Scholastics had understood time and history in the abstract. But they were more interested in "eternity" than in human history. They distinguished *sacred* from *secular* time in order to give absolute priority to the former. The Scholastics lacked a sense centered on natural and human history. Herder provided a new mode of thinking for the German public by offering just such a historical sense.

Most important, this historical sense was lacking in the orientations of Leibniz and Kant. The thought of Leibniz was almost entirely nonhistorical. Mathematics and physics (monadology) do not require a developed notion of human history, because they are largely theoretical, abstract sciences. Perhaps Kant sought to sidestep the question of history (and phenomenology) by focusing largely on the transcendental structures of the ego. His thinking was strikingly nonhistorical, even transtemporal. The reader of Kant is often left with a sense of timelessness and of rationality at rest, a sense of the structure, but not the content, of time.

Herder and Hamann

Herder had a noteworthy German cohort in developing a historical sense, Johann Georg Hamann. Herder's senior and friend, Hamann had also embraced history and sought to instill the German nation with it. Hamann's thought was largely a reaction to the Enlightenment and to the ideas of Benedict Spinoza. The German public, in general, had reacted to the twin specters of Spinoza and Enlightenment science with fear and loathing. Spinoza's ideas were considered deterministic and covertly antireligious. The advance of English science, especially chemistry, was threatening because the "soul" could then perhaps be explained by the chemical processes of the mind within the brain.

Hamann sought to guarantee religion a safe respite. Faced with the advance of Spinoza's ideas, Hamann retreated into irrationalism and a sort of counter-Enlightenment orientation. He sought to link biblical history with modern history, and to prevent any juncture between history and science, materialism, or atheism. Again, Hamann had a sense of history, but it was of a religiously orthodox orientation; history is the story of the divine will. Above all, he viewed *history as entirely sacred;* a strictly secular history could not be conceived of. In short, Hamann viewed history as an irrational but divine phenomenon.

Confronted by the twin specters of Spinoza and Enlightenment, Herder chose a different strategy from Hamann. While Herder still viewed history as a divine phenomenon, he claimed that history is *rational and an object of science.* As the third element of his strategy, Herder chose to incorporate a broad range of natural sciences into his scientific account of history.

Like Hamann, Herder saw history as divine, but in a very different sense. Hamann had sought to develop history out of revealed scripture; Herder, on the other hand, attempted to delete all religious or mythological content from his history. He would accept no divine intervention, no miracles, as a cause in history. In short, Herder's explicit project was to develop a *scientific (non-theological) account of history.* He saw the inevitability of considering time not as entirely sacred but also as secular (natural scientific), and even as an agency of human free will (the historical sense). And so he set off on the project of authoring a coherent history of the universe, nature, man, society, and God, all strictly within a natural scientific methodology.

Herder was not concerned with Spinoza, since his natural science was strongly influenced by Roger Joseph Boscovich, in that his theory of atoms was a point-particle theory rather than the Newtonian corpuscular theory—meaning that instead of Newton's extended atoms, Herder insisted on Boscovich's dimensionless force-points. (Recall that Herder, like Kant, was in the generation of thinkers immediately after Newton, Leibniz, and Boscovich. It seems most likely that Herder discovered Boscovich's ideas when Kant was struggling with them in the early days at Königsberg. Kant finally resolved his struggle with Boscovich by incorporating the theory of force-points into his own natural philosophy in his late work *Metaphysics of Natural Sciences.*) Force-points completely subverted Spinoza's monism and determinism so abhorred by the German sense of piety. Boscovich's natural science (*Theory of Natural Science,* 1776) was a theory of *force,* not *substance.* By adopting an atomism most closely related to Kant and Boscovich, Herder was able to sidestep Spinoza's "deterministic" metaphysics entirely. But Herder could not deny the advance of Enlightenment science, so he chose to concentrate on the natural within history. He began with an explanation of the universe and the origins of nature, and then worked his way through the rise of animal life. Finally, he gave a historical account of human civilizations.

Herder saw that the sciences had to be given a sense of time. Previously the sciences were taught as a hierarchy rather than as having a shared history. The static classifications of Aristotle and Linnaeus could no longer do justice to the advance of science. To account adequately for the development of the universe, Earth, and man, a timeline was necessary with which to distinguish developmental stages. This meant discovering the unidirectionality of time and of a sequence of events following that linear model. The sciences could no longer burst forth fully grown from the brow of Zeus.

Developing an interdisciplinary history of the sciences was a daunting challenge for Herder. Writing well before Charles Darwin, Herder did not have a theory of evolution. Nor did Herder have the advantage of the theory of developmental history authored by Karl Ernst Ritter von Baer. Yet he traced history from the origins of the universe to the rise of humankind across a panorama of the sciences. Breaking completely with Hamann, Herder saw in history not the divine will of God, but instead the actualization of divinity in man.

Herder's Legacy

Herder's contributions to the historical sense—religion, rationality, and science—proved fateful to subsequent German thought on history. After Herder, the first on the scene, Johann Gottlieb Fichte, adopted the religious interpretation of history provided by Herder; history was indeed the actualization of the "absolute ego" in man. Fichte also adopted the notion that there is a deeper logic and rationality to history; it is the gradual actualization of the Idea (logos). Fichte returned to Hamann's tradition of pietism and developed his system of ethical idealism entirely within a quite preachy, harshly moralistic tone. But Fichte ignored the natural sciences in favor of his own principle of ego. Next in the line of succession came F. W. J. Schelling. In keeping with these predecessors, Schelling viewed history as sacred time. He also saw history as the gradual unfolding of an Idea, and so all things are ultimately knowable by reason. In contrast to Fichte, Schelling greatly esteemed the natural sciences. If Fichte's system may be called "ethical idealism," Schelling's system could be called an "idealism of nature" or "objective idealism." Thus Schelling was a more complete replication of Herder's historical sense even if in the strange new idiom of systematic idealism. After Schelling, Hegel advanced the historical sense to perhaps its most extreme form. In fact, Hegel adopted Herder's three contributions, mediated through Fichte and Schelling, as his own. Religion had its eternal safe haven in Hegel's Absolute Spirit, and history was the moving image of eternity. Also, history had its own rational process for Hegel, which may be captured in a system of logic. Rationality is indeed only the standard of knowledge for a historical stage of development. Hegel's Absolute sought to subsume Nature and man entirely within itself. Finally, Karl Marx reworked Hegel's absolute idealism, mediated by Ludwig Feuerbach, the utopian socialists, and later Darwin, into an all-inclusive theory of historical materialism. Directly or indirectly, these thinkers traced several fundamental elements of their historical sense back to Herder.

It should be observed that each stage within Herder's history was equal to every other; there is no progress or higher spirituality as history proceeds (in contrast to Hegel's notion). He did believe, though, that human history should be narrated ultimately in terms of races and nations. Thus Herder truly set the scene for Fichte's entrance onto the stage.

Greg Whitlock

See also Baer, Karl Ernst Ritter von; Boscovich, Roger Joseph; Fichte, Johann Gottlieb; Hegel, Georg Wilhelm Friedrich; Idealism; Kant, Immanuel; Leibniz, Gottfried Wilhelm von; Marx, Karl; Schelling, Friedrich W. J. von; Spinoza, Baruch de; Time, Sacred

Further Readings

Beck, L. W. (1965). *Early German philosophy: Kant and his predecessors* (chap. 15). Bristol, UK: Thoemmes Press.

Herder, J. G. von. (1963). *God, some conversations.* Indianapolis, IN: Bobbs-Merrill. (Original work published 1787)

Herder, J. G. von. (2002). *Philosophical writings.* Cambridge, UK: Cambridge University Press.

Herodotus (c. 484–c. 420 BCE)

Herodotus of Halicarnassus in Caria (currently Bodrum on the Aegean coast of Turkey) was one of the earliest known historians and is most renowned for his chronicling of the Greco-Persian Wars. His approach to the writing of history set a precedent for later historians' concern with establishing causal links between past and current events.

At the time of Herodotus' birth, Halicarnassus was a Greek city located on the fringes of the Persian Empire and thus subject to their monarchal control. Herodotus examined the root of the great conflicts between the Greeks and non-Greeks; from its origins in the Lydian kingdom, to the failed revenge of the Persian King Darius at Marathon, to the final unsuccessful efforts of his son King Xerxes at the Battle of Thermopylae and the famous naval battle at the Straits of Salamis off the coast of Attica. Very little is known of the author's life except that he never claims to be an eyewitness to the events he describes, although he often makes mention of conversations with those who were present, as if he were there. It is thought that many of these conversations might have come secondhand via the exiled grandson of the Persian Zopyrus.

Herodotus' *Histories* is composed of nine books starting in Book 1 with the story of the Lydian king Croesus in the mid-6th century BCE and ending with the founding of the Athenian empire following the second Persian War in approximately 480 BCE. The books of Herodotus' *Histories* cover a significant period of time in which the author attempts to uncover how the enmity between Greeks and Persians began. In doing so, Herodotus' main narrative is composed of many smaller narratives in great detail. These narratives focus on a set of chronological events surrounding the reason for the conflict between Greeks and non-Greeks. Often these narratives temporarily move backwards or sideways to describe circumstances or details that Herodotus feels shed light upon the main thrust of the story. Herodotus' historical method not only is one of enquiry into his narrative but also integrates a line of questioning of his sources, not all of which he trusted. Often he writes that while he is under a moral obligation to report what was said to him, he is under no obligation to believe it, and this inability to decide is cited. While Herodotus never disguises the fact that the Greek-speaking world is the geographic and cultural center of his perceptions, he is often very open-minded when describing the cultural details of foreign societies, although he inevitably must compare them to his own, by relating what he views as a rationale for diverging modes of life and religion.

For Herodotus and his historical reasoning, actions are the result of a prior cause or reciprocity. A wrong inflicted in the past will be returned in equal measure in the future; this is especially the case in regard to kinship relations. And this is why Herodotus' work extends to the distant past to find the answer to the events that caused Greeks and non-Greeks to go to war. In addition, the historian does not focus his energies upon the logical explanation of events and human actions, as they often seem random or too easily assigned to fate or the intervention of divine will, and to some readers this will lend to a sense of storytelling. Although some accounts by Herodotus have been charged with straining the reader's credulity, one should not discount all he records. It is believed that Herodotus died sometime in the 420s while living in the Athenian colony at Thurii in Southern Italy.

Garrick Loveria

See also Greeks, Ancient; Homer; Peloponnesian War

Further Readings

Kapuscinski, R. (2007). *Travels with Herodotus.* New York: Knopf.

Strassler, R. B. (Ed.). (2007). *The landmark Herodotus: The histories* (A. L. Purvis, Trans.). New York: Pantheon.

Hesiod (c. 700 BCE)

Hesiod was a probable contemporary of Homer, and the writings of both represent one of the earliest phases of Greek literature. Although a variety of writings are attributed to him (his only known complete works are *Theogony* and *Works and Days*), Hesiod stands out not only as an important source of Greek mythology but also as one of the first Western philosophers to conceive of and elaborate a cyclical view of time. He is said to have been a simple shepherd who was rewarded with deep insight by the Muses, who came to him in a mist. His views reflect a more practical understanding of life than do those of many of the ancient Greek philosophers who were from an elite class.

Theogony, which likely includes a compilation of oral traditions and myths, tells of the origins and chronology of the gods as well as the creation of the heavens and the earth. Like nearly all creation myths, it sets up a view of time and a relationship between humans and the world of the gods. However, it is in his *Works and Days* that Hesiod elaborates his views on the topic of time.

Hesiod divides existence into five stages, or "Ages of Man." The first, the Golden Age, is ruled by Cronus, the god of time. Under his control, it is an era of truth, justice, and peace. People are filled with wisdom, are of a contented nature, and, though considered mortal, are ageless. Zeus, the son of Cronus, takes over at the dawn of the Silver Age. It is one in which differences and discord begin, and morality declines. Human existence is shortened and the necessity to work is born. During the Bronze Age, strife and violence continue to increase, as do human passion and frailty. The fourth stage is the Heroic Age and is connected with the Trojan War. While human attributes seem to improve somewhat, violence dominates. The last is the Iron Age and is associated with the present. In many ways, it is similar to the present age of degradation as conceptualized in Hindu (Kali

Yuga) and Buddhist mythology. Human life is at its shortest, while truth, justice, and peace are suppressed by crime, greed, deceit, and violence. The gods forsake humanity and evil rules.

Ultimately, however, Hesiod uses *Works and Days* to provide a vindication for morality and the righteous. In an eschatological vision, goodness and justice (*dike*) prevail and triumph, while injustice (*hubris*) loses out. A subsequent section of the text is devoted to teachings on morality and ritual propriety.

Hesiod's influence can been seen in the Pythagorian concept of a cyclical reality and possibly in the works of Homer as well. His elaboration of a golden age, his views on the necessity and value of physical work, and his prescriptions on morality have all had a continuing impact on Western thinking.

Ramdas Lamb

See also Herodotus; Homer; Mythology; Presocratic Age; Time, Cyclical

Further Readings

Hesiod. (1991). *The works and days; Theogony: The shield of Herakles* (R. Lattimore, Trans.). Ann Arbor: University of Michigan Press.

Hesiod. (1999). *Theogony, works and days* (M. L. West, Trans.). New York: Oxford University Press.

Hibernation

The term *hibernation* comes from the Latin word *hibernare,* which means to pass the winter. The term is commonly used for a type of deep winter dormancy in some animals (i.e., mammals, birds, reptiles, and amphibians). A similar type of dormancy that occurs in summer is referred to as estivation. Some insects and snails also exhibit a similar state of dormancy in winter that is often called winter estivation or diapause.

Hibernation (winter dormancy) involves a periodic (seasonal) drastic reduction of the animal's metabolic rate and body temperature for an extended period of time. Generally, in hibernation, the body temperature drops almost to that of the surroundings (the bear being a well-known exception). Basic body processes, like breathing and the rate at which the heart beats, are drastically reduced so that the hibernator appears almost to be in a comatose state. The hibernating animal then lives through the winter on a reserve of body fat and/or externally stored food until it awakens in the spring.

Hibernation is therefore considered to be a fundamental adaptation to the environment since it is essentially an adjustment in the animal to allow its survival during those periods of the year when environmental conditions are so severe that exposure to the elements will be fatal. To survive in these harsh conditions, some animals can temporarily migrate from the hostile environment (e.g., birds). Some animals can remain active but adapt by

- growing thicker fur (e.g., weasels and snowshoe rabbits),
- storing extra food for the coming winter months (e.g., beavers, some squirrels and mice),
- changing their diet to that which is available during the winter (e.g., the red fox normally eats fruit and insects but eats small rodents in the winter),
- using shelters to protect them from the cold.

For others (e.g., reptiles, ground squirrels, and certain bears), migration is impossible and other adaptations are either impossible and/or ineffective as protection. These animals then adapt by means of hibernation.

Hibernation is seen mainly in small animals such as marmots, ground squirrels, chipmunks, dormice, and northern bats whose food supply is very limited or nonexistent in winter. Cold-blooded hibernators include such amphibians as frogs and toads and such reptiles as lizards and snakes that pass the winter with body temperatures that are the same as the ambient temperature (i.e., near freezing). More recently, a tropical primate, the fat-tailed dwarf lemur (*Cheirogaleus medius*), endemic to Madagascar, has also been found to hibernate.

Outside of winter, natural hibernators lacking food and facing other adverse conditions cannot *choose* to hibernate to survive. Indeed, unless an animal is genetically disposed to hibernation or other states of dormancy, it cannot become dormant at will to escape inhospitable conditions.

Some refer to hibernation as "time migration" since hibernation allows the animals to skip over the entire period of the inhospitable winter season and truly "live" only in the periods of plentiful food and higher temperatures that occur outside of winter. Indeed, to the hibernating animal, time passes unnoticed since it is insensible and unaware during winter and becomes active or "alive" only at the end of hibernation.

Interestingly, according to an American legend, the hibernating groundhog (woodchuck) can be used to predict a period of time, namely, the length of the winter season. It is said that the groundhog creeps out of its hole on February 2 of each year (Groundhog Day) and that if the day happens to be sunny and the groundhog sees its shadow, there will be 6 more weeks of winter. On seeing its shadow, the groundhog supposedly returns to hibernation again. If, on the other hand, the groundhog does not see its shadow, it stays above ground, cutting short its hibernation. This is supposed to indicate the imminent arrival of spring, or the end of winter. Of related interest is a 1993 comedic movie by the same name, *Groundhog Day*. It describes a television reporter (a weatherman) who, while waiting to report on the arrival of the hibernating groundhog, becomes "stuck in time" on Groundhog Day. In the movie, time stands still since the events of that day are repeated without end, making it impossible to move on to the next day!

Hibernation is very different from sleep since it involves an extreme reduction in metabolism. For example, a black bear's heart normally beats about 60 times per minute but this is reduced to as little as 5 times per minute during hibernation (about a 90% reduction). The blood pressure of an active mouse varies between about 80 and 120 millimeters of mercury (mm Hg). By contrast, blood pressure in a hibernating mouse varies between about 30 and 50 mm Hg. A hibernating marmot may reduce its breathing from 16 to 2 breaths a minute and its heartbeats from 88 to 15 per minute. Hibernating turtles reduce their metabolic rate by as much as 95%.

Terms like adaptive hypothermia, or the lowering of body temperature to adapt to changes in the external environment; torpor (state of dormancy and inactivity); and suspended animation (temporary suspension of vital functions) are often associated with hibernation.

The Hibernaculum

To hibernate, most animals go into shelters such as burrows, caves, and dens as a buffer from the lethally low ambient temperatures. The temperature of these shelters is normally slightly above the ambient temperature and does not fluctuate as much. This shelter is called a hibernaculum. There is no standard type of hibernaculum. For example, most rodents hibernate in underground burrows. Bog turtles (*Clemmys muhlenbergii*) use hibernacula that include abandoned animal burrows, mud cavities, and the base of tree stumps. Tortoises use underground sites such as burrows excavated in soil or rocky caves. Bats choose a hibernaculum that is not only cold but humid. For example, in winter, the Indiana bat (*Myotis sodalis*) hibernates in limestone caves and also in some manmade shelters, like underground mines. Adders (*Vipera berus*) in cold areas hibernate in clusters in underground areas made by other animals, for example, in abandoned mammal or tortoise burrows. Animals that are not in the wild will even use crawl spaces and basements in buildings.

Most hibernacula must remain at temperatures slightly above freezing (not too cold or too warm), must have a relatively high humidity, must be safe from predators and other intruders, and in some cases must have space for the animal to store food. Indeed, hibernacula are usually dark, protected, and secluded to keep the hibernating animal safe. A suitable hibernaculum is key to the survival of the animal when it is so vulnerable during hibernation.

Length of Hibernation

The period of hibernation may vary, lasting several days or weeks depending on the animal's species, the ambient temperature, length of winter, and the animal's latitudinal location. Woodchucks hibernate in underground burrows from September or October until March (about 5 to 6 months). The fat-tailed dwarf lemur hibernates for around 7 months of each year in tropical winter temperatures that are relatively high.

Typically, whatever the exact length of hibernation, it is generally characterized by periods of sporadic arousals. The intervals between these arousals depend on the animal's size, body temperature, and other internal factors. For example, in one study, hibernating hedgehogs were observed to have 12–18 arousals. The average duration of these arousals was 34 to 44 hours. Some arousals are initiated by external stimuli like noise and are called alarm arousals. Other arousals do not appear to be initiated by any external trigger and seem to be under the control of signals internal to the animal (endogenous signals). Final arousal from hibernation always occurs at the end of winter, and it is believed that this may be controlled by a combination of exogenous (environmental) and endogenous signals that work to prevent the animal from reentering hibernation.

What Initiates/Triggers Hibernation

Entry to hibernation may be triggered by environmental factors such as temperature, day length, and shortage of food. Indeed, animals such as hamsters and chipmunks are sometimes called *facultative* hibernators because they hibernate in response to environmental conditions.

For *obligate* hibernators (e.g., ground squirrels, marmots, and white-tailed prairie dogs), on the other hand, the cycle of storing food, hibernating, and arousing seems to be controlled mainly by a signal or cue that originates internally from the animal itself (an endogenous signal). For example, when the golden-mantled ground squirrel (*Spermophilus lateralis*) is kept in the laboratory under constant environmental conditions (i.e., in the absence of environmental triggers), hibernation occurs just as if it were in its natural habitat. This indicates that environmental triggers (exogenous cues) are not essential for hibernation in this animal and so the golden-mantled ground squirrel can be considered to be an obligate hibernator.

On the other hand, some birds, like the common whippoorwill, enter a hibernation-like state once their food is removed (environmental trigger). Similarly, food availability has been shown to play a major role in the start of hibernation in the Japanese dormouse (*Glirulus japonicus*). At the same time, another environmental trigger, low ambient temperature, does not appear to have significant influence on the hibernation of the Japanese dormouse once there is an adequate supply of food. Hibernation in the desert tortoise (*Gophents agassizii*) also appears to be only weakly influenced by exogenous environmental cues.

Thus, for some animals some type of endogenous signal or cue is the main determinant of entry to and exit from hibernation. It is believed that this internal signal is regulated by a biological rhythm based on a 24-hour cycle (called a circadian rhythm). The onset and end of hibernation in most animals appear to be triggered by both endogenous signals and exogenous environmental cues.

Biological Clocks

The endogenous cue that triggers hibernation is believed to be based on an internal or biological clock. The biological clock is believed to control not only the hibernation of animals but also many repetitive physiological functions in humans, as well as the migration of birds and the flowering of plants. It is also believed that the periodic arousals characteristic of some hibernators are regulated in part by the biological clock. Also, final arousal from hibernation in spring may be triggered partly by the biological clock.

Although it is endogenous, it is believed that the biological clock responds to several external environmental cues, in particular day length or more accurately the photoperiod (the ratio of daylight to darkness). The photoperiod is an exogenous environmental cue that is "sensed" by the animal from certain sensory organs, like the eyes and also from light-sensitive receptors in its brain. The location of the biological clock in animals, its properties, and the mechanisms by which it operates are only now beginning to be discovered. Indeed, the study of biological clocks and their associated rhythms has led to the scientific discipline called chronobiology.

Thus, the timing of the entry into and exit from hibernation may be due primarily to an endogenous circannual (about a year) or circadian (about a day) rhythm with the timing being reset annually by environmental factors such as day length.

Effects of Global Warming/Climate Change

There has been some concern that higher global temperatures may have led to shortened winters that could have a negative effect on animals that hibernate. For example, some research done in Italy suggests that climate change is indeed bringing animals out of hibernation early. This is believed to put their feeding and breeding habits out of synchronization with the environment, causing unusual weight loss and stress in the animal. For instance, one study observed the effect of annual temperatures in southern England (between 1983 and 2005) on the body condition of female common toads (*Bufo bufo*). This study suggested that "partial" hibernation (early emergence from hibernation) occurred in the female toad as a result of the mild winters experienced during those years. It is thought that this contributed to a decline in the bodily condition of the toad, which negatively affected its survival.

Another study of the effect of temperature change on several traits of over 1,000 temperate-zone animal species worldwide revealed that springtime events (e.g., blooming of flowers, laying of eggs, and the end of hibernation) now occur about 5.1 days earlier per decade on average. Some studies on marmots in Colorado also show that they are ending their hibernations about 3 weeks earlier than they used to in the late 1970s. It has been reported that dormice now hibernate 5–1/2 weeks less on average than they did 20 years ago. Reports indicate that some bears in Spain (slightly different genetically from bears in other parts of the world) have stopped hibernating altogether, remaining active throughout recent winters. Research and debate continues on climate change and its effects on animals' biology and behavior, including hibernation.

Artificially Induced Hibernation

There has always been interest among medical researchers in the chemical induction of hibernation. The potential benefit of putting trauma patients into hibernation and giving doctors more time to repair severely damaged tissue is just one advantage of artificially induced hibernation in humans. Science fiction writers have also explored the possibility of preserving human life in a reversible state of so-called suspended animation. Supposedly, one could exist in such a state for several years without aging, without the need for food and the elimination of waste, and reawaken unaffected by a journey through space of many years to another planet. Inducing artificial hibernation in humans (even in fiction) is therefore potentially a way of "buying time."

Hibernation cannot be induced in nonnatural hibernators by manipulating the environment, so researchers have searched for a possible blood-borne chemical that could induce hibernation. It has been discovered that a serum extracted from hibernating animals such as the woodchuck, when injected into active animals, can induce a hibernation-like state. This serum contains a substance called the Hibernation Induction Trigger (HIT). Experiments with hydrogen sulfide have also shown potential for inducing a hibernation-like state in nonhibernating animals.

Other Types of Hibernation

The term hibernation is now used widely outside of the life sciences to refer to any period or state of temporary dormancy or inactivity. For example, it is used in computer technology to refer to the state of dormancy that can be achieved by powering down a computer while retaining data about running programs and the status of the input/output devices. When the computer resumes from the state of hibernation (on powering up), it reads the saved state data and restores the system to its previous state.

Companies have been said to go into hibernation as a survival strategy to avoid closure (analogous to animal hibernation, the business operations might be reduced drastically to maintenance levels until, for example, brighter market conditions emerge). Projects that are temporarily dormant have sometimes been said to be placed "on ice" and "in hibernation." The term hibernation is also sometimes used for any temporary loss in function as seen, for example, in organs in the human body.

Jennifer Papin-Ramcharan

See also Clocks, Biological; Cryonics; Ecology; Global Warming; Heartbeat; Seasons, Change of; Sleep; Solstice

Further Readings

Dunlap, J. C., Loros, J. J., & DeCoursey, P. J. (Eds.). (2004). *Chronobiology: Biological timekeeping.* Sunderland, MA: Sinauer Associates.

Harder, B. (2007). Perchance to hibernate. *Science News, 171*(4), 56.

Reece, W. O. (2005). Body heat and temperature regulation. In *Functional anatomy and physiology of domestic animals* (pp. 369–378). Baltimore, MD: Lippincott Williams & Wilkins.

Roots, C. (2006). *Hibernation. Greenwood guides to the animal world.* Westport, CT: Greenwood Press.

Roth, M. B., & Nystul, T. (2005). Buying time in suspended animation. *Scientific American, 292*(6), 48–55.

Whitrow, G. J., Fraser, J. T., & Soulsby, M. P. (2004). Biological clocks. In *What is time: The classical account of the nature of time* (pp. 31–48). New York: Oxford University Press.

Hinduism, Mimamsa-Vedanta

Among the six principal schools that developed in the Brahmanical traditions in India, Mimamsa and Vedanta form the most recent group, the others being Nyaya, Vaisesika, Samkhya, and Yoga systems. These two schools, that is, Mimamsa and Vedanta, are also known as Purva-Mimamsa and Uttar-Mimamsa. Founded by Maharshi Jaimini and Maharshi Samkara, respectively, these schools develop their theories of time in a unique way. Remaining faithful to the core thought of the Upanishads, namely the reality of the immutable Brahma, they explain the phenomena of change in diverse ways.

The Purva-Mimamsa school has two sects, one known as Bhatta or Kumarila Mimamsa, propounded by Kumarila Bhatta, and the other known as Prabhakara Mimamsa, established by Prabhakara Mishra. The two sects differ somewhat in their treatment of time.

In the Jaimini Sutra the concept of time is discussed in connection with action. Some element of time is associated with all actions. Action is in a way determined by time. The Mimamsa school bases its philosophy on the Vedic hymns. According to these, while interpreting Vedic hymns, if the hymns are correctly pronounced, they refer to the laws of time or life. But it is also said in the Jaimini Sutras that time is not the cause of the result of action. Result is due to the effort.

The Vedas never refer to time as the sole cause, for a result never comes simply due to the passage of time. But it is true that sometimes if after several sincere efforts, one fails to achieve results, one can attribute the failure to time. In ancient thought time is associated with decay, death, and failure. Time changes our position. But the Mimamsa text clearly points out that everything happens due to the impelling force, and time is not connected with it. The concept of change is associated with both nature and time. The concept of time is associated with any new being. But time itself is immeasurable as it is without beginning or end. It is through the instrument of intellect that we receive the idea of time. For the purpose of our understanding, the seers have divided time into *yugas, manvantaras,* and *kalpa.* These are nothing but the exercise of our mental faculties. When we analyze the etymological meanings of these terms, the point becomes clearer. The word *yuga* is derived from *yuj,* which means "to fix or concentrate the mind." The word *manvantara* is derived from *manu,* which is the same as mind. Similarly, one of the meanings of kalpa is research or investigation. And the word *kala* for time is itself derived from *kal,* which means to perceive or consider. So, it might be said that the idea of time involves a series of functions of the intellect or the mind.

It is mainly the orderly arrangement of objects or the succession of events that gives us an idea of time. The appearance of a new object in nature reminds us about the role of time.

Though the Mimamsakas are known as theists and God plays a supreme role in their systems, still time cannot be personified as God, as God creates and sustains the universe. But time does not play the role of creator and sustainer, although sometimes it is treated as the greatest destructive force.

After life begins on this earth, time starts acting upon everything of the universe, but time itself remains unaffected. Therefore it is said to exist for other things and not for itself. Like the *prakriti,* or nature, of the Samkhya system, time remains unaffected throughout its functions.

For our practical purposes we divide time into segments but time itself is a continuous whole.

Time is like butter that is beginning to separate out of cream; it partakes of the nature of both the butter and the cream, yet the two can be separated. Like the endless chain of desires, time flows continuously. It has been said that when the mind begins to function, desire arises, hence the idea of time is associated with desire.

The Mimamsa school follows the sacred text known as the Gita. In the Gita it is stated that the number 12 refers to time. This number constitutes the basis of the calculation of the time of day and night, the months of the year, as well as the great ages of time, all of which are multiples of the number 12. The Mimamsa text refers to *dvadasa-satam,* which may mean both 112 and 1,200. These two numbers refer to the passage of time. There are four yugas: *Satya, Treta, Dvapara,* and *Kali.* The entire duration of these yugas make a Manvantara, and a thousand yugas constitute a kalpa. The duration of these four yugas is said to be 1,728,000 years, 1,296,000 years, 864,000 years, and 432,000 years, respectively, of human time. The above figures are nothing but multiples of 1,200, which may be said to refer to prakriti, or time. In the Mimamsa text, the number 12 plays an important role, as it represents time or prakriti.

Bhatta and Prabhakara Schools Compared

Both of the schools of Mimamsa (Prabhakara and Kumarila) admit that substance is that in which quality resides, and they treat time, like space, as substance.

Kumarila, in his commentary titled *Slokavartika on Sabara-bhasya,* an important text of the Mimamsa school, says that time is one, eternal, and all-pervasive. But although time is all pervasive, it is conditioned by extraneous adjuncts for the purpose of empirical usage. And this is the reason that there are various divisions and subdivisions within all-pervasive time. *Tantravartika,* another commentary in prose on *Sabar-bhasya,* describes time as eternal like the Veda. Not only Kumarila but Prabhakara, another philosopher of the Mimamsa school, also treats time as one of the eternal substances.

Regarding the perceptibility of time, the two schools of Mimamsa (Bhatta-Mimamsa and Pravakara-Mimamsa) differ in their viewpoint. The Bhatta school held that time is perceived, whereas the Pravakaras joined the Nyaya-Vaisesika system in maintaining inference to be the valid means for knowing time. Parthasarathi Misra, a noted Bhatta-Mimamsaka philosopher, in his work *Sastradipika* points to the fact that all perceptual cognitions also include the duration of the perception. In our everyday dealings we express such statements as "we are perceiving the pot from the morning . . ." And this statement clearly reveals the fact that time is perceived. While recording a statement, we always use such expressions as "now," "then," and the like. Such expressions would be meaningless if time were not also perceived along with the perception of the object. These words refer to the time-content of the perception itself. The Bhatta-Mimamsaka wants to stress that the temporal dimension of all our perceptions would remain unaccounted for without the contention that time is amenable to perception. The Nyaya-Vaisesika philosophers deny that time can be said to be perceived, as it has no perceptible qualities such as colors and form. It can only be inferred.

But the Bhatta-Mimamsaka retort that absence of color in time does not prove that time is imperceptible. Possession of a quality like color is not a criterion of perceptibility. Thus the contention of the opponent suffers from the fallacy of false generalization. However, the Bhatta-Mimamsakas acknowledge that time as such is never an object of perception, but it is perceived always as a qualification of perceptible objects. Therefore time is perceived as a qualifying element and never independently of the perceptible object. This is why we perceive events as slow, quick, and so forth, which involves a direct reference to time.

In *Manameyadaya,* a notable Mimamsa work, one of the exponents of Bhatta-Mimamsa, Narayana, refutes the claim of Nyaya-Vaiseseka that time is inferred on the basis of such notions as simultaneity. Narayana expands his view with the following example: In our everyday life we experience such events as "Devadatta and Yajnadatta arrived simultaneously" or "Ram arrived late." Here the question is whether such awareness, received through perception, can be denied to have time as its essential component. The point that emerges is that such direct, immediate awarenesses as simultaneity or lateness cannot be ascertained, as in the above examples, through anything but sense-perception. "Devadatta," "Yajnadatta," and "Ram" are related directly to time, and the latter is perceived along with the substances that it qualifies.

Prabhakara also treats time as one of the eternal substances. According to Prabhakara, space and time cannot be perceived as they cannot be seen or touched. Prabhakara considers some degree of magnitude along with color and touch and other such qualities as a necessary condition of proper sense perception. Time, though large, is not perceptible because it is devoid of color and touch.

Vedanta

The Brahmanical tradition finds its full expression in the system of Vedanta. The system follows the literal teaching of the Brahma Sutra, where a complete identity or oneness is shown between the world and the Brahman, hence it denies the reality of the finite world of time and space. The two main commentaries of the Brahma Sutra are the following: Samkara-Bhasya, propounded by Maharshi Samkara, and Sribhasya, propounded by Maharshi Ramanuja. Both of them believe that Reality is non-dual, but their interpretations differ. The school of Samkara is known as *Advaita-vedanta,* and the school of Ramanuja is known as *Visistadvaita-vedanta.* In the school of Advaita-vedanta, the problem of Creation is reduced to a problem of appearance, whereas in the Visistavaita school of Ramanuja, the world is conceived as the body of God. The concept of time is discussed here mainly from the Advaita-vedanta standpoint of Samkaracharya. The conceptual framework of this school has the unique characteristic of expressing its position with the single dictum, "Reality is the-one-without-another." Here the concept of being is treated as timeless, involving the notion of time as appearance. This nondual nature of Brahman (reality) reduces all duality (difference) to a problem of appearance. In this metaphysical system there is no place for the concept of objective real time. To accept time as real is to invite pluralism. But this view cannot be accommodated in the system of Advaita-Vedanta because "Reality is one-without-a-second." So the question arises, What will be the status of time here?

To understand the status of time, it is necessary to discuss the notion of causality of the Vedantins that is known as Vivartavada. In this theory, the effect is characterized as "no-other," that is, *ananya,* to the cause. In his commentary on the Brahma Sutra, Samkara brings out the importance of the term ananya as the effect being no-other in relation to the cause. This is the kernel of the vivartavada of the Advaita Vedanta. In the Vedantaparibhasa, an important work on the Vedanta glossary of technical terms, the significance of the term *vivarta* is explained. The term vivarta is to be understood by contrasting it with the term *parinama.* When the cause and the effect are taken to be realities with the same status, the effect is designated as parinama of the cause; for example, prakriti and its different aspects evolve in Samkhya system. But if the cause and the effect cannot be given the same ontological status, it is termed vivarta, as in the case of the world as effect in relation to its cause, Brahman, the Real, in Advaita Vedanta. Here the effect is indeterminable, related to neither sameness nor difference. The effect is nothing but the *appearance* of the cause—the effect has no separate existence of its own other than a difference of name and form, just as the earthen jug cannot have any reality apart from the clay, its material cause. The Advaita Vedantin asserts that no change or modification can be ascribed to that which is real. In other words, change can never be granted a separate ontological status under any circumstances.

The process of Creation as discussed in the Brahma Sutra shows that the world is in time. And time is characterized by succession and sequence. The steps shown in the creative process involve succession. This succession implies that the world must be in time. The creation and the dissolution of the world are a temporal series.

The Vedantins believe in the rebirth of the soul. The experiences gathered by the finite soul in each birth and death are events in time. The Vedanta schools agree unanimously on one point: that the liberated soul is beyond the domain of the temporal world. Therefore, finite souls live, move, and have their being in time. But here a question arises: How can an eternal and changeless Brahman be held to be preceding the world? Precedence, succession, sequence—these are all time relationships. If Brahman precedes the world and brings the world into existence at a certain point of time, Brahman cannot be called eternal and cannot be beyond time. In the Vedanta texts the word *eternal* is used in different senses. Sometimes the word *eternal* means existing for an unending span of time. Sometimes this word is used in the sense of being beyond time. In other places it is taken as transcending time yet somehow including it. There is a distinction between self-contained Brahman and Brahman-who-creates-this-world in the Vedanta texts. But this distinction is apparent, not real. The temporal order of the world is in Brahman. Thus Brahman is beyond and yet somehow includes time.

In fact, the Advaita Vedanta reduces the problem of creation to a problem of appearance. Just as the earthen jug, which has a distinct name and form, does not have an existence independent of the "clay" from which the jug is made, similarly the world as effect has no independent existence without its cause: Brahman. The Advaita Vedantins do not accept Brahmaparinamavada, a theory where the world is conceived as a real transformation of Brahman. The Vedanta school maintains Brahman as the cause, but cannot accept any transformation or modification of the cause. The empirical world is entirely dependent on the self-existent Brahman, having no reality of its own. Brahman is the indispensable cause of the world-appearance. *Maya,* or illusion, explains the existence of this temporal and spatial world. But it should be pointed out in this connection that the idea of appearance does not imply that the world is merely a subjective dream. Samkara, in his commentary on the Brahma Sutra, refutes the Buddhist school of subjective idealism by saying that the world is empirically real. The idea of the false appearance of the world is intended to disclose its entirely dependent status, that is, it is not real per se. Brahman as real per se as cause of the origin, sustenance, and destruction of the world points to the formulation of the idea of being as timeless as the very presupposition of the temporal. There are two types of definition of Brahman as cause of the world in the Advaita Vedanta texts: one is *tatasha-laksana,* the accidental or modal definition; the other is *svarupa-laksana,* or the nonrelational definition. The modal definition indicates Brahman as the ground of the experienced world as it appears to be. But the Advaita-Vedanta advocates a nonrelational definition of Brahman as *sat-cit-ananda swarupa* (being–consciousness–bliss). The temporal world as effect is not ontologically real, but Brahman as its foundation is self-existent. In Advaita-Vedanta, Brahman is characterized as nontemporal, nonspatial, and noncausal. Time, space, and causality are all categories of the empirical world. These categories presuppose pluralities that are the products of our ignorance or *avidya.* It is because we are veiled with ignorance that we experience this plural world. Brahman transcends these categories and is the timeless Reality. Thus Brahman is beyond the spatiotemporal sequence of multiple events and things.

Sri Harsa, an ardent follower of Adaita Vedanta, raises the question whether such time distinctions as past, present, and future are integral or not integral to time as such. If these distinctions are integral to time as such, then the notion of the unity of time has to be given up. On the other hand, if one upholds the unity of time as essential and the differences of time-forms as inherent in it as its threefold nature, then the designations of past, present, and future could be made at random for want of a strict basis for such usages. This is to say that one cannot accept the differences of time-forms as integral to one time, either. But if it is said that it is due to the sun's motion that time divisions arise, Sri Harsa also finds that to be equally untenable. He holds that the solar motion cannot account for the determinations of the different time forms, since it is used as the common criterion for the divisions of all three time forms—past, present, and future. It is said that the same qualifying factor—the solar motion that distinguishes the past day from the present and also distinguishes the present day from the future—is not a basis for valid argumentation. In this way Sri Harsa rejects the reality of time.

Not only Sri Harsa but also Citsukha, another great exponent of Advaita Vedanta, in his important work *Tattvapradipika,* has discussed the problem of time thoroughly. While refuting the existence of time, Citsukha argues that neither perception nor inference can establish the reality of time. The ontological reality of a substance can be established through perception only if the substance in question can be said to be perceived by vision or touch. But as time is without form and is intangible, sense perception cannot be the source of knowledge. Time cannot be perceived even though the mind works only in association with the external senses. It cannot even be inferred in the absence of perceptual data.

Again it is possible to point to the existence of such mental states as happiness or sadness, which are known through introspection and need not be first known through sense perceptions. Could the existence of time be established in a similar way? But Citsukha rules out this possibility, as it is possible only for facts that are subjective, but time as an objective category of existence cannot be thus established.

Time cannot be inferred as the substratum of the notions of priority and posteriority, succession and simultaneity, quickness and slowness, as these notions are not associated always and in all places. It may be argued by the Naiyayikas that the entity that establishes relations between the solar vibration and worldly objects is called time. But Citsukha replies that all-conscious self is the cause of the manifestation of time in things and events; therefore it is unnecessary to assume the existence of a further entity called time. The notions of priority and posteriority are associated with a larger or smaller quantity of solar vibrations. So here also it is not necessary to admit time as a separate category.

Concluding Remarks

In conclusion, it may be noted that the Brahmanical tradition expresses its conception of being in its radical form in the Vedanta system, rejecting the pluralistic model. The system advocates a philosophy of nondualism and focuses on the reality of the unchanging eternal Brahman as the one and only Reality. The concept of time plays a phenomenal role without any ontological reality.

Debika Saha

See also Becoming and Being; Change; Eternity; Hinduism, Nyaya-Vaisesika; Hinduism, Samkhya-Yoga; Idealism; Materialism; Metaphysics; Ontology; Time, Measurements of

Further Readings

Balslev, A. N. (1999). *A study of time in Indian philosophy.* New Delhi: Munshiram Manoharlal Publishers.

Jha, G. (1942). *Purva-Mimamsa and its sources.* Banaras, India: Hindu University.

Mandal, K. K. (1981). *A comparative study of the concepts of space and time in Indian thought* (2nd ed.). Patna, India: A. Mandal.

Radhakrishnan, S. (Ed.). (1952). *History of philosophy: Eastern and Western.* London: Allen & Unwin.

Hinduism, Nyaya-Vaisesika

The Nyaya-Vaisesika schools of thought founded by Maharshi Gautama and Kanada, respectively, advocate a plural, realistic worldview. These schools focus on the reality of time as vital to their total conceptual framework. Their philosophical outlook regarding the role of time is distinctly different from that of other schools of the Indian tradition.

Time, in the context of the Nyaya-Vaisesika philosophy, has been studied in its various aspects. The question of its existence and how it is related to different ontological issues like causality, motion, and space are all highlighted in the Nyaya-Vaisesika system. This system believes in the theory of Creation. Kala, or Time, is considered as the eternal, unique, all-pervading background of the creative process. All events derive their chronological order through time. But time possesses no specific physical quality like color and thus cannot be an object of external perception. Neither is it perceivable internally, as the mind has no control over external or internal objects independently of a physical sense organ. But the question of its existence is arrived at by a series of inferences. The notions of priority (*paratva*) and posteriority (*aparatva*), of simultaneity (*yaugapadya*) and succession (*ayaugapadya*), and of quickness (*ksipratva*) and slowness (*ciratva*) constitute the grounds of inference for the existence of time.

It is a recognized fact that priority and posteriority are related with the movement of the sun. When using the term *now,* it is natural to refer to the motion of the sun above or below the horizon by so many degrees. But a pertinent question is raised by the Nyaya-Vaisesika philosopher: How can any object be related at all with the motion of the sun? It is not possible to establish any direct relation of inherence, as the motion of the sun inheres in the sun alone, and not between the object in question and the sun, which are far apart. There must be a connecting medium that joins the two and that is capable of transmitting the quality of one to the other. This fact leads to the inference of time. Regarding the question of whether time is perceived or inferred, there is a philosophical debate among the Indian realists. The main voice in denouncing the view of the Nyaya-Vaisesika is the Bhatta-Mimamsaka school. In reply to the Nyaya-Vaisesika view that time cannot be said to be perceived as it has no sensible qualities such as color and form, the Bhatta-Mimamsakas maintain that sensible qualities are not the criteria of perceptibility. However, they do acknowledge that time as such is never an object of perception, though it is always perceived as a qualification of sensible objects. It is for this reason events are perceived as slow or quick, which involves a direct reference to time. The Nyaya-Vaisesika philosophers answer that a directly perceived time would point to only a limited, divided time, which is not absolute time but conventional temporal time. Sridhara in his *Nyaya-Kandali* and Jayata Bhatta in his *Nyaya-Manjari* take up the issue in a brilliant manner. According to them, time cannot be established as objects of ordinary perception, as time does not possess any finite dimension.

There are certain typical terminologies in every Indian philosophical system that are unique within their field; the Nyaya-Vaisesika system is no exception. Besides normal perception, extraordinary perceptions are also accepted by this school. One of the extraordinary perceptions is known as *Jnanalaksana.* This is the type of perception where the perceived object is given to the senses through a previous knowledge of it. For example, when someone says, "I see the fragrant sandalwood." Here one visualizes fragrance due to a prior knowledge of it from past experience. Similarly, the perception of "the present pot" is composed of two elements. "The pot" is an object of ordinary perception, whereas the time element indicated by the word *present* is derived from previous prior knowledge of time, resulting in the extraordinary perception of "the present pot."

So the sense perception of time as a qualifying element of the object is analyzed by the Nyaya-Vaisesika philosopher as based on the previous inferential knowledge of time. Moreover, the Nyaya-Vaisesika system believes that time must be of unlimited magnitude as it has to determine the priority and posteriority of all finite substances of the world. In other words, it must be a ubiquitous substance. Time is also conceived as *nimitta karana* or the instrumental cause of all objects of the world. This causation also leads to the point that time is eternal.

Kala (time) is treated in the Nyaya-Vaisesika system as the substratum of motion. The statement "I am at present writing" clearly points out that time is the substratum of objective motion. Not only is time the substratum of objective motion, it is the cause of the origination, maintenance, and destruction of all objects of the world.

Time and Causality

In Nyaya-Vaisesika literature, the role of time comes to the forefront in discussions of causality. For any causal operation the idea of absolute time is accepted as a necessary presupposition. The distinction between the eternal and the contingent is clearly mentioned in their writings. An object that is ever-present cannot be amenable to creation or destruction. The contingent is defined as that which does not exist prior to its creation and ceases to exist after its destruction. An object like *akasa* (i.e., ether) is never nonexistent whereas an absolute nonentity like a hare's horn is ever nonexistent. Any causal operation is necessarily an event in time. No event is conceivable without time as its *adhara* or receptacle. According to Nyaya-Vaisesika philosophy, to deny time an objective reality is to face a static universe where there is no room for any change or movement.

The theory of causality known as *asat karyavada* presupposes time in the Nyaya-Vaisesika system. The theory is so called because the effect is conceived as *asat,* that is, nonexistent prior to the causal operation. Cause is defined as that which is an invariable and necessary antecedent to an event. Here, the idea of antecedence has an immediate temporal reference. The idea of time always follows when it is said that the effect succeeds the cause.

Another characteristic of time is that, though time is a *dravya* (substance) it does not presuppose any substratum for its existence, as it is something undivided and eternal. Time is *svatantra,* or independently real. As time is all-pervasive, it has no form. The main argument that the Nyaya-Vaisesika proposes is that time, being a necessary condition of all movement and change, must itself be free from general attributes.

Divisions of Time

But here a question arises: If time itself is free from general attributes, are divisions of time real? The Nyaya-Vaisesika does accept the conventional time divisions. Generally, time is divided into past, present, and future. This division does not affect the inherent nature of time. These conventional time divisions can be explained with the help of limiting adjuncts, which are of finite duration. It is only with relation to events that time takes on these distinctions. Saying that something is happening refers to an event that has already begun but has not yet ended. Similarly, the past is the time related to an event that completed its cause of action, and the future is related to an event that has yet to begin. It is only the actions that are labeled as past, present, and future. Real time is never affected by these *upadhis,* or external adjuncts, namely, different solar motions.

This conception of absolute time in the Nyaya-Vaisesika system has been criticized by all the schools of Indian philosophy. Here is the view of one of the major systems: The Advaita Vedanta school maintains that this argument is not plausible because all the three divisions of time are related in the same way with the same solar motion. For example, a particular day is understood as the present with reference to a particular solar motion. But it must be admitted that the same solar motion is responsible for past and future days. The day referred to above is understood as present on the day itself, as past on the days that follow, and as future on coming days. Here the three time determinations, namely present, past, and future, have the same conditioning factor, namely, solar motion.

The Nyaya-Vaisesika answer is that the above position has come about due to the absence of certain necessary qualifications regarding relationship. When the time of an event is in "actual" relationship to solar motion we cognize it as present. A thing is cognized as past when the particular relationship "has been" and "is no more," and when the relationship in question "will be" and "is not yet," it is recognized as future. But the vedantin does not accept this position and says this argument is merely tautologous. The terms *actual, has been,* and *will be* are synonyms of present, past, and future, respectively.

In answer to his opponent, the Nyaya-Vaisesika suggests a modified definition of time division. The time that is conditioned by a particular action is present in respect to that action alone and not in relation to other actions.

In fact, the criticisms that the opponent offers are applicable only to the time division of the Nyaya-Vaisesika system. But time division is not inherent in the nature of time; hence it does not at all affect time-in-itself. Thus the criticisms do not really affect the reality of absolute time. These time divisions belong only to the events that are in time. The Nyaya-Vaisesika does not agree that the time series is the "schema" of the understanding and is formed by the conceptual fusion of discrete movements. The reason behind this view is the following: It is not possible to encounter the objective existence of a momentary entity. "The jar exists for many moments." This perceptual judgment cannot prove the objective existence of moment. It is a common experience shared by us all that a moment is supposed to vanish just after its contact with the sense organs; it cannot synchronize with perceptual judgment. A question, therefore, arises: Is not the use of the term *moment* superfluous? The Nyaya-Vaisesika system maintains that it is not superfluous, it is part of the whole cognition procedure. Moreover, every cognition presupposes a determinant datum, hence a particular datum in this case should be pointed out. It is said that the datum in this case is a particular motion and the pre-nonexistence of disjunction caused by it as the determining factor of motion. Each of the moments covers a definite duration; hence they cannot separately be the datum. These two, taken together, form the datum of cognition.

Moment, Time, and Space

In the *Prasastapadabhasyam,* one of the major authentic books of the Nyaya-Vaisesika system, the unique character of time is brought out in relation to space. Time has only one form of mensuration. The point is elaborated by saying that the temporal order of events is such that it does not allow for reversibility, whereas the spatial order does allow that. Thus, in the realistic framework of Nyaya-Vaisesika, another question crops up: How are they related? It is said that every cognition acquires them at the same time; hence cognition itself is the connecting link. The moment itself is the datum of such cognition. Therefore the concept of moment arrives indirectly in the framework of cognition. A moment is the point of time that refers to the final phase of cause and the initial phase of effect. Vallabha, a famous commentator, in one of his books defines moment as that intervening span of time that comes between the completion of the totality of caused conditions and the production of effect. This discussion reflects the realistic attitude of the Nyaya-Vaisesika system. These two—space and time—are characterized as *niravayava* (noncomposite), *avibhajya* (indivisible), and *vibhu* (ubiquitous). All objects are in space and time, but space and time cannot be said to be contained in anything.

This traditional view of absolute time is questioned by the later school of the Nyaya-Vaisesika system. Raghunath Siromoni, one of the famous philosophers of the Nabya-Nyaya system, challenged the entire conceptual framework of the pluralistic metaphysics of Maharshi Gautama. In his book *Padarthatattvanirupanam,* he criticized the traditional understanding of time. He does not find enough justification for accepting these two all-pervasive substances that require adjuncts in order to account for spatial and temporal divisions. He wants to identify space and time with God, who is also characterized as an all-pervading substance. Regarding time, he gives up the notion of absolute time and introduces a new category called *ksana* (moment). But he rejects the notion of moment in relation to motion as inadequate. Instead he sets up a new category consisting of things that are not motions but that share the universal repeatable property of momentariness.

This version of Raghunatha Siromani was not approved, however, by the later adherents of the school. In fact, the idea of absolute time is integral to the whole fabric of Nyaya-Vaisesika thought as all the concepts are woven together against the background of this particular concept; that is, absolute time. Therefore, in the Nyaya-Vaisesika system, time is a reality per se, the very presupposition of the reality of change.

Debika Saha

See also Causality; Cognition; Eternity; Hinduism, Mimamsa-Vedanta; Hinduism, Samkhya-Yoga; Materialism; Metaphysics; Perception; Time, Absolute

Further Readings

Bhadhuri Sadananda. (1947). *Studies in the Nyaya-Vaisesika metaphysics* (Series 5). Poona, India: Bhandarkar Oriental Research Institute.

Chatterjee, S. C. (1939). *The Nyaya theory of knowledge.* Calcutta, India: University of Calcutta.

Jha, G. (1916). *Padarthadharmasangraha of Prasastapada: With the Nyayakandali of Sridhara.* (English Translation). Beneras, India: E. J. Lazarus.

Niyogi Balslev, A. (1999). *A study of time in Indian philosophy.* New Delhi: Munshiram Manoharlal Publishers.

Hinduism, Samkhya-Yoga

The Samkhya-Yoga system is one of the oldest Brahmanical traditions of India. Founded by Maharshi Kapil and Maharshi Patanjali, respectively, these two schools share a basic ontological structure, with some subtle differences in their treatments of time.

The Samkhya system rests on a dualistic metaphysics, tracing the whole course of the universe to an interplay of two ultimate principles—*Purusa* and *Prakriti.* The dichotomy of matter and consciousness is expressed by these two terms; Purusa is the unchanging principle of consciousness, whereas Prakriti is the ever-changing principle of matter. The Yoga system shares this metaphysical dualism of Samkhya but whereas the latter is atheistic, the former is theistic. Both of them, however, deny the Nyaya-Vaisesika view of absolute time.

Time, Causation, and Change

The concept of time in the Samkhya-Yoga system can be grasped only in the framework of their metaphysical thought, along with their theory of causation termed as *Satkaryavada*. This theory explains the relation between Purusa and Prakriti in a unique way. Everything that is manifest is dependent on a cause. This causal dependence shows the contingent character of the objects. It is not possible to posit any entity without a cause. In the Samkhya system the understanding of the manifest as caused leads ultimately to the postulation of Prakiti as the unmanifest, uncaused principle. Prakriti is dynamic; it is matter that evolves ceaselessly. It is the formal, material, and instrumental cause of the world. This Prakriti consists of three qualities, or *gunas: sattva,* as something illuminating; *rajas,* as the active component; and *tamas,* as the restraining forces. These three gunas, through mutual cooperation, create this objective world.

The *Samkhya-Karika* of Isvarakrisna contains a detailed discussion regarding the change and origination of the 25 categories, but what is interesting is that nowhere is time recognized as a separate category of existence *(tattva).* In fact, time is taken as an aspect of concrete becoming. The Samkhya system does not recognize any conception of time as independent of change. Change is not understood in abstraction, but as a concrete becoming.

The Samkhya recognizes that the principle of change is ultimate, and so this principle cannot be derived from any fixed principle. The heterogeneous categories of creation with all its varied aspects presuppose always a primordial movement that remains even in the state of cosmic dissolution. It is because of the interplay of the gunas (qualities) that Prakriti is always in motion. In his *Samkhya Karika,* Vacaspati Mistra, the famous commentator, remarks on this ever-changing nature of Prakriti. There are two types of manifestations of Prakriti, *Sadrisaparinama* (homogeneous manifestation) and *Visadrisaparinama* (heterogeneous manifestation). Though no unique object arises from the homogeneous manifestations of the gunas, as they do not mix with one another, yet the process of self-production that is sattva giving rise to sattva, raja giving rise to raja, and so on, continues. The gunas never remain unmodified, even for a moment, as the very characteristic of the gunas is dynamicity. Creation of varied objects is possible due to the heterogeneous movements of Prakriti. This unceasing, ever-active nature of Prakriti points to the principle of change. And this principle of change is nothing but time. Some scholars, such as Madhavacharya in his commentary on *Parasara Samhita,* have characterized Prakriti as the very personification of time.

Here it may be asked whether the process of change is enough to account for time-experience. The Samkhya replies in the positive. Time-experience is totally dependent on the duality of Purusa (consciousness) and Prakriti (matter).

This system rejects the idea of time as a receptacle, advocated by the Nyaya-Vaisesika philosophers. Here the process of becoming as a movement from the potential to the actual is identical with the causal process. The notion of empty time is rejected as merely a thought construction. In other words, the reality of change does not necessarily imply a conception of time apart from the changing material.

In *Samkhya Karika,* Vacaspati Misra observes that it is not necessary to postulate an absolute time for the explanation of action, solar motion, and so on. The temporal phases can be explained with reference to the creative movements of Prakriti.

The concept of concrete becoming is primary. In the state of cosmic dissolution, Prakriti is to be taken not only as a principle of unconscious matter but also as the principle of time itself in its transcendental aspect. In the conception of dynamic Prakriti, the Samkhya system combines both time and matter in the same principle.

Vijnanabhiksu, a noted Samkhya commentator of the 16th century, advocates a different view. In his *Samkhya Pravacana Bhasya,* he maintains that time is nothing apart from the ubiquitous ether or akasa, which is an evaluate of Prakriti. Limited space and time are ether or akasa itself perceived as space and time by limiting adjuncts. The same limiting adjuncts, which are said to account for time-divisions, can do so with reference to akasa, making it unnecessary to postulate time as a separate ubiquitous substance called time. Aniruddha, another noted Samkhya commentator, in his *Samkhya-Sutra Vritti,* also advocates that limited time and limited space are really ether (akasa) conditioned by upadhis (limiting adjuncts) and hence are included in the unified entity of the ether.

In fact, when Vijnanabhikshu consigns the status of space and time to the level of Prakriti itself, he is merely pushing the realistic dualism of the Samkhya to its logical limit. In the Samkhya system, Prakriti is taken as the rootless root of all, and this is the reason for including everything, even time, as part of Nature or Prakriti itself.

In the *Yoga-Sutra* of Patanjali, time is seen as instant. The Yoga school shares the basic ontological structure of the Samkhya. The Yoga system develops its specific stand on the reality of the instant (*ksana*) and the merely subjective construction of sequence (*krama*).

"Sequence" or krama is the temporal succession of forms of the same object. This sequence is broken up into infinitely short instants (ksana). If the flow of time were absolutely continuous, no change could ever happen. But at each moment in time, a subtle, imperceptible change takes place; this imperceptible change is perceptible only to the Yogin, and it is the accumulated effect of these minute changes of which common people become aware. This theory has its modern parallel in the conception of time not only as objective and relative but also as a continuous phenomenon.

The Yoga-Sutra focuses on the three modes of immanent change as the *dharma parinama, lakshana parinama,* and *avastha-parinama.* The Yoga system, like Samkhya, holds that change is incessant in the dynamic Prakriti—the ultimate substance, or *dharma.* When the change is perceptible as a charge of *guna* (quality) in an object, it is called *dharma-parinama. Laksana parinama* directly involves the notion of temporal phases (past, present, and future). The potential is the future state, the manifested is the present, and what has happened is the past phase of the thing. *Avasta-parinama* is that aspect of change that deals with the notion of new and old. This can be explained with an example. An earthen jug is made from a lump of clay. The change of quality through the different stages of the processing of the clay constitutes its dharma-parinama. When it is said that the clay that is at present transformed into a jug was in its past only a lump, it refers to the laksana-parinama. This jug, whether new or old, is known as *avastha-parinama.* The clay remains the same, though the shape changes. This theory of change of the yoga system coheres perfectly with the Samkhya Yoga metaphysical framework. The point that emerges here is that the main substance (*dharmin*) remains the same throughout the three stages of past, present, and future, and the temporal changes are ascribed to the external aspects or qualities of the substance. How does this change in time occur? This is answered by the third type of change, that of the *avastha* (condition) of a thing. Time is a succession of individual moments that imperceptibly alter the condition of the jug; this is the well-known process of decay or aging.

If we substitute consciousness for clay in the above example, we find that emergence and restriction are the forms or dharma of consciousness, each being connected with the three segments of time—past, present, and future—and experiencing its own growth and decay.

The concept of instant (ksana) plays a very important role in this system. As discussed earlier, the concept of time is explained as instant here. The *Yoga Bhasya,* a notable commentary on the *Yoga-Sutra* by Vyasa, holds that the whole universe undergoes change in a single moment. It is said that just as the atom is the minimal position of matter, so the moment is the minimal duration taken by an atom to change its position.

In the Yoga metaphysical structure, a discrete view of time is advocated, over against the concept of time as unitary and as an independent category of existence. This system holds that the idea of time as instant is objectively real, but the idea of time as duration, (e.g., day, month, year) is a mental construct. These are ideal representations. In the *Yoga-Bhasya,* the idea of the reality of instant is recognized. In fact, the idea of time as a continuum or as a substance is unreal for yoga. Time is *vastu-sunya* (unreal) as a substance but time as instant is real (*vastu-patita*). According to the Yoga system, the idea of sequence can be formed only when two moments exist simultaneously. But in reality, two moments cannot and do not occur other than successively. What we experience is always the present, because the past (earlier) and the future (later) moments have no objective reality. In the *Yoga-Sutra,* Vyasa clearly states that the moment is real, and that the idea of a unitary objective time, whether as a collection of moments or as an objective series, is a subjective representation.

Being and Becoming

The idea of time as instant does not hold true of being. In this system, being is not instantaneous. As it was remarked earlier, the substance or being remains the same, persisting through the changes. In the Yoga system, the concept of becoming presupposes the substratum of Prakriti that continues and supports the process of evolution. Not only that, in the Yoga system, the root nature of consciousness is unchanging and unchangeable. The Yoga system recognizes the self as timeless.

Language

From the perspective of philosophy of language, the concept of time plays a distinctive role in the Yoga system. Patanjali in his *Mahabhasya* points out that time is the basis for the tense distinctions of verbs. Here a very pertinent question arises: What accounts for the tense distinctions of verbs? Despite the same verbal root, a word may be used in various ways in the sense of past, present, and future. An analysis of the nature of verbs is always associated with the concept of time. In the *Mahabhasya* of Patanjali one comes across the idea of time as that in association with which bodies grow and decay. Furthermore, the concepts of day and night are described as possible due to the movement of the sun in connection with time.

From the above discussion, it becomes clear that though the Samkhya-Yoga systems share a basic common metaphysical structure, their treatments of the concept of time are not similar. Both of them reject the concept of absolute time, yet in dealing with time, the Samkhya system accepts Prakriti as time personified, whereas the yoga philosophy explicitly advocates a discrete view of time.

Following Professor K. C. Bhattacharya, an eminent Indian philosopher, an explanation of the above fact may be given. It is true that in both the systems time is not admitted as a separate *tattva,* or category. Time in both is concrete becoming, and this becoming is not an event in time. Here becoming, or *Parinama,* may be understood as the cause turning up its own limitation or as the limited effect having got manifested. In the former view, the real objective fact is the act of the cause, its act of production, the succession of cause and effect being only a retrospective construction. In the latter view, the real fact is the antecedence of the cause to the effect, the prior becoming of the effect, but the causal act as the initial moment of this becoming is but an abstraction. Therefore to Yoga that presents the former conception, time is real as the causal act, the initial moment of this becoming. To Samkhya, the effect being preexistent in the cause is never turned up, so that the time relation of sequence alone is real time.

In conclusion, it may be said in the Samkhya-Yoga system the notion of time is beautifully illustrated through Nature, or Prakriti, in its undifferentiated, transcendental state and its differentiated, multiform condition. The reality of time is the reality of the ever self-modifying primary substance.

Debika Saha

See also Becoming and Being; Causality; Change; Consciousness; Duration; Hinduism, Mimamsa-Vedanta; Hinduism, Nyaya-Vaisesika; Materialism; Metaphysics; Ontology

Further Readings

Bhattacharya, K. C. (1956). *Studies in philosophy* (Vol. 1). Calcutta: Motilal Books.

Dasgupta S. N. (1974). *Yoga philosophy.* Delhi, India: Motilal Banarasidas.

Feuerstein, G. (1989). *The yoga-sutra of Patañjali: A new translation and commentary.* Rochester, VT: Inner Traditions.

Sinha, B. M. (1983). *Time and temporality in Samkhya-yoga and Abhidharma Buddhism.* New Delhi: Munshiram Manoharlal.

Histories, Alternative

The literary category of alternative histories is also known as alternate histories, uchronias, allohistories, or counterfactual histories. The last term is often used for serious studies from a historical perspective. Historians run experiments of a sort in complex areas such as economic history by imagining all events that shape an outcome are static except for one change. For example, consider a counterfactual study in which Calvin

Coolidge and a conservatively controlled and business-favoring Congress did not lead the way to a huge difference in wealth between the very rich and the middle class during the 1920s. Economic historians believe this was one major factor that led to overspeculation and then a Depression. In this situation, would the Great Depression have occurred, and if it did not, what major changes would this cause? Would Roosevelt have been elected in 1932?

Besides historical counterfactuals, a good deal of alternative history falls into the fiction genre, and in particular, science fiction. These stories begin with the premise "What If" some single significant event in history either did not happen or happened differently and then go on to explore possible developments. This genre is found in many different languages besides English.

Examples of alternative history appeared in Western literature as early as 1836 with the publication of Louis-Napoléon Geoffroy-Chateau's *Napoléon et la conquête du monde, 1812–1832, Histoire de la monarchie universelle.* This novel tells the story of Napoleon crushing all his enemies and ending as emperor of the whole world. Far earlier in time, Herodotus, in the 5th century BCE, looked at the consequences had the Persians defeated the Greeks at the Battle of Marathon in 490 BCE. Livy's *Ab Urbe Condita* (history of Rome, c. 29 BCE) is considered to have some alternative history aspects.

There were sporadic additional writings now considered to be alternative history during the 19th and early 20th centuries. Examples include a story by Edward Everett Hale in the March 1881 *Harpers* titled "Hands Off." This tale imagined Joseph was not sold into slavery in Egypt, leading to an eventual conquest of the Mediterranean by the Phoenicians.

In 1899, Edmund Lawrence published *It May Happen Yet: A Tale of Bonaparte's Invasion of England.* H. G. Wells's *A Modern Utopia* (1905) falls into one of the subgenres of alternative history, the parallel or alternative Earth plot. In this work, the point of divergence from actual history is the complete skipping of the Dark Ages. In 1926 Charles Petrie published "If: A Jacobite Fantasy," in the *Weekly Westminster,* exploring the idea the Hanoverians fled from Bonnie Prince Charlie in 1745 and the Stuarts were restored to the throne.

More stories and books appeared during the 1920s and mid-1930s. In December 1939, a short story by L. Sprague de Camp titled "Lest Darkness Fall" appeared in the magazine *Unknown.* This story, later published as a novel, is generally credited with the beginning of alternative history as a distinct part of science fiction. "Lest Darkness Fall" involves an archaeologist, Martin Padway, who, while visiting the Pantheon in Rome in 1938, is suddenly transported to the 6th century BCE. There Padway recreates some 20th-century inventions and eventually succeeds in averting the Dark Ages. Although the plot is similar to Mark Twain's *A Connecticut Yankee in King Arthur's Court* (1889), what distinguishes the two is Twain's focus on social inequities and de Camp's exacting historical accuracy in his details. An alternate candidate for this honor is the 1934 short story "Sidewise in Time" by Murray Leinster. However, Harry Turtledove, who has written a number of alternative history novels, as well as some other authors, credits de Camp with getting him interested in this genre.

In the past 20 years there has been a dramatic upsurge in alternative history, not only in novels and short stories but also in all forms of entertainment media. A brief list would include *Timestalkers* (1987), *Sliding Doors* (1998), the *Quantum Leap* television series (1989–1993), *Time Cop* (1994), and *Fatherland* (1994).

An idea that is at the heart of much of alternative history is the "butterfly effect." This is a concept connected with chaos theory and was described by Edmund Lorenz in a 1972 talk titled "Predictability: Does the Flap of a Butterfly's Wings in Brazil Set Off a Tornado in Texas?" A traveler into the past may alter one tiny thing, perhaps just by his or presence alone, which leads to significant changes in the future. This was the basis of one of the most popular classics of the genre, Jack Finney's *Time and Again* (1970), in which a present-day New Yorker joins a secret government project and travels back to 1880s New York. By preventing one chance meeting between two people in the past, the entire project in the future never happens.

One of the largest publishing areas of alternative history is Nazism, the Third Reich, and the Second World War. Gathering interest since the 1970s, this subarea has sparked more novels, short stories, films, television shows, plays, and even comic books than any other topic in alternative history. There have been at least 50 novels written with plots ranging from the Nazis winning World War II to something happening to prevent Hitler from ever coming to power. One of the best known is Philip K. Dick's *The Man in the High Castle* (1962) in which the Axis powers won World War II. More recently, Philip Roth's novel *The Plot Against America* (2004) is based on the famous aviator and America-first isolationist Charles Lindbergh defeating Franklin D. Roosevelt in the 1940 election, leading to America's neutrality in the war against Hitler's Nazi Germany.

The increasing popularity of alternative history, particularly in the popular culture, may owe something to an increase in imagination on the part of filmmakers, disenchantment with the current social and political events in everyday life, and a wish for alternatives. The late-20th-century proliferation of alternative histories was made possible in part by developments in atomic physics. Quantum mechanics, and in particular the Heisenberg uncertainty principle, contradicts age-old beliefs in absolute determinism—the doctrine that all future developments in the whole universe, from the largest galaxy to the smallest particle, are predetermined. If true, this would mean history could never change.

Present-day physics, however, generally accepts a less deterministic view of the cosmos. In the past half-century or so, as physicists and mathematicians attempt to link classical and quantum physics with approaches such as string theory, the possible existence of parallel universes is being explored. Parallel or alternate universe concepts are explored in a number of science fiction novels that may or may not be linked to alternative histories. For example, novels in which the South won the Civil War may predicate an alternative universe in which this is true. Other alternate universe stories with no historical basis involve simple differences such as Jack Finney's *The Woodrow Wilson Dime,* in which the simple slip between parallel universes of a dime with Wilson's image on it causes the protagonist to move into a significantly different, but in some ways similar life.

Charles R. Anderson

See also Eternal Recurrence; Experiments, Thought; Multiverses; Novels, Time in; Sin, Original; Progress; Time and Universes; Worlds, Possible;

Further Readings

Hellekson, K. (2001). *The alternate history: Reconfiguring historical time.* Kent, OH: Kent State University Press.

Nedelkovich, A. B. (1994). *British and American science fiction novels 1950–1980 with the theme of alternative history.* Belgrade: University of Belgrade.

Rosenfeld, G. D. (2005). *The world Hitler never made: Alternate history and the memory of Nazism.* New York: Cambridge University Press.

History, End of

The end of history usually does not mean the end of time. In fact, it means that history has a definite goal in mind, one it is attempting to attain or one it has already reached. These expectations placed on human development depict an image of a meaningful history and comply, in an academic sense, with the school of thought we attribute to the philosophy of history. Classically speaking, when we view history from a philosophical perspective, we don't perceive things as just singular occurrences. We consider them in their historical development. Who determines the course of history? Where do the laws that advance history actually come from? Can the goal of history be judged as positive or negative? All these questions imply how diverse the range of opinions is in regard to this matter. Yet, at the same time, it is possible to identify some of the criteria that are characteristic of end-time thinking: First, people commonly contemplate the end of history during periods of change, times that give them reason to think about the present and the future, because they are either afraid of the unknown or hoping for better times. Second, a primary prerequisite for believing that history has a certain goal that it is heading toward is based on the conception that occurrences are part of a progressive process. Third, this historical process must, however, be understood as human-made; the goal of history must be attainable within the life of human beings and it should not be reserved only for those of divine stature.

Historical Process

There is a basic difference between the way ancient and modern historians view the idea of progress. In early antiquity, history was generally seen as a decline, as the never-ending distancing from the golden ages of mythic times. So, in that sense, the historical process is firmly in divine hands, if it is of any significance at all: History is initiated by the gods and is therefore principally beyond human influence.

According to ancient thought, time flows constantly, without being directed, in the stream of eternal sameness. Certainly, during the course of antiquity, a cyclical picture of history is developed that has been documented in the cyclical theories dealing with transitional political policies. But change occurs essentially along the known paths of human existence. Compared with the anthropological constants that characterize humankind, social and cultural factors are of no particular significance. This even applies to a thinker like Heraclitus (c. 535–475 BCE), whose philosophy centers around the cosmic idea of becoming and passing away. But this thought has nothing to do with progress; change will not necessarily produce any improvements, so it is nothing to place your hopes on. That is also a reason why there are no relevant drafts of the future from that period of time. History is fundamentally the rendition of stories that illustrate human existence—as exemplified by Thucydides (460–c. 396 BCE) in his study on the Peloponnesian War.

Not until Judaism appears on the scene does a new awareness of time evolve: The belief in a future Messiah becomes a historic expectation that defines life itself. The anticipation of a redeemer gives an extraordinary meaning to the future and a structure to the historical process that has a redeemer as its goal. Consequently it is possible to imagine an end of history, even though this development is not historically manmade. Christianity's development is similar: Although world history is divided into a time before and a time after the birth of Christ, historical appreciation remains focused on the Second Coming. Particularly in times of crisis, the promise of redemption has rekindled the specter of doomsday and transformed the fear of Judgment Day into the hope that all misery will come to an end. So it's no wonder that throughout history believers have tried to accelerate salvific history. A vivid illustration of that is offered in the so-called Paupers' Crusade of 1096. But the popular doomsday calculations can also be seen in that connection, because they include the wish to make the Redeemer's Second Coming predictable. The Catholic Church has always eyed these number games with suspicion, in part because they spark a discussion about the expiration date of their own reign. Augustine (354–430) clearly rejected doomsday calculations: First, this knowledge is reserved for God. Second, however, there could not be a qualitative improvement of humanity, historically speaking, because, according to Augustine, Original Sin is the reason why humankind is never purely good or bad; it will always remain in that tense relationship where it depends on redemption. In a similar way, Martin Luther also rejected the thought of humankind actively promoting the Second Coming.

Nevertheless, this belief still finds support within the Christian community, and, in the course of time, it has been partly responsible for the evolution of two complementary structures: While the wish continues to grow that Christ's Second Coming should be "prepared" within the worldly community, the meaning of Original Sin has been losing its significance. The transition is recognizable, for example, in Joachim von Fiore (1130–1202): Joachim differentiates among three realms, those of the Father, the Son and the Holy Ghost, and accordingly teaches that the third realm is immanent, born from the simple strength of monastic contemplation without divine assistance. These teachings are an early example of the budding secularization of doomsday expectations that gained further momentum during the Renaissance. During this era, the vision of history changes drastically. History no longer serves as a voice of divine revelation. It takes on its own significance: History becomes a forum of humankind's ascent to its own Creator who endows it with his divine image, thus distinguishing humankind and setting it apart. With humankind's increasing (self-)confidence, which also includes a worldly portion of truth and goodness, it develops a willingness to accept evil as a worldly matter for religious reasons, as did Thomas Müntzer (1486/1490–1525). According to Müntzer, the teachings of original sin begin to fade, and the battle between good and evil that was carried out within humanity itself now becomes a battle of good people versus evil people. That is why Müntzer felt obligated to "purge" the Christian community of the false-believers, in order to herald the Kingdom of God.

Meaningful History

Human beings make history. This understanding of history becomes more and more widely accepted until it is universally acknowledged in the French Revolution. During the revolution, history itself becomes the subject. Time draws a line. It becomes the modern age, leaving the ancient regime behind. Contrary to prior experience, traditions no longer have any intrinsic value of their own. They are outdated; the past is actually seen as the past. People are no longer embedded in existing structures. There are now unforeseen possibilities, and the future appears to be open and manageable. In antiquity it was the anthropological constants that determined thinking; now it is social circumstances. Conditions that are hostile to freedom and contradict the ideals of equality are removed by means of revolution and reform. The end of history now refers to a situation where human capabilities are optimally developed and established by people themselves.

For Immanuel Kant (1724–1804) this is a regulative idea that shows lawmaking reformers which way to go; for other representatives of German Idealism, for example Johann Gottlieb Fichte (1762–1814), this demand sounds "more final." But it was especially the early socialist currents of the 18th and early 19th centuries that portrayed a happy end of history with their utopias; among them were names like Charles Fourier (1772–1837), Henri de Saint-Simon (1760–1825), and Wilhelm Weitling (1808–1871). They all shared a belief in the power of the Enlightenment and the goal of making the system of political dominance and economic exploitation superfluous.

August Comte (1798–1857), one of the founding fathers of French sociology, developed one of the most mature programs of this nature: Like many of his socialist comrades-in-arms, Comte was inspired by the innovative capabilities of science and technology and regarded the social sciences as the foundation for a controlled "reconstruction" of society, quite similar to the way the natural sciences controlled nature. To him, progress is no longer a question of faith or hope but one of positive knowledge, "positive" in the sense of real, useful, and precise knowledge.

Comte based his Law of Three Stages on this belief, claiming that knowledge first passes through a theologically fictional stage, then a metaphysically abstract stage before finally reaching a scientifically positive state formed by a spirit of order and progress. As soon as it is possible to subject the social changes to an intellectual authority, consisting of the most capable scientists, happiness on Earth will be assured. An earthly paradise under the political leadership of the intellectual elite sounds like a "positivistic" Plato—and it is no wonder that Karl Marx (1819–1883) rejected this and other programs as idealistic.

According to Marx, it was naïve to believe that the effective powers of capitalism could be replaced by humanistic reason. Marx was convinced that the only way bourgeois society could be overcome was through its own contradictions. For instance, Marx's well-known quote about capitalism digging its own grave says that the bourgeois relations of production themselves will reproduce the class struggle between capital and labor without initializing the instruments of compensation. Therefore, this conflict, according to Marx's analysis, has an antagonistic structure: More and more workers will be enslaved by economic necessity and forced to carry their own skin to market, while more and more capital will be amassed in the hands of very few who will become richer and richer. It is this law of development for history itself that determines the course of affairs, no matter what the players want to do or should do. So the end of a certain history is predetermined. In this case, it is the history of bourgeois society.

End of Modern Society?

As is commonly known, bourgeois society not only failed to come to an end, it also successfully outlived its rival political systems in the context of the East–West conflict. The collapse of communist rule in the Soviet Union in 1989/1990 enabled the spread of liberalism in Central and Eastern Europe and inspired the hope of a "new world order," supported by a consensus of the community of states and based on human rights and human dignity. This optimism is reflected in the analysis published by American political scientist Francis Fukuyama titled *End of History,* first as an essay in 1989 where it was tagged with a question mark, and then as a book in 1992. Fukuyama claims that political reason has reached its goal, because liberal democracy has succeeded in becoming the adequate form of government for people in comparison with other political systems. Significant occurrences should not be excluded, but they will not alter the basic situation that—seen from a universally historical perspective—political development must result in a marriage between democracy and the market economy. There are no other alternatives.

Fukuyama bases his historical considerations on an interpretation of G. W. F. Hegel by Alexandre Kojève (1902–1968). In Kojève's version of Hegel's *Phenomenology of Spirit,* the end of history is reached when the contradictions that had previously existed in the various struggles for recognition are resolved and shared by everyone. Contrary to Marx, Kojève specifically assigns capitalism the competence not only to spread affluence and global standards of values and rights but also to balance the inequalities of the market and to redistribute prosperity. He envisions a state of universal freedom and the successful emancipation from all pressures for modern society.

Kojève's interpretation, which is especially influential in French philosophy, should not be understood as a true-to-the-original reconstruction of Hegel's philosophy. Hegel (1770–1831) himself does not assume that the dialectical logic of the struggle for recognition will be overcome. The state may restrain the struggle between individuals, but the intergovernmental world remains in its natural state and thus in the paradigm of the struggles for recognition. It has always been this point of view that others have used to accuse Hegel of having glorified the nation-state. For Hegel, however, it is more a matter of the plurality of concrete life forms. The Age of Enlightenment's universal demand should not seduce us into overlooking the fact that particular "national characters" have lives of their own regarding specific political communities; in Hegel's opinion, that would be an example of apolitical thinking denying itself the insight into the bourgeois era's cultural bondage. The tendencies of bourgeois society to expand into a world community is based on the success of the economy's division of labor. But it is Hegel's conviction that "no state can be established" on the interplay of diverse egoistic interests, calculated economic decision making, and formal lawmaking alone.

From the perspective of Hegel's philosophy of freedom, there is no end in sight for history, either. In Hegel's opinion, freedom and consciousness have a dialectic relationship: History advances in an awareness of freedom when its consciousness becomes conscious of itself and propagates itself as self-consciousness, as in the case of the French Revolution, which marked a decisive breakthrough toward general and political freedom. Yet this new understanding of freedom also introduces new demands that Hegel discusses in the context of the Jacobin terror. In view of the demands of the modern globalized world, Hegel would hardly speak of the end of history. Instead he would speak of the need to understand the ambivalences of freedom. This not only requires that the processes that brought about a lasting impulse for integration following the end of the East–West conflict in Europe are taken into consideration. It also requires that the "opposing movements" of regionalizing systems of order are understood, movements that stir up new conflicts and transform old systems. So it remains to be seen whether the effective cultural diversity, for example, between Islam and the West, will be consolidated in a new systematic rivalry, a "Clash of Civilization" (S. Huntington). And the future of liberal democracy is just as uncertain, considering that the majority of nations will hardly accept it as the universal and ultimate goal of history.

This "realistic" point of view of an uncertain future has also led, among other things, to a greater theoretical humility within political philosophy and the theories of international relations. On the one hand, it has promoted historico-philosophic reflections on end time that are connected to postmodern and post historical concepts. Even Kojève is ambivalent about his envisioned final stage of history: In the general state of happiness, man's "negativity" fades, Kojève claims, and with it his will to overcome himself, his ability to transcend his own nature. This view also absorbs Friedrich Nietzsche's (1844–1900) thoughts on the "last man," who tries to numb the pointlessness of his existence by satisfying his "little desires." One of Nietzsche's central thoughts consists in enlightening the world about the end of all illusions. Human reason set forth to retrace Christianity's promise of truth and morality; but, according to Nietzsche, it found—in the truest sense of the word—nothing. Consequently, man's own self-enlightenment had to end in the "death of God," making man responsible for the meaning of his own life. In Nietzsche's opinion, it is a task only the strongest individuals are able to cope with.

Postmodernism celebrates this disillusionment—in the words of Jean-François Lyotard (1924–1998)—as the "end of the great tales." It opens the world to a pluralistic diversity of views and attitudes that also allow criticism of the outdated ideas of a self-dependent subject. According to the views of postmodern theorists, it doesn't have as much to do with the end of history as with the end of the totality of every historical interpretation based upon reason. They don't see the open future as uncertain. They understand it as a form of liberation, opening the world to the chance of new ideas. Specifically, this belief in the power of innovation, particularly in the areas of the arts and culture, is branded by the post-histoire representatives as an illusion. Not diversity but simplicity characterizes the personality of the modern age. Yet this hectic business, as Arnold Gehlen (1904–1976) claimed, cannot deceive us from seeing that it is essentially just cultural redundancy. Only remakes and recycling, but no real progress anymore, neither in the arts nor in other social areas. The post-histoire view says that history has come to its end, because society has lost the strength to set new goals.

By reviewing the history of philosophy (of history) and considering the various schools of thought, it seems that any reflections on the end of history are a question of mood that can be compared to a theory of colors: ranging from hopeless black to gloomy gray to a happy colorfulness and a confident rosy-ness. So the course of history is, to quite a relevant extent, a question of what people themselves are willing to risk.

Oliver W. Lembcke

See also Augustine of Hippo, Saint; Clock, Doomsday; Comte, Auguste; End-Time, Beliefs in; Enlightenment, Age of; Fichte, Johann Gottlieb; Gehlen, Arnold; Hegel, Georg Wilhelm Friedrich; Heraclitus; Kant, Immanuel; Luther, Martin; Marx, Karl; Nietzsche, Friedrich; Postmodernism; Teilhard de Chardin, Pierre; Thucydides; Time, End of; Universe, End of; Zeitgeist

Further Readings

Cohn, N. (1970). *The pursuit of the millennium: Revolutionary millenarians and mystical anarchists of the Middle Ages.* Paladin: London.

Fukuyama, F. (1991). *The end of history and the last man.* New York: The Free Press.

Kojève, A. (1980). *Introduction to the reading of Hegel: Lectures on the phenomenology of spirit* (A. Bloom, Ed., J. H. Nichols, Trans.). Ithaca, NY: Cornell University Press.

Koselleck, R. (2004). *Futures past: On the semantics of historical time* (K. Tribe, Trans.). New York: Columbia University Press.

Lyotard, J.-F. (1984): *The postmodern condition. A report on knowledge* (G. Bennington & B. Massumi, Trans.). Minneapolis: University of Minnesota Press.

Niethammer, L. (1992). *Posthistoire: Has history come to an end?* (P. Camilier, Trans.). Verso: London.

Scholem, G. (1995). *The messianic idea in Judaism and other essays on Jewish spirituality.* New York: Schocken Books.

Hitler, Adolf (1889–1945)

German dictator Adolf Hitler (born Adolf Schicklgruber) was the leader of the Nazi party and Führer of the Third Reich of Germany from 1934 to 1945. He and his National Socialist movement were one of several nationalistic and racist mass movements that arose from the political and economic problems that followed World War I. Hitler and his followers were responsible for many of the events that led to World War II, most of which shaped the latter 20th century. His promotion of racial supremacist doctrines and anti-Semitism culminated in the Holocaust, the systematic murder of more than 6 million Jewish civilians. Consequently, Hitler left a tragic and indelible mark on recent human history.

With the outbreak of war in 1914, Hitler volunteered for action in the German army. He fought on Germany's Western front for 4 years. Injured and gassed, Hitler won the highly respected Iron Cross. After the German defeat, he became convinced that his country was not defeated on the battlefield but was stabbed in the back by traitors, Jews, and Marxists who sabotaged Germany with the Treaty of Versailles, by which the victorious Allied powers inflicted humiliation and ruinous reparation payments on a defeated nation.

After the war, Hitler attended a German Workers' Party meeting, joined, and became one of the party's best speakers. The party was one of many political groups formed after Germany's defeat that called for national unity and promoted anti-Semitism. These militant and nationalistic parties were similar to those fascist parties in Italy and Spain that were radical, anti-liberal, anti-Marxist, and authoritarian. When Hitler became leader of the party in 1921, it was renamed the National Socialist German Workers' Party (the official name of the Nazi party).

Runaway inflation and the French occupation of the Ruhr Valley contributed to the growth of the Nazi party in Germany. Hitler hoped to end the Ruhr occupation by staging the "Beer Hall" Putsch of November 8–9, 1923, to march against the government in Berlin. The attempt failed and Hitler was tried for treason and sentenced to prison. During his imprisonment, he dictated *Mein Kampf* (1925), an autobiographical account of his life and political philosophy in which he described Germany's need to re-arm, become self-sufficient, suppress Marxism, and rid itself of the Jewish minority.

The conditions of the Great Depression were especially severe in Germany during the postwar period of the Weimar Republic; widespread poverty and public demoralization aided the growth of the Nazi party to the point that, in January 1933, President Hindenburg called Hitler to be chancellor of a coalition government. Hitler consolidated his power by persuading the government to suspend the constitution and to grant him emergency powers. At the death of President Hindenburg in August 1934, Hitler merged the offices of Chancellor and President to become the single leader of both the Nazi party and the Third Reich. Consequently, Germany became a one-party state as other political parties were banned and the unions were destroyed.

Hitler began massive rearmament and a building program to ready Germany for war. This stimulated the German economy, making him a popular leader. He included a propaganda campaign promoting German nationalism, stressing themes from the 19th century that focused on military losses to the French and called for a return to the glories of the Prussian state.

His power secured in Germany, Hitler took advantage of the appeasement policies of the French and English governments. Germany walked out of the Geneva Disarmament Conference and withdrew from the League of Nations. Hitler militarized the Rhineland in 1936, contrary to the disarmament clauses of the Treaty of Versailles. Next, Austria was annexed to Germany in the 1938 Anschluss, followed by the annexation of the Sudetenland of Czechoslovakia with the 1938 Munich Agreement. These annexations led Hitler to be named *Time* magazine's Man of the Year.

On September 1, 1939, Hitler's armies invaded Poland. Alarmed at Germany's full occupation of Czechoslovakia, a violation of the Munich Agreement, France and Great Britain honored their agreement to keep Poland independent and declared war against Germany. By 1940, Hitler had gained control of Norway and Denmark. Next, he focused on France, Holland, and Belgium. Only Great Britain, in the Battle of Britain, repelled German advances.

Hitler invaded Russia on June 22, 1941. The Soviets held the Germans at Stalingrad in 1943. After the United States declared war on Germany and the Allied efforts intensified, the German army lost momentum and the tide of war began to turn against Germany. Hitler gradually withdrew from public view and ignored advice from his military generals. He also intensified his execution of Jewish people in conquered territories; Jewish ghettos and villages were emptied and survivors sent to concentration camps for extermination. But a series of Allied victories continued to drive back the German armies, and an invasion of Germany became inevitable. As Soviet troops converged on Berlin, Hitler escaped to an underground bunker and committed suicide on April 30, 1945. Germany surrendered in May.

Hitler's dream of a greater Germany existed a little over 12 years and caused the death of approximately 50 million people globally. It took a coalition of world powers to defeat the German military in a war that lasted nearly 6 years. To prevent future conflicts, the Allies formed the United Nations, whose first act was to ratify the creation of the new Jewish state of Israel, in response to the Holocaust.

The legacy of Hitler's actions continued for nearly half a century as a weakened Europe became a battlefront in the Cold War between the Soviet Union and the United States. Germany itself was divided in half; though defeated, both sides of Germany were coveted by the superpowers for military support. To some degree it was the political instability in postwar nations that allowed anticolonial movements in Asia and Africa to form independent states, and the resistance to these movements led in turn to conflicts such as the Vietnam War. Former German territories in central Europe, liberated by the Soviets at the end of the war, had Communist regimes installed. The Soviet occupation plus the memory of the failed Munich Agreement made political appeasement unattractive as a foreign policy; it helped to establish the policy of communist containment by the United States that lasted until the reunification of Germany in 1990 and the dissolution of the Soviet Union in 1991.

Laura Sare

See also Judaism; Marx, Karl; Stalin, Joseph

Further Readings

Bullock, A. (1964). *Hitler: A study in tyranny* (Rev. ed.). New York: Harper & Row.

Fest, J. C. (2002). *Hitler* (R. Winston & C. Winston, Trans.). New York: Harcourt Brace Jovanovich.

Hitler, A. (1943). *Mein Kampf* (R. Manheim, Trans.). Boston: Houghton Mifflin.

Shirer, W. L. (1990). *Rise and fall of the third Reich.* New York: Simon & Schuster/Touchstone.

Toland, J. (1991). *Adolf Hitler: The definitive biography.* New York: Anchor.

Homer

Homer is the conventional name of the supposed unitary author of the famous Greek epics *Iliad* and *Odyssey* that were composed shortly after written language came into use again in the first half of the 8th century BCE. The *Iliad* dates to the second half of the 8th century BCE and is the oldest written literature in the West that has come down to us. The *Odyssey* seems to have been written slightly later, possibly by a different author. Marking the end of a long oral tradition, both poems show characteristics of oral improvisation, as in the use of formulaic phrases typical of extempore epic traditions, including the repetition of entire verses. At the same time the poet is held in highest esteem because of his ingenious and creative composition.

There is not a single contemporary document about Homer's life, person, and time, since the first sources from the 7th and 6th centuries mentioning him do not seem to know an official calendar and refer to him merely as a poet of past times. The question of authorship, date, and transmission, the so-called Homeric question, was discussed by modern scholars for about 200 years. Whereas F. A. Wolf (1795) and his school of "Analysis" tried to identify different poets emphasizing the inconsistencies in both epics, the "Unitarians" put stress on their artistic unity. Since the 20th century, the mechanisms and effects of oral transmission, as well as the poetic and aesthetic quality of the poems, have been the focuses of attention.

In both the *Iliad* and the *Odyssey* time is referred to in a highly elaborate way. Times of the year and the day are expressed by verbal pictures; for example, the morning is frequently described by the appearance of the rosy-fingered goddess Dawn (Eos) and by a variety of terms, such as the frequent use of *etos* for a certain year and *eniautos* for the general course of a year with its regular natural phenomena that recur every 12 months. The meaning of some of these terms is still under discussion: looking *prosso kai opisso* is usually translated as looking "forward and backward," that is, "into the future and into the past," although it seems to mean looking "forward and beyond" at times. In this latter case, one anticipates the near future and the more distant future that hides behind the immediate future.

Modern scholarship has proved wrong H. Fränkel's theory of Homer's indifference toward time and Th. Zielinski's opinion that Homer was not fully able to describe parallel plots. Scholars have underlined the difference between the time of narrating and the narrated time (the described actions do not happen as fast or as slowly as the narration goes on) as well as the fine technique of looking back and forward that creates a complex view of time instead of a simple chronological account of the events. Thus both poems consist of a selection of scenes essential for the plot, necessarily focusing on certain periods of time within the set time span. For instance, the plot of the *Iliad,* whose theme is the anger of Achilles, extends over 51 days within the 10th year of war between the Achaeans and Trojans. It does not include the end of the war, because Achilles' anger had ended before the conquest of Troy. The events of only 15 days and 5 nights are described in detail, while those of the remaining days are merely mentioned (e.g., 9 days of plague, absence of the gods for 11 days, Achilles' ill-treatment of the corpse of Hector for 11 days). Aristotle praised Homer for concentrating on the plot in a way that would befit tragic writers, by leaving out all scenes that would need to be included from a historical point of view but that are not an essential part of the plot, which evolves from the characters of the protagonists and their interaction with the gods according to Zeus' plan.

Anja Heilmann

See also Herodotus; Hesiod; Mythology; Presocratic Age; Thucydides

Further Readings

Morris, I., & Powell, B. (1997). *A new companion to Homer.* Leiden, New York, & Köln: Brill.

Hominid–Pongid Split

During the past 2 centuries, curious scientists have laid the foundation for answering the question of human origins using advances in technology. Fossil evidence and genetic studies have made this quest possible. Patient accumulation of evidence has led to gradual acceptance by the general public of the conclusion that our species has evolved, during 5 million years, from an ancestor in common with the four great apes in the remote past.

Aristotle, considered the founder of biology, was the first to effectively place animals in a specific order of taxonomy. By organizing species into common groups, he established a foundation for understanding the similarities among life forms on this planet. Even so, he taught the eternal fixity of all species.

Linnaeus, the father of modern taxonomy, took the science of classification to the next level. A species is seen as a group of individuals that share enough biological similarities that they can interbreed with one another successfully. The primate order, which was formed about 70 million years ago, is divided into two suborders: Prosimii and Anthropoidea. Being hominoids, humans and pongids (apes) are placed into the same taxonomic family due to their biological similarities.

Linnaeus's writings were available in the century before Darwin's *On the Origin of Species* (1859). If it were not for Linnaeus's work in biological classification, it would not been possible to speculate on how some species die off and new species are formed. This was later explained by Darwin with his theory of evolution. Darwin realized that species are not in a static state, since everything is in a dynamic process that results in constant change. With the uses of relative dating techniques, Darwin came to the conclusion that fossils in rock strata demonstrate a continuous organic evolution that led to a species' survival or extinction.

Advances in radiometric techniques have allowed scientists to date fossil evidence more accurately. The modern technology of chronometric dating, through the use of radioactive decay, has greatly strengthened the hypothesis that our species has evolved from a common ancestor close to the four living great apes (orangutans, gorillas, chimpanzees, and bonobos). Darwin argued that our evolution took place in Africa. Scientists realized that, to find the truth about our evolutionary past, they had to gather empirical evidence to solve the mysteries of our origin and history. For more than a century, they have searched for the so-called missing link between the great apes and humans.

The first major discovery that established our descent from a common ancestor with the apes was made in Africa, and known as the "Zinj" skull. This unearthing in 1959 by Mary D. Leakey shed light on the hypothesis that our species is more similar to the great apes than had been thought. Several decades ago, anthropologists thought that the split between fossil hominids and fossil apes had occurred during the Miocene epoch, about 12 million years ago. However, more recent hominid species found in southeast Africa are also pongid-like. This evidence from the Pliocene, about 5 to 7 million years ago, determines that our split from the pongids was more recent than had been thought.

During the Miocene, hominoids emerged in many varieties. Before the end of the following Pliocene epoch around 3 million years ago (mya), the environment had changed; the tropical jungles of Africa had become scattered woodlands and open savannas. This change directly resulted in the split between early hominid-like hominoids and pongid-like hominoids. Based on fossil evidence, this split from Pliocene pongid-like hominoids happened roughly 6 mya. The species that were able to adapt to living a terrestrial life survived to form our more recent ancestors. Bipedality is the true separation between the fossil apes and more human-like forms. Many other specializations like tool-making, articulate speech, and a far more complex brain where all later developments during the past 3 million years.

An early ancestor that was successful in this transition from quadrapedal movement to bipedality is known as *Australopithecus afarensis;* the features of *afarensis* are from the neck up chimpanzee-like and from the waist down human-like. The discovery of *afarensis* was only the beginning; scientists are still trying to uncover more clues from our distant past. Interpreting the evidence in order to determine the evolutionary advantage for becoming bipedal continues.

Owen C. Lovejoy, an anthropologist and an expert in anatomy, was the person whom Donald C. Johanson trusted in studying his discoveries at Hadar. Lovejoy has been an influential contributor to the understanding of our past; he believes that our evolution was based on a social advantage that leads to our survival. Hominids created an evolved breeding strategy that enabled humans to raise fewer children, but also allowed them to focus on the success of a limited number of offspring. Lovejoy believes that hominids' opportunity for success revolved around living in large monogamous groups. This is supported by the secondary characteristics that have formed to create the modern human; these phenotypes control the amount of competition between males. In the animal kingdom, an obvious way for species to determine if a female is in heat is to observe the inflammation of sex organs and the growth of breasts. These characteristics are clear signs that lead males to fighting, which would lead to the disbanding of groups. In hominids, it is difficult to make this distinction when a woman is not pregnant. Many criticize this theory of a monogamous relationship, because it goes against Darwin's evolutionary logic. Without the most dominant males passing on their genes to multiple partners, how would hominids survive? Lovejoy answers this argument with the theory that by males appealing to a singular mate, using their time during the day to gather food for not just themselves but also their partner, they allow the females to spend all their time raising the offspring. This allows for a successful passing on of their genes. Another major distinction between the pongids and hominids is that we evolved from being simply insectivores or vegetarians to being primarily omnivores; this allowed for a broad diet that provided a better opportunity for survival in an environment where grasslands where slowly replacing forests in Africa. Through cooperation, *afarensis, Homo habilis, and Homo erectus* allowed humans the opportunity to exist. The creation of a more complex brain came later in the development of *Homo sapiens sapiens.*

From close genetic similarities and observations of human and pongid social behavior, many scientists now recognize that, as mentioned earlier, the split between the hominids and pongids did not occur in the Miocene epoch, but more recently in the Pliocene. In recent decades, biologists have come to the conclusion that our species is more similar to the great apes than had been imagined even by Darwin.

Ryan J. Trubits

See also Aristotle; Darwin, Charles; Evolution, Organic; Fossil Record; Fossils, Interpretations of; Fossils and Artifacts; Haeckel, Ernst; Huxley, Thomas Henry; Olduvai Gorge; Scopes "Monkey Trial" of 1925

Further Readings

Birx, H. J. (1988). *Human evolution*. Springfield, IL: Charles C Thomas.

Foley, R. A., & Lewin, R. (2004).*Principles of human evolution*. New York: Wiley.

Johanson, D. C. (1994). *Ancestors: In Search of human origins*. New York: Villard.

Hourglass

The hourglass, also known as the sandglass, is a timekeeping instrument that has been used for centuries. The sandglass is constructed using two conic reservoirs that are joined at their apexes by a small hole. One reservoir is filled with a finite volume of sand that flows into the bottom reservoir over a set duration of time. When the sand has run completely from the top reservoir through the small opening, the previously determined time period can be recorded and the timepiece inverted to begin running again. The hourglass has been made in various time denominations ranging from 30 seconds to longer than an hour.

History

The actual date of the hourglass's emergence cannot be definitely placed. There are, however, many theories as to when the hourglass emerged as a common timekeeping instrument. The most significant clues to the invention of the hourglass deal with the dates of portrayals and the emergence of the technology needed to create the glass to make the reservoirs. No record exists of the sandglass's original form, so researchers have to work with the modern hourglass shape. Items were made from glass as far back as 2500 BCE, but the art of glass-blowing was not perfected until around 70 BCE.

The earliest definite depiction of an hourglass dates to between 1337 and 1339 CE. This depiction can be found in an Italian fresco on the walls of Palazzo Pubblico in Siena and was painted by Ambrogio Lorenzetti. Another early depiction of an hourglass can be found at the Mattei Palace in Rome. The palace contains a Greek bas-relief depicting Morpheus, the god of time, holding an hourglass. German artist Albrecht Dürer (1471–1528) profiles the hourglass prominently in many of his artistic works. The sandglass holds a prominent place in all three of his most famous copper engravings. *Knight, Death, and the Devil* (1513), *St. Jerome in His Study* (1514), and *Melancholia I* (1514) depict the hourglass just to the left of the central character. The first textual reference can be found in the Receipt of Thomas de Statesham, which dates to around 1345. This receipt references Thomas's acquisition of 12 hourglasses ("*pro xii orologiis vitreis*") in Flanders. The second textual reference dates to slightly after this, in 1380, and can be found in a furniture inventory of King Charles V of France: "*ung grant orloge de mer, de deux, grans fiolles plains de sablon en ung grant estuy de boys garny d'archal*" (a large sea clock with two large phials filled with sand, in a large wooden brass-bound case). The hourglass also finds reference in classical literature, specifically in the works of William Shakespeare. Specifically, Henry V (Act 1 Scene I) refers to an hourglass in the opening Chorus.

One theory credits the invention of the hourglass to an 8th-century monk of Chartres named Luitprand, but most historians believe that if there is any truth to this theory it would find credence as a reinvention or an improvement on the ancient sandglass after the Dark Ages. Many historians believe the hourglass came into use about the same time as, or slightly after, the clepsydra (water clock). Since both timekeepers use the same basic principle, emergence during the same approximate time period could be likely. For a long time no distinctions were made between the two, and they were simply called *horlogues*, leaving only context to determine which of the two was being described. This theory places the invention of the hourglass at around the 3rd century BCE in the Egyptian city of Alexandria. The sand for the hourglass would have been more accessible in regions with drier climates. In countries such as Egypt, water would have been scarce and not a resource that could be needlessly wasted. The more probable date of emergence coincides with the inception of sea travel, as the technology would aid in nautical navigation.

The hourglass has many advantages over the clepsydra. Sand, unlike water, flows at a consistent pace when pressure is applied from above. For this reason, sandglasses kept a more consistent and steady time. Moreover, unlike water, the sand would not freeze or evaporate, so no replacement filler was ever needed. Most high-quality sandglasses did not actually use sand, however, but finely crushed eggshells. The eggshells provided a smoother flow than sand and therefore provided a more accurate method of keeping time.

Several factors contribute to an hourglass's ability to accurately measure time. The type and quality of sand is important, and it must have a rate of flow that does not fluctuate.

Source: Patricia E. Green/Morguefile.

Ancient Variations

The hourglass has had many different variations since its invention. Although the hourglass and sandglass have become synonymous in modern meaning, ancient variations of the hourglass used different fillers as the timepiece evolved. Some ancient variations used mercury instead of sand. These variations required an extremely fine hole between the reservoirs to keep the mercury from flowing too quickly. Another mercury variation used a small bubble of mercury inside a narrow tube containing air. The resistance from the internal air pressure allowed the mercury to fall slowly through the tube and therefore allowed the measurement of a useful time period.

Since the original form of the hourglass is not known, it would be fair to consider the clepsydra and early sand clock (*clepsammia*) among the very early forms of the hourglass. Both of these early timepieces use a set amount of water or sand, respectively, and measured time by the volume of water and sand in a bottom reservoir, but could not be reset without refilling the top.

Ancient and Modern Uses

The hourglass found use in various places throughout history. The most notable of these is in the field of nautical navigation. In the early era of sea travel, it was crucial for ancient mariners to know the speed of their ship in order to know their location at sea. The hourglass provided a timepiece that would remain stable and reliable even in turbulent weather. Thirty-second sandglasses were used to calculate speed. Tying a rope to a heavy piece of wood and throwing the wood overboard behind the ship calculated speed. The wood created resistance with the water and dragged the rope out behind the boat. The rope would have knots tied 47-foot, 4-inch intervals so that each knot that ran through the fingers of the navigator in the 30-second time period indicated one nautical mile in an hour.

Sandglasses were also used as timers for all sorts of activities. They became the first industrial timers for early factories. *Das Feuerwerkpuck* (1450), a German essay on manufacturing, contains an illustration of a stamping mill being timed by an hourglass. Both politicians and clergy also used the sandglass as speech timers. The hourglass was introduced into the church during the 16th century as a tool to restrain preachers from speaking too long. The hourglass also served as a symbol during olden funerals that the sands of life had run out. In today's era, the sandglass has all but disappeared as a timer except for use in kitchens as the common 3-minute-egg timer.

Matthew A. Heselton

See also Archaeology; Clocks, Atomic; Clocks, Mechanical; Sundials; Time, Measurements of; Timepieces

Further Readings

Balmer, R. T. (1978, October). The operation of sand clocks and their medieval development. *Technology and Culture, 19*(4), 615–632.

Bruton, E. (1979). *The history of clocks and watches.* New York: Rizzoli International.

Cunynghame, H. H. (1970). *Time and clock: A description of ancient and modern methods of measuring time.* Detroit, MI: Singing Tree Press.

Macey, S. L. (1980). *Clocks and the cosmos: Time in Western life and thought.* Hamden, CT: Archon.

Humanism

While the word *humanism* is recent, the idea of humanism is one of the oldest and most transnational worldviews in human history. Most worldviews are defined in terms of the distinctive beliefs they hold, but the principal feature of humanism is not so much its core articles of belief, but the method by which inquiry into the world is undertaken.

The American philosopher Paul Kurtz provided what is perhaps the simplest understanding of humanism when he defined it in terms of its four constituent features. First and foremost, humanism is a method of inquiry; second, it presents a cosmic worldview; third, it offers a set of ethical recommendations for the individual's life stance; and, fourth, humanism expresses a number of social and political ideals. It is important to note the order in which these characteristics have been listed. Indeed, this order is a result of the seriousness with which humanism takes cosmic and evolutionary time. Number 4 is the least important of them, not because social and political ideals are unimportant, but because the nature and orientation of those ideals have been more susceptible to change over time and between continents. Numbers 2 and 3 are more important because the details of the worldview and the ethical recommendations have greater commonality over time and across cultures.

But the most important single characteristic of humanism is the one that is most impervious to the passage of time. The most constant defining feature of humanism is that it is a method of inquiry. From the Carvakas and Ajivikas in ancient India, from Kongfuzi and Wang Chong in China, from Thales and the Greek thinkers, through the Renaissance to the Enlightenment and on to the 21st century, humanism is best understood as a method of inquiry. The *conclusions* of that inquiry may change between cultures and between centuries, but the *method* of inquiry has remained essentially the same.

Symptomatic of the uniqueness of humanism is that for most of its life it has functioned very well without the word. Humanism as a *concept* has its origins in ancient India, China, and Greece—each one arising independently—but the actual *word* was not coined until 1808, in Germany. To make things more complicated, *humanist* existed as a word long before *humanism,* originating in the Renaissance, although traceable to the Roman word *humanitas.* But as humanism is defined principally by its method rather than by its conclusions, the lack of a word to act as a catch-all for those conclusions is a trivial issue.

Other philosophers, while agreeing that the method of inquiry remains the single paramount feature, have tried to articulate some general idea of what humanist philosophy would actually subscribe to. The best thought-out was by the American philosopher Corliss Lamont (1902–1995). In an influential account of the philosophy of humanism, one that went through several editions, Lamont outlined what he called the 10 central propositions of humanist philosophy.

1. Humanism believes in a naturalistic metaphysics or attitude toward the universe that considers all forms of the supernatural as myth.

2. Humanism believes that humankind is an evolutionary product of the nature of which we are a part; it has no survival after death.
3. Having its ultimate faith in humanity, humanism believes that human beings possess the power of solving their own problems, through reliance primarily upon reason and scientific method applied with courage and vision.
4. Humanism believes that all humans possess genuine freedom of creative choice and action, and are, within certain objective limits, the masters of their own destiny.
5. Humanism believes in grounding all human values in this-earthly experiences and relationships, and it holds as its highest goal this-worldly happiness, freedom, and progress—economic, cultural, and ethical—for all humankind.
6. Humanism believes that the individual attains the good life by harmoniously combining personal satisfactions and continuous self-development with significant work and other activities that contribute to the welfare of the community.
7. Humanism believes in the widest possible development of art and awareness of beauty, including the appreciation of nature's loveliness and splendor, so that the aesthetic experience may become a pervasive reality in the life of man.
8. Humanism believes in a worldwide democracy, peace, and a high standard of living on the foundations of a flourishing world order.
9. Humanism believes in the complete social implementation of reason and scientific method; and thereby in the use of democratic procedures, including full freedom of expression and civil liberties, throughout all areas of economic, political, and cultural life.
10. Humanism, in accordance with scientific method, believes in the unending questioning of basic assumptions and convictions, including its own.

The 10th proposition is significant. As with Kurtz's preference for seeing humanism as a method of inquiry, Lamont is clear that humanism is just as rigorous in questioning its own presuppositions as those of any other system. We may quibble over this or that point in Lamont's list, but the 10th proposition helps ensure the need for ongoing inquiry as the core humanist process.

Another account of humanism, by the philosopher Mario Bunge, seems at first sight to come from a different perspective than that of Kurtz or Lamont, but in the end it is the similarities that are more significant. Bunge speaks of humanism as involving concern for the lot of humanity. This concern he spells out in what he calls the seven theses of humanism, which go in this order:

1. *cosmological thesis:* whatever exists is either natural or manmade;
2. *anthropological thesis:* the common features of humanity are more significant than the differences;
3. *axiological thesis:* there are some basic human values, like well-being, honesty, loyalty, solidarity, fairness, security, peace, and knowledge, and these are worth working, even fighting, for;
4. *epistemological thesis:* it is possible to find out the truth about the world and ourselves with the help of experience, reason, imagination, and criticism;
5. *moral thesis:* we should seek salvation in this life through work and thought;
6. *social thesis:* liberty, equality, solidarity, and expertise in the management of the commonwealth;
7. *political thesis:* while allowing freedom of and from religious worship we should work toward the attainment or maintenance of a secular state.

Bunge is an advocate of what he calls *systemism,* which postulates that every thing and every idea is a system or a component of another system. In this way, the first thesis covers the broadest territory, and provides the foundation for the other theses. Looking at it another way, the theses run from the hard sciences through to the social sciences. Interestingly, in the first thesis, which addresses the individual human being, Bunge says what Kurtz also says, namely, the importance of the method of inquiry. From the point of view of time, the focus of inquiry for this encyclopedia, the crucial point of Bunge's systemism is that it proceeds from sectors in which time operates on a macro scale, to sectors operating on much more limited time scales, well within the human ability to grasp.

Humanism in the Ancient World

Humanism is a collective name that can be given to some of the oldest and most transcultural worldviews on Earth, beginning quite independently in India, China, and Greece. Looking to India first, humanist elements can be found in four of the six classic schools of thought (*darshana*) that make up classical Indian philosophy: the Samkhya, Mimamsa, Yoga, and Vaisesika. They either doubted the existence of any God or gods, or valued the acquisition of knowledge of the world as a worthwhile end in itself, or saw harmonious living within the world as freed from illusion as possible, as a worthwhile goal.

The oldest of the Indian *darshanas* is the Samkhya school, which was mentioned first in the 4th century BCE, although it had probably been flourishing for a century or so by then. The earliest known Samkhya work is the *Samkhya Karika,* written by Iswara Krishna about 200 CE. Samkhya thinking developed an elaborate metaphysics that was, broadly speaking, atheistic, recognizing only two ultimate realities: *purusa* (sentience) and *praktri* (matter). *Praktri* is uncaused, eternal, and in a state of constant evolution. It is composed of three essential substances: essence, energy, and inertia. Cosmic history began with these elements being in total equilibrium. Evolution began with the arrival of *purusa,* and is the principle for which evolution continues. *Purusa* is itself entirely uncreated and is neither God nor some sort of prime mover, although some strands of Samkhya thinking tend to promote *purusa* as a variation of universal spirit. It would be a mistake, however, to equate this notion of a universal spirit with God in the Western sense. What divided Samkhya thinkers was whether the notion of God was capable of any proof, or whether it was a mistaken belief. The Samkhya system advocated liberation from the bondage of *praktri* by knowledge, a liberation best equated with philosophical wisdom.

Standing alongside the classical Indian darshanas was the materialist school of thought known as the Carvakas, traceable to the 6th century BCE. Carvaka philosophy shared with other Indian schools the belief that the universe is interdependent and subject to perpetual evolution. But at that point they parted company, holding instead a range of unorthodox beliefs: that sacred literature should be regarded as false; that there is no deity, immortal soul, or afterlife; that karma is inoperative and illusory and matter is the fundamental element; and that only direct perception, and no religious injunctions, can give us true knowledge. The aim in life is to get the maximum amount of pleasure from it. This had various interpretations, from unalloyed hedonism to an altruistic service for other people on the principle that this will maximize one's own happiness as well as that of others. It will be seen that the humanist variants in Indian thought involved to no small extent a differing view of time, being more linear and embedded in nature.

Other Indian movements of a broadly humanist orientation in India include the Ajivikas and the Sumaniya, about which little is now known. The Ajivikas flourished between the 6th and 3rd centuries BCE, and their influence can be traced for more than 1,500 years. Like some early Jains and Buddhists, the Ajivikas went about naked to indicate their contempt for worldly goods. In the main, they upheld a principle of nonaction, denying that merit accrued from virtuous activity or demerit from wicked activity. Coupled with this was a thoroughgoing determinism and skepticism regarding karma and any sort of afterlife.

Humanist elements can be seen in some elements of contemporary Hinduism. For instance, the Rahda Soami group, which is engaged in building a magnificent temple in Agra, not far from the Taj Mahal, is broadly humanistic in outlook. They disavow caste barriers and undue emphasis on ritual and ceremony, and encourage secular and honest living.

In contrast to India, the humanist strand in China became the principal strand of thought. This is largely, though by no means exclusively, due to the influence of Kongfuzi (551–479 BCE), who is known in West by his Latinized name, Confucius. Chinese humanism is placed squarely in temporal time. Confucius's genius involved transforming the naturalism and humanism latent in Chinese thinking into the strongest forces in Chinese thought down to this day. He said that maintaining a distance from spiritual beings was a sign of wisdom; he expressed no opinion on the fate of souls, and never encouraged prayer.

Confucius realigned the concept of *junzi* (*chun-tzu* in old spelling), which had traditionally meant "son of the ruler" into "superior man," with the effect that nobility was no longer a matter of blood or birth, but of character. This was a very radical departure from the customary Chinese thinking preceding him. Along with this, he radically transformed the notion of *jen* from meaning kindliness to the more general "man of the golden rule," or perfect *junzi. Jen* was expressed in terms of *chung* and *shu,* or conscientiousness and altruism. Confucius spoke of the doctrine of the mean, which essentially meant doing things "just right," which in turn meant doing things in harmony with the Dao.

The clearest articulation of Confucian values was given in *The Great Learning,* the classical work of Confucian ethics. Originally part of a large Confucian work called the *Book of Rites, The Great Learning* was later extracted from it and made a standalone Confucian work. It was probably not written by Confucius himself, but by Zengzi (a disciple) or Zi Si (Confucius's grandson). The reference to great learning is to distinguish it from small learning, which children are in need of. The power of *The Great Learning* is that it encapsulates the Confucian educational, moral, and political program in a simple format. First are the Three Ways, or Three Aims: clear character; loving the people; and abiding in the highest good. These Ways are achieved by the Eight Steps: the investigation of things; extension of knowledge; sincerity of the will; rectification of the mind; cultivation of the personal life; regulation of the family; national order; and world peace.

The Three Ways and the Eight Steps encapsulate the Confucian emphasis upon reforming one's character and building social cohesion and the intimate links between the two notions. Another development of what could be called humanist thinking was developed by Mengzi (371–c. 289 BCE), known in the West as Mencius. The Great Morale (*Han Jan Chih Ch'i*) involved the morale of the individual who identifies with the universe. There are two ways of cultivating the Great Morale: understanding the Dao, or the way of principle that leads to the elevation of mind; and the accumulation of human-heartedness, or the constant doing of what one ought to do because one is a citizen of the universe. Through the constant accumulation of understanding of the Dao and cultivation of human-heartedness, the Great Morale will gradually develop within oneself. Mengzi also believed the Great Morale was achievable by everyone, because we all have the same nature, one that, at heart, is good. It is important to see how both processes begin with the individual, working on the basic premise that one can play no useful role in society if one's own personality and family are in disorder. At no point is there any reference to any supernatural order or command morality. Like the Indian humanist systems, Chinese humanism is located within natural time and subject to natural rhythms. It also has important points in common with Mario Bunge's systemism.

In Greece, humanism can be traced back to the philosophers known as the Presocratics; that is, philosophers active before Socrates. Many historians and philosophers have had reason to admire the general attitude of the Presocratics, even if their actual opinions are of less value than the attitudes that underlie them. Bertrand Russell praised them for not only having an open, scientific attitude of inquiry but for their creativity and imagination. In fact, most of the basic categories of philosophy began with them. The first of them was Thales (c. 624–548 BCE), the man credited as being the father of philosophy. Thales' claim to fame rests on being the first person to try to explain the world not in terms of myths, but by observation of the world as he actually saw it. Where Homer attributed the origin of all things to the God Oceanus, Thales taught that water was the prime element in all things.

One of the core foundational concepts of humanism as articulated in Greece was the notion of *Paideia.* The term comes from classical Greek philosophy and is derived from the Greek word *pais* or *paides,* which, translated literally, means "boy," but in its broader meaning can be defined as "education for responsible citizenship." It was understood by the Greeks that this sort of education was an effective democracy. *Paideia* had four characteristics: it offered a unified and systematic account of human knowledge; it provided a technique of reading and disputation based upon mastery of language and intellectual precision; it worked on the assumption that the human personality can be improved by education; and it valued the qualities of persuasion and leadership as important for the vital task of taking part in public affairs. The ideals of Paideia were given their best voice by Socrates who, during the trial for his life, declared that the unexamined life is not worth living. These ideals were imitated by the Romans, from whose language the Latin word *humanitas* comes.

The humanist tradition of Greek thought received a check in the philosophy of Plato, and never fully recovered. Several Hellenistic movements, in particular Epicureanism and Stoicism, reflected important elements of humanist thought, but all had been effectively countered or crushed by Platonism and then by Christianity.

Renaissance Humanism

The collapse of the Roman Empire was followed by what might still justifiably be called the Dark Ages. In these long centuries of faith and ignorance, the spirit of inquiry was almost lost in the West. It revived only with the Renaissance. In its broader sense, the Renaissance began with the life of Petrarch (1304–1374) and ended with the death by execution of Giordano Bruno for heresy in 1600, with the most productive years of the Renaissance being the century before the sack of Rome in 1527.

Medieval thinking had been dominated by cultural pessimism: weariness with this world, and a suspicion, derived mainly from Saint Augustine, that human affairs are inevitably tainted with corruption and selfishness. Renaissance thinking, by contrast, was positive and optimistic. While Renaissance humanism differed on many points, it was pretty unanimous in its condemnation of the monastic life with all its implications of defeatism and withdrawing from one's civic duties.

Humanist thought in the Renaissance is also responsible for some very basic units of periodization of history as developed in the West. The very notion of "Renaissance" came from the idea that this age was the first to rediscover and fully appreciate the cultural grandeur of the ancient world. What happened in between then and now was slightingly dubbed the Middle Ages—those ages of interest only insofar as they stood between the glories of antiquity and the glories of today. We still speak of the Middle Ages, and they have retained in popular imagination the gloomy picture drawn of them by the Renaissance humanists.

The thinkers of the Renaissance saw themselves as the scourge of medieval dogma and the pioneers of a new cultural and intellectual orientation centering on the majesty of the ancient world. The new style of thinking was stimulated by the rediscovery of ancient thinkers, in particular Lucretius, Cicero, and Plato. The Renaissance humanists were optimistic about the power of culture to effect positive social change. At its least helpful, what can be called Renaissance humanism encouraged too backward-looking a stance. The tremendous enthusiasm for the ancient scholars deteriorated into a conservative climate where simple quotation of an ancient authority was sufficient to bring a dispute to an end. Only after the sack of Rome in 1527 did this optimism decline and a return to contemplation and an escape from the world once more became a discernable trend in Renaissance thought.

Very few Renaissance thinkers were atheists: almost all were theists, the majority of them remained Christians, though of heterodox sympathies. Belief in personal immortality came to be questioned, and the understanding of God changed: God was something that could be understood through our learning. Indeed, learning was a deeply pious activity, in that learning about nature meant, ipso facto, learning about God. The influence of antiquity inevitably meant a renewed interest in the pre-Christian ideas of the ancients. Few people declared an overt preference for the pagan ideas, and a rather sophistical justification for interest in the subject developed. The triumph of Christianity, so this argument went, had expunged the danger in paganism, so there now would be little harm in studying it.

The Renaissance was a period of significant historical and scriptural research. For instance, Lorenzo Valla (1405–1457) was responsible for exposing the fraud known as the Donation of

Constantine, upon which the papal claims to own its large territories in Italy was based. Valla also wrote *Dialogue on Freewill,* a frankly skeptical work that is a very sympathetic account of Epicurus. Valla was influential in teaching people the need to read scriptures with a skeptical frame of mind. This religious scholarship of Renaissance scholars was usually undertaken with a mind to reform religion and purge it of its recent and harmful additions. This drive led directly to developing some of the first ideas since the ancient world of religious toleration. This trend came to an end only when the religious reformer Savonarola was burned at the stake in 1498.

Contemporary Humanism

The humanism of both the ancient world and the Renaissance were stifled by religious reaction. It slowly rose again to the surface in the 18th and 19th centuries for a complex array of reasons. The Renaissance and Reformation had broken forever the monolithic idea of a single Christian belief that covered the known world, excepting only a few heathens around the periphery. The discoveries of other civilizations by European explorers also showed that people could live quite happily without Christianity. This, along with greater knowledge of the civilizations of India and China, from which Europe might learn, had dramatic consequences for the monoculturalism of Christian Europe. The Enlightenment in Europe (roughly, the years between the 1680s and the 1780s) was characterized in part by the taking to heart of this realization.

It was during these changes in perception that the word *humanism* was coined, in 1808. Friedrich Immanuel Niethammer (1766–1848) was a German educational reformer who wanted to develop an educational philosophy avoiding the excesses of the Roman Catholic reactionaries and the radicals of his day, known as the Philanthropinists. Niethammer posited humanism as combining the respect for tradition stressed by the conservatives with the innovative education of the whole child, as advocated by the reformers. Niethammer was a close friend of the philosopher George Wilhelm Friedrich Hegel (1770–1831) and it was from him that subsequent writers, thinkers, and reformers took up the idea of humanism and adapted it to their purposes. By the 1870s humanism was known and used in the English-speaking world. It retained its general use as a catch-all idea for the values of the Renaissance or of the classical world until about the end of the 19th century when the philosopher F. C. S. Schiller (1864–1937) took the word up as the title for his brand of subjectivist pragmatism. From there, American religious progressives took humanism up as the word to denote a brand of nonsupernatural religion as personal commitment in the context of a common humanity. This position came to be known as religious humanism.

It was only after the Second World War that humanism became the most widely used word among secularists, rationalists, and freethinkers. Chief articulators of this kind of secular humanism were Corliss Lamont in the United States, referred to earlier; M. N. Roy (1887–1954) in India; and Hector Hawton (1901–1975) and H. J. Blackham in the United Kingdom. Religious humanists and secular humanists do not differ so much in what they believe. Both groupings are fundamentally naturalistic and reject supernaturalist interpretations of religion. They are also generally opposed to ecclesiastical authority being exercised in society and see science as having constructive contributions to make in science, philosophy, and society. Where they differ is more in their general perceptions of what religion is and what their response to it should be.

Returning to Paul Kurtz's outline of humanism, we can trace the changes in two of the points and the continuities in others. Item 2, the cosmic worldview, has become more modest and less geocentric, as required by developments in astronomy and cosmology. Item 4, the political ideals, has become inclusive of more groups than earlier humanisms would have endorsed. Toleration of slavery in ancient Greece and the widespread misogyny found in Confucianism and Greek thinking no longer play a part in humanist thinking.

In contrast to these changes, item 3, the ethical ideals, has remained fundamentally the same. The values identified by Solon, Pericles, Confucius, the Carvakas, or the Epicureans still resonate today. Enjoyment of the moment, resistance to the transcendental temptation, love of nature, impatience with display, imperviousness to materialism, respect for learning, civic values, and family responsibility; all were recommended by the ancient humanists and all find enthusiastic support among their contemporary successors.

But it is the first item of Kurtz's outline that has remained the most constant. The endorsements of skeptical, open-minded inquiry given by Socrates, Wang Chong, and the Ajivika thinker Upaka find direct parallels in the work of Bertrand Russell, Albert Einstein, and all the other humanist philosophers and scientists who have shaped the modern world. Humanism is first and foremost a method of inquiry, and it is the conclusions of that inquiry that furnish us with our ideas and beliefs. Humanists have always rejected static formulas of thought, or bowing to arguments from authority, or accepting command moralities. In contrast to systems that are founded on acceptance of bodies of thought, humanism assumes that as knowledge comes from humans, it is bound to include some element of error and therefore will be in need of revision as our learning grows. In this way, contemporary humanism can reject the anthropocentric conceit of the Renaissance humanists, the hedonism of the Carvakas, the humorlessness of the Stoics, and still honor their role in the long stream of humanist thought. And as our scientific and philosophic knowledge grows, doubtless some important features of 21st-century humanism will be replaced. But the primacy of humanism as a method of inquiry will remain.

Humanism and Time

We can conclude this general account of humanism by returning to its philosophy of time. Humanist outlooks differ from supernaturalist outlooks in many important ways, and these differences have important implications for their respective beliefs about time. In the centuries before the Copernican revolution, classical Western religions posited a geocentric universe that was both very young and very small and that revolved around one's own religious heartland and was geared to one's own needs. This viewpoint became untenable after the 17th century, along with the conceptions of time that attended it. The outlooks that have developed since then, including the humanist ones, have accepted the need for a cosmic perspective, which involves recognizing the relative unimportance of our galaxy, our planet, our species, and ourselves in the wider scheme of things. Baruch de Spinoza spoke in these terms when he extolled the virtue of *sub specie aeternitatis,* or "under the aspect of eternity." This cosmic perspective has been reiterated from humanist viewpoints by people as diverse as George Santayana, H. G. Wells, Bertrand Russell, Steven Weinberg, and Richard Dawkins. It is also similar to Mencius's idea of the Great Morale, mentioned above.

The next major shift in perception came as a result of the breakthroughs in geology and biology in the 19th century; in particular, Charles Darwin's articulation of natural selection as the means by which evolution takes place. Evolution has demonstrated that human beings must extend their newfound cosmic modesty to the animal kingdom. Ernst Mayr determined four basic beliefs central to theism that were overthrown by evolutionary thinking: belief in a constant world; belief in a created world; belief in a world created by a wise and benign creator; and belief in the unique position of humanity in that creation. Each of these beliefs required a radically different notion of time than is compatible with a scientific account. Once again, this involved a serious recalculation of our understanding of time and the relative position of humanity to time. The ongoing dialogue between evolutionary theory and humanist thinking is testimony to this dynamic relationship.

Finally, the falling away of generally accepted externally imposed goals and purposes in life has stimulated among many humanist thinkers an appreciation of living each moment fully and joyfully. Here some of the more religiously and poetically inclined humanists have helped provide the language necessary to appreciate and savor each moment for the joy it can bring, if only we can look openly at it. This has involved some very creative and significant new understandings of what something like "life eternal" can mean.

Bill Cooke

See also Bruno, Giordano; Christianity; Confucianism; Darwin, Charles; Einstein, Albert; Ethics; Farber, Marvin; Hinduism, Samkhya-Yoga; Marx, Karl; Materialism; Presocratic Age; Russell, Bertrand;

Sagan, Carl; Santayana, George; Spinoza, Baruch de; Values and Time; Wells, H. G.

Further Readings

Bullock, A. (1985). *The humanist tradition in the West.* London & New York: Thames & Hudson.

Bunge, M. (2001). *Philosophy in crisis.* Amherst, NY: Prometheus.

Chan, W.-T. (1973). *A source book in Chinese philosophy.* Princeton, NJ: Princeton University Press.

Cooke, B. (2006). *Dictionary of atheism, skepticism, and humanism.* Amherst, NY: Prometheus.

Fowler, J. (1999). *Humanism: Beliefs and practices.* Brighton, UK: Sussex Academic Press.

Fung, Y.-L. (1960). *A short history of Chinese philosophy.* New York: Macmillan.

Goicoechea, D., Luik, J., & Madigan, T. (1991). *The question of humanism: Challenges and possibilities.* Amherst, NY: Prometheus.

Hawton, H. (1963). *The humanist revolution.* London: Pemberton.

Hiorth, F. (1996). *Introduction to humanism.* Pune, India: Indian Secular Society.

Kurtz, P. (1986). *The transcendental temptation: A critique of religion and the paranormal.* Amherst, NY: Prometheus.

Kurtz, P. (1989). *Eupraxophy: Living without religion.* Amherst, NY: Prometheus.

Lamont, C. (1965). *The philosophy of humanism.* New York: Frederick Ungar.

Mayr, E. (1993). *One long argument: Charles Darwin and the genesis of modern evolutionary thought.* London: Penguin.

Riepe, D. (1961). *The naturalistic tradition in Indian thought.* Seattle: University of Washington Press.

Thrower, J. (1980). *The alternative tradition: A study of unbelief in the ancient world.* The Hague, The Netherlands:: Mouton.

Walter, N. (1997). *Humanism: What's in the word.* London: RPA.

Wilbur, J. B., & Allen, H. J. (1979). *The worlds of the early Greek philosophers.* Amherst, NY: Prometheus.

Hume, David (1711–1776)

David Hume, a Scottish moral philosopher, historian, and public economist, is considered among history's most important British men of letters. A major representative of British Empiricism, Hume influenced generations of thinkers, in particular Immanuel Kant, John Stuart Mill, Adam Smith, David Ricardo, and Friedrich August von Hayek. His writings on perception, causality, history, time, morality, and economics prepared the ground for many important subsequent philosophical and economic schools, for example, Utilitarianism, Rationalism, and the Austrian liberal economic school of the 20th century.

Life and Work

Hume was born into a noble albeit no longer prosperous family in Edinburgh on April 26, 1711. After the death of his father in 1713 he grew up with his two elder siblings under the care of his mother at Ninewells in the Scottish lowlands.

At the young age of 12 he followed his older brother to Edinburgh University. There he first studied mathematics; later on, he studied law but never completed the degree. Soon Hume turned his concentration to the great ancient philosophers, especially Cicero and Seneca. Beyond them he read the major contemporary British writers like Joseph Butler, John Locke, and George Berkeley; he admired them as fathers of moral science that was based on an experimental approach.

In 1734, after several years of intense reading, he felt sick and exhausted. In an effort to recover his health through a change of habits, Hume relocated to Bristol to take up work for a sugar importer. Because this way of life did not suit him at all, he quickly decided to resign and, still in the year 1734, took up his former studies again—but this time in France. Declining to accept gainful employment forced him to live a very modest life for years.

In France he stayed mainly in La Flèche, where roughly a century before Descartes had received his education in the Jesuit College, which still existed in Hume's time. There he read the continental philosophers' works and soon gained back his mental strength. Between 1734 and 1737 Hume composed his initial work, *Treatise of Human Nature.* To finalize the editing and for the sake of managing the book's release he moved to London. The first two volumes—*Of the Understanding* and *Of the Passions* were finally published in 1739.

Portrait of David Hume by Allan Ramsey, 1776. Hume was one of the greatest philosophers in Western history, as well as an accomplished historian and economist.

Eventually the third volume, *Of Morals,* appeared. *The Treatise* included Hume's attempt to introduce a "science of human nature."

But the *Treatise's* tepid reception and miserable sales left Hume disappointed since his ambition from his youngest age onward had been to become a reputable man of letters. Having experienced this frustration, he returned to Ninewells to carry on his studies. In 1741/1742, Hume published his *Essays Moral and Political.* In contrast to the previous work, this one met with wide success.

In search of a stable income and space for further free development, Hume applied in 1745 for a vacant chair of Ethics in Edinburgh. But the "murmur among the zealots" about Hume's alleged skepticism and atheism as voiced in the *Treatise* gave rise to successful opposition against his appointment. For the same reason he failed again 6 years later in Glasgow in his second and final effort to obtain a professorship.

After several short engagements, he followed the call of General James St. Clair—a cousin of his—in 1746 to accompany him as a secretary on a campaign against France, and 2 years later on a diplomatic mission to Vienna and Turin. During his absence his *Philosophical Essays Concerning Human Understanding*—today denoted as his first *Inquiry*—were published; these were succeeded by *Inquiry Concerning the Principles of Morals* and his major work as political economist, *Political Discourses.* The latter attracted international attention.

In 1752 Hume began engagement as Librarian to the Advocates' Faculty in Edinburgh. His primary benefit was being able to pursue his independent studies, allowing him to elaborate and release his six-volume *History of England,* which, again, met with great success. His belief in the necessity of historical classification of current issues and his view on history as immeasurably valuable "collections of experiments" motivated this body of work. In 1757 his *Four Dissertations* were published.

Fortune led him to Paris again, this time (1763) as secretary to the English ambassador. He enjoyed his opportunities to associate with noble Parisian society, with its stimulating and convivial salons. Returning to Britain in the company of Jean-Jacques Rousseau, with whom he shared a brief friendship, Hume soon became undersecretary of state. As a newly prosperous man, he retired to Edinburgh in the late 1760s.

David Hume died in Edinburgh on August 25, 1776, after having spent some years of comfortable life. His most controversial literary work, *Dialogues Concerning Natural Religion,* was published posthumously in 1779.

Moral Philosophy and Method

Hume emphatically distinguished "ought from is"—in today's scientific language to speak from normative versus positive analysis.

Along with Berkeley and Locke, Hume is regarded an important representative of British Empiricism; Hume's theory of knowledge follows the distinction of the perceptions (sensations vs. reflections) of the human mind into impressions and ideas; according to Hume, any idea, however elaborate, is based on a bundle of simple impressions. Hume therefore stated that any theory not empirically based must be rejected categorically. He presented the copy principle (that simple ideas come from simple impressions of the outside world) and thought about the principles of natural connections of ideas in mind—the process of association. These findings had implications for his judgments of the status of the outside world and free will.

Hume bewailed the primacy of passions over sanity in human behavior, causing him to state,

"Mankind are so much the same, in all times and places, that history informs us of nothing new or strange. . . ." Some authors took this as an indication of Hume's disregard of time and history; others believed he grasped the understanding of history and reason being central to philosophical advance.

Hume challenged the widespread unconscious supposition of causality between two events closely following another in time. Historically constant conjunction of two events should not be misunderstood as a proof of natural coincidence through causation and must not lead to the firm expectation of conjunct occurrence in the future.

Hume described the custom and habit of induction based on humankind's natural tendency to form expectations of a general circumstance out of single observations of incidents—that is to say, inductive generalizable deductions. He denied that this way of generalizing in order to build expectations for the future was rational, because he did not detect the slightest evidence to confirm such generalization or to believe in even the prospectively constant conjunction as a form of stationary behavior over time. On the contrary, there may occur significant inconsistencies between past and future events. Some traces of parallelism to John Stuart Mill's thought may be discerned here.

Hume doubted the moral rationalists' (in particular Locke's) and the "selfish schools" (e.g., Thomas Hobbes's) behavioral assumptions and thought about the origins of humankind's natural sympathy (benevolence) and morality. Hume tended to be a skeptic, especially in regard to religion.

Political Economy

Beyond his philosophical and historical thoughts, Hume made relevant contributions to the new science of economics. By means of progressive empirical inquiries, he disproved the theories of the strong contemporary school of mercantilists who believed in the achievements of protectionism. Hume spoke in favor of free trade and more dynamic currency regimes and hereby set a cornerstone of classical economics; in this respect he strongly influenced his close friend and later famous economist Adam Smith, as well as constituting a source of Ricardo's monetary theory. Furthermore, he was convinced that in the absence of economic freedom any real political freedom would be unreachable, an idea that later inspired Hayek and others.

Matthias S. Hauser

See also Berkeley, George; Causality; Ethics; Kant, Immanuel; Morality; Rousseau, Jean-Jacques

Further Readings

Mossner, E. C. (1954). *The life of David Hume.* London: Nelson.

Stewart, J. B. (1992). *Opinion and reform in Hume's political philosophy.* Princeton, NJ: Princeton University Press.

Stroud, B. (1977). *Hume.* London: Routledge.

Husserl, Edmund (1859–1938)

Edmund Husserl, a German philosopher, was the founder of the phenomenological tradition. He studied mathematics, science, and philosophy in Leipzig, Berlin, and Vienna. In Vienna (1881–1882) he studied with Franz Brentano, who inspired him to establish the phenomenological method, which later became one of the most important philosophical movements in 20th-century continental Europe, influencing among others Martin Heidegger, Jean-Paul Sartre, and Maurice Merleau-Ponty. Husserl taught at Halle, Göttingen, and Freiburg. Before he turned his interest to the development of phenomenology, he worked on the philosophical foundations of mathematics (*Philosophy of Arithmetic,* 1891). In this connection, his discussion with Gottlob Frege, who established analytical philosophy, was important.

Husserl's concept of time was initially inspired by what Brentano had taught in Vienna. Two crucial insights of Brentano became the foundation of Husserl's phenomenology: First, the psychic contents of an experiencing consciousness can be the subject of an unmediated inner perception; and, second, consciousness is always related to something by intention, meaning that there is no consciousness without an intended object. In short, the phenomenal "givenness" of psychic contents and "intentionality" as an irreducible structure of consciousness form the starting points of phenomenology. Time is the system that orders the contents of consciousness first of all, giving rise to before-and-after relations and making the past experiences qua memory accessible to the subject. This means that past experiences become recognizable as one's own experiences; thus, time is a function of self-identity. Until this point, Husserl's notion of time remains quite Kantian.

The Problem of Subjectivity

Specific to Husserl's time theory is the problem of a subjectivity that constitutes time but is also related to an original temporally distributed flow of experiences. That means consciousness constitutes time and is itself temporal. In order to explain this, Husserl introduces the "absolute consciousness," a unifying transcendental function of consciousness that is in itself atemporal. This theoretical construct is contentious not only among his interpreters but also for himself because it cannot become the subject of phenomenological analyses. Phenomenology as Husserl has developed it is a philosophy of description in a strictly scientific sense. The goal of philosophy should be the precise description of how things appear to us. It is not about how things are in themselves, but rather how we apprehend them. Our experience is a source of apodictic truths for Husserl. But the individual contents of consciousness do not belong to these truths; it is the way or the structure in which they are given that can be an apodictic truth. The Kantian distinction between the thing-in-itself and how it appears is to be found in *Logical Investigations* (1900–1901), but in his analyses of time Husserl gave up that sharp distinction. The thing-in-itself is then defined as the identity that can be intuited through manifold apprehensions or perspectives of it. Therefore it no longer belongs to the realm of transcendent things if transcendent is understood as being beyond the reach of our recognition.

The analyses concerning time-consciousness are fundamental to Husserl's phenomenology because time is not an object like other objects; it is rather the way in which objects are given to and apprehended by consciousness. That is why time is so fundamental to phenomenology, and its analysis stretches over the whole period of Husserl's investigations. He focuses on subjective time, the time-consciousness and not on an objective "world-time." In general it can be said that phenomenology is always concerned with how things appear to a consciousness. The question of how things are in themselves misses the point that there is no object without a subject. This does not mean that Husserl is an idealist. He thinks that the attempt to describe something while leaving out the condition that it must be described by someone cannot be a sufficient description and, therefore, cannot be true. Husserl distinguishes two perspectives: the "natural attitude" and the "phenomenological attitude." The former is the perspective we adopt in everyday life, in which we naturally believe in the existence of the world. In the latter, the philosopher suspends all convictions and intentions that belong to the "natural attitude," especially convictions about existence. This suspension is called *phenomenological epoché*, and is gained by a reductive method that leads through different levels back to pure intention. Intention is the way in which a consciousness is directed toward an object. There are many ways in which an object can be given to a consciousness (in perception, imagination, representation, etc.). In order to describe the nature of these modes, Husserl uses variation, the so-called eidetic variation, which is a method of distinguishing between contingent and necessary features of the intention. Intentionality is the subjective structure of consciousness, which apprehends an object. In an analysis, a distinction is made between the *noema* (object of consciousness) and the *noesis* (corresponding mental activity). They can be analyzed separately, but they are not reducible to one another.

Consciousness and Time

The central question concerning time is how temporal objectivity can be constituted by a temporal consciousness. Analogous to this question, the phenomenological analysis splits at first into two subjects: the temporal object (e.g., a melody) and time-consciousness (the succession of now-moments and their becoming more and more past and the anticipation of the impressions yet to come). The constitution of a temporal object, an object that endures identical to itself over time, needs time. That means the time the consciousness takes to constitute a temporal object needs to be just as long as the object endures. While the intuited identity of the object appears as an integral whole, the object is in fact given only in successive perspectives or temporal parts. Husserl takes a melody as an example of how time-consciousness constitutes such a temporal object. He asks why we not only hear the successive tones while listening to a melody, but also grasp them as a whole. His explanation is that time-consciousness is not only consciousness of the present moment ("primal impression"), but rather encompasses moments already past ("retention" or "primary memory") and anticipates others still in the future ("protention"). The now-point is fundamental as a source of the present since consciousness is defined as a continually flowing stream of contents, but the present moment is ideal only in the sense that it has no duration. It is a kind of border between the past and the future. The temporal field, the duration of the present composed of past, primal impression, and anticipation, is necessary for the connection of the different tones and the recognition of the melody. The succession of primal impressions causes the backward-shift of the former primal impressions, and, therefore, the temporal field of what is recognized as present is continually changing. The function of "secondary memory," which is distinguished from retention, is to keep all continually changing apprehensions in mind, so that the whole continuum of the temporal phases of the melody can be apprehended as a whole. The content of this secondary memory is not present, but remembered or, as Husserl puts it in *On the Phenomenology of the Consciousness of Internal Time* (1893–1917), the "re-presented present." This kind of memory is also distinguished from what is commonly called memory, the reproduction of an impression in the mind. Time-consciousness is clearly distinguished from and not reducible to any other kind of mediated consciousness like imagination, sign-reading, or consciousness of pictorial images; it is a consciousness sui generis since it apprehends something as present that is not present as a whole, but that is held in consciousness as if it were present as a whole; therefore, it does not reproduce its contents as, for example, imagination does.

Time-consciousness not only accompanies the constitutions of temporal objects of these specific kinds, such as melodies. It is a type of consciousness that is always at work and necessary for all kinds of object-constitution. For example, if we look at a house, we do not see it as a whole, although we intend it as a whole. In fact, what is given is a manifold of perspectives on that house, which we mentally synthesize. At no stage of this perception is the sum total of perspectives ever given; there is always something absent, but we apprehend the absent perspectives simultaneously with the present ones. Perception of an object is always a process of synthesizing different perspectives or sense data; therefore it is spread over a certain period of time. The perception of things exhibits the same necessary temporal structure as the perception of objects that are distributed only in time, like melodies. The problem is to clarify how a consciousness, whose contents are successively and that means temporally distributed, can constitute enduring temporal objects and an objective time.

How Time Is Constituted

Husserl attempted to describe the constitution of time in countless manuscripts. As with all phenomenological analyses, he tries to elaborate the structures and processes that are at work in every act of consciousness but that do not appear in the normal course of objective events within the natural attitude. The constitution of time has three different stages. The first is constitution of the objects or things in objective time; they are constituted in the duration of the present by apprehension of past, present, and future together. The second stage is the basis for the first, the object-constituting sense data or the manifold impressions in their wholeness within a so-called pre-empirical or prephenomenal time. This time is called prephenomenal because its succession is not apprehended. The third and most fundamental stage is the wholeness of the "absolute time-constituting flow of consciousness." All these stages represent forms of reflective acts of consciousness that constitute different kinds of objectivity, such as temporal objects, objective time (i.e., measured time), and personal experiences as objects of consciousness. The problem with this model and that Husserl himself was aware of is that of constitution via reflexivity. This concept may lead to a vicious circle: If the constituting reflexivity is not itself atemporal (which is also problematic), it would need to be constituted at a higher-level reflexivity or consciousness, and so on.

Time-consciousness consists according to Husserl of two different but inseparably united kinds of intentionality: The first one he calls "transverse intentionality" (*Querintentionalität*), which has the function of keeping the past phases of an experience in mind in order to preserve its duration and objective identity. This intention is directed onto the duration and process of the experience. The second kind of intentionality is called "horizontal intentionality" (*Längsintentionalität*) and, this being the fundamental function, it is directed to the whole continuum of the inner flow of time. It functions as self-reflection, or in other words, *the retention of retention*. This intentionality is the necessary condition of possibility for an awareness of time-consciousness, which Husserl calls "absolute consciousness." The contentious question is whether this transcendental form of consciousness is to be understood as temporal or not. If it is understood as temporal, the theory ends up in a vicious circle because the transcendental instance itself would have to be constituted by something, and so on. Husserl tends to define it as nontemporal, thus avoiding this vicious circle. But having just shown that objectivity depends on temporality, how can it be consciousness if it is not temporal? If the absolute consciousness were not temporal, it would not be consciousness of something. Husserl saw this problem in his constitutional theory and did not settle the question definitely.

Later Analyses

In his later manuscripts, which are not yet available in translation, the *Bernauer Manuskripte über das Zeitbewußtsein* (1917/1918) and *Späte Texte über Zeitkonstituition* (1929–1934), he constantly revised his theory. There Husserl shifts the emphasis of his thought from analyses of the inner time-consciousness to its intersubjective conditions. He follows the idea of a universal time-structure, which is the necessary condition for both history and the human life-world (*Lebenswelt*). With the new topic he develops a new method. The former analyses were dedicated to the intentional structure. They described the static relationship between subject and appearances of objects, which he called static phenomenology. Later on he searched for a way in which phenomenology could also explore the conditions for the dynamic process of the flow of consciousness and the constitution of objects. This next step he called genetic phenomenology. The later manuscripts on time exhibit this new direction, as, for example, in his *Die Welt der lebendigen Gegenwart und die Konstitution der ausserleiblichen Umwelt* (1931), in which Husserl follows the thought of constitution beyond the subject into the life-world. With the ontology of the life-world, Husserl examines both the meaning of and necessary conditions for history more closely. From *Ideas*, in which Husserl maintained that the possibility for objective experience lies purely in the subject, in the transcendental function of consciousness, it was a long way until his later work *Cartesian Meditations*, in which he speaks of transcendental intersubjectivity. This shift from idealism to ontology of the life-world was accompanied by the analyses of time. The phenomenology of time therefore is one of the most important topics on the way to a phenomenological philosophy.

Husserl's Legacy

Among Husserl's students were Martin Heidegger and Maurice Merleau-Ponty, who also worked on the problem of time. His influence extended beyond these prominent students, though. He influenced Rudolf Carnap's theory of logical empiricism; Marvin Farber, who introduced Husserl's phenomenology to the United States; Roman Ingarden, who focused on phenomenological aesthetics; Jean-Paul Sartre and Emmanuel Levinas, two of the most important phenomenologists in France; and Edith Stein. Husserl was also a source of inspiration to Jacques Derrida, the analytic phenomenologist Hubert Dreyfus, and many phenomenological philosophers in Germany and North America.

Yvonne Foerster

See also Consciousness; Derrida, Jacques; Farber, Marvin; Frege, Gottlob; Heidegger, Martin; Idealism; Intuition; Merleau-Ponty, Maurice; Metaphysics; Ontology; Time, Phenomenology of; Time, Subjective Flow of

Further Readings

Bell, D. (1999). *Husserl.* London: Routledge.

Bernet, R., Welton, D., & Zavota, G. (Eds.). (2005). *Edmund Husserl: Critical assessments of leading philosophers.* London: Routledge.

Brough, J. (2004). Husserl's phenomenology of time-consciousness. In D. Moran & L. E. Embree (Eds.), *Phenomenology: Critical concepts in philosophy* (pp. 56–89). London: Routledge. (Original work published 1928)

Husserl, E. (1973). *Cartesian meditations* (D. Cairns, Trans.). The Hague, The Netherlands: Nijhoff.

Husserl, E. (1980). On the phenomenology of the consciousness of internal time (1893–1917). In R. Bernet (Ed.), *Edmund Husserl: Collected Works, Vol. IV* (J. B. Barnett, Trans.). Dordrecht, The Netherlands: Kluwer Academic.

Husserl, E. (1982). *Ideas pertaining to a pure phenomenology and to a phenomenological philosophy* (F. Kersten, Trans.). The Hague, The Netherlands: Nijhoff.

Smith, B., & Smith, D. W. (Eds.). (1995). *The Cambridge companion to Husserl.* Cambridge, UK: Cambridge University Press.

HUTTON, JAMES (1726–1797)

James Hutton was a Scottish geologist, chemist, and naturalist noted for formulating the uniformitarianist doctrine and the Plutonist school of thought. After short careers in law and medicine at the University of Edinburgh, he followed his interest in chemistry and the nascent science of geology. He analyzed metamorphic and igneous rocks at Glen Tilt in the Cairngorm Mountains in the Scottish Highlands and layers of sedimentary rock at Siccar Point on the Berwickshire coast, east of Edinburgh. After these studies, he noted that the origins of sedimentary and igneous rocks are different and formulated theories on the earth's origin that paved the way for modern geological science.

Hutton began a dispute with the popular Neptunist school, which suggested that all rocks developed by precipitating out of a single great flood. After studying the Devonian Old Red Sandstone along Scotland's coast, he realized sedimentary rocks originated not from a single flood but a series of successive floods. He established what became known as Hutton's Unconformity. Hutton reasoned that there must have been several cycles of depositing sedimentary layers, with each cycle involving deposition on the seabed, uplift and erosion, and new deposition on the seabed. He suggested that the stratigraphic record clearly indicated gradual geomorphologic processes throughout the earth's history.

Moreover, Hutton found granite penetrating metamorphic schists at Glen Tilt. This indicated that the granite had been molten in the past, suggesting that granite formed from the cooling of molten rock and not precipitation out of water. The suggestion that plutonic and volcanic activity were the sources of rocks on the surface of the earth replaced Abraham Werner's Neptunism theory, which claimed that rocks had originated from a great flood and were basically sedimentary in origin.

Hutton concluded that the history of Earth can be explained by observing the geological forces now at work. This is the basis of uniformitarianism, the doctrine that assumes that the natural processes of the past are the same as those that can be observed operating in the present: "the present is the key to the past." The theory of uniformity is one of the most basic principles of modern geology. It contrasts with catastrophism, which states that Earth's surface features originated suddenly, through catastrophic geological processes (e.g., the biblical flood) that were radically different from current processes. Because of these hypotheses, Hutton was accused of atheism and poor logic, especially by Richard Kirwan, an Irish scientist who supported the catastrophist theory. Note, however, that many catastrophic events (e.g., earthquakes, glaciations, hurricanes, tsunamis, volcanic eruptions, meteoritic impacts, etc.) are perfectly compatible with uniformitarianism.

In 1795, Hutton summarized his views in a major work, *The Theory of the Earth,* where he affirmed: "Earth is very old and present-day geological structures formed slowly by processes observable today, such as erosion and deposition." Hutton's theory was first presented at meetings of the Royal Society of Edinburgh in 1785, and published in Volume I of the *Transactions of the Royal Society of Edinburgh,* 1788. Two years before his death, Hutton published *The Theory of the Earth* in two volumes, consisting of the 1788 version of his theory with slight additions. In this work, Hutton compiled material on various subjects previously published, such as the origin of granite. His ideas influenced Charles Lyell's principles of geology, which in turn influenced Charles Darwin's theories of adaptive evolution.

Ignacio Arenillas

See also Chronostratigraphy; Catastrophism; Erosion; Geological Column; Geologic Timescale; Geology; Lyell, Charles; Paleontology; Plate Tectonics; Sedimentation; Uniformitarianism; Wegener, Alfred

Further Readings

Bailey, E. B. (1967). *James Hutton: The founder of modern geology.* New York: Elsevier.

Baxter, S. (2004). *Ages in chaos: James Hutton and the true age of the world.* New York: Forges Books.

Repcheck, J. (1987). *The man who found time: James Hutton and the discovery of the earth's antiquity.* London & Cambridge, MA: Simon & Schuster.

Huxley, Thomas Henry (1825–1895)

Within a short time of its 1859 publication, Charles Darwin's *On the Origin of Species by Means of Natural Selection* evoked a wide variety of reactions. Many of his readers enthusiastically embraced natural selection as the elusive mechanism that explained the process of evolution, while others recoiled from Darwinian naturalism, which seemed to obviate divine involvement and purpose. Responses came from all quarters—scientists, philosophers, theologians—and the concept of evolution soon made its way into almost every academic discipline. Because of his temperament and bad health, Darwin shunned the limelight, especially public confrontation. Thomas Henry Huxley, one of Darwin's closest friends and confidants, entered the debate about evolution with a more combative spirit and quickly earned the nickname "Darwin's Bulldog." No epithet was given more deservedly. Though he did not agree with every aspect of Darwinism, Huxley advanced the arguments that related to evolution—its tempo and mode—through his knowledge of comparative anatomy and paleontology. In Charles Lyell and the writings of other geologists, he and his peers found the long stretches of geological—or evolutionary—time needed to produce the vast, fascinating array of extinct and living species. As Huxley conducted research on several broad fronts, he promoted an agenda of change and made a significant impact in science, education, and society at large.

Thomas Henry Huxley (usually referred to as T. H. Huxley) was born in 1825 in Ealing, a small village west of London, and grew up under humble circumstances. Like Dickens, Huxley obtained most of his early education through voracious and wide reading. After a medical apprenticeship, he received a scholarship to study medicine at Charing Cross Hospital (London). Huxley gave special attention to anatomy and physiology and completed this preparation in 1845. In a manner that has some parallels with Charles Darwin's experience on the *Beagle,* Huxley entered the Royal Navy as an assistant surgeon on the *HMS Rattlesnake,* which sailed to Melanesia and surveyed Australia's coast. On a cruise that lasted nearly 4 years (1846–1850), he—also taking on the duties of amateur naturalist—observed a wide range of wildlife and human cultures. While on this journey, Huxley sent detailed studies of invertebrates back home and, upon returning to England, was elected a Fellow of the Royal Society. In 1854, he left the navy and became a lecturer in Natural History at the School of Mines, in London, where he launched his lifelong study of and writing about various topics in comparative anatomy, paleontology, and related disciplines. Over the next 4 decades, T. H. Huxley held many positions in educational and scientific organizations and institutions (including the Anthropological Institute and the British Museum); he also accepted a number of government appointments that drew upon his expertise. Huxley won many awards and promoted scientific research (especially in the lecture hall and laboratory) and publications (and helped start the journal *Nature*). Throughout his career, he identified enemies (scientists, churchmen, and politicians) and attacked them vigorously through the spoken and written word. Along the way, Huxley, known in his intimate circles as "Hal," was devoted to his family, loyal to his scientific colleagues, and committed to improving the situation of the working class.

As was true for many scientists of his era, T. H. Huxley was interested in—and later obsessed with—the lively discussion concerning evolution, even before Darwin published *On the Origin of Species*. Many participants in this dialogue influenced Huxley, one way or the other, at some point in his own intellectual development (e.g., Chambers, Cuvier, Etienne Geoffrey Saint-Hilaire, Haeckel, Hooker, Lamarck, Lyell, Marsh, Spencer). K. E. von Baer, the father of embryology, helped Huxley bring order to his biological worldview through the study of fixed types. Even though he became Darwin's ardent defender and champion, Huxley never fully accepted his friend's fundamental claim that the slow process of natural selection could transform one species to another. Huxley, the scientist, insisted that he needed experimental proof that this mechanism could produce such a powerful transmutation. He also thought that Darwin restricted himself too much by insisting on gradual changes, since Huxley believed that changes could occur "by jumps" (*per saltum*). (Though based on considerably less data, this debate anticipated in some ways the modern discussion concerning "punctuated equilibrium," itself a form of natural selection.)

Although he argued with scientist and churchman alike, Huxley saved some of his most potent venom for the church. He rejected any form of special creation and insisted that a logical, scientific approach would not allow any sort of divine involvement or plan (as held by Adam Sedgwick, Louis Agassiz, Asa Gray, and many others). As a student and admirer of Hume, Huxley's rational, scientific naturalism—supported by his self-professed "agnosticism" (a term usually attributed to him, though others defined it differently)—brought mixed reactions from Rome and the Anglican Establishment. Huxley successfully devoted himself to the task of deflating the church's power and influence in certain aspects of British society, including the universities (Oxford and Cambridge, in particular) and scientific education and research. Indeed, one of the most famous events in the history of science took place at Oxford's Museum of Natural History in the summer of 1860, when Darwin's Bulldog briefly debated Bishop Samuel Wilberforce about evolution. Much (though not all) of Huxley's (and Darwin's) disdain for Richard Owen, a leading comparative anatomist and paleontologist at the British Museum, resulted from the latter's stubborn reference to a divine plan—what some might call intelligent design today.

Whatever course the 1860 Huxley–Wilberforce debate took—and accounts vary a bit—it is clear that sparks flew with reference to the subject of human descent from apes. This subject remained pivotal for T. H. Huxley, as evident in the 1863 publication of what many regard as his most important book, *Evidence as to Man's Place in Nature*. In this same year, Charles Lyell published a volume on this topic, *The Geological Evidence of the Antiquity of Man*. Both Huxley and Darwin counted on Lyell's earlier discussion of "geological time" (though Lord Kelvin objected to Lyell's uniformitarianism) but, of course, nobody in that era had access to the inventory of fossil specimens known today—much less the modern science of genetics. Huxley's study antedated Darwin's 1871 book, *The Descent of Man and Selection in Relation to Sex*, though the former writer said little about the causes of change. Huxley defeated Owen in a debate concerning human and gorilla brain anatomy and insisted, as he did in his encounter with Wilberforce, that the ape's human descendants should feel no shame because of their common ancestors. He held this view at a time when Europeans had limited knowledge of the gorilla. Nevertheless, from Huxley's perspective, human beings were part of the animal kingdom and shared a "pedigree of prodigious length," and the frontispiece in *Man's Place in Nature* (which depicts a human skeleton followed by a gorilla, chimpanzee, orangutan, and gibbon) secured its place in the iconography of science. For Huxley, and his contemporaries, whose outlook was formed, in part, by the place they occupied in the British Empire, the "savages" who lived in remote corners of that empire served as more recent, albeit stone-age, specimens in the same human family tree.

Huxley's study of paleontology led him to accept the transmutation of species, though he did not believe that Darwin's gradualism was reflected in the fossil record. Darwin also recognized this weakness in his argument. The *Archaeopteryx* and *Compsognathus* fossils offered intriguing illustrations of the transition from reptiles to birds, and other aspects of the fossil record helped Darwin and Huxley identify similarities and delineate the continuity between species. Othniel C. Marsh, paleontologist at Yale's Peabody Museum, introduced Huxley to *Hesperornis,* another bird fossil that fueled the latter's speculation about dinosaur–bird evolution. Huxley's knowledge of fossils (including other dinosaurs) and a close reading of Ernst Haeckel's two volumes on *Morphologie* allowed him to see family trees—a genetic connection between the past and the present. T. H. Huxley found special significance in his proposed sequence of horse fossils, also contained in the Peabody collection.

In his 1859 letter to Charles Darwin, Huxley expressed his willingness "to go to the stake" (figuratively speaking by their day!) in support of evolution. He remained loyal to his good friend and continued to promote the cause after Darwin's death, in 1882. Indeed, Huxley's entire career reflects the curiosity and tenacity that allowed him to achieve greatness from such a humble start in life. For Darwin's Bulldog, even the piece of chalk (formed from the remains of countless microorganisms) that a carpenter carried in his pocket reflected the earth's antiquity; his enthusiasm for learning about that past provides a good example even today.

Gerald L. Mattingly

See also Darwin, Charles; Design, Intelligent; Evolution, Organic; Haeckel, Ernst; Lyell, Charles; Saltationism and Gradualism; Scopes "Monkey Trial" of 1925

Further Readings

Barr, A. P. (Ed.). (1997). *Thomas Henry Huxley's place in science and letters: Centenary essays.* Athens: University of Georgia Press.

Desmond, A. (1997). *Huxley: From devil's disciple to evolution's high priest.* Reading, MA: Addison-Wesley.

Di Gregorio, M. A. (1984). *T. H. Huxley's place in natural science.* New Haven, CT: Yale University Press.

Huxley, T. H. (1863). *Evidence as to man's place in nature.* London: William & Norgate.

Huxley, T. H. (1970). *Collected essays (1893–1894)* (9 vols., Reprinted ed.). New York: Georg Olms Verlag.

Lyons, S. L. (1999). *Thomas Henry Huxley: The evolution of a scientist.* Amherst, NY: Prometheus.

Weiss, K. M. (2004). Thomas Henry Huxley (1825–1895) puts us in our place. *Journal of Experimental Zoology, 302B,* 196–206.

White, P. (2002). *Thomas Huxley: Making the "Man of science."* New York: Cambridge University Press.

I

Ice Ages

The most recent ice age can be considered the last million years of geologic time—the Cenozoic era, the latter part of which comprises the Pleistocene and Holocene epochs, which include the most recent ice age. There is evidence, however, that other ice ages occurred much earlier in time, although their existence and extent are more difficult to determine. Evidence of ancient climate change, which covers a range in temperatures from tropic to arctic, is mostly found with fossil evidence. For example, a reef complex and marine invertebrates tend to indicate a tropical area, while mastodon fossils show a cold environment. Thus, knowledge about the ice ages depends to some extent on the study of fossils.

At present, we know of six ice ages. The oldest two known occurred during the Precambrian era, more than 570 million years ago; one of these may have extended into the early Cambrian era. The next oldest ice age occurred during the Permian period and had a time span of about 55 million years, beginning about 280 million years ago. The next ice age came in the early Cretaceous period and another in late Cretaceous, but both were rather limited in extent. The span of time between the early and late Cretaceous ice ages is about 71 million years, beginning about 136 million years before the present (BP). The present ice age began about 1 million years ago, and because glaciers still exist but are receding, *it* can be considered an interglacial stage, presuming another glacial stage is forthcoming.

The Scandinavian Peninsula is a landscape that was largely shaped by glaciers over the last ice age. The moderate resolution imaging spectroradiometer (MODIS) instrument aboard NASA's Terra satellite captured this image of the Scandinavian Peninsula on February 19, 2003. Along the left side of the peninsula one can see the jagged inlets (fjords) lining Norway's coast. Many of these fjords are well over 2,000 feet (610 meters) deep and were carved out by extremely heavy, thick glaciers that formed during the last ice age.

Source: Jacques Descloitres, MODIS Rapid Response Team, NASA/GSFC.

The ice ages of the Cryptozoic eon might be difficult to imagine, but glacial till or tillite with a thickness of more than 500 feet has been identified, as have areas with grooved, striated, and faceted boulders, dating from middle Huronian time during the Proterozoic era in Canada, north of the present Great Lakes, extending more than 1000 miles in diameter. Also, ancient tillites have been found in Manitoba, eastern Greenland, and northern Utah, where layers of tillite with other formations have a thickness greater than 12,000 feet and could be a part of the Cambrian. This type of ice age evidence is also found in southwest Africa, as well as the Transvaal, the Katanga, Griqualand, and South Africa. In Australia, the Flinders Range shows tillite more than 600 feet thick. Ancient tillites are also found in northeast China, and northwestern and eastern India, dating from the end of the Proterozoic. With only tillite to show the location of the glaciation, and without fossil evidence from the Precambrian, it is impossible to estimate the time periods for these occurrences, except that these ice ages, with perhaps others not yet known, occurred approximately from 500 million to 2 billion years BP. So, at least two ice ages are presumed to have occurred in the Precambrian eras.

The ice age of the Permian period was mostly in the southern hemisphere. Primarily consisting of ice sheets, the glaciers covered a large part of southern Africa, portions of Nigeria, Uganda, and the southern tip of Madagascar. Three small portions in India, three areas in Australia, and six sections of South America were also covered with glacial ice. These areas show striated, grooved, and polished rock, with adjacent tillite formations sometimes greater than 100 feet in thickness. These locations appear to have been glaciated repeatedly, all moving in a northerly direction lying within 20° to 35° of the equator. It is likely that these land masses were much farther south than they are at the present time, giving rise to the acknowledgment of continental drift. That is, ice sheets in areas of India and Nigeria that are now north of the equator were well south of the equator in Permian times. Fossils formed during the Permian have been found near the glaciated areas. Especially noted were the small, hardy tongue ferns, *Glossopteris* and *Gangamopteris*, surviving the increasingly harsh climate. These fossils tend to confirm the specific time of the Permian ice age.

The ice age during the Cretaceous period, again, was mostly in the southern hemisphere during Aptian time or in the Late Cretaceous. This is when the dinosaurs ceased to exist, during the Laramide disturbance. Thus, the end of the Mesozoic era and the beginning of the Cenozoic era, about 70 million years BP, is adequately marked. It is noted that the plateau of eastern Australia was ice-capped, with glaciers flowing toward the west into the sea. Icebergs also calved into an inland sea to the east.

During the Cenozoic era, the most recent and well-known ice age occurred. The Pleistocene epoch had four ice ages, or in some accounts, one great ice age with four separate ice advances separated by interglacial stages. Using North American terminology, beginning with the Nebraskan stage about 1 million years ago and lasting about 100,000 years, global warming, or the Aftonian interglacial stage, was about 200,000 years in length. Then, the Kansan stage of glaciation began and lasted for about 100,000 years, followed by the Yarmouth interglacial stage, which lasted for 310,000 years. The Illinoian stage was only about 100,000 years in length, but it appears to have the greatest ice coverage of the four advances during the Pleistocene. The Sangamon interglacial stage was about 135,000 years long; this was the shortest of the three interglacial stages. The Wisconsin stage began about 55,000 years ago and ended a mere 11,000 years ago with the advent of the present interglacial stage, which is known as the Holocene epoch. It is known that the Wisconsin glacial stage had four distinct ice advances and recessions, making a complex landscape across the Midwest of the United States with the same being true in Europe and Siberia.

It is generally accepted that during the last million years or so, crustal movement has been poleward in the northern hemisphere, and that the continents whose shapes and location we know today were different long ago. These are all, of course, geological concerns involving time. Also to be considered are meteorological concerns, which involve the cyclical changes in temperature of land, sea, and atmosphere. At present there is a warming trend that began an estimated 11,000 years ago. It is questioned whether we are at the end of the ice ages or merely between two glacial ice advances. If the global atmospheric temperature decreased a mere 5° C, another ice advance would be likely to begin. How long is this warming trend to last? How has humankind affected this warming trend? What effect will it have if we are still in the Pleistocene ice age or at the end of it? There are more questions than answers. Perhaps significant, however, is that the occurrences of the six known ice ages are spaced closer in time toward the present.

Approximately 4,500 to 5,000 years ago, ice covered all of Minnesota; it had been receding from Iowa since about 2,800 years earlier, although ice had reached as far south as northern Missouri. Today the last vestige of this ice is the arctic polar ice cap, which is continuing to recede. As the ice receded, vegetation and wildlife followed, with aborigines migrating northward from the southern climes. So it is today with nations looking for environmental resources where the ice is melting. Since the shortest known interglacial stage is the Sangamon, which lasted 135,000 years and ended 55,000 years ago, it can be expected that a phase of global warming is just beginning and will be here for tens of thousand of years more, whether affected or not by the activities of humankind.

Richard A. Step

See also Cretaceous; Erosion; Geology; Glaciers; Pangea; Wegener, Alfred

Further Readings

Ausich, W., & Lane, N. G. (1999). *Life of the past* (4th ed.). Upper Saddle River, NJ: Prentice Hall.

Chernicoff, S., & Whitney, D. (2007). *Geology* (4th ed.). Upper Saddle River, NJ: Pearson Prentice Hall.

Martini, I. P., Brookfield, M. E., & Sadura, S. (2001). *Principles of glacial geomorphology and geology.* Upper Saddle River, NJ: Prentice Hall.

Weiner, J. (1986). *Planet earth.* Toronto, ON, Canada: Bantam Books.

Idealism

The term *idealism* refers to any philosophical system or thesis that emphasizes the mental (idea) or the notion of very high or preeminent value (ideal). Because minds evaluate, it is possible to imagine a deep relationship between these two very different criteria. There are some connections, but clarity demands that they be understood separately. Only the emphasis on the mental yields a distinctly idealist conception of time.

When emphasizing value, *ideal, idealistic,* and *idealize* serve many functions. For example, norms and goals are often called ideals. On that basis, some moral attitudes such as optimism, commitment, and cheerfulness are deemed idealism, as in "the idealism of youth." Similarly, in international relations, idealism refers to a policy or posture obligating a state to promote higher aims among nations, such as peace, justice, cooperation, open borders, or democracy, while *practical* idealism refers to a position between that kind of idealism and what is known as *realism,* which refers to a state's selfishness in international affairs.

A second line of usage takes off from the fact that whatever would fully satisfy a desire is called ideal. Hence, in aesthetics, idealism has to do with representing things as we would like them to be (idealized) rather than as they are. With a little pejorative shading here, *idealist* calls out a vice. Thus, in ethics, the idealist is one whose standards are unrealistically high, or who is too confident about the virtue of a person or line of action, or perhaps altruistic to a fault. A system of ethics might be called idealistic if it sacrifices too much for the sake of a particular principle; or if it elevates any aspect of moral life too high, for example, sympathy at the expense of autonomy, but especially the spiritual or rational at the expense of the sensual or immediate; or if it believes too much in the goodness, or perfectibility, of human nature; or if it overestimates the moral efficacy of a line of causation, such as instruction, role models, prayer, meditation, sobriety, free markets, self-denial, the individual, the mass, the ego, the id, or rationality. Slightly more pejorative shading renders the idealist an idle dreamer of utopias, or an impractical adherent of some perfect, best, or ultimate in some domain. At the deepest levels of pejoration, *ideal* gives way to *idea,* in the sense of imaginary, and the idealist is primarily a fantasist.

Philosophy and Idealism

When the emphasis falls on mind, rather than value, there are both broad and narrow uses.

Most broadly, idealism is the negative of naturalism, materialism, or realism. Both *naturalism* (which abjures supernatural explanations) and *materialism* (which asserts that reality is ultimately physical) relegate the mental to a nonbasic, nonfundamental ontological status. Idealists do the reverse. Realism holds that the world exists independently of thoughts about it. Idealists typically deny some or all of that. However, there are systems of idealism that grant a good deal of realism, and so the most common and broadest ways of thinking about idealism treat it as the negative either of materialism or naturalism, or both.

For Karl Marx, it is the negative of materialism. If materialism is built upon the particular, the describable, and the sensuous, idealism stresses the universal, the indescribable, and the supersensuous. Where materialism posits the mechanistic, idealism introduces the teleological, and where it abjures evaluation, idealism inserts appraisals. Taken in this way, idealism includes all theses and systems that ultimately ground reality in the animate or the mental, whether taken as idea, mind, will, ego, nous, logos, word, text, absolute, God, life force, or earth turtle. Every kind of theism in which something animate is the source, ground, or essence of nature is idealism. Every account of causation that introduces teleology is idealism. Every ascription of a value to the essence of things is idealism. In this sense, the history of philosophy is very much a history of idealism, if only because it has been rare for thinkers to adhere to strict materialism and to forgo teleological and normative language in their descriptions of the world. When they have deviated into materialism, many have settled on a dualistic view in which it inevitably became all but impossible to allow the material equal footing with the ideal, making them realists who maintain a form of idealism.

Such views are often called *realistic idealism,* which refers to any doctrine that recognizes the existence of nonmental, nonideal entities but relegates them to a subordinate status as compared to the ideal. Theistic schools of this sort allow some category of nonmental existence independent of the divine mind but also hold that the divine created it, perfected it, controls it, or is coeternal with it. Nontheistic schools include Plato's system, in which timeless, disembodied ideas are organically united in the idea of the Good. Insofar as the world of particular things is downgraded into a mere imitation of these transcendent universals, Plato's system is an instance of realistic idealism. Another example is psychological idealism, which is the doctrine that ideas or judgments cause thoughts or behaviors. Like Plato's system, this doctrine is understood as denying either naturalism or materialism, or both, but not realism.

Epistemological idealism is the view that all entities other than minds are exclusively noetic objects, meaning that they have no reality apart from being perceived or thought by a mind. Edmund Husserl attempted to systematically think through a hypothetically couched form of epistemological idealism. In his phenomenology, the essences of objects, including the self, the other, time, and causation, are methodically treated *as if* they were exclusively noetic, apart from any other status they might actually have. A nonhypothetical system is found in George Berkeley's doctrine of immaterialism, where to be is to be perceived, and all that exists must be either perceiver or perception. Another is the transcendental idealism of Immanuel Kant, in which knowledge is wholly a product of the logical self or of transcendental unity of apperception.

In the existentialism of Martin Heidegger, learning is giving oneself to oneself. In postmodernism, there is nothing outside the text, and there are signifiers but no signified, both of which directly imply that our minds can encounter only thought. Semantic idealism is the thesis that our descriptions refer only to ideas rather than to things, so that, for example, the tautology, "When I speak of my right hand, it is my right hand of which I speak" is false, and instead, "When I speak of my right hand, it is an idea of my right hand of which I speak" is true. The implausibility of semantic idealism has been a frequent challenge to forms of epistemological and psychological idealism that rely on or imply it.

In its narrowest senses, *idealism* is variously qualified so as to sort and organize metaphysical outlooks. For example, impersonalistic idealism grounds the world in unconscious, spontaneous mental energy, whether as life force, precognitive urge, or primordial will. Personalistic idealism, also known as personalism, grounds the world in a self-conscious or purposive principle, such as nous, logos, or ego. The relationship between the mental world ground and individual, finite minds is crucial.

Monistic idealism holds that each finite mind is a mode, aspect, or projection of the One. Pluralistic idealism grants varying degrees of freedom, autonomy, privacy, uniqueness, and causal independence to the thoughts and acts of finite minds. With regard to their means of accounting for the natural world, schools of idealism are either subjective or objective. *Subjective idealism* holds that the natural world is a projection of our minds and appears to imply solipsism. *Objective idealism* identifies the natural world with the thoughts or acts of the world ground. Given these distinctions, Arthur Schopenhauer's philosophy, which grounds the world in blind will and affirms that every lion is one lion, can be understood as an impersonalistic, monistic, objective idealism. Schopenhauer's primary influence, the Vedic philosophy, which teaches that the finite self is a moment of the impersonal *Brahman,* is also an impersonalistic, monistic, objective idealism.

Schopenhauer's antipode, G. W. F. Hegel's philosophy, which treats thought as the ground of the world, freedom as the condition of thought, and self-consciousness as the condition of freedom, obviously grounds the world in a self-conscious principle, and it is thus a personalistic, pluralistic, objective idealism. Henri Bergson's philosophy of *elan vitale* is an impersonalistic, pluralistic, objective idealism, a status it shares with Friedrich Nietzsche's philosophy of will to power.

George Berkeley's immaterialism is often described as "subjective idealism," but that cannot be the case as the term is defined here. Berkeley denies only matter in nature, not nature's independence from finite minds. Objects in nature turn out to be exclusively noetic in his system, but they are not projections of our minds, but rather thoughts perceived in the mind of God, a view echoing prior work by Arthur Collier and John Norris. Because objects in nature are accounted for in terms of the acts of God's mind, rather than the acts of finite minds, Berkeley is an objective idealist with a personalistic world ground and a pluralistic outlook on finite minds.

A version of subjective idealism is found in the "windowless monads" of Gottfried Leibniz, which perceive the real world by perceiving only themselves. A more recent example is the decidedly antirealist school of postmodernism. Given its denial that our minds can encounter anything other than thought, the question becomes whose thought, our own or that of another, is encountered in the objects of nature. The postmodernist cannot offer us an encounter with the thoughts or acts of an independent world ground, because that would mean encountering the signified rather than the signifier. With the only route to objective idealism thus blocked, the postmodern idealist must be a subjective idealist for whom the natural world is a mental projection conditioned by our backgrounds and differences. Because these backgrounds and differences condition the world, postmodernism is an impersonalistic, notably monistic, subjective idealism, in which an unmistakable degree of monism among finite minds is achieved through the notion of impersonal, collective construction of reality—to be is to signify as a group member.

In the philosophy of mind, *hylozoism,* the view that matter is alive, and *hylopathism,* the view that matter is sentient, or that whatever is ontologically more basic than consciousness, such as neurons and their building blocks, already contain the essentials of consciousness, are impersonalistic idealist views. Alfred North Whitehead's *panexperientialism,* which holds that the fundamental constituents of reality are experiences, is somewhat more personalistic than these, as is *panpsychism,* which attributes mind to matter. *Panentheism,* which holds that the One both transcends and is imminent in the universe, and *pantheism,* which identifies the natural world with the One, are yet more personalistic. The *occasionalism* of Nicholas Malebranch, in which natural causation is an illusion and God is the sole cause of all events, including all human acts and cognitions, is an example of a highly monistic causal pantheism, the omnisufficiency of a personal One, in the philosophy of mind.

Time and Idealism

Because idealist thinking is found in all ages and languages, the term *idealism* is frequently restricted so as to sort schools according to historical criteria, for example, as 19th century, modern, or ancient idealism; German or British; Vedic or Buddhist; philosophical or religious. By the same token, there is no single idealist attitude pertaining to time. History's various gods and goddesses of time, such as Chronos, Aion, Kali, and the figure of Father

Time, express personalistic views, while a World Tree and a World Snake express impersonalistic views. Personalistic creationism in this area has often stumbled over the fact that the notion of a creator of time looks incoherent, because a creator of time must change from a state of not creating time to a state of creating time, and thus time must exist (as the medium of change) before it is created.

In philosophy, there has been one characteristically idealist thesis about time, known as the *ideality of time* thesis, which holds that time is not to be found outside of mind, or that it is exclusively noetic, or phenomenal, a view expressed as early as Antiphon the Sophist. Plotinus held it as well. Its modern proponents, in their various ways, include Leibniz, Baruch Spinoza, and most of the school of 19th century German idealists, including Schopenhauer, Hegel, Johann Fichte, and Friedrich Schelling. Kant held that space and time were necessary, a priori preconditions of understanding. The external world is necessarily represented in space and time, while the inner world of thought is necessarily represented temporally. In both cases, time is empirically real but transcendentally ideal, meaning that it is an actual datum of experience, though we can have no reliable conception of it outside of experience. In contrast, Nietzsche probably denied the ideality of time in favor of a real eternal return. More recently, many of the existentialists who were inspired by Husserl's work entertained phenomenological versions of the ideality of time, including Heidegger and Karl Jaspers. In the school of British Idealism, F. H. Bradley and J. M. E. McTaggart embraced the ideality of time thesis. The latter famously argued for the unreality of time and produced a system that was so pluralistic that there was no God or One in it. Instead, McTaggart grounded the world in the thoughts of a community of immortal spirits.

Belief in the ideality of time has often been held as a consequence of the belief that all things are ideal, but it has also occurred, for example in Kant and McTaggart, that belief in the ideality of time thesis has been the basis on which a more general idealism was erected and justified.

Idealism's Critics

Critics of idealism have been many, beginning at least with Aristotle. The most notable German critics are Kant and Nietzsche. Kant argued against all forms of thought, including all forms of idealism, that move beyond the limits of categorical experience to predicate the thing in itself. Nietzsche suggests that Kant's transcendental idealism is based in the faulty argument that a priori synthetic truths are possible by virtue of a faculty, and he describes Hegelian idealism and Kantian skepticism as delaying tactics employed in a largely theological struggle to prevent the emergence of a naturalistic world view.

In Britain, G. E. Moore and Bertrand Russell led a revolt against the breed of Hegelian idealism that was then dominant in English philosophy. It was the seminal event in the birth of the school now known as analytic philosophy.

Bryan Finken

See also Berkeley, George; Hegel, Georg Wilhelm Friedrich; Kant, Immanuel; Marx, Karl; Materialism; McTaggart, John M. E.; Nietzsche, Friedrich; Plato; Schopenhauer, Arthur; Solipsism; Whitehead, Alfred North

Further Readings

Ameriks, K. (Ed.). (2001). *The Cambridge companion to German idealism.* Cambridge, UK: Cambridge University Press.

Ewing, A. C. (1934). *Idealism: A critical survey.* London: Methuen.

Foster, J. (1982). *The case for idealism.* London: Routledge.

Franks, P. (2005). *All or nothing: Systematicity, transcendental arguments, and skepticism in German idealism.* Cambridge, MA: Harvard University Press.

Muirhead, J. (1931). *The platonic tradition in Anglo-Saxon philosophy: Studies in the history of idealism in England and America.* London: Allen and Unwin.

Neujahr, P. J. (1995). *Kant's idealism.* Macon, GA: Mercer University Press.

Stove, D. (1991). *The Plato cult and other philosophical follies.* London: Blackwell.

Ides of March

In the complex world of the Roman calendar, each month had a day called its *ides.* In March, May, July, and October, the ides fell on the 15th of the month. In the remaining months, the ides fell on the 13th. Ides comes from the word *iduare,* which means, "to divide" and literally signified the division of a month into two equal parts. The Roman calendar divided each month into *kalends, nones,* and *ides,* each celebrated in their own ways, indicating lunar events, and serving as a calendar for the common citizens. The ides was a day traditionally marked for ritual offerings and sacrifices.

Religious and political leaders kept calendars private and were often adjusting the length and structure to remain in sync with the lunar and seasonal changes. In the Roman and Julian calendars, the term *ides* was commonly used, equivalent to using such modern terms as *next week, today, tomorrow,* or *yesterday.*

The ides of March took on a whole new meaning in the year 44 BCE. It was on this day that Julius Caesar, emperor of Rome, was assassinated by a group of nobles in the Roman Senate. William Shakespeare captured this historic assassination in his play *Julius Caesar.* In Act One, Scene Two, a soothsayer proclaims a warning to Caesar in the infamous line, "Beware the ides of March."

After the Julian calendar, inspired by Julius Caesar, displaced the Roman calendar in 46 BCE, the term *ides* was used only colloquially to denote the middle of the month. Those living under the traditional Roman calendar most accurately interpreted the ides as the 15th of the month. With the passing of time, the expression "beware the ides of March" came to signify the prediction of doom, much like the aura of superstition that associates bad luck with Friday the 13th.

Debra Lucas

See also Caesar, Gaius Julius; Calendar, Gregorian; Calendar, Julian; Calendar, Roman; Rome, Ancient

Further Readings

Evans, G. B. (1974). *The Riverside Shakespeare.* Boston: Houghton Mifflin.

Feeney, D. C. (2007). *Caesar's calendar: Ancient time and the beginnings of history.* Berkeley: University of California Press.

Michels, A. K. (1967). *The calendar of the Roman Republic.* Princeton, NJ: Princeton University Press.

Immortality, Personal

Personal immortality is best understood as the belief in the actual survival beyond physical death of the core element of our personality or consciousness—often called the soul—for an indefinite period. The primary assumption of this belief is that the core element of consciousness, or soul, is entirely distinct from our body, and so can be removed from it on the body's death with no compromise in quality or essence. The consequences of belief in personal immortality on one's view of time are profound and involve a comprehensive rejection of temporality. It is also considered of fundamental importance in the development of religion.

Beginnings

It seems apparent that some of our predecessors' earliest speculations revolved around death and its consequences. The existence of burial sites among Neanderthals from 60,000 years ago and early *Homo sapiens* from about 35,000 years ago suggest that death was believed to be a transitional state. The practice in primal societies of killing off aging people was due, in part, to the supposition that their bodies, not being yet decrepit, would be useful to them in the afterlife. And the mutilation or eating of enemies was done with the same view in mind; the destruction of their bodies so as to cripple their ability to exact revenge from beyond the grave.

Similar views carried on in the early civilizations. The Egyptians, for instance, held high store on immortality, but their extraordinary efforts to preserve the physical body with processes like mummification suggest they could not conceive of immortality without a physical body. The coronation of the pharaoh was held to coincide with either the rising of the Nile in the early summer or receding of the waters in autumn when the fertilized fields were ready to be sown. As part of the coronation the pharaoh would reenact the deeds of Osiris, who represented the life-giving waters of the Nile. Beliefs about Osiris embodied the cycle of birth, death, and rebirth around which Egyptian life revolved. And from this cycle came the promise of immortality, which for many centuries was the preserve only of the pharaohs but eventually became available to anyone who could afford the expensive rites and observances.

Asian Traditions

Indian notions of immortality resonate to a different beat, bound up as they are with the idea of rebirth. It is one thing for the soul to survive death, but immortality is another thing altogether. The Vedas spoke simply of an afterlife presided over by the god Yama, but ideas of rebirth developed later. In what has become understood as the quintessential Hindu view, the soul will undergo an almost countless number of rebirths and, along the way, gradually rid itself of the life-clinging vices of greed, hate, and delusion. The final aim in Hinduism is *Moksa,* or liberation. Here and only here is true immortality achieved, but at the cost of having shed all traces of existence apart from the universal whole into which it has merged. Much the same is true of the Buddhist notion of *nibbana,* or in Sanskrit, *nirvana.*

In China, Confucianists and Taoists alike shared the Indians' suspicion of the egotism thought to underlie a desire to live forever, but they had no thought of rebirth, thinking instead that we have but one life to live. One tradition of thought expressed this well by speaking of the "three establishments." Instead of yearning for immortality in heaven, they advised we focus more on being remembered well here on earth, which could be done in any of these three ways: *establish virtue* to be remembered as an upright and honest person; *establish achievement* to be remembered by what one achieved in life; and *establish words* to be remembered by any written legacy one may have left. The Asian traditions, then, do not value personal immortality. In fact, their core insight is to rid oneself of the delusions that would presuppose such a goal to be worthwhile.

The Middle East and Judeo-Christian Traditions

As long as the physical body was bound up with the afterlife, as in Egypt, thinking on personal immortality could only go so far. It was the Persian sage Zoroaster who is credited as the first to associate personal immortality with the beliefs and actions of one's life. As part of a rebellion against the polytheism of the Persia of his day, Zarathustra posited a universe characterized by a cosmic struggle between good and evil. The good was exemplified by Ohrmazd and the evil by Ahriman. While Ahriman owed his existence to God, he had of his own free will chosen the evil path. Zarathustra taught that human beings can also choose to follow the righteous or the evil path and that our immortal destiny depends on the decisions we make.

As is now widely recognized, Jewish, and later Christian, religious concepts owe a heavy debt to Zoroastrianism. Jewish eschatology was primarily concerned with the fate of the nation, but the collapse of the Jewish nation and the Jews' exposure to Zoroastrian beliefs while in exile in Babylonia in the 6th century BCE introduced new ideas of personal immortality linked to the passionate belief in the justice of God. It is likely that the development of personal horoscopes in Babylonia in the 5th century BCE was due to the influence of Persian doctrines of personal immortality.

Until the exposure to Persian ideas on personal immortality, there was little discussion in the Jewish tradition of what happens after death beyond mention of a dim underworld called Sheol, to which all departed go, whether righteous or wicked. *Sheol* originally referred simply to the collective graves of the tribe. Sheol was a sad place where the soul wanders aimlessly, an enfeebled remnant of the living person. After the Persian ideas became available, some evidence exists that views developed to include the possibility of intervention from Sheol into worldly affairs. These ideas were resisted by the more learned sections of society, and more skeptical ideas became prevalent.

The other important influence on Jewish thought on immortality was Platonism. Before Plato, little attention had been given to a life after death. The funeral oration of Pericles, for example, makes no mention of an afterlife for those who had fallen in defense of Athens. Popular notions of Hades were much like Sheol: dark, gloomy, and unwelcoming. So when, in the *Phaedo,* Plato argued for personal immortality as experienced by an immortal soul, he was going against the grain of Greek thinking.

As a result of these different influences, it is hardly surprising that Jewish thinking on immortality was not consistent. On the one hand it was believed that at the end of time the soul would be reunited once again with the body and that the righteous and the wicked would receive rewards or punishments accordingly. But alongside this was the idea that the soul, being immortal, survived bodily demise and received its punishment or reward immediately upon death. And then there is the eloquently phrased skepticism of any sort of afterlife, as expressed in Ecclesiastes (9: 3–10).

For first century Judaism, then, the question of immortality was hotly contested, with the Sadducees arguing against it on the grounds that such an idea has little or no scriptural warrant. Against them stood the Pharisees, who were less traditional and more influenced by the Greeks in this respect, and their case won in the end. The influence of the later Greek thinkers, with their notions of an immortal soul for which the body was simply the vessel, can be seen in the thought of Philo of Alexandria (c. 20 BCE–c. 50 CE).

Christian teaching on immortality inherited some of the ambiguities of Judaism, but Saint. Paul's strong commitment to personal immortality smothered them. Saint Paul was quite clear about the central role immortality had in his theology. "And if Christ be not risen, then is our preaching vain, and your faith is also vain. . . . If in this life only we have hope in Christ, we are of all men the most miserable." (1 Cor 15:14, 19) His strong interpretation of immortality was followed up by Saint Augustine (354–430 CE), whose works were vastly influential in establishing Christian dogma for more than 1,000 years.

But even in Saint Paul there was a certain ambiguity. It is not certain, for instance, that eternal life can be understood straightforwardly as surviving beyond death. Elsewhere Paul speaks of our mortality being clothed with immortality, but once again this is as likely to refer to the eternal life of salvation with no suggestion of personal survival. This more sophisticated argument has not found favor among the vast majority of Christian believers, who continue to believe in their own personal immortality.

It is true, however, that there are two quite different conceptions of immortality in the early church, neither of which was clearly articulated. On the one hand, the belief that with the Second Coming of Christ all the dead will be resurrected requires that they lie dormant until then. But on the other hand, there is the belief that our souls depart from our bodies on death and immediately go to whichever point of repose is allocated to them. In the early years of the church, this distinction did not matter a great deal, because the return of Christ was believed to be imminent. But as time passed and Christ did not return as expected, the problem of this intermediate period between death and resurrection took on a greater urgency, particularly as the promise of immortality constituted its principal point of difference with other religions and with paganism.

Renaissance and Modernity

The importance of the idea to the Christian appeal notwithstanding, major areas of confusion remained for centuries to come. The individuality of the soul—an idea essential to the notion of personal immortality—only became established dogma at the fifth Lateran Council of 1513. But Pope Leo X's effort to enshrine belief in immortality was indication that it was under attack. In the general spirit of inquiry that characterizes the Renaissance, belief in personal immortality was one of the first supernaturalist nostrums to be queried. Of particular interest here is Pietro Pomponazzi's (1462–1525) *On the Immortality of the Soul.* Published only 3 years after Leo's edict, Pomponazzi's treatise looked at the discrepancies between Aristotle's view of the soul and that of official Catholic teaching. In particular, it explored the difficulties in reconciling the status of the perished soul in the twilight zone between death and resurrection. It ended up saying that personal immortality could not be established by reason and could only be believed as an article of faith. His work was publicly burned and he narrowly escaped more serious persecution.

Over the next centuries, thinking on immortality diversified a great deal. Immanuel Kant (1724–1804) devoted much energy to placing immortality as a postulate of the moral law, but almost completely ignored any description of what immortality might be like. A landmark in thinking on personal immortality came from Ludwig Feuerbach

(1804–1872), who was one of the first to articulate a contemporary understanding of immortality. In *Thoughts on Death and Immortality* (1830), Feuerbach argued that the ancient world conceived of immortality as achieved through identification with the state and the people, and it felt no consequent need to denigrate the limitations of a life lived in the here and now. But the specifically personal immortality before a personal transcendent God was a relatively recent development of Christianity, this not having been the main understanding of immortality during the Middle Ages. But, he argued, in the face of the "double nothing" of living people without essences and essences without meaning, the traditional dualism of Christian dogma was no longer viable. God, by this way of thinking, became an alien "other," outside the framework of human love. Instead, the earlier notions of immortality as a social memory and as part of a larger community were now returning to their rightful position.

In the wake of Feuerbach, a variety of naturalistic accounts have been offered. Following on from the ancient Indians and Chinese, 19th and 20th century freethinkers were suspicious of the motives that underlie the popularity of the idea of personal immortality. Nowhere was this more forcefully put than by the American pragmatist thinker William James, who wrote, "The pivot round which the religious life, as we have traced it, revolves, is the interest of the individual in his private personal destiny. Religion, in short is a monumental chapter of human egotism." Many freethinkers of the time thought the same. For example, George Anderson (1824–1915), an English businessman and philanthropist, wrote a poem called "Immortality," which was published in the *Agnostic Annual* in 1897. Anderson understood well the attraction behind claims of personal immortality: "So man's weak vanity was touched and flattered,/And thus he listened to the wondrous tale,/That all creation might be whelmed and shattered,/While He o'er death itself would prevail."

Several attempts were made in the 19th century to reconcile belief in immortality with the naturalistic worldview, the most influential of which was the phenomenon of spiritualism, which attracted considerable support between the 1850s and the 1920s. Through séances run by mediums, it was widely believed that contact could be made with the deceased. The more devoted spiritualists were convinced that they had achieved a major scientific breakthrough. Personal immortality was no longer a matter of faith or dogma, but of testable fact. The last great wave of interest in spiritualism happened during the First World War, as grieving family members sought to establish contact with loved ones killed in the fighting. But spiritualism declined after the war, as greater scrutiny of the mediums' practices revealed them to be tricks that magicians could reproduce or even simple fraud.

Other thinkers tried to avoid the question of whether personal immortality is actually true or not. For instance the pragmatist philosopher F. C. S. Schiller (1864–1937) acknowledged that dogmas about immortality were accepted because they were what people want to believe. But he specifically sidestepped the question of whether personal immortality was true and confined himself solely to the value of the belief on a person's conduct and well-being. He went further when he insisted that the ethical argument for immortality remains independent from whatever science may discover. This style of thinking has remained influential among religious progressives.

The most thorough critique of personal immortality from the 20th century was *The Illusion of Immortality* (1935) by Corliss Lamont (1902–1995), an American philosopher who studied under John Dewey. This work went through three editions and remained in print for 40 years. Lamont was not unsympathetic to the motivation behind the wish for immortality, but it was clear to him that immortality was incompatible with a naturalistic account of things. This said, he did more than simply deny the traditional conception of immortality; he also broadened the range of ways we could think about immortality, which involved many more perspectives than simply personal immortality, which relied upon a heavy dualism. Lamont identified *ideal immortality,* or the eternal moment that Spinoza and Santayana spoke of; *material immortality,* where the material that makes us up is subsumed after death back into nature; *historical immortality,* or the simple fact of our existence in time; *biological immortality,* or our continued existence through our children; and *social immortality,* or survival through the memory of our achievements.

An interesting development over the past half century has been an increase in interest in the beliefs of the general citizen on matters such as this. For example, the World Values Survey, which polled people in 74 societies around the world between 1981 and 2001, found a general decrease in belief in personal immortality as societies become more prosperous, although with an interesting anomaly. In agrarian societies, belief was measured at 55%, in industrial societies at 44%, and in postindustrial societies at 49%. The slight rise of reported belief in personal immortality in postindustrial societies may be accounted for by the inclusion of the United States, where the figures are out of alignment with all other postindustrial nations, with its growth of various "new age" spiritualities. Interestingly, this was the only anomaly in an otherwise even decline in all recognized forms of religious belief and expression from agrarian to postindustrial societies.

More recently a wide range of Christian thinkers have reconciled themselves with the naturalistic account of the world and have abandoned the more supernaturalistic interpretations of personal immortality. Many have developed variations of Lamont's ideal immortality to reposition the argument. These, along with the more naturalistic interpretations, mean that immortality has a wider range of interpretations now than at any time in history. In general, however, personal immortality has given way to variations of the nonpersonal immortalities Lamont and others have outlined. And in so doing, time has been handed back its absolute dominion.

Bill Cooke

See also Aquinas, Saint Thomas; Augustine of Hippo, Saint; Christianity; Feuerbach, Ludwig; Judaism; Kant, Immanuel; Plato; Egypt, Ancient

Further Readings

Anderson, G. (1897). Immortality. In C. A. Watts, (Ed.), *The agnostic annual* (pp. 47–48). London: Watts.

James, W. (1908). *The varieties of religious experience.* London: Longmans, Green.

Lamont, C. (1959). *The illusion of immortality.* New York: Philosophical Library.

Norris, P., & Inglehart, R. (2004). *Sacred and secular: Religion and politics worldwide.* Cambridge, UK: Cambridge University Press.

Schiller, F. C. S. (1912). *Humanism: Philosophical essays.* London: Macmillan.

Whitrow, G. J. (1988). *Time in history.* Oxford, UK: Oxford University Press.

Incubation

Incubation refers to the amount of time required in a developmental period. It stems from the Latin root *incubare,* which means to lie upon. This correlates with the term's most common use today, avian incubation. However incubation can also be applied to chemistry, microbiology, medicine, and even ritual.

Avian incubation denotes the time when the parent birds, usually female, sit upon their eggs. The body heat from the parent provides warmth to the growing eggs and maintains a consistent environment, providing constant humidity, shade, and temperature. Some species of bird, such as megapodes, incubate their eggs by burying them, with the eggs receiving warmth from geothermal heat and organic breakdown of vegetable matter. The incubation period varies from species to species, ranging from only 11 days up to 85 days. Some species may also begin incubating at different times of development to either stagger the brood's growth or create a simultaneous hatching.

In chemistry and biochemistry, incubation is the period of time and set of conditions required to maintain a chemical reaction. These condition constants can include temperature, pressure, concentration of reactants and products, and the presence and concentration of a catalyst. By utilizing the proper chemical incubation period, chemists are able to synthesize the highest yield of products in the shortest time.

In medicine, or more specifically pathology, incubation is the amount of time before a patient begins showing symptoms after a pathogen or disease-causing bacterium enters the body. This is often also referred to as the latency period. This incubation period varies for each disease and can range from a few minutes to several years. Generally the incubation period in children is shorter than the incubation period in adults.

Because these diseases are caused by pathogenic organisms, a valuable tool for diagnosing these illness are cultures, including blood, stool, urine, and sputum cultures. These cultured samples are actually grown in optimal conditions in machines called incubators, which can range in size from small tabletop units to those as large as rooms. These microbiological incubators can be programmed to simulate highly specific conditions, including carbon dioxide and oxygen levels, humidity, and temperature. Most incubators are programmed to simulate the internal environment in a human, because pathogenic bacteria usually experience optimal growth in these surroundings.

Another form of medical incubation is done in neonatal intensive care units (NICUs). Here, newborns in critical condition are placed in incubators, which are essentially large open warming units. These provide a constant temperature and oxygen level and a relatively clean environment. More advanced incubators may feature monitoring equipment and pressure capabilities, which help keep premature infants' airways from collapsing. However, this technology has its drawbacks, as newborns in the NICU often experience high noise and light levels, reduced physical contact with humans, and separation from their parents.

In a less scientific sense than the other definitions, incubation may also be applied to the mystical practice of sleeping in a sacred or divine area in hopes of achieving a spiritual enlightenment, whether it be through a dream, vision, experience, or cure. In ancient times, this practice was very common, especially for followers of the Greek deity Asclepius, the demigod of healing and medicine. Several present-day Christian sects and Greek Orthodox monasteries still practice this method of incubation today.

Christopher D. Czaplicki

See also Chemistry; Medicine, History of; Thanatochemistry

Further Readings

Bergtold, W. H. (1917). *A study of the incubation period of birds: What determines their length.* Denver, CO: Kendrick-Bellamy.

Hamilton, M. (2006). *Incubation: The cure of disease in pagan temples and Christian churches.* Whitefish, MT: Kessinger.

Industrial Revolution

The Industrial Revolution was a period in modern history when the production of goods by hand was gradually converted to methods of manufacture using large machines and assembly lines. This shift is seen by many historians as the force chiefly responsible for the birth of the modern era and for the ongoing phenomenon of globalization that emerged in the 19th century and now continues into the 21st.

The Industrial Revolution began in Great Britain. In 1650 the population of England was approximately 10 million, of which 90% earned a livelihood through farming of one kind or another. In 1821, Thomas Malthus wrote *Principles of Political Economy,* in which he discussed the concern that a lack of moral restraint by the British working class might lead to unchecked population growth. This behavior on the part of some people would create a fall in production, and the world would face a famine. Malthus gave a very gloomy prediction that worldwide mass starvation would eventually occur. In just 200 years, 1650 to 1850, the English population soared to more than 30 million, with less than 20% at work in fields, barns, and granaries. During this time England experienced the first Industrial Revolution, as entrepreneurs and capitalists began procuring the natural resources and labor power to produce goods for profit on a scale hitherto unknown.

The Industrial Revolution succeeded the longstanding agrarian age, which had begun millennia before when much of humankind developed agriculture and gave up nomadic lifestyles. Agrarian societies allowed people to escape from dependence on food sources over which they had no control. In such societies people produced surpluses that could be used to feed new classes of non–food-producers. At the same time, these societies required increasing amounts of land, which led to conflicts over territory. The need to store and defend food supplies and to house non–food-producers resulted in the growth of villages and small cities. Agrarian societies developed extensive division of labor and interdependence.

In *Principles of Political Economy and Taxation,* economist David Ricardo furthered the explanation for the changes that were occurring during the

Homestead Steel Works, *by B. L. H. Dabbs, 1893–1895. Workers watch as a foundry ladle prepares to pour molten iron into ingot molds at Carnegie Steel Company's Homestead Steel Works.*

Industrial Revolution. Ricardo explained the notion of cooperation as a consequence of prosperity in his theory of comparative advantage. Ricardo concerned himself with explaining the factors that caused growth in an agrarian economy. At the time, modern science and technology had not been applied to agriculture, and capital tools such as hoes and plows were of relatively minor importance as productivity inputs. Ricardo argued that the average productivity of labor would eventually decrease. This decrease in productivity would lead to falling wages for workers.

The Industrial Revolution in the simplest terms can be described as the advent of a decline in household production. It started in Britain for several reasons. The increased population provided labor for factories and markets. Britain had a rich supply of raw materials. Britain had many small shop owners that knew how to run a business. The change in production was an economic one, yet over time it became a political one as well as a social one.

Political Thoughts on the Industrial Revolution

The first Industrial Revolution was the impetus that led to the science of political economy. A new way of thinking was needed to explain and justify the social conditions. This began as philosophers in Europe struggled with the fundamental problem of how the individual pursuit of self-interest would lead to the highest social good. According to Arild Saether, theology and the power of the universal church as an explanation for human behaviors were replaced by new theories independent of church doctrines. There was a new distinction between positive and natural laws. Natural law was the manifestation of divine law. This law was revealed through nature based on reason. On the other hand, humankind created the concept of positive law. People needed to create laws to secure peace. New laws were required, because in a natural state everyone has the same equal and unlimited right to everything. With the advent of the Industrial Revolution, social structures changed and continued to develop specialized production of material goods. A continued propagation of the natural law would have resulted in a war of all men against all men, and self-destruction would have resulted. Hence, it was necessary for people to seek an agreement of cooperation and common wealth with each other. Works were written to explain positive laws. Such positive law acquired dominance in the 18th century and was firmly established in the 19th century.

The origins of the early teachings of political science are widely agreed upon. The Industrial Revolution had created a wealthy class of industrial capitalists who had to pay workers at least enough to live on. England at that time had a living wage that depended greatly on food prices. Ricardo advocated trade between countries. Consequently, the Corn Laws were passed. The Corn Laws placed tariffs on grain imports and created export subsidies that kept food prices high and increased the wages that capitalists had to pay. The basis of David Ricardo's theory, accepted in Parliament, was that specialization and free trade were beneficial to all trading partners. Adam

Smith and Ricardo had successfully explained the need and advantages for material cooperation and the acceptance of economic inequalities.

Adam Smith, a Scottish philosopher, wrote *The Wealth of Nations* in 1776; the book was widely accepted as a blueprint for economic activity and has been called the "bible of capitalism." Smith's teaching underscored the way humans reconciled self with others during the first Industrial Revolution. Smith discussed the division of labor as a consequence of social prosperity and people's desire to barter and exchange one thing for another. He explained how people managed to coexist by offering goods and services for sale in quantities that satisfied each other's private wants. Smith described how the invisible hand of the marketplace led to a cooperation that provided productivity and distribution that met the needs of society, and he argued that people would act in their own self-interest to provide goods and services of the greatest possible value. That in turn would allow for one to trade a good or service of a proportional value for another that one otherwise would not be able to receive. Those who produced the best would thrive, while those who provided the wrong good or provided an inferior good would lose out. Smith explained how the Industrial Revolution shaped what became known as the free market.

The free market paradigm provided an incentive for productivity and trade and provided a rationale for people to accept the idea of economic inequalities. Smith in essence justified the unequal distribution of wealth, arguing that in the long run everyone would gain. This notion of inequality was not new to the human race, but the free market explanation was an extension of the new concept of positive law. At the time of Smith's publication, England's Industrial Revolution saw the new manufacturing technologies, the development of more efficient forms of transportation, and the increase in productivity of agriculture that led to a massive movement from the countryside to the city. People in search of work migrated from rural areas to crowded cities and toiled long hours.

In 1875, Karl Marx completed work on the book *Capital* (*Das Kapital*). Marx's treatise developed an alternative understanding of the value of human labor. Marx's work served as an intellectual basis for a different vision of society, one that originated in class conflict, a struggle between labor resources and owners of capital. The labor theory of value, according to Marx, was the value of any commodity that resulted from the labor needed to produce the commodity. Labor provided more value in a day than it was paid. Capital owners were able to extract a profit when labor was used for the production of commodities. This notion proved to be a basis of the controversy as to how cooperation among humans over material value was organized. The philosophy of Marx pointed to the inevitably antiprogressive results of the economic and social inequality that the Industrial Revolution produced. Marx's view of progress was not supportive of the Industrial Revolution and the decline of family production. His philosophy was not based on incentives for people to be productive in order to get their needs and wants met. Marx's work in the field of political economy promoted an egalitarian economic and social structure, unlike the theories of other political economists, which were concerned with explaining the Industrial Revolution.

These three scholars—Smith, Ricardo, and Marx—are considered the progenitors of modern economics; in due course their ideas were exported across the Atlantic and promoted in the United States.

The Industrial Revolution in the United States

The Industrial Revolution that began in the mid-19th century in the United States must have seemed to others to have come out of nowhere. Europe had been the undisputed economic and technological leader in the world until, suddenly, the United States appeared in the vanguard. By 1926, the United States was producing about 45% of the world's industrial output, including 80% of the world's automobiles and 50% of its steel, electricity, and crude oil. America's experience of the Industrial Revolution had a decisive effect on the role of human resources in the economy and on social conditions for the American family. The Industrial Revolution in the United States thus mirrored Europe, as production changed from the manufacture of goods in small workshops to making goods with machines in factories.

The Industrial Revolution in the United States placed the family squarely in the capitalist mode of production, a fact well documented by the stories of urban working-class families' struggles to survive the Industrial Revolution. Changes in social relationships in the workplace and in households as a result of the Industrial Revolution have been viewed as both progressive and antiprogressive. The limits of capital's extensive mode of consuming labor in the first phase of industrialization, it has been argued, created a false consciousness among urban working-class families. The forces inducing conversion to a more intensive regime of separate spheres between work and home in the latter half of the 19th century were crystallized. Changing forms of family occurred as a result of the Industrial Revolution, as modern industry created the social construction of the "male breadwinner wage."

Thus, the economic structure of modern industry led to significant changes for American families. Prior to the industrial revolution, families had been largely self-sufficient work units. Fathers and mothers worked side by side, and children joined in the work as soon as they were able. Almost all of the values and skills needed for life were learned in the family setting. A sense of self-worth came naturally as did a sense of one's place in society.

The Industrial Revolution and the growth of cities changed all that. Fathers left home to work for money wages, and real wages rose quickly. Many families left their small farms and moved to the city. Children became more of an economic liability and less of an economic asset as they became unable to contribute to family maintenance. During the Industrial Revolution, work changed from requiring physical strength to requiring the ability to manipulate factory machinery, and what adults did was separate from children. This prompted the social need to limit the size of the family during the early years of the 20th century. The modern image of the family consisting of mother at home, a father at work, and children playing at home was born during the Industrial Revolution. Some scholars see the current crisis in family life as a long-term effect of the Industrial Revolution, one that poses a serious challenge for the 21st century.

In the United States as elsewhere, the Industrial Revolution greatly enhanced the economic fortunes of the middle classes and brought economic and social improvement for many people owing to the increase in earned wages. The lower classes, however, gained far less economically and socially. As the nation advanced, it failed to build adequate support systems for families in which both parents were employed. Indeed, the United States has not put national policies in place to provide for children's needs, as many European countries have done. Much of the energy behind popular demands for political change arises from the perception that the socioeconomic system that resulted from the Industrial Revolution and its aftermath has expanded social inequalities. As some economists and sociologists have pointed out, economic power in the 21st century has remained concentrated in the hands of those who own and control the means of production.

Marianne Partee

See also Economics; Evolution, Cultural; Fossil Fuels; Global Warming; Marx, Karl; Technology Assessment; Timetables; White, Leslie A.

Further Readings

Adam, L. (1965). *Agricultural depression and farm relief in England 1813–1852.* New York: A. M. Kelly.

Buchholz, T. G. (1989). *New ideas from dead economists.* New York: Plume Books.

Malthus, T. (1821). *Principles of political economy.* Boston: Wells and Lilly.

Marsh, R., & Tucker, M. (1992). *Thinking for a living.* New York: Basic Books.

Marx, K., & Engels, F. (1982). *Collected works.* New York: International.

Pfeffer, R. (1979). *Working for capitalism.* New York: Columbia University Press.

Reich, R. (1992). *The work of nations.* New York: Vintage Books.

Smith, A. (1937). *The wealth of nations.* New York: Random House. (Original work published 1776)

Wirth, A. G. (1992). *Education and work for the year 2000: Choices we face.* San Francisco: Jossey-Bass.

Infinity

Infinity refers to that which has no end. Inherently filled with paradoxes and contradictions, it is a concept found in math, science, and philosophy, and it can refer to time, space, and numbers.

Mathematical Infinity

Infinite Sets

In mathematics, infinity is not a number itself but a construct to refer to a sequence of numbers with no ending. The common symbol for infinity, ∞, was introduced into mathematical literature by the English mathematician John Wallis in the 17th century. Positive integers are an example of a set of infinite numbers, because there is no last number. For any number that one can consider, there is always a higher one.

Nineteenth century mathematician Georg Cantor stated that a collection is infinite if some of the parts are as big as the whole. This can seem to create paradoxes. For example, if one has an infinite number of objects, he or she can add or remove objects from the group and still have the same quantity. In the above example, positive integers make up an infinite series, but so do a set of just positive *even* integers. The second set comprises only a part of the first set, which would appear to make it only half as large. But in both cases, a larger number always follows, so they are both infinite.

Aristotle described two ways of looking at an infinite series. In actual infinity, one conceives of the series as completed. In potential infinity the series is never completed, but it is considered infinite because a next item in the series is always possible.

Other Uses of Infinity in Mathematics

The concept of infinity was introduced into geometry in the 17th century by Gérard Desargues. He developed projective geometry, in which "infinity" is the point where two parallel lines meet, a concept that was already being used in art.

In the algebraic function $y = 1/x$, y approaches infinity as x approaches zero. The expression y could never actually be infinity, since x could never be zero, because an expression of division by zero has no meaning in algebra. However, the smaller that x gets, the more times it could be divided into 1.

Paradoxes of Infinity

The Greek philosopher Zeno of Elea wrote of the dichotomy paradox. In order to travel a certain distance, we would first need to travel half of the total distance. In turn, before we reach that point, we must first have gotten halfway there (1/4 of the total distance), and so on. The distance we would need to travel first becomes infinitely smaller, so that it is impossible even to get started—for every distance to travel, there is always a smaller distance to travel first. These circumstances obviously would not have the same result in a real-life occurrence. However, the paradox demonstrates the idea of a distance being infinitely divisible. In other words, there are an infinite number of fractions between zero and 1 (or any two numbers).

The same idea is demonstrated in Zeno's paradox of Achilles and the tortoise. In the story, Achilles, a swift runner, races the slow tortoise but allows him a head start of 100 feet. If we imagine that each racer is going at a constant speed (one fast, one slow), then after some period of time, Achilles would have reached 100 feet—the tortoise's starting point. But by that time, the tortoise would have advanced a further distance. By the time Achilles catches up to the new point, the tortoise would have gone still farther. Zeno states the paradox: Achilles would never catch up to the tortoise, because the more distance he travels, the further ahead the tortoise is. To catch up to the tortoise in the story, Achilles would need to complete an infinite number of actions. Aristotle responded to this paradox by using the ideas of actual and potential infinity. Like the dichotomy paradox, this paradox assumes an actual infinity of points between Achilles and the tortoise.

Neither of these paradoxes would happen in a real-life occurrence, because a person can travel a given distance without having to take account of each point between the start and finish. Therefore, the traveler would be able to reach the destination, and Achilles would catch up to (and surpass) the tortoise in a finite time.

Bertrand Russell described the Tristram Shandy paradox, which is based on the title character of the novel by Laurence Sterne. In the novel, Shandy attempts to write his autobiography but finds that it takes him 2 years to write about the first 2 days of his life. Shandy complained that he would never be able to complete the writing, because the more he wrote, the more material there would be to write about. Russell suggested that if Shandy lived for an infinite number of years, then every day of his life would eventually be written about.

The paradox lies in the fact that there would never come a day when Shandy could finish the book, because with each day of writing, he became a year farther away from his goal.

Infinity of Time

Time and the Universe

In some schools of thought, time is inextricably linked to the universe itself. Thus, many theories dealing with the infinity of time are actually referring to the infinity of the universe. Saint Augustine wrote that time exists only within the created universe, with God existing outside of time in the "eternal present." While science theorizes that the universe began with the big bang, we cannot know if time began at this point as well or if there was time before there was a universe. According to some theories, the universe will end with a big crunch, but again, it is unknown if time would end at that point as well. Richard Tolman, in his oscillatory universe theory, hypothesized that the universe is in an infinite cycle of big bangs followed by big crunches.

An Infinite Past

The possibilities of an infinite past or infinite future have been discussed by philosophers for millennia. In 1692, English theologian and scholar Richard Bentley rejected the idea of an infinite past while accepting the idea of an infinite future. He wrote that the revolution of the earth, occurring in the present, could continue into infinity, because the future was inexhaustible. However, an infinite number of revolutions in the past was not possible, because each revolution would have once been part of the present and therefore finite. Immanuel Kant also rejected an infinite past, saying that an infinity cannot be completed, and if the past were infinite, we would never have arrived at the present moment.

Modern Perspectives

Debate about various aspects of infinity has continued into the 20th and 21st centuries. Albert Einstein wrote that our universe was finite but had no boundaries. Physicist Stephen Hawking built upon this idea, proposing that time and space together were finite but boundless (similar to the surface of the earth) and therefore had no beginning.

Galileo stated that infinity, by its very nature, was incomprehensible to human minds. Nevertheless, it is likely that infinity, with its possibilities and paradoxes, will continue to be a source of theory and discussion.

Jaclyn McKewan

See also Aquinas, Saint Thomas; Augustine of Hippo, Saint; Bruno, Giordano; Eternity; Gödel, Kurt; Kant, Immanuel; Russell, Bertrand; Zeno of Elea

Further Readings

Clegg, B. (2003). *A brief history of infinity.* London: Robinson.

Craig, W. L. (1978). Whitrow and Popper on the impossibility of an infinite past. *The British Journal for the Philosophy of Science, 30,* 165–170.

Lavine, S. (1998). *Understanding the infinite.* Cambridge, MA: Harvard University Press.

Whitrow, G. J. (1980). *The natural philosophy of time.* New York: Oxford University Press.

Information

The theory of information is fundamental to a rational understanding of the temporal fabric of our world. The principal reason for this is that information and time cannot be separated reasonably—a fact that will become evident from a short consideration of how we know about time and information. Only if information produces effects on objects whose changes in time are measurable physically is there a chance to test hypotheses about information objectively. Time, again, is measurable by a physical system only on condition that this system has information about its internal changes. To measure time, information about change is necessary, and to measure information, change in time is necessary.

Motivated by concrete problems of empirical research and engineering, the theory of information is steadily developing into a coherent system of mathematical models describing the complex abstract structure of information. On the one hand, information is an entity difficult to grasp, because the interactions between its syntactic, semantic, and pragmatic components constitute a complex structure. On the other hand, information is something supposed to exist almost everywhere: Its abstract features are claimed to be discernible in the most different kinds of systems such as quantum entanglements, cells, humans, computers, and societies.

The need for a unified theory of information is, thus, felt at nearly all frontiers of science. Quantum physicists are discussing whether an interpretation of their experimental data in terms of information could resolve some of the paradoxes that haunt their understanding of the subatomic world. Cosmologists are calculating the information content of black holes. Geneticists are describing the hereditary substance as containing information that is decoded in cells by means of the genetic code. Evolutionary biologists are assuming that the most important steps in the history of life on earth consisted in establishing increasingly efficient ways of information processing. Neurophysiologists are talking about the brain as the most complex information processor known to us. Computer scientists are constructing and programming machines that process information more and more intelligently without being continuously assisted by humans. Communication engineers are building information networks that connect human brains to computers in a new kind of symbiosis. Economists are proposing mathematical theories of economic behavior based on the distribution of information among people who buy and sell goods in a market. Sociologists are characterizing developed countries as information societies in a globalized world.

If we do not want to get lost in the jungle of information concepts used in different sciences, and if we want to make the relation between time and information clear, we are well advised to orient ourselves with the help of a rough classification of the manifold aspects of information and to distinguish its syntactic, semantic, and pragmatic components. The scientific disciplines of syntax, semantics, and pragmatics are known to anyone who studies language or other sign systems. In short, *syntax* is the study of signs in their relation to other signs; *semantics* is the study of signs in relation to their conceptual and referential meaning; *pragmatics* is the study of signs in relation to the agents using them. That all three disciplines are also of utmost importance for an information theorist should not come as a surprise, because sign systems basically are means of transmitting information. Each sign is, as it were, a package of information: a syntactic unit that transports a meaning from a sender to a receiver.

This entry presents some fundamental characteristics of the syntactic components of information and reflects on the relation between syntactic and semantic features of information. The reason for restricting ourselves to these topics is that the general semantics and pragmatics of information still constitute a mostly uncharted continent. To explore it further, not only sophisticated mathematical models but also new concepts for analyzing the close relation between time and information will have to be developed.

Syntactic Features of Information in Shannon's Communication Theory

Syntactic features of information are described by a variety of mathematical models. The most important one resulted from Claude E. Shannon's understanding of communication as transmission of information. Shannon (1916–2001), researching on communication systems first as an engineer at the Bell Laboratories and then as a professor at the Massachusetts Institute of Technology, has rightly been called "the father of information theory." The scheme of a general communication system he introduced in his classic paper, *A Mathematical Theory of Communication* (1948), still is the basis for most research in information theory.

According to Shannon, a general communication system consists of five components: an information source, a transmitter, a channel, a receiver, and a destination. The information source generates a message that is to be transmitted over the channel to the destination. Then the transmitter encodes the message in a signal that is suited for being transmitted via the channel. In the channel, there normally exists a certain probability that noise distorts the signal. The receiver decodes the transmitted and possibly distorted signal. Finally, the message is delivered to the destination. This general communication system can easily be exemplified by telephony. A person (the information source) speaks into a telephone (the transmitter) that encodes sound waves in a sequence of analog or digital signals. These signals are transmitted via fiberglass cables, air, satellites, or other channels. The physical structure of the medium of transmission, atmospherics, defective electronic devices, jamming stations, and other noise sources might distort the signal. A telephone at the other end of the channel (the receiver) decodes the transmitted signals in sound waves that some person (the destination) can hear.

Shannon introduced the general communication system in order to solve two main problems of information transmission. First, how many signals do we minimally need on average to encode a message of given length generated by an information source? Second, how fast can we reliably transmit an encoded message over a noisy channel? Both problems ask for principal spatial and temporal limits of communication, or in other words, for minimum code lengths and maximum transmission rates. To answer these questions, Shannon defined an information-theoretical analogue to entropy, the statistical measure of disorder in a thermodynamic system.

Omitting formal technicalities, Shannon's crucial idea goes as follows. For any message that is generated by an information source, the information content of the message is equal to the amount of uncertainty that the destination of the message loses on its receipt. The less probable the receipt of a message is, the more information it carries to the destination. Shannon's measure of the entropy of an information source, in short: Shannon entropy quantifies the average information content of a message generated by an information source.

Shannon's noiseless, or source, coding theorem shows that the entropy of an information source provides us with a lower boundary on the average length of signals that encode messages of the information source and are transmitted over a noiseless channel. If the length of these messages goes to infinity, the minimum expected signal length per message symbol goes to the entropy of the information source. Shannon entropy defines, thus, in terms of signal length, what a sender can optimally achieve in encoding messages.

In his noisy, or channel, coding theorem, Shannon proves the counterintuitive result that messages can always be sent with arbitrarily low, and even zero, error probability over a noisy channel—on condition that the rate (measured in information units per channel use) at which the message is transmitted does not exceed an upper limit specific to the particular channel. This upper limit is called *channel capacity* and can be quantified by an ingenious use of Shannon entropy. If we subtract from the entropy of the information source the conditional entropy of the information source given the messages that are received by the destination, we get the mutual information of the information source and the destination. *Mutual information* measures how much the uncertainty about which message has been generated by the information source is reduced when we know the message the destination has received. If the channel is noiseless, its capacity is equal to the Shannon entropy of the information source. The noisier the channel is, the more signals we must transmit additionally in order to correct the transmission errors. The maximum mutual information of an information source and a destination equals the capacity of the channel that connects both. That a sender who wants to transmit a message reliably tries to achieve channel capacity, regardless of how noisy the channel is, seems to be a vain endeavor, because any correction signal is subjected to distortion, too. Yet Shannon could show that for any noisy channel, there do exist codes by means of which a sender can transmit messages with arbitrarily small error at rates not above channel capacity—alas, he did not find a general procedure by which we could construct such codes, and up to now no information theorist has been able to perform this feat.

Shannon's coding theorems prove, with mathematical exactness, principal physical limits of information transmission. His channel coding theorem shows that if we want to reliably transmit a message over a noisy channel—and any realistic channel is noisy—we must respect the channel capacity as an upper limit on the transmission rate of our message. If we want to be sure that another person receives our message in its original form, we must take the properties of the medium of transmission into account and make the encoding of the message as redundant as necessary. To make an encoding redundant means to make it longer than required by the noiseless coding theorem. It means that we need more time to transmit a message over a noisy channel than over a noiseless one. Shannon's information theory implies, thus, an economics of information transmission: Given the goal of reliable information transmission and knowing the noise in a channel, we must respect a spatial lower limit on the length of encodings and a temporal upper limit on their transmission rate.

Shannon entropy measures only a syntactic property of information, more precisely: a mathematical property of statistical distributions of messages. Its definition does not explicitly involve semantic or pragmatic aspects of information. Whether a transmitted message is completely nonsensical or very meaningful for its destination, Shannon entropy takes into account just the probability that a message is generated by an information source. Since Shannon published his mathematical theory of communication, further statistical measures of syntactic aspects of information have been defined. For example, the theory of *identification entropy*, developed by the German mathematicians Rudolf Ahlswede and Gunter Dueck at the end of the 1980s, refers to Shannon's general communication system yet introduces a decisive pragmatic difference as regards the purpose of communication. In Ahlswede and Dueck's scenario, the information source and the destination are not interested in the reliable transmission of all messages that the information source can generate. The destination just wants to be sure as fast as possible that one particular message has been sent, which might have been encoded in different signals by the sender. It is the situation of someone who has bet money on a horse and only wants to know exactly whether this horse has won the race. Such a relaxation in the goal of communication allows an enormous increase in the speed of information transmission.

Semantic Features of Information in Shannon's Communication Theory

The semantic and pragmatic features of information are much more difficult to formalize than its syntactic features. Some approaches to semantic aspects of information try, therefore, first to identify syntactic properties of signals that may be correlated to the fact that these signals have a meaning for both the information source and the destination. When we speak about meaning, sense, reference, and other semantic concepts in an information-theoretical context, we do not suppose that information sources and destinations have complex psychological qualities like those of human beings. Access to semantic aspects of information is, thus, not restricted to self-reflective agents who associate, with signs, mental representations as designations and who refer consciously to objects in the real world as denotations. In this sense, "the semantic component of information" just means that at least a set of messages and a set of signals are interrelated by means of a convention.

A code, as a system of conventional rules that allow encoding and decoding, is the minimum semantic structure *par excellence*. It is normally not possible to infer which message is related to which signal, and vice versa, if we know only the elements of both sets (i.e., the messages and the signals), and the natural laws that constrain the encoding of messages in signals and the decoding of signals in messages. Thus, the most characteristic feature of the semantics of information is the conventional nature, or contingency, of the relation between messages and signals.

From this perspective, Shannon's theory of communication as information transmission does say a lot about semantics implicitly, because it is also a theory of the encoding of messages in signals and the decoding of signals in messages. Shannon's coding theorems express information-theoretical limits on the syntax of signals if the latter semantically represent messages under pragmatic constraints on the compressibility of encodings and on the reliability of transmissions.

Let us now focus our discussion on the channel coding theorem and the measure of mutual information. The higher the channel capacity—that is, the higher the maximum mutual information of an information source and a destination—the more is known about the statistical properties of the information source given the destination, and vice versa. We can express, for each channel, the information transmission distance between a given information source and a given destination in terms of time minimally needed by the fastest receiver for interpreting transmitted signals correctly as syntactic units that represent other syntactic units, namely messages. The noisier the channel between sender and receiver is, the less certain the semantic relation between a received signal and a transmitted message is for the receiver. Because the most general pragmatic function of communication is, for Shannon, the loss of uncertainty, the gain of uncertainty due to noisy channels must be counteracted by the use of longer signals. Then the actual rate of information transmission over the channel decreases and the transmission time increases. The more effort has to go into making a signal a reliable representation of a message, the longer the receiver needs to infer the transmitted message from a received signal.

Conclusion

We started our investigation into the relation of time and information by observing two very general facts: To measure time, information about change is necessary; and to measure information, change in time is necessary. We ended up describing an important example of the latter fact in semantics: The less certain the semantic content of a signal is, the more time is required to receive further signals needed for getting to know the message. Shannon's theory of communication contains, thus, a quantitative insight into the context dependence of information: Interpreting signals is a process that must obey temporal constraints depending on the media used for information transmission. In-depth analysis of the semantic, and also the pragmatic, features of information will arguably need more insight into the interdependence of time and information.

Stefan Artmann

See also DNA; Entropy; Logical Depth; Maxwell's Demon; Quantum Mechanics

Further Readings

Arndt, C. (2001). *Information measures: Information and its description in science and engineering.* New York: Springer.

Cover, T. M., & Thomas, J. A. (1991). *Elements of information theory.* New York: Wiley.

Pierce, J. R. (1980). *An introduction to information theory: Symbols, signals, and noise* (2nd ed., rev.). New York: Dover.

Shannon, C. E., & Weaver, W. (1998). *The mathematical theory of communication.* Urbana: University of Illinois Press. (Original work published 1949)

Von Baeyer, H. C. (2003). *Information: The new language of science.* London: Weidenfeld & Nicolson.

Intuition

Intuition is a concept with a number of meanings. It derives from the Latin *in-tueri:* knowing from within, that is, by contemplation. Intuition is a mode of perceiving objects, concepts, and ideas through direct apperception, as when people refer to having had an insight or illumination. In modern psychological approaches to various processes of judgment, intuition, in contrast with analytical thought, is described as a fast and instant thought process that does not demand mental effort, is not necessarily conscious, and allows neither description nor explanation. One illustration is constituted by heuristic processes, which are a type of unexplained rules of thumb that allow us to make various judgments, including perceptual judgments, rapidly. Another common description of intuition is "just having a hunch."

Intuition and time are related in both directions: On one hand, we can examine the role of intuition in the processes that lead to time perception and perception of duration, while on the other, we can observe the role of time in intuitive processes themselves.

Time is a key dimension to which every living creature must relate in order to adapt to its environment. Nevertheless, in the absence of any specifically defined perceptual apparatus for duration perception, on one hand, and, on the other, of any specific stimulus that represents temporal duration, the question arises of how time apperception occurs. Though this question is relevant in relation to any living creature, it is especially so to human beings due to their awareness of time. This is why philosophers have always wondered about the nature of time and about how the sense and consciousness of time and duration emerge.

An early example of this can be found in Saint Augustine of Hippo (354–430), who deals with the question of time in the sixth book of his *Confessions.* He argues that, when not asked about the nature of time, he knew the nature of time; however, when asked to explain, he was unable to explain. His text suggests that time is a type of primary experience that we cannot explain. That is to say, time is not experienced as the product of higher-order cognitive activity. Indeed, some major philosophers share a notion of time as a primary intuitive concept.

Immanuel Kant (1724–1804) held that time and space are two major intuitions of the human mind. He rejected doctrines like Gottfried Leibniz's, according to which experience is characterized by an inherent rationality. Kant argued that time and space are a priori categories necessary for human experience to occur. As such, he thought time and space to be enabling conditions of the human mode of experience. Time is not a real feature of things-in-themselves but rather a structure of the knowing mind.

Henri Bergson (1859–1941) rejected Kantian idealism but still associated time with intuition. He distinguished between an intuitive and an analytical mode of thinking. Intuition, for Bergson, yields metaphysical knowledge of reality as time, change, and creative evolution, which is certain knowledge. Reason, on the other hand, does not render absolute knowledge about reality but only relative knowledge about the material objects of science and experience. If reality is in flux—that is, not static—then for Bergson it is intuition that gives us this awareness of time and change as creative evolution.

Modern psychological theories have been dealing with the question of whether intuition forms the substrate of perception on the whole and of time perception in particular. But while most of them consider sensory input from the external environment a key component in a person's image of the world, some acknowledge that this sensory component is not sufficient. One example is the *constructive perception* approach (e.g., that of Jerome Bruner). *Gestalt psychology* (J. P. Kotter, Kurt Koffka, Max Wertheimer) holds that perception does not consist of a direct reflection of physical features of external reality. The perceptual system imposes organizing principles (*gestalt rules*) on the incoming stimuli, and these eventually determine the resulting image in the human brain. These organizing principles could be considered a priori, intuitive rules.

Developmental cognitive research, to which Jean Piaget (1896—1980) was a crucial contributor, has dedicated much attention to the development of the ability to grasp time in its various respects, e.g., the ability to understand the sequential nature of the time in which external phenomena occur. According to Piaget's findings, the ability to grasp time evolves in stages and gradually. A case in point is the development of the ability to integrate information according to the sequence and order in time of phenomena or actions. Children below age 12, it was found, are affected by intuitive thinking when making judgments about time, and this causes misperceptions. Abstract thought and logical reasoning evolve only in the formal operations stage, between ages 12 and 16. The ability to fully grasp time concepts matures only at a relatively late stage, and this testifies to the complexity of time perception. On the basis of this research, Piaget concluded that human understanding of time is not intuitive and depends on the acquisition of complex mental abilities.

Contemporary models of psychological time, too, consider the experience of time as the product of a cognitive process based on elaboration and integration of various types of information. This, it should be noted, does not necessarily clash with the Kantian approach: It could be that the concept of time as such—rather than the experience of duration—is intuitive and a priori.

Modern psychological approaches to cognitive processes and judgment consider intuition as a mode of thinking with unique features, which is unlike analytical thinking. This distinction refers to time as one of its parameters. First, intuitive thinking is very rapid, in contrast to analytical thinking. Second, intuitive thinking relies on past experience and on tacit knowledge (H. A. Simon), that is, on accumulated information that has been encoded in the memory over time.

Daniel Kahneman and Shane Frederick (2002) suggested that judgment can be achieved in two manners. The first is by means of a rapid intuitive, associative, and automatic process that demands no mental effort—this is also known as System 1. Alternatively it can result from a slower process that requires mental effort and that is controlled by rules and laws: System 2. Intuitive errors, accordingly, occur under two conditions: System 1 errs (usually due to applying past experience that is not relevant to the judgment in question) while System 2 fails to get activated in order to correct the error. Intuitive thinking, it must be noted, can produce effective and adaptive judgments if it is based on relevant experience, as happens in the case of expertise in the field of judgment. It is not easy to remove bias in the case of intuitive errors. Many perceptual illusions are the result of the activation of intuitive heuristics: Mere awareness of this does not suffice to avoid the perceptual illusion. Time illusions, too, which are expressed in erroneous experience of the clock duration of various events, are the result of heuristic intuitive thinking. Thus, for instance, in the case of the "filled time illusion," a time interval "filled" with stimuli and events is perceived as longer than a time interval identical in duration but "empty" in terms of stimuli and events (see also Time, Illusion of).

In part, time affects heuristic thinking as it does due to organizational processes in the memory that occur involuntarily in the course of time. As a result, a person is not always aware of the correct order of events in objective time, so that it may occur that later information may be perceived as—or combined with—earlier information. This may be reflected in intuitive biases like false memories or hindsight bias.

Often errors in eyewitness evidence are actually in part the result of such processes. And so we observe that the very dependency of intuition on the accumulation over time of information in the memory may cause intuitive bias regarding the dimension of time itself.

Dan Zakay

See also Augustine of Hippo, Saint; Bergson, Henri; Consciousness; Kant, Immanuel; Memory

Further Readings

Al-Azm, S. J. (1967). *Kant's theory of time.* New York: Philosophical Library.

Kahneman, D., & Frederick, S. (2002). Representativeness revisited: Attribute substitution in intuitive judgment. In T. Gilovich, D. Griffin., & D. Kahneman (Eds.), *Heuristics and biases* (pp. 49–81). New York: Cambridge University Press.

Lacey, A. R. (1989). *Bergson.* London: Routledge.

ISLAM

The religion of Islam originated in 622 CE, inspired and guided by the teachings of Muhammad ibn Abdullah, who lived in Mecca, Arabia, and who (in the Islamic belief tradition) received revealed truths directly from God and recorded them in the holy text called the Qur'an. As a religion, Islam has its roots in the same Middle Eastern monotheistic faith as Judaism and Christianity, and they share many of the same beliefs, practices, and sacred texts. Followers of all of these Abrahamic religions believe themselves to be descended down through time from the first man (Adam) and to have been chosen as God's special people through a covenant made between God and the Hebrew patriarch Abraham in about 1800 BCE. However, while Judaism teaches that the Messiah has not yet been born, and Christians believe that Jesus Christ was the divine Messiah, followers of Islam believe that Christ and Muhammad were both prophets of God in the tradition of Abraham, Moses, Samuel, and others but were not themselves divine.

Muslims believe that Muhammad was the last messenger or prophet sent by God to restore his people to the original, uncorrupted version of the Abrahamic faith. Muslims ("those who submit to God") therefore seek to return to the earliest and purest forms of worshipping God through the constant practice of the five pillars of Islam: professing faith, fasting, prayer, alms giving, and making the *hajj* pilgrimage to Mecca. The word *Islam* means "submission," and Muslims seek to demonstrate their obedience to God's will in all aspects of daily living, holding the teachings of the Qur'an to be a literal and uncorrupted version of God's laws, in contrast to the practices and texts of Judaism and Christianity, which have been distorted and corrupted by the long passage of time since the covenant was made between Abraham and God. In contrast to what they believe about the Qur'an, Muslims consider all earlier holy texts of the Abrahamic tradition (the Old and New Testaments) to be only partially "revealed" versions of God's will, because they were not recorded directly by the Hebrew and Christian prophets themselves but were embellished and altered by the followers and descendants of those prophets.

Time is also an important criterion in determining the *Sunni* (the "trodden path") practices that Muslims use to regulate religious practice and beliefs: *Sunnah* are defined as those religious actions and practices that Muhammad introduced during the 23 years of his ministry and that were subsequently passed down as tradition to subsequent generations by his closest companions. Adhering to tradition, by ensuring that religious holy days are observed at the correct times and for the correct durations of time, is therefore an extremely important aspect of Islamic culture. And, the Islamic world has evolved intricate and highly formalized ways of marking the passage of time and the timing of religious observances and holy days or months through use of a lunisolar calendar.

Origins of Life

The Islamic version of the origins of life and humanity has many parallels with Judaic and Christian beliefs. For Muslims, the creation of the universe is proof of the all-encompassing and absolute power of Allah (God). Allah is the beginning and end of all things, and the universe itself is an expression of his will, subject to his laws. For Muslims, the creation story derives from the Qur'an, which describes the division of the heavens and earth by Allah (God) from one solid mass into their present states over six long stages of time (not literal days as in the Judaic or Christian versions of the creation myth). Allah next created all of the living creatures and vegetation that inhabit the earth and sky, as well as the angels and heavenly bodies. Last, Allah used clay, earth, sand, and water to make and breathe life into the first man (Adam).

By breathing his own spirit into Adam, God distinguished him from the animals and endowed him with the ability to think and freely choose to submit to Allah or not, as well as to use his intelligence to seek to understand the workings of God in the rest of Creation. After Allah took Adam to live with him in paradise, Allah also created the first woman (Eve, or Haw) from Adam's side, and placed the two humans in a beautiful garden in paradise, forbidding them to eat the fruit of a certain tree. Ibis, an evil jinn (a disobedient angel, or Satan), tempted Adam and Eve to disobey God and eat the fruit of that tree. When they did, God cast them out of paradise and sent them to live on earth. In Islamic belief, all human beings are descended from Adam and Eve.

End of Time

As with the other Abrahamic religions, Islam depicts the end of time as a period of great disruption and trouble. This period will directly precede the Second Coming of the Messiah, who will bring an end to evil and suffering and will establish a new era of the kingdom of God. The Qur'an does not provide specific details about the end of time (Allah will not reveal these until that time arrives), but it does describe many signs that will signal its imminent arrival. The signs associated with the end of time will include such dramatic environmental events as drought, flood, the falling of the stars, and the cleaving of the heavens. As well, the earth will be purified of nonbelievers, either through mass conversion or by death. Most of humanity will have begun to worship the *Dajjal* or Antichrist, who opposes God's law and is aided by two leaders (Gog and Magog) who head nations opposed to God. The Dajjal will be ultimately defeated by the *Mahdi* ("the guided one"), a great leader descended from Muhammad who will become the final Muslim caliph. He will unify all of the various sects within Islam and will lead Muslims to a period of great prosperity and equality.

The prophet Jesus Christ will return to earth to aid the Mahdi in establishing Islam as the one true religion and to help defeat all of the unbelievers. He will then live on for 44 years until he dies and is buried next to the prophet Muhammad. Christ's death and burial will be followed by the resurrection of all humankind in preparation for the final judgment day. The sequence of events leading up to the judgment day will include a first trumpet blast, which will crush the mountains and kill all living creatures on earth (including the angels); a second trumpet blast, which will mark the resurrection of all humans; the descent of Allah to earth in order to pass judgment on all humans; the passing of thousands of years while humanity awaits judgment; and the final judgment or division by Allah of humankind according to whether or not their good deeds have outweighed their bad deeds during their lifetimes. Those who have submitted to Allah's will and followed his commands will enter paradise and then be placed to the right of his throne; those who rebelled against Allah by practicing evil ways will be placed to the left of his throne and condemned to live in hell for all time.

Helen Theresa Salmon

See also Adam, Creation of; Calendar, Islamic; Qur'an; Time, End of

Further Readings

Aslan, R. (2006). *No God but God: The origins, evolution, and future of Islam.* New York: Random House.

Denny, F. M. (2006). *An introduction to Islam.* Upper Saddle River, NJ: Pearson Prentice Hall.

Grieve, P. (2006). *A brief guide to Islam: History, faith and politics.* New York: Carroll and Graf.

Neusner, J. (Ed.). (2006). *Religious foundations of western civilization: Judaism, Christianity, and Islam.* Nashville, TN: Abingdon Press.

Riddell, P. G., & Cotterell, P. (2003). *Islam in context: Past, present, and future.* Grand Rapids, MI: Baker Academic.

JAINISM

Jainism is a non-Brahmanical philosophical tradition of India. Enriched by the deep insights of 24 spiritual gurus or *tirthankaras,* from Risabha (the first) to Mahavira (the last), Jainism has its own distinctive flavor that distinguishes this philosophy from others. To understand the concept of time in Jainism, it is necessary to have an overall idea of the metaphysical system of the Jainas. Jainas believe in a many-sided view of reality, known as *anekantavada,* or the theory of pluralism. This concept of reality can accommodate identity and difference, permanence and change. There are two main important terms in Jaina philosophy, *dravya* and *astikaya,* which pervade Jaina writings. The concept of *kala* (time) is to be understood in terms of these two concepts.

Umasvati, a famous Jaina philosopher, defines *dravya* or substance as that which possesses qualities and modes. Substance is always characterized by qualities and modes. Jainism offers six substances as ultimately real, and time is one among them. They are *jiva* (the soul), *pudgala* (the matter), *dharma* (the principle of motion), *adharma* (the principal of rest), *akasa* (space) and *kala* (time).

Amongst the six substances, five are known as *astikaya,* a term which, negatively speaking, helps to understand the concept of time. In the *Dravya-Samgraha,* a famous book of the Jainas written by Nemichandra, the term *astikaya* is explained in the following way: This term consists of two parts, *asti,* meaning "to exist" and *kaya,* meaning "body." Therefore this term refers to those five substances that exist and occupy space. The term *kaya* also technically is to be understood as consisting of many indivisible particles (*pradesa*). So, five out of the six substances apart from time can be characterized as astikayas. This notion of indivisible particles points to the fact that the Jainas believe in the general atomic conception of the universe. The five substances are called astikaya (extended) because the particles of which they are made are not separate. This is also the reason why time is not classified as an extended substance, even though it has existence.

According to Jaina philosophy, time is real, as it possesses origination, decay, and permanence, the characteristics of all real things. Time is as real as the five others, as it is the accompanying cause or condition of the modification of substances. This reality of things is explained by the Jaina philosophers by recognizing the two aspects of things, *parjaya* (the mode) and *dravya* (the substance). The former is the series of temporary modes with coming and going, and the latter is the core thing that remains constant. Time as a real substance possesses all these states. The above can be explained by taking an example from our everyday life. When a man clenches his fingers into a fist, then this phenomenon—clenching of the fingers—occurs (origination), and the previous state of the fingers necessarily is at end (decay), yet so far as the fingers are concerned, they continue to be substantially the same (permanence).

In the famous Jaina text entitled *Pancastikayasara,* the atomic conception of time is discussed. The

This 15th-century Jain temple at Ranakpur represents a culmination of Jaina temple building in western India.
Source: Tonis Valing/iStockphoto.

characteristics of time atoms are such that each atom is distinct and can never be mixed with other time atoms. This is the reason for advocating the theory that time has one *pradesa* only, or in other words, the time atoms can never be combined. And this also leads to the point that the time atoms can only constitute a unidimensional series that is unilateral. In Jaina terminology, this is known as *urdha-pracaya* (unilateral) as opposed to the *tiryak-pracaya*—a multidimensional series or horizontal extension. All other substances except time possess extensions in both these dimensions. As kala (time) consists of infinite *samaya* (instants), it possesses only unilateral extension. This unilateral extension of kala points to the fact that the world is constantly progressing with the help of time.

The Jainas held that there are two kinds of kala, *vyavahara kala* (conventional time) and *niscayakala* (transcendental time). Vyavahara kala is also known as samaya in Jaina philosophy. In our daily life we use the concepts of moments, hours, and days to indicate the beginning and end of a particular event. This conventional or practical time is indicated by modification, activity, distance, and proximity. The Jainas describe modification as the self-variation in a thing without a change in its substance. Activity is caused by the movement of a body. Distance and proximity are to be understood here in relation to time. In our everyday life we label a thing as new or old by observing its state of modification. On watching the sunrise in the east, we say we have the daybreak, and so we gather the ideas of midday, evening, and night by observing the movements of the heavenly bodies. A thing under actual observation is said to be proximate in time, otherwise it is distant in time. In this way, the external objects indicate and refer to the conventional or empirical time denoted by moments, hours, and days, as we call them.

In the Jaina literature, *pancatikaya-sara,* the conventional or empirical time periods, are discussed elaborately. The minutest unit of kala is *samaya.* This unit is defined as a period of time taken by an *anu* (atom) in traversing to its consecutive anu. *Vyavahara kala,* or conventional time, consists of infinite samaya. The time undertaken by an adult person in inhaling and exhaling is termed a *prana.* Seven pranas make one *stoka.* Seven stokas compose one *lava.* Thirty-eight and one-half lavas form one *nali.* Then comes *muhurta,* which is equal to two nalis. Thirty muhurtas form one *divaratra* (day and night). Fifteen divaratras form one *paksa.* One month comprises two paksas. Two months equal a *ritu* (season). Three ritus make one *ayana.* A year is formed by two ayanas. These divisions and subdivisions of time are linked to the different units of measurement pertaining to *vyavahara-kala* or conventional time.

The above divisions are dependent on some external factors, such as the motions of the heavenly bodies. But in addition to the motions of the heavenly bodies, vyavahara kala or conventional time is dependent on niscayakala, or transcendental time, for the determination of its measure. This niscayakala is variously designated as *paramarthkala* or *dravyakala. Vartana,* or continuance, is the main characteristic of niscayakala. Brahmadeva, a commentator on Jainism, holds that vartana is the accompanying cause of the modifications of things, like the stone beneath a potter's wheel or like fire in the matter of studying in winter. This vartana is the characteristic of real time. The stone underneath the potter's wheel does not directly impart motion to the wheel, but still the stone is indispensable in the matter of the movement of the wheel. Similarly, fire is not directly responsible for one's study, but study is impossible in winter without fire for heat and light. The case of time is comparable.

Therefore the changes that can be observed in things point to something more than those changes and the changing things. Every change is accompanied by a definite point of time, called samaya, or conventional time, while the continuing reality underlying this conventional time is the dravyakala, or transcendental time. The transcendental time is conceived as having both a beginning and an end. This point brings an important feature of the Jaina understanding of time. The samaya, or instants arranged unilaterally, are conceived of as permanent—beginningless and endless. This is indicative of the fact that the transcendental time can never be made an object of sense perception. But one can arrive at knowledge by inference. Another characteristic is that time is *amurta* or formless; in this respect it is similar to the concept of soul. Time or kala is described as *niskriya,* or devoid of activity, and it is different from soul, which is *sakriya* or active.

The Jaina conception of transcendental time is to be contrasted to the idea of time as appearance. In fact the concept of time plays a very important role in the pluralistic metaphysics of the Jaina system. Nothing can be conceived in this universe without time.

Jainism with its realistic viewpoint of thinking accepts the reality of change, which in turn indicates the reality of time. It is not possible to deny the reality of time, as every change involving birth, growth, and decay happens within the framework of time.

Thus time is confined within the limits of *lokakasa,* or the space that is occupied by other substances; the *aloka,* which is beyond this, has not time within it. But as aloka is a substance, it must have modifications also. Now how are these modifications possible in the aloka, signifying void which has no time within it? This point is stated in the following way. Aloka, although it is beyond the loka, is still a part of the akasa, or space. Time being within the akasa, it brings about modifications in any part of it, just as a pleasant object coming in contact with a particular part of a body causes pleasant feeling all over the body, or a potter's stick moves the whole wheel by striking at a particular point of the pot. Innumerable and partless atoms of time fill the lokakasa. In Alokakasa there is absence of matter, and so there is absence of *kalanus,* or instants.

Anu is the smallest part of a substance, and when these anus are combined inseparably, the substance constituted of them is an extended substance. The five substances—jiva, pudgala, akasa, dharma, and adharma, are extended substances, because their smallest constituents are mixed together, and they remain inseparably combined with one another. But the case is different with time. Time has also minute kalanus, or instants, but these anus or ultimate particles are not capable of mixing up with one another. Therefore, these ultimate particles remain always separate. It is for this reason that Jaina thinkers compare time substance with a heap of jewels. The jewels embedded in a garland remain separate; similarly these anus do not form a kaya; that is, they do not aggregate as the other five substances do. In addition, each kalanu occupies only one pradesa , or space point. Therefore, it does not possess an aggregate or volume space point comparable to that portion of space that is obstructed by an anu or atom. These time atoms are imperceptible, formless, and inactive. Thus they have no extension, but they possess existence.

There is again another principle of classification, on the basis of which substances are classified as conscious or unconscious. Following this line, jiva alone is singled out as conscious, and all other substances, including time, are labeled as insentient.

In conclusion, it may be pointed out that the Jainas make change the essential character of all reality, which is intelligible only in terms of time. Jaina thought, with its law of pluralism accompanied by time as a constant variable, stands unique among the philosophical traditions of India.

Debika Saha

See also Christianity; God and Time; Hinduism, Mimamsa-Vedanta; Hinduism, Nyaya-Vaisesika; Hinduism, Samkhya-Yoga

Further Readings

Mandal, K. K. (1968). *A comparative study of the concept of time and space in Indian thought.* Varanasi, India: Chowkhamba Sanskrit Series.

Mukherjee, S. (1944). The Jaina philosophy of non-absolutism. Calcutta: Bharati Jaina Parisad.

Potter, K. H. (Ed.). (1970). *The encyclopedia of Indian philosophies* (Vols. 1–2). Delhi: Motilal Banarsidas.

Rao Venkateswara, R. (2004). *The concept of time in ancient India.* Delhi: Bharatiya Kala Prakashan.

Tatiya, N. M. (1951). *Studies in Jaina philosophy,* Banares, India: Jain Cultural Research Society.

Janus

In Roman mythology, Janus was the god of doorways and of transitions in time. He was depicted as a double-faced head with the two faces looking in opposite directions. As the god of transitions, Janus was deeply associated with the notion of time.

Janus was one of the most ancient gods worshipped by the Romans. Unlike many other Roman gods, Janus has no Greek counterpart. The Romans invoked Janus at the beginning of each month, as well as at the new year. They prayed to Janus during important events such as birth, marriage, and harvest time. To the Romans, Janus signified transitions such as from primitive society to civilization, childhood to adulthood, and especially from war to peace and vice versa. The Romans built numerous temples in honor of Janus, and the name January for the 1st month of the year is a modern vestige of the deity.

As the god of beginnings, the Romans believed that Janus predated even Jupiter, the primary Roman god. According to tradition, Janus was the first king of Latium. His rule witnessed a golden age in which laws, coinage, and agriculture were introduced. The Romans believed that Janus had protected Rome from the legendary attack by the Sabines. In memory of this protection, the Romans left the gates of Janus's temples open when Rome was at war and closed them during times of peace. In addition to temples, Rome also had many gateways called *jani* (the plural form of *janus*) which often bore the symbol of Janus. The most important janus was the Janus Geminus located in Rome. It had a gateway at each end through which the Roman legions would parade before departing for war. The coins of the Roman Republic also carried the double-faced image of Janus. The Roman Empire later often depicted Janus with four faces looking in opposite directions, known as Janus quadifrons. Two of these faces were often bearded and the others shaven.

Janus was known as the inventor of religion, festivals, temples, and coins. The Romans held him in very high esteem as the custodian of the universe. In this role, he transcended many other gods in importance. Janus was present in all aspects of time: beginnings and endings, sunrise and sunset, the progression of life, and during rites of passage. He was invoked on a wide variety of important occasions in Roman cultural life.

In the Roman religion, Janus was believed to have come from Thessaly in Greece. He was invited to rule Latium with Camese. The two became the parents of the god Tiberinus, who was the Roman river god. The Tiber River in Rome is his namesake. Janus was also the father of Fontus, the Roman god of springs. Worship of the god Janus in effect ceased with the decline of the Roman religion. Janus is often cited in popular culture in connection with personalities or ideas that have two opposing sides.

James P. Bonanno

See also Calendar, Roman; Rome, Ancient

Further Readings

Beard, M., North, J., & Price, S. (1998). *Religions of Rome.* New York: Cambridge University Press.

Dumezil, G. (1966). *Archaic Roman religion.* (P. Krapp, Trans.). Chicago: University of Chicago Press.

Turcan, R. (2001). *The gods of ancient Rome: Religion in everyday life from archaic to imperial times.* New York: Routledge.

Jaspers, Karl (1883–1969)

Karl Theodor Jaspers was a German existential philosopher whose initial profession had been psychiatry. In later life he attained an international reputation owing to his highly publicized statements as a political philosopher in postwar Germany and received several honorary doctorates and prestigious awards honoring his philosophical work for peace, freedom, and humanity.

Jaspers was born into a middle-class family in Oldenburg, northern Germany, on February 23, 1883. The liberal democratic mindset in his family deeply shaped his way of thinking. Jaspers first studied law, but partly owing to pulmonary disease contracted in his youth, he switched to the study of medicine. Due to his illness Jaspers was forced to live a calm and retiring life. Notwithstanding, he passed his exams in Heidelberg with distinction. In 1910 he married the nurse Gertrud Mayer, daughter of a Jewish merchant and sister of his fellow student Ernst Mayer.

Jaspers received his medical degree in 1908. He obtained his second doctorate from the philosophy faculty of Heidelberg University for his book *General Psychopathology,* still considered an important psychological work today. Beginning around 1910 Jaspers gradually had moved toward philosophy—mainly inspired by his readings of Baruch Spinoza, Immanuel Kant, Georg W. F. Hegel, Friedrich Nietzsche, and Søren Kierkegaard. His first predominantly philosophical book was *Psychology of World Views* (1919). By 1922 he had became full professor of philosophy in Heidelberg University.

In 1909 Jaspers made the acquaintance of Max Weber, one of the most important liberals, economists, and sociologists of prewar Germany. Weber was to play a guiding role in Jaspers's personal and academic development. During the same period Jaspers also met Ernst Bloch, Georg Lukács, Friedrich Naumann, and Georg Simmel.

Jaspers dissociated from the neo-Kantian philosophy that prevailed at this time and was one the first researchers who fought with insistence for a widening of the scope of philosophical sciences. Jaspers's main work is chastely titled *Philosophy* (1932) (*Philosophical World Orientation, The Illumination of Existence, Metaphysics*). Some further prewar works are *Man in the Modern Age* (1931), *Reason and Existence* (1935), and *Philosophy of Existence* (1938). Jaspers investigated the influences of borderline situations on the human being's attitudes and described a transcendental theory of the encompassing (*das Umgreifende*). Jaspers's attitude toward religion, however, was critical, and he did not consider it essential to philosophy.

Jaspers became deeply involved in the history of philosophy, which he understood as a facet of existential philosophy. He was among the first to think of history beyond the notion of a path of time or a path of ideas. Thus, when considering the heritage of the great philosophers, he replaces the link through time with the link through reason. One could say Jaspers widened the dimensions of history and time by ignoring their strict chronological order and interpretation.

Jaspers also dealt with Friedrich Nietzsche's thoughts about the eternal return or recurrence of time and the necessary overcoming of weakness and nihilism through such recurrence. Jaspers esteemed Nietzsche's work, although he disagreed with some parts of Nietzsche's legacy, particularly in Nietzsche's interpretations of the existence and role of transcendence. Furthermore, Jaspers was opposed to any thinking of the concept of the superior man.

Another outstanding person in his life was Martin Heidegger, with whom he shared an intense discourse. Nonetheless they quite frankly took very opposite philosophical positions. Jointly they are considered the most important representatives of German existential philosophy—clearly demarcated from Jean-Paul Sartre's existentialism. Their relationship abruptly ended the day Heidegger declared his accordance with the aims of Hitler. More satisfying was Jaspers's lifelong friendship with his former student, the political philosopher Hannah Arendt.

As a consequence of Jaspers's marriage to a Jewish woman and his declining to collaborate with the Nazi regime, he was subjected to certain restrictions, including the prohibition to teach and to publish. After the end of the war, however, with his reputation intact, he was among the well-regarded academics who helped to refound Heidelberg University, but in 1948 he chose to relocate to the University of Basel in Switzerland out of his dissatisfaction with the political and academic developments in postwar Germany.

Induced by his experiences during the Nazi tyranny, Jaspers had become an increasingly politically thinking person and fought against the quick rehabilitation of former Nazi collaborators like Heidegger. His well-known and controversial book

The Question of German Guilt appeared shortly after the war's end (1946). Jaspers assumed the role of a strong advocate for freedom and purer democracy. Some explicitly humanistic works, such as *The Atom Bomb and the Future of Mankind* (1961), followed. At this stage he pronounced the importance of an enlightened and educated bourgeoisie to the stability of democratically ruled societies.

Jaspers's circumscribed influence in modern philosophical theory, especially outside the German sphere, is partly due to the fact that more famous philosophers such as Heidegger and Theodor W. Adorno did not take issue with the greater part of his perceptions and ideas.

Karl Jaspers became a Swiss citizen in 1967. He died in Basel on February 26, 1969.

Matthias S. Hauser

See also Hegel, Georg Wilhelm Friedrich; Heidegger, Martin; Kant, Immanuel; Morality; Nietzsche, Friedrich; Space and Time; Weber, Max

Further Readings

Ehrlich, L. (1975). *Karl Jaspers: Philosophy as faith.* Amherst: University of Massachusetts Press.

Josephus, Flavius (27–100 CE)

Flavius Josephus, a Jewish native of Jerusalem and later a Roman citizen, achieved lasting significance for his writings about Judaism and the Roman Empire (*The Jewish War, Antiquities of the Jews, Against Apion,* and *The Life of Josephus*). His own life and writings illustrate the 1st century clash of the Jewish and Greco-Roman worlds.

Although his historical works follow the methods of Greek historians like Thucydides, Josephus communicated the Jewish tradition of ancient history and the beginnings of time to the Greco-Roman world. As with Jewish histories in general, his observations of the passage of history emphasize a linear progression of connected events driven by divine providence. Though Josephus showed great interest in the accurate record of factual history, his ordering of events focused on the quality and readability of the story, often to the sacrifice of exact chronology.

Born to an aristocratic family within the Jewish priesthood, Josephus rose to prominence as a Pharisee and political leader in Judea, later receiving a military command during the Judean revolt. Josephus had been convinced of Rome's glory and power since he visited the empire's capital at age 26, and his own writings indicate that he felt neither convicted about the cause for revolt nor confident of any sort of victory for the Jews. Josephus was defeated and captured by the Romans at Jotapata in 67 CE, but he quickly made a powerful acquaintance with Emperor-to-be Vespasian by predicting his rise to power. Vespasian later secured Josephus's freedom and his Roman citizenship and provided for him in Rome with a residence and pension. Once he was in Rome, Josephus began his writing. Little else is known about the latter part of his life.

Major Works

In Rome, Josephus wrote *The Jewish War,* a commissioned account of the Judean revolt and the destruction of Jerusalem, which received imperial endorsement from Vespasian. For this task Josephus drew upon his own experiences in the war, Roman military records, and his many contacts. Josephus was politically sensitive in his handling of the conflict, generally pointing all blame toward the Judeans who had stirred up the revolt.

In his next undertaking, *Antiquities of the Jews,* Josephus composed his own version of Jewish history for explanation to the Greco-Roman world, covering a span of some 5,000 years. As an appendix to this account, he attached *The Life of Josephus,* a short autobiography that focused on his ambiguous role in the Judean revolt. *Antiquities* was based on the Hebrew scriptures as well as other rabbinic sources and traditions, but it also contains interpretations and inferences that were uniquely his. When dating events, he often used Jewish reckoning alongside that of other civilizations such as Egypt or Rome. As in *Genesis,* in *Antiquities,* time was said to begin with the 7-day divine Creation of the world, which became the model for a 7-day week. Josephus also recorded the origins of the feasts and celebrations that make up the Jewish year.

In his last work, *Against Apion,* Josephus defended his own accounts of Jewish history and offered an apology for the Jewish religion. Against Roman accusations that Judaism was a young and unimportant religion, Josephus argued that Judaism was in fact quite ancient and possessed a rich antiquity equal to that of the Greeks and Romans. In this work, Josephus sharply criticizes Greek histories initially but then uses them to help validate many aspects of Jewish history. Throughout his writing he confirmed his belief that God guided the events of history and that all things happened at the time appointed them by God.

Adam L. Bean

See also Creation, Myths of; Judaism; Philo Judaeus; Thucydides

Further Readings

Cohen, S. J. D. (2002). *Josephus in Galilee and Rome.* Boston: Brill.

Fieldman, L. H., & Hata, G. (Eds.). (1987). *Josephus, Judaism, and Christianity.* Detroit, MI: Wayne State University Press.

Hadas-Lebel, M. (1993). *Flavius Josephus* (R. Miller, Trans.). New York: Macmillan.

Josephus, F. (1999). *New complete works of Josephus* (W. Whiston, Trans.; P. Maier, Comm.). Grand Rapids, MI: Kregel.

Joyce, James (1882–1941)

James Joyce, widely regarded as the most important novelist of the 20th century, was born in Dublin to a family of modest means and was educated at Belvedere College, Dublin. He began his writing career within the then-dominant mode of literary naturalism but soon outgrew those conventions and produced, over several decades, a remarkably original body of work that, for sheer inventiveness and virtuosity of prose style, gained him international acclaim. Joyce's novels virtually defined modernism in literature, influenced several generations of writers, and continue to exert a fascination for serious readers.

The early work *Dubliners,* a collection of penetratingly observant short stories, was followed by the autobiographical novel *Portrait of the Artist as a Young Man,* which introduced experimental techniques that Joyce later explored fully in the radically innovative novel *Ulysses.* Finding the atmosphere of Ireland to be stiflingly parochial and repressive, Joyce resolved to live as an expatriate in order to maintain fidelity to his vocation as an artist. Composed in Paris, Trieste, and Zurich, where Joyce and his wife Nora had taken up successive residences after leaving Ireland, the novel *Ulysses* was completed in 1921. This book, Joyce's masterpiece, focuses on the lives of several residents of Dublin whose paths intersect over the course of a single day. In planning *Ulysses* Joyce took enormous pains to follow a rigorous outline that, beneath the surface of the novel, incorporates detailed parallels between the day's events and the major episodes of the *Odyssey,* a classical Greek epic work attributed to the poet Homer. Published in France, Joyce's novel was immediately recognized for its masterful use of the technique that became known as *stream of consciousness.* This literary tool, afterward widely imitated, became a mainstay in modernist literature, using a constant flow of subjective observations, reactions to events, reflections, and ideas as they pass through a character's mind, providing a uniquely intimate view of a character's mentality. Most famously, the final chapter of *Ulysses,* commonly referred to as Molly Bloom's soliloquy, is a lengthy, labyrinthine, richly associative interior monologue composed of only three sentences. As if to remind his readers that time presses on even at the close of a story, Joyce allows for the final lines of the novel to linger on a slight and momentary notion.

Owing to chronic poverty and the pressures of family, specifically the deteriorating mental health of his daughter, Lucia, Joyce found himself exhausted upon completing *Ulysses.* After a year's hiatus he began writing again in 1923. A work in progress that would eventually become *Finnegans Wake* occupied Joyce for more than 6 years and involved the use of narrative techniques and multilingual word play that left even the highly inventive idiom of *Ulysses* far behind. A complex amalgam of myths and folktales, archetypal themes, and timeless stories told in a unique style of language filled with puns, wordplay, and allusions to events that span world history, the novel posits the idea that history is not linear but cyclical and is bound to repeat. In this Joyce echoes the philosopher Friedrich Nietzsche's idea of eternal recurrence; in this way the *Wake* opens onto a vista that embraces all of human endeavor, encompassing past, present, and future. The final passage of the book ends in midsentence, continuing with the first phrase on the first page, thereby manifesting the very cyclical nature that Joyce came to see as a fundamental not only to the act of storytelling but also to the universe itself.

In the last decade of his life, internationally famous and with his literary reputation secure, Joyce became nearly blind, undergoing a series of eye operations that were only partially successful and were complicated by alcoholism. He died in Zurich in 1941 following surgery for a stomach ulcer.

Christian Austin

See also Homer; Mann, Thomas; Nietzsche, Friedrich; Novels, Time in; Proust, Marcel

Further Readings

Ellmann, R. (1982). *James Joyce: A biography* (Rev. ed.). New York: Oxford University Press.

Joyce, J. (1967). *A portrait of the artist as a young man.* New York: Viking Compass. (Original work published 1917)

Joyce, J. (1976). *Finnegans wake.* New York: Viking Penguin. (Original work published 1939)

Joyce, J. (2003). *Ulysses.* Ann Arbor, MI: Borders Classics. (Original work published 1922)

JUDAISM

Judaism is the religious, ethnic, and cultural heritage derived from the ancient people of Israel and the Hebrew scriptures. While the practice of Judaism is centered around the books of the Hebrew Bible, and among those primarily the five books of the Torah, its developed form, often referred to as a civilization, represents the accumulation of custom and tradition over centuries of thought and writing.

Intriguingly, many of the most central aspects of Judaism are connected to unique ways of measuring time and marking its passage. Judaism follows a nuanced calendar that prioritizes the cycles of the moon over the revolution of the earth around the sun. Within this calendar, the Jewish year is given distinct color and texture by numerous holidays and festivals that make up the liturgical schedule for worship. Through this continuous liturgical cycle that celebrates milestones of national and historical significance, Judaism captures both the cyclical and linear elements of time. Each year moves forward into history but also repeats the same journey as those before it—for example, the complete reading of the Torah, the celebration of special days like Pesach, Shavuot, or Sukkoth, and the blast of the *shofar* to greet the new year on Rosh Hashanah. Throughout its long history, Judaism has also contributed significantly to the discussion and philosophy of time. The ancient Hebrews solidified their religious and cultural identity in part by their beliefs concerning the beginning of time and the creator God who moved it forward through history. Centuries later, medieval Jewish thinkers grappled with influential philosophies of time originating outside their tradition and tested their compatibility with Jewish theology. In the European Age of Enlightenment, with the relaxation of strictures forbidding Jews from participating in civic culture, Jewish thinkers and philosophers engaged more freely in dialogue and debate with their Christian counterparts. Today, as full participants in secular society, observant Jews still preserve ancient customs revolving around the Jewish calendar and religious year.

Jewish Calendars and Chronology

Throughout most of its history, Judaism has used a calendar based primarily on a lunar year with necessary intercalations made to avoid excessive divergence from the solar year. This year is divided into twelve lunar months: Nisan, Iyyar, Sivan, Tammuz, Av, Elul, Tishri, Heshvan, Kislev, Tevet, Shevat, and Adar. In leap years, an extra intercalary month called Adar II is added. The names of these Jewish months closely resemble Babylonian month names, a probable result of the 6th century BCE exile of the Hebrew people in Babylon.

Each month begins at the new moon, or *molad* ("birth" in Hebrew). Since the average lunar month is approximately 29.5 days, the Jewish months alternate between 30 and 29 days in length. This results in a typical year of around 354 days and a leap year of around 384 days. The present system of leap years is based on a 19-year cycle in which the 3rd, 6th, 8th, 11th, 14th, 17th, and 19th years are intercalated with a 13th month. Originally, the commencement of new months and insertion of leap months were physically observed rather than calculated. New moons were observed and announced to Jewish communities by the religious leadership, and leap months were inserted when it was observed that the calendar and festival cycle were deviating too much from the agricultural season. In the 4th century CE, patriarch Hillel II formulated a new mathematical foundation for the calendar that introduced the 19-year cycle of leap years and diminished the need for physical observation. His calendar method was refined for the remainder of the 1st millennium and has remained essentially unchanged since that time.

For Judaism, the day begins and ends at dusk, rather than midnight. Thus, the Sabbath begins at sundown on Friday of the secular calendar and ends 24 hours later. For the reckoning of times for daily prayers and observances, 24 uneven hours (proportionate divisions of the daytime and the night) are used, while 24 even hours are used for most other purposes. Rather than minutes and seconds, Jewish custom divides the hour into 1080 *halakim,* which equal about 3.5 seconds each.

A potentially confusing aspect of the Jewish calendar is the starting point of a new year. Originally, Nisan corresponded to the "1st month," as was the case in the Babylonian ordering of the months. Likewise, it is between Adar and Nisan that the leap month Adar II is inserted. Despite this fact, the 1st day of Tishri is actually considered the start of the new year, called Rosh Hashanah, approximately corresponding with the autumnal equinox. Though Rosh Hashanah is the 1st day of the month of Tishri, it does not always occur on the day of the *molad.* In the interest of avoiding certain scenarios, such as Yom Kippur being a Friday or Sunday, Rosh Hashanah is sometimes delayed by one or two days. Because of this, the year always begins on a Monday, Tuesday, Thursday, or Saturday. An individual year is succinctly described by three facts that determine liturgical schedules: the day of the week of Rosh Hashanah, the number of days in the year (353, 354, or 355 for regular years; 383, 384, or 385 for leap years), and the day of the week on which Passover begins.

Standard Jewish chronology numbers the years from the supposed date of the Creation of the world, sometimes indicated by the abbreviation AM for *anno mundi,* "year of the world." Though various calculations have been made by Jews and Christians based on biblical genealogies and histories, Jewish dating accepts the conclusion attributed to 2nd century rabbi José ben Halafta that the Creation occurred in the year 3761 BCE. By this chronology, the year 5768 corresponds with 2007–2008 CE.

While the calendar system described above reflects the mainstream of Jewish tradition throughout its history, there have been dissidents. Most noteworthy is a group of solar calendars described in the 2nd century BCE book of *Jubilees* and in multiple versions of the book of *Enoch.* The main calendar of this type was distinctively solar, describing the year as 364 days or exactly 52 weeks. These calendars are generally regarded as sectarian within the Judaism of their time, and there is little evidence to determine the extent of their observance. Materials referring to such calendars have been discovered at Qumran, but references to other systems make it unclear exactly what calendar the Qumran sect observed. Clearly, the 364-day calendar would drift from the actual 365.5-day solar year at a steady pace, but no intercalation method is mentioned in literature connected to these calendars, a fact that causes some to doubt their sustained observance.

The Sabbath

Throughout history, a primary distinguishing characteristic of Judaism has been the observance of the Sabbath (or Shabbat), a day of rest observed on the seventh and final day of each week. The Sabbath has been central to Jewish identity since ancient Israel, and it figures prominently throughout the Hebrew Bible. Biblical passages connect the institution of the Sabbath with God's 7-day Creation of the world, the Israelite wanderings in the wilderness, and the Decalogue. Prophetic texts sternly rebuke neglect of the Sabbath in Israel, while later rabbinic writers further delineate what would be acceptable and unacceptable on the Sabbath. Jews observe the Sabbath (which actually runs from Friday evening until Saturday evening) as a time of relaxation, prayer, and worship. Food is prepared ahead of time, and most activities classified as "work" are prohibited. Most of the Sabbath is spent together by families at home, but synagogue services are typically held on Saturday morning.

While the concept of a 7-day week is taken for granted in modern culture, in the ancient world it was a cultural novelty of Judaism. Because it does not directly follow any natural phenomenon (as days, months, and years all do), the week is one of the more puzzling chronological developments in human civilization. While the Jewish week appears to have been a unique development, there are interesting parallels from the ancient Near East, such as Assyrian "unlucky days," which occurred on the 7th, 14th, 19th , 21st, and 28th days of the month (the 19th is understood to equal seven cycles of seven from the start of the previous month). The 7-day week in its modern form is a combination of the Sabbath-based week of Judaism and the Roman planetary week, which emerged around the turn of the common era. A further development was a Christian adoption of the 1st day of the week, Sunday, as its day of worship and rest after the day of Jesus' resurrection, contributing to the development of the 2-day weekend now widely observed.

Jewish Holidays

The Jewish year is given shape by a number of festivals and holidays that are central to Jewish religious life. These observances provide a liturgical cycle of worship, a continuous revolution that commemorates annual occurrences and special events from Jewish history. Because of the original uncertainty or delayed information concerning the fixing of new years and months, especially for diaspora Judaism, a second day was added to most festival days, a convention that is still maintained despite more precise methods of calculation.

Of special importance for Judaism are the 3-week-long pilgrimage festivals (Passover, Shavuot, and Sukkoth), which are each tied both to events from the biblical history of Israel and to agricultural seasons. Passover (or Pesach), beginning on Nisan 15, is connected with the Israelite exodus from Egypt and the beginning of the spring harvest. The most widely celebrated Jewish holiday, Passover, is commemorated with a special seder meal. Shavuot (the Feast of Weeks or Pentecost) follows Passover by 7 weeks and celebrates the giving of the Torah at Mount Sinai, as well as the end of the barley harvest and the beginning of the wheat harvest. On the 15th of Tishri, Sukkoth (the Feast of Booths or Tabernacles) recalls the Israelite wilderness wandering and celebrates the ingathering of grain from the fields into barns. Observers of Sukkoth construct a *sukkah,* a simple booth, in which they are to dwell during the 7 days of the festival.

In addition to the three central festivals, many other special days mark the Jewish year. As explained above, Rosh Hashanah ("head of the year") occurs at the beginning of Tishri, in autumn. The Jewish new year is ushered in with the blasts of a *shofar,* a ram's horn, and marked with a day of introspection and special synagogue worship. Shortly after the start of the new year on Tishri 10, Jews observe Yom Kippur, the Day of Atonement. This is the most somber of Jewish holidays, marked by fasting, abstinence from work, and lengthy synagogue services of confession and repentance. Other significant occasions include Shemini Atzeret ("assembly of the eighth day") and Simhat Torah ("joy of the Torah"). These immediately follow the 7 days of Sukkoth in Tishri, and Simhat Torah marks the completion of the annual cycle of readings through the Torah, the first five books of the Hebrew Bible. Purim is a celebration of Jewish survival from Persian oppression, revealed in the book of Esther, and is held on Adar 14 or Adar II 14 in leap years. Hanukkah, the Festival of Lights or of Dedication, is well known for its proximity to the Christian Christmas holiday and commemorates the miraculous burning of a menorah lamp for 8 days at the rededication of the Jerusalem temple in the 2nd century BCE. This holiday is celebrated for 8 days, beginning on Kislev 25 with the lighting of candles in a menorah.

Time in Jewish Thought and Literature

Though the ancient people of the Hebrews do not seem to have developed any extensive philosophies of time, among ancient civilizations they were pioneers of the concept of historical identity. In some sense, the entire development of the Hebrew Bible and subsequent Jewish tradition centered around the conviction that historical events that made up the progression of time had meaning and purpose. It was the activity of humanity (guided by the hand of God) that made up history or time for the ancient Hebrews rather than purely abstract measurements or theories. Modern scholarship does assert that the Hebrew scriptures are not "historical" in the modern sense; they represent a mixing of mythology, historical reality, and theological-historical interpretation. Still, it is precisely through these mediums that they communicated the narratives that formed the Hebrew identity and gave meaning to the progression of time. This begins with the very first passage of the Hebrew Bible, which asserts that time began with the Creation of "the heavens and the earth" by God. This affirmation became, for both Jewish and Christian thinkers, a frequent point of contrast with Greek philosophies, which often argued for the eternality of matter and a cyclical nature of history.

In addition to prescribing a beginning of time at Creation, the ancient Hebrews developed beliefs concerning the end or consummation of time, generally classified as "eschatology." The seeds of Jewish eschatology are present even in the Torah, as the concept of Israel as God's chosen people becomes fully developed, but it is the prophetic writings that contain the bulk of Jewish eschatological language. As the course of Israelite history grew darker and the prophets struggled with what they considered unfaithfulness to God among the Israelites, the focus turned increasingly toward concepts of future divine judgment. Terms such as "that day" and "the day of the Lord" became common indicators of a coming time when God would intervene on a universal scale, judging all peoples and setting right the injustices of the world. Within this realm of thought, the concepts of a messiah and of life after death developed and grew, particularly in late prophetic and intertestamental literature. Increasingly, the hope of resolution of injustice in the world was pushed to a future time, when all people would be resurrected and face judgment.

By the medieval period, many Jewish philosophers interacted with classical Greek traditions about the subject of time. Moses Maimonides accepted a primarily Aristotelian concept of time, in which time was an "accident" of the motion of physical bodies. Hasdai Crescas, on the other hand, rejected the Aristotelian conception and connected time to the consciousness of the mind, not motion, a concept more akin to that of Plotinus. For these and other philosophers, numerous questions of God's relationship to time became pressing: Does God exist within time or outside of it? Was there time before Creation? If so, what was God doing during that time? Can "eternity" be understood in the normal sense of time? Such questions were discussed among Jewish thinkers as well as theologians of the Christian and Islamic traditions with diverse and varying answers. For both monotheistic faiths, the philosophical questions of time, eternity, and cosmogony have been a constant component of theological discussion.

Judaism in the modern world is very diverse and difficult to characterize as a whole. For many, Judaism is more a cultural than religious identifier. Despite this, observing time by the traditional calendar and customs of Judaism remains a central aspect of faith for many. Keeping the sabbath is fundamental to being a Torah observer, and the Jewish calendar system is still the framework of worship. The richness of Jewish customs regarding time and their persistence throughout history, even in the face of modern secular assimilation, testifies to the remarkable character of Judaism as a religious, cultural, and ethnic entity that has preserved a genuinely unique identity.

Adam L. Bean

See also Adam, Creation of; Bible and Time; Creation, Myths of; Genesis, Book of; Moses; Noah

Further Readings

Beckwith, R. T. (1996). *Calendar and chronology, Jewish and Christian: Biblical, intertestamental and Patristic studies*. Leiden, The Netherlands: Brill.

Beckwith, R. T. (2005). *Calendar, chronology and worship: Studies in ancient Judaism and early Christianity*. Leiden, The Netherlands: Brill.

Brin, G. (2001). *The concept of time in the Bible and the Dead Sea Scrolls*. Leiden, The Netherlands: Brill.

Richards, E. G. (1999). *Mapping time: The calendar and its history*. Oxford, UK: Oxford University Press.

Skolnik, F., & Berenbaum, M. (Eds.). (2006). *Encyclopedia Judaica*. New York: Macmillan Reference.

Steel, D. (2000). *Marking time: The epic quest to invent the perfect calendar*. New York: Wiley.

Stern, S. (2001). *Calendar and community: A history of the Jewish calendar, second century BCE—tenth century CE*. Oxford, UK: Oxford University Press.

Stern, S. (2003). The rabbinic concept of time from late antiquity to the Middle Ages. In G. Jaritz & G. Moreno-Riano (Eds.), *Time and eternity: The medieval discourse* (pp. 129–146). Turnhout, Belgium: Brepols.

Talmon, S. (2005). What's in a calendar? Calendar conformity, calendar controversy, and calendar reform in ancient and medieval Judaism. In R. L. Troxel, K. G. Friebel, & D. R. Magary (Eds.), *Seeking out the wisdom of the ancients* (pp. 451–460). Winona Lake, IN: Eisenbrauns.

Wacholder, B. Z. (1976). *Essays on Jewish chronology and chronography*. New York: KTAV.

Zerubavel, E. (1985). *The seven day circle: The history and meaning of the week*. New York: The Free Press.

Kabbalah

Kabbalah, derived from the Hebrew root meaning "to receive," is the name given to a Jewish mystical tradition that originated around 100 BCE. Owing to variations in transliteration from the Hebrew, it is sometimes rendered as Kabalah, Qabalah, or Qabbalah. The name Kabbalah is attributed to the scholar Isaac the Blind (c. 1160–1236), sometimes called the Father of Kabbalah, although the practice of Kabbalah predates him. As a form of Jewish mysticism, it is primarily concerned with directly experiencing God through meditation, spiritual exercises, or interpretations of scripture (particularly the Torah) and other writings. Kabbalah has both a traditional and a distinct beginning.

Traditionally Kabbalah is believed to date from the relationship between Adam and God. According to the first three chapters of Genesis, Adam enjoyed a unique experience with God: Adam conversed directly with God without any mediation from another person or text. He was able to know God face-to-face. This is the kernel of mysticism in general and Kabbalah specifically: to know God directly.

Viewed this way, many persons from the *Tanakh* (Hebrew Bible; in Christianity, the Old Testament) are viewed as having kabbalistic experiences. According to Jewish teaching, the events in the book of Genesis occurred prior to God giving the Law of Moses. Within Genesis, people speak with God directly. While Genesis makes no mention of the type of relationship between Melchizedek and God, Melchizedek (Gen 14: 18–20) is mentioned as a priest of God prior to the establishment of the Torah (Law). Therefore (the reasoning goes), because there was no mediation, Melchizedek must have had a direct experience with God. Abraham spoke directly with God (Gen 12–22) as did Moses (Exodus through Deuteronomy).

In other books of the Tanakh, prophets are mentioned as having direct experiences with God. This is important for understanding Kabbalah, because the prophets had experiences with God after the Law of Moses was given. To a Kabbalist, therefore, the Law of Moses does not stand between a follower of God and God, such as in the role of a mediator, but instead the Law of Moses becomes a gateway into communication with God.

Kabbalah has a distinct beginning that can be traced to a wealth of Jewish mystical literature from the period 100 BCE to 1000 CE. Over time, the literature and understanding of Kabbalah increased. Yet the central Kabbalah text is the *Sefer ha-Zohar* (Book of Splendor), which is usually given simply as Zohar. The Zohar contains commentary and stories. Its preeminence among Kabbalah is borne out by the number of commentaries that have been written about the Zohar. This book is attributed to Simeon bar Yohai (2nd century CE), although the first record of the book is from the 13th century in Spain as the work of Moses ben Shem Tov de Leon (1240–1305). He either edited the writings of Simeon bar Yohai or based his work on Simeon; in either case the work originated with Simeon and was completed by Moses de Leon.

Soon after the Zohar, *Ma-arekhet ha-Elohut* (The Order of God) appeared, which attempted to systematically explain Kabbalah doctrine. In addition, *Sefer ha-Bahir* (Book of Light) was another early Kabbalah text that explained the *sefirot* (the emanations of God).

A central tenet of Kabbalah is the tree of life. This symbol contains a diagram (sometimes represented as a tree) with ten sefirot (emanations): sovereignty, foundation, endurance, majesty, beauty, loving kindness, judgment, wisdom, understanding, and the crown. The sefirot exhibit different aspects of God. They are to be viewed as a unit and individually. As a unit, these represent the way that God communicates or interacts with the world. Individually, these are steps along which the kabbalist works or progresses in an attempt to draw closer to God.

Prominent teachers of Kabbalah include Moses Cordovero (1552–1570), who wrote *Tamar Devorah* (Palm Tree of Deborah), which explains Kabbalah, and *Or Yaqar* (Precious Light), which is a commentary on the Zohar. Isaiah ben Abraham ha-Levi Horowitz (1565–1630) wrote *Shenei Luhot ha-Berit* (Two Tablets of the Covenant). Isaac the Blind (c. 1160–1235) is believed to have written the *Sefer ha-Bahir* (Book of the Brightness). Over time, texts and commentaries on Kabbalah have been written and collected, and interest in the texts has increased both within and outside Judaism.

Similar to other mystical traditions, Kabbalah considers time and space as mundane realities. Within the spiritual realm, therefore, concepts such as time, space, and place are nonexistent. Understanding God entails understanding his actions apart from time; God's qualities exist beyond time and cannot be limited or perceived by temporal constraints. This in itself tends to restrict interest in Kabbalah to truly serious adherents, because practitioners must remove themselves from the physicalness of their surroundings in their efforts to fully understand God.

Among Jewish religious groups, the Kabbalah greatly influenced the ultraorthodox Hasidim. This influence can be seen in the Hasidic theology of working toward a union with God and in their concept of *devekut* (from the Hebrew for "cleaving," the cleansing or preparation of prayer and mediation toward God).

In understanding the intent toward the Torah of Kabbalah, the teaching of Bahya ben Asher is instructive. He held to four different interpretations of scripture: the literal, the homiletic (preaching), the moral, and the mystical. The Kabbalah is not solely interested in the mystical meaning of scripture, yet it is concerned with utilizing the former three while focusing on the mystical. For the kabbalist, the Torah does not solely teach people about God and how to live before God; instead it starts with those and brings the person into communication with God.

Kabbalah texts can be difficult to understand, many being written in esoteric language. Additionally, some kabbalists in the past believed that no one under the age of 40 should practice Kabbalah. Kabbalah was understood to be reserved for those willing to devote themselves to a mystical relationship with God.

Recently, more accessible forms of Kabbalah have become popularized and are currently practiced by people from various backgrounds, both Jewish and gentile. Kabbalah centers can be found in major cities worldwide, and a number of Web sites offer instruction and guidance in teachings that are based on elements of the kabbalistic traditions.

Mark Nickens

See also Bible and Time; Genesis, Book of; Judaism; Moses; Mysticism

Further Readings

Carmody, D. L., & Carmody, J. T. (1996). *Mysticism: Holiness east and west.* New York: Oxford University Press.

Epstein, P. (1978). *Kabbalah: The way of the Jewish mystic.* Boston: Shambhala.

Scholem, G. (Ed.). (1995). *Zohar: The book of splendor: Basic readings from the Kabbalah.* New York: Schocken Books.

Kafka, Franz (1883–1924)

Franz Kafka, among the most highly acclaimed writers of the 20th century, was a Jewish

Czechoslovakian who wrote allegorical stories in the German language. As literary critics have pointed out, Kafka's meticulously constructed tales seem to defy categorization, containing elements of the comedic and satirical as well as the tragic. He has been described as a realist, an absurdist, a Marxist, a sociologist, an existentialist, and a comedic theologian. Now widely considered the paradigmatic artist of his time, Kafka developed a personal idiom of expression that imposed a unique literary order on the disorder of the emerging modern world and gave universal form to his own personal nightmares and obsessions. Many later writers have acknowledged Kafka's influence, including Jorge Luis Borges, Milan Kundera, Gabriel García Márquez, Carlos Fuentes, and Salman Rushdie.

A striking aspect of Kafka's work is in its uncanny foreshadowing of the nightmare of bureaucratic control and mechanized murder that, more than a decade after Kafka's premature death from tuberculosis, would take shape as the Holocaust. In his novels *The Trial* and *The Castle,* both published posthumously, the central characters undergo persecution by petty officials: Joseph K. is anonymously accused of a crime whose nature remains unspecified; his attempts to clear himself only draw him deeper and deeper into a labyrinthine legal system that ultimately issues a death sentence carried out in an impersonal and humiliating manner. The protagonist of *The Castle,* who is identified only by the initial "K.," is repeatedly thwarted by a succession of bureaucrats and office holders in his efforts to gain access to the forbidding, enigmatic structure and ultimate source of power that dominates the town where the story is set.

A sense of oppression and anxiety characterizes much of Kafka's literary output. In the novella *The Metamorphosis,* which was published during Kafka's lifetime and received considerable acclaim, the main character, Gregor Samsa, a traveling salesman and the primary breadwinner for his mother, father, and sister, awakens one morning to find himself transformed into a giant insect. His efforts to confront the reality of this existential nightmare are described in such a way as to convey the dark humor as well as the horror of his situation.

Other acclaimed stories include "The Judgment," "In the Penal Colony," "A Country Doctor," and "A Hunger Artist," as well as the unfinished novel *Amerika,* the first chapter of which is the acclaimed story "The Stoker."

Kafka was born in the cosmopolitan city of Prague, Bohemia, which is now part of the Czech Republic. His father Hermann, a businessman, was physically imposing, overbearing, unsympathetic, and intimidating, and their relationship, always difficult, grew more strained as the years passed. The primordial theme of father-son conflict would emerge in Kafka's writing again and again in the form of confrontations with remote and sometimes hostile authority figures. Kafka refused to work in the family business, intent instead on a university education. After several years of changing majors at Ferdinand-Karls University in Prague, he went on to earn a doctoral degree in law.

As a student Kafka made acquaintances with other intellectuals and aspiring artists. He befriended Max Brod, a lifelong friend and confidant, to whom he would later entrust his literary estate. Throughout Kafka's life, he would continually battle bouts of physical, emotional, and mental exhaustion and would seek rest and treatment at sanatoriums. His first sanatorium stay occurred before he finished his law degree in 1905. In 1906, just after graduation, the pressures of family, professional, and scholarly pursuits overcame him, and again he sought the refuge of the sanatorium.

Two years after graduation, he took employment as an attorney with the Worker's Accident Insurance Institute for the Kingdom of Bohemia, where he worked diligently until shortly before his death. While in this employment, Kafka wrote several articles on the prevention of industrial accidents. In 1909, he published his first stories, excerpts from an abandoned novel, *Description of a Struggle,* in the journal *Hyperion.*

Kafka's complex inner life and his ambivalence concerning intimate relationships caused him to break off several affairs and marital engagements. As a young man, he would confess to his friend Brod that he was unsuitable for marriage and that he could never love a woman if she returned that affection.

Kafka obtained an early retirement from his long-term employer, and in 1922 he wrote "Investigations of a Dog'" and *The Castle.* By this time his health was seriously affected by the chronic tuberculosis that would eventually take his life. He lived transiently between sanatoriums and his sisters' homes. In 1922, he befriended a woman named Dora Dymant, with whom he lived briefly and to whom he eventually proposed marriage. Early in 1924, however, increasingly ill and near death, he traveled back to Prague, where he lived briefly with his parents and wrote "Josephine the Singer." He died on June 3 in a sanatorium near Vienna.

Although Kafka had directed Brod, his literary estate executor and lifelong friend, to destroy all his unfinished manuscripts, Brod salvaged the unfinished manuscripts instead. In the1920s three novels were posthumously published, and in 1931 a collection of incomplete tales was printed, together establishing Kafka as a 20th century literary master who, against the background of a decaying empire, had prophetically glimpsed the outlines of the world that was soon to come and had responded with a sense of dread and foreboding.

Debra Lucas

See also Dreams; Existentialism; Novels, Time in

Further Readings

Brod, M. (1947). *Franz Kafka* (G. H. Roberts & R. Winston, Trans.). New York: Schocken.

Kafka, F. (1948). *The penal colony: Stories and short pieces* (W. Muir & E. Muir, Trans.). New York: Schocken.

Kafka, F. (1995). *The trial* (W. Muir & E. Muir, Trans.). New York: Schocken. (Original work published 1925)

Sokel, W. (2002). *The myth of power and the self: Essays on Franz Kafka*. Detroit, MI: Wayne State University Press.

Kant, Immanuel (1724–1804)

Immanuel Kant, like Plato and Aristotle, counts as one of the most influential philosophers of all time. He developed a new theory of time and space that combined insights from rationalism and empiricism. In his practical philosophy he developed a completely new approach, which makes Kant one of the most distinguished scholars of the European Enlightenment.

Kant was born in Königsberg (formerly East Prussia, since 1945 renamed Kaliningrad, Russia) on April 22, 1724, as the son of a saddler. He was brought up in the spirit of Pietism, a Lutheran movement that focused on love, which is realized through duty and good deeds. He never left Königsberg. It was there where he went to school and attended university, where he received permission to teach as a university lecturer (*Privatdozent*) in 1755, and again where he was appointed full professor of logic and metaphysics in 1770 after declining an offer from Jena in 1769. He held this position for the rest of his life and died on February 12, 1804.

The tone of the German intellectual landscape in which Kant started his work was largely set by the so-called "school philosophy" of Leibniz and Wolff, a thoroughgoing form of rationalism. Wolff, continuing Leibniz's work, believed that every truth of reason could be deduced from the principles of noncontradiction and sufficient reason. When Kant encountered Hume's empiricist philosophy, he felt awakened from his dogmatic slumbers. In 1770 he consolidated many of the intellectual gains he had made during the 1750s and 1760s in his Latin inaugural dissertation (*Habilitationsschrift*). He introduced an important new theory about the epistemology and metaphysics of time (and space). His fundamental new philosophical insight, which he later elaborated in his famous work *Critique of Pure Reason*, goes as follows:

> 1. The idea of time does not arise from the senses but is presupposed by the senses. . . . 5. Time is not something objective and real (*tempus non est obiectivum aliquid et reale*); it is neither an accident, nor a substance, nor a relation; it is the subjective condition, necessary because of the nature of human mind, of coordinating everything that we experience (*quaelibet sensibilia*) by a certain law, and is a pure intuition. For we coordinate substances and accidents alike, as well according to simultaneity as to succession, only through the concept of time.

The same holds for space, which he maintains is also neither objective nor real. Kant combines

Hume's empiricist point of view with the rationalistic assumption of an ordered universe. Hume believes that with sense-knowledge, the given consists of impressions and sensations, and, for this reason, we can only discover post hoc a succession in time without causal relation. Rationalists assume that there is a fundamental causal connection between events (a post hoc principle of sufficient reason). Kant explains that time and space, as our subjective forms of intuition, are the reason why we have the experience of causal connections without holding that these causal connections are real in an absolute sense. But they are real to us, because all human beings have the same experience, which results from the notion that all human beings have the same subjective forms of intuition; that is, of time and space. Kant's position on time and space is of great philosophical interest. For example, Einstein's theory of relativity shows that it is not possible for us to have absolute time or space, so it seems to be true that our experiences are just of appearances and do not tell the whole story.

In his *Critique of Pure Reason* (1781/1787) Kant developed these important insights to encompass all the categories used in thought. However, Kant admits that this work does not describe the whole system but instead is the blueprint for a new transcendental philosophy, thereby destroying the metaphysics of the past. Kant's central idea is to argue that the unity of consciousness, the so-called transcendental unity of apperception, categorizes all of the sensations according to the forms of intuition (time and space) and the forms of intellect; that is, the pure concepts (categories). This explains why there are universal and necessary laws for every being with consciousness and intellect, in other words, for every human being. Kant calls this knowledge synthetic a priori, because it is necessary and universal, on the one hand, but is not inferable through the analysis of concepts, on the other. From that follows that "things" that are not explicable in space and time, for example, God or an immortal soul (if they exist), cannot be assessed by human beings without leading to an endless series of irresolvable disputes. The previous system of metaphysics is now null and void. As a result, we cannot derive ethical conclusions from theological (or metaphysical) "knowledge" of the good. We are left without a solid foundation for a consequentialist account of ethical reasoning.

Kant proposes an alternative theory of practical reasoning that can be found in *Fundamental Principles of Metaphysics* (1785), *Critique of Practical Reason* (1788), *The Metaphysics of Morals* (1797), and numerous sections of other works. He assumes that only a will that acts for the sake of duty and not for the sake of inclination is a morally good will. The will is good when, for the sake of duty, it follows the fundamental commandment he calls the "categorical imperative," a principle of action that is a universal moral law. All rational beings endowed with freedom, regardless of their particular interests and social background, have to adopt this principle. Kant formulates it in the *Groundwork for the Metaphysics of Morals* in four different ways. The most famous formulation is as follows: "Act only according to the maxim (the determining motive of the will) through which you can at the same time will that it become a universal law." By *universal law* Kant means that this principle does not allow for exceptions, in contrast to the laws of nature. Another formulation of the categorical imperative demands the impartial respect of humanity in every human being: "Treat humanity . . . never simply as a means, but always at the same time as an end." Here Kant introduces the concept of human dignity, which remains among the most discussed of all ethical concepts.

In his third critique, the *Critique of Judgement,* Kant tries to unite theoretical and practical reason. Nature can be understood as a determinate system only under the regulative idea of being a system serving an ultimate end. This idea is regulative because it helps us to understand nature according to our capacities of reason. Consequently, we do not know whether nature serves an ultimate end at all; instead we are just left to postulate it.

Many aspects of Kant's theoretical and practical philosophy have been questioned, in particular his assumption of synthetic a priori propositions and his ethics of the categorical imperative. His theory of time and space, however, will likely remain a cornerstone of philosophical thought for generations to come.

Nikolaus J. Knoepffler

See also Ethics; Hume, David; Intuition; Metaphysics; Morality; Space; Space and Time; Time, Perspectives of

Further Readings

Guyer, P. (Ed.). (2006). *The Cambridge companion to Kant and modern philosophy.* Cambridge, UK: Cambridge University Press.

Kant, I. (1992ff). *The Cambridge edition of the works of Immanuel Kant* (P. Guyer & A. W. Wood, Eds.). Cambridge, UK: Cambridge University Press.

Kühn, M. (2001). *Kant: A biography.* Cambridge, UK: Cambridge University Press.

Kierkegaard, Søren Aabye (1813–1855)

Søren Aabye Kierkegaard was a Danish literary author, philosopher, and theologian whose work has had an abiding influence on philosophy, especially existentialism and postmodernism in the 20th century. His wide-ranging thought focuses largely on points where theological explication of the Christian faith and philosophical investigation of concrete human existence intersect and inform one another. Throughout his authorship, Kierkegaard proposed different understandings of existential (and theological) human temporality—how the human individual dwells and relates to himself or herself and the eternal order in the midst of time.

For Kierkegaard, time itself is a fluid and infinite succession of discrete points that are passing by. This succession is one of ceaseless quantity that spreads out extensively. Temporality, taken as signifying the temporal domain, is characterized in Kierkegaard's work as a realm of plurality (difference) and becoming (movement). This becoming is not a necessary one. All finite, temporal actuality for Kierkegaard is contingent; it could have been another way if another possibility had been actualized. (It is for this reason that human understanding or knowledge of events in time, in history, involves an irreducible uncertainty due to the contingent status of its objects.) In the midst of this fleeting impermanence, any given moment in time—from the perspective of time alone—is an infinite vanishing, a nothing. Considering time as such, there is no present and thus no past or future. There is only the infinite succession. The differentiation and distinction into present, past, and future only arises in the relation of time to another order—to the eternal.

The temporal, for Kierkegaard, is understood in contradistinction and dialectical relation to the eternal. Generally, the eternal is presented as an eternal present—a present in full in which the relative succession of time is annulled. (For Kierkegaard, in a sense, the present is eternal before it is temporal.) "The eternal," while it is used in various ways throughout Kierkegaard's authorship, refers more specifically to the transcendent God and that which has to do with one's relation to him (such as one's moral ideals and one's future life beyond death). Kierkegaard presents God as the source of ethical-religious stability in human existence that, in himself, transcends the temporal order (see below). As opposed to the ancient Greek intellectual conception (in Kierkegaard's estimation) of eternity in the form of abstract ideas, Kierkegaard affirms the Judeo-Christian "eternal" of a personal creator God who transcends history and yet actively relates to it—in its origin (creation), its progression (guiding and interacting), and its consummation. It is the divine eternal that makes time into something like unified narrative.

Following the Judeo-Christian understanding, Kierkegaard understands the eternal God as creating the temporal-finite order out of nothing. This created temporal realm is that of "existence"—of what has come to be in contingent finite-historical becoming, for the temporal realm as a whole, as freely created by God, could have been otherwise. In Kierkegaard's writings, existence is primarily discussed as the realm of ethical coming-to-be, of human existence and self-becoming.

The existing human being, for Kierkegaard, is a synthesis of the temporal and the eternal. The human at once dwells in time and relates to the eternal in *the moment*—a point in the succession of time at which an individual comes into contact with something of an order other than the finite-temporal. The moment is where the "present" of the eternal relates to time (where there is no present as such) in human consciousness. In the moment, time is meaningful (is more than mere succession) inasmuch as it relates to a transcendent order. It is only now that temporality—in terms of present, past, or future—is truly posited.

In Kierkegaard's anthropology, the existing human self is an identity that is actively produced in and through time. Human living and temporality are understood in terms of *repetition*—the project of trying to make of one's life something of the eternal, the stable, the abiding. For Kierkegaard, "aesthetic" repetition, the attempt to make pleasurable experiences permanent, fails because of the discontinuous and contingent nature of time. In ethical-religious repetition, however, one attempts to forge an identity through the discontinuity and becoming of time. In moral commitment, one seeks to establish a continuity over time—a continuity of temporal moments relating to the eternal—in which one becomes a self by relating to the eternal, ultimately to God. Here the unity of the self comes to be in the midst of the instability of time relative to something stable transcending time. The willing of one thing (God, the good, the eternal) in the midst of the plurality of one's temporal life makes one into one—into one thing—into a self.

The problem in establishing this self as a continuity over time is that of sin, of moral failure. When one considers the moral task of repetition along with the possibility of freedom, one encounters a psychological state of *anxiety* regarding the possibility of sin, of moral failure. Anxiety, for Kierkegaard, specifically relates to dwelling in time. For anxiety has to do with the human spirit's awareness of itself as free, as being able to actualize various non-necessary possibilities in the future—namely with the possibility of moral failure. Anxiety has to do with the tension in the synthesis of the self between our dwelling in time as finite beings and our "spiritual" relating of our finite lives to the eternal. Our freedom to act (in relation to the eternal, rightly or wrongly) and our responsibility for our actions incites anxiety over the very real possibility of failure—a kind of dizziness of freedom. In such a failure, one fails to relate to God and to relate to one's self as a dependent being-in-relation to God—not wanting to be what one is, failing to be a self.

Kierkegaard, as a kind of Christian theologian, affirms the *incarnation* of the eternal God in the historical person of Jesus Christ. In Jesus Christ, the incarnate God-man, there inheres a paradoxical unity of the eternal and the temporal, the divine and the human. The eternal entered into time in order both to save humanity by reconciling individuals to the eternal as a gift and to provide a prototype of the eternal moral character of God in a human life. The human individual comes into contact with this event of the eternal in time *in the moment* of Christian salvation. As at once a supreme if not archetypical instance of the ethical moment (see above) and an analogue of the incarnation, the moment of Christian salvation is a moment in time that has eternal significance. Kierkegaard observes that in this intensive moment of relating to Christ in salvation inheres the paradox that something eternal (one's eternal salvation) can be decided in time. Kierkegaard writes that it is quite understandable that this is something that does not "make sense" in the order of temporality alone—that a moment can be more than merely one moment among moments. In the moment of the possibility of one's salvation, one meets the moment of the incarnation with faith or offense. In *faith,* one believes and accepts that in Christ the eternal to which one has failed to attain a proper and consistent relationship in one's temporal life has come into time and given itself to one as a gift. In this moment of receiving the gift of salvation, there is a *contemporaneity* with Christ—for, in the moment of faith, the moment of the incarnate Christ's being in history is "present" to one. Through the gift of Christ and the example of his life, one can attain an identity that is not a mere aggregate of equivocal points of time in succession—one can become a self in relation to God.

Christopher Ben Simpson

See also Becoming and Being; Christianity; Eternity; Existentialism; God and Time

Further Readings

Evans, C. S. (1983). *Kierkegaard's* Fragments *and* Postscript: *The religious philosophy of Johannes Climacus.* Atlantic Heights, NJ: Humanities Press.

Kierkegaard, S. A. (1980). *The concept of anxiety* (R. Thomte, Trans.). Princeton, NJ: Princeton University Press. (Original work published 1844)

Kierkegaard, S. A. (1985). *Philosophical fragments* and *Johannes Climacus* (H. V. Hong & E. H. Hong, Trans.). Princeton, NJ: Princeton University Press.

Knezević, Bozidar (1852–1905)

As a science-oriented philosopher and historian living in Yugoslavia during the emergence of Darwinism, Bozidar Knezević was greatly influenced by the theory of evolution. He extended Darwin's process view of life on earth to include the future of organic evolution on this planet within a cosmic perspective. Knezević's philosophy of time and evolution was presented in his two-volume work *The Principles of History* (1898–1901). His incredible vision was a mixture of facts, concepts, and rational speculations that embraced both the ascent and subsequent descent of life forms on earth (including our own species) over a vast period of time.

Knezević acknowledged the awesome immensity of this dynamic universe in terms of both time and space. He claimed that humankind is merely a fleeting aspect of cosmic reality. Focusing on our planet, he saw three major stages of material development: inorganic, organic, and psychic levels of evolution. Within this sweeping perspective, our species recently emerged on the earth. Like all other forms of life, Knezević maintained that humankind is subject to the same finite history of evolution and then devolution that will eventually engulf all species on this planet. He referred to this history as the semicircular model of cosmic reality in general and organic evolution on earth in particular. This geometrical scheme is an engaging approach to understanding and appreciating the vulnerability of life forms throughout time.

According to Knezević, the history of our planet has been a material evolution of organic forms from simplicity to ever-increasing complexity, from the first appearance of the earliest life forms and then the fishes, through amphibians and reptiles, to mammals and the recent emergence of humankind. In the future, he speculated that organic evolution will result in the disappearance of life forms: Humans will vanish first, this will be followed by the extinction of the other mammals, then reptiles and amphibians will vanish, and finally the fishes and simplest organisms will become extinct. As the last species to emerge in evolution, humankind will be the first form of life to disappear; our species is linked to the fatal destiny of the earth and this solar system. Likewise, the first life forms to appear on earth will be the last forms of life to vanish.

For present environmentalists, this semicircular model of organic evolution may be relevant to a meaningful degree. If our planet changes significantly, then it could be the fact that the more complex life forms will become extinct first, while the fishes and invertebrates will survive much longer, far beyond the duration of mammals, reptiles, and amphibians. Eventually, perhaps even bacteria and viruses will vanish from the earth.

For Knezević, time is both creative and destructive. Through evolution, time created the enormous diversity of organisms on this planet, as well as life forms and intelligent beings on other worlds. Through devolution, time will destroy all life among the galaxies. Knezević also envisioned all other planets, galaxies, and entire universes passing through such a semicircular history. Collectively, all these semicircular histories represent God as an infinite series of creations and destructions within the endless flux of cosmic reality. Only matter itself endures throughout time, while time creates and destroys all objects and events in a process that is both unending and beyond human comprehension.

H. James Birx

See also Cosmology, Cyclic; Darwin, Charles; Evolution, Organic; Extinction; Extinction and Evolution; Materialism; Spencer, Herbert; Time, Cyclical

Further Readings

Birx, H. J. (1982). Knezević and Teilhard de Chardin: Two visions of cosmic evolution. *Serbian Studies 1*(4), 53–63.

Knezević, B. (1980). *History: The anatomy of time (the final phase of sunlight)*. New York: Philosophical Library.

Kronos

See Cronus (Kronos)

Kropotkin, Peter (1842–1921)

Peter (Pytor) Alekeyevich Kropotkin, born in Moscow into the Russian aristocracy, was a geographer, revolutionary anarchist, libertarian communist, zoologist, anthropologist, economist, philosopher, and sociologist. His writings were published in several languages and widely discussed, exerting considerable influence, especially among political thinkers and activists of the period preceding the Russian Revolution of 1918.

In the course of his geographic work Kropotkin demonstrated that eastern Siberia was affected by post-Pliocene continental glaciations. As a result of his animal life studies, Kropotkin theorized that mutual aid was the key to understanding human evolution, contrary to the notion of the natural world as shaped entirely by ruthless competition. He suggested that science and morality must be united in the "revolutionary project." Education should be global and humanistic and should empower everyone equally. Children should learn not only in the classroom, but also in nature and in living communities. To him, education and science should be based upon mutual aid and serve as a revolutionary activism with which to transform the entire world. This was not only morally correct, but the only life worth living.

Evolution is influenced and shaped by adaptation to a changing environment and adverse circumstances. Among many animal societies, competition between individuals, though important, is secondary to intraspecies cooperation in the survival of the species. Adaptation to the environment and the struggle against adverse conditions lead to an evolutionary theme, resulting in individuals working in partnership to protect their offspring.

In the animal world, most animals live in societies. Kropotkin felt that the survival strategy of safety was a concept that needed closer examination. This was not just a struggle for existence but a protection from all natural conditions that any species may face. Each individual increases its chances by being a member of a group. Mutual protection allows certain individuals to attain old age and experience. Among humans, these collective groups allowed for the evolution of culture. In the earliest band societies, social institutions were highly developed. In later evolution of clans and tribes, these institutions were expanded to include larger groups. Chiefdoms and state societies shared mutual identities in groups so large that an individual did not know all members. The idea of common defense of a territory and the shared character of nationalism appeared in the growth of the group sharing a collective distinctiveness.

Solidarity gives the species as a whole a better opportunity to survive. Thus, supportive attention for the well-being of relationships of the group is selected for. The manipulative and sly individuals are cleansed from the pool of ancestors, and those most supportive of mutual social life are selected to survive. Along with this, safety in numbers allows for increased chances of survival. Cooperation increases chances of an intelligent response to a threat to the group or the offspring by a coordinated effort. Be it hunters in a pack or herbivores cooperating, these efforts develop social skills and an awareness of comrades. The needs of the progeny and for continued existence bring together groups for reproduction and protection of the young. Because of this, security of all individuals is enhanced through mutual aid.

Ethics are thus a part of natural evolution. The strongest section of social companionship, which is the attraction of all major religions, is learned through the observation of nature. As animal behavior becomes more complex, association becomes less instinctual and more learned. Conscious awareness of the needs of others becomes increasingly important to the survival of the group and all its members. Communication becomes central to sociability. Vocal communication, exchange of ideas, and replication all help to teach collected knowledge to other individuals. Being companionable is an adaptation to the environment, and struggle and friendliness are the main foundations of evolution.

Through cooperation, less energy is expended for gathering food or fighting off danger. Cooperation, observed Kropotkin, favors the evolution of intelligence. Moreover, social feelings are central to development of societies, justice is part of the natural world, and ethics have a biological origin. The struggle for survival most often has a collective origin.

Carrying capacity determines population density, not by the most favorable circumstances but by the most server-limiting factors. Competition between individuals adds little to the survival of the species. Species become extinct not because they kill each other off but because they fail to adapt to a changing environment. Intraspecies competition is secondary.

According to Kropotkin, peace and reciprocal support are the rule within the group. Because societies or bands were the first social organizations in human evolution, they are the origin of consciousness, the source of social conscience, the innovators of ethics, the originators of the best of religion, gods, and our humanity. The sharing of resources necessary for life precedes all other systems of distribution. Reciprocity is the foundation of our common identity; the community precedes the individual. The life of the community becomes the validation for the life of the individual. Self-sacrifice in all major truth-seeking or sacred systems is the source of the greatest joy for the individual.

Sympathy and self-sacrifice are positively vital to human shared advantage for improvement. Sense of suitability versus hard-heartedness is the starting point of human unity. Mutual aid everywhere can be seen as the dawn of the human spirit and the soul of our humanity. The quest for power is also part of the human condition. With intelligence, conscience needs to be learned. The will to power and the quest for wealth by individuals everywhere undermines community and solidarity. Democratic communism is replaced by a stratified society based upon coercion, exploitation, and oppression. Popular democratic uprisings are as old as class society.

Social instincts are based on the pleasure of companionship; the collaboration of others is the primary basis of ethics. Compassion is the starting point of public service. The public good is expressed in the dialogue of mutual aid. The guide to action is always moral self-control and reciprocal support.

Kropotkin begins by asking the origin and meaning of social ethics. To him, exploitation is caused by unacceptable wealth based upon the poverty of others and is intolerable. Poverty is the direct result of wealth. It is the poor who support the rich through their hard work. This planet has the capacity to feed, clothe, and house every human comfortably. But, the problem must be attacked. It is important to know what is possible and to understand the right thing to do. Where science and ethics come together is the realm of sociology (anthropology).

Freedom and necessity join, and agency becomes action founded upon information. Liberty, based upon knowledge of the essential principles of nature, is the source of sovereignty. Needs and determinism are the sources of free will; people have true choices only if they have the proper knowledge of how to act to lessen unexpected consequences.

In Kropotkin's view, ethics ultimately does lead to fulfillment, for satisfaction comes from the abandonment of exploitation and oppression. Harmony between the individual and the community is this instinct of sociality. Through imagination, we are able to feel what we have never experienced and identify with the joys and sorrows of others. This strengthens our individual identity by morally connecting us with others. Individual initiative grows out of belonging to a community. Each unique individual works together as part of a group to develop strategies for the welfare of all. This is the starting place for the expansion of the individual personality. Moral courage is the first step to overcome passivity—breaking with the narrow philosophies and religions of the past. Fulfillment comes through mutual aid and egalitarian self-restraint. Conversely, economic individualism and personal salvation are the bedrock of narcissism and existential isolation.

Michael Joseph Franciscon

See also Darwin, Charles; Ethics; Evolution, Organic; Humanism; Marx, Karl; Materialism; Morality

Further Readings

Kropotkin, P. (1967). *Memoirs of a revolutionist.* Gloucester, MA: Peter Smith.

Kropotkin, P. (1967). *Mutual aid.* Boston: Extending Horizons.

Kropotkin, P. (1968). *Ethics: Origin and development.* London: Benjamin Blom.

Kropotkin, P. (1968). *Fields, factories and workshops tomorrow.* London: Benjamin Blom.

Kropotkin, P. (1970). *Revolutionary pamphlets.* New York: Dover.

Kropotkin, P. (1989). *The conquest of bread.* Montreal, QC, Canada: Black Rose Books.

Woodcock, G. (1971). *The anarchist prince: Peter Kropotkin.* New York: Schocken.

K-T Boundary

The Cretaceous-Paleogene (K-Pg) boundary, dated about 65 million years ago, marks the base of the Paleogene period and consequently of the Cenozoic era. ("K" is traditionally used as an abbreviation for the Cretaceous period (from the German term *Kreide*) in order to avoid confusion with the abbreviation "C" of the Cenomanian epoch and the abbreviation "C-T" of the Cenomanian-Turonian boundary.) It is more popularly known as Cretaceous-Tertiary (K-T) boundary, but the term *Tertiary* has become informal and is not used in the geologic timescale. From the chronostratigraphic point of view, the K-T (or K-Pg) boundary was formally defined at the base of a dark clay bed at the El Kef stratotypic section (Tunisia). This bed is commonly called the *K-T boundary clay,* and its basal part is characterized by a millimeter-thick reddish layer, called *K-T airfall layer,* containing meteoritic impact evidence such as anomalous-concentrated iridium, siderophile trace elements in chondritic proportions, microdiamonds, nickel-rich spinels, shocked quartz, and altered microtektites.

The K-Pg or K-T boundary is widely known, because it represents one of the five major mass extinction events recorded in the earth's history, which affected the most famous paleontological group, the dinosaurs. However, the dinosaurs were not the only victims of this biotic catastrophe. It is considered that 75% to 80% of then extant species were extinguished; this included the total extinction of dinosaurs (*Tyrannosaurus, Triceratops,* and *Ankylosaurus* among others), plesiosaurs, mosasaurs, ichthyosaurs, primitive birds (*Enantiornithes, Hesperornithiformes*), ammonites, belemnites, rudists, and orbitoid foraminifers. Many other groups were severely affected, such as mammals (~80% extinguished), osteichthian fishes (~96%), nautiloids (~50%), gastropods (~80%), bivalves (~60%), brachiopods (~70%), scleractinian corals (~80%), thermospheric ostracods (~75%), planktic foraminifers (~96%), radiolarians (~85%) and calcareous nannofossils (~75%). Marsupial mammals became extinct except for those in Australia and South America. Remarkably, insects, amphibians, lepidosaurs (lizards, snakes, etc.), crocodilians, turtles, and insectivore mammals and birds were not very affected by the K-T extinction. Moreover, although their populations were initially decimated, many species of terrestrial plants and of marine phytoplankton (diatoms or dinoflagellates) also tended to survive, due surely to their capacity to form resistant cysts, spores, or seeds.

The K-T extinction was therefore selective, with a species' survival depending on its position in the food chains. In the open ocean, the food chain is and was based on the microscopic phytoplankton (unicellular algae), such as coccolitophorids (calcareous nannoplankton), dinoflagellates, or diatoms. The marine protozoons and animals at successively higher levels in this food chain (phytoplanktivorous and carnivorous) were very strongly affected (planktic foraminifers, ammonites, carnivorous fishes, and marine reptiles such as plesiosaurs, mosasaurs, and ichthyosaurs). However, those animals whose diet was suspensivorous or detritivorous—or those who lived on these—tended to survive, at least partially (e.g., benthic foraminifers and many bivalves, bryozoans, brachiopods, and fishes). In the terrestrial environment, the food chain is and was based on plants, so the herbivorous and carnivorous animals directly or indirectly dependent on this vegetation—all dinosaurs and many birds and mammals—became extinct. On the contrary, those terrestrial animals whose diet was detritivorous (e.g., insects), who were potentially scavengers (e.g., cocodilians, turtles) or insectivorous (ancestral mammals and many birds, amphibians and lepidosaurs) tended to survive.

The most widely accepted theory for explaining the K-T mass extinction event is the meteoritic impact theory proposed by Luis Alvarez (1968 Nobel Prize winner in physics), Walter Alvarez, Frank Asaro, and Helen Michels in 1980. They discovered a reddish-greenish sedimentary layer at Gubbio (Italy) that marked the boundary between Cretaceous and Tertiary sediments, and it contained an anomalous concentration of iridium—hundreds of times grater than normal. Iridium is extremely rare in the earth's crust but abundant in chondritic and iron meteorites. For this reason, Alvarez's team proposed that a collision with a large asteroid (about 10 kilometers in diameter) caused the K-T extinction event 65 million years ago and the deposition of the reddish-greenish layer or airfall layer at Gubbio. Almost at the same time that Alvarez's team proposed this collision, Jan Smit suggested the same theory after studying the Caravaca section (Spain), which was a more expanded and continuous section than that at Gubbio. He proved that the airfall layer coincides with the largest, most sudden planktic foraminiferal mass extinction in evolutionary history, suggesting a clear relation of cause and effect between the impact event and the K-T catastrophic mass extinction. Since then similar airfall layers have been found at sections worldwide, such as the El Kef section, suggesting the totality of the K-T event.

The discovery of the ~180-kilometer-diameter Chicxulub crater on the northern Yucatan Peninsula (Mexico) supported strongly the Alvarez group's theory, because such a crater was coincident in age with the K-T boundary and has just the size to have been formed by the predicted 10-kilometer-diameter meteorite. In addition, strange and thick K-T clastic sediments were discovered around the entire Gulf of Mexico, suggesting the destabilization of continental margins in the region as result of giant tsunamis and earthquakes generated from the impact point at Yucatan.

Most geologists and paleontologists agree with Alvarez's impact theory, although alternative scenarios have been suggested for explaining the K-T extinction. They include sea-level regressions, increases of the volcanic activity linked to the Deccan Traps (an extensive basalt province in India), or multiple causes, that is, all the aforementioned causes operating at the same time. However, these hypotheses have not been accepted by the geologic and paleontologic communities, because they would be more compatible with uncorroborated gradual extinction patterns occurring hundreds of thousands of years before and after the K-T boundary.

According to Alvarez's theory, the impact raised a vast dust cloud of fine ejecta in the atmosphere and triggered global firestorms due to the fall of incandescent melted fragments worldwide (tektites and microtektites). The dust and soot cloud generated a global atmospheric darkening, sudden climatic cooling and acid rain, a phenomenon known as "impact winter" (similar to a nuclear winter). The blockout of sunlight caused the cessation of all photosynthesis and the severe disruption of food chains, initiating the catastrophic mass extinction event at the K-T boundary. The dust and fine ejecta that covered the atmosphere after the Chicxulub impact were deposited slowly, probably over months or a few years, forming the K-T airfall layer worldwide.

Atmospheric carbon dioxide content increased rapidly as result of the emissions from the impact site and the widespread fires as well as the temporary cessation of photosynthesis and primary productivity. After the dust cloud settled, the high concentration of carbon dioxide initiated a greenhouse effect period that lasted about 10,000 or 15,000 years. The K-T boundary clay was deposited during this period of global decrease in primary productivity and of increased greenhouse effect after the impact winter. All postimpact paleoenvironmental aftermaths of longer term (decrease in primary productivity, low biodiversity, greenhouse effect, disruption in the water column stratification, etc.) have been recorded in the K-T boundary clay worldwide. Only the most cosmopolitan, generalist, and opportunistic species were able to survive the environmental damage of both the impact winter and greenhouse effect periods.

Terrestrial plants and marine phytoplankton, and consequently primary productivity, recovered progressively over thousands of years, reestablishing the food chains and decreasing the concentration of carbon dioxide in the atmosphere. Many niches remained vacant after the K-T mass extinction, so the opportunistic surviving species began to occupy them. Small and new marine and land species appeared in the earliest Tertiary, following a model of "explosive" adaptive radiation. For instance, mammals diversified, evolving from the small insectivorous mammals that had survived the K-T event and becoming dominant on land. Similar adaptive radiations occurred in other marine and terrestrial groups. Thanks to the K-T extinction of the dinosaurs, the mammal groups could diversify during the Tertiary, including those that formed the evolutionary lineage of the primates that culminated with the appearance of the species *Homo sapiens.*

Ignacio Arenillas

See also Chicxulub Crater; Cretaceous; Dinosaurs; Evolution, Organic; Extinction, Mass; Foraminifers; Fossil Record; Geologic Timescale; Geology; Nuclear Winter; Paleogene; Paleontology; Permian Extinction

Further Readings

Alvarez, L. W., Alvarez, W., Asaro, F., & Michel, H.V. (1980). Extraterrestrial cause for the Cretaceous-Tertiary extinction. *Science, 208,* 1095–1108.

Alvarez, W. (1997). *T. rex and the crater of doom.* Princeton, NJ: Princeton University Press.

Ryder, G., Fastowsky, D., & Gartner, S. (Eds.). (1996). *The Cretaceous-Tertiary event and other catastrophes in Earth history.* Geological Society of America Special Paper 307. Boulder, CO: Geological Society of America.

Smit, J., & Hertogen, J. (1980). An extraterrestrial event at the Cretaceous-Tertiary boundary. *Nature, 285,* 198–200.

KUHN, THOMAS S. (1922–1996)

Thomas S. Kuhn is recognized as one of the foremost historians of science to have emerged in the 20th century. He is credited with a controversial approach to understanding scientific development, one that emphasizes rare but significant revolutions that challenge and even overthrow traditional ways of observing the world. Such revolutions, he claimed, change our views so profoundly that it may be unreasonable to compare theories that originated in different times. It is unfair, for example, to judge the writings of Aristotle (384–322 BCE) according to the standards of proof required today, not only because Aristotle could not have been aware of the scientific discoveries since his time, but also because the ancient Greeks understood worldly phenomena according to a different set of guiding principles. The scientific revolutions of more than 2000 years have rendered our perspectives beyond comparison; Kuhn calls this the "incommensurability thesis."

Paradigm Shifts

Kuhn studied physics at Harvard before teaching the philosophy of science at a variety of schools (most recently Princeton and MIT). His book *The Structure of Scientific Revolutions* (1962) inspired intense controversy by calling attention to the limitations of accepted scientific methodology. Kuhn wrote that science, or rather scientists, proceed through time operating according to a collection of accepted theories. There are few challenges to these theories, because they are assumed to reveal truths about the world. Because the scientific method has already exhaustively tested such explanations, any new work by scientists is expected to conform to accepted laws and principles. Occasionally, however, the conventional knowledge base is unable to answer a particular problem, or worse, it is at odds with a new theory or observation. When a solution emerges that resolves such a crisis by overthrowing previously accepted elements of scientific knowledge, then what Kuhn calls a *revolution* has occurred.

The term *paradigm shift* originated with Kuhn's vision of a scientific revolution that interrupts the normal thought paradigm. The traditional view of science presumes continual progress as advancements contribute to an ever-growing base of knowledge. When theories are revised and formerly held principles abandoned, then science simply has witnessed a correction that brings us closer to an underlying truth. Kuhn rejected this traditional view. Instead, he claimed that there are long periods with negligible progress, what he called *normal science* (operating within conventional knowledge) that are occasionally punctuated by revolutions. There *is* monumental progress in science, Kuhn would claim, but only through the process of alternating eras: convention, revolution, convention, revolution, etc.

Scientists might pursue monumental innovations (e.g., inventions, discoveries, cures for diseases), but they are held back, Kuhn claimed, by the prevailing wisdom of the time. In other words, the tendency is to think *inside* the box. This urge for innovation, restrained by conventional thinking, Kuhn referred to as *the essential tension.* In his view, revolutions are not actively sought. Instead, changes in thinking occur as a matter of necessity when accepted principles are replaced by the new paradigm. Advancement has less to do with individual accomplishments than with new modes of thinking that are so profound that they render obsolete some aspect of the former status quo. If science consists of games or puzzles, then there are familiar rules under which we attempt to solve problems. A true paradigm shift introduces not just a new puzzle to be solved, but new directions for answering old questions. The new theory not only resolves some inadequacies of past thinking, but also launches a new set of problems and guidelines for proceeding. Kuhn was not suggesting that individuals strive to overthrow established consensus. Rather, he was describing a cycle of convention, crisis, revolution, and new convention.

Evolutionary Science

To Kuhn, this theory was not just for the history and evolution of frameworks in science; science itself is evolutionary. When science meets with the hostile environment of an unsolvable problem, adjustments are made to address the challenge. In the traditional view of science, competing theories battle against one another; through the trial and error of the scientific method, the strongest theory emerges victorious and progress is claimed. For Kuhn, however, the competition pits the existing body of scientific knowledge against conflicting ideas; challenges to the standard way of thinking are discredited by conventional scientists, because their operating principles are regarded as truths. If a sufficient number of scientists present the same new idea over time, or especially if a new idea successfully resolves a major problem that has plagued science for a very long time, then that new idea might be paradigm shifting.

In embracing the revolution, the evolutionary change is dramatic. Where traditional science would (reluctantly) accept new information as an addition to the growing body of knowledge, however, Kuhn emphasized that some old theories, instead, are discarded to make way for the new. In organic evolution, fins are replaced with limbs, not augmented by them; it is absurd to imagine a creature that collects and maintains every feature its ancestors have ever possessed (the platypus notwithstanding). To continue the analogy, Kuhn saw revolutionary science as changing significantly enough that appendages are no longer recognizable. Traditional science, like Darwin's original conception of evolution, envisioned a series of continual changes over time. To recognize biological evolution as a series of periodic responses to catastrophes, however, is more akin to Kuhn's vision: Scientific revolutions are reactions to intellectual crisis, but they are rare.

The conventional view of science, according to Kuhn, is mistakenly teleological; it presumes that scientific inquiry is progressing toward a particular goal. The principles originally developed through the observation of patterns in nature have been elevated to such esteem that they are taken as truths about the world or, in any case, as steps toward eventually knowing an objective and unchanging Truth. Just as evolutionary adjustments might be arbitrary, though, responding as they do to challenges posed by the external environment, in Kuhn's view scientific advancement is vulnerable to sociological influence. History has shown that political bias sometimes keeps science at bay: Followers of Copernicus were convicted of heresy for daring to support his theory of heliocentrism (e.g., Giordano Bruno was burned at the stake; Galileo Galilei was placed under house arrest). Indian astronomers as early as the 7th century BCE (and ancient Greeks 300 years later) had believed the earth to rotate around the sun, yet religious influence prevented acceptance for centuries. Like the Connecticut Yankee, regarded as a magician for his demonstrations in King Arthur's court, scientific knowledge is contingent upon a particular era in time.

Kuhn's view of the history of science recognizes that what passes for "knowledge" is subject to revision and conventional science restricted by its own parameters. The scientific timeline is neither static nor perpetually expanding as we endlessly acquire new knowledge. Paradigm shifts are so radical it is inappropriate to hold theories from one era to the standards of another. According to Kuhn, true scientific advancement replaces obsolete principles with new ways of thinking that revolutionize our view of the world.

Elisa Ruhl

See also Aquinas and Augustine; Aristotle and Plato; Bruno and Nicholas of Cusa; Copernicus, Nicolaus; Darwin and Aristotle; Einstein and Newton; Galilei, Galileo; Hegel and Kant; Nietzsche and Heraclitus

Further Readings

Fuller, S. (2000). *Thomas Kuhn: A philosophical history for our times*. Chicago: University of Chicago Press.

Horwich, P. (Ed.). (1993). *World changes: Thomas Kuhn and the nature of science*. Cambridge: MIT Press.

Kuhn, T. (1959). The essential tension: Tradition and innovation in scientific research. In E. C. Taylor (Ed.). *The third (1959) University of Utah Research Conference on the Identification of Scientific Talent* (pp. 162–174). Salt Lake City: University of Utah Press.

Kuhn, T. (1962). *The structure of scientific revolutions*. Chicago: University of Chicago Press.

La Brea Tar Pits

The La Brea Tar Pits, or Rancho La Brea, are a famous group of tar pits located in Hancock Park in the heart of downtown Los Angeles, California. This location contains one of the richest and best preserved assemblages of fossilized Pleistocene life in the world. The plant and animal remains found represent hundreds of species that populated southern California from almost 40,000 to 10,000 years ago. Rancho La Brea provides a wealth of information on life during the most recent ice age.

Tar pits form in places where crude oil seeps to the surface of the earth through cracks in the bedrock. A portion of the oil evaporates, leaving thick, viscous pools of asphalt behind. Local natives used this resource for thousands of years to waterproof baskets and canoes. As Caucasian settlers moved into the area, the tar was collected and used for roofing in the nearby town of Pueblo de Nuestra Señora la Reina de los Angeles.

A family named Hancock owned and worked a 4,400-acre ranch that encompassed the tar pits in the late 1800s and continued to mine the pits for tar to use as a sealant. Occasional bones unearthed were assumed to be those of cattle that strayed, got stuck in the tar, and perished. Eventually paleontologists heard about the bones and performed the first scientific excavation in 1901. Work intensified as the number of finds increased. From 1913 to 1915, digging at over 100 pits produced more than 1 million bones. In 1924, businessman G. Allan Hancock donated the 23-acre plot containing most of the tar pits to Los Angeles County. Early excavators tended to keep only the large, complete bones for museum display. When digging resumed in 1969, more care was taken in cleaning and analyzing over 40,000 additional specimens retrieved.

Tallying and cataloging the bone and plant remains provided insight into the activity around the tar pits so long ago. Wood, plant, and pollen remnants illustrate a cooler and wetter environment than exists today. The park is best known for the large mammals found in the pits, including mammoths, camels, sabertooth cats, dire wolves, and giant ground sloths, but many other species have been revealed. Since excavations at Rancho La Brea began, evidence of more than 650 species has been recovered: 231 vertebrates, 159 plants, and 234 invertebrates. More than 95% of the mammal bones found belong to only seven species. Four were carnivores: the dire wolf, the sabertooth cat, the coyote, and the North American lion, and three were herbivores: the ground sloth, the bison, and the western horse. The most common bones found are those of the dire wolves, with over 2,000 sabertooth cat specimens ranking second. Of the mammals found, 80% were predators, and 60% of the birds were birds of prey. These percentages reflect the exact opposite of what you would expect to find in the wild.

In a stable ecosystem, herbivores always outnumber carnivores by about 10 to 1. The leading theory to explain this discrepancy is called *entrapment.* Asphalt is very sticky, especially when it is warm. Only one inch of this tar can hold a horse or cow until it dies of starvation or exposure. In dry weather, dust can blow across the surface, making the ground appear solid. A trapped animal thrashing around trying to escape would attract the attention of many predators. In attacking the prey, some of the attackers would also get stuck, providing additional food for other carnivores. This cycle repeated for 30,000 years to create the accumulation seen at Rancho La Brea today. Although the collection seems to represent a huge number of animals, it averages out to one herbivore being trapped every decade, and several scavengers attempting to retrieve the meat and joining their prey.

Although thousands of individual animals have been found in La Brea, only one human skeleton has been discovered. The skull and partial skeleton of a Native American woman was recovered in 1914. She was in her early 20s when she appears to have been killed by a blow to the head. Her bones have been dated to about 9,000 years ago.

The La Brea Tar Pits provide a treasure trove of information about life during the late Pleistocene. While studying the bone collection reveals information about ancient food chains, the sophisticated chemical tests available today reveal a surprising amount of detail about how animals became trapped in the tar pits and what happened to them in their final hours of life. The Page Museum in Hancock Park opened in 1977 to house the heritage of over 1 million specimens recovered from the tar. The museum features more than 30 exhibits of reconstructed skeletons and robotic sculptures, a glass-walled working laboratory where visitors can watch the curators work, hands-on displays, films, and many wall murals. Rancho La Brea sheds light on a fascinating time period when glaciers last covered much of the hemisphere.

Jill M. Church

See also Dating Techniques; Extinction and Evolution; Fossil Record; Fossils, Interpretations of; Museums; Paleontology

Further Readings

Harris, J. M., & Jefferson, G. T. (1985). *Rancho La Brea: Treasures of the tar pits*. Los Angeles: Los Angeles County Museum.

Perkins, S. (2004). L.A.'s oldest tourist trap. *Science News, 165,* 56–58.

Stock, C., & Harris, J. M. (1992). *Rancho La Brea: A record of Pleistocene life in California*. Los Angeles: Los Angeles County Museum.

Laetoli Footprints

In 1976, a trail of fossilized hominid tracks was found in Laetoli, Tanzania. These footprints were determined to be at least 3.6 million years old, and some of the earliest evidence ever found to illustrate upright, bipedal walking in early hominids. Thanks to a chance series of events including a volcanic eruption, a light rainfall, and a second deposit of ash, scientists today can observe firsthand the preservation of a specific moment in prehistoric time.

Laetoli was first surveyed by anthropologists in 1935, when Louis S. B. Leakey and his wife Mary D. Leakey evaluated the area. They were excavating at Olduvai Gorge 2 days to the north when Maasai tribesmen told them there were many bones at Laetoli. The team investigated when the Olduvai season ended. After making a few scattered, fragmented finds, they left. Later, during the 1938–1939 season, a German team surveyed the Laetoli area. The Leakeys returned a couple of more times, with few results. In 1974, Mary Leakey found better preserved hominid remains, resulting in a more systematic evaluation of the area. One night in 1976, team member Andrew Hill was joking around with colleagues, throwing dried elephant dung in a mock battle. As Hill ducked to avoid a hit, he noted what appeared to be animal prints preserved in the exposed volcanic tuff. Further investigation revealed a trail of hominid footprints.

The majority of the Laetoli footprint site was excavated in 1978. The exposed trail is about 80 feet long. The prints are distinctively different in size and in two parallel tracks. Despite the size disparity, the strides are the same length, indicating that the two hominids were walking together and compensating to match their strides. Scientists have disagreed as to whether the tracks represent two or three individuals. The smaller prints in the westernmost trail have sharp, clear impressions. The larger prints are blurred and more indistinct. Some argue that a third hominid was walking in the prints made by the larger individual, while others claim the larger, heavier hominid was simply sliding more in the wet ash. The footprints are almost indistinguishable from those of modern humans except for their small size. They have a deep heel impression, a distinct arch, and push off from the toes at the end of the stride. The big toe is in line with the other toes like the toes of a modern human, rather than being an opposing toe like that of a chimpanzee.

Regardless of whether the prints were made by two or three hominids, the importance of this find cannot be overstated. It is, literally, rock-solid proof that early hominids were bipedal long before large brains evolved and stone tools were used. The only hominid fossils found at Laetoli belong to *Australopithecus afarensis,* indicating that individuals of this species made the trail. Despite years of searching, no stone tools have been found in the Laetoli beds, indicating for now that hominids had not yet developed into the tool-making stage.

Study and debate about the Laetoli footprints is ongoing, but no one can deny the uniqueness of this find. Approximately 3.6 million years ago, the Laetoli area was savanna that supported a varied animal population. At the beginning of a rainy season, Sadiman, a volcano to the east, erupted. The shower of ash was not severe enough to drive the animals away. The season brought intermittent showers, turning the ash to a fine mud that the local fauna continued to walk through. Several times over the course of a few weeks additional light layers of ash covered the area. The scattered rains continued to make a mud that left distinct impressions. Finally a heavy ash fall buried the area deeply, protecting the Footprint Tuff from erosion. Over time, the ash cemented and became rock. Weathering eventually brought the Footprint Tuff layer to the surface again, giving modern man a unique view of a specific event in the life of early hominids. A brief moment in prehistoric time has been preserved.

Jill M. Church

See also Anthropology; Dating Techniques; Evolution, Organic; Fossils, Interpretations of; Olduvai Gorge; Paleontology

Further Readings

Hay, R. L., & Leakey, M. D. (1982). Fossil footprints of Laetoli. *Scientific American, 246,* 50–57.

Leakey, M. D., & Harris, J. M. (Eds.). (1987). *Laetoli: A Pliocene site in northern Tanzania.* Oxford, UK: Clarendon.

Stringer, C., & Andrews, P. (2005). *The complete world of human evolution.* New York: Thames & Hudson.

Lamarck, Jean-Baptiste de (1744–1829)

During the latter half of the 18th century, a handful of science-oriented enlightened thinkers took time and change seriously; they saw history in terms of process and progress. Jean-Baptiste de Lamarck extended this perspective to become the first serious evolutionist. He argued for the mutability of animal species over vast periods of time, in an age when most other naturalists still maintained the fixity of life forms on a planet held to be only several thousand years old. As an invertebrate paleontologist, he took the fossil record in the geological column as empirical evidence demonstrating the evolution of species, into life forms of increasing complexity and greater perfection, as seen in their preservation up through the rock strata. Lamarck presented this controversial view of organic history in his major work, *Zoological Philosophy* (1809).

Lamarck's evolutionary interpretation of life was in sharp contrast to the one offered by the contemporary vertebrate paleontologist Georges Cuvier, who held that the fossil record (with its gaps) represented a history of periodic creations and extinctions. Cuvier's vision suggested a series of divine creations, each special creation eventually followed by worldwide extinction due to a planetwide catastrophe. Thus, Cuvier was not an evolutionist. However, Lamarck saw the biological continuity of species (without any extinctions) throughout organic history. Because evolution challenged the entrenched beliefs in the biblical story of Creation in the Book of Genesis, with its 6 days of Creation, religious naturalists were not open to Lamarck's new worldview, despite those ongoing discoveries in geology and paleontology that clearly supported an ancient earth and the mutability of species on it, respectively.

Lamarck held that spontaneous generation would explain the sudden appearance of the first forms of life on earth. But, the fossil evidence to the contrary, he needed to account for the evolution of species over considerable time. Lamarck offered two explanatory mechanisms: the inheritance of acquired characteristics through use and disuse, and a form of vitalism that manifested itself only in complex animals. Lamarck's famous but now notorious example to demonstrate his first explanation involves the evolutionary history of the giraffe. He held that, eons ago, short-necked giraffes fed on the leaves of eye-level trees. Over time, these trees continuously became taller. Consequently, in order to reach and eat the ever-higher leaves, the giraffes needed to constantly stretch their necks. This acquired characteristic of a longer neck (due to continuous stretching) was inherited by the offspring over numerous generations. The ongoing accumulation of this useful and specific characteristic has resulted in the long-necked giraffe of modern times.

Furthermore, Lamarck speculated that complex animals with consciousness could actually will those characteristics (structures and functions) that they needed or desired in order to adapt to, and survive in, changing environments; he even thought that, through willing, new organs and behavior patterns could emerge over time. Not surprisingly, because of a lack of convincing experimental evidence, naturalists have not accepted as true Lamarck's two explanatory mechanisms for the evolution of life on this planet.

Concerning the origin of our own species, Lamarck boldly claimed that the human animal had emerged from an orangutan-like primate somewhere in Asia. Through slow evolution, he maintained, our human species gradually acquired those mental faculties and biological characteristics it has today. Moreover, Lamarck argued that humans differ from the living apes and monkeys merely in degree rather than in kind; the superiority of human reason is grounded in the larger and more complex brain of our species. However, the present fossil record points to Africa as the cradle of humankind. Even so, modern primatology does clearly demonstrate the striking similarities between the human animal and the great apes.

Lamarck died blind and poor, unaccepted and unappreciated by most of the naturalists of his time. Nevertheless, he was a significant link between the ideas of earlier natural philosophers, who held to the fixity of species, and those pivotal scientific writings of Charles Darwin that presented the mutability of life forms in terms of organic evolution by natural selection. In fact, Darwin's *On the Origin of Species* (1859) appeared exactly 50 years after the publication of Lamarck's *Philosophy of Zoology* (1809).

H. James Birx

See also Darwin, Charles; Evolution, Organic; Haeckel, Ernst; Huxley, Thomas Henry; Lysenko, Trofim D.; Spencer, Herbert

Further Readings

Birx, H. J. (1984). *Theories of evolution.* Springfield, IL: Charles C Thomas.

Lamarck, J.-B. (1984). *Zoological philosophy: An exposition with regard to the natural history of animals.* Chicago: University of Chicago Press. (Original work published 1809)

Packard, A. (1980). *Lamarck: The founder of evolution—His life and work.* North Stratford, NH: Ayer.

Language

We look back at the thoughts of our predecessors, and find we can see only as far as language lets us see. We look forward in time, and find we can plan only through language. We look outward in space, and send symbols of communication along with our spacecraft, to explain who we are, in case there is anyone who wants to know. (David Crystal)

Among the most popular areas of study that involve explorations to characterize humankind is the study of language. Throughout history, individuals have been fascinated about how language works and even more by the unique characteristics of each language as well as the common ones among the languages of the world. The phonology of different languages is especially interesting to some researchers; others are intrigued by grammar and morphology that show similarities in structure among languages. Some find great satisfaction in studying expressions of idiom, metaphor, and humor; others look to how societies function and get along as they adjust their language use for communicating in specific environments (e.g., formal or casual). Whatever the intentions or goals of those who study language, that this field of endeavor has endured for thousands of years attests to the changing nature of language over time, wherein there is always something new on the horizon to try to understand.

Definitions of Language

Although there are some distinct characteristics that are used to identify languages (e.g., phonology, grammar), definitions of *language* and *languages* vary among scholars depending upon their particular curiosity and the questions for which they want to find answers. Joel Davis comments that there is somewhat of a dilemma for linguists to capture all of the characteristics into one definition. How does one describe language incorporating the terms of the language he or she uses? It is like defining a word by using that word in the definition. Thus, many researchers attempt to compare human languages in order to find common components and to identify differences, hoping such pursuits will add to the specificity of the term. Language definition is somewhat like trying to hit a slowly moving target, because there are many factors of language that gradually change as societies change. Abram de Swaan proposes that a language may be defined by the capacity of any two speakers to understand each other and incidentally notes that the variety and complexity of languages is such as to be comparable to that of life itself.

Studies about language can be separated from studies about the history of humankind only by an imaginary line that has to be crossed quite regularly. Definitions of language have to allow for distinctions that influence it from other areas of human evolution, such as the development of communities, trade, music, and art (in themselves kinds of language).

The Generative Nature of Language

If one were to converse with thoughtful individuals about their native tongues, one would probably find out that they agree that their own language is not static but rather is creative and permits new and adapted forms of words and expressions as the times require them. As societies and communities progress and as educational opportunities grow, so does the mother tongue. But, in some cases where there is little progress, languages become endangered and sometimes extinct. Technological terms, for example, not occurring in the first few decades of the 20th century are now part of many languages of the world: the word *computer,* for example, appears in French as *l'ordinateur,* in Japanese as *keisanki,* and in Maori as *rorohiko*. Contrastively, languages among hunter-gatherer tribes in Africa, as explained by Lenore Grenoble and Lindsay Whaley, are more threatened by extinction than those spoken by groups that are involved in socioeconomic reorganization and growth.

Spoken Language Over Time

Linguists recognize that amidst the intricate web of factors involved in language change, modifications in sounds and vocabulary seem to lead the other stages, particularly grammar, even though the total process may be extremely slow, occurring sometimes over centuries. John McWhorter discusses how the "erosion" of sounds over time leads to the development of new words, or even of new languages. Consider, for example, McWhorter's description of the movement from the Latin word for woman, *femina* (FEH-mee-nah) to *femme* (FAHM) in French. The accented syllable in the Latin word for woman remained, while the other two weak syllables eventually disappeared. Latin evolved into several new languages, including French, and although it is no longer a spoken language, Latin endures in written form.

Throughout human history, individuals have modified their indigenous languages and have created new words and expressions. For example, William Shakespeare (1564–1616) left an enduring influence on the English language with hundreds of words commonly used today that he created by using nouns as verbs, using verbs as adjectives, and adding prefixes and suffixes (e.g., *countless, bet, laughable, excitement, torture).* Words are also coined or invented when there is a need for new ones. Among his many malapropisms, U.S. President Warren G. Harding's *normalcy* (for *normality*) continues in common American English usage today. Some novel words endure, and others disappear (e.g., *defidate,* to pollute, circa 1669; *yuppie,* attributed to Bob Greene of the *Chicago Tribune; guillotine,* renamed for Dr. Guillotine who urged the machine's use in the French Revolution).

Language Mixtures

Although the grammars of spoken languages are rather stable, bound by rules, and slow to change, they may be affected by the incorporation of new expressions, especially in those societies whose members may have come from nations with different mother tongues. Such is the case in the United States, considering the large waves of immigration from the mid-1800s to the 1920s. As first and second generation children of immigrants from Western Europe and Asia learned English, they adopted accommodation behaviors, or *interlanguage* for communicating appropriately in places in their communities when their English was not yet fluent (e.g., at the store, in mixed ethnic neighborhoods). An interlanguage is defined in one of two ways. It may be the way that one individual, still learning the language of a society different from one's own, creates novel terms or mixes terms between one's language and the target language. A Polish immigrant might say something like, "*Ja będę iść do marku*" ("I will go to the market"), substituting the English word, *market,* for the Polish word, *rynku,* but retaining the final letters from the Polish word.

A second characterization of interlanguage regards the reciprocal relationship in communication between two individuals with distinct mother tongues. In this case, each individual accommodates the other in clever verbal manipulations. The use of interlanguage may continue among speakers for considerable time, or it may eventually lose its utility once the speakers become fluent in a single new language. Larry Selinker provides an example of an Israeli who continued to produce English sentences of the type, *I bought downtown the postcard,* even as a fluent bilingual in Hebrew and English.

Quite often, when individuals become bilingual, they switch between the two languages in their attempts not only to be understood but also to clarify for themselves what they mean. This behavior, called *code switching,* has been studied extensively by sociolinguists for spoken language as well as written language where it also might occur. Over time, accommodations such as interlanguage and code switching influence the development of new words and expressions in the main target language.

A special case where factors such as code switching and interlanguage are especially helpful for understanding language growth is manual language, particularly sign language that is used by deaf communities. Of over 100 such sign languages used throughout the world, the most studied is American Sign Language (ASL). Because of the efforts of William C. Stokoe in the 1960s, ASL gained recognition as a true language. Researchers have since explored all facets of the language to support its identity. Sociolinguistic research into ASL continues into the 21st century to identify characteristics that are equivalent to those found in spoken languages. Besides the areas that concern formal linguistics (i.e., phonology, grammar, semantics, pragmatics), research is now replete with studies regarding code switching and pidginization. Research about social interactions within deaf communities and between deaf individuals and hearing individuals has been most fruitful.

Code switching and interlanguage are relatives to a language dynamic known as *diglossia.* This occurs when members of a community recognize that there are varieties of their mother tongue that have to be selected for particular situations, such as a more formal variety for the business world, or for worship or governmental meetings. These varieties may differ in grammar as well as word usage. Joshua Fishman explains that diglossia can also involve two distinct languages (e.g., Latin in some Roman Catholic Church services, but a native language outside of church).

Language varieties are special forms of language that endure within speech communities, and they may have taken a long time to establish (e.g., formal French and vernacular French). Often, a more formal variety is one that is used in written literature, while the less formal variety will never be used in writing. What is interesting about diglossia is the existence of two or more varieties of a language, side by side, that do not eventually mix. This is not the case, however, with pidgins and creoles.

Pidgins and Creoles

Pidgins are formed when individuals who speak two different languages create a means for immediate communication that involves the integration of characteristics of each of their languages. Most pidgins endure only as long as they are needed for a particular purpose, but some, such as *Tok Pisin* of Papua New Guinea, remain as a common vernacular. Historically, pidgins have developed throughout the world, with many used predominantly in areas where trade necessitated communication between European traders or colonizers and native people. These "half-languages," as John McWhorter calls them, can occur any time when conditions warrant it between speakers. Although pidgins do have limited rule systems, they do not qualify as natural languages even though some pidgins endure and are expanded over time so that they have all the typical characteristics of a true language.

In some societies, pidgin languages have undergone a process of change from one generation of speakers to the next so that they become the mother tongue, or more appropriately, a vernacular. These new languages are called *creoles,* taken from a Portuguese term that describes persons of European descent who were born and have lived in a colonial territory. The process by which a pidgin becomes a creole is known as *creolization.* It involves the stabilization of grammatical rules and vocabulary that allows for language generativity and novel use, similar to that in natural languages. McWhorter explains that that there can be varieties of a creole language, somewhat as on a continuum. Individuals may show flexibility for using a creole language in more than one variety depending upon the circumstances and needs of a particular communication.

More Evidence of Language Change

Another set of characteristics that are evidence of language change include dialects, jargon, slang, and colloquial expressions.

Dialects are forms of language that involve pronunciation, vocabulary, and grammar that differs from what is designated as the standard form of a language. The origins of dialects vary. The language, Icelandic, for example, began as a dialect branching from Proto-Germanic, the precursor of the Germanic line of languages in the Indo-European language tree. Sometimes dialects have developed within groups of individuals that live apart from a larger society; sometimes they have developed over time in a geographic region that has been influenced by the merging of linguistic elements from two or more varieties of languages. People of the Appalachian Mountains speak a dialect that was influenced by varieties of English from Scotland as well as England.

The term *dialect* may also be used as an equivalent term to describe a language. Many Chinese languages are called dialects (e.g., Oirat dialect, a Mongolian language), as are some creoles.

Dialects are many times erroneously identified solely by speech patterns. They are likewise mistaken in many instances as having to do with one's social status. Historically, Americans in the northern states have judged individuals from the south with significant bias due to their dialects. As access to communication media becomes more common, bringing people of all walks of life together on a daily basis, these biases are weakened in America.

Slang and jargon provide evidence that languages can change with the influence of individuals who might be assumed to have little or no effect upon how the majority will continue to speak (e.g., adolescents). Robert MacNeil and William Cran, recounting what they learned about Californian *surfer dude* speech from the writer George Plomarity, explain that it is not actually uncommon for slang expressions to take on formal use over time. For example, someone who works in a law office might comment that he is *caught inside,* meaning in a situation that cannot be avoided where a great amount of information and interactions are coming in all at once (just like waves of the ocean).

Language Adaptations and Language Continuance

Because one of the main goals of humans is to interact with other humans, it might seem obvious that language adaptation would be relatively common. And, individuals frequently do adapt their language to facilitate communication, sometimes creating pidgins or using interlanguage, for example. Over time, what started out as an adaptation may show novel and somewhat permanent changes within existing natural languages, or new languages such as creoles might evolve. Yet, considering the tendency of societies to grow and change, access to certain language forms might also be limited to specific groups of people within a community. Sociolinguists have documented barriers to communication in thriving as well as developing nations and societies in the world.

Abram de Swaan, a political sociologist of language, envisions the world as a complex language system, a galaxy constituted of a global constellation in which there are major star patterns determined by factors identified in social science theories (e.g., trade, economics). As do many sociolinguists, de Swaan points out that language can be used as a tool for isolation and for insulation of groups of people. It can be manipulated to keep individuals from participating in specific areas within a society as well as enabling individuals to bond with others. For example, literacy continues to be prohibited to women in certain societies. Contrastively, interlanguage or pidgins can open up communication between peoples with different indigenous languages for the sake of commerce.

In some nations it may take a concerted act of the leaders within a society to establish venues that will help to keep a native language viable and to generate new words and expressions that are necessary for communication in subjects and categories that may influence a nation's growth. It may be a matter of pride among the people themselves to try to maintain their native tongue. Fearful of the demise of their lingua franca, Ewe, the people of the small country of Togo took measures to develop terms such as those needed for technology and new forms of commerce. Once a French colony, the Togolese Republic uses French as an official language, but it strives deliberately to keep its indigenous tongue, Ewe, alive. There is also a considerable body of literature written in Ewe, and this is helpful for sustaining the language.

Written Language and Literacy Over Time

The provision of a written form of a spoken language tends to support the viability of that language. Although this is so, forms of written language and literacy have not always been available to all classes or groups of people in societies from the time of their invention. Literacy was used as a tool for economic and political power in ancient civilizations, and this continues even to the present. Fishman describes the nature of literacy as "a sociocultural phenomenon in its own right, thus subject to many of the same political and self-interest concerns which come into play when making a language/dialect decision."

Alice Joan Metge documents this situation in her history of the Maori of New Zealand. Up until the mid-1800s, most of the Maori had maintained their native tongue and had established a base of literature. Those in power at that time, however, developed language-planning goals insisting upon the use of English in the community and as the only language in schools. Children were punished for using the Maori language even into the 1960s. At the beginning of the 20th century, some 90% of Maori children knew their indigenous language, but by 1953–1958, this number had dropped to 26%. Families also contributed to this loss. They recognized opportunities for their children if they used English. Thus, instead of creating terminology for industry and commerce in the Maori language, communities used English words and spoke for the most part in English. In the early 1970s, the Maori Youth Movement was successful in convincing the government to support efforts to revive the Maori language. The Ministry of Education adjusted the school curricula for children, and courses were offered in Maori up through the university level. Maori became integrated in many venues in society. Now, Maori terms and expressions necessary for daily life have increased in number and are not treated as novel. In the 21st century, New Zealand continues to celebrate Maori Language Day and has week-long activities to encourage pride in the native tongue. Literature is thriving and programming in Maori is available on the Internet as well as on local television.

The situation of the Maori is not unique. Fishman explains that as communities strive to modernize, the development of literacy involves their statement of economic, political, social, and cultural power. The leadership elite who may have had exclusive command of literacy will not necessarily be satisfied to permit the common person the capability to acquire it and join them in acts of literacy. Yet, literacy, just like spoken language, is growing relatively rapidly in developing as well as established nations.

Considering literacy in Europe, Fishman cites research that indicates there were only six languages of literacy by the year 950 (i.e., Latin, Greek, Hebrew, Old Church Slavonic, Arabic, and Anglo-Saxon). At the beginning of the 20th century, there were 31; by 1937, 56. By the late 1980s in Western Europe only, there were 67 different languages used in schools and adult education to teach literacy. In all Europe by 1989 there were at least some 200 languages. Fishman estimates that in the 20th century alone, the languages of literacy in Europe increased sixfold. As of 2003, he sees the same kind of dramatic growth in countries in Africa and Asia.

It appears to make sense that a spoken language will survive if its speakers have access to that language in writing. As in the case of Maori, that may be true. There is another direction, however, that some languages take where they appear to be on the verge of extinction in spoken form. A literary tradition of the language may develop and survive beyond its conversational utility.

One language that falls into this group is Picard, a language recognized by Belgium as an indigenous language and acknowledged as a language by the European Bureau for Lesser Used Languages. Although it is very similar to French in many respects, France considers it separate from French, and as with other minority languages, does not particularly care to recognize Picard as a "good" language. Picard is spoken in an area near the English Channel, in a northern region of France and in southern Belgium. Linguist Julie Auger explains that by all appearances spoken Picard should be dying, because most of the speakers in larger towns are older individuals, and they do not seem to see the necessity for concerted transmission of the language to the young. In actuality, the French government has given very little support to groups such as the Picards.

On the other hand, Auger has observed that written Picard is, in her words, "varied and dynamic." There were hints of a literary tradition as far back as the 16th century, but it was in the latter half of the 19th century that there were significant numbers of written works in Picard. By the year 2000, literature and a new magazine were thriving, and these works were being read in print as well as on the Internet. In the final analysis, Auger is hopeful that the expanded literary tradition will continue to get readers more involved in Picard and that the French government may enable it to thrive as a spoken language by giving it recognition and support.

Human Languages and Artificial Languages

Early in the 20th century, with the growth of cognitive psychology, the idea of the generative nature of human language aroused the curiosity of scientists regarding the possibility of creating languages with computers that might mimic human languages. If this were possible, then perhaps the computer could act as a bridge among peoples with different spoken languages. The challenge was particularly provocative for individuals such as Marvin Minsky, who, in an important work titled *A Framework for Representing Knowledge,* proposed structures that stimulated thought regarding the duplication of human language in machines.

In order to function as human languages, artificial ones would have to have grammatical rules that a machine could apply at the level of a sentence, and the machine would have to be able to apply inferential reasoning to capture the real, singular meaning of ideas or the multiple meanings of ideas, expressions, and words. For example, how might a computer interpret words for emotions and feelings and distinguish between them as well as indicate variations by degree (e.g., love of a friend versus love of a picture)? How might it deal with the ambiguity that is so prevalent in human spoken languages (e.g., *Bill told John that he loved Mary)?* In other spoken languages, including English, how might a computer distinguish among polysemous words?

Since the 1970s, the field of artificial intelligence has provided several powerful paradigms that in the 21st century are often taken for granted. Minsky's ideas and those of others like himself regarding frames of knowledge have been translated in such schema and models as that of Vinton Cerf's Arpanet, a precursor of the Internet, as well as in scientific applications such as those of Ray Kurzweil, considered a second Albert Einstein by many. Among his myriad accomplishments, Kurzweil has become well known for his print-to-speech reading machine for the blind as well as for his work with voice activated word processing. Cerf, the vice president of Google since 2005, has been called the father of the Internet. He has had a special interest in the field of communication and deafness because he himself is hard of hearing and his wife, Sigrid, is deaf.

Developments in artificial intelligence have enabled mutual understandings among humans beyond that of their distinct spoken languages. These new means for communication are realized in the scientific world as well as in commerce and even in music, another area studied by Kurzweil. Such developments point to the potential for humans to use "invented" forms of language in ways that can add to the growth of civilization and to enable a convergence of knowledge and sharing of cultures among diverse societies around the world.

Patricia N. Chrosniak

See also Anthropology; Consciousness; Language, Evolution of; Languages, Tree of; Memory

Further Readings

Aboba, B. (1993). *The online user's encyclopedia.* New York: Addison-Wesley.

Bartlett, J. R. (1859). *Dictionary of Americanisms* (2nd ed.). Boston: Little, Brown.

Cooper, R. (Ed.). (1982). *Language spread: Studies in diffusion and social change.* Bloomington: Indiana University Press.

Coulmas, F. (Ed.). (1989). *Language adaptation.* New York: Cambridge University Press.

Crystal, D. (1997). *The Cambridge encyclopedia of language* (2nd ed.). New York: Cambridge University Press.

Davis, J. (1994). *Mother tongue: How humans create language.* New York: Carol.

de Swaan, A. (2001). *Words of the world.* Malden, MA: Blackwell.

Fishman, J. A., Hornberger, N. H., & Putz, M. (2006). *Language loyalty, language planning and language revitalization: Recent writings and reflections of Joshua A. Fishman.* Bristol, UK: Multilingual Matters.

Garry, J., & Rubino, C. (Eds.). (2001). *Facts about the world's languages.* New York: H. W. Wilson.

Grenoble, L. A., & Whaley, L. J. (Eds.). (1998). *Endangered languages: Current issues and future prospects.* New York: Cambridge University Press.

Joseph, B. D., DeStephano, J., Jacobs, N. G., & LeHiste, I. (Eds.). (2003). *When languages collide: Perspectives on language conflict, language competition, and language coexistence.* Columbus: The Ohio State University Press.

Kurzweil, R. (1999). *The age of spiritual machines: When computers exceed human intelligence.* New York: Viking Press.

Lucas, C. (Ed.). (1995). *Sociolinguistics in deaf communities.* Washington, DC: Gallaudet University Press.

McNeil, R., & Cran, W. (2005). *Do you speak American?* New York: Nan A. Talese.

McWhorter, J. (2001). *The power of Babel.* New York: Henry Holt.

Metge, A. J. (1976). *The Maoris of New Zealand: Rautahi* (2nd ed.). London: Routledge & Kegan Paul.

Minsky, M. (1975). A framework for representing knowledge. In P. H. Winston (Ed.), *The psychology of computer vision.* New York: McGraw-Hill.

Selinker, L. (1992). *Rediscovering interlanguage.* New York: Longman.

Language, Evolution of

The study of the origin and evolution of human language is an essential one to characterize what distinguishes humans from other species on earth. Yet this noble and challenging endeavor has been a source of much speculation as well as research as far back as 3,000 years ago. The search for answers to such questions as why language began and how it evolved has provided researchers with a particularly grand puzzle. In spite of recurring fascination about the topic, without sources of direct evidence it has been virtually impossible to verify many theories, to know precisely when language originated and what the first spoken language was like. At some points in modern history there has been great frustration about the directions of the search, and this even led to a moratorium on scientific investigations in 1866 that slowed down research until the late 20th century.

Shortly after the publication of Darwin's *On the Origin of Species,* ideas and speculations about language evolution were so rampant and frequently quite strange that the primary authority regarding language study, the Société de Linguistique de Paris, put a ban on all discussions about the origins of language. It was not until the last decade of the 20th century that renewed vigorous study emerged. Many believe that the presentation in 1990 of a research paper, *Natural Language and Natural Selection* by Steven Pinker and Paul Bloom, inspired the rapid spread of study into the 21st century.

Although one might assume that researchers in subfields aligned with linguistics would have long been in the forefront of the study of language origins and development, this has not necessarily been the case. Many of the most curious have been in fields such as psychology, anthropology, biology, neuroscience, computer science, and archaeology. At the beginning of the 21st century, a number of researchers, including Marc Hauser, Noam Chomsky, Morten Christensen, and Simon Kirby, called for interdisciplinary research to help define the faculty of language and subsequently posit some answers regarding its evolution. The complexity of the subject requires collaborative efforts among specialists from distinct fields who can deal with the important questions together that otherwise in isolation lead these specialists down a narrow tunnel, so to speak, where there appears to be little light at the end.

Defining Language

The term *language* may refer to one of several constructs (e.g., language of music, language of love, computer language, spoken language, body language). Language is a major component in the larger category of communication among humans as well as among nonhuman primates, birds, insects, and water mammals. Characterizing human language, Noam Chomsky wrote,

> The human faculty of language seems to be a true "species property," varying little among humans and without significant analogue elsewhere. . . . Furthermore, the faculty of language enters crucially into every aspect of human life, thought, and interaction. It is largely responsible for the fact that alone in the biological world, humans have a history, cultural evolution and diversity of any complexity and richness, even biological success in the technical sense that their numbers are huge. (Chomsky, 2000)

Studies of communication among species other than *Homo sapiens* have helped to develop a clearer understanding of the unique nature of human language. For example, scientists studying the dance of the bees have identified two characteristics of this signaling system—distance and direction—that help other bees find a source of nectar. In their dance, bees map distance and direction into particular angles and speeds, some greater and smaller, more or less. There is no variation, no modification, of this behavior otherwise.

Joseph Greenberg notes that language systems such as the dance of the bees are iconic. In other words, there is always a one-to-one correspondence between the purpose of alerting for nectar and the specific movements in the dance. Human language, by comparison, is a complex symbolic system with infinite possibilities for generating meaning. Grammar and syntax distinguish human language from the languages of other creatures. Thus, trying to identify the steps through which human language evolved is much more challenging than researching the dance of bees. The search for the origins of human language is a search for a generative and growing system of human exchange between minds and external events, as James Hurford puts it.

Origins and Evolution

The question of where and when the earliest language occurred requires studying the evolution of modern humans from their ancestors, the first hominids (i.e., beings within the taxonomic family, Hominidae). For explorations of human language evolution, it is necessary to consider two realizations of language, gestural language and spoken language.

Gesture

The majority of study regarding the evolution of language focuses upon verbal communication, but a second area of research provides many answers about human language, that is, gestural language. In the latter case, scientists may be concerned with the communicative function of gestures in themselves, or they may be concerned, as is Michael Corballis, with the evolutionary processes of gestural language to speech. He proposes that language emerged from manual gestures rather than from vocalization.

Although there is no documented evidence that early hominids used gestural language, Corballis provides a set of arguments supporting the practicality and effectiveness of gestural communication that may have been used as far back as 25 to 30 million years ago. For example, as an essentially spatial mode of communication, gesturing would provide a silent visual means to alert others about predators or to support the hunting of game.

Somewhere around 6 to 7 million years ago (mya), hominids and apes diverged into distinct groups. At that time, archeological findings indicate, both groups were assumed to have used simple gestures, and both probably vocalized for emotional needs. Hand signals for purposes other than just emotional release are believed to have been used by hominids designated as the genus *Australopithecus,* who are identified by two important characteristics: nonprojecting canine teeth and bipedal locomotion. Because they lacked dental defenses, they needed to develop some way of surviving. Thus, C. Loring Brace postulates that they must have created primitive cultures. This idea is backed up by the discovery of rudimentary stone tools from around 2.5 million years ago. Scientist Ralph Holloway, among others, comments that the early bases for language may have had roots at the time of tool creation and use. Hominids at this time would have had a reason to establish groups for hunting, scavenging, and cooperative living and thus have had a need to develop forms of communication within these groups. Because the hands were no longer needed for locomotion, they could be used for gestures and making tools. These activities could have stimulated the evolution of the brain and especially those parts of the brain necessary for speech.

It seems logical that the early *Homo erectus* would use hand signals and body gestures to communicate. Corballis notes that, given the fossil evidence, by 2 mya the gestures would have been fully syntactic and vocalizations would have been progressively increasing. By 100,000 years ago, early *Homo sapiens* would have used more speech than gestures, but gestures would still have been a characteristic of communication.

Theories about early communication gestures have been most affirmed by those who study manual languages such as American Sign (ASL) or any of the other modern sign languages used by deaf persons in countries other than the United States. Just as research about language evolution and gestures has been slow in developing, it was not until the 1960s that sign language research began to build with studies of the grammar of ASL. The growth of research was due to the work of William C. Stokoe, who provided the foundation for identifying sign languages as true, formally structured languages and not merely cryptic means for communication. Since his time, sign language research has included study within several areas of linguistics, including sociolinguistics and neurolinguistics.

In the field of sociolinguistics, researchers such as James Woodward have established a large body of knowledge of interlanguage communication, studying areas of the world where deaf persons use sign language within their own established communities and/or with hearing individuals. This research is similar to that regarding the development of pidgins and creoles in spoken languages.

What of gestural language among hearing, speaking individuals? Gestures supply visual–spatial support to conversations, enabling a speaker to use emphasis or to demonstrate information that otherwise might take a longer period of time to describe verbally. When there is little or no opportunity to speak, gestures can take on a form of syntax designating meaning. For example, individuals who try to communicate with someone who speaks a language other than their own use hand gestures accompanied by facial expressions and body language in an orderly fashion that parallels their spoken language.

Considering neurolinguistics, much has been learned from the last decade of the 20th century into the 21st regarding language function areas in the brain and sign language. For example, just as do hearing listeners of spoken language, deaf persons show increased activity in the left side of the brain for receptive communication in sign language. Both of the language-mediating areas of the brain, Broca's area and Wernicke's area, are activated when deaf signers watch sentences presented in ASL. Even so, deaf persons show more activity in the right side of the brain than do hearing persons. This has frequently been attributed to the fact that ASL has a distinct spatial component where facial, head, and body movements complement the hand signs.

Philip Lieberman explains that the hominid brain and hand evolved in a precise way that facilitated manual dexterity. He notes that studies in the 20th century comparing the brains of normal individuals to aphasics and those with mental retardation show that there is a link between speech and syntax and motor activity. If that is true, then it makes sense that as the human brain developed, and as the skull and throat developed in the earliest hominids, so too would the use of gestures have developed as well as the growth of spoken language.

Spoken Language

Scientists document that during the Late Pleistocene period that began around 130,000 years ago, brain size in hominids, classified as archaic *Homo sapiens,* had reached modern levels. The increased expansion of the brain is taken as a sign of brain reorganization for language. At the same time, another major indicator of the evolution of humans, the manufacture of tools, shows multiple traditions of stylistic differences across specific regions of the world to which these hominids had migrated. C. Loring Brace explains that these geographic areas are coincidental with the areas of the world that have been mapped for major language families. Thus, he posits that there may be a connection between the development of tools and the development of verbal communication in these geographic areas.

Fossils show that the necessary mouth and throat anatomy for speech were in place even 150,000 years ago. The structure of the human brain and the larynx have evolved so that humans are the only ones who have a low larynx that facilitates speech but that can also interfere with swallowing and be obstructed by food. With this kind of physical structure, it would seem that hominids of the middle Paleolithic era would necessarily cultivate vocalization. Yet, this is a quandary, because other evidence (e.g., hints from unearthed art, burial and living sites, and jewelry) would signal that verbal language did not appear until about 40,000 years ago. This time of social organization and growth is known as the Upper Paleolithic explosion. Most archaeologists would agree that at that time there must have been liberal use of language.

Some scientists believe that there was a sudden cultural shift; others believe the development of speech use was gradual. Just because the hominid anatomy was available for speech, the evolution of the neural pathways may not yet have been ready earlier in time, that is, before 50,000 to 40,000 years ago. Another missing piece of the puzzle is the fact that it is hard to infer without fossil evidence the ways that soft tissue was connected to the anatomy of the vocal tract.

Lieberman believes that speech should have been possible between 150,000 and 100,000 years ago, because there is evidence that the African hominid living during that time had a cranial structure identical to that of modern humans. The tongue was rooted in the throat and not restricted to the vocal cavity, and the larynx had descended. This kind of structure is necessary for rapid and complex speech, and there would be little reason to think that speech might not exist among these hominids. However, their vocal tracts were not completely formed, nor were their abdominal and thoracic muscles that, among modern humans, receive stimulation for speech through the thoracic region of the spinal cord.

Many scientists indicate that there is little evidence about the development of human culture that would support the argument for the necessity of speech before 50,000 years ago. Alternatively, some scientists, such as John Yellen and his colleagues, have discovered signs of a kind of trading network among the Katanda people of Africa, as well as barbed bone, obsidian, and stone used for spear points, and the use of pigments such as red ochre that can be dated between 80,000 and 90,000 years ago. Yellen and his colleagues explain that these archaeological traces may indicate the existence of early societies whose members had modern behavioral potential and hominids with anatomies that resembled those of modern humans. Thus, there may have to be a change regarding the target year estimate of 40,000 years ago for the emergence of speech.

Clues From Written Language

Observations of written language provide clues regarding the existence of grammar and syntax of spoken language as well as changes in these components that signal the difference between human language and that of other species.

Formal Written Language

There is some agreement among scholars that the oldest known formal written language was in use about 5,000 years ago and was probably Sumerian, an extinct isolate language that shows few characteristics common to any of the other spoken language families. There are over 30,000 preserved cuneiform writings in this language as well as evidence from remnants of Sumerian monuments. These show accounts of commerce and politics, culture, poetry, myths, and epics about the origins of civilization in the Fertile Crescent. As a spoken language, Sumerian must have existed prior to being put down in writing. It was spoken in Mesopotamia, in part of what is now modern Iraq, until 2,000 BCE and was supplanted by Akkadian, a Semitic language that like Sumerian is now extinct. The Akkadians, a nomadic people, lived peacefully among the Sumerians until the conquest by Sargon of Akkad, who insisted upon language assimilation. The geographic region continued to use Semitic languages through the time of the Babylonian empire, whose people spoke a dialect of Akkadian. Subsequently, another Semitic language, Aramaic, predominated in the region.

Informal Written Languages

Writing systems that are characterized as representations of informal language lack the consistent use of grammar or syntax. Informal written languages existed as far back as 30,000 years ago. Evidence is seen in cave and other drawings and engravings at burial sites and in artifacts such as jewelry. These forms are found in several key archaeological sites around the world. Although explorations have been made predominantly on the African continent and in the Near and Middle East, one land that has been a source of curiosity about language evolution is Australia. The Aborigines, living in Australia for perhaps 40,000 years or longer, are one people whose ancestors have left a pictorial record across the continent in carvings and paintings representing symbols of life around them, some of these over 23,000 years old. Many of the Aboriginal drawings are in geometric shapes, and they appear to be accounts of numbers of people and events within designated time frames.

It has been estimated that at one time there may have been over 500 spoken languages among 200 distinct groups of Aborigines. These languages show no clear connection to any of the other language families of the world and have been classified by anthropological linguists into 28 of their own families. Currently, in a population of some 400,000, there are fewer than 30,000 who speak one of the remaining 263 original languages. And, it has only been a little over 100 years since outsiders have helped the Aborigines establish a formal written language based upon the visual material that has survived and developed over the centuries. However, there is still much to be learned about these people to put one more piece in place in the puzzle of language evolution.

Language and Future Changes

Recent research in anthropological linguistics as well as genetic linguistics has supported our understandings of the origins, evolution, and current status of languages around the world. Coupled with studies by allied and growing fields such as computer science, biology, and neuroscience, the horizon appears bright, revealing ever more information about language change over time. Linguist Salikoko Mufwene remarks that there is a need for researchers to be open to a variety of alternative assumptions that might advance and enrich current and future thought about language evolution. He states,

> A language is more like a bacterial, Lamarckian species than like an organism. A subset of innovations/deviations in the communicative acts of individual speakers cumulate into the "invisible ecological hand" that produces evolution. . . . There are internal and external factors that bear on language evolution, but they apply concurrently in all cases of language evolution.

Given what is already known about the development of language in humans, it is highly probable that scientific explorations, especially by researchers of several fields collaborating, will continue to reveal subtle as well as vivid evolutionary processes that enable creative and distinct language use among all persons across the earth.

Patricia N. Chrosniak

See also Anthropology; Evolution, Cultural; Hammurabi, Codex of; Language; Languages, Tree of; Rosetta Stone

Further Readings

Asher, R. E., & Moseley, C. J. (Eds.). (2007). *Atlas of the world's languages* (2nd ed.). New York: Routledge.

Brace, C. L. (1995). *The stages of human evolution.* Englewood Cliffs, NJ: Prentice Hall.

Chomsky, N. (2000). *New horizons in the study of language and mind.* Cambridge, UK: Cambridge University Press.

Christiansen, M. H., & Waller, S. (2003). *Language evolution.* New York: Oxford University Press.

Corballis, M. C. (1999). The gestural origins of language. *American Scientist, 87,* 138–145.

Corballis, M. C. (2002). *From hand to mouth: The origins of language.* Princeton, NJ: Princeton University Press.

Crystal, D. (1997). *The Cambridge encyclopedia of language* (2nd ed.). New York: Cambridge University Press.

Greenberg, J. H. (1968). *Anthropological linguistics: An introduction.* New York: Random House.

Hauser, M. D., Chomsky, N., & Fitch, W. T. (2002). The faculty of language: What is it, who has it, and how did it evolve? *Science, 298,* 1569–1579.

Holden, C. (1998). No last word on language origins. *Science, 282,* 1455–1458.

Lieberman, P. (1998). *Eve spoke: Human language and human evolution.* New York: W. W. Norton.

Mufwene, S. S. (2001). *The ecology of language evolution.* New York: Cambridge University Press.

Stokoe, W. C. (2002). *Language in hand: Why sign came before speech.* Washington, DC: Gallaudet University Press.

Yellen, J. E., Brooks, A. S., Cornelissen, E., Mehlman, M. J., & Stewart, K. (1997). A Middle Stone Age worked bone industry from Katanda, Upper Semliki Valley, Zaire. *Science, 268,* 553–556.

Languages, Tree of

It is estimated that there are between 6,000 and 10,000 living languages in the world. Not all of these languages have been identified and not all have names; not all of these languages will survive in the 21st century. The languages that prevail have done so because of evolutionary processes over vast periods of time. These processes have been characterized and documented by scholars, and all identified languages have been represented using the metaphor of *language families* in schematics known as *language trees.*

Among human languages, there are spoken languages that are indigenous to a geographic area, those that are spoken in places other than the area of origin, and those that are called *official* languages as designated by an official body within a geographic community. There are creole and pidgin languages that developed from contact between two particularly different language-speaking groups, typically a result of trade between Europeans and peoples on other continents, such as Africa. There are language dialects that are sometimes designated as languages, as occurs in China. There are also gestural languages, such as sign languages used by deaf persons. Each sign language has a distinct grammar, syntax, and phonology according to the country in which it is used. It is estimated that there are at least 112 sign languages in the world.

There are many reasons for knowing about the organization and life of human languages. One reason is to support understanding of the history, the current dynamic, and the future of specific cultures and nations. Another is to assist in understanding the role that diversity plays among peoples in the world.

In 1993, the United Nations Educational, Scientific, and Cultural Organization (UNESCO) embarked on a project to find out which languages were viable, endangered, close to extinction, and extinct. As a result of this process, classification schemes for the world's languages were modified, and maps were collected into an atlas that was in its second edition in 2001.

Besides the UNESCO project, the study of the origins of human languages has been especially fruitful from the last decade of the 20th century up

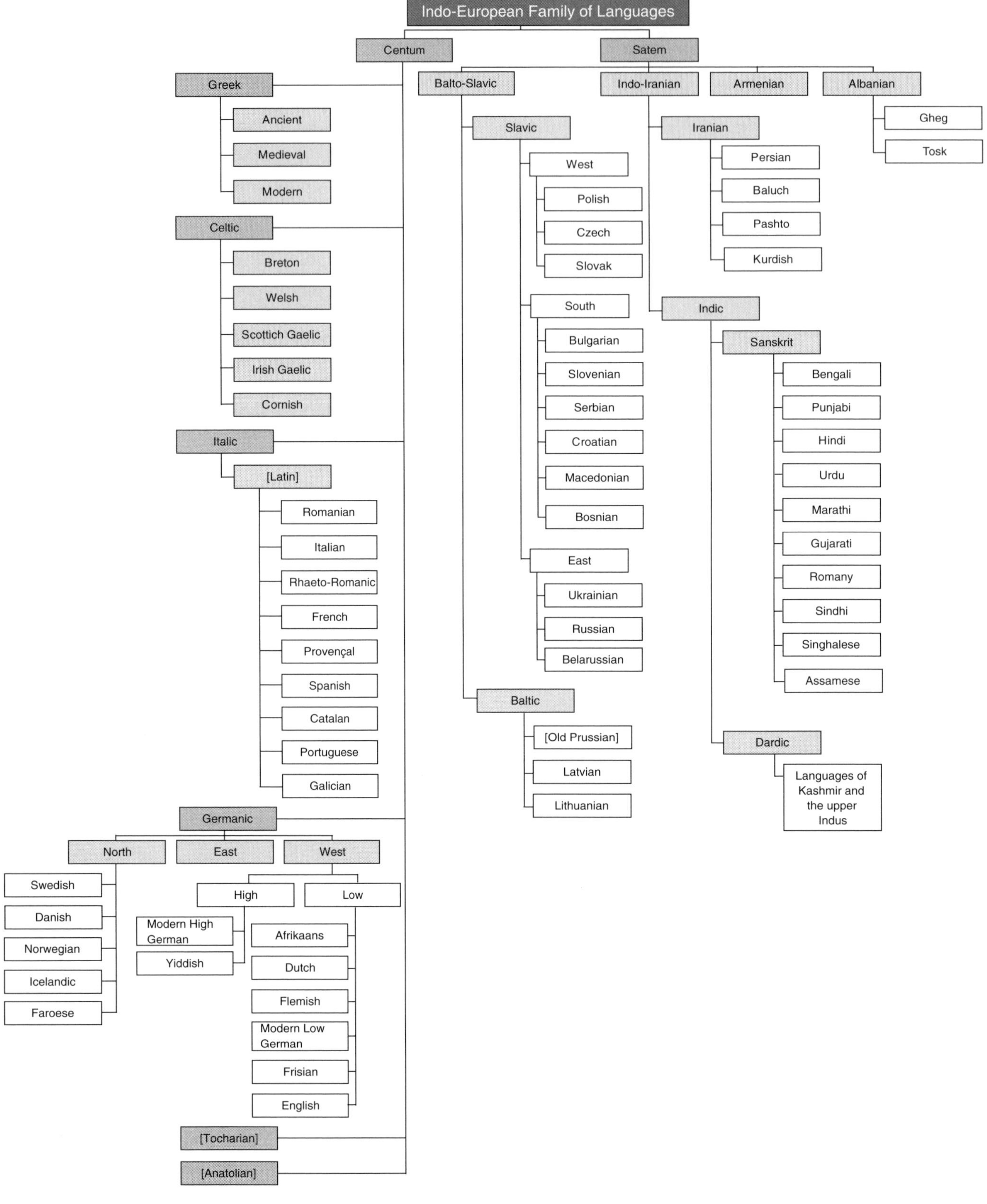

Figure 1 The Indo-European family of languages

to the present, and this research has helped to create schematics for the classification of languages into their particular families. Currently there are 94 different language families in the world whose member languages have been identified by a process called *genetic classification,* where linguists and philologists looked for similarities that showed descent from the same parent language.

Lyovin explains that languages are compared to one another especially for recurring sound correspondences between and among the words of languages that have roughly the same meaning and belong to the basic vocabulary. Each language can be further characterized in relation to other families, constituting distinct *language trees.* The schematic for these trees is usually reversed in a hierarchical format with the main, central family name at the top and the branches extending out systematically below it. For example, there are over 2.5 billion speakers of languages in the Indo-European family, the most studied of language families. This number constitutes 44.78% of the world's people. The Indo-European family tree is divided into two major components, the Centum languages (Western European) and the Satum (Eastern European and Asian). Subsequently, there are branches, such as West Slavic, that then are subdivided into the actual spoken languages of each branch.

Although the Indo-European language family predominates for number of speakers in the world, there are only 430 living languages in this tree. The Niger-Congo tree, however, is constituted of 1,495 living languages that are spoken by 6.26% of all the world's people. Six of the 94 language families account for about five-sixths of the world's population. Among the other 88 language families, there are *isolates:* individual languages with no known connections to any other group of languages. Gilyak, for example, is an isolate language spoken by about 1,000 people on Sakhalin Island and along the Amur River in East Asian Russia. Not every language that is spoken by few individuals is an isolate. Yukhagir is a separate language family of East Asian Russia that is spoken by only 120 people, mostly the elderly, and it is expected to become extinct once these speakers die.

As stated above, each language tree is composed of several subcategories, or *branches* of languages that share certain common characteristics. Languages in the same branch may or may not share commonalities of grammar and syntax, vocabulary and phonology. Some may use the same scripts and alphabets, while others show no similarity in this regard. For example, French and English have comparable alphabets, although French has diacritical marks not used in English. Russian uses a Cyrillic alphabet that has little resemblance to the variation of the Roman alphabet used for Polish. There are, however, words that sound quite similar in Russian and Polish, such as the word for *opportunity* (in Polish, *okazja [oh kahzh'yah];* in Russian, transliterated from Russian Cyrillic to English, *okaziya [oh kah zy'yah]*). Russian is an East Slavic language in the Indo-European family tree; Polish is a West Slavic language.

Some languages have an exact one-to-one correspondence between each letter of their alphabets and each sound in words; others have several sounds that are associated with the same letter or combination of letters. For example, in English /g/ can be "hard" or "soft" as in the words, *gorilla* and *cage.*

Languages vary in the number of consonants and vowels and how they are produced. There are tonal languages such as Chinese where words are pronounced with obligatory changes in pitch, or tones, to each stressed syllable. In Mandarin Chinese, the national language, a syllable might look like this example from Lyovin:

(C)(G)V (N or G) + Tone

where C represents a consonant, G a glide (nonsyllabic vowel), V a full vowel, and N a nasal consonant. There are 235 living languages in the Republic of China. Chinese is in the Sino-Tibetan languagc family; the languages of Myanmar and Thailand are also members of this family.

One of the oldest language families in Africa is the Khoisan. There are estimates that languages in this family existed some 60,000 years ago, which might place them beyond the commonly proposed time for the appearance of speech at 40,000 years ago. Khoisan languages are recognized by the incorporation of clicks in speaking; the grammars of the 22 languages are rather distinct. Most speakers are found in southwestern Africa around the Kalahari Desert regions, particularly in Botswana and Namibia. Some language scientists, such as Abram de Swaan, are concerned that the neglected study of such small language families as Khoisan might mean the eventual loss of information about the origins of language and the development of cultures prior to colonization by the Western world.

Patricia N. Chrosniak

See also Anthropology; Evolution, Cultural; Language; Language, Evolution of

Further Readings

Asher, R. E., & Moseley, R. E. (2007). *Atlas of the world's languages* (2nd ed.). New York: Routledge.

de Swaan, A. (2001). *Words of the world.* Malden, MA: Blackwell.

Gordon, R. G. (Ed.). (2005). *Ethnologue: Languages of the world.* (15th ed.). Dallas, TX: SIL International.

Lyovin, A. V. (1997). *An introduction to the languages of the world.* New York: Oxford University Press.

Wurm, S. A., & Heyward, I. (2001). *Atlas of the world's languages in danger of disappearing* (2nd ed.). Paris: UNESCO.

Laplace, Marquis Pierre-Simon de (1749–1827)

Marquis Pierre-Simon de Laplace, French mathematician and cosmologist, sought to explain the origin of the universe in terms of mathematics, physics, and philosophy. Although his conceptual reference for the temporal framework of the universe is not evident, other than under the general heading of orbital computations, Laplace's speculation within the nebular hypothesis poses a unique and interesting account for the origin of our solar system. Laplace was well known for his publications, *Systems of the World* (1796) and *Celestial Mechanics* (1825).

The existence of the solar system, with all the stars, planets, and comets, retains the appearance of regularity and stability. Laplace, with observations and mathematical computations, understood the governing factors and influence of both gravitation and motion among celestial objects. The vastness of our solar system, as evident by cometary orbits, was suggested as going beyond the then-known limits of our system. It was postulated that the solar system was a product of the expansion and contraction of a superheated sun whose atmosphere existed beyond the known planets. These newly formed planets possess similarities in directional rotation (including satellites), plane, and projection. Such commonalities were evidence of a common origin, albeit only a probable origin.

As for temporal facets of the known solar system, Laplace's concepts of time are held within the mechanical operations of a solar system in particular and the universe in general. The totality of time or age of the solar system or universe became secondary to operational understanding. Perhaps in Laplace's view, time measured within human consciousness becomes irrelevant when compared with the vastness and mechanistic system of the universe. Essentially, the birth of our solar system, like that of the universe, is indifferent to humankind's existence or to life in general.

It is rather interesting to note that before modern science and evolutionary theory, Laplace had mentioned, though briefly, that life may exist on other planets, or in his words, this is extremely probable. This assumption concerning the probability of the existence of life on other planets can be seen as a logical outcrop of the commonalities that are shared among the planets. This novel idea of extraterrestrial life is devoid of both human and divine interaction. Should life forms exist elsewhere in the universe, their existence becomes as irrelevant to our species as the universe is to human existence. The mechanistic nature of the universe, both in implementation and adjustments, becomes a set of interconnected subsystems in which individual life, assuming the existence of life on other planets, becomes insulated. Within Laplace's cosmology, there is no need for a divine essence that permeates the universe. Probability and mechanics replace the deity or force in both planetary and human existence.

In Laplace's substantial contributions, inferences regarding time become subtle statements without regard for theology and philosophy. The replacement of the Son with the Sun, as it were, has remarkable connotations for human existence and placement within nature. Laplace advised extreme caution to readers regarding his conjectures. Nevertheless, the materialistic explanation and justification seems to support this view. It is unknown if he held any theistic views before his death in 1827. However, the contradiction between science and traditional religious beliefs would surely become problematic. The concept of time as supported by science would be greater than human experience would allow. Timelessness or eternity would only be inferred as applying to the planets and the mechanical operations of solar systems and the universe.

David Alexander Lukaszek

See also Cosmogony; Kant, Immanuel; Nebular Hypothesis; Planets; Time, Planetary

Further Readings

Laplace, P.-S. (1966). *Celestial mechanics.* Bronx, NY: Chelsea. (Original work published 1825)

Laplace, P.-S. (1976). The system of the world. In M. Bartusiak (Ed.), *Archives of the universe.* New York: Vintage Books. (Original work published 1796)

Lascaux Cave

Lascaux Cave, like many archaeological sites, was discovered by chance. Four young boys stumbled across the cave in 1940 in southern France near the town of Montignac. Inside, these adventurers encountered what later researchers further explored and analyzed: a cavern filled with paintings and engravings on limestone walls depicting an array of images, which were determined to have been created millennia before. Today, Lascaux Cave's contents provide a glimpse of a people and a time in prehistory of which little else is known.

Lascaux is a multichambered cave extending approximately 250 meters in total length when considering all its areas. Chamber widths vary considerably so that the height of Lascaux's ceiling is equally diverse in size, going from less than one meter to several meters high, with variations in dimensions sometimes happening within a single chamber. As for when the cave was occupied, researchers conducted the dating of Lascaux's occupations through the use of multiple techniques including carbon-14 dating and the observance of animal species depicted in the wall art. The results of these tests have generated a range of corresponding dates, particularly between 18,000 and 16,000 years ago.

The actual paintings and engravings within the Lascaux Cave include a myriad of designs ranging from realistic and abstract images of animals to images and designs whose meaning remains undetermined. The images include horses, aurochs, bison, and felines. In addition to the wall art discovered in the Lascaux Cave, researchers discovered tools such as projectile points, miscellaneous flint tools used for engraving, and pottery sherds. As with other caves discovered in Europe with similar artwork, the paintings and engravings within the Lascaux Cave demonstrate the use of multiple techniques to generate images on stone. The materials used to generate colors for the wall art at Lascaux were predominately iron and manganese oxides.

Questions remain as to the importance and function of these images to their creators. Some researchers have postulated that the images were to serve as magical aids in hunting, perhaps in rituals; others have argued that the images were created as a form of decorative art. While answers to these questions remain elusive, interest in the images of Lascaux Cave remains strong. However, further investigation of the cave and its contents has been suspended and access has been restricted. This was considered necessary to protect the cave and its contents from continued deterioration caused by multiple contaminants that were destroying Lascaux's art. When and if access will be granted to researchers in the future remains to be seen. Fortunately, researchers have at least acquired a glimpse into the lives and activities of those who occupied and decorated Lascaux Cave, thereby providing us with some understanding of an era that remains relatively unknown.

Neil Patrick O'Donnell

See also Anthropology; Altamira Cave; Chauvet Cave; Evolution, Cultural; Olduvai Gorge

Further Readings

Aujoulat, N. (2005). *Lascaux: Movement, space, and time.* New York: Harry N. Abrams.

Tattersall, I., & Mowbray, K. (2006). Lascaux Cave. In H. J. Birx (Ed.), *Encyclopedia of anthropology* (Vol. 4, pp. 1431–1432). Thousand Oaks, CA: Sage.

Last Judgment

The Last Judgment refers to an event that, according to religious tradition, will occur after the world has ended and God pronounces his final verdict on the human race. For Christians, the Last Judgment is the stage in which all people are judged according to their behavior when they were still alive. The righteous will receive their reward and will spend an eternity in fellowship with God; however, the evil will spend an eternity in hell, acknowledging they had the opportunity for heaven but chose otherwise.

Some Christians believe that there will be only one judgment, because one's soul is asleep or unconscious between its demise and the Last Judgment (a view held by the religious reformer Martin Luther and others). Most Christians believe, however, that souls are not asleep and that they receive their punishment or reward after death. This first judgment is called the *particular judgment,* and it is different from the Last Judgment, where individuals are sentenced for their belief or lack thereof. This acceptance regarding one's final judgment is considered dogma by Roman Catholic believers, but the Church feels that the Last Judgment is not an actual trial, because the individuals already deceased are either residing in heaven or hell or working off their sins in purgatory—all resulting from their particular judgment at the moment of their death. Protestant believers in millennialism regard the two judgments as describing separate events, one at the moment of death and one after the end of the world.

According to this belief, the Last Judgment will take place after the deceased have been resurrected and undergone a complete reunion of the body and soul, where their evil acts are judged and their eternal sentence will be known to all before their fate before the resurrection is continued. At that specific moment, the joys of heaven and the sorrows of hell will be evident, because everyone present will be able to feel both pleasure and pain. This scenario appears most directly in "the sheep and the goats" section in the Book of Matthew. This belief regarding the afterlife can also be found in the books of Daniel, Isaiah, and Revelation.

Christianity is not the only religion that deals with the end-times. In Islam, there is *Yawm al-Qīyāmah* (the day of the resurrection). At a time preordained by Allah but unknown to humanity (at a time when people least expect it), Allah will consent for the Qiyâmah to commence. The archangel Israfel, named "the caller" in the Qur'an, will puff mightily into his horn, and out will come a "blast of truth." This specific occasion can also be seen in Jewish eschatology, where it is called "the day of the blowing of the shofar," found in Ezekiel 33:6. The Qur'an states that the Qiyâmah will last 50,000 years. Some Islamic scholars believe that this period refers to the vastness of man's spiritual advancement (one day is equal to 50,000 years), or that the day may signify the final triumph of Truth in the world, from the time when revelation was first granted to man.

In the Qiyâmah as envisioned, *Alameen* (humanity, the jinn, and all other living beings) are gathered upon a vast, white plain under a blistering sun. Each is completely unclothed, uncircumcised, and standing so close together that some are submerged in their own sweat. The depth of one's submergence in sweat depends upon how good or pious one was. Those that practiced good *adab* (following good etiquette) by following the Five Pillars of Islam daily are considered *nadirah* (shining and radiant). However, the expressions of the disbelievers are called *basirah* (miserable and grimacing). Although they are all nude, anxiety and fear are so great that no one thinks to look at another's nakedness. The creatures wait to be brought before Allah for their judgment, but they are terrified. The prophets plea to Allah with the phrase "sallim, sallim," translated as, "Spare your followers, Oh God." Those that followed Muhammad while he was alive, but then left

Islam after his demise, are apostates and are engulfed in fire. The angels are afraid, as state several *hadiths* (actions and utterances the prophet made while on earth), because Allah is livid with anger, more so than before or after.

In the Hindu religion, *pralay* is the specific time when the earth will be completely annihilated. Garuda Purana is one of the *puranas* (means "belonging to ancient time" and is a genre in Sanskrit literature) that are part of the Hindu body of texts known as the *smriti*. This sacred text discusses in vivid detail precisely what to expect after someone dies, specifying the different torments one can expect for the evil committed while alive, including being scorched in hot, boiling oil and given to bloodsuckers as prey. Some Western theologians see this scenario as being comparable to the Christian idea of Judgment Day.

Garuda Purana are directions given by Lord Maha Vishnu to Sri Garuda (also known as the king of birds—a *vahana* of Lord Vishnu). This specific purana looks at a wide range of subjects, including cosmology, various remedies for sickness, language syntax, and information regarding jewelry. The Garuda Purana is deemed the most respected Vedic book discussing the Nine Pearls (the nine sacred gemstones), and the second half of the tome deals with life after death. The Hindus residing in northern India customarily read this purana as they cremate the bodies of their loved ones.

Cary Stacy Smith and Li-Ching Hung

See also Apocalypse; Bible and Time; Christianity; Eschatology; God and Time; Gospels; Immortality, Personal; Luther, Martin; Michelangelo Buonarroti; Nirvana; Parousia; Qur'an; Religions and Time; Revelation, Book of; Time, Sacred

Further Readings

Braaten, C. E. (2003). *The last things: Biblical and theological perspectives on eschatology.* Grand Rapids, MI: Eerdmans.

Schwarz, H. (2001). *Eschatology.* Grand Rapids, MI: Eerdmans.

Smith, J. I. (2006). *The Islamic understanding of death and resurrection.* London: Oxford University Press.

Upton, C. (2005). *Legends of the end: Prophecies of the end times, antichrist, apocalypse, and messiah.* Ghent and New York: Sophia Perennis et Universalis.

Latitude

For the purpose of finding an relative location on the earth's surface with precision, geographers have drawn certain circles to which the position of any point may be referred. The art of navigation may have begun approximately 6,000 to 8,000 years ago; eventually it became a science of vast importance, particularly during Europe's age of discovery, when knowing one's exact position at sea saved both ships and the lives of sailors. With longitude being determined by time, and latitude by the angle of the sun at noon, mariners were able to find their exact position.

The earth, not a perfect orb, is an oblate spheroid, as discovered by A. R. Clarke in 1866. It is known that the earth rotates on its polar axis, or a line between the north and south poles. From this axis great circles can be contrived, such as the equator or 0° latitude, which has a 90° angle to the axis, and divides the earth in half, making the equator the largest or great circle. With the equator at 0°, the north and south poles are at 90° north and south latitude, respectively, with various angles from the axis in between. Latitude, then, is a line or parallel on the earth's surface that runs east and west at various angles from the axis and measures distance north and south on the earth's surface. Thus, the only latitude being a great circle is the equator with all other latitudes being smaller circles.

But, a number of great circles can be drawn on the earth depending on the angle from the axis, such as longitude. Longitude is a line or meridian that runs north and south but measures distance east and west (see the entry on John Harrison in this encyclopedia). Using an accurate timepiece such as a ship's chronometer, longitude can be calculated, and using a sextant the latitude can be found so that an exact position on the earth's surface can be plotted. Depending upon the accuracy required, latitude and longitude are measured by degrees, minutes, and seconds of an angle. The distance between lines of longitude decreases as one proceeds from the equator to the poles or as latitude increases.

It was Pythagoras, around 529 BCE, who first advanced the concept of the earth being a sphere with a moon evolving around the sun along with five known planets at that time. Later, Eratosthenes, around 300 BCE, was the first to place parallels on a sphere, but he named them after places. He also calculated the earth's circumference very accurately for that time, using the distance between two latitudes in Egypt. It was Ptolemy who first wrote about latitude and longitude in his *Geographia,* about the time that Alexander the Great was expanding Greek civilization.

The tools of navigation are a chart, a chronometer, a compass, and a sextant. Probably the first tool was the chart. The first charts, perhaps appearing as early as 2,700 BCE, were quite inaccurate. But, as time passed and technology improved, charts have become very accurate, so that contemporary charts actually a margin of error of less than one tenth of 1%. For measuring latitudinal position, the common quadrant was invented. From this crude instrument the astrolabe developed, first appearing about the 3rd century BCE. The astrolabe, like the common quadrant, was a device to measure the angle of the sun and various stars from the zenith. It was a cumbersome procedure involving several mariners. Later, the cross-staff was developed, with which the visible horizon was used rather than the zenith. Then, the back-staff or sea quadrant was invented by John Davis in 1590. Finally, in 1730, John Hadley perfected the sextant that is still in use today, commonly to measure latitude at noon using the angle between the horizon and the sun.

Another important tool is the compass. Crude though it may have been at first, it certainly decreased the chance of being lost at sea or in the desert. Perhaps the first compasses came into use about around the 11th century and were used by Norsemen. But, it wasn't until almost the 20th century that an acceptable compass was developed by Lord Kelvin for the British Admiralty. An accurate timepiece for use at sea, the chronometer, was developed by John Harrison in the 18th century for measuring longitude; this was the last tool to be developed for accurate navigation at sea. Today, the global positioning system (GPS) is used to find accurate location or position.

Latitudes of significance, besides the equator and the poles, include the Arctic and Antarctic Circles at 67.5° north latitude and 67.5° south latitude, respectively, and the Tropic of Cancer and the Tropic of Capricorn at 23.5° north latitude and 23.5° south latitude, respectively. These smaller circles represent the relationship of the sun's illumination at the summer and winter solstices. During the vernal equinox and the autumnal equinox, the sun's rays are directly on the equator, but during the solstices the sun's direct rays are over the tropics of Cancer and Capricorn, while the poles are in either total darkness or sunshine. These relationships between the earth and the sun are directly related to our weather patterns, seasons, and climate.

Richard A. Stephenson

See also Astrolabes; Chronometry; Harrison, John; Presocratic Age; Timepieces

Further Readings

Bowditch, N. (2002). *American practical navigator.* Washington, DC: National Imagery and Mapping Agency. (Original work published 1802)

Christopherson, R. W. (2004). *Elemental geosystems* (4th ed.). Upper Saddle River, NJ: Prentice Hall.

Dunlap, G. D., & Shufeldt, H. H. (1972). *Dutton's navigation and piloting* (12th ed.). Annapolis, MD: Naval Institute Press.

Law

Law is a cultural achievement of humankind. It is subject to an evolutionary process in which a sense of right and wrong is developed, and the legal system as the end of this evolution is connected and continuously adapted to the conditions of human existence within a complex framework of social interaction. Essentially the main function of law consists in judging and sanctioning human behavior according to the judicial code of legal/illegal. It is possible to differentiate systems of law from an historical point of view depending on whether this function is intended to uphold and continue past customs or to steer the future.

Ancient and medieval systems of law are structurally characterized by a strong connection between law, politics, and morals; a case in point is the historical trial against Socrates (469–399 BCE). The law of those eras is deeply rooted in social customs and traditions. It is not so much the will of the lawmaker that determines what is right. What is right is primarily established by the authority of the judge, often personified by rulers themselves, who legitimize the law through their decisions. His point of view is principally directed toward the past as a time when conditions were right. And their verdict must uphold this continuity—old law is good law.

According to Georg Wilhelm Friedrich Hegel (1770–1831), legal thinking changed with the "invention" of subjectivity during the eras of the Renaissance and the Reformation. Individuals came to be seen as subjects who make themselves the persons they are. No longer did the demands of the past determine the life of an individual. Instead, individuals came to be defined by the tasks they choose themselves and their dedication to fulfilling them. But the joy of being the creative sculptor of one's own biography, the "plastes et factor," as Pico della Mirandola (1463–1494) fittingly calls it in his *Oratio de Dignitate Hominis,* is joined by worries about the future, a future that is essentially open and uncertain. Consequently, even today, the main task of law is to steer the future.

Temporality of Law

It has been obvious since the Enlightenment, or beginning of the Modern Era, that differentiations are made between law, politics, and morals. Living an upright life has increasingly become a question of outwardly obeying laws with no particular regard to moral considerations. The difference between making and enforcing laws has been institutionally reinforced. In general, the judge's verdict has been given a legal framework and thus become much less dependent on the center of political power—due in part to the condition that decisions must be made promptly. Law will always be power oriented, but, thanks to its own normative nature, it also establishes a bond with the future in that general expectations of an individual's behavior are stabilized—even though these expectations may be disappointed in individual cases. An essential factor that fosters this function of law is the construction of its own temporality.

Self-reflexivity is a prerequisite for this construction of law's inherent time: Law exists, because it is valid—specifically as long as it is valid. Differentiating between natural and legal time is therefore relevant for the validity of law. For example, as long as law is carried out under the auspices of a particular ruler, the validity of the regulations is bound to the "lifetime" of that ruler. As a result, the effectiveness of a legal system decisively depends on whether it is possible to create and change norms according to its own legal rules, in other words, to separate them from a specific time period. This process is particularly evident, because there is usually no mention made about when the validity of legal norms ends. They are, in a certain sense, "timeless." In some cases possible changes are even explicitly excluded, for instance by rhetorical expressions, such as human rights are "inalienable," human dignity is "inviolable," and the core of a constitution is valid "forever." These examples make it clear how the normative nature of law tries to immunize itself against social changes, especially in the field of politics. From a legal perspective, it may not be possible to prevent rulers from violating human rights; but by no means should they claim to act in accordance with the law when violations are committed.

Positivity of Law

The construction of a legal time provides for law's ability to structure its relationship to natural time: Legal validity means the present, which is differentiated from life's unchangeable past but also from the open (and, in that sense, more or less uncertain) future. In this corridor of the present, legal norms can develop their own "history" but are also latently subject to becoming history by a stroke of the legislator's pen as a result of their positive nature (i.e., by being recorded in a fixed, written form). Yet, seeing as their "time" is left to the will of the legislature, their history must be differentiated from the historical course of events. In effect, legal norms represent a particular moment in history that is highlighted in the course of time and conserved. Their changeability takes place in accordance with the given rules of law, in other words, in accordance with those norms that are higher and that determine the changeability of the individual legal norms. If that were not the case, norms could be changed at any time and would never develop a "timeless" effect.

The positivity of legal norms structurally guarantees a somewhat anachronistic character by welding past, present, and future under the conditions of law and creating a context that detaches itself from natural time. That is immediately obvious in the power of legal norms to call forth a specific future. They determine which behavior will be right and which will be wrong in the future. Of course, it is neither possible to know what the future will bring nor how individuals will behave, but it has already been established how their behavior will be judged legally. Following Immanuel Kant (1724–1804), this does not relieve individuals of their freedom to select their own goals and determine how they want to behave, but it requires everyone to respond to the law. Only with this common reference, which can be taken into consideration by evaluating the consequences of one's own behavior, is it possible to establish a common present.

Law also reacts in a similarly selective way in regard to the past; for all that remains, from a legal point of view, is what the legal norms specifically deal with. Everything else is nothing but the preface to a case that plays no role in determining the actual verdict. On the one hand, the past is permitted only in accordance with legal proceedings and admissible evidence. On the other hand, the unchangeable nature of the historical past does not touch the legal past, because legislation can changed retrospectively.

Legal Time

Legal time is expressed in a number of regulations. The calendar, for example, can be of individual nature in creating time divisions or in differentiating between standard time and daylight savings time. Typically, law leaves the measurement of time up to the laws of time, but it also connects it with value judgments about the quality of legal persons (e.g., being of legal age) or legal relationships (e.g., obtaining possession by prescription). Furthermore, appointments or deadlines are also objects of legal value judgments, for instance, when "immediacy" is mentioned, meaning that the person in question is required to act without culpable delay.

A perfect example of an individual construction of legal time measurement is the "legal second," a second that has no counterpart in natural time. It is purely an invention of law. Nevertheless, legal time is not completely unrelated to natural time, especially in regard to the omnipresent social problems of time. If law is to do justice to its function of binding and steering the future, it should always keep the conditional possibilities of legal behavior in mind: Things should be done on time. And that requires a worldly adaptation, a synchronization of law in connection with worldly situations, so that the rules can be followed within an appropriate period of time. Law itself should be "just in time."

Oliver W. Lembcke

See also Democracy; Ethics; Evolution, Cultural; Evolution, Social; Hammurabi, Codex of; Hegel, Georg Wilhelm Friedrich; Kant, Immanuel; Morality; Scopes "Monkey Trial" of 1925; Statute of Limitations; Time Management; Values and Time

Further Readings

Greenhouse, C. J. (1989). Just in time: Temporality and the cultural legitimation of law. *Yale Law Journal, 98*(8), 1631–1651.

Kirste, S. (2002). The temporality of law and the plurality of social times: The problem of synchronizing different time concepts through law. In M. Troper & A. Verza (Eds.), *Legal philosophy. General aspects: Concepts, rights and doctrines* (pp. 23–44). Stuttgart, Germany: Steiner.

Luhmann, N. (2004). *Law as a social system.* Oxford, UK, and New York: Oxford University Press.

Leap Years

The calendrical convention of the leap year is important to how we keep track of time. In the Gregorian calendar, adopted in 1582, a leap year is a year to which an extra day (February 29) is added. This extra day occurs in every year evenly divisible by 4. There is one exception to this rule.

For years marking the turn of a century (e.g., 2000, 2100), if the year can be divided evenly by 400, it is considered a leap year. The year 2000 was a leap year, whereas 2100 will not be.

Leap years correct for the difference between a solar year and our calendar year. A solar year is the time (approximately 365.25 days) it takes the earth to complete a single orbit around the sun. Counting 365 days from January 1 to December 31 means that in four Earth orbits around the sun, the calendar difference (4 times .25) would have the New Year starting on January 2.

Early in history, the Egyptians, with sophisticated astronomical skills, realized the calendar would get out of date with the solar year. They therefore began adding a day every 4th year. Later, the Romans used a calendar with alternating months of 30 and 29 days. This produced a calendar year of 354 days, which came up short by 11.25 days a year, meaning in just 3 years the calendar became out of synchronization by more than a month. Their imperfect solution was to add a month every 2 or 3 years, thereby causing more confusion.

The emperor Julius Caesar solved the problem in 45 BCE by establishing a calendar of 365 days and adding 1 day every 4 years. In order to bring the calendar into alignment with the solar year, he added extra months to 1 year. This worked for a while, but the remaining inaccuracy led to a difference between the calendar beginning of spring and the vernal equinox, which defines spring.

The first day of spring occurs when the sun is directly above the equator while moving from south to north. This day, when the lengths of day and night are equal, is the vernal equinox. Similarly, in fall, the autumnal equinox occurs when the sun is again directly overhead but moving from north to south. The slow changes in calendar year versus the solar year meant that by the 16th century CE, the beginning of spring had moved from March 23 to March 11. Pope Gregory XIII ordained the change the calendar in 1582 to move the 1st day of spring to March 21. To do this, 10 days were eliminated, so the day after October 4 was October 15. Non-Catholic countries changed as well but not until 1752 or later. Other cultures use leap years but with different methods to calculate when to insert days or even months.

Even with the new adjustments, there is still a very small difference in the length of a calendar day and a solar day. However, it will take about 8,000 years before the difference will amount to a full day.

Charles R. Anderson

See also Calendar, Egyptian; Calendar, Gregorian; Calendar, Julian; Calendar, Roman; Earth, Revolution of; Time, Measurements of

Further Readings

Duncan, D. E. (1998). *Calendar: Humanity's epic struggle to determine a true and accurate year.* New York: Avon.

Steele, D. (2000). *Marking time: The epic quest to invent the perfect calendar.* New York: Wiley.

Leibniz, Gottfried Wilhelm von (1646–1716)

Gottfried Wilhelm von Leibniz was an outstanding German philosopher, mathematician, scientist, historian, and diplomat. He tried to develop an

Gottfried Wilhelm von Leibniz, German philosopher and mathematician.

Source: Bildarchiv Preussischer Kulturbesitz/Art Resource, New York.

adequate metaphysical system to answer the fundamental question, "Why is there something rather than nothing?" and examine the worldview resulting from his answer. After all, he observed, nothing is simpler and easier than something. But there *is* something, so why is that so? This entry discusses Leibniz's core ideas on metaphysics, including his concepts of space and time.

Two of Leibniz's guiding basic principles were the principle of contradiction (whatever implies a contradiction is false; thus, every true proposition is a tautology—always true) and the principle of sufficient reason (nothing exists without a sufficient reason why it is so and not otherwise). In an argument analogous to one of Saint Thomas Aquinas's five ways to prove the existence of God, he argued that the reason for the existence of the world cannot lie in the contingent things (they might have all gone out of existence at the same time, and then what?); it must lie in a necessary being—God. God is the metaphysically perfect substance, containing all perfections, such as omniscience and omnipotence. Leibniz's idea of God seems to be like a kind of supermathematician; God examined all the infinite number of possible worlds and chose the best, that one that is "simplest in hypotheses and richest in phenomena." In an effort similar to work done by Johannes Kepler in determining the optimum dimensions of a wine cask to contain the most liquid with the least material required, Leibniz's God might have applied a kind of minimax procedure to come up with the optimum combination of characteristics for the world. If there had been no best combination, there would not have been a sufficient reason for God to create a world, and there would be none; thus this must be the best possible world. When God *thought* the best possible world, his thoughts *became* the world. (Leibniz argued that thoughts require signs, and the world is the sign of God's thoughts.)

René Descartes (1596–1650) had thought that the essence of (physical) substance was extension (as contrasted with mental substance, whose essence was thinking), and so a geometrical account could explain all properties of bodies. Leibniz saw that this was inadequate, that a geometrical account of motion failed to take inertia and momentum into account. Thus Leibniz concluded that the essence of substance is *action,* it has a force (*conatus*), a striving to change (for the better). The basic building block of the world, thought Leibniz, must be indivisible, a formal atom—a *monad.* Leibniz's monads are like little minds; it is as if bodies are a byproduct of these minds (he sometimes refers to a body as a momentary mind). These monads are "windowless," unaffected by anything other than their own programs. It is as if each monad is driven by its own program (something like a computer program), its internal tendency to move from one state to the next. This "program" has existed as long as the monad and determines its current state and what state it will be in next. It is like a finite-state automaton, with its initial state and the function that describes how it changes; these totally define it. Its principle of change Leibniz calls "appetition"—it is a striving for perfection. The world itself cannot *be* perfect, for then it would be God, and there could not be two Gods (there is not a sufficient reason for *any* two identical things, according to Leibniz); but the world strives toward perfection, and it is this striving that underlies change. Each monad, since its program totally defines it, contains its future and the traces of its past: "The present is big with the future, the future might be read in the past, the distant is expressed in the near."

Each monad has perceptions, which are its perspective on the universe, and Leibniz says that each monad reflects the whole (universe), though imperfectly. Thus it may contain the whole, but not be aware of all its aspects, or some of them may be less "in focus" than the nearer ones. A monad's perceptions are its properties; thus the monads form a plenum, because otherwise a monad could perceive nothingness, but nothingness cannot be one of its properties. A monad is a "concentration and a living mirror of the whole universe, according to its point of view." Thus monads make up the world and perceive the world at the same time; they are all part of the world harmony created by God's thought. Thus there is hope for knowledge of the truth regarding objective reality (of that world). The dominant monad of the aggregation of monads that gives rise to a body is its *soul* monad, which is dominant over the others. It has memory and reason and is farther up the hierarchy of monads. A rational soul, in discovering truth through science (or mathematics or logic), imitates what God does in the world as a whole; it is a kind of image of God. It tries to know God, but since God is infinite, it can never achieve full knowledge, but it can continue striving in "perpetual progress" (and never get bored!).

Leibniz argues that there are two realms: the realm of Nature and the realm of Grace. Because perceptions (and feelings) cannot be explained by mechanical causes (like trying to find your perception/idea of red by surgically examining your brain), there must be another realm (of final causes) that governs the soul monads (in their striving toward the best, to maximize good over evil). In the realm of nature, bodies follow the laws of efficient causes; these two realms will, according to Leibniz, always be in harmony, as God thought them. There will be two possible explanations, one following final causes, the other following efficient causes in the realm of nature, for every event; Leibniz says they are "equally good" explanations. Thus the mental and the physical realms are in sync because of God's preestablished harmony; they are like two clocks created and started off by the same watchmaker that continue to tell the same time even though there is never any interaction between them, or like members of a chorus who sing the same music even though they never touch each other or directly interact. They run in perfect parallel (which is his explanation to Descartes' problem of how mind and body affect each other, without Descartes' desperate solution that they interact in the pineal gland).

Leibniz views space and time as relative (see the entry on Newton and Leibniz). Space is created by the positioning of bodies, bodies are not placed in an absolute preexisting container space, and time is the succession of states (physical or mental) of bodies or minds. Without bodies, there is no space; without change, no time. Space and time are thus, for Leibniz, relations; they are ideal (based in ideas), phenomenal. Time is the more fundamental, because monads have their internal programs to move from one state (condition) to another, in their striving toward perfection. The difference between one state and another state creates their temporal relation (first one state, then the next, creates the temporal relation of earlier and later time, of past and present, or present and future). This would be time in the realm of grace, from one perceived state to another. Monads that congregate together as a body guided by a soul monad act and change in the realm of nature, so time in that realm would measure the difference between different bodily states. Derivatively, physical space would be defined by the relative position of bodily monadic aggregates.

Stacey L. Edgar

See also Descartes, René; God and Time; Idealism; Metaphysics; Newton and Leibniz; Ontology; Space; Spinoza, Baruch de; Theodicy; Time, Relativity of

Further Readings

Leibniz, G. W. (1989). Monadology. Principles of nature and grace. Discourse on metaphysics. In R. Ariew & D. Garber (Eds.), *G. W. Leibniz: Philosophical Essays.* Indianapolis, IN: Hackett. (Original works published 1686 and 1714)

Leibniz, G. W. (1996). *New essays on human understanding* (P. Remnant & J. Bennett, Trans.). Cambridge, UK: Cambridge University Press. (Original work published 1764)

Lemaître, Georges Édouard (1894–1966)

Georges Édouard Lemaître, a Belgian engineer, astrophysicist, and Catholic priest, was known for his theoretical contributions to the understanding of the origin of the universe. Theologically unorthodox and based on available scientific evidence, Lemaître's theories postulated that the universe was expanding from an unknown point of origin. This expansion, as evidence suggested, would entail that the origin of the universe would require vast differences in space and time from the condition of the universe today. In what is now commonly referred to as the big bang theory, Lemaître speculated that the universe was a product in the breakdown of the primeval atom, whereby the stars, planets, and nebula exist in an expanding universe. Furthermore, he attempted to bridge the gap between science and religion by suggesting that there are two distinct metaphysics, one for science and another for religion. Lemaître was known for his publications *Discussion on the Evolution of the Universe* (1933) and *Hypothesis of the Primeval Atom* (1946).

Contrary to previous concepts of the universe as being both infinite and static, scientific contributions from Edwin Hubble (1889–1953) and Albert Einstein (1879–1955) suggested that the universe was not static but expanding and relative. This would have serious consequences for both the traditional concepts of time and space. Although Lemaître held that the mass within the universe is constant and distributed in a nonuniform manner, the expansion of the universe would indicate that the velocity of expansion will change in time and space. Insofar as the origin of the universe, Lemaître suggested, was based on quantum theory, the quanta would decrease in number and increase in the amount of energy. When reduced to a few quanta or a single quantum whose atomic weight would reflect the total mass of the universe, the concepts of both time and space would be devoid of meaning. This primeval atom would contain the undetermined nature of the present universe; whereas spatiotemporal reality did not begin until the disintegration of this primeval atom.

Lemaître's theory, though speculative, had a profound impact on traditional cosmology. Philosophically and theologically, the beginning of the universe would call into question previous human ontology and teleology. The existence of God, the concept of the void, the stability of the universe, and humankind's place in nature would conflict with traditional perspectives. Lemaître, a proponent of dissonance, never acknowledged this conflict between science and religion, for he held that science and religion were two separate and different types of knowledge. However, an exact epistemological synthesis—understanding and blending the temporal nature depicted by science and what is stated in scripture—was never stated, and thus the two accounts remain irreconcilable.

Although problematic, Lemaître's contribution to science and cosmology remains as important as it is controversial. According to Lemaître, the universe, like the planets, had a definite beginning. Time and space, in relation to the newly "created" universe, would provide the unwritten basis for the development of preexisting matter in the universe. Humankind, a minute part in this cosmological unfolding of time and space, would not only ponder the evolution of our species but also the evolution of the universe.

David Alexander Lukaszek

See also Big Bang Theory; Black Holes; Cosmogony; Einstein, Albert; Gamow, George; Singularities; Universe, Contracting or Expanding; Universe, End of

Further Readings

Lemaître, G. (1931). The beginning of the world from the point of view of quantum theory. In M. Bartusiak (Ed.), *Archives of the universe.* New York: Vintage Books.

Lemaître, G. (1931). A homogeneous universe of constant mass and increasing radius accounting for the radial velocity of extra-galactic nebulae. In M. Bartusiak (Ed.), *Archives of the universe.* New York: Vintage Books.

Lemaître, G. (1949). *Beaumarchais.* New York: Knopf.

Lemaître, G. (1950). *The primeval atom.* New Jersey: Van Nostrand.

Lenin, Vladimir Ilich (1870–1924)

The Russian revolutionary Vladimir Ilich Ulyanov, who later adopted the name V. I. Lenin, was born to Mariya Blank, a doctor's daughter, on April 10, 1870, in Simbirsk, where his father was an educator, schoolmaster, and inspector of primary schools. He was one of six siblings.

The second child and oldest son, Alexander (Sasha), went to St. Petersburg University in the fall of 1882, where he majored in natural science. While there, he joined an organization known as Narodnaya Volya, or the People's Will. This terrorist wing believed in a socialist party led by the working class. Socialism was the logical extension of democracy, but czarists did not allow for peaceful development of democracy. Legal opposition was not allowed. The only option for purging the official hatred of any popular movement toward socialist democracy was terrorism. In February 1887, the People's Will planned the assassination of Alexander II, czar of Russia. The police discovered the plot, and as a result, Alexander Ulyanov was executed in 1887 for being a part of the group that attempted to kill Alexander II.

Because of this experience, young Vladimir was propelled into revolutionary activity. As such,

Vladimir would become a leader in the Russian Revolution. A diligent scholar, Lenin attained a mastery of social theory and a sound understanding of the history of social movements and historical trends. As a leading activist, Lenin wrote a series of articles that would not only serve as a guide for revolution in Russia but also lay the intellectual foundation for what would become Marxism-Leninism. These articles included *The Development of Capitalism in Russia.* In this work, Lenin tried to prove that that by 1900, Russia had already been incorporated into the world capitalist system. As a result, the peasants were rapidly being divided into capitalist farmers and the rural proletariat. Russian industry was divided between traditional handicrafts, backward manufacturing, and a modern machine industry in which a great deal of foreign capital was invested.

In 1902, Lenin wrote the pamphlet, *What Is to Be Done?* "Without revolutionary theory there can be no revolutionary movement," Lenin stated. He wrote that unions by themselves would only lead to struggles over wages and working conditions; a socialist consciousness needed to be introduced from outside the working class. This would require an organization of professional revolutionaries. The organization would need to be disciplined, conspiratorial, and centralized. The "party of a new type" would become the vanguard of the working class and would lead, but remain separate from, the broader democratic workers' movement.

In 1904, Lenin wrote *One Step Forward, Two Steps Back.* At this time, Lenin's writing was based upon the assumption that the workers' party needed to be taken over by those workers who were educated with a socialist awareness. The intellectuals were undependable, obsessed with eccentricity, anarchism, and egoism, and had an intense horror of discipline.

Materialism and Empiriocriticism was written in 1908 as a rebuttal to the physicist Ernst Mach, who had stated that reality was only the actuality of our experience, and therefore science could record only our subjective experiences. In *Materialism and Empiriocriticism,* Lenin affirmed that science is the observation of the material universe, existing independently of the observer. Physical sensations are the direct connection to the external world. Every ideology is conditioned by its historical setting. Science is no different, yet science is valid. It is independent of the observer to the degree it corresponds externally to tangible nature.

Sensations, wrote Lenin, are our obvious link between our consciousness and the external world. Energy affects our bodies, which are excited by stimuli and provoked by the external world. This in turn excites a chemical response in our nervous system and is transformed into consciousness by the mental activity of our brains. The energy of the exterior excitation changes physical and chemical corporeal activities and is transformed into sensations. Sensations are changed into consciousness. Consciousness is cast and grows. Consciousness interprets sensations. This implies a dynamic and interactive process in mental development. We learn through actively interacting with both the social and physical environment.

Physical and social realities, Lenin noted, are constantly changing. There is an eternal process of conflict and fusion of conflicting parts leading to the death of the old and the birth of the new. This simple logic is basic to anything that can be studied. It is more than a part of mental processes, or the way that the world is studied and understood. This logic is broadly similar to the way the universe is constantly evolving. The opposition and combination of components forming ever-greater wholes is an approximation of what is really happening to the universe outside the mind. When these opposing and interacting parts are fused, there is something basically different being formed that will replace what went before. The new thing soon develops its own tensions as it begins to break down. These new tensions not only lead to the extinction of this entity but also give birth to its replacement.

Science, according to Lenin, is a specialized form of logical practice, founded upon the certainty that there is an external reality that is independent of our consciousness. Through careful observation and the use of specific scientific method to analyze observations, this external reality is more closely revealed than by any other technique of understanding. Because all scientific observations are approximations, science is a constantly growing discipline. Every generation will get progressively closer to this external reality. The limitations of the categories used are always qualified, modest, changeable, provisional, and approximate. Each scientific breakthrough is built on earlier breakthroughs. Because every ideology is historically conditioned, science itself occurs in a specific historical setting. This is the basis of materialism; the fact that political states are a reality can be understood in terms of matter in motion.

In 1908, Lenin wrote *The Agrarian Program of Social Democracy in the First Revolution*. Most peasants in Russia were downgraded to farm workers and tenant farmers. Only a small number of peasants had enough land to endure as farmers. A minority of the farmers were rapidly becoming more like American capitalist farmers. The large feudal *latifundium* (large estates) remained. The *latifundiums* slowly developed into large farms modeled on the German Junker type. With the breaking down of feudalism, capitalism developed. The market economy merely meant that the state was becoming a major landowner.

Lenin contributed to the Marxist theory of imperialism with his text called *Imperialism: The Highest Stage of Capitalism*. According to Lenin, imperialism was "the final stage of capitalism," a sign of the breakdown of capitalism and the transition to socialism. Competitive capitalism of the early 19th century had become increasingly centralized, with fewer competitive firms surviving the increasingly intense competition. As a result, capital became concentrated in larger firms. In this way, "monopoly capitalism" was able to create ever-larger surpluses by limiting competition. Markets at home became glutted, and investment opportunities in the industrialized nations declined. This meant that corporate capitalists were forced to export capital to ensure future profits. This became, for Lenin, a major distinction between the earlier competitive capitalism and its later descendent, monopoly capitalism. Competitive capitalism exported finished goods in exchange for raw materials produced in the poor areas of the world. Monopoly capitalism exported its capital to these areas. Capital was invested to create modern ways of extracting those same raw materials. Instead of mines being owned by the local traditional elite, capitalists in the rich industrial nations owned them. This caused an increase in overall capital on a world scale while arresting development in the main capitalist countries.

The principal feature of modern capitalism was the domination of monopolist consolidations by giant capitalist firms. The monopoly control was most firmly established when all sources of raw materials were jointly controlled by several large surviving firms. Monopoly capitalism, with a few highly centralized firms effectively dominating the economy, created imperialism out of its own needs. In order to find continued profits in an already overdeveloped economy at home, investments flowed to less developed areas of the world, where the capitalist economy had not reached a saturation point, therefore making profits much higher. Export of capital was the fundamental principle of imperialism. The export of capital greatly influenced and hastened the growth of capitalism in those countries to which it was exported.

The next two most important of Lenin's works, which outlined his sense of history, were *State and Revolution* and *"Left-Wing" Communism: An Infantile Disorder*. Both were models for the direction revolution would take. In *State and Revolution*, Lenin maintained that the workers could not merely take over control of the existing state; they needed to smash it. Then, when the returning bourgeoisie was overthrown, the state would wither away. In *"Left-Wing" Communism: An Infantile Disorder*, Lenin disapproved of leftists in the West for disregarding parliamentary tactics and legal opportunities to create their own socialist revolution. Lenin also summarized the significance of all communist parties that had become centralized and disciplined and were following the lead of the Russian party.

Michael Joseph Francisconi

See also Consciousness; Dialectics; Economics; Engels, Friedrich; Mach, Ernst; Marx, Karl; Materialism

Further Readings

Lenin, V. I. (1939). *Imperialism: The highest stage of capitalism*. New York: International.

Service, R. (2001). *Lenin: A biography*. Cambridge, MA: Harvard University Press.

White, J. D. (2001). *Lenin: The practice and theory of revolution*. Hampshire, UK: Palgrave.

Leonard da Vinci

See Fossils, Interpretations of

Libraries

A library is a repository for recorded information. The concept of a library has evolved but has existed in some form for thousands of years. The ability to record information in a format more permanent than oral tradition added another dimension to human communication time. Information that has been transcribed into a solid, visual format can be accurately remembered over generations. While libraries have physically changed over the course of history, their cultural role has not. Libraries and librarians are responsible for acquiring, maintaining, and providing access to information. They keep business, legal, historical, and literary records of a civilization. Libraries could be considered the memory of a society.

The word *library* is derived from the Latin *liber,* meaning "book." Traditionally, library refers to a collection of books, or a room in which such a collection is kept. In the earliest days of recorded information, however, there was no distinction made between a record room (archive) and a library, so it could be said that the concept of a library has existed for as long as writing.

The Ancient World

The emergence of large, complex civilizations in Egypt and Mesopotamia 6,000 years ago necessitated the development of writing to provide the record keeping needed to maintain a stable and orderly society. Ancient civilizations in East Asia indicate similar development, with collections of official records dating to as early as the 12th century BCE in China.

Hieroglyphs emerged from the artistic history of Egypt, starting a tradition of literacy and recorded knowledge. This writing system used pictures to reflect words and ideas. As writing became more widespread and vital in Egyptian society, simplified glyphs were developed, resulting in hieratic (priestly) and demotic (common) scripts. These cursive writing forms were better suited to writing texts on papyrus, greatly increasing the speed in which scribes could write business or literary texts.

In Mesopotamia, the most abundant material for information storage was clay. Cuneiform writing, done with a wedge-shaped tool, was impressed into tablets of damp clay. After the tablets dried, they were permanent. A temple dating to the 3rd millennium BCE in Nippur had several rooms filled with tablets, indicating the existence of a well-developed archive. Large collections of tablets also appear to have been deliberately organized with a catalog listing the contents of the collection.

In East Asia, records were first inscribed on bone, stone, bronze, or tortoise shell and date to the Shang dynasty (approximately 1200 BCE). By the 8th century BCE, less durable materials such as bamboo or wooden slats were used. Lengths of silk appear in the historical record during the 4th century BCE. Paper was invented in China in the 1st century CE, but it would take several centuries for it to completely replace the cumbersome bamboo or the expensive silk. In China, records were often destroyed during war or purged when new rulers assumed the throne, so history would appear to begin during their reigns. The Han dynasty, which succeeded in 206 BCE, ended the repression. They recovered earlier works that were hidden, encouraged writing and record keeping, and developed formal methods to classify information.

In the Western world, libraries had their origins in classical Greece and Rome. Most of the larger Greek temples appear to have had libraries and record rooms. As early as the 6th century BCE, Greek leaders were constructing large public libraries, and a sophisticated book trade existed in Athens by the 5th century BCE. Many large book collections were owned by wealthy citizens. One of the finest private collections of ancient times was owned by Aristotle. His collection was seized and taken to Rome by Sulla in 86 BCE as war booty. From there, copies of his texts became the basis of the greatest library in antiquity at Alexandria.

The Alexandrian library was planned by Ptolemy I in the 3rd century BCE and established by his son Ptolemy II. The founders of this library intended to collect the finest copies available of every piece of Greek literature and arrange it systematically to facilitate research. It was staffed by many famous Greek writers and scholars and was estimated to contain 700,000 scrolls of parchment and the first use of vellum.

Private libraries were common in classical Rome. It was considered fashionable to have a library in your home. Interestingly, imperial Rome is known for military skill, not intellectual life, but Romans revered books and transported hundreds of collections virtually intact back to Rome when plundering foreign lands. Most of the captured libraries were incorporated into the private libraries in officers' homes. Eventually, Romans decided that placing these collections in large public buildings would display the glory of Rome. Julius Caesar is credited with actually planning the first public library facility, but he died before the plans could be completed. Tiberius, Vespasian, Trajan, and many later emperors created libraries throughout the city.

The Middle Ages

Books were thought to be essential to spiritual life. As monastic communities were established in the 2nd century, monasteries included a library for reading and study. They also incorporated a *scriptorium,* a room where manuscripts were copied. Strict rules existed for the use of the books, but at the same time, monasteries often lent materials to other monasteries or to the public. The contents of these libraries included scriptures, religious histories, philosophical writings, and some secular literature. Unknown scribes copied texts repeatedly for generations, preserving cultural transmission through the Middle Ages. Important developments during this time include the codex, or book, and the use of vellum (animal skin) for the pages of books. This new format for recording information was far more durable than papyrus. After universities were founded in the 11th century, students who were also monks returned to their monasteries and deposited their lecture notes in the monastery collection, expanding its contents. The libraries of newly founded universities and monasteries were the main places to study until late in the Middle Ages. Books were very expensive, and only the wealthiest individuals could purchase copies.

The Renaissance

Private book collections developed again through the 13th to 15th centuries. The growth of commerce, widespread education and literacy, the new learning of the Renaissance, and the invention of the printing press by Johannes Gutenberg widened the circle of book collectors. Paper also became the preferred writing surface in the 15th century. Paper and the printing press made books far cheaper, faster, and easier to produce for the rapidly growing reading market. Along with a sophisticated book trade, many fine public libraries opened in Italy and throughout Europe.

17th and 18th Centuries

In the 17th and 18th centuries, book collecting everywhere continued to expand. In Europe and North America, many fine private collections were formed that eventually became the core of national and university collections. National libraries today receive one free copy of every book and periodical produced in that country and are maintained with national resources. They attempt to collect and preserve the nation's literature while keeping an international scope. A new form of library also developed at this time, the circulating library for popular literature. Circulating libraries were operated by booksellers for profit, but they made a large body of literature available to the general public for a nominal cost.

20th Century

The look and feel of libraries has changed drastically since the 19th century. A leading figure in transforming library service was Antonio Panizzi, an Italian refugee who began working for the British Museum in 1831 and was its principal librarian from 1856 to 1866. He revolutionized library administration and developed a code of rules for catalogers. He also recognized the potential of libraries being open to all for research and reflected this notion in his planning of the British Museum reading room. With the emergence of a large reading public and an enormously expanding stock of books and periodicals, libraries had to expand their organization and storage capacities.

The paradigm for libraries shifted again in the 20th century with the advent of new information technologies. An information explosion after World War II created a need for new methods to store information. Compact movable shelving, microfilming, and remote storage evolved to address this need. Today, the vast quantity of information available through databases and other online resources requires computers and highly qualified professional librarians to help the public navigate through the overwhelming amount of information freely available to everyone. Many types of libraries and services are available today. There are great national libraries, university and research libraries, and public libraries that are generally open to everyone. In addition, there are special libraries founded to meet the research needs of a specific group, school libraries for the students of that institution, and private libraries that reflect the interests of the collector. Archives exist as separate entities today and are generally collections of papers, documents, and photographs preserved for historical reasons.

The Future of Libraries

Libraries have always been driven by the needs of their users. That will not change. Today, most libraries open to the public are fairly neutral in their collections—they have something for everyone. This may have to change given the exponential growth of information available. To be able to manage diverse subjects and formats, while also preserving existing collections with limited funds, is becoming increasingly difficult. Libraries of the future will likely have to specialize and tailor their collections to specific groups of users. Resource sharing between libraries will continue to grow in importance.

While the appearance of libraries and the methods of doing research will continue to change, the main purpose for their existence is unchanging. Libraries and librarians will continue to be responsible for acquiring, maintaining, and providing access to the collective knowledge of a civilization, no matter what form that knowledge may take in the future.

Jill M. Church

See also Evolution, Cultural; Hammurabi, Codex of; Information; Museums; Rosetta Stone

Further Readings

Battles, M. (2003). *Library: An unquiet history.* New York: Norton.

Casson, L. (2001). *Libraries in the ancient world.* New Haven, CT: Yale University Press.

Harris, M. (1999). *History of libraries of the Western world.* Lanham, MD: Scarecrow Press.

Lerner, F. (1998). *The story of libraries: From the invention of writing to the computer age.* New York: Continuum.

Maxwell, N. K. (2006). *Sacred stacks: The higher purpose of libraries and librarianship.* Chicago: American Library Association.

Rubin, R. (2004). *Foundations of library and information science.* New York: Neal-Schuman.

Staikos, K. (2000). *The great libraries: From antiquity to the Renaissance.* New Castle, DE: Oak Knoll Press.

Life, Origin of

The origin of life remains today one of the most challenging puzzles to science. The challenge is twofold: (1) qualify the essence of life and (2) explain its appearance on Earth. Although both aspects have been subject to much scientific investigation, no satisfactory explanation has been formulated so far.

Life is often defined as the distinctive property of particular physical systems: living organisms. One way of defining life is to say that a physical system is alive if and only if it can replicate with variation and therefore be submitted to natural evolution. From a temporal perspective, the origin of life can therefore be defined as the point in time when, for the first time on Earth, a particular physical system simultaneously displayed a set of given properties, namely replication and variation.

How life is defined has a strong impact on the question of its origin: Taking a more restrictive definition of life, one that would for instance require that such replicating systems with variation be able to complete at least one thermodynamic cycle, would move forward the point in time when the origin of life might be traced back to. In addition, defining life as a collective set of properties of living organisms also raises the question of the temporal order of appearance of each of these properties: For instance, which appeared first, replication or variation?

Roughly speaking, life most likely appeared on Earth some 3.8 billion years ago. This dating is framed, on the one hand, by the formation of the planet some 4.5 billion years ago, with still some heavy meteoritic bombardment until about 4 billion years ago; and, on the other, by the oldest cellular fossils dated 3.6 or potentially even 3.8 billion years ago. As such, the time frame available for life to appear is about 200 to 400 million years.

The first modern hypotheses of the origin of life were formulated in the 1920s by Alexander Oparin and John B. S. Haldane, separately: The first living systems would have appeared in a primitive ocean of organic molecules, all of them resulting from prebiotic chemical processes, that is, chemical processes compatible with the physicochemical conditions believed to be those of the primitive Earth, before the appearance of life. The first scientific experiments supporting these hypotheses are those of Stanley Miller in 1953, who demonstrated the possibility of synthesizing certain organic molecules—amino acids—under prebiotic conditions. Since then, the prebiotic synthesis and chemical behavior of numerous other organic molecules in prebiotic conditions have been investigated, including those of proteins, lipids, and nucleic acids. Today, the scientific field of research on the origin of life draws upon a large number of disciplines: molecular biology, biochemistry, prebiotic chemistry, and theoretical biology but also planetology, geology, or even astronomy, in order to define the environmental conditions of the primitive Earth or to search for primitive life forms on alternative planets.

Far from having occurred at a particular moment or point in time, as a sudden event, the origin of life is likely to have been the result of a long and gradual process. The question remains whether this gradual process is truly continuous or might otherwise display some sorts of sudden steps akin to phase transitions, for instance. In the latter case, such steps could be used as particular landmarks for defining the origin of life, even if these are still speculative today. Due to this specific time frame of several hundred million years during which life is thought to have appeared on Earth, experimental research on the origin of life has to consider a broader scope of physicochemical possibilities than usual, for instance chemical reactions with longer characteristic times than in usual laboratory experiments, or environmental conditions that would be much different from current ones (temperature, pressure, pH, etc.). Over such long periods, the role of chance also might come to play an important role, enabling unlikely molecular encounters to happen or precellular components to assemble and disassemble in many different ways as in a "tinkering" process. Therefore, life as we know it on Earth would somehow also keep the trace of numerous specific events or "frozen accidents," all of them highlighting the very historical nature of life.

Life as we know it on Earth might be one particular instance of a more generic phenomenon, the characterization of which is among the goals of artificial life research. What is more, not only could life appear under different forms, but it might also appear in different places in the universe. Exobiology research focuses in particular on the quest for extraterrestrial signs of life. Such a discovery would not only be an astounding discovery, but it would also shift the question of the origin of life and raise the question of its potential perennial existence over the known history of the universe, an old theory known as panspermia. In addition, life as we know it today might not be life as it has been in past, or, as a matter of fact, life as it will be in the future. Life as we know it today might not be life wherever nor forever.

Could there be, as well, multiple parallel pathways leading to the origin of life? If life is defined as a set of collective properties of particular physical systems, its origin might thereby be traced back to the chains of events that led to the appearance of each property separately and, as such, of those specific components enabling the physical systems to display such properties. Hence, for instance, the origin of life on Earth might be traced back to the origin of prebiotic proteins, prebiotic nucleic acids, or even prebiotic lipidic compounds, all of them being basic components of current living systems. In such a case, the origin of life could dissolve into as many origins as there are components necessary to build a living system.

Could life still be originating on Earth as we speak? This would make the origin of life a continuously recurring event, continuously generating new origins, so to speak. Such a hypothesis appears unlikely: It would require not only abiotic chemical processes, that is, chemical processes able to produce organic molecules without appealing to compounds of existing living organisms, but also physicochemical processes that could lead to the organization of these molecules into living systems. However, it appears extremely likely that most abiotically synthesized organic molecules today would end up being immediately assimilated by existing living organisms, without any possibility to further assemble themselves into new living systems, life thereby preventing any new origination of itself.

Christophe Malaterre

See also Creationism; Evolution, Chemical; DNA; Evolution, Issues in; Evolution, Organic; Genesis, Book of; Gosse, Philip Henry; Materialism; Oparin, A. I.

Further Readings

Bernal, J. D. (1951). *The physical basis of life.* London: Routledge & Kegan Paul.

Brack, A. (Ed.). (1999). *The molecular origins of life: Assembling pieces of the puzzle.* Cambridge, UK: Cambridge University Press.

De Duve, C. (1995). *Vital dust: The origin and evolution of life on earth.* New York: Basic Books.

Eigen, M. (1992). *Steps towards life: A perspective on evolution.* Oxford, UK: Oxford University Press.

Fry, I. (2000). *The emergence of life on earth: A historical and scientific overview.* New Brunswick, NJ: Rutgers University Press.

Popa, R. (2004). *Between chance and necessity: Searching for the definition and origin of life.* Heidelberg, Germany: Springer-Verlag.

Life Cycle

All living organisms have a life cycle beginning with the initiation of life and continuing until death. Scientists generally consider a life cycle to encompass the changes that an organism experiences from the beginning of a specific developmental stage until inception of the initial developmental stage in the next generation. The stages an organism will experience and the length of the life cycle varies widely across species. Species survive only when sufficient offspring complete the life cycle to continue the species. Such cycles vary from minutes to multiple years.

The time to complete a life cycle is generally correlated to the size of the organism. Bacteria complete their life cycle in 30 minutes by dividing to form another generation. Higher organisms take much longer. The giant sequoia, for example, produces its first fertile seeds after 60 years. Some plants and animals die after reproducing, with reproduction symbolizing the end of the life cycle. Other animals, notably humans, live for many years after their reproductive cycle is complete.

Most simple organisms complete their life cycle in one generation with an initial splitting in two of an adult, the growth of the new organisms, and upon reaching maturity, the division of each one into two new individuals. Higher animals can also have a single generation with the fusion of male and female sex cells to create the embryo and the growth of the offspring to reproductive maturity when they produce sex cells that create new offspring. Plant life cycles, however, are usually multigenerational. A spore will germinate and grow into a gametophyte. At maturity, the gametophyte forms gametes. With fertilization, the gametes develop into sporophytes. Once reproductive maturity is reached, the sporophytes produce spores and a new life cycle begins. This process is frequently referred to as "alternation of generations" and occurs in some fungi and protists as well.

The time the young animals spend developing prior to being born (the gestation period) or hatched varies across species. Small marine animals frequently produce large numbers of tiny eggs that hatch at a very early developmental stage, while others produce fewer, larger eggs with the young exiting at a more advanced developmental stage. The smaller young will need longer to develop before becoming adults, while the more fully developed young move quickly into adulthood. This effect is common in higher organisms as well. Generally the more developed the young, the fewer produced in one reproductive cycle, as there is more chance of the young surviving. Larger animals take longer to bear their young, as the young must be ready for the hardships of life. Elephants have the longest pregnancy period of any mammal at 22 months. Smaller animals have shorter gestation periods. For example, the domestic cat has a pregnancy term of approximately 9 weeks.

Hibernation

Hibernation is a state of dormancy that some animals enter into during cold months of the year. Bears are well known for this; however many other animals hibernate, including rodents. Hibernation allows the animal to exist during times of food shortage and unfavorable conditions. In the spring, the animal rises and is generally ready to reproduce. Plants and seeds also experience dormant periods until conditions improve to allow them to continue their life cycle.

Metamorphosis

A number of animal species include metamorphism in their life cycle. Metamorphosis is a process of great change between stages in the cycle. The stages provide a way for the organisms to live separately from other stages in order not to compete for food or habitat. For example, frogs lay eggs in the water. The embryo quickly grows and hatches into a tadpole. The tadpole lives in the water and breathes through gills. As the tadpole grows, its tail shrivels, and legs and lungs develop. When the gills disappear, the tadpole comes to the surface and is now considered to be an adult frog. Most species of frogs grow into adults within one season and reach reproductive maturity within a year. Thus, mature frogs and tadpoles live in very different habitats, lessening the competition for food and enhancing other aspects of survival.

Butterflies also use metamorphosis in their life cycle. Each one moves quickly from a caterpillar (larval) stage to a pupal or chrysalis stage, when the larva is dormant, and then emerges from the pupa as a full adult ready to reproduce, with one cycle normally taking a year to complete. Fruit flies also undergo metamorphosis, changing from one form to another completely different form, moving from an egg to larva to pupa to adult. Each phase is a natural phase of the life cycle. The whole process normally takes about a week. Other animals that use metamorphosis are amphibians (toads, newts, etc.) and a wide variety of insects.

Invertebrate animals have a variety of life cycles. Metamorphosis, a series of rapid physical changes, is common. Most insects lay eggs. Incomplete metamorphosis, or simple development, involves the eggs hatching into nymphs and moving to the adult stage. Complete metamorphosis, or complex development, includes a pupal stage between the larval and adult stages. As immature larvae and nymphs grow, they shed their exoskeleton, which is a tough exterior skin. Each time the exoskeleton is shed, the new stage is called an *instar.* The final instar develops into a pupa in complex life cycles or into an adult in simple life cycles. Most common insects generally have three to six instar stages; thus a complete insect life cycle includes the egg, larval, three to six instar, and adult stages. Ladybugs provide a good example of complex development. The adults lay eggs that hatch into wormlike larvae, which go through several instar stages, and the final instar covers itself in a hard pupa and emerges as a new adult. The life cycle takes 4 to 7 weeks. Most insects have a 1 year life cycle. However, wide differences exist. For example, the life cycle of a fruit fly is 1 week, while one species of cicada has a life cycle of 17 years.

Reptiles

Most reptiles are egg laying (oviparous). There is great variety in reptile mothers. Some eggs are laid and left on their own, although usually in well-protected and hidden nests. Sea turtles lay their eggs on sandy beaches and return to the water, leaving the young to fend for themselves. Other species stay with the eggs and aggressively defend the nests. Examples of these include crocodiles and pythons. Some reptiles are ovoviviparous, including many species of snakes and lizards, with very thin shelled eggs remaining in the mother's body. These species appear to have live births, but in reality the eggs hatch just before the young are born, making it seem that the young are born rather than hatched.

Reptiles produce hatchlings that are exact replicas of the parents. The hatchlings must quickly grow and learn to defend themselves. Many reptiles grow through a process of molting, where the outer layer of their skin is shed to allow for physical growth. The molting times will vary. Most reptiles will molt shortly after hatching. During periods of rapid growth, molting can occur four or more times a year. Adults still molt but more infrequently, generally one or two times a year.

Fish

Fish are primarily egg laying. Many species lay the eggs and leave them alone to survive. Most eggs are soft shelled. Some species of sharks lay eggs with a tough outer coat to help protect the eggs. A few species protect the eggs through burying them in the sand or by being "mouth brooders," where one parent will hold the eggs in his or her mouth until they hatch. Male seahorses have a special pouch where they hold the eggs. Baby fish are called "fry." The young are generally able to reproduce within a year after birth.

Birds

Birds lay eggs on land even if, like ducks and loons, they spend more time on the water. Most birds stay with the nests (brood) to keep the eggs warm and safe from predators. Eggs vary in size, from the very small robin egg, approximately 1 inch around, to the large ostrich egg, which is 5 to 6 inches around. Hatching time is 12 to 14 days for robins and 35 to 45 days for ostriches. Ostriches will reproduce at 2 to 4 years of age, while robins reproduce within a year. Comparing the robin and the ostrich provides further evidence that the size of the organism affects the time taken for each stage of the life cycle.

Humans

Humans are generally considered to have four stages in their life cycle: infancy, childhood, adolescence, and adulthood. The gestation period is 9 months. Infancy is the stage between newborn and 2 years of age. Many changes occur in this stage, and the infant learns to walk and talk. Childhood occurs between 2 years of age and puberty (adolescence). Many physical changes occur in childhood, such as the rapid development of teeth and bones. The time of puberty varies but usually starts at 8 to 12 years of age. Adolescence is a period of many changes, including the sexual maturing of the human body. Males continue to grow physically until they are 21 years old, while females usually reach peak physical growth by 18 years of age. During adulthood, considered to be 18 years of age and over, the body begins to slow down. Changes occur, including thinning hair and body shape modifications. Sometimes old age is added to the life cycle stages. In developed countries the average life expectancy is 77 to 80 years. A few people live for more than 100 years. Life cycles are thus complex and show immense variation across species.

Beth Thomsett-Scott

See also Birthrates, Human; Clocks, Biological; DNA; Dying and Death; Extinction and Evolution; Fertility Cycle; Gestation Period; Hibernation; Longevity; Malthus, Thomas; Maturation; Metamorphosis, Insect

Further Readings

Kalman, B., & Langille, J. (1998). *What is a life cycle?* New York: Crabtree.

McGraw-Hill encyclopedia of science & technology. (2002). New York: McGraw-Hill.

Light, Speed of

The physical qualities of light have always been important, but its speed has been difficult to measure. Light is the fastest known physical phenomenon, so it took the development of modern timing equipment to be able to measure it. The early civilizations, such as that of the ancient Greeks, tried to quantify light but found its qualities elusive. It was only in 1973 that the speed of light in a vacuum was accepted at the current value of 299,792,458 meters per second (m/s).

One of the first observed facts about light was that it was faster than sound. When something produced both a loud sound and a bright light, such as a cannon firing, the light always reached the observer before the sound. The difficulty trying to quantify this difference was the huge difference in the speed of the two phenomena and the lack of good clocks. Sound has a speed of 331.3 m/s (at 0^0 C, 0% humidity, and 1 atmosphere of pressure) while light goes at 299,792,458 m/s. To early observers light seemed to travel instantaneously. For many centuries the speed of light was a philosophical concept, and there were debates between those who believed it had an infinite speed and those who felt it had to be finite.

It was the invention of more accurate clocks that help generate the first realistic calculation of the speed of light. One of the first experiments to yield a rough estimate was made by Ole Christensen Roemer in 1676, who used the orbital motion of Jupiter's moon Io to make his measurement. By recording the exact moment Io entered Jupiter's shadow when Jupiter was closest to Earth and comparing that time to the time many months later when Jupiter was a known distance farther from Earth, he calculated a speed of 220,000,000 m/s for light. In 1728, this observed estimate was improved by James Bradley, who measured the apparent motion of stars at different times as compared to the speed of Earth in its orbit. His calculation of 298,000,000 m/s was accepted as the more accurate value.

As clocks and clockwork mechanisms improved, it became possible to divide time into smaller and smaller segments. The first earthbound measure of the speed of light came in 1849, when Hippolyte Fizeau used a mechanical apparatus to measure the speed of light. He used a beam of light focused on a mirror several thousand meters away. By forcing the light to pass through a rotating cogwheel, he found that at a certain rotation, light would pass through the cogwheel on both its forward and return journeys. A calculation based on a combination of distance of the light source from the cogwheel, the number of teeth on the cogwheel, and the rotational speed of the cogwheel yielded a speed of 313,000,000 m/s. This type of experiment was later refined using rotating mirrors and prisms to estimate a speed of 299,796,000 m/s in 1926. It took the development of oscilloscopes with time resolution in the subnanosecond range to further refine this number by measuring the delay of a light pulse from a laser or LED.

One of the interesting behaviors of light is that its observed speed lowers when it passes though media other than a vacuum. The refraction of light as it passes from air to water is a visible manifestation of the slowing of light as it moves into a denser medium. This has led scientists to explore the concept of "slow light" by increasing the index of refraction of various media. It is possible to increase the path that photons must take through a medium by using specialized conditions such as a Bose-Einstein condensate. Scientists have succeeded in slowing the measurable apparent speed of light to less than 1 m/s.

Since 1983 the speed of light has been treated as a defined constant, with refinement of its present value by further experimentation not needed. As a constant it functions as an absolute measure of distance (1 meter = the path traveled by light in 1/299,792,458 of a second), which has allowed for finer and finer descriptions of the velocity of objects as time measurement tools have improved.

John Sisson

See also Black Holes; Clocks, Atomic; Einstein, Albert; Experiments, Thought; Newton, Isaac; Observatories; Planetariums; Time, Relativity of; Singularities; Space Travel; Stars, Evolution of; Time, Measurements of; Twins Paradox

Further Readings

Clegg, B. (2008). *Light years: An exploration of mankind's enduring fascination with light.* London: Macmillan.

Nimtz, G. (2008). *Zero time space: How quantum tunneling broke the light speed barrier.* Weinheim, Germany: Wiley-VCH.

Logical Depth

The time necessary to generate an object, given a concise description of it, is a factor that intuitively plays an important role when we assess the value of this object. We are inclined to consider an object valuable if we believe that a long sequence of rather difficult actions has led to its existence. It is, thus, a promising idea to use the length of the process in which an object has been generated as an indicator of its complexity, on which indication we then base our assessment of its value. The measure of logical (or algorithmic) depth brings this reasoning into a mathematical form. It was introduced, in the 1980s, by Charles H. Bennett, a physicist and computer scientist doing research at IBM. Along with being one of the founders of quantum computing, Bennett is also among the most important contributors to the structural science of complexity.

Bennett's definition of logical depth is based on central ideas of algorithmic information theory. The algorithmic information content of a string of binary digits (bits) is, broadly speaking, equal to the bit length of the shortest computer program that, if run on a certain type of computing machine (a universal Turing machine), outputs this string and then halts. Such a program is the most concise algorithmic description of the bit string. If this string can, again, be interpreted as the digital representation of an object (e.g., as containing a detailed account of parameters of a physical system at a certain time), then the shortest program outputting the string encodes the algorithmically most economical hypothesis about the causal history of the digitally represented object.

Bennett defines, in its simplest version, the logical depth of an object represented by a binary string as the run time of the shortest program that outputs the binary string and then stops; this run time is measured in execution steps of the program on a universal Turing machine. A long time that elapsed in the real-world generation of an object does not correspond necessarily to a logically deep time: If, in a long period of real time, there did not occur many events that contribute to the generation of the object, this stretch can be summarized by a few steps in the simulated causal history of the object. Thus, a logically deep object is one whose simulation on a computer proceeds slowly when the shortest program outputting its digital representation is run. In contrast, logically shallow objects are either very regular objects that can be simulated by short and fast programs or very irregular objects that can be simulated by incompressibly long and fast programs. Under the general constraint, stated by the second law of thermodynamics, on the possible growth of order in physical systems, a fast increase of order, measured by logical depth, is very improbable.

From an information-theoretical perspective, the logical depth of an object is not equal to the Shannon entropy of its digital representation, given a probability distribution of the binary digits. It is equal to the number of operations necessary to generate an object starting from its most concise description. These operations can, at least principally, be reconstructed if we know the digital representation of the object and search for the most concise program that outputs this representation and then halts. Logical depth is, thus, correlated with the redundancy in the causal history of the object, that is, with the difference in the number of bits between an explicit digital representation of the generation of the object and the shortest program that outputs the digital representation of this object and then stops.

The simple definition of logical depth given above is, however, not robust enough. If, for example, a program that generates the digital representation of an object much faster than the shortest program is just a few bits longer than the latter, the exclusive focus of logical depth on the minimum length program leads astray. The simplest solution to this problem is to define a mean logical depth of an object on the base of a rate of exchange between run time and program size. Such a rate of exchange must guarantee that, whereas short and fast programs outputting the digital representation of the object contribute most to its logical depth, long and slow programs contribute least. In calculating the average run time of all programs that output the digital representation of an object, the shorter and the faster the program that generates the desired output, the higher its run time must be weighted.

Stefan Artmann

See also Entropy; Information; Maxwell's Demon; Time, Asymmetry of; Time, Measurements of

Further Readings

Bennett, C. H. (1988). Logical depth and physical complexity. In R. Herken (Ed.), *The universal Turing machine* (pp. 227–257). New York: Oxford University Press.

Bennett, C. H. (1994). Complexity in the universe. In J. J. Halliwell, J. Pérez-Mercader, & W. H. Zurek (Eds.), *Physical origins of time asymmetry* (pp. 33–46). Cambridge, UK: Cambridge University Press.

Gell-Mann, M. (1994). *The quark and the jaguar: Adventures in the simple and the complex*. New York: Freeman.

Longevity

Longevity can be defined as either a duration or a life span. Although the term might bring to mind the time spent by a living being in a situation, such as a job or living, the term can also be seen as much more. The length of an occurrence, or its longevity, can be measured in various ways. It can be measured in clock time, which measures all occurrences according to the same scale, such as a 24-hour day, separated into durations of equal size; or it can be measured relatively, as compared with the longevity of another event. It can be accurate according to the clock, or it can be seen in terms of perceived duration, also known as social time.

Measuring longevity, or the duration of an event or occurrence, has changed as the methods of accurately depicting the passage of time have enabled its division into smaller and smaller intervals. Time was once measured by passages of the sun or recurrences of events, such as seasons or the appearance of a star in the sky. Time now can be measured in increments far shorter than a second.

Measuring time in relation to another occurrence, such as a natural event, enables one to better comprehend longevity. However, recording and communicating such a duration to others requires their awareness of the relative measuring stick. The advent of the calendar, whether cyclic or linear, solar or lunar, enabled the consistent measuring of longevity according to a predetermined scale. Once this became possible, one could accurately date to the year, as defined by that calendar, a person's life span or the duration of an event. However, because a variety of calendric measurements coexisted, the definition of a year could vary greatly, causing inaccuracies in comparing longevities over the eons based on the generic term *year.*

The advent of the clock, which first determined time by the hour (the shortest interval that could then be measured with reasonable accuracy), enabled one to define the longevity of shorter events, such as meetings or the time of prayers, with the result that events could be more closely defined. However, again there was the difference in the length of hour, whether it was solar or "of the clock." Therefore, the hour of the day would have to be defined as clock time when stating a time; this is the basis for the term *o'clock,* meaning "of the clock," as opposed to a time based on another measuring system of the hours.

When examining longevity, one must look not only at the number but also the measuring system used at the time the number was created. In Biblical times, for example, longevity of life, if considered by the same time spans used today, might seem extremely long. However, when examining the length of a year in the times written about, one might determine that a year's length was much shorter than that measured today by the Gregorian calendar.

Without a standard measure, duration is notoriously subjective. An hour spent, for example, at a baseball game can seem to be much briefer than an hour spent in a dentist chair, and thus the longevity of these events, while they appear the same when measured by the standard measurements, seem to differ greatly to those experiencing them

The ability, and desire, to measure longevity is an essentially human trait. By knowing how long it takes for a plant to grow, or the length of time a dog lives as compared to a fish or hamster, humans gain a degree of control over the environment. A growing dependence on clock and calendar time has led to increased desires to measure more by these standards. Although society has benefited from a standardized method of measuring longevity, the existence of the method has also led to more of a dependence on such measurements. Actual time is becoming more important than perceived or social time.

Sara Marcus

See also Dying and Death; Fertility Cycle; Gerontology; Life Cycle; Medicine, History of; Time, Measurements of

Further Readings

Kurtzman, J., & Gordon, P. (1976). *No more dying: The conquest of aging and the extension of human life.* Los Angeles: J. P. Tarcher.

Newton, T. (2003). Crossing the great divide: Time, nature and the social. *Sociology 37*(3), 433–457.

Trivers, H. (1985). *The rhythm of being: A study of temporality.* New York: Philosophical Library.

Whitrow, G. J. (1972). *What is time?* London: Thames & Hudson.

Longitude

Longitude is a geographical measurement that describes the position east or west of the prime meridian for a given location. On maps, longitude is illustrated by vertical lines called meridians of longitude. Unlike parallel latitudinal lines, meridians of longitude all intersect at the north and south poles. Using longitude measurements for east and west along with latitude measurements for north and south, the location of any place on the earth can be precisely described. While the equator provided the logical starting point for latitudinal measurements, there was no such obvious place for measuring longitude. Eventually the Greenwich Meridian—on which the original British Royal Observatory was located—was internationally accepted as the prime meridian. In relation to time, longitude corresponds directly to the local time of any location and is the basis for time zone divisions.

The concept of longitude originated in ancient Greece, but the quest for accurate measurement of longitude continued until modern times. While many methods were proposed and attempted for calculating longitude, the most significant development came with the production of the first highly reliable mechanical clocks. These enabled travelers to compare the time of a known location with the observed local time in order to calculate their longitude on land or at sea. Modern satellite technology now enables precise latitude and longitude to be calculated using the global positioning system (GPS).

Longitude is closely related to time, because the local time of any place on the earth is directly relative to its longitude. Each of the earth's 24 time zones represents a section 15° wide in longitude, the distance over which local time changes by one hour (1/24th of 360°). The relationship is also historical; the need for an accurate way of finding longitude, especially while navigating at sea, was a primary driving force in the advancement of timekeeping technology.

Longitude was used at least as early as 300 BCE in ancient Greece. In the 2nd century CE, Claudius Ptolemy published *Geography,* one of the earliest developed mathematical works concerning latitude and longitude, in which he discussed an earlier theory of Hipparchus for figuring longitude. Recognizing that longitude could be calculated by observing the same event in two places and comparing the local time of each, Hipparchus proposed that lunar eclipses could be used for this purpose. The soundness of this theory was outweighed by the fact that sundials, the only available timekeeping devices, would not work well when observing something that occurs only at night.

After centuries with no scientific advancements, in 1514 Johann Werner surmised a new theory for finding longitude: the lunar distance method. Werner proposed that longitude could be found by precise comparisons of the moon's apparent position relative to the stars. This method had its problems, but with necessary observation data it had potential. Precisely for the gathering of such data, the Royal Observatory was founded in Greenwich, England, in 1676.

In an age that depended upon maritime travel, the need for more accurate navigation became urgent, prompting several European countries to offer generous rewards for anyone who made breakthrough discoveries toward calculating longitude. John Harrison received the largest of such rewards ever given by Great Britain in 1770 CE for his invention of a highly accurate portable timekeeper called a chronometer. The chronometer enabled mariners to use a simpler, more accurate method of finding longitude. A traveler could compare the observed local time at noon with the time of a known location kept on a chronometer; the difference in time told the difference in longitude between the two places. With so much data already available in the almanacs published by the Greenwich Observatory, it was the natural location to which chronometers were set, and the longitude at which it was located later became the prime meridian for the entire world. The chronometer method for identifying longitude remained standard until more advanced technology replaced it in the 20th century—first radio based systems, then GPS satellites.

Adam L. Bean

See also Astrolabes; Earth, Rotation of; Harrison, John; Observatories; Time Zones

Further Readings

Harrison, L. (1960). *Sun, Earth, time, and man.* Chicago: Rand McNally.

Howse, D. (1997). *Greenwich time and the longitude.* London: Philip Wilson.

Sobel, D. (1995). *Longitude: The true story of a lone genius who solved the greatest scientific problem of his time.* New York: Walker.

Lucretius (c. 99–55 BCE)

Titus Lucretius Carus was a Roman poet and philosopher who is best remembered for his comprehensive epic *On the Nature of Things.* This remarkable work offered a dynamic worldview that departed significantly from the Aristotelian interpretation of this universe, life forms on earth, and the place our human species occupies within nature. Because his ideas differed greatly from those of the Greek thinker, Lucretius was not taken seriously by his contemporaries. In fact, his provocative thoughts on time, change, and reality would not be appreciated by scholars until his lost manuscripts were revived from philosophical oblivion during the Renaissance.

Aristotle had presented a geostatic and geocentric model of the universe. He separated the ethereal heavens from our material planet and claimed that the finite but eternal spherical cosmos is enclosed by a fixed ceiling of stars, each star equidistant from the earth. For him, both terrestrial linear motion and celestial circular motion are caused by the existence of the Unmoved Mover beyond the stars; this perfect entity of reflecting thought is the ultimate object that all desiring things attempt to emulate in terms of their development from potentiality to actuality. Within his philosophical system, Aristotle taught that species are eternally fixed, with the human animal occupying the highest position in a static ladder of planetary organisms. Lucretius boldly challenged each of these basic Aristotelian assumptions.

Rejecting religious beliefs and ignorant superstitions, but indebted to the earlier thoughts of Epicurus, Lucretius presented a strictly naturalistic interpretation of and explanation for the existence of this eternal and infinite universe. He argued that endless reality consists only of material atoms and the void, there being no difference in makeup between celestial and terrestrial objects; thus, his natural philosophy taught the cosmic unity of all existence. Furthermore, Lucretius saw this dynamic universe as a creative process within which, over time, material atoms combine by chance to form an infinite number of stars, planets, and organisms. As a thoroughgoing naturalist, he taught that all objects and events are a part of an ongoing material reality that has no center, design, purpose, or goal.

For Lucretius, if there are immortal gods somewhere in this universe, then they have no interest in human existence. Nevertheless, with his powerful imagination, he speculated that life forms and intelligent beings (perhaps beings even superior to humans) exist elsewhere on other worlds throughout the cosmos. Within this dynamic worldview, Lucretius glimpsed the forthcoming evolutionary framework. He held that, over time, the material earth itself gave birth to those plants and animals that now inhabit this planet, including our own species. Furthermore, for him, organic history is full of both creativity and extinction.

With exceptional insight, Lucretius outlined the sociocultural development of the human animal. In their prehistoric state, our naked but robust ancestors lived in caves and subsisted on pears, acorns, and berries. They wandered through the forests and woodlands searching for wild beasts (e.g., boars, lions, and panthers) with clubs and stones. Later, our ancestors wore animal skins, learned to use fire, and lived in nomadic hunting/gathering societies that waged war. Over time, humans even developed the use of symbolic language as articulate speech. With the emergence of agriculture, they learned to cultivate plants and domesticate animals. Eventually, civilizations appeared and flourished, with people living in cities, using metals (first copper, then bronze, and later, iron), and developing art, law, and religion.

Lucretius offered penetrating insights into the nature of our universe and the history of life on earth that far surpassed the cosmology of Aristotle. He enlightened later thinkers with his bold speculations, influencing several recent major philosophers from Herbert Spencer and Henri Bergson to Alfred North Whitehead and Pierre Teilhard de

Chardin. This is an impressive testament to Lucretius, whose astounding vision was first presented over 2,000 years ago.

H. James Birx

See also Aristotle; Bergson, Henri; Bruno, Giordano; Cosmogony; Materialism; Ovid; Poetry; Presocratic Age; Rome, Ancient; Spencer, Herbert; Teilhard de Chardin, Pierre; Whitehead, Alfred North

Further Readings

Birx, H. J. (1984). *Theories in evolution.* Springfield, IL: Charles C Thomas.

Gale, M. R. (Ed.). (2007). *Oxford readings in classical studies: Lucretius.* Oxford, UK: Oxford University Press.

Gillespie, S., & Hardie, P. (Eds.). (2007). *The Cambridge companion to Lucretius.* Cambridge, UK: Cambridge University Press.

Lucretius. (1995). *On the nature of things: De rerum natura* (A. M. Esolen, Trans.). Baltimore, MD: Johns Hopkins University Press.

Luther, Martin (1483–1546)

Martin Luther, a German priest, monk, and theologian, was among the leading figures of the Reformation. As part of the exceptional role he played in the history of religion during his lifetime, Luther added some new thoughts to Christian thinking about time.

He was born to a burgher family. From approximately 1490 onward, Luther attended a variety of schools. In 1501, he began his basic philosophical studies at the University of Erfurt; these basic studies were required for further studies in theology, medicine, or law. In the spring, Luther decided to follow his father's wishes and began to study law. On July 2, 1505, he experienced a severe thunderstorm and, believing himself to be in mortal danger, vowed to become a monk if he survived. Fifteen days later, he joined an Augustinian monastery in Erfurt. In spring 1507, he received his ordination to the priesthood. In the same year, also in Erfurt, he began his studies of theology, which were to continue later in Wittenberg. From November 1509 to April 1510, he traveled to Rome on behalf of the brotherhood. In 1512, he received his doctorate in theology, was appointed as professor of theology at the University of Wittenberg, and became subprior of the monastery in Wittenberg.

On October 31, 1517, he sent 95 theses about indulgences to archbishop Albrecht of Mainz and other theologians with the demand to discuss them. In these theses, Luther proclaimed that the selling of indulgences as it was practiced at this time was not the right way to salvation. The theses were followed by turmoil in the church. Luther was asked to abjure his theses, but he insisted on having a disputation. In 1518, he wrote to Pope Leo X to inform him better of the situation. The answer was the issuing of a Papal Bull in 1520 in which 41 of Luther's theses were called heretical and which contained a warning that Luther risked excommunication.

In consequence of his refusal to recant, Luther was excommunicated from the church on January 3, 1521. In May 1521, he was ordered to the Reichstag zu Worms (Diet of Worms) with the promise of free passage on the way. Luther refused once again to deny his claims without their falsification by arguments from the Bible. On his way back to Wittenberg, Luther was kidnapped by men of his own sovereign and supporter, Frederick III of Saxony, and brought to the Wartburg in Eisenach to be kept there hidden in safety. In this exile, Luther translated the New Testament into German.

Luther stayed in Eisenach until March 1522, when he returned to Wittenberg to moderate the movement for religious reform, which was initiated by Andreas Karlstadt and which exceeded Luther's goals. In the following years, the reform movement quickly spread, first to Saxony and later to other parts of Europe. From 1524 to 1526, peasants whose discontent had been fueled by Luther's writings revolted. After attempting to pacify the peasants, he took a position on the side of the sovereigns in this evolving war. In 1525, he married Katharina von Bora, a former nun. Over the years, there were many debates, especially with Müntzer, Karlstadt, and Zwingli, on the right course of the reformation and its theological contents.

At the Reichstag zu Speyer (Diet of Speyer) in 1529, the Catholic members of the diet wanted to cancel the provisional acceptance of the reformed members of the diet. The protestation against this decision established the name "Protestants." During the Reichstag zu Augsburg, the reformed parties presented their confession, the so-called Confession of Augsburg (*Confessio Augustana*), which was composed by Luther's friend Philipp Melanchthon. In 1534, Luther's translation of the Old Testament was published.

During the following years Luther's state of health deteriorated, although he stayed head of the German reformation and was often called upon in case of conflicts in theological disputes as well as in worldly matters. In February 1546, he traveled (although seriously ill) from Wittenberg to Eisleben to mediate in a conflict threatening the livelihood of his siblings' families. He died there, after successfully settling the matter, on February 18, 1546.

Luther's view of time is always eschatological. Influenced by Pauline theology, two of the bases of his view of time are the virtues of *fides* (belief) and *spes* (hope). The difference between these two virtues is practically irrelevant for Luther's concept. The Christian hope and belief aim for a future in which the realm of God is present. The suffering in this world for the Christian has two parts: first, the suffering every human has to endure due to illness and misfortune, and second, the worry about life, because every real Christian knows himself to be a sinner. In Luther's view, the life of the Christian is justified because he endures the second suffering. In his view, Christ gives the promise that the realm of God is open for the people who are worrying about their sins.

But the object of Christian hope, the realm of God, cannot be perceived during one's lifetime but stays hidden from the physical world. According to Luther's belief, the exact design of the realm of God cannot be imagined by a human being. It will always stay hidden from the searcher's mind and eyes. But in his view, it is possible that hell and heaven are not worldly places in real time and space but exist only within the soul. According to this line of thinking, every believer's conscience might be the place of final judgment and hell. Therefore, resurrection might take place directly after every individual's death, because it would not be bound to the dimension of time as humanity understands it. Here, the influence of Saint Augustine of Hippo is obvious. For Augustine, time exists only within the soul, because neither past nor future exist in the present, while the present itself has no extent.

Luther does not subscribe to the idea, however, that to believe is a pure decision of the individual. Following the tradition of Paul and Augustine, he is convinced that God manifests the true belief in people. In this way, he pleads for a kind of predestination. Belief is to be understood as a gift given by God to humanity. This gives humankind also a kind of freedom of conscience and belief, because its belief is not made by itself. But in the end, rational thinking is not an adequate means to fully understand the working of God.

Markus Peuckert

See also Augustine of Hippo, Saint; Bible and Time; Christianity; Eschatology; Genesis, Book of; God and Time; Gospels; Sin, Original; Time, Sacred

Further Readings

Bainton, R. H. (1995). *Here I stand: A life of Martin Luther.* New York: Meridian.

Kittelson, J. M. (2003). *Luther the reformer: The story of the man and his career.* Minneapolis, MN: Augsburg Fortress.

McKim, D. K. (Ed.). (2003). *The Cambridge companion to Martin Luther.* Cambridge, UK: Cambridge University Press.

Lydgate, John (c. 1371–c. 1449)

John Lydgate was a poet and monk of the Benedictine abbey of Bury St. Edmunds in Suffolk, England. Adept at writing for a variety of patrons and purposes and in a wide range of styles, he penned short devotional lyrics as well as vast moralistic tomes running to tens of thousands of lines each. Although after the Protestant Reformation, his reputation waned, during the 15th century Lydgate was the most popular didactic poet in England. Some of his works survive in more late-medieval manuscript copies than even certain poems by Geoffrey Chaucer, whom modern readers generally consider to be Lydgate's superior as a poet. The longwinded didacticism of "the Monk of Bury" has left him open to the frequent charge of tediousness. This and the apparent irregularity of his meter (in comparison with Chaucer's) drew widespread critical disparagement from the 19th to well into the 20th century. In the context of discussions about time, however, Lydgate is interesting both because of his attitudes toward the history of civilizations and because of his place in the history of literature.

Born around 1371, the poet apparently followed customary practice among medieval monks by taking his surname from his place of birth, in this case what is now the modern village of Lidgate in the county of Suffolk. He entered St. Edmund's Abbey as a teenager and was sent to the University of Oxford for further training in theology. Although he was a Benedictine monk, he spent plenty of time in the world outside the cloister and, because of his skill as a moralistic versifier, garnered the patronage—essential to premodern poets—of some of the most illustrious figures of his day. These included Henry V (whom he evidently met at Oxford), Henry VI, and Humphrey, Duke of Gloucester, among others; they were either the recipients of or the guiding influences behind some of Lydgate's longest productions. For example, *The Troy Book* (c. 1420–1422) addresses Henry V's victory over the French at Agincourt, while *The Lives of Saints Edmund and Fremund* was written to commemorate Henry VI's Christmastide sojourn at St. Edmund's in 1433–1434. The immense *Fall of Princes* (c. 1431–1439), Lydgate's longest work at over 36,000 lines, was commissioned by Duke Humphrey.

Both in his so-called courtly works, written for royal or aristocratic patrons, and in his less ambitious poems, Lydgate was chiefly concerned to praise God and his saints and remind his readers of their own dependence upon them. As creatures living out their lives in earthly time, Lydgate's readers were expected to conform their wills to Christ's in eternity and to prepare their souls for final judgment at his hands.

Until late in the 20th century, literary critics dazzled by Chaucer and other contemporary poets like William Langland and the anonymous (presumed) author of *Pearl* and *Sir Gawain and the Green Knight* found little to commend in the writings of a derivative, digression-prone moralist. Since the 1980s, however, scholars have returned to literary history with a keen interest in the relationship between writers and the sociopolitical conditions of their times. Much has been done of late to salvage modern understanding of his importance by situating Lydgate in the contexts of contemporary court politics and religious devotion.

To broach the topic of Lydgate and time, however, is to return to older critical discussions that dismissed 15th-century English literature as little more than an artistically arid interim between the ages of Chaucer and Spenser (or, alternatively, Shakespeare). In some respects, Lydgate does indeed seem self-consciously traditional. He affirmed pious orthodoxy at a time when the church in 15th-century England sought to curtail speculation on religious matters. Moreover, he appears to confirm modern scholars' suspicion that medieval writers lacked a sense of historical change, a basic awareness that customs, religious rituals, and dress did not remain the same over the course of 2,000 years. Finally, Lydgate was seemingly unaware of, or hostile to, the advances in humanistic thought characteristic of the Renaissance taking place in the Italy of his day.

Even so, the Benedictine poet had clear notions about time and history, whether or not those notions were reactionary. His interest in classical antiquity was real. It is evident in his long historical romances, *The Siege of Thebes* (early 1420s) and *The Troy Book,* as well as in *The Fall of Princes,* a series of didactic vignettes about illustrious biblical, classical, and medieval personages who rose to great heights of prosperity only to suffer abrupt downfalls. Yet Lydgate was no archaeologist: He was not concerned to investigate the distant past as it "really" unfolded. On the contrary, in his work he often rails against those pagan customs that he judges to be irreconcilable with Christianity.

To say that he therefore lacked a historical sensibility, however, is to assume that he should have shared the modern historian's dispassionate interest in learning exactly what happened and where and why it did. His reasons for studying antiquity were very different. He used ancient narrative materials such as the Troy and Thebes legends partly because he had a bookish interest in old stories, partly to find evidence for what he believed were eternally valid moral principles, and partly for the purpose of addressing political concerns that were current in the England of his own day. In unfolding the tragedies of Thebes with unstinting attention to their violence, he may have wished, as some scholars have argued recently, not so much to condemn the "ancient" and the "pagan" out of hand as to warn English kings and princes against the recurrent perils of empire building in all times and places.

Admirable attempts were made in the 1950s (by Walter Schirmer) and 1960s (by Alain Renoir) to treat Lydgate as a harbinger of English Renaissance humanism. Derek Pearsall's landmark study of the poet (1970) instead argued, without apology, that the Monk of Bury was much the product of his time and place. Debate continues apace, with ever more energetic scholarship focusing on the very meaning of formerly unexamined terms like *medieval, conventional,* and *humanistic.* To this ongoing conversation scholars like Lee Patterson, David Lawton, Paul Strohm, James Simpson, and others have made valuable contributions. All agree that Lydgate was a more active thinker and a more purposeful commentator on history than earlier scholars suspected.

Joseph Grossi

See also Alighieri, Dante; Chaucer, Geoffrey; Christianity; Humanism; Novels, Time in; Poetry

Further Readings

Nolan, M. (2005). *John Lydgate and the making of public culture.* Cambridge, UK: Cambridge University Press.

Pearsall, D. (1970). *John Lydgate.* Charlottesville: University of Virginia Press.

Scanlon, L., & Simpson, J. (Eds.). (2006). *John Lydgate: Poetry, culture, and Lancastrian England.* Notre Dame, IN: University of Notre Dame Press.

Schirmer, W. (1961). *John Lydgate: A study in the culture of the XVth century* (A. Keep, Trans.). Berkeley: University of California Press.

Simpson, J. (2002). *Reform and cultural revolution: 1350–1547. The Oxford English literary history* (Vol. 2). Oxford, UK: Oxford University Press.

Lyell, Charles (1797–1875)

Charles Lyell is widely regarded as one of history's most notable and influential scientists. His work in historical geology helped cement our understanding of the earth's formation. Yet, for all Lyell's efforts to gain acceptance and recognition of geological theory and the emerging science of historical geology, it is his impact on Darwinian evolution and the understanding of humanity's place in time that is arguably his greatest achievement.

Born in Scotland in 1797, Lyell took an early interest in nature studies, eventually focusing on geology for his career path. Inspired by geologist James Hutton's idea of *uniformitarianism,* Lyell supported the idea that Earth's geological elements are continually affected by slow processes of change (erosion, uplift, etc.) and always have been. Between 1830 and 1833, he summarized many of his emerging ideas in his chief work, *Principles of Geology,* which ultimately influenced

ENGRAVED FOR THE ECLECTIC BY GEO E PERINE N.Y.

SIR CHARLES LYELL.

Lyell's version of geology came to be known as uniformitarianism because of his insistence that the processes that alter the earth are uniform through time.

scientists not only in geology but also in the fields of archaeology and biology as well. Lyell continued to publish his findings, which were widely read and admired by his scientific peers. He was knighted in 1848 in recognition of his many contributions to science.

Charles Lyell's primary legacy revolves around uniformitarianism and the earth's age. Whereas Hutton formalized uniformitarianism, Lyell arguably became its most influential proponent. Lyell's research and argumentation helped cement the idea that the earth's landscapes took eons to form, a thought in stark contrast to Archbishop James Ussher's Bible-based estimate that the planet was formed approximately 6,000 years ago. The premise that billions of years instead of thousands were required to form Earth, the solar system, and the entire universe was radical in the extreme and challenged doctrines whose authority lay in religious scripture. Lyell helped revolutionize not only geological thought but scientific thought in general by helping scholars look beyond conventional ideas of time and change.

Lyell's *Principles of Geology* made an impression on countless scientists, including Charles Darwin. For Darwin, Lyell's arguments supported the idea that organisms, including humans, needed as much time to change as rivers needed to carve out canyons and mountains needed to rise and fall. Such a conceptualization made Darwin's evolutionary ideas appear all the more plausible. Consequently, Lyell, though initially hesitant to embrace the ideas of evolution of organisms, provided Darwin with direction. The ensuing shift in humanity's understanding of time, the pace at which events unfold in the natural world, and our own place in the natural order is due in part to Lyell's contribution, which cannot be overvalued.

Neil Patrick O'Donnell

See also Darwin, Charles; Earth, Age of; Geological Column; Geology; Hutton, James; Huxley, Thomas Henry; Lamarck, Jean-Baptiste de; Materialism; Paleontology; Smith, William; Spencer, Herbert; Steno, Nicolaus; Stratigraphy; Uniformitarianism

Further Readings

Matson, C. C. (2006). Uniformitarianism. In H. J. Birx (Ed.), *Encyclopedia of anthropology* (Vol. 6, pp. 2239–2241). Thousand Oaks, CA: Sage.

McKready, T. A., & Schwertman, N. C. (2001). The statistical paleontology of Charles Lyell and the coupon problem. *American Statistician, 55*(4), 272–278.

Recker, D. (1990). There's more than one way to recognize a Darwinian: Lyell's Darwinism. *Philosophy of Science, 57*(3), 459–478.

Young, P. (2006). *Lyell, Charles*. In H. J. Birx (Ed.), *Encyclopedia of anthropology* (Vol. 4, pp. 1503–1505). Thousand Oaks, CA: Sage.

LYSENKO, TROFIM D. (1898–1976)

A biologist and agronomist, Trofim Denisovich Lysenko was director of the Institute of Genetics in the Academy of Sciences in the Soviet Union under the regime of Joseph Stalin. To the public he was portrayed as a heroic example of the self-educated peasant. Although the theories he espoused were based largely on misunderstandings of genetics and scientific principles, the exercise of political influence and power enabled Lysenko to hold sway over biological research in the Soviet Union for decades. Lysenko's genetic theories rejected the theories of Gregor Mendel; they were the result of blending a superficial understanding of the theories of Jean Baptistede Lamarck with specific selections from Charles Darwin's theories that would support Lysenko's latest interpretations of inheritance.

Born near Poltava in the Ukraine, the son of a poor Russian farmer, Lysenko's first employment was as a gardener. In 1921, he studied at the Uman School of Horticulture. Shortly thereafter, he was chosen for the Belaya Tserkov Selection Station, and in 1925 he received a doctorate at the Kiev Agricultural Institute. Lysenko was highly interested in the theories of Ivan Vladimirovich Michurin, who taught that the environment is directly responsible for the development of hybrids that are very different from their parents. By controlling the environment, a breeder can select the type of hybrid to be developed. Individuals are highly plastic and not limited to the genetics of their parents. The results are then under the authority of the breeder. Although Michurin claimed many successful results, no one else seemed able to replicate his outcomes.

While working at an agricultural experimental station in Azerbaijan in 1927, Lysenko came up with the idea that fields could be fertilized by planting a winter crop of field peas. The field peas would then provide livestock with forage through the winter. Lysenko claimed that the use of chemical fertilizer did little to improve crop yield. Chemical fertilizers were successful the first year, but failed in succeeding years in areas of poor soil fertility.

This was the beginning of a career that would last until 1964. Each of Lysenko's failures would be rapidly followed by a new stunning success by the "peasant genius," as reported in the Soviet press.

In the Soviet Union, the long winters required that seeds survive long cold periods. Lysenko claimed that cooling the seeds before planting increased their strength and therefore the yield of the next crop. Lysenko selected spring wheat with a short "stage of vernalization" (exposure to cold) and a long "light stage," which he then crossed with wheat of a longer stage of vernalization and a short light stage. This led to increased yields and new varieties of grain. Largely as a result of the adulatory reports of this experiment by the Soviet press, Lysenko became editor in 1935 of his own agricultural journal, called *Vernalization*. His failures were not reported but led rather to new claims of success. His only real success, however, was his popularity among Soviet farmers, who were unenthusiastic about Soviet agriculture in the early 1930s. While most of his experiments ultimately hurt the Soviet farmer, because he was one of their own, his popularity remained intact. This served the needs of the Soviet government.

Lysenko was most noted for his criticism of the genetic theories of Mendel, Morgan, and Weisman, which he saw as antiscientific, decadent, and metaphysical. Lysenko set out to save science founded upon dialectical materialism from religious dogmatism masquerading as empirical science. His claim was that these genetic theories made the same errors as those of Thomas Malthus, who was exposed by both Karl Marx and Friedrich Engels for using deductive logic instead of empirical science.

Lysenko made the following argument: Intraspecies competition would only weaken the species in its adaptation to a changing environment. Morgan-Mendel genetics would be too slow to allow any species to survive in the real world; fixity of species was the core of the metaphysical superstition of the Morgan-Mendel school. Darwin proved that species do adapt to their environment and, in the process, species evolve into new species. Morgan-Mendel's ilk destroyed the insights of Darwin. According to them, genes and chromosomes that were fixed and passed on from generation to generation could not change fast enough to prevent extinction. This is why genetic theory would be metaphysical and not scientific. It created a dilemma that could only be avoided by introducing mutations or genes that would randomly make mistakes in the next generation. Furthermore, they said that most mutations were lethal, making all life on this planet impossible. Their only solution was divine intervention. Thus Morgan-Mendel genetics introduced religious dogmatism as a replacement for honest science. Lysenko claimed to merge Marx and Darwin; in fact he was closer to Lamarck than either Marx or Darwin.

By 1927, the Soviet Union had survived the First World War, revolution, civil war, invasion by 18 powerful nations, an economy in ruins, famine, isolation, and a desperate need for solutions. Simple remedies for complex problems were more attractive than more complicated solutions. The Soviet Union emerged from chaos with a political structure that was both rigid and bureaucratic. The Communist Party's tight control over all aspects of society led to fear of innovative ideas out of step with party directives among the top party leadership. The scientific community was particularly controlled in an inflexible and mechanical way. Science was defined as part of a larger global class struggle. There were two kinds of science: bourgeois capitalist science and revolutionary proletarian science guided by dialectical materialism.

The Communist Party used hero worship as a way to gain support for its policies. Lysenko, being a peasant himself, was a hero figure. When Russian farmers were resisting forced collectivization, both Stalin and the resisting peasants loved Lysenko. Lysenko was the son of a poor farmer who used commonsense arguments. His lack of understanding of genetics proved to most farmers that he had more connection to the land than most professors who had been "corrupted by Western ideas." With this support, he could make the case for genetics as part of an imperialist conspiracy to destroy true science.

Lysenko used anti-intellectual values and the support of a coercive government to secure his tight control over agricultural science and biology for many years. By early in the Cold War, his control was complete. The resolution in 1948 of the Lenin Academy of Agricultural Sciences of the USSR made any disagreement with Lysenko's findings illegal. Textbooks at all levels of education did not even mention Mendel's genetics. As a result, many serious biologists were exiled, tortured, imprisoned, murdered by the government, or committed suicide. With Stalin's death in 1953, there was a slight thaw in Soviet biology, and in 1956, more criticism of Lysenko's theories became possible. Nikita Khrushchev offered some protection. However, Lysenko resigned as president of the Academy of Agricultural Sciences in 1954, and in 1956 his resignation as president of the All-Union Academy of Agricultural Sciences was announced. In 1965 he was removed as director of the Institute of Genetics, and Lysenko was officially blamed for much of the failure of farms in the Soviet Union. Trofim Lysenko died in 1976.

Michael Joseph Franciscon i

See also Darwin, Charles; DNA; Evolution, Organic; Lamarck, Jean-Baptiste de; Materialism

Further Readings

Carroll, S. B. (2006). *The making of the fittest: DNA and the ultimate forensic record of evolution.* New York: Norton.

Joravsky, D. (1986). *The Lysenko affair.* Chicago: University of Chicago Press.

Lecourt, D. (1977). *Proletarian science? The case of Lysenko* (B. Brewster, Trans.). Atlantic City, NJ: Humanities Press.

Lewontin, R., & Levins, R. (1976). The problem of Lysenkoism. In H. Rose & S. Rose (Eds.), *The radicalisation of science: Ideology of/in the natural sciences* (pp. 32–64). London: Macmillan.

Medvedev, Z. A. (1969). *The rise and fall of T. D. Lysenko.* New York: Columbia University Press.

Roll-Hansen, N. (2004). *The Lysenko effect: The politics of science.* Amherst, NY: Humanity Books.

Safonov, V. (1951). *Land in bloom* (J. Fineberg, Trans.). Moscow, USSR: Foreign Languages Publishing.

Soyfer, V. N., Gruliow, L., & Gruliow, R. (1994). *Lysenko and the tragedy of Soviet science.* New Brunswick, NJ: Rutgers University Press.

MACH, ERNST (1838–1916)

Perhaps best known as a founder of the field of philosophy of science, in which he held a chair at the University of Vienna, the Austrian physicist, mathematician, philosopher, and science historian Ernst Mach also had a major influence on the emerging discipline of physiological psychology as well as on the development of physics itself. Albert Einstein would later credit Mach's critique of Newtonian concepts of absolute time and space as a decisive influence on the development of relativity theory.

A prominent element in Mach's thought is antimechanism, or the refusal to accept the doctrine that reality, including psychic phenomena, consists essentially of matter in motion. Along with naturalism, the principle that nothing exists beyond nature, it was largely Mach's embrace of Darwinian evolution that shaped his ideas. According to Mach, human culture, science, mind, and the senses have an evolutionary history, and indeed knowledge itself is a product of biological evolution. The earliest organisms responded to simple experience, thereby constructing an elementary picture of the primordial world; out of these first interactions, more complex understandings emerged, forming innate capacities in our remote ancestors that gradually developed through adaptation into increasingly elaborate constructions. The acquisition of memory permitted greater scope for awareness of spatiotemporal relations than what is given directly to the senses; much later on, memory was greatly extended by the capacity to communicate culturally. For Mach, scientific activity is not only the product of biological evolution; it also serves to advance the evolutionary process by giving rise to further adaptations as new data are confronted and understood. From among all available ideas, whether derived logically or from dreams or fantasy, scientists select those theories that best fit the data. In this way science proceeds, as does biological evolution, by a process of selection.

After earning a degree in physics, Mach undertook studies in anatomy, physiology, and chemistry at the medical school of the University of Vienna, where he later designed and taught a course in physics for medical students. From the pioneering work of Gustav Fechner, the founder of experimental psychology, Mach learned that there are quantifiable thresholds of perception, in other words, that sensations can be measured. Thus, a mathematical relationship exists between the psychological realm and the physical. Here lay the key to an experimental methodology that would produce, in studies by Mach and his contemporaries and up to the present day, significant advances in the psychology of perception, including color, sound, space, and time.

Today Mach's name is commonly associated with measurements of the velocity of sound. The development in the mid-19th century of more powerful guns and cannons had led to the production of bullets and shells that traveled at speeds greater than that of sound vibrations. Mach's research into supersonic motion, published in 1877, helped to establish the field of modern aerodynamics; the Mach number, still in constant use by engineers, is the ratio of the speed of a projectile to the speed of sound.

Mach's investigations into optical phenomena included the discovery of so-called Mach bands, an effect of contrast perception that creates the illusion of narrow light and dark bands at the boundaries of contrasting areas. Of greater significance is Mach's more fundamental insight that *perception itself is always relational;* that is, we perceive not the world itself, but relations between sensations. Our senses have evolved to perceive contrasts between stimuli. It is the interaction of a new experience with the residue of a previous experience, or the difference between successive sensations, that forms the basis of perception.

Sanford Robinson

See also Darwin, Charles; Epistemology; Evolution, Organic; Memory; Perception; Psychology and Time; Space, Absolute; Time, Absolute; Time, Relativity of

Further Readings

Blackmore, J. T. (1972). *Ernst Mach: His work, life, and influence.* Berkeley: University of California Press.

Mach, E. (1984). *Analysis of sensations and the relation of the physical to the psychical* (C. M. Williams, Trans.). Chicago: Open Court. (Original English publication 1897)

Mach, E. (1986). *Popular scientific lectures* (T. J. McCormack, Trans.). Chicago: Open Court. (Original English publication 1898)

Machiavelli, Niccolò (1469–1527)

Niccolò Machiavelli was a Florentine diplomat, political theorist, historian, and poet. He was politically active in the courts of Louis XII of France, Cesare Borgia, Maximilian I, and Pope Julius II. After the return to power of the Medici in Florence in 1512, Machiavelli underwent banishment and withdrawal to Sant' Andrea, where he wrote his two major political works: *The Prince* (*Il Principe,* 1513), and *Discourses on the First Ten Books of Titus Livy* (*Discorsi sopra la prima deca di Tito Livio,* 1513–1522). These political writings present themselves as political counsel. They deal mainly with the possibilities for the lasting stabilization and self-preservation of polities. The theme of time emerges against the background of this practical problem. Machiavelli's analysis of history focuses on the way different factors are persistent or variable with time. He formulates advice for different time frames and deals with the correct handling of the opportunities and dangers of time as a factor of political action.

The Prince

Machiavelli's project can be understood as an answer to the then current political crisis of an Italy that was splintered into city-states. *The Prince* is his most influential work. Formally the work follows in the Middle Ages tradition of "mirrors for princes" that deal with the presentation of the kingly virtues. But Machiavelli breaks with the tradition in content. Instead of a normative orientation around Aristotelian virtue ethics and the Christian natural law tradition, he lays out a series of rules for political cunning. Not the ideal, but rather the actual determines Machiavelli's advice. Most of this advice relates in particular to the acquisition and maintenance of power by a new prince. Machiavelli's achievements in *The Prince* testify thereby to his efforts to recommend himself to the Medici for reinduction into the service of the state.

The virtues (*virtù*) Machiavelli recommends for princes should not be understood as classical virtues. These would even be harmful. The prince must appear to have those virtues that are considered good, but he must also have the ability to contravene mercy, humanity, and religion. For Machiavelli, justice and political success are not connected.

This counsel, directed to the achievement and preservation of the power of the autocrat, gave Machiavelli the reputation of a "teacher of wickedness." This emancipation of politics from morality was influential in the history of ideas. The directives, focused on the preservation of the prince, provided a basis for later thought on questions of national interest and are to this day a point of reference for political realism.

The *Discourses*

The *Discourses* are a commentary by Machiavelli on the Roman histories of Titus Livy. But at the same time, he unfolds a republican theory of the state. His analysis of the history of Rome has the goal of enabling a revival of Rome's political success. Since the basic structure of the world is invariable, history can serve as a teacher in current political questions. The imitation of ancient Rome could thus be the solution to the political crisis of contemporary Italy.

Machiavelli claimed that there are particular, necessary rules in history that hold for all time. These *necessità* are not interventions into history by a god (*providentia dei*) but rather regularities comparable to natural laws. Throughout time, political events follow necessarily from particular preconditions. This compulsion can result from natural circumstances or from the actions of humans.

For Machiavelli, human nature is a constant. Humans always tend to the bad rather than to the good, and they ceaselessly follow their appetites and ambitions (*ambizione*). They are not political beings by nature but must be domesticated and cultivated by institutions. Only in the well-ordered state can they develop the necessary powers for the preservation of the polity.

Politics, Government, and Time

The connection between *The Prince* and the *Discourses* has always created difficulties for interpreters. The techniques aimed at the retention of autocratic power and the reviving of republicanism are two different, situation-dependent proposals for solving the problem of the stabilization of a polity. A consistent reading presents itself against the background of Machiavelli's conception of political time.

Human nature imprints a determinate structure on the course of history. Following the ancient historian Polybius, Machiavelli formulates a cyclical model of the forms of government. States change from a condition of order to a condition of disorder. They thereby pass through various forms of administration, from autocracy to popular government. Monarchy degenerates to tyranny, aristocracy to oligarchy, and democracy to anarchy. For Machiavelli, all of these forms of government are to be rejected, the good ones because stability is ephemeral, the degenerate forms because of their badness.

According to Machiavelli, periods of political decline require an autocrat to bring new order to the polity, because the people are not in a position to do so. The advice in *The Prince* is directed in particular to this politically effective agent (*uomo virtuoso*). But the function of the *uomo virtuoso* seems to be temporally limited. If the polity is able to maintain itself after the establishment of laws and institutions, then the republican mixed constitution is for Machiavelli the better form of government. The considerations in the *Discourses* apply to such a government.

Republics are best able to maintain their inner stability and external capabilities of expansion. In them, the prince, the nobility, and the people can govern and oversee themselves together. Republican freedom is the result of orderly conflict between the nobility and the people.

Machiavelli makes temporal continuity the criterion of success in politics. The other goals of political action are subordinated to it. The key to defense against the permanent dangers of decay and corruption is *virtù*. In Machiavelli's usage, this denotes a category of accomplishments that lead to political success. *Virtù* can be found in individuals, in a people, or in the military.

Since *virtù* cannot be inherited, the competence of a people is better than that of an autocrat as the starting condition for a stable polity. The mere continuity of a republic over generations is grounds for its precedence as a form of government. Republican freedom can be a means of achieving enduring political stability.

Virtù represents power in the fight against *Fortuna*. Opposed to *necessità*, *Fortuna*, often personified as a female deity, stands for unpredictability in politics. She is the irrational moment in time. Her temper can help a polity to greatness or bring about its fall. She predestines the path of human action, but not absolutely. Machiavelli sees about half of the action as being left to human skill.

Even if the arrival of *Fortuna* is uncertain, there are ways to take precautions against her. Machiavelli compares her with a raging torrent that in its times of calm allows dams to be built. But the human tendency is to not think about changing times. This idleness is a sign of lacking *virtù* and offers *Fortuna* easy prey.

On the other hand, correct *virtù* can harness *Fortuna* for political success. The goddess of fortune can appear as the bringer of favorable opportunities (*occasione*). If they are recognized and exploited, then the agent has a share in luck. Machiavelli talks of the *occasione* as hurrying by, with the hair brushed forward covering the face so as not to be recognized. If the opportunity passes by, then one tries in vain to grab it by the bald back of its head.

The contest with *Fortuna* requires a deep sense of situation and adaptiveness to actual temporal circumstances (*qualità dei tempi*). Time, writes Machiavelli, drives everything before it, and it is able to bring with it good as well as evil. But time waits for no man, and only the one who adapts to it will have luck in the long run. Republics offer here the best conditions because of the diversity of their citizens; they are more flexible and adaptable than an autocracy. The *virtù* of one person fit for one time is not likely to change when circumstances change.

However, even the degeneration of a republic cannot be stopped, only slowed. Machiavelli calls time the father of truth. The enduring badness of humanity becomes manifest and in time spoils goodness. Thus, for Machiavelli, there is a natural limit to the life of all things in the world. Decay is inherent to time.

Robert Ranisch

See also Ethics; Law; Magna Carta; Morality; Rome, Ancient; Time, Cyclical; Values and Time

Further Readings

Machiavelli, N. (1983). *The discourses.* London: Penguin Classics. (Original work published 1513)

Machiavelli, N. (2003). *The Prince.* London: Penguin Classics. (Original work published 1513–1522)

Orr, R. (1972). The time motif in Machiavelli. In M. Fleisher (Ed.), *Machiavelli and the nature of political thought.* New York: Atheneum.

Skinner, Q. (2001). *Machiavelli: A very short introduction.* Oxford, UK: Oxford University Press.

Magdalenian Bone Calendars

A number of authors (e.g., André Leroi-Gourhan) have examined the possibility that Paleolithic engravings may have constituted a form of notation—certainly this is the case with the Azilian painted pebbles, but the name most associated with Magdalenian bone calendars is that of Alexander Marshack of Harvard's Peabody Museum. Marshack determined that certain objects of art from the late Stone Age Magdalenian, Solutrean, and Aurignacian periods of the Upper Paleolithic may have been not simply *objets d'arts* or hunting tallies but may have served as lunar calendars. The Upper Paleolithic period stretched from roughly 30,000 years ago to 10,000 years ago, with some range of variation. This was during the height of the Würm glaciation, when much of the classic "cave art" is believed to have been produced.

Although polychromatic paintings of ice age animals found on cave walls (parietal art) in southwestern France and northern Spain, most notably those of Lascaux and Altamira, are the best-known examples of Upper Paleolithic art, portable art also exists from this time. Venus figurines often are displayed as examples of Upper Paleolithic portable art, but there also are objects of unknown use. Among these are pieces that Marshack claims represent calendars or seasonal notations. These are exemplified by a piece carved in the round on an antler. This carving, when rolled out on a flat matrix, represents reindeer, snakes, and salmon. The carvings show scratches that have been taken to be arrows. Marshack showed them to locals in the area where the piece was found, and they proclaimed the scratches to be *farin sauvage,* or wild wheat. Marshack determined that because of the antlers on the deer, the exposed genitalia on one of the snakes, the appearance of a hook on the jaw of a salmon on the carving, and the wild wheat in seed head, that the carving represented a short period of time in the spring when all of these factors co-occur.

Further, Marshack believed that he had found evidence of thousands of engraved and painted notational sequences from Spain to Russia and an actual lunar calendar showing phases of the moon carved into a piece of mammoth ivory from the Ukraine. For some, the most convincing evidence comes from a 10-centimeter-long ovaloid Aurignacian antler plaque from Blanchard, Dordogne. The Blanchard plaque shows 69 round and "paisley" pocks in a complex line that snakes back and forth across its flat surface. Marshack argues that when the pockmarks are microscopically examined, it is apparent that they were made over an extended time with multiple tool points and engraving pressures. The paisleys and circles correspond to the waxing, full, and waning moons over nearly two and a half months. The snakelike arrangement of the marks appeared to be the moon's rising and setting positions to an observer facing south. Some of Marshack's claims have been seconded in recent years by Francesco d'Errico by use of a scanning electron microscope. Although a number of people remain unconvinced by Marshack's discoveries, it is undeniable that he has made a significant contribution to our understanding of Upper Paleolithic observational abilities.

Michael J. Simonton

See also Anthropology; Time, Prehistoric; Timepieces

Further Readings

Elkins, J. (1996). On the impossibility of close reading: The Case of Alexander Marschack. *Current Anthropology, 37*(2), 185–226.

Leroi-Gourhan, A. (1993). *Gesture and speech* (A. B. Berger, Trans.). Cambridge: MIT Press. (Original work published 1964)

Marschack, A. (1964). Lunar notation on Upper Paleolithic remains. *Science, 146*(3645), 743–745.

Marschack, A. (1972). Cognitive aspects of Upper Paleolithic engraving. *Current Anthropology, 13*(3–4), 445–477.

Marschack, A. (1991). *The roots of civilization* (rev. ed.). Mt. Kisko, NY: Moyer Bell. (Original work published 1972)

Magna Carta

The Magna Carta (Latin for "Great Charter") is considered one of the most significant documents in history. First issued by King John of England in 1215, it formalizes the covenant whereby the king was compelled by the barons and the authorities of the church to make a series of promises regarding rights and privileges, thus imposing constraints on the powers of a monarchy hitherto regarded as absolute. Over the following centuries, the Magna Carta has had a strong influence not only on British constitutional history but also on the democratic development of other nations.

The signing of the Magna Carta has long been lauded as a defining moment in human history and has been cited as a touchstone of English common law by those seeking the guarantee of rights and freedoms in nations throughout the world. It is important, however, to note that the charter did not grant rights and privileges to all people in 1215. The English barons, or noblemen, and the church drafted the Magna Carta to assert their own rights and thereby check the power of the monarchy. The common people of England initially gained little. The document's prestige increased over time with the development of parliamentary government. It became a symbol for liberty and individual rights against oppressive rule.

Earlier English kings had made agreements with their barons. The Magna Carta differs in that it is the first important example of the king's subjects demanding rights and forcing him to concede. John became king in 1199 and lost little time in angering his subjects. He waged war unsuccessfully in France, which required him to raise taxes and conscript more nobles for military service than had his predecessors. He broke with feudal tradition by failing to consult with his barons on important issues. John also clashed with Pope Innocent III over the appointment of the archbishop of Canterbury. This quarrel resulted in John's excommunication from the Roman Catholic Church in 1209. Reconciliation with the church was immediately followed by additional military defeats. In 1213, the disaffected barons and clergy met and outlined a list of articles. John dismissed these twice before the barons raised an army. The king then reluctantly arranged to meet with them at Runnymede, south of London, and agree to terms.

The Magna Carta was concluded on June 19, 1215, following much squabbling. Written in Latin, as was the custom during the age, it was essentially a peace treaty. It contained a preamble and 63 articles, which were for the most part reactionary rather than revolutionary. The articles were divided into separate groups dealing with issues such as law and justice, conduct of royal officials, trade, the independence of the church, and the royal forests. The concluding articles concerned King John's loyalty to the Magna Carta and the right of the barons to challenge the king should he ignore the charter. The Magna Carta was reissued and amended a number of times in subsequent years. The vague wording of many of the articles has left considerable room for interpretation over time. Later generations interpreted clauses 39 and 40 in particular as protections of habeus corpus and trial by jury, but the reasoning has been much debated.

The stature of the Magna Carta increased over time as English lawyers and members of parliament cited it in arguing the rule of law. In addition to English common law, the charter has influenced constitutions throughout the world, including that of the United States. Four of the original copies are known to remain.

James P. Bonanno

See also Hammurabi, Codex of; Law; Morality

Further Readings

Holt, J. C. (1965). *Magna Carta.* Cambridge, UK: Cambridge University Press.

Turner, R. V. (2003). *Magna Carta: Through the ages.* London: Pearson Education.

Maha-Kala (Great Time)

In the tradition of India, the creation of the universe is the purpose and significance of Maha-Kala, or Great Time. The Great Time is personified by the god Siva and his alter-ego Maha-Kala. Siva is the artistic part of creation. Maha-Kala is the power (called Pralaya or the Great Dissolution) to dissolve the universe. Destruction is necessary for creation; reproduction requires both creation and destruction. This interaction is the reproduction of nature and the cycle of birth and death. Both male and female principles are at work: The female is destructive by dissolving what is ceasing to exist into herself; the male is creative by being the source of new existence continually being created in the universe; this interaction is represented by the concept of Sankara.

Siva Maha-Kala represents the continual rebirth of the universe, which is going on continuously. Also, it is the life cycle of the universe, which is born, matures, and dies. The universe is created by Siva and destroyed by Maha-Kala. New universes are created out of the destruction of the previous universe. This then brings about the Mahadaeva, who is the deity without comparison—the Great God who is Siva.

Unity is the restorer, which is symbolized by Linga the Phallus, or Global Oneness. The linga is the erect penis of Siva, which represents the respect and meaning of life. When combined with Yoni, the symbol for the female genitals, it becomes the reproduction of the universe. Yoni is the center of all that is spiritual. Through this union of the material and the spiritual, the universe is formed.

Maha-Kala is the destroyer essence of Siva; but she is more than that. She is the force that absorbs all of creation unto her. From this material, which Maha-Kala allocates into her womb, Siva can fashion the universe once more. The destroyer is the regenerator. Through destruction, rebirth and evolution are possible. Through understandings of the nature of destruction and creation, union with true enlightenment becomes possible.

Mahayogin is the knowledge of the secrets of the universe. Because of this, Maha-Kala is the destroyer of human passions. With the annihilation of all excitement, true enlightenment is possible. With illumination, it is possible to achieve ultimate understanding and tranquility in the endless shifting cosmos and escape from the affliction and suffering of life, existence, death, and fate.

Siva Maha-Kala is also the Nataraja, the lord of dancers and the dance of creation, destruction, embodiment, liberation, and maintenance. With the invention of the cosmos, the devastation of creation is realizable. The personification of the interaction of contrary forces brings release and preservation from stagnation. Nataraja is the artistic representation of Siva Maha-Kala through dance. The dance expresses the continuing creation and destruction of the universe as a single interactive and ongoing process. The dance makes clear that creation is a movement in which the universe creates itself out of its own destruction. This is why Nataraja is important in this vital understanding.

In different regions, people find different ways to express this insight. Different stories were developed to understand Siva Maha-Kala nature. In South India, in the woods of Maharashtra, lived a group of sages who had strayed from the Way. Siva and Vishnu tried to win the mystics back and restore rule over the area. Vishnu took the form of a beautiful woman, and the anger of the monks increased. Siva Maha-Kala danced to break down the priests' control over the region. The philosophers then created a tiger out of fire to pounce and kill Siva Maha-Kala. Siva Maha-Kala very gently removed the pelt of the tiger and covered himself with the fur of the tiger as if it were a robe. Then a venomous serpent was formed. Siva placed the snake around his neck, which became a most beautiful garland. The heretics then shaped together a giant who looked like an overgrown and deformed dwarf. Siva Maha-Kala broke this monster's back, returning substance to the earth. With song and dance, malevolence was overcome, and everything foul was recycled to create beauty.

Nataraja the dancer produces ecstasy, through which the divine embraces human life. Through the dance, life embraces the four directions. Siva Maha-Kala holds the hourglass drum, which is creation. The drum becomes the pulse of the universe. Sound is the first element created by the universe. From the sound and songs of the universe came the Sanskrit language—the second element and the carrier of wisdom.

Siva Maha-Kala then holds up the tongue of flames that is the next element. Fire is the destroyer. It is the element of destruction, annihilation, and extinction that generates the raw materials for creation to begin again. From this, Abahaya, or protection, is born. Protection is needed for life. Through this dance, Siva Maha-Kala gives birth to the son he sired. Ganesha, the son, removes obstacles to enlightenment, leading to the escape from birth and death.

Original beginnings and inspired strength are possible when slothfulness, apathy, and preoccupation with self are overcome. The soul of the universe rests within the spirit of each individual. Through wisdom, the understanding of this relationship can be appreciated.

Eternity and Time embrace each other. The mountain streams feed the oceans of the world. Siva is both Kala (fleeting time) and Maha-Kala (the Great Time or eternity). Mahayugas, or Great Eons, are but flashes of time. Eternity and Time stand in continual unity, conflict, disintegration, and reunity at every point from the eternal past to the eternal future. Time is the tension between destruction and reproduction.

Buddhism and Maha-Kala

With the rise of Buddhism, Maha-Kala became the realization of the eternity of ever-changing time. Ngawang Drakpa founded the Dhe-Tsang monastery. While traveling in the region of Eastern Tibet, a large crow flew down to the monk and pilfered his scarf. Days later and some distance away, the monk discovered his scarf draped over a juniper tree. This became the spot the monastery was built. To this day in Tibet, Maha-Kala means the "great black one."

The local Bonpo (the indigenous animistic masters) feared the coming of Buddhism. They used magic to prevent the building of the Buddhist monastery. What the Buddhists built during the day would collapse at night. When the crow, Maha-Kala, saw this, he carried a correspondence between Ngawang Drakpa in Dhe-Tsang and the Most Holy Master Tsongkhapa in Lhasa. From this communiqué, the solitary hero Bhairava Sadhana was gathered. This enhanced and intensified the decisive factors of Buddhism. Due to this improved Buddhism, the Bonpo monks became Buddhists.

With the construction of the monastery, it was agreed upon that there was a need for a guarding statue to protect the abbey. That very day, three black men from India showed up and offered their services. These sculptors were contracted to complete the sculpture. When work began on the figure, there was only one black man left. When the icon was only half-completed, a rite was planned and implemented to celebrate the holiness of the site. Tibetan dancers were asked to perform a dance of rejoicing for the ceremony that celebrated the founding of Buddhism in this region of Tibet. With dancing in progress, the black Indian began to dance. No human ever saw a dance more wild or beautiful. Everyone stopped what she or he was doing to watch the untamed magnificence. At the height of the performance, the sculptor disappeared and the image of Maha-Kala was completed. The same mysterious event occurred at the same time at other sites in which two other icons were constructed in exactly the same way. This was the work of none other than Maha-Kala, protector of the holy site and guardian to the Great Time.

Concluding Remarks

In the tradition of India, Maha-Kala represents time, eternity, destruction, and creation. This Great Time leads us to realize that our lives are transitory flashes in the eternal ocean of change. Because of this, our ignorance creates wisdom, pride becomes humility, desire leads to detachment, jealousy gives support to secure accomplishments, and anger gives way to inner peace. Eternity is forever and changes constantly, being destroyed and reborn. The universe also returns to its beginnings and starts over. What happens to every individual happens to the universe.

Michael Joseph Francisconi

See also Cosmogony; Cosmology, Cyclic; Eternal Recurrence; Eternity; Nietzsche and Heraclitus; Time, Cyclical

Further Readings

Many forms of Mahakala, protector of Buddhist monasteries. (2005, January). Retrieved July 3, 2008, from http://www.exoticindiaart.com/newsletter

Miller, B. (1986). *The Bhagavad-Gita: Krishna's counsel in time of war.* New York: Bantam Classics.

Olivelle, P. (1998). *The early Upanishads. Annotated text and translation.* Oxford, UK: Oxford University Press.

Ramanujan, A. K. (1973). *Speaking of Siva.* New York: Penguin Classics.

Rig Veda. (2005). New York: Penguin Classics.

Shiva as Nataraja—Dance and destruction in Indian art. (2001, January). Retrieved July 3, 2008, from http://www.exoticindiaart.com/newsletter

Tulku, U. R. (2004). *As it is.* Berkeley, CA: North Atlantic Books.

Valenza, R. (1994). *Maha Kala in the center.* Occidental, CA: Nine Muses.

Malthus, Thomas Robert (1766–1834)

Thomas Robert Malthus was an English political economist of the classical school and a representative of utilitarianism as well as of early demographic science. He is best known for his *Essay on the Principle of Population* (1798), in which he assesses the problems of the growth of the human population in respect to their supply of food and other vital products. Malthus is considered an economic pessimist, an economist of crisis whose theories were of high impact despite their being partially erroneous.

Life and Writings

Thomas R. Malthus was born on February 13, 1766, in Surrey, southern England, as the sixth child of a prosperous family. His father, Daniel Malthus, is said to have known the philosophers David Hume, James Mill, and Jean-Jacques Rousseau personally. Little Thomas was educated by his father and private tutors; starting in 1784, Malthus attended Cambridge University's Jesus College, where he studied mathematics, classical languages, and literature. He obtained a master's degree in 1791 and became a fellow of Jesus College only two years later. In 1797 Malthus was ordained in the Anglican Church and decided to officiate as country clericalist in his home county.

In 1804, however, Malthus resigned from the priesthood to marry Harriet Eckersall, with whom he had three children. Soon after, in 1805, he became the first professor of political economy at the college of the East India Company at Haileybury, Hertfordshire. Already in 1798, *An Essay on the Principle of Population, as it Affects the Future Improvement of Society, With Remarks on the Speculations of Mr. Godwin, M. Condorcet, and Other Writers* had been published anonymously. In this, Malthus's main work, which he revised continually until his death, he showed his deep economic and social pessimism, as he was convinced that any increase in productivity and wealth would inevitably be outrun by the evoked growth in the population's number. Consequently, he doubted the benefit of a high birth rate for an economy.

Around 1810 Malthus met one of the most important economists of the time, David Ricardo; an active correspondence and a close friendship developed. In 1819 Malthus obtained fellowship in the Royal Society; moreover he became a member of the Political Economy Club, and in 1834, he was one of the founders of the Statistical Society of London.

In his *Principles of Political Economy Considered With a View to Their Practical Application* (1820) Malthus wrote about the motivating parameters of individual economic decisions as individually rational cost/benefit considerations.

Population Theory

Following David Hume's empirically scientific attempt, Malthus generated his insights with respect to the constitution and the prospect of the society in a completely nonidealistic way, based solely on direct observation. He thereby formed one contradiction of significance to the French Revolution's idealists' and in parts even anarchists' (e.g., Rousseau, William Godwin, and M.-J.-A.-N. de Caritat marquis de Condorcet) theoretic idea of a perfect society.

Malthus did not believe in any kind of utopia. He instead came to the conviction that a steady state of a society's economic prosperity would be utterly impossible due to an inevitably geometrical growth of population combined with at best an arithmetical growth of production of food. In consequence, at any time, the population will be too numerous to live in comfort. War, diseases, and famine are considered natural regulators and necessary to constrain the excessive growth of population and following misery. This bad fate Malthus deemed divine destiny to keep morality and to evade profane idleness. However, later in his life Malthus came to regard self-imposed moral restraint as an alternative check on population-food equilibrium.

Malthus sometimes is criticized for his unfounded usage of statistical methods and subsequent generation of his arguments. On the basis of a sophisticated empiricism, Malthus unfortunately built somehow arbitrarily theoretical constructs of ideas.

Impact

So-called Malthusianism had a sudden and deep impact, especially on English social policy. Malthus thereby vigorously turned the balance against public generosity toward the poor. On the other hand, in stating public investments as a means to resolve economic slumps by stimulating the aggregate demand in his *Principles of Political Economy,* Malthus preceded or at least influenced John Maynard Keynes.

For decades, Malthus's economic doctrine was crucial to the economic and especially welfare-related politics of several European nations. Because of his influence on Ricardo and Keynes, for example, Malthus's thought remains, to some extent, still vivid today.

In evolutionary theory Malthus left his mark as well; Charles Darwin admired him all his life and followed Malthus's thoughts about the struggle of existence and the implication of the fitness of a species' individual representatives for the evolution of the species.

Despite their popularity, wider parts of Malthus's thoughts proved false. As one of numerous examples, he did not anticipate the technological advances leading to the agricultural revolution, which resulted in discharging growing parts of the population from the necessity of agricultural work, thereby paving the way for industrialization. Industrialization in turn finally led to a considerable increase in prosperity.

Matthias S. Hauser

See also Darwin, Charles; Dying and Death; Economics; Extinction; Extinction and Evolution; Extinctions, Mass; Hume, David; Rousseau, Jean-Jacques

Further Readings

Bonar, J. (2000). *Malthus and his work.* Boston: Adamant. (Original work published 1885)

Hollander, S. (1997). *The economics of Thomas Robert Malthus.* Studies in Classical Political Economy/IV. Toronto, ON, Canada: University of Toronto Press.

Malthus, T. R. (1998). *An essay on the principle of population.* New York: Prometheus. (Original work published 1798)

Mann, Thomas (1875–1955)

Thomas Mann was a highly regarded German novelist and social critic of the 20th century. He won the Nobel Prize for literature in 1929, and his works were considered classics by the end of his life. His novels and essays combined philosophy, psychology, and political insights with his literary craft. His themes often centered on dualism: the coexisting physical and spiritual human natures, the life of action and the life of thought.

Mann's writings detail the complexity of reality and time. *The Magic Mountain,* his novel published in 1924, most clearly explores the inner time-consciousness of the main character, Hans Castorp, as he seeks knowledge and adjustment to life in a tuberculosis sanatorium. This novel is full of references to understanding time. Mann insists that time cannot be narrated and is not linear. Hans remarks about time being a turning point in a circle. The daily routines and seasons circle around the characters' lives.

Thomas Mann was born Paul Thomas Mann into a prosperous middle-class family in Lübeck, Germany, on June 6, 1875. He was baptized as a Lutheran. He was one of five children of a prominent merchant and city councilman. When his father died, the family moved to Munich, where he received his early education. He began the daily habit of writing in his personal diary when he was a schoolboy in the 1890s. In 1905, he married Katia Pringsheim, an educated woman and the only daughter of a professor in Munich. She devoted herself to him, his career, and their six children. He traveled on the lecture circuit and vacationed around Europe. They had a life of culture, order, and comfort. He had many famous acquaintances in the fields of literature, music, psychology, and politics of the time.

In 1933, while Mann and his wife were vacationing in Switzerland, they were advised not to return to the political turmoil of Germany. Adolf Hitler's actions forced Mann into a reluctant exile. Mann was very concerned about getting his private diaries back. On July 7, 1935, Mann received Harvard University's honorary doctor of letters degree with Albert Einstein. In 1938, Mann and most of his family settled in the United States, where he continued his writings in the German language. His children grew up to succeed in a variety of literary and scholarly endeavors. In 1944, he became a U.S. citizen. Mann moved back to Switzerland and died there on August 12, 1952.

Major Novels and Essays

After writing several essays and journal articles, Mann published his first novel, *Buddenbrooks,* in 1901. This novel thoroughly detailed the story of three generations of a family as they declined physically but grew to include several failed artists. The values and attitudes of the middle class were in conflict with those of the artists. These internal conflicts of opposing forces leading to change are associated with the philosophy of dialectics. Mann read the classics, and his writings reflect his thinking about the nature of Western middle-class culture as well as his version of his own family.

His short novel *Death in Venice,* published in 1912, also is considered a mirror of his own life and his psychological issues. This novel details a writer's moral conflict and collapse through a humiliating, uncontrollable, and unfulfilled passion for a young boy.

During the period from 1914 to 1918, Thomas Mann supported Germany's slide in World War I. He wrote a lengthy essay published in 1918 as *Reflections of a Non-Political Man.* This was part of a disagreement with his older brother, Heinrich Mann, who was also a published author but who was very opposed to the military buildup in Germany. At this point, Thomas Mann praised Nietzsche for supporting the acceptance of ambiguity as a great personal strength. His essay claimed the romantic view that art would not surrender to the system but could remain isolated from politics. Later he realized that this position was political itself, and he finally reconciled with Heinrich in California in 1942.

The publishing of *The Magic Mountain* in 1924 marked the end of 12 years of work. Just as he had come to believe that there is no turning point in politics, this work incorporated the notion of time's circular nature in both the novel's form and its content. The reference to time not being linear seems to refer to Hegel's idea about bad infinity being linear but real infinity being circular and dialectical. The protagonist Hans Castorp notes that the longest day of the year, June 21st· is called the first day of summer, yet the days start getting shorter at that point, so it truly is the beginning of winter. Joy and melancholy can exist at the same time in a dualistic philosophy.

Mann used techniques of the composer Richard Wagner to tell this story: multiple themes with variations and the exploration of the characters' emotional lives. This novel is based on the setting of the Davos Sanatorium, where his wife spent 6 months in 1912 for a lung condition. *The Magic Mountain* is an allegory of Western civilization's sickness as well as a sympathetic telling of the story of individuals dealing with personal sickness, real or imagined. Man is the master of contradictions, and the human goal is not to decide but to reach harmony with the human condition.

Through the years Mann wrote essays on Freud, Goethe, Nietzsche, Tolstoy, and Wagner. These essays detail his intellectual struggles, which shaped his fiction writing. He wanted to understand people completely and took great interest in understanding himself and his world, as is minutely detailed in his diaries.

In 1943, Mann published a four-novel series on the biblical Joseph, *Joseph and His Brothers*. The project took him 10 years to complete. This epic begins in the timeless tribal existence of the desert and moves into the historical timeline of Egyptian civilization. Mann wanted to restore a belief in the power of humane reason. These novels show the individual as a reflection of his epoch as well as of his personal story.

Doctor Faustus, published in 1947, tells the tale of a great German composer who bargains with the devil and rejects love and moral responsibility in favor of artistic creativity. Mann connects the 12-tone musical system to totalitarianism. He writes that a chord has not one key but is all about relationships. This became a novel about commitment and the failure of accepting ambiguity. At the end of his career, he finally chastised the dialectic and recognized the importance of fighting evil. He aided the Allies through his writing and saw Franklin D. Roosevelt as saving the world. He chose to leave the United States when the anticommunists and Senator Eugene McCarthy held power and influence.

Ann L. Chenhall

See also Dostoevsky, Fyodor M.; Goethe, Johann Wolfgang von; Hegel, George Wilhelm Friedrich; Hitler, Adolf; Joyce, James; Nietzsche, Friedrich; Novels, Time in; Proust, Marcel; Time, Cyclical; Tolstoy, Leo Nikolaevich; Wagner, Richard

Further Readings

Kesten, H. (1982). *Thomas Mann diaries, 1918–1939* (R. Winston & C. Winston, Trans.). New York: Abrams.

Kurzke, H. (2002). *Thomas Mann: Life as a work of art: A biography* (L. Wilson, Trans.). Princeton, NJ: Princeton University Press.

Mundt, H. (2004). *Understanding Thomas Mann.* Columbia: University of South Carolina Press.

Maritain, Jacques (1882–1973)

Jacques Maritain, French Catholic philosopher, composed numerous influential works on topics including epistemology, metaphysics, moral philosophy, sociopolitical philosophy, philosophy of art, and mysticism. In Maritain's metaphysical philosophy he discussed the experience of the progression of time by finite beings in contrast to the divine, eternal perspective of time. In the latter, all moments of time are simultaneously known in a single instant, which has neither beginning nor end. Maritain also held that the human soul, which he associated primarily with intellectual activity, was eternal, existing always in the thoughts of the creator. Maritain's Christian-humanist perspective has been called Thomist, owing to Maritain's great affinity with the perspectives of 13th century philosopher and theologian Thomas Aquinas.

Born in Paris in 1882 and baptized into the French Reformed Church, Maritain studied philosophy and science from 1901 to 1906 at the Sorbonne, where he met Raïssa Oumansoff, a Jewish Russian student whom he would marry in 1904. Jacques and Raïssa Maritain became disillusioned with the rationalist scientism at the Sorbonne, and they resolved together to end their lives in suicide if they could not find a satisfactory understanding of truth. In their subsequent quest the Maritains were influenced first by the metaphysical perspectives of Henri Bergson and then by the writings of Léon Bloy, which led them to convert to Roman Catholicism in 1906.

Maritain began to study Thomas Aquinas's immense work, *Summa Theologiae,* in 1910. Aquinas's perspective appealed to both the philosopher and the Christian in Maritain, and its impact on him was acute and enduring. Maritain soon began publishing, and he lectured in philosophy at the Institut Catholique de Paris from 1913 to 1933 and at the Pontifical Institute of Medieval Studies in Toronto from 1933 to 1945. During these years he wrote without ceasing, becoming the dominant philosophical voice for Catholics in France and the United States. After World War II Maritain served as the French ambassador to the Vatican from 1945 to 1948, personally befriended Pope Paul VI, and taught philosophy at Princeton until retiring in 1956. After Raïssa's death in 1960, Maritain joined a Dominican order, The Little Brothers of Jesus, and lived with them in Toulouse as a hermit until his death in 1973.

In *Existence and the Existent,* Maritain discussed the relationship between time—the experience of finite beings—and eternity—the divine perspective of all existence. From the perspective of divine eternity, all moments of time—past, present, and future—are present and tangible in a single instant, Maritain reasoned. This eternal divine perspective does not imply absolute determinism, according to Maritain, but included acknowledgment of the free choices given to created beings, which rightly operate within the measurable succession of time. Along similar lines, Maritain viewed the human soul or intellect as eternal and timeless, presupposing that the intellectual activity observable in human beings must always have been and must always be existing, even before its creation in the thoughts of the creator.

In reflections and observations about history, Maritain—though highly modern in many ways—despaired of much of the course of the modern world. Maritain saw in history a pattern of contrasting progresses. He observed that while modern civilization had clearly achieved many advances in scientific knowledge and humanitarian causes, these gains were mirrored by moral and spiritual depravity, political totalitarianism, and wars with unprecedented cost of human life. The answer for Maritain lay in neither capitalism nor communism but in a renewed Christendom characterized by justice, truth, love, and tolerance. Seen in many ways as a liberalizing influence in Catholicism, in 1967 he shocked many by writing very critically of Vatican II in one of his last works, *The Peasant of Garonne.*

Adam L. Bean

See also Aquinas, Saint Thomas; Bergson, Henri; Christianity; Eternity; Immortality, Personal; Metaphysics; Mysticism

Further Readings

Barre, J.-L. (2005). *Jacques & Raïssa Maritain: Beggars for heaven* (B. Doering, Trans.). Notre Dame, IN: University of Notre Dame Press.

Dunaway, J. (1978). *Jacques Maritain.* Boston: Twayne.

Kernan, J. (1975). *Our friend, Jacques Maritain.* Garden City, NY: Doubleday.

Marx, Karl (1818–1883)

Karl Marx is often named as one of the two greatest intellectual innovators of the 19th century, the other being Charles Darwin. He helped to redefine the fields of sociology, history, economics, and anthropology. Much of what followed in these disciplines is a response to the theories he outlined in his writings. Known as the father of Marxism, a revolutionary socialist movement worldwide, he also strongly influenced and in part defined such topics as social stratification, historical sociology, materialist anthropology, cultural ecology, social history, and social economics, just to name a few areas. For example, sociologists since Marx have tried to disprove, defend, or reform his theories. Few can ignore the writings of either Marx or his followers.

Karl Marx was born in the German Rhineland city of Trier on May 5, 1818. Both his mother and father were Jewish by birth. However his father, who was well read in the humanist writings of the Enlightenment, converted to the Lutheran faith to secure employment opportunities at a time when many occupations were closed to Jews. Marx's mother and the rest of his family converted later. Having been born into a Jewish family and raised as a Protestant in a Catholic city helped mold the character of young Karl.

At age 17, Marx enrolled in law school at the University of Bonn. At Bonn he became engaged to Jenny Von Westphalen, the daughter of a baron who was also a professor at the Friedrich-Wilhelms University in Berlin. The next year, Marx transferred to the University of Berlin. There he became interested in philosophy. He associated with the Young Hegelian movement, which was a radical humanist movement. Because a university career was closed to all Young Hegelians, Marx took up journalism as editor of the radical journal *Rheinische Zeitung*. This would eventually result in Marx's exile to Paris.

In Paris, Marx made contact with the French socialists. As a result of pressure brought to bear by the Prussian government, who feared the anti-Prussian underground in France, Marx was labeled a radical and an undesirable foreigner and was forced by the authorities to leave Paris; he moved to Brussels, where he lived for three years. There Marx dedicated his time to an intensive study of history and expanded the materialist conception of history. He developed what later would become known as historical materialism. In 1848, Marx moved back to Paris in support of the revolutions in France and Germany. In 1849, he moved to Britain, where he died in 1883. Marx developed his methodology of historical materialism in the early years, and it served as a model for his later work in political economy. It is important to analyze the evolution of his life from the days when he was influenced by the Young Hegelians until he wrote *German Ideology* in 1847 as he moved from philosophy to historical sociology.

Karl Marx, the father of modern communism. Marx believed that the downfall of capitalism by revolution and its replacement with a society based on socialism was inevitable.
Source: Library of Congress, Prints & Photographs Division.

Marx the Scholar

According to Irving Zetlin, Karl Marx is credited with establishing sociology as a discipline. Since then, sociology has been defined by a debate with the ghost of Marx. Karl Marx was a true heir of the Enlightenment. Marx's sociology was historical, materialist, and dialectical, and it was part of a social, political, and economic revolution. In his early years, Marx was interested in philosophy, encouraged by his association with the Young Hegelians. These philosophers wore their radical atheism as a public badge of honor. Particularly influential in the life of young Marx was Ludwig Feuerbach. Feuerbach taught that God is a human creation in which human characteristics, requests, requirements, and promise are projected onto an entity that is creative fiction. We then worship that entity as if it were real. He believed that ideas actually come to pass from the lives of real people. Therefore, only when people come to realize this can they end their alienation and restore their species-being. This is our shared common humanity. From this, Marx arrived at the idea that we would need to look at how real people live in their environments in order to study properly the content of their cultural ideas, ideology, and religion. Political beliefs, values, ways of life, and religious convictions all develop and change within human communities. These communities are embedded in an ever-changing environment. The environment is rooted in a constantly evolving set of conditions.

In the following sections, the principles Karl Marx espoused are briefly outlined.

Early Marx

Following is a summary of the young Karl Marx's views on democracy, power, the state, religion, bureaucracy, and law. According to Marx, democracy makes the assumption that all people are equal, even if they are not. The nature of any state is the specific historical circumstances that reflect particular social relations. The state is the design of unequal amounts of power that competing groups use to control the administrative system of political institutions.

Power, within the politics of a state society, often appears to operate independently of individuals within that society. In point of fact, groups compete with vastly different amounts of political power. Power cannot exist as a force independent of individuals within these competing groups. Economic classes and social groups compete over control of resources that are necessary for wealth and power. The state appears to operate externally to and autonomous of these struggles. This is a hallucination. The state is constantly changing, reflecting changes in these power struggles. While the political appears to be the cause of these changes, it is more like the effect. The state as a neutral arbiter of these conflicts is reflective of power, wealth, and class interests. Not all classes, families, or individuals are equal in their influence on the state. Being a neutral arbiter is an illusion.

Social interaction between real people, like families and the surrounding community, can be observed empirically. This includes how people provide for their material needs collectively in that community through working together. This is called interaction in civil society. Civil society is an abstraction that reflects authentic social relationships. Civil society is the private satisfaction of personal desires in a social setting. The state reflects the public expression of power; it is also an abstraction that illustrates genuine social relationships. The state is the public manifestation of the private yearnings expressed in civil society; it is made up of real individuals interacting and competing for political power. These human beings relate mutually to nature and each other through work, forming groups to cooperate in meeting their real physical and social needs. The state is the public expression of this, and civil society is the private expression.

Because ideology is used to justify one group's control over economic resources necessary for political power, it is usually declared that the dominant group represents all of society, including the powerless.

The state, according to Marx, cannot be separated from real individuals. There are real people doing what they must to survive. In order to live, people need access to economic resources. Each group competes for power with unequal political resources. Sovereignty is understood as the abstract reflection of who has more real power. This abstraction is often confused by the fact that appearances mask reality. Sovereignty appears as that which gives people power. Individuals define their citizenship within this struggle for sovereignty. However, with control over basic resources, political freedom masks real oppression and exploitation.

Humans are human only in a social context. The "social" is made up of definite individuals. The individual is a social product. The state and civil society are interconnected abstractions, and the separation of the two is an intellectual tool that makes the science of society possible. The government (state) only appears as outside of and above individuals of civil society. The monarch is a real person who abstractly stands for the state. The state is an abstraction of power used to coerce the people. The monarch uses real power supported by other individuals, the military and police, to enforce his will. Others who work for the monarch use implements of coercion to force people to obey the will of the monarch. This threat of violence is reinforced by ideology, doctrine, and religion. The monarch is the personification of the sovereignty of power. People are excluded from the use of power. The monarch represents the unity of the people, a people without power.

Sovereignty of the people is a concept that stands opposed to the sovereignty of the monarch. The monarch speaks for God and not the people. In a republic or democracy, the government is the imagination of an abstraction called the people.

Instead of sovereignty defining citizenship, it is the belief of the citizen that gives sovereignty its perceived reality. Each type of government reflects real power relationships between groups of individuals. Thus, each government defines sovereignty to meet the interests of the more powerful group. The military, the courts, the church, and the media define what most people believe to be real about sovereignty. Political consciousness and political culture are learned in institutions that are largely controlled by the power elite. These learned explanations justify and hide real power in society. In order to operate more smoothly, the established power relations are often falsely represented as being in everyone's best interest. Alternative views are learned, in opposition to the established political culture. Each form of government is part of a cultural completeness, which everyone is taught to see as true. Imperial democracy is no contradiction. These abstractions reflect the interests of those who control the political resources of the state. Political constitutions operate on faith. A republican body of laws reflects the development and evolution of commerce and private property. Government based on a constitution is a bourgeois artifact. During the Middle Ages, property, economy, and society were embedded in the political structure. In capitalist society, the separation of the economic from the political is possible because of political action and reflects the interests of the economically powerful.

Religion in a monarchy is the foundation of the political constitution. A republic or democracy uses the private lives of its citizens as its justification. Private property and commerce are the material groundwork for the republic. In a monarchy, private property and commerce are political gifts of the sovereignty.

Bureaucracy is the practical structure of the state. It is the social relationship in which the abstract state becomes real. There are hierarchies of power that used specialized knowledge as their justification. The imaginary state functions through a real bureaucracy.

Bureaucracies specialize in carrying out the mundane details of the state. Formal goals are translated into action, coming into conflict with real goals. Specialization creates a hierarchy of knowledge. The upper echelons of the hierarchy are in charge. The lower levels carry out the mundane details. The bureaucracy operates outside of government and is a real part of the state. In carrying out the law, the law is changed. This is a daily process, continually causing the constitution to become obsolete; therefore, it must be updated when the discrepancy becomes too great.

The law exists only in people's imagination. The law becomes real because people, including people who enforce the law, believe it to be real and behave accordingly. Laws must first be interpreted, and the interpretation is constantly subject to change. When people in power replace old laws with new ones, change is amenable. Individuals must inevitably give up their interpretations and accept the official one.

The private realm in a republic is established through economic inequality. In public, the political lives of the citizens appear to be equal. Inequality requires the illusion of equality to make it seem acceptable. Freedom is established through a lack of freedom. This means that persons are free to work for anyone who will hire them. If they do not work for someone and accept that individual's authority, they die. The republic has freed them from the feudal lord only to enslave them to the owners of business.

Marx: Jews in a Christian State

In Marx's view, the Christian state oppresses everyone. The issue of emancipation of the Jew requires the emancipation of the Christian at the same time. The Jew is a Jew because he says no to Christ. The Christian cannot free the Jew and still be a Christian. The Christian state, because it represents Christianity on some level, must exclude Jews from being treated as fully equal. Religious constraints undermine political emancipation. The official recognition of one established faith as being more important than others oppresses everyone. People must insist on the complete separation of church and state. The legal abolition of any established religious privilege is a prerequisite for popular sovereignty or democracy. Religion must remain a private decision. If one chooses to have no religion, that decision must be protected. Not any one religion should be necessary for political power.

Freedom from religion and freedom of religion are two sides of the same struggle. The ultimate religious freedom is when the state recognizes no religion as superseding any other. Only when religion is recognized as a private decision, and only when one is free not to believe in any kind of higher power, is there freedom of religion for all.

The state must become secular. Humanism is a belief in all religion as superstition. Humanists believe than humans can be ethically decent and richly fulfilled without a higher power. Because secular humanists do not believe in a god, their religious freedom is the test case. If they are free to not believe, all others are free to worship in their own way. Religion becomes a private decision and is kept out of the public discourse over policy. In a secular state, people become the universal abstraction, replacing God as the political explanation of the state. The secular state then unites the believer and nonbeliever. The public realm protects the private realm, allowing for a great deal of diversity in society.

There cannot be much freedom of diversity in a Christian state. Religion limits choices. The state remains incomplete, because it continues to exist only by being attached to religion. Faith supplements coercion. In a democracy, the paramount hypothetical excuse for the state is "the people" and not "God." It is necessary for a democratic state to stay away from any religious commitment to God, or the philosophy of democracy is compromised. Religious accountability must remain independent of the commonly shared political culture of a nation. The foundation of democracy is the secular state. A Christian state, an Islamic state, or a Jewish state cannot be democratic.

Human rights struggles are historically a contest against tradition. Traditional religions are the heritage of the chosen. Even if a Christian state guarantees religious freedom, it still favors Christians over Jews. The confinement of privilege divides. Equal rights to diverse opinions in a matter of faith can exist only in a secular community. The state protects the religious rights of all by restraining the religious power of the favored group.

Liberal democracy requires egoism, individualism, private property, and freedom of expression in matters of faith. The liberal will claim that the rights of one individual are limited only when they threaten to injure the rights of someone else. This assumes equality, yet because of private property the society is founded upon enormous inequalities. Freedom of egoistic individualism was the result of political revolution. The civil society of religion was replaced by a secular egoistic civil society.

When the individual Christian is no longer Christian, individualism replaces the Christian community. The commitment to "other" as the basis of the Christian community is replaced by the selfish crusade of greed. The new state protects the individual citizen who is atomized in his greedy quest for power and material things.

Beyond Capitalism

According to Marx, people create their religious beliefs from their own imaginations. Religious beliefs reflect the real world and the lives of real people. Faith is the hope of the powerless. The only real power that can overturn the attraction of faith is the power of people who have real democratic control over their lives. The lack of collective control over the political power and the economy of a society make religion a necessity to help people get through each day. Happiness in this life requires people to be the authors of their lives and not victims. Political struggles need to deal with social and economic conditions of life, not merely to provide philosophical debates over religious issues.

Political action and philosophy cannot be disconnected without making both negligible in changing the lives of the poor. People cannot have genuine political power without first having collective control of their actual lives. This means they must have cooperative control over the economy. This can happen through joint action of a social movement. The capitalist came to power by way of a liberal political revolution. Liberal philosophy and capitalism are tied together and cannot be rent asunder. Private property and individualism are represented in each person's self-interest. Only when it is questioned whether some groups benefit more than other groups does liberal philosophy begin to unravel. The capitalist now replaces the feudal aristocrat as the major oppressor of the direct producers, the poor.

The liberator soon becomes the new oppressor. The capitalist class fights for liberty, human rights, and private property against the ancient feudal order. When capitalists gain power, human rights cannot threaten their private property. The working class must be kept in tow. The poverty of the worker becomes a necessary condition for the wealth of the capitalist. Workers need their own revolution to go beyond the liberal society of capitalism.

Alienation

Workers, wrote Marx, create wealth; the wealth belongs to the capitalist. The capitalist becomes rich and powerful because of the wealth created by the workers. The workers become weak and poor as a result of their labor. This is because the labor of a worker is sold to the capitalist like any other commodity, and the capitalist will try to buy it as cheaply as possible. By selling the products of labor, the capitalist gains his wealth. This wealth is the source of the capitalist's power over the worker. Workers sell the source of their own slavery. The products made by workers become their chains of bondage.

Nature, which is necessary for life, must first be changed through labor into the means by which we are able to live. Humans, being a part of nature, are free only as a part of nature. Because the resources of production belong to the capitalist, nature becomes an unavoidable condition for the enslavement of the worker. They can live only if they get jobs. Their physical survival is possible only if they can find work.

The true nature of work is art. Work is the creative relationship between the worker and the rest of nature. Only through labor can we develop our full creative potential as humans. Yet the planning and design of labor is taken away from the worker. For the capitalist who owns the resource, the final product, the labor power, and the labor process, work becomes a source that cripples the worker. This turns workers into an adjunct to a machine. They can live only at the pleasure of this stranger who has plundered their humanity. Workers, once a vital part of nature, are now foreigners. Nature, their human essence and their birth, is now an unfriendly centaur prepared to devour their lives. The artistic celebration of life through labor is forever shattered. Our kinship with nature is eternally vanquished.

Workers stand in open competition with their coworkers. Their community has been plundered from them. Private property separates workers not only from nature but also from their own community. Greed and envy, as well as lewdness and ill will, become the glue that holds together a society founded upon accumulated wealth, individualism, and private property. The wealthy are always in awe of the wealthier but are also spiteful and afraid of the very rich. Wealth requires the cunning to live off the income looted from the workers. Until everyone becomes a worker and all wealth is held in common, the rich can only survive by stealing wealth from the poor.

The abandonment of private property is only the first step in returning the wealth to the poor who originally created it. We can democratically control our own society only after everyone becomes a public employee. Democracy grows only by empowering everyone through democratic control over the process of production. This process reintroduces the aesthetic and harmonious connection between the workers and nature. The alienation of all workers from nature, from the products they make, and from the community in which they live amidst the very process of creation will come to an end. Only then will the estrangement of one's work that has been plundered be returned. Democracy, socialism, and communism are our rediscovery of nature. Humans are not only a part of nature; they are human only in a natural and social setting.

It is through labor that this unity between the individual, the community, and nature becomes real. Humans are as much a product of social labor as they are a part of nature. We not only produce, we create ourselves. We write our own history through our social interactions in a natural and social setting.

Logic and philosophy become separated from the practical. Intellectual work becomes separated from physical work. This is a counterfeit way to live. Nature becomes alienated from nature, and our lives from reality. Living nature becomes a lifeless fact. Workers can regain their souls only if they are able to bring together abstract thought with the natural setting through their joint activities with other workers.

Equality is achieved through the celebration of nature and our common humanity. This is communism. Communism occurs where nature and humanity meet through democracy. Socialism is the path to communism. The mind and the body are reunited. Political democracy is transcended by economic democracy. Alienation is replaced by a communal relationship with nature. Thought, action, and creation are brought together as people are reintroduced to nature. We, as humans, live life as a ceremony to be fully indulged and as a burden to be endured. Work becomes entertainment and not drudgery.

Labor can become fulfilling only when private property has been eliminated. When owners possess nature, those who do not own anything become objects for sale. The life of workers becomes sold to the highest bidder; where slaves are abundant, their price is low. Capitalism cannot be reformed, and it can never be democratic. Workers are not human but objects of trade. Their lives are those of slavery and alienation. The ethics of communism require that all people should be the composers of their destinies and not their blood offering. People must be reunited with themselves.

Marx's Method

The real study of history, in Marx's view, begins with the material formulation of real people living their everyday lives, with people's relationship to nature. Through these relationships, humans produce their own means of subsistence. Each generation inherits and reproduces this means of subsistence and then changes it to fit their changing needs. This happens in the context of a historically and culturally specific setting and shapes individual human nature. Production determines how people are organized and interact.

Production molds all other social relations. This includes the relations of one nation to another as well as the internal social structure of a single nation. With every new change in the forces of production, there exists a corresponding change in the relations of production. These changes lead to changes in the division of labor. With changes in the division of labor, there are changes in the property relations of the nation. Ultimately, this means ideology changes as well.

The earliest division of labor is between the town and country. Industrial and commercial interests are separated from agriculture. With these changes in the division of labor, there are changes in property relations. When private property restricts access, the resources of subsistence become constrained. Each type of stratified society is founded upon this unequal access to needed resources. Each society has its own type of ownership and its own type of property relations. Historically, specific relations then develop between groups of people.

The first type of property is tribal or communal. This undeveloped stage of production has simple technology. The social structure is based upon the family and the extension of family called kinship. This evolves into ancient communal or state ownership. Private property develops but remains subordinate to the communal property of the state. With the development of an economic surplus, the town or administrative center stands opposed to the countryside that supports its life.

Feudal ownership begins with estate ownership. Peasant serfs are the economic foundation. Property is organized through hierarchical land ownership. Nobles are an armed body of retainers. In the city, the guilds of master, journeyman, and apprentice copy the feudal relations of the country. Property ownership changes to meet the changes in production relations; this causes changes of the status of the serfs and in relations between the town and the rural aristocracy. This occurs because social relations continuously change. The ideas of the age are the direct result of the real material life of the people. People produce ideas through their productive lives.

The first historical act is production to satisfy material life. Following the first historical act is the production of new needs that are the practical result of satisfying the needs of material life. People reproduce themselves, their families, and their culture daily. These acts of production and reproduction are exhibited by the historical past of a people; but this very activity also changes both the people and their culture. Old needs are changed and new needs are created. With expanding needs, production of life is both social and natural. Humans are both the animal creations of nature and the social creations of society. Each society creates its own social organization based upon its own historical mode of production. The nature of society is based upon the mode of production and consciousness. People's relation to nature molds their relation to each other. People's relation to one another affects their relation to nature. Production, human needs, population pressure, and change will follow.

The division of labor begins within the family. After that, within the rest of society, the division of labor continues to evolve. It subdivides between mental and physical labor. With this division of labor, an unequal distribution of property (both quantitative and qualitative) and its resources occurs. This leads to the development of private property. Private property results from the activity of the property-less and grows as a result of that activity. Because of the development of the division of labor, there is a concurrent development of the contradiction between the individual and the community.

With these divisions between ownership and work, mental and physical labor, and the community interests and particular deeds of labor, property becomes an alien force. The alienation of the worker occurs within society. Tools, resources, and human activities appear to control people rather than being controlled by the producing people. Through increasing specialization, labor is imposed upon the individual as a source of exploitation. The job also appears to own the individual. What we produce becomes an objective power over us. These illusions take on a reality that frustrates our best-laid plans.

During the evolution and development of alienation, the state develops as a community divorced from the realities of the individual. The state becomes a community unto itself. It is important to remember that the struggles within the state are class struggles. These struggles are class wars for mastery of the political powers of the state, whether peaceful or violent. Each class tries to conquer power in order to best represent the interests of that class. Any cooperation that exists is determined by the division of labor of that particular mode of production and by the class that controls the state. This cooperation is not for the benefit of all. The goal of cooperation is to benefit property owners who control the power of the state. Property relations, cooperation, and the state all change as a result of these class struggles. Society changes and the culture itself changes because of events brought about by this class struggle. Any kind of property relation that restricts access to resources causes resistance among the people who have limited or no control over the property that they work with to produce a surplus, which the nonproducers accumulate. Any indigenous class struggle is always an international class struggle. The world market economy exacerbates this struggle, and international struggles are then expressed locally.

Because property ownership restricts people's access to needed resources, direct producers become estranged from themselves. Material and intellectual production, as well as the producers themselves, do not belong to the producers but to the nonproducing minority. Until people are reunited with their creativity, work becomes a painful experience. When people become united with their own creative activity, they then achieve the capability to experience a joy that life holds dear. Cooperation must be transformed from cooperation for the benefit of the few into cooperation for the benefit of all, so that work and joy may be reunited. With the universal development of the productive forces of a market economy, a contradiction between the worldwide interdependent social economy and the private control of that economy for the benefit of the few is established. Only a world revolution can resolve that contradiction.

The universal development of production is a precursor to most people becoming property-less on a global scale. At the same time, world history replaces local history. Civil society is the result of the historical development of a new global system of production. Civil society is then seen as the material relation between people, people and nature, and the forces of production. This civil society exists only because of the rise of the bourgeoisie along with the evolution of the modern state, industrial production, world commerce, and professional bureaucracy.

History can be defined as a succession of economic systems along with changing ideological traditions. History modifies old circumstances by changing activities. The products of consciousness are the products of social life. Ideas reflect material production and its social relations that are historically inherited. Circumstances create people the same way that people make circumstances.

Humans need each other for survival. People cannot be free if they are hungry. Freedom is a historical action and not a state of mind. Freedom, if it is to exist, must have historical and technological foundations. Because the world is altered by the changes in industrial production, it can either increase alienation and exploitation of the direct producers, or it can increase freedom, depending on who controls the means of production. In the end, society changes according to the changing needs of social production.

Human unity with nature exists through industry. Social science must reflect this if it is to understand the deeper underlying connections between specific social actions and global trends. Industry, commerce, production, and exchange establish distribution, which in turn gives birth to ideological possibilities. Along theses lines, socioeconomic classes are determined by the mode of production. Every class society creates its own ideological support. Bourgeois society develops science to meet the needs of its mode of production. This is possible because the ruling ideas of any class society are those of the ruling class. Those who control the material forces of society rule the ideas of that society. Workers are subject to those ideas. The dominant ideologies reflect the dominant material relations.

Division of labor begins with the separation of physical and mental labor. Class antagonism soon develops. The exploited classes become the revolutionary classes. Each new ruling class presents its particular interests as the interests of society as a whole. This means the class making the revolution speaks as the new leader for the entire society. In the beginning, the revolutionary class leads the opposition against the old ruling class. The revolutionary class at this time is connected with the oppression of other exploited classes. Once the new class gains control of the state and the new means of production, opposition to the new class fully develops from yet other exploited classes. The old ruling ideas die with the new ruling class, and their new ideas become the dominant ideology. New opposition develops a new alternative ideology for a new struggle against the current ruling class.

Nature is constantly altered through human labor. This causes nature to become a product of human labor. The separation between town and country develops with the separation of mental and physical labor, and this leads to the development of state society. The class of nonproducers controls the coercive powers of the state and the means of production. When this happens, private property develops out of the surplus created by the producers who now have no property.

In the Middle Ages, the urban rabble controlled no resources; thus they were the most oppressed. The journeymen and apprentices were organized to meet the interests of the guild masters. Peasants remained isolated and weak and were controlled by the lords. Fear of the rabble united the nobles, the masters, the journeymen, and the peasants against this element in the towns.

Separation of production from commerce arrived with trade. Merchants became the new class. Manufacturing grew out of this marriage of merchants and guilds. Merchant capital became movable capital; the guilds became increasingly independent from the merchants. The merchant could then hire workers outside the guilds for manufacturing. With new types of manufacturing, unemployment became common.

With the rise of manufacturing, nation-states increasingly competed with each other. Trade wars became common. Within the nation, the capitalist and the worker related and competed with each other. The big bourgeoisie came to dominate the means of production but not the state. Commerce and navigation expanded rapidly, making the national bourgeoisie international in scope. Navigation and colonial monopolies went together. Protective laws sheltered the older bourgeoisie, making them dependent upon the state. The colonial monopolies controlled the market, and at home the market was administered and protected. Free trade was banned.

Competition came with big manufacturing. Movable property evolved into real private property. Competition separated the bourgeoisie and the workers from their own classes. Through private property, the state became independent of other forces in society. This was done even though the state organized society for the general interests of the bourgeoisie. Through the state, individuals of the ruling class asserted their common interests in spite of internal conflicts among the capitalist class. This was done because the state mediated the larger common interests of the capitalist class. There was a disintegration of natural communities with the evolution of private property, and civil law grew to define private property and natural interests.

Civil law defines property as if it were the general world of the people, not the property owners. The bourgeoisie, as a class, slowly absorbs the other propertied ruling classes. Its mode of production becomes dominant. The proletariat without property develops at the same time as this capitalist class. Industrial, financial, and commercial property becomes the dominant theme that relates to all forms of property. One class has conflicting antagonisms with another class. Class position limits life choices and defines the limitations and potential of every individual. This division of labor creates a reality independent of the will of the parties involved. Freedom can only be established on material grounds in a community in which the division of labor has been outgrown. In the past, economic reality acted independently of the will of the individuals of that society. Class is defined as a community that shares a common interest. Class is a condition of life and lifestyle.

Freedom is possible in resistance to oppression. Communism overturns all earlier forms of relations of production. Control of necessary resources returns to a community of individuals. This struggle is shaped by the material life and the history of a people. The conditions of real activity, which are the preconditions for the movements of a society, become a fetter to further movements of a people at a certain point. The resulting effect is that one type of material activity replaces another.

Historical conflicts grow out of contradictions between coexisting productive forces and between those forces and the rest of a society. The industrial capital of an advanced nation exports those conditions all over the world. Big industry equals social production while it is privately owned. This type of private property is the result of the accumulated labor of others. The division between ownership and labor becomes complete under capitalism. Forces of production do not belong to those who work with the tools of production but rather to the nonproducers. In modern industry, the workers are separated from the tools, their work, the products they make, themselves, and their coworkers. In addition, the state looms over them like an alien power opposed to the workers' class. The ruling class sets forms of distribution in motion that reproduce this inequality. In Marx's view, revolution becomes the only hope of the oppressed and exploited.

Michael Joseph Francisconi

See also Christianity; Dialectics; Economics; Engels, Friedrich; Evolution, Cultural; Evolution, Social; Feuerbach, Ludwig; Hegel, Georg Wilhelm Friedrich; Humanism; Judaism; Lenin, Vladimir Ilich; Materialism; Religions and Time;

Further Readings

Althusser, L. (1969). *For Marx.* New York: Vintage Books.

Berlin, I. (1996). *Karl Marx.* Oxford, UK: Oxford University Press

Marx, K. (1964). *The economic and philosophical manuscripts of 1844.* New York: International. (Original work published 1932)

Marx, K., & Engels, F. (1970). *The German ideology.* New York: International. (Original work published 1932)

Novack, G. (1971). *An introduction to the logic of Marxism.* New York: Pathfinder Press.

O'Malley, J. (Ed.). (1994). *Marx: Early political writings.* Cambridge, UK: Cambridge University Press.

Materialism

Philosophical materialism maintains that all things can be understood in terms of matter in motion. The only things that exist are matter and energy. Because of this, there is an association between different varieties of materialism and the scientific method. Many scholars credit Greek philosophers such as Democritus and Epicurus as the intellectual predecessors to philosophical materialism.

During the 17th and 18th centuries, materialism emerged as a philosophical tradition. Pierre Gassendi (1592–1655) raised the question of consciousness as integrated into the physical world and known through the senses. Julien Offray de La Mettrie (1709–1751) and Baron Paul Heinrich Dietrich von Holbach (1723–1789) taught that consciousness is simply the consequence of the biological structure and activity of the brain. Dialectical materialism and physicalism have existed since the 19th century as the modern expression of philosophical materialism. Physicalism is the point of view that any observed study can be articulated as a record of visible physical objects and events.

The main principle of dialectical materialism is that everything consists of matter briskly in motion, and everything is constantly changing, breaking down, and dying, while constantly being renewed and reborn. This is "the struggle of opposites."

Physicalism, or logical positivism, states that most things can be understood through science and mathematics. The pronouncements made by metaphysics, ethics, and religion are pointless, because their proposals cannot be verified by observation and experimentation or by logical deduction.

Three themes relate historical materialism to social action. They are materialism, action, and freedom. Action within nature is central to movement. Freedom through action is central to liberation and sovereignty. By way of action, we continuously alter the orchestration we have with nature. Given this, preexisting but changing boundaries limit freedom. Frontiers we cannot transgress include the physical universe, biology, ecology, social arrangements, technology, populations, organization, social design, and the mode of production. Theory leads to action; from action comes theory, freedom, determinism, and moral choice. This interaction cannot be separated. Natural history, which consists of geology and biology, is in inseparable unity with human history, which includes history, sociology, anthropology, and psychology.

Methodological Materialism

Karl Marx

The German economist Karl Marx (1818–1883) subscribed to the concept that there are real regularities in nature and society that are independent of our consciousness. This reality is in motion, and this motion itself has patterned consistencies that can be observed and understood within our consciousness. This material uniformity changes over time. For Marx, tensions within the very structure of this reality form the basis of this change; this is called dialectics. These changes accumulate until the structure itself is something other than the original organization. Finally, a new entity is formed with its own tensions or contradictions.

Human interaction in a natural setting is a given, because people are, at their core, a part of nature. It is because of this interaction that people are able to live. Through cooperation and labor, people produce what they need to survive. People live both in a community and in a natural environment. Any study of history cannot separate people from either the social or the natural environment.

According to Marx, the interaction between a social organization, called *relations of production*, and the use of technology within an environment, called *forces of production*, can be used to understand many particulars about the total culture. The evolution from band-level society to tribal-level society, tribal to chiefdom, and chiefdom to state-level society, has to take into consideration changes in the organization of labor, including the growing division of labor and ultimately changes in the technology people use.

With changes in the organization of labor, there are corresponding changes in the relationship to property. With increasing complexity of technology and social organization, societies move through these diverse variations to a more restrictive control over property; eventually, in a state society, restrictions develop around access to property based upon membership in economic classes.

A social system is a dynamic interaction among people, as well as a dynamic interaction between people and nature. The production required for human subsistence is the foundation upon which society ultimately stands. From the production of the modes of production, people produce their corresponding sets of ideas. People are the creators of their ideology, as people are continually changed by the evolution of their productive forces and of the relationships associated with these productive forces. People continually change nature and thus continually change themselves in the process.

Julian Steward

Julian Steward (1902–1972) is credited with the twin concepts of *multilinear evolution* and *cultural ecology.* Multilinear evolution is the exploration of recurring themes in cultural change. Cultural laws are then described in ways that make these changes clear. What become apparent are patterns of historical change that explain arrangements of the interaction between parts of a society and the larger environment. Cultural traditions are made up of basic characteristics that can be studied in context. Similarities and differences between distinct cultures can be studied in a meaningful way, and cultural change becomes more understandable. The evolution of recurrent forms, processes, and functions in different societies has similar explanations. However, each society has its own specific historical and evolutionary movement.

Cultural ecology is the adaptation of a unique culture, modified historically in a distinctive environment. This provides for observation of recurrent themes that are understandable by limited circumstances and distinct situations. The importance here is to discover specific means of identifying and classifying cultural types. "Cultural type" serves as a guide in the study of cross-cultural parallels and regularities. This allows investigation into the reasons for similarities between cultures with vastly different histories. This, of course, depends upon the research problem. But for problems related to historical change, economic patterns are important, because they are more directly related to other social, cultural, and political arrangements. This is the "cultural core." Cultural features are investigated in relation to environmental conditions. Unique behavioral patterns that are related to cultural adjustments to distinctive environmental concerns become more understandable. The cultural core is grouped around subsistence activities as demonstrated by economic relationships. Secondary features are related to historical possibilities and are less directly related to historical change. Changes are, in part, evidenced by modification in technology and productive arrangements as a result of the changing environment. Culture is a means of adaptation to changing environmental needs. Before specific resources can be used, the necessary technology is required. Social relations reflect these specific technological adaptations to the changing environment. These social relations organize specific patterns of behavior and its supportive values. A holistic approach to cultural studies is required to see the interrelationship of the parts.

Leslie A. White

Leslie A. White (1900–1975) looked at culture as a superorganic unit that was understandable only in cultural terminology. The three parts of a culture were the technological, the social, and the ideological. All three parts interact, but the technological was the more powerful factor in determining the formation of the other two. Thus, cultural evolution has all three parts playing important roles, with the technological influencing the sociological to the greater degree, and the sociological ultimately determining the ideological. Culture becomes the sum total of all human activity and learned behavior. It is what defines history. Through technology, humans try to solve the problem of survival.

To this end, the problem arises of how to capture energy from the environment and use this energy to meet human needs. Those societies that capture more of this energy and use it most efficiently are in a more advantaged position relative to other societies. This is the direction of cultural evolution. What decides a culture's progress is its capability of "harnessing and controlling energy." White's law of evolution, simply stated, says that a society becomes more advanced as the amount of energy harnessed per capita per year is increased, or as the efficiency of the activity of putting the energy to work is increased. This is cultural evolution.

Marvin Harris

Cultural materialism is based on the concept that human social existence is a pragmatic response to the realistic problems that are the consequence of pressures of the interaction between populations, type of technology, and the environment; with the economy as ever-important. Marvin Harris is the major spokesperson for this model, in which the social scientist investigates the basic relationship between particular social activity and overall tendencies.

Human communities are connected with nature through work, and work is structured through social organization. This is the foundation of the production of all societies. The way people come together to provide for their necessities, how these goods are distributed within the population, and networks of trade and exchange establish what is possible for the social organization. This relationship between environment, technology, population pressure, and social organization sets up the potential alternative ideologies within any culture.

How the basic needs are met within a society affects all members of that society, though often not equally. Ideology reflects not only the interaction between culture and nature but the understanding of this relationship. The model used is one that begins with the infrastructure, which includes environment, technology, and population pressure, Infrastructure is roughly similar to the Marxist concept of forces of production.

The structure or social organization is similar to the structure in the Marxist theory of relations of production or social organization. Structure is one step removed from the human interface with nature, and therefore the infrastructure has more influence on the structure than vice versa. The superstructure is what would be called ideology, or the symbolic and the ideational by other theories. The superstructure is twice removed from the human interface with nature and thus influenced more by both the infrastructure and the structure. The economy is the interaction between the infrastructure and the structure. This would be the mode of production of Marxism.

Unlike Marxism, cultural materialism emphasizes the primacy of the infrastructure over the structure in the formative relationship between the various parts of society. Changes in technology that are adaptive, given the environment and population pressure, are likely to be selected for and kept. This, in turn, will create long-range changes in both the structure and superstructure. Marxism, because it is dialectical, explains the relationship between forces of production and relations of production and is more reciprocal than cultural materialism. In Marxism, the forces and relations of production together make up the mode of production or, roughly, the economy. Again, cultural materialism would show, more than Marxism would, that the economy or infrastructure and structure more closely influence the substance of the superstructure. Only in cultural materialism do the forces of production determine the relations of production, and only ultimately does the mode of production control the superstructure using a Marxian model. The materialism of the Marxist is founded upon Hegelian dialectics; the philosophical foundation of cultural materialism is logical positivism.

Marxism, cultural ecology, and cultural materialism all begin with the first premise that the study of any social system is the dynamic interaction between people, as well as the dynamic interaction between people and nature. Because people come together in groups for human subsistence, they are social animals. This is the foundation upon which society ultimately stands. In producing what people need to live, people also produce their corresponding sets of ideas. In this way, it can be said that people are the creators of their history and ideology, though usually not in ways they are aware of. The process is historical, because people are continually changed by the evolution of their productive forces, and they are always changing their relationships associated with these productive forces.

Cultural core is the central idea of cultural ecology. This core is made up of economic patterns, because they are more directly related to other social, cultural, and political arrangements than are interactions between populations, types of technology, or the environment. The cultural core sets the limit of what is possible rather than directly determining what other theorists would call superstructure. Current scholars in the field add the use of symbolic and ceremonial behavior to economic subsistence as an active part of the cultural core. The result of cultural beliefs and practices continuing sustainability of natural resources become more likely. Symbolic ideology is as important as economics in defining the cultural core. Through cultural decisions, people continually become accustomed to a changing environment. Cultural ecology is closer to Marxism than it is to cultural materialism.

Finally, in the debate between Hegelian dialectics and logical positivism as the philosophical foundation for methodological materialism, the radical behaviorism of B. F. Skinner stands closer to the cultural materialism of Marvin Harris.

B. F. Skinner

The radical behaviorism of B. F. Skinner (1904–1990) begins with the idea that psychology is the science of behavior and not the science of the mind. The ultimate source of behavior is the external environment, not the world of ideas. Skinner maintained behavioral explanations of psychological observable facts as physiological influences. Behavior includes everything that an organism does. Thinking and feeling are other examples of behavior. All behavior is what psychologists try to explain. Skinner promoted the explanation that environmental characteristics are the correct causes of behavior. Environmental factors are external and separate from the behavior being studied, and one can influence behavior by manipulating the environment. Conditioning is caused by the influence of the total environment, including physiology. Conditioning is also influenced by culture and the ability of the organism to learn its own history. Each individual observes private events, like thinking, which is also a behavior. Introspection is also a behavior that is affected by the environment.

Radical behaviorism claims that behavior can be studied in the same manner as other natural sciences. Animal behavior is similar enough to human behavior that comparisons can be made. Ultimately, for all animals including humans, the environment is eventually the cause of the behavior that is studied; and an inclination for operant conditioning, or the modification of behavior. The occurrence or nonappearance of rewards or punishment is conditional upon what the animal does. This is achieved through reinforcement of already existing behavior, either by introducing a stimulus to an organism's environment following a response, or by removing a stimulus from an organism's environment following a response. Reinforcement will cause a behavior to occur with a greater rate of recurrence. Punishment or removal of the stimulus, also called negative reinforcement, will lead to a decrease in frequency, leading to extinction of such behavior. When an aversive stimulus is inflicted, a subject learns to avoid the stimulus. The avoidance learning may still be in place for a time.

Consistent with the theory of operant conditioning, any behavior that is repetitively rewarded, without error, will be more rapidly changed than when behavior is reinforced sporadically. This will lead to a more constant occurrence of a particular behavior and is comparatively more resistant to extinction.

Michael Joseph Francisconi

See also Darwin, Charles; Dialectics; Engels, Friedrich; Farber, Marvin; Feuerbach, Ludwig; Harris, Marvin; Lenin, Vladimir Ilich; Marx, Karl; Presocratic Age; White, Leslie A.

Further Readings

Afanaslev, A. G. (1987). *Dialectical materialism.* New York: International.

Afanaslev, A. G. (1987). *Historical materialism.* New York: International.

Bakunin, M. (1970). *God and the state.* New York: Dover.

Cameron, K. N. (1995). *Dialectical materialism and modern science.* New York: International.

Engels, F. (1975). *The origin of the family, private property and the state.* New York: International. (Original work published 1884)

Engels, F. (1977). *The dialectics of nature.* New York: International. (Original work published 1883)

Engels, F. (1978). *Anti-Dühring: Herr Eugen Dühring's revolution in science.* New York: International. (Original work published 1878)

Foster, J. B. (2000). *Marx's ecology: Materialism and nature.* New York: Monthly Review.

Harris, M. (1980). *Cultural materialism: The struggle for a science of culture.* New York: Vintage Books.

Harris, M. (1998). *Theories of culture in postmodern times.* Walnut Creek, CA: Rowman & Littlefield.

Marx, K., & Engels, F. (1970). *The German ideology.* New York: International. (Original work published 1845)

Materialism. In *Columbia electronic encyclopedia* [Electronic edition]. Retrieved February 27, 2007, from http://www.reference.com/browse/columbia/materialsm

Skinner, B. F. (1971). *Beyond freedom and dignity.* New York: Vintage Books.

Skinner, B. F. (1974). *About behaviorism.* New York: Vintage Books.

Steward, J. H. (1955). *Theory of culture change: The methodology of multilinear evolution.* Urbana: University of Illinois Press.

Vitzthum, R. C. (1995). *Materialism: An affirmative history and definition.* Amherst, NY: Promethus Books.

White, L. A. (1949). *The science of culture: A study of man and civilization.* New York: Noonday Press.

Maturation

Maturation, the growth and transformation of a single-celled zygote to a multicellular organism, has fascinated humans for centuries. In only a matter of months, a single cell can develop and mature to a complex organism. Although the maturation period for each species is different, the result is the same: a complex living organism.

No matter the species, the early stages of maturation are common to almost all animals. The first step is fertilization, in which the male and female sex cells or gametes fuse, creating the single-celled zygote. In species living in an aquatic environment, external fertilization is the usual method, in which the female deposits eggs into the environment to be immediately fertilized by the male. This method of fertilization usually requires courtship and environmental cues to be sure a male is present for fertilization, as well as to prevent the destruction and drying out of the eggs. In dry environments, the only way for sperm to reach the egg is by internal fertilization. By this method, sperm are deposited in or around the reproductive tract of the female. After fertilization, the newly formed zygote creates a fertilization envelope to prevent polyspermy, or union with more than one sperm cell.

Two distinct development modes begin after fertilization, protostome development and deuterostome development. Examples of common organisms that feature protostome development are molluscs and arthropods, while chordates and echinoderms are common examples of organisms that feature deuterostome development, among many others. From this point, maturation begins with the first cell divisions, known as cleavage. Unlike normal cell division, cleavage divisions are rapid, with no time for cell growth between each division. This process virtually partitions off the single-celled zygote into smaller cells called blastomeres, each complete with its own nucleus. In deuterostome cleavage, the cells divide in a radial pattern, in which the planes are parallel or perpendicular to the vertical axis of the embryo. Cleavage in deuterostomes is also indeterminate, meaning these new cells are not yet fated and can form an entire organism if isolated. In protostome development, the cleavage pattern is spiral, in which the planes of division are diagonal to the vertical axis of the embryo. Cleavage in protostomes is also determinate, in that these new cells are already fated and cannot form a whole organism when isolated.

This cleavage continues, forming a multicellular ball of blastomeres, known as a morula. Eventually, the number of blastomeres grows to form a single-layered ball of cells known as a blastula, featuring a hollow cavity in the center, known as a blastocoel. In humans and mammals specifically, this stage forms a blastocyst, with the key difference being the presence of an inner-cell mass within the blastocoel, as well as an outer epithelial lining to the blastocyst, known as the trophoblast. Also, in many other species, different concentrations of yolk, stored nutrients for maturation, tend to offset cleavage patterns. This is due to high concentrations of yolk found at the vegetal pole of the blastula and low concentrations of yolk at the animal pole. Thus the blastomeres at the animal pole tend to be smaller then the blastomeres at the vegetal pole. Yolk concentration can be so high that cleavage is hindered and even incomplete, a phenomenon known as meroblastic cleavage. Alternately, cleavage that is unimpeded by yolk and continues to completion is known as holoblastic cleavage. This uneven distribution of yolk is very distinct in different species and characterizes how they develop.

Although at these stages of maturation we start to see characteristic differences in each species, gastrulation generally follows cleavage blastulation. Gastrulation is the rearrangement of the blastula to form a gastrula consisting of two or three germ layers: the ectoderm and endoderm, with the mesoderm as the third and middle layer. Gastrulation also forms a primitive gut, known as the archenteron. This rearrangement allows the cells to interact in new ways and causes changes in cell shape, motility, and adhesion. One process of rearrangement is invagination, in which the single-celled blastula buckles into the blastocoel to form a second layer. This invagination causes the direct formation of the archenteron as well as the blastopore. This blastopore becomes the primitive mouth in protostomes and the primitive anus in deuterostomes. A second mechanism is involution, in which cells "roll" over the lip of the blastopore into the interior of the embryo. The combination of these two processes forms the gastrula. The archenteron and coelom, the primitive body cavity, differ in deuterostomes and protostomes. In deuterostomes, the body cavity formation is described as enterocoelous, in which the mesoderm forms outpockets from the archenteron and forms the body cavity. In protostomes, the formation of the coelom is described as schizocoelous, in which mesoderm near the blastopore split and outpockets to form the coelom.

Organogenesis is characterized by the development of primitive organs, vessels, and body systems in the embryo. Each of the three germ layers of the gastrula gives rise to the beginnings of very specific organs throughout the body. The ectoderm, the outermost layer of the gastrula, develops into the epidermis (skin) and sense receptors, as well as a majority of the nervous system. The mesoderm, the middle layer of the gastrula, gives rise to the skeletal and muscular system, as well as the reproductive system, circulatory system, and excretory system. The endoderm, the innermost layer of the gastrula, forms the epithelial lining of the digestive and respiratory tracts and the liver pancreas, as well as several glandular organs. One of the first organs to develop is the neural tube. The ectoderm thickens and pinches inward, forming the primitive spinal cord.

In mammals, all of these maturation processes occur during pregnancy, or gestation. The length of the gestational period directly correlates to the size of the growing organism. The typical human gestational period is about 40 weeks. However, the typical gestation period of a rodent may be only 21 days, while the gestation period of a cow averages about 270 days. Gestation in elephants can last as long as 600 days. In humans, pregnancies are usually split into three trimesters of about 3 months each. Organogenesis is usually completed by the end of the first trimester. At this point, the embryo is no longer considered a gastrula, but a fetus; fetuses average only 5 cm in length. In the second trimester, the fetus becomes more active within the womb and grows to approximately 30 cm. The final trimester results in birth of the child, or parturition. It is believed that hormone levels within the blood control labor, the process of birth. However, the mechanism behind this is not yet fully understood. Labor consists of three phases: the thinning and dilation of the cervix, the delivery of the baby, and the expulsion of the placenta.

Modern science has provided this insight into these complex processes, but this has not always been the understanding of maturation. As far back as 2,000 years ago, Aristotle proposed the idea of epigenesis, in which the animal develops from a relatively formless zygote. This theory, which was more accurate than most then believed, contrasted with the theory that prevailed to the 18th century, that of preformation. In preformation, it was believed that within the egg or sperm was a preformed miniature infant or homunculus and that this homunculus simply grew and matured within the womb until it was born. It wasn't until the 19th century, with the invention of light microscopy, that scientists were able to get a more accurate understanding of the complex process of maturation.

Christopher D. Czaplicki

See also DNA; Fertility Cycle; Gestation Period; Life Cycle; Metamorphosis, Insect

Further Readings

Campbell, N. A., & Reece, J. B. (2005). *Biology* (7th ed.). San Francisco: Pearson Education.

Ulijaszek, S. J., Johnston, F. E., & Preece, M. A. (1998). *Human growth and development.* Cambridge, UK: Cambridge University Press.

Maximus the Confessor, Saint (c. 580–662)

Born in Constantinople (present-day Istanbul, Turkey), Saint Maximus (Maximos) the Confessor influenced Eastern Christianity through his writings, debates, and personal witness as a Byzantine monk, spiritual writer, and opponent of monothelitism. Maximus's primary writings consist of Biblical commentary (*Quaestiones et Dubia, Quaestiones ad Theopemptum Scholasticum, Quaestiones ad Thalassium),* Christological debates (*Opuscula Theological et Polemica*), explanations of the liturgy (*Mystagogia*), and ascetic practice (*Liber Asceticus*). From his writings, it is evident that Maximus showed a great interest in the concept of timelessness and its relationship to the temporal world.

Maximus came from a wealthy household in Constantinople and received a good education. From 610 to 613, he worked as a secretary for Emperor Heraclius I, but he retired to the monastic life. Maximus resided in two monasteries in Turkey: Philipikos in Chrysopolis and Saint George in Cyzcius. Threatened by the Persian invasion of 626, Maximus moved to North Africa, making stops in Crete and Cyprus before reaching Carthage in 628. From North Africa, he defended the Christological teachings of the Cappadocian and Chalcedonian fathers against the Christological heresy of monothelitism, which taught that Jesus Christ had two natures but only one will, in contrast to the orthodox view that Jesus had two natures and two wills, human and divine. Maximus disputed this heresy with expatriarch Pyrrhus I, and later condemned monothelitism at the synod of the Lateran, in Rome, in 649.

Refusing to sign the conciliatory declaration of Emperor Constans II, Maximus was arrested in 653, charged with treason, and exiled to Bizye (Byzia), in Thrace, in 655. Tried and convicted again for treason in 662, Maximus endured torture and amputation of the tongue and right hand. This time the Byzantine monk was exiled to Lazica (on the eastern coast of the Black Sea) where he remained until his death in 662. Following his death, the Eastern Church called Maximus "the Confessor," for his faithfulness under torture in defending the orthodox teaching on the nature of Christ.

Maximus studied the classical Greek philosophies and labored to find their relation to or compatibility with the Christian worldview. Particularly interested in Neoplatonism, Maximus looked at the connection between Plotinus's "one" and the "many" (or creator and created). This examination made Maximus believe in the concept of timelessness, in which the individual achieves a supratemporal state or existence. The Eastern doctrine of deification complemented his view of timelessness, because the individual works to become one with God through prayer, meditation, and other ascetical practices. The "one" (God) became like the "many" through the incarnation (*logos*), and the "many" possess the ability through grace to become divine like the "one" because of the resurrection and ascension.

Although Maximus thought individuals reached a supratemporal state, he also recognized that the individual lives in a temporal world and experiences a beginning and an end, following a sequence of events in history. In his writings, Maximus makes this distinction using the term *aion* in reference to an endless amount of time and the term *chromos* in reference to the daily passage of time and events in history. Maximus, like other Neoplatonist Christians, had to reconcile the Greek understanding of a cyclical calendar with the Judeo-Christian understanding of linear history. For Maximus, the Christian scriptures focus on the history and chronology of past events, as well as point toward a fixed goal in the future (the eschatology). Both views of time share the concept of process and movement. Maximus acknowledges that God created the universe and interacts in history, but time does not restrict God. Therefore, the individual follows the movement of time and events in a linear sense, but through the process of deification, time does not restrict those who achieve union with God.

Because he was recognized as a saint, Maximus's life and works greatly influenced Eastern orthodox theology, and this Byzantine monk worked diligently to defend the nature of Christ. In his writings, Maximus the Confessor played a key role in bringing together Greek philosophy and a Christian worldview and contributed to the discussion or study of timelessness in a temporal world.

Leslie A. Mattingly

See also Christianity; Eternity; God and Time; Mysticism; Plotinus

Further Readings

Maximus the Confessor. (1985). *Maximus Confessor: Selected writings* (G. C. Berthold, Trans.). New York: Paulist Press.

Plass, P. (1980). Transcendent time in Maximus the Confessor. *The Thomist, 44*, 259–277.

Maxwell's Demon

Maxwell's demon is the name of a thought experiment that has been intensely discussed by physicists and philosophers investigating the relation between energy, information, and time. In this experiment, an imaginary but not supranatural being is using his extraordinary sensorial and intellectual powers to violate the second law of thermodynamics and, therefore, to falsify the fundamental principle by which modern physics explains the irreversible direction of time. The demon is named after one of the greatest physicists of all time, James Clerk Maxwell (1831–1879), who developed the theory of the electromagnetic field and conceived, in 1867, the experiment with the fictitious being.

The problem Maxwell wanted to solve by his thought experiment is whether a physical process might be conceived that, though being consistent with all known laws of physics, breaks the second law of thermodynamics. This law states that the entropy of a thermodynamically closed system never decreases so that it is possible to objectively measure the irreversible flow of time by comparing the entropy of different states of such a system. How did Maxwell suppose his demon to perform the feat of violating the second law? The demon, who has superhuman abilities but must act according to the laws of physics, is sitting in a closed container that is filled with gas and divided, by an impermeable partition, into two halves. In the partition there is a door that can be opened and closed at will. The demon observes the gas molecules approaching the door in both halves and sorts them by deciding which molecules are allowed to move through the door so that, for example, the faster ones will be gathered in the left half and the slower ones in the right half. Because the left half gradually becomes warmer and the right one cooler, the resultant temperature difference can be used to do work. Suppose that Maxwell's demon and the door could operate in a reversible manner, giving up some amount of energy when opening the door without friction and taking in the same amount of energy when closing it without friction. Then the container would constitute, for all practical purposes, a *perpetuum mobile* that, thanks to the demon, could completely transform heat flowing from a reservoir at lower temperature to a reservoir at higher temperature into work.

Even a short description of Maxwell's demon shows that information plays a crucial role in reasoning about the thought experiment. The demon must know the positions and the velocities of the molecules that are approaching the door to decide reasonably whether it is to be opened or closed. To bring order into a disordered system (thermodynamically speaking, to lower its entropy), information about the components of the system is needed. Therefore the physical nature of information must be taken into account if we want to find the weak spot of Maxwell's demon and to show that the thought experiment is not a counterargument against the universal validity of the second law of thermodynamics. The history of the discussion about Maxwell's demon mirrors the difficulties encountered in the development of a physics of information. Arguments for the possibility of Maxwell's demon, and against it, revolved around the information-gathering activity of the imaginary being: How does the demon come to know the location and speed of single molecules, and is the thermodynamic cost of gathering this information high enough to rescue the second law of thermodynamics because it produces at least the same amount of entropy that it makes possible to consume?

In 1982, a quantum physicist and computer scientist, Charles H. Bennett, brought forth the conclusive argument against the possible existence of Maxwell's demon. Bennett showed that it was not the gathering, as everyone had thought before, but the erasure of information that is decisive for an adequate understanding of the demon. The reason for this is that the demon needs a memory in which information about the molecules can be stored. Because the demon is, though fictitious, a physical being, this memory is finite. At some time the demon must begin to forget—that is, to erase information about the locations and velocities of molecules he has observed; otherwise, he could not memorize new information about molecules approaching the door. "Erasing information" means here that it is impossible for the demon to gather the deleted information once again. Bennett calls this kind of irreversibility that concerns the inability to infer a former state of a system from information about its given state, *logical irreversibility*.

Because different pieces of information must be represented by different physical states, logical irreversibility is combined with thermodynamic irreversibility. The higher the number of different equiprobable pieces of information a system can have, the higher its entropy is. Erasure of information thus decreases the entropy of the demon's memory: At first, the memory unit containing the information could have been in a number of different equiprobable states ("molecule x is in state y at time t"); then the erasing process resets it to a predefined state ("unknown molecule is in unknown state at unknown time"). Because the erasure of information in the demon's memory shall be logically irreversible, this process must emit an amount of heat into the environment of the demon that increases the environment's entropy to a higher extent than the entropy of the demon's memory is decreased. Bennett shows that this net increase at least counterbalances the decrease in the environment's entropy that happens thanks to the information-gathering activities of Maxwell's demon. Altogether, the entropy of the whole container is at best constant, and the second law of thermodynamics is rescued. By the power of information-theoretical and thermodynamical reasoning, Maxwell's demon thus shows us that the irreversibility of time is the same as the loss of information that we, as finite beings, must inevitably experience.

Stefan Artmann

See also Entropy; Experiments, Thought; Information; Logical Depth; Time, Arrow of

Further Readings

Bennett, C. H. (1973). Logical reversibility of computing. *IBM Journal of Research and Development, 17,* 525–532.

Bennett, C. H. (1987). Demons, engines, and the second law. *Scientific American, 257,* 108–116.

Leff, H. S., & Rex, A. F. (Eds.). (2003). *Maxwell's demon 2: Entropy, classical and quantum information, computing.* Philadelphia, PA: Institute of Physics.

McTaggart, John M. E. (1866–1925)

John M. E. McTaggart, a British Hegelian philosopher and one of the most independent minds of his generation, produced, among other things, a logical and coherent argument for the essential unreality of time. He was the son of Francis and Caroline Ellis; the surname McTaggart was added by his father in order to fulfill a condition for an inheritance. The now prosperous family sent young McTaggart to the prestigious Clifton School and Trinity College, Cambridge. While visiting his widowed mother in New Zealand in 1892, he met Margaret Elizabeth Bird. The two married during his next visit to New Zealand, in 1899. His entire academic career, between 1897 and 1923, was at Cambridge, where he developed his brilliant, though idiosyncratic, blend of quasi-Hegelian idealism and atheism.

A genial man, McTaggart was a longtime friend of G. E. Moore (1873–1958), despite the latter's role as the most influential critic of British Hegelianism. And along with Moore and Bertrand Russell (1872–1970), McTaggart was a member of the irreverent Cambridge club known as the Apostles. But his friendship with Russell came to an end during the First World War, when McTaggart led a campaign to have Russell thrown out of Cambridge University for his vocal opposition to conscription. McTaggart died suddenly and unexpectedly in January 1925.

One of the many paradoxes of McTaggart's life is that he produced no disciples and yet generated some of the most exhaustive commentary of any British philosopher of his generation. His importance among philosophers was given graphic illustration in C. D. Broad's massive three-volume *Examination of McTaggart's Philosophy* (1933, 1938), which remains one of the most comprehensive expositions of a 20th-century philosopher's body of work. Broad succeeded McTaggart at Trinity College. McTaggart's views on time were defended as recently as 1960 by the British antirealist philosopher Michael Dummett. McTaggart was a philosopher's philosopher, and he devoted little time toward engaging the interest of nonspecialists. But among professional philosophers he is best remembered for his work in logic, which remains influential to this day. An important example of McTaggart's logical power is his theory of the unreality of time.

Aspects of McTaggart's Philosophy

McTaggart's earlier career was spent articulating a comprehensive though idiosyncratic interpretation of the philosophy of Hegel. *Studies in Hegelian Dialectic* (1896) reworked the notion of proceeding with successive stages of thesis, antithesis, and synthesis. *Studies in Hegelian Cosmology* (1901) was more radical in its reexamination of Hegel's concept of the absolute idea, and *A Commentary on Hegel's Logic* (1910) dissected Hegel's argument from pure being to the absolute idea.

The closest McTaggart ever came to writing a popular work was with *Some Dogmas of Religion* (1906, with a second edition in 1930). Once again, being an atheist with respect to questions of the existence of God or gods while also maintaining a highly individual conception of immortality, he came to conclusions that were characteristically idiosyncratic. His justification for religion was dauntingly rigorous. Any religious belief, he argued, required the prior belief that the universe is good. But there is no reliable method by which one can believe this other than dogmatically. And dogmas, in turn, require a metaphysical investigation, for which most people lack the time or inclination. Therefore, regardless of whether the religion is actually true, the vast majority of people accept their religion on false grounds. This in turn will lead to a larger number of people living without religion, but also without its consolations, and who are therefore unhappy. This said, McTaggart was no more convinced that there was a link between religious belief and happiness.

But as against this line of argument, McTaggart advocated a mitigated version of immortality. After criticizing most arguments against, as well as many of those for, personal immortality, he advocated a disembodied mind that linked with a universal spirit that was composed principally of love. His contribution to the development of 20th-century atheism is more substantial than he is given credit for, although the fault for this lies with McTaggart himself.

McTaggart on Time

Many of these arguments came to rest on McTaggart's core belief in the unreality of time, which were first given serious expression in *Mind* in 1908 and developed in the 33rd chapter of his main work, the two-volume *The Nature of Existence* (1921, 1927). Like much of his work, this book was broadly Hegelian in outlook rather than proceeding specifically from a particular argument of Hegel's. He worked along Cartesian lines, postulating the existence of any one thing to existence of pluralities of things, to the existence of "the Absolute," which is the sum total of all the various substances without being anything more of itself than any of its constitutive parts.

Attempts had been made before McTaggart to construct arguments along these lines, though none with anything like his attention to logical detail. For instance, it has been claimed that a series of paradoxes by Hui Shi (c. 380–c. 305 BCE) was an argument for the unreality of time as part of a general program of problematizing the distinctions between space and time. And arguments for the unreality of time were advanced by the Sarvastivadin school of Buddhism about 500 CE. Only with McTaggart, however, was a concerted and deliberate aim made to argue for the unreality of time. His argument began with the observation of two types of temporality. There are events (which he called the A series) that figure either as past, present, or future, while others (the B series) operate either as earlier or later. Only the A series of events are essential to the idea of time, because only those sorts of events require a distinction between past, present, and future. Consequently, any difficulty in regarding the A series as real means an equal difficulty in regarding time as real. Past, present, and future can, more or less, be described, McTaggart admitted, but they cannot be defined.

The next step in the argument is crucial. McTaggart then argues that past, present, and future are clearly incompatible, and yet the A series needs each one at every event. To the objection that past, present, and future happen successively rather than simultaneously, he replied that any one moment still has its past and future and, as such, remain incompatible, and insofar as time depends on this series, time cannot be real.

He then infers that if a B series without an A series can constitute time, then change must be possible without an A series. A change of this sort means that an event (a position in time in McTaggart's usage) ceases to be an event while another one begins to be an event. But this cannot be, as nothing can cease to be an event or begin as an event. So without the A series there can be no change, because the B series is not sufficient in itself for change. And as events in the B series are time-determinations, it follows that there can be no B series where there is no A series, because where there is no A series, there is no time.

Now McTaggart does allow for events having an order—this he calls the C series—and events in order may become relations of earlier and later, in which they would become a B series. But this order does not necessarily imply that they must change, because change must be in a particular direction.

Having demonstrated that there can be no time without an A series, McTaggart then goes on to prove that the A series cannot exist. The characteristics of A series—the supposed sequence of past, present, and future—are either a relation or a quality. Either way, a fatal contradiction exists. Each event is the same, whether in the past, present, or the future, and its relation to each event's past and future must also always be same. McTaggart also argues that past, present, and future are incompatible. Each event has a past and future, and is in this way predictable, and yet events are also incompatible with each other. It presupposes the existence of time to erase the incompatibility, and yet the existence of time is what this argument sets out to demonstrate.

McTaggart conceded that it may well be possible that the realities we perceive as events in time are part of some nontemporal series in the manner of the C series. This seemed compatible with Hegel, who argued for a timeless reality, of which the time-series is but a distorted reflection we have of it. But this did not affect the core argument that the A series is "as essential" as the B series in that the distinctions of past, present, and future are essential to time, and that, if these distinctions are never true of reality, then one cannot include time as part of reality.

As mentioned above, McTaggart's theory of the unreality of time has not found general acceptance. It is not coincidental that his most loyal defender after Broad was Michael Dummett (1925–) the British exponent of antirealism. Others, like Roy Bhaskar (1944–), have found value in McTaggart's distinction between the A and B series without endorsing his conclusions about time's unreality. Opponents of the theory claim it amounts to little more than a play on tenses. J. J. C. Smart (1920–), for instance, argues that the idea of change can be expressed in the language of the B series by speaking of points in time *differing* from each other, which does not require us to say that events change. In effect, McTaggart's nonuse of tensed verbs with respect to the B series and use of them with respect to the A series is what sustains the apparent contradiction his argument rests upon. A simple reversal in the distribution of tenses, and the problem disappears.

McTaggart's denial of the existence of time is the best known of his arguments, although he did not stop there. *The Nature of Existence* also featured arguments that denied the existence of material objects, space, and a range of mental processes. These claims rested on a quite different argument, however. These entities could not exist by virtue of not meeting the requirements of a relation he called determining correspondence, a complex relation of any substance to the almost infinite range of divisible parts it could possibly be divided into.

Conclusion

Notwithstanding the solid support of Broad and Dummett, McTaggart's arguments have not found wider favor. The very strength of the argument—its logical power—was also its weakness, because the argument rested on logical grounds alone. Even if we overlook Smart's powerful objection to those logical grounds, McTaggart's argument for the unreality of time is fatally undermined by virtue of having taken too little account of the facts of science. At much the same time McTaggart was working out his theory of the unreality of time, developments in physics were establishing that time was very real indeed. The second law of thermodynamics and its corollary in entropy makes it clear that time is a fundamental part of the universe and that it is unidirectional. The fate of McTaggart's theory of the unreality of time is an object lesson in the need of scientific understanding, or at least of a multidisciplinary approach, when doing serious philosophy. Though not a contender as an explanation of the universe, McTaggart's argument for the unreality of time is an impressive intellectual achievement.

Bill Cooke

See also Hegel, Georg Wilhelm Friedrich; Humanism; Idealism; Nietzsche, Friedrich; Russell, Bertrand; Time, Nonexistence of

Further Readings

Berman, D. (1990). *A history of atheism in Britain.* London: Routledge.

Broad, C. D. (1933, 1938). *Examination of McTaggart's philosophy.* Cambridge, UK: Cambridge University Press.

Mace, C. A (1957). *British philosophy in the mid-century.* London: Allen & Unwin.

McTaggart, John M. E. (1921, 1927). *The nature of existence.* Cambridge, UK: Cambridge University Press.

McTaggart, John M. E. (1969). *Some dogmas of religion.* New York: Kraus Reprint. (Original work published 1906)

Passmore, J. (1967). *A hundred years of philosophy.* London: Penguin.

Westphal, J., & Levenson, C. (Eds.). (1993). *Time.* Indianapolis, IN: Hackett.

Media and Time

In the context of 19th and 20th century modernity—its technical revolutions and specific ways of experiencing time—the insight has grown that our ideas of time are decisively shaped by our interaction with media: Media influence our understanding of time. Particularly in investigations in cultural and media theory, there have been attempts to work out the dependencies between media and conceptions of time. The following cultural-historical transformation can be considered as paradigmatic for the assumption of media-time dependency: the change in our experience of time through the invention of the clock.

Thanks to a uniform mechanical process, time became precisely measurable, independent of subjective impressions of time and independent of sequences of natural events (e.g., day/night). The cultural establishment of clock time since the 13th century has not only changed social processes but also changed and extended our consciousness of time. In contrast to the experience of event time, cultural awareness of qualitatively indifferent, infinitely divisible time has grown through clock time, which made possible a massive economization of time with all the known accompaniments, such as lacking time and acceleration. In a media-theoretical perspective, this historically indubitable finding serves as evidence for the assumption that a certain conception of time (idea of abstract linear time) depends on the invention of a medium (the clock). But whether the clock, as a "time machine" (Marshall McLuhan), can form a model for proving a media-time dependency is problematic. The question is whether a clock is a medium at all. A clarification of the relationship between time and media, and a differentiation of ideas of time using media parameters, cannot take place without a preceding differentiation of the concept of media.

Both in everyday language and in the sciences, the ways in which the term *media* is used are very varied and not to be subsumed under one category. If, along with the media philosopher Marshall McLuhan, one pursues a very broad media concept, according to which media are technical inventions (artifacts), indeed artificial extensions of the human, then the clock (as a technically optimized form of time measurement) can also be treated as a medium. The advantages of such an approach lie in the possibility of treating very different technical inventions (e.g., the car and the telephone) in their cultural context. The disadvantage of such a broad understanding is conceptual imprecision. The differentiations between medium and tool and between medium and machine remain largely unclear. How broadly one grasps the media concept ultimately depends on which questions and aims one has and which facts are relevant in an academic context. The connection between medium and time thus cannot be described generally but only by considering the respective perspectival character. This entry distinguishes four perspectives in which the connection between temporality and mediality is respectively outlined.

The Media-Theoretical Perspective

Although the definition of the media concept is still quite controversial in media studies, there is nonetheless widespread agreement as to what counts among the central determinations. The material and technological basis of information and communication processes belong essentially to mediality. Relevant to this, however, are not only media in the sense of equipment and its technological changes and progress but their consequences for human information and communication processes. That is, mediality encompasses both the equipmental aspects of communication (in the broadest sense) and communicative practice as such.

In a media-historical perspective, significant changes in our ideas of time become clear. From a macroperspective, three phases can be distinguished above all. In a first phase, in which communication is primarily shaped by nonwritten media (gestures, voice, etc.) and ritualized oral tradition, cyclical ideas of time dominate. In the context of a culture of orality, events and series of events are largely considered as occurrences within larger cycles. In a second phase—starting with the development of phonetic writing (alphabet) in the 11th century BCE, via the invention of movable type printing in middle of the 15th century, through to writing as a mass medium since the 18th century—the idea of linearity develops into the dominant conception of time. Time is comprehended less according to the structure of cyclic recurrence and more as a linear order of the succession or sequence of events. By contrast, in a third phase—starting with the invention of new communications technologies (telegraphy and the telephone) and representational media (photography and cinematography) in the 19th and 20th centuries (radio and television in the first half of the 20th century, the computer and the Internet in the second half)—the ideas of simultaneity and nonlinear temporality develop. The transitions between these three phases are quite fluid, with overlap and interference being the norm. Already on historical grounds it is not convincing to assume, as is sometimes suggested, that different media paradigms like orality, scripturality, or digitality imply a strict difference in conceptions of time.

Whether the connection between media development (from a culture of orality, via written culture, to the digital world) and a changing conception of time (from the cyclic idea of time to the idea of linearity and finally to simultaneity) is to be considered as monocausal dependency or merely as a (theoretically difficult to grasp) correlation is controversially discussed. There are different arguments—according to which media concept is taken as basic—for there being a connection.

1. If one first considers media as *information and communications media,* then two lines of argument can be highlighted as examples using developments within writing culture. The first relates above all to the performative aspect, or the type of activities, in media use. The cultural techniques (writing and reading) linked with writing are shaped by the primarily linearly shaped *performance* of information processing. Information is here initially ordered not pictorially or sculpturally but in a strict linearity. The cultural dominance of the medium of writing and the performances linked with this forces the priority of the linearity of temporal succession (over the temporally indifferent division of space) in the sense of a temporal schema that regulates both the production and the reception of writing. That writing and reading are not only linear but also highly complex cognitive feats shaped by forward and backward references is not a compelling objection to the linear conception of time; this complexity is rather the condition for developing an idea of linearity (cf., Edmund Husserl's concept of retention/protention).

The second line of argument for a time-media correlation results from the observation that certain media practices encourage the emergence of certain epistemic rules and systems of order. In the context of writing culture, media practices interact with the *order systems* of knowledge, which are characterized by essentially linear ideas of temporality. For the ideas of open, transpersonal, and cross-generational development processes (i.e., not only the tradition of knowledge but also growth and progress in knowledge) are reinforced by the media production, storage, distribution, reception, and reprocessing (reinterpretation or change) of knowledge. Furthermore, the media practice of writing culture is one factor in the dominance of a model of perfectibility that is based on generally accessible and examinable (i.e., also methodically reproducible) knowledge and aims at an open, infinite progress in knowledge.

The perfectibility here ultimately lies less in what has in fact respectively been attained than in the possibility of constant improvement. Further, it is a factor in the development of a modern self-consciousness, which not only addresses its temporary and local environment in writing but nourishes its self-consciousness not least from the fact that as an individual it simultaneously understands itself as an expression of humanity and addresses all of posterity ("writing for eternity"). Linear conceptions of time gain in importance both through the knowledge orders (cf., in particular the idea of encyclopaedic knowledge pursued by Diderot and others in the spirit of Enlightenment), which developed in writing culture (above all after the invention of movable type printing) and aimed at general comprehensibility and infinite progress, and through the individual that, so to speak, in writing comports herself toward the whole of humanity. These linear conceptions of time are not restricted to local conditions and events, or to the mere recurrence of events, but take account of open processes and events that build upon one another. The dominance of a linear conception of time in the context of writing culture manifests itself particularly in the historical thinking that is typical of modernity and that has been spreading into practically all areas of knowledge since the late 18th century.

2. Alongside the first line of argument for media-determined changes in and shaping of ideas of time, which is based on the aspect of media *practices* (performances) and the forms of discourse and *order systems* shaped by media practices, a further line of argument results when one treats media primarily as *representational media* and not as *information and communications media.* Representational media, which are distinguished less technologically than functionally from communications media, are more precisely characterized by not only transmitting given (physical, biological, technological, social, or historical)

information but by generating meaning or semantic units. In this perspective media are considered not with regard to their transmission function but with regard to their function in the production of certain contents (media contents).

Representational media (e.g., dance, music, written text, film) produce semantic units, that is, they generate individuable and reproducible complexes of meaning. If one focuses on the level of semantic units articulated by representational media (instead of the paths of transmission for given information and media practices), then different (both fictional and nonfictional) temporal conditions, in part media-specific ones, can be recognized. Examples are, say, the distinction between *narrative time* and *narrated time* in literature, and the possibilities for extreme time contraction and time extension, or—partly analogous to literature—the extreme leaps in time in film though jump-cut or flashback methods. Thus well-known films such as *The Matrix* also show in exemplary fashion how specific forms of imagination and temporal conditions can be generated through the specific representational possibilities of a medium. Think, say, of the so-called bullet time, in which natural speed conditions are suspended and a normally invisible bullet is slowed down extremely in its motion, becomes visible, and is overtaken or moved around by naturally much slower bodies (e.g., humans). What such examples make particularly clear is the possibility of using representational media to generate temporal conditions that must be considered to be naturally highly unlikely, if not even impossible. That is to say: Representational media articulate temporal relations in a manner largely independent of the basic physical beliefs that are indispensable in our everyday life.

At this level of consideration—the level of media-dependent and, so to speak, physics-independent semantics of imagined time constellations—two aspects should be emphasized: (1) First, focusing on representational media shows that certain ideas of temporal processes are generated in a way dependent on representational methods and possibilities. At the semantic level, not only are established patterns of temporal information processing (according to the model of "natural" sequences of events) presented, but new "naturally" "meaningless," but semantically "meaningful" ideas of temporal arrangements are (also) generated. (2) Furthermore, the representational capacities of representational media permit humans a particular relationship to temporality. Using representational media, we can become aware of dimensions of time that are naturally inaccessible to us (past sequences of events or possible future scenarios). They enable us on the one hand to submerge in a past (either historical or fictive) beyond our own biographical horizon, but on the other hand they also enable us to imagine our future and to exchange views about possible future scenarios.

The particular cognitive ability of humans to process information with the help of time concepts and to comport themselves toward nonpresent events finds its adequate expression in the use of representational media: Released from what is respectively present, meaningful patterns of events are articulated in representational media, the meaning of which reaches beyond the time field of individuals. It is only through the production and storage of such semantic units that we can comport ourselves toward the past and future independently of the narrow horizon of our own experiences (cf., for example, the mythical past articulated in the Homeric epos or in aborigines' songs).

3. A third line of argument for the time-media correlation relates less to systematic connections, starting from relatively broadly conceived and functionally differentiated media concepts (information and communications media, representational media), and is based rather on a primarily historical and quantitatively defined class of media. Central to this are modern mass media (newspapers or the press, posters, radio, television, Internet) through which information can be spread in quantitatively high numbers (in "masses") and within a very short time. Four different aspects should be emphasized here:

- A first temporal aspect of mass media consists in the particularly fast, tendentially synchronous informing of a large number of information receivers.
- A further temporal aspect of mass media becomes recognizable when one considers complex communications situations (primarily not dialogical,

but marked by feedback processes). The understanding of information spread by mass media is essentially shaped by the demands of being up-to-date and the high frequency and fleetingness of the associated importance of information. The value of a piece of information (but also of a transmission format) is here measured in terms of its up-to-dateness. The rhythm of information determined by this (e.g., periodic appearance of newspapers or magazines, continuous transmission in radio or television) and the respective program structure (e.g., news or entertainment programs) have an effect on the audience's everyday life and feeling for time. Overall it holds that the semantics of information in the context of mass media is shaped by its specific communicative functions. The importance of information is here temporally indexed in a particular way. It is coupled to the relatively short intervals of time during which information, subjects, fashions, or program formats are perceived as being up-to-date.

- A third aspect is seen in the context of electronically based mass media (radio, television, Internet): News about real events, but in particular television pictures of events, is made accessible to the audience not only close to the time the events occur, as with newspapers, but instantaneously ("live") and "in real time." Being able to observe spatially distant events almost instantaneously through mass media is part of everyday experience these days. This fact encourages the idea (which is misleading, because it abstracts from the medium's selective mechanisms) of being able to adopt a quasi-divine standpoint in which all events in the world can be made simultaneously accessible, independently of their spatial position.
- A fourth aspect results from the *internal temporality* of the basic technical processes of information transmission, which have changed significantly through progress in the history of technology. The time interval between information production and reception has constantly become shorter; information transmission is so accelerated that the transmission time is converging on zero. An example of technical progress is the contrast between the sending of letters made possible by stagecoaches and the digitally based information transmission in an e-mail. The duration of information transmission diverges considerably. Instantaneous information transmission, live programs and real time, and the ideas of time shaped by their mass-media presence here appear to be determined in equal measure by the internal temporality of the underlying technology. This fact seems to confirm a thesis that has been much discussed in media studies, namely that ideas of time might be determined by the speed of technical processes. This thesis, which ultimately amounts to a medium a priori (the assumption of a medium basis that precedes, makes possible, and structurally determines our cognitive abilities), should however be rejected, because it starts with a very one-sided and abstract basic idea (i.e., from a media concept that lacks evolutionary and conceptual embedding; see point 2 below).

The Social-Pragmatic Perspective

In this perspective, the media-theoretically propounded thesis of a medium a priori is largely retracted and relativized. As a consequence, the assumption that experience and ideas of time are primarily determined by the history of media or the respectively guiding media is no longer dominant. For within this perspective, the accent lies not on the technological aspect but on the aspect of the human practices in which media are functionally defined and embedded. Here primordial connections between human actions are the starting point for focusing on the dependencies between media and ideas of time. Two approaches can be distinguished here: (1) a socioeconomic approach and (2) a communications-theoretical approach.

A brief general comment beforehand: For a theory that examines human practice and the social forms shaped by this, time plays a central role—and not only because of the historical nature of these phenomena. Humans must possess a concept of time so as to be able to act at all (e.g., so as to make end-means distinctions); beyond this they must possess a particular degree of time concepts (such as the differentiation of past, present, future) and be able to communicate these in order to act socially, that is, to be able to coordinate their actions with other agents.

Against this background, the extent to which media are an important factor precisely in a social-pragmatic perspective can be better understood: In order to coordinate actions socially, not only is the cognitive competence to be able to deal with temporal processes needed, but also an exchange of information between agents, in part over large temporal and spatial differences, is required. The functioning of communication between agents also depends essentially on the functioning of communications media and the speed of information transmission. Which actions are possible and are realized also depends on how, and above all how quickly, the communicative routes run.

1. Within a socioeconomic approach, the media-time correlation is treated above all in the context of an acceleration theorem. The sociologist Hartmut Rosa argues that we are today living in an age in which all social processes are being shaped by a ubiquitous tendency toward acceleration. This line of argument can initially draw support from the media-theoretical finding that especially through the introduction of digital media, and that means through the detachment of information transmission from bodily media (such as postmen, for example), a maximal acceleration of data transfer has come about. But the fact is now interpreted in the context of complex social systems and leads to the thesis that technical acceleration is embedded in different acceleration processes (acceleration of life processes as well of social and cultural changes) with in part paradoxical consequences (e.g., processes slowing down as a result of maximal increase in complexity).

2. If one advocates a stronger communications-theoretical approach, there is a shift in the lines along which problems of the media-time correlation are developed. Here too the question of mediality is embedded in a superordinate reference system. Yet it is not so much social systems that are central, but rather anthropologically founded discourse structures that regulate communicative actions and within which meaning contexts are produced and handed on. In relation to modern media and multimedia methods of representation (especially on the Internet), the thesis of a fundamentally altered experience of time is advanced. Various authors (the communications theorist Vilém Flusser, the sociologist Jean Baudrillard, the speed theorist Paul Virilio, the German literature and media theorist Götz Grossklaus) agree that the altered experience of time is characterized by a new primacy of the present.

There are three arguments above all that support the thesis of a primacy of the present:

- The first (media-epistemological) argument is general. (It gets by without closer differentiation of media and usually occurs in combination with other arguments.) It emphasizes that the experience of time within media communication is fundamentally relative to the present. In particular, the dimension of the past (e.g., certain events) as such loses importance in the context of digital media and is primarily experienced and perceived as a mode of current execution.
- Alongside that, so to speak Augustinian and rather non-medium-specific argument, a media-ontological argument is also discussed. Starting with the media-specific network structure of communicative processes (cf., digital media, particularly the Internet with its almost instantaneous and worldwide transmission of sound, images, and/or text signs), the aspect emphasized here is that there is a tendency to synchronize spatiotemporally distant processes within such forms of communication (Niklas Luhmann, Götz Grossklaus). The primacy of the present appears in this line of argument as the *primacy of simultaneity:* For the audience and agents within the network structure, the present extends into a field of the present in which remote events are synchronized within a time window (cf., the preceding argument) into a media reality.
- The third argument for the primacy of the present also has media-ontological implications. However, it focuses not on the synchronization effects of media communication but on the *modality* of media reality for the media user (sender, receiver; information producers and consumers). What matters here is the conviction that in the conditions of global networking of mutually independent information producers and storage, as well as communicative agents, the dominant impression for the individual agent is that of stepping into an open space of possibilities. The experience of presence here has nothing, or hardly anything, to do with the idea of a progression of transient present points but rather culminates in the idea of a field of the present generated

by selective individual decisions and which are surrounded by an open horizon of possibilities. That is, the experience of time leads to an idea of time in which the classical directional vectors (e.g., a line of progression from the past through the present to the future) do not dominate. Rather the accent is here on an idea of time in which a field of presence is surrounded by a variable horizon of possibilities, that is, by the future in all directions, and in which the dimension of the past appears as a residue (lost possibilities) at most (cf., Vilém Flusser).

The Media-Philosophical Perspective

The media-philosophical treatment of the time-medium connection, the discussion of dependencies and fundamental changes in time experience, and the ideas of time that build on these, converge (also in authors) in part with the preceding perspectives and findings, but in part they also compete with these. In methodical and systematic respects, three media-philosophical approaches can be distinguished above all. It is true of all these media-philosophical approaches that they also pose the question of mediality in a broad frame of reference and thus develop the subject of time at a problem level that is to be respectively specified.

1. One media-philosophical approach that is much discussed and widespread, above all in Europe, operates with fundamental assumptions about the philosophy of history and tends to teleologize media-historical changes (cf., Jean Baudrillard, Paul Virilio, Vilém Flusser). Modern and postmodern experiences of time play a prominent role in this context. Starting with the basic assumption about the philosophy of history that the historical processes aiming at progress and humanity (in the spirit of Enlightenment and modernity) are being dissolved, the focus for culture-critical and time-diagnostic purposes is above all on the experiences of time that speak for a negative eschatology. Here again the phenomenon of acceleration affecting all areas of life is to be mentioned in particular. In this perspective too, acceleration does not prove to be a desirable effect in reaching certain ends (e.g., acceleration of information exchange up to the speed of light) but is regarded as the signature of an overall cultural process in which the realization of human ends is structurally undermined by excessive acceleration. In conscious contrast to the assumption that a media age might bring about a free exchange of information conducive to the project of a humane world (McLuhan), here the thesis is advocated that information is obliterated by the accelerating speed of information transmission and by the excess of information in electronic media. Paul Virilio puts it pithily in an interview: "In contrast to what we are told, information in real time is not real information, but an action—like a slap in the face."

2. The second media-philosophical approach operates meta-theoretically and with conceptual criticism—in a highly cross-disciplinary perspective—analyzing the scope and coherence of media-studies theories and theses. Related to the problem of time, the central question is whether one can really infer fundamentally altered concepts of time from the media-determined development of new ideas of time. What is discussed above all are the facts that (a) time is a freely available and manipulable representational parameter, and (b) the respectively generated media reality depends on speed conditions (at the representational level or at the reception level). This approach challenges attempts to understand technical innovations and experiences of time that are in part interpreted as revolutionary (and the undoubtedly linked social changes gradually taking shape in the context of action of the Internet and cyberspace)—as the basis for a "revolution" in the time concept.

Against such attempts (which mostly go along with the thesis of a medium a priori) it should initially be objected, with a *transcendental argument,* that precisely the diagnosis of new experiences of time presupposes, and precisely does not dissolve, a relatively stable concept of time (i.e., a complex of certain categorical distinctions such as before/after, succession/simultaneity, etc.) in order to be able to register variations or changes in ideas of time. That is, as suggested, not an objection to the specifics of a media reality in which time is a freely available representational parameter leading to certain ideas of reality—in part independently of the sequence of events actually observed. Thus with TV weather forecasts, for instance, the (future) movement of weather fronts is calculated and visually simulated on the foundation of a certain data basis. The future's being visually brought to mind by simulating a complete sequence of events is a phenomenon that is just as everyday as it is interesting for the theory of science. It is, however, not yet any indication of a fundamentally altered consciousness of time.

Alongside the transcendental argument, a *genealogical or evolutionary argument* should also be asserted from a media-philosophical perspective. The assumption of a respectively media-determined, uniquely new time consciousness gets caught up in inconsistencies when it is supposed to explain transitions from one paradigm medium to the next. Only in a genealogical and evolutionary consideration (i.e., one incorporating both biological and cultural developments) does the origin become explicable of the foundation of cognitive abilities that are built on by the ideas of time that are respectively varied or extended by technological conditions. And only in this way can it be made plausible why, despite epochal changes in media history, we are quite able to understand past representations and time experiences—which on the assumption of strict epochal breaks would remain an explanatory gap.

In the framework of a *topological analysis,* in which media-theoretical theses are discussed in comparison with results from other disciplines and examined with regard to their coherence, it must further be pointed out that our media practice has until now not significantly changed anything about evolutionary and neurally based timescales of our time window. (Neurologically speaking, perception of the present spans only about three seconds.) The assumption that media-generated time windows, or experience of the simultaneity of remote events, directly correspond to our neurologically based time window should be corrected to the extent that the concern here can be only with a structural analogy. (Also, so far it has yet to be clarified what exactly the relata—media and mental processes—are and how they are connected). In media-philosophical terms, one cannot speak of a simply quantitative extension of the time field but of qualitatively, not quantitively, extended time windows having become possible within an evolved timescale. The comparative temporality of media (cf., primacy of the present), the extension of the present (the tendency to blank out the past dimension, and the time inherent to events before their media processing) is not to be equated with a basic change in our perception of time. "Extension of the present" means more precisely "conceptual extension of the present with relatively constant perception of time."

3. The third media-philosophical approach differs from the preceding ones—the (1) culture-critical and (2) concept-theoretical approaches—in not understanding the concept of mediality in terms of the sender-receiver model. The foundation is a deep-set (quasi-metaphysical) understanding of mediality that accentuates the significance of the "in between." The guiding conviction in this is that the structure of mediality is to be considered as an interval structure, which cannot be reduced to the structures of subjectivity or intersubjectivity, but which rather precedes these. This conviction has been developed through engagement with the tradition of phenomenology (Husserl) and structuralism (Ferdinand de Saussure), particularly in postmodern French philosophy (Jacques Derrida, Gilles Deleuze, Jean-François Lyotard). In the context of this—altogether nonanthropocentric—concept of mediality, time is a key theorem: Time is here conceived as a dynamic structure that cannot be reduced to the sphere of cognitive competences, subjectively available parameters, or the communicative exchange of information.

This conception of time was most radically advocated by Derrida, who characterized media dynamics as "dead time" (*De la Grammatologie*), as distinct from the (metaphysical and phenomenological) paradigms of subjectivity. Following on from this, many attempts have been made to comprehend time completely in terms of the underlying technologies (i.e., those underlying information processes; cf., Friedrich Kittler), which leads, however, to the inconsistencies of media-aprioristic approaches mentioned above. On the other hand, however, attempts have also been made to deploy this altered conception of time for the ontology of individual media without relapsing into a media a priori. Deleuze has carried this out in exemplary fashion for the medium of film by classifying the semantic potential of film on the basis of differentiating subject-dependent (or subject-centered) representations of time and subject-independent representations of "time itself." In so doing, he has created a thesis that the medium of film is particularly suited to the representation of temporal processes. Using the history of author cinema, he has drawn attention to a transformation in the direction of those temporal processes that are no longer to be comprehended via subject-centered conceptions of time.

An Outlook: Time and Media in the 21st Century

It is difficult to deny that experiences of time and ideas of time stand in a close connection with media and media practice. Yet the thesis that time must be specified using different media, and time concepts correspondingly revolutionized, is today—at the start of the 21st century—just as unconvincing as was the thesis that electricity must be differentiated into types of electricity according to the underlying substances (so that there might be, for example, a specific "resin electricity") at the beginning of the 19th century. Time is not a function of media technology. Nonetheless it is true that experiences of time vary, in part considerably, according to which media practice is individually or socially dominant. To put it in the way the social psychologist and cultural historian Robert Levine—who in view of culturally differing rules of social time talks of the "language of time" having different "accents"—does, the language of time also has different idioms and accents in the history of media. But different idioms are not yet different languages. Epochal misunderstandings are likely, but the ability to understand remains fundamentally possible even with a change in media paradigms.

Since the end of the 20th century—that is, since the introduction of the Internet and the development and use of computer animations, simulators, digital movement control, and cyberspace—and increasingly with the beginning of the 21st century, new experiences of time have been taking shape that have not been empirically researched much so far. In connection with the aspects already mentioned (cf., above all the primacy of the present discussed above), the idea of virtual time presents a particular theoretical challenge. This idea is linked with the suspicion of a problematic development. For in the context of digital media, the idea of virtual time seems to be establishing itself as dominant, both in the private realm (e.g., with PC gamers) and in the socioeconomic realm (in the "global player's" field of activity), at the cost of the idea of the time of real developments or a (biographical or sociopolitical) historical time. The much-discussed media-determined loss of reality here exhibits its time-theoretical implications: Virtual time seems to contain the loss of historical time.

Virtual time means here not so much that time is a freely disposable representational parameter but rather that time is an ordering of available events. That is, the idea of virtual time is marked by the image of events being located in a possibility space and standing freely at one's disposal. Here events do not stand under the dictate of irreversibility but are considered in the mode of "as if" and are, according to the aim, reversible. An example of this is the "different lives" in computer games. When your own figure in the game fails ("dies"), the game doesn't end; you simply don't yet reach the next level. To begin with, then, an event within virtual time has quite different consequences for your own actions to an analogous event in real time. That is, in the extreme case, where no more action is normally possible and everything is finished, in virtual time everything can start again anew.

The flipside of such media practice is that actions within virtual time can come at the cost of actions in real time. One should not only think here of the mostly young, intensive player of online games with an endless format, whose gaming passion leads him to neglect to attend sufficiently to his "real" life. Rather one should also think of possibilities for media actions of a global player who pursues abstract profit intentions, largely unburdened by local political or social circumstances, and must pay hardly any or no attention to the "real" life of a company or a social group on the ground. The specific profit intentions lead to a strategic set of actions that is guided by a complex play of information in a global framework. Thanks to the media networking of the world, the global player acts in an endless field of options (i.e., one not limited by local conditions) in which partial failure can be compensated for, or even itself transformed into elements of a win-win situation, by strategic turns. One of the problematic consequences of the action space created by media detachment, or the entry into virtual time, is a peculiar dissonance. The actions made possible by media (in the case of the global player, the global connection of information flows largely independent of local circumstances; in the case of the online gamer, submergence in a virtual world and an attractive role independent of the gamer's own social role or physical constitution) adhere to a different timescale and a different future, one less rooted in the past, than actions in real time.

The tensions between actions in virtual and real time are thus also structurally determined. One of the chronopolitical tasks for the 21st century is to bring geopolitically differentiated, local timescales on the one hand and global connections (of economic processes) made possible by media on the other into a balanced relationship. Admittedly no media-philosophical solution to this problem is to be anticipated; rather it should be recalled what the examples already show: Media-based virtual time and the complex structure of real time are not two spheres that are completely independent of each other. They are conceptually linked with each other (for the virtual time of media practice can only be comprehended in contrast to certain aspects of real time, with some aspects of real time being retained, others suspended) and factually connected with each other (an action in virtual time—e.g., "occupying and defending a country"—is always a certain action in real time too—e.g., "sitting at the computer").

Experiences at the beginning of this century have in the meantime shown that the entry into virtual time (reversible events) made possible by media does not have to mean an exit from real time (irreversible events) but is embedded in contexts of action in which different transitions in both directions and countless functional connections are to be registered. Thus in the realm of online games, it can be observed that the media practice of casual gamers or users of online forums (chat rooms) by no means have to lead to the oft-attested losses of reality. Actions in virtual time often function only as normal moments of relaxation in everyday working life; sometimes they even provide possibilities to try out communicatively and to form one's own identity beyond the often restrictive local conditions. That is, entry into virtual time does not have to come at the cost of time for individual development but can even benefit this. In addition, the media practice of intensive players ("heavy gamers"), for instance, shows that the entry into virtual time is by no means always completely detached from and unburdened by real economic processes.

That entry into virtual time also costs real time is a factor that is meanwhile represented within games in the form of certain advantages in the game and that leads to virtual time and real time standing in an economic relationship to one another that is no longer hidden. Players who lack the real time to work for the desired game advantages (the attainment of which is time-intensive) can nonetheless acquire these advantages by buying these from other players who have invested enough real time to work for the advantages (at the level of virtual time).

However one evaluates these different relations (and in view of some morally and politically problematic developments, one will not want to endorse them in every respect), in any case they show that theories are fundamentally too short-sighted that on the one hand reduce media-determined experiences of time to the logic of the underlying media technologies and on the other hand want to capture the specificity of these experiences of time with strict conceptual oppositions. The undeniable connection between time and media cannot be explained through simple certainties but only by starting with sufficiently complex descriptions of the respective media practice.

Ralf Beuthan

See also Film and Photography; Information; Language; Music; Time, Perspectives of; Timepieces; Virtual Reality

Further Readings

Baudrillard, J. (1994). *The illusion of the end.* Cambridge, UK: Polity Press.

Crary, J. (2001). *Suspensions of perception: Attention, spectacle, and modern culture.* Cambridge: MIT Press.

Deleuze, G. (1986). *Cinema 1. The movement-image* (H. Tomlinson & B. Habberjam, Trans.). Minneapolis: University of Minnesota Press.
Deleuze, G. (1989). *Cinema 2. The time-image* (H. Tomlinson & R. Galeta, Trans.). Minneapolis: University of Minnesota Press.
Derrida, J. (1967). *Of grammatology* (G. C. Spivak, Trans.). Baltimore: Johns Hopkins University Press.
Innis, H. A. (1991). *The bias of communication.* Toronto, ON, Canada: University of Toronto Press. (Original work published 1951)
Levine, R. (1997). *A geography of time.* New York: Basic Books.
McLuhan, M. (1964). *Understanding media.* New York: McGraw-Hill.
Postman, N. (1982). *The disappearance of childhood.* New York: Delacorte Press.
Virilio, P. (1984). *Negative horizon: An essay in dromoscopy* (M. Dregener, Trans.). New York: Continuum.

Medicine, History of

Medicine is as old as civilization itself; humankind has always made attempts to heal, cure, and prolong life. Even in the nascent phases of our development, our hominid ancestors recognized disease and sickness, and they made attempts to combat this with what was available at that time, mostly herbs and rituals. In our current phase of evolution, technology and applied intelligence have improved our understanding of what disease, sickness, and even death actually are. Now with the completion of the Human Genome Project and the possibility of genetically tailored treatments, an unprecedented chapter in medicine is about to begin. This entry provides a brief account of medicine's evolution and speculates on its future direction.

Primitive, Ancient, and Modern Medicine

The word *medicine* is derived from the Latin word *medicus,* meaning "physician," and the feminine declension *medicinus,* which means "of a doctor." However, the actual definition of what medicine *is* has changed as our civilization itself has developed through time.

Medicine in its primitive form could be defined as a *ritual* practice, sometimes involving a sacred object that a society believes capable of controlling natural or supernatural powers that act as a form of prevention or remedy for physical ailments. This could involve the use of herbs, potions, prayers, or incantations and was usually performed by a specialized member of a society—a shaman or medicine man. These individuals held special status in their societies for their apparent ability to heal. It should also be noted that even though herbs and potions may have been believed to have supernatural effects by those giving and taking them, some of these herbs and potions did have potent pharmacological effects that were not understood as such until the modern era.

Evidence of prehistoric surgical procedures has been found, most notably for trepanation, a process in which a hole is drilled into the skull, evidently for therapeutic purposes. Trepanation is believed to be one of the oldest invasive surgical procedures, and without any doubt it is one of the first neurosurgical procedures. Evidence on prehistoric human skulls and in cave paintings of the Neolithic era (8500 BCE) establishes that this procedure was in fact practiced widely, most likely with the intent to cure aliments such as headaches, seizures, and possibly psychiatric disorders.

Some of the first evidence of ancient medical information and texts can be found in ancient Egypt in the so-called Edwin Smith papyrus, which has been dated to around 3000 BCE. This scroll is also believed to be one of the first ancient textbooks that illustrates in detail the examination, diagnosis, treatment, and prognosis of a large number of physical ailments. There is evidence of what are believed to be the first surgeries performed in manuscripts from 2750 BCE (250 years after the Edwin Smith papyrus is dated).

In ancient Greece and in early Europe, a natural system of medicine was devised that was called *humoral medicine.* Hippocrates is often credited with innovating this system, but it was actually derived from Pythagoras's idea of humoral medicine, which was based on the treatment of a patient by balancing what were called the four humors: blood, phlegm, black bile, and yellow bile. An imbalance in any of these four humors is what was believed to cause physical ailments. (Pythagoras's four humors were influenced in turn by Empedocles's "four elements.")

A modern definition of medicine is that it is an objective *science* of diagnosing, treating, and in some cases preventing disease and other insults to the body or mind. This type of science implements treatments with pharmacological drugs, diet, exercise, and other surgical and nonsurgical interventions. These therapeutic interventions are performed by a specialized member of society who has extensive education and training. In addition, modern medicine is now based increasingly on scientific evidence and clinical trials, or what is presently called "evidence-based medicine." More specifically, according to the definition provided by the Centre for Evidence-Based Medicine in 1996, evidence-based medicine is the conscientious, explicit, and judicious use of current evidence in making decisions about the care of individual patients.

Comparing and Contrasting Primitive and Modern Healers

There are several similarities between primitive medicine and modern medicine. First, the healers themselves share many similarities. They are in all accounts possessors of some form of specialized knowledge, be it of the spirit world or of anatomy and physiology. In order to perform healing, they all have to undergo some form of training or rite of passage that entitles them to special recognition within their society as a person who can heal. In the past, it was an apprenticeship or a ritual. Today there are formal study, graduations, clinical training, and certifications. In addition, healers hold a high status in their respective societies and command special recognition for what they do.

Second, healers, primitive and modern, have a similarity in their methods of healing, and that is the administration of medicines. Even though the actual types of medication given by the healers differ drastically, the intention is the same, which is to put something external into the body to elicit an improvement of physical symptoms. The universal expectation among patients, primitive or modern, is that the healer "make the pain go away."

Third, healers of any era must inspire the patients' belief. Modern physicians understand that patients' belief that the doctor can help them get well is a large part of successful treatment. The patient needs to have confidence in the physician's knowledge of medicine and experience of practicing it successfully. For ancient medicine as well, the patient had to believe that the shaman's connection with the spirit world was real and that the rituals and herbs would cure aliments.

Healers and Their Methods: Then and Now

Although ancient medicine and modern medicine are separated by thousands of years and have become very different in how they are practiced, another notable similarity is that, as previously mentioned, the function of healing and curing is and was performed by a specialized member of society; not just anyone could do it.

Evidence of shamanism has been dated back to the Neolithic period (8500 BCE), which would predate all organized religions. Presently, some forms and remnants of shamanism are still seen in some societies in Africa and South America. It is also seen in regions of Asia such as Korea, the Ryukyu Islands, and sparsely populated rural areas in Japan.

Many disciplines define a shaman and his function, in a basic sense, as an intermediate between the physical (or what is call the natural) world and the nonphysical realm (or what is sometimes referred to as the spirit world). It was believed that the shaman had the ability to travel between these two worlds and was able to commune with spirits and ancestors to assist them in the healing (or in some cases the harming) of another person. Essentially, the core of shamanism is based on the belief that the physical world is somehow interconnected with the nonphysical world and that the nonphysical world can have a profound effect on the physical world.

Contemporary physicians have long since been divested of their powers of enchantment and connection with the spirit world. They are equipped instead with scientific knowledge of disease and disorders of the human body, and they have access to a vast body of medical knowledge and research. They also have the benefits of modern technology, such as diagnostic imaging and laboratory tests, and thousands of pharmacological agents to choose from. Along with primary and family care physicians who treat a wide variety of general conditions, there are more specialized physicians who treat one specific bodily system or one type of disease or disorder—for example, cardiologists, dermatologists, and psychiatrists.

Medieval Physicians

In medieval Europe, a typical physician was neither well versed in the rituals of the spirit world nor educated in topics such as pharmacology or physiology. Rather, he studied humoral medicine, alchemy, astronomy, and dogmatic textbooks that were sometimes centuries old. These were his tools for diagnosing and treating patients, mostly on the basis of conjecture. Historically the medieval physician did very little to cure rampant outbreaks of diseases or plagues. In fact, most were unwilling to actually touch their patients for an examination. The transitional period between the medieval era and the Renaissance was the time of Paracelsus (1493–1541), who has been credited with pioneering the use of chemicals and minerals to treat sickness. He believed that the human body (and nature) should be studied and understood and that alchemy, whatever its purported ability to make gold and silver from base metals, was unsuited as a tool for medicine.

Sickness: Then and Now

How did our earliest ancestors view sickness and disease? Primitive humans would most likely have perceived sickness and disease as something unnatural or even as a supernatural event. Lacking knowledge of infectious diseases or microorganisms, sick persons could be perceived as being adversely affected by the spirit world or a curse, because they were being affected by something that was invisible or poorly understood. Therefore, only an individual capable of communicating with, and able to influence, the spirit world could abate these supernatural events and thus heal.

In mainstream Western culture today, when someone becomes sick, it now automatically comes to mind that "she has a virus" or "he needs an antibiotic for his infection." No longer do most people blame evil spirits or retribution from the spirit world for illness. Spells, curses, and the "evil eye" have been discounted as causes of illness; the average person today has a better understanding of what "being sick" is and what can be done to cure various aliments.

Pharmacology: From Plants to Human-Made Pharmaceuticals

A change has taken place in *what* is used to cure illness and disease. Early humans had to be very resourceful and use what was abundant and available to them at the time, mostly plants. Today, medical botany or herbalism, the use of plants and plant extracts to cure and heal aliments, has largely been supplanted by modern pharmacology, and most drugs are now manufactured in high-tech laboratories. They are distributed in exact doses, taken for prescribed periods of time, and have extensively studied effects.

In 1960, a Neanderthal skeleton (determined to be over 60,000 years old) was uncovered from a burial site in what is known today as the Shanidar Cave (in Iraq). This specimen was buried with eight species of plants that are believed by medical anthropologists to be used medicinally all over the world. However, the earliest record of the use of plants for medicinal purposes is found in paintings in the Lascaux Cave in France, which have been dated to between 13,000 and 25,000 BCE.

It is believed that early tribal societies eventually created a small semireliable repertoire of medical knowledge based on generations of trial and error experience. The observed effects of specific plants and herbal preparations were transmitted from generation to generation and used therapeutically by a specialized member of society. Ancient societies and early European physicians would use potions and tinctures to treat patients, although they lacked the knowledge of what these potions were doing in a pharmacological or physiological sense.

Penicillin, a byproduct of a fungus, was one of the first antibiotics discovered and used therapeutically. Originally noted by a French medical student, Ernest Duchesne, in 1896, it was rediscovered by Alexander Fleming in 1928. However, the internal use of penicillin as an antibiotic did not begin until the 1940s. Since that time, thousands of medications and hundreds of classes of medications, from blood pressure medication to antidepressants, have been manufactured and administered to patients.

Eventually, with the continued growth and application of knowledge gained from the Human Genome Project, medications may be developed that are genetically tailored for optimum effectiveness with a given patient's genetic makeup. With further research, some genetic diseases will be treated directly by genetic therapies, that is, by actual alteration or manipulations of the genome. It is believed that this will improve the efficacy of treatments and reduce side effects and adverse reactions. The treatment of a patient with therapeutics engineered to correct underlying genetic causes is call "gene therapy" or "genetic medicine." The concept of genetic medicine will challenge the manner in which modern medicine treats life-threatening illness such as cancer, which currently utilizes the surgical extraction of tumors, radiation treatments, and chemotherapy. Our current approach to treating cancer with chemotherapy and radiation inevitably involves some toxic effects for the patient as physicians attempt to localize these effects to the actual cancer.

Modern medicine, despite some shortcomings, has provided our present-day population with several tremendous benefits. With the advent of new medications, treatments, and vaccines, people are now living longer and more productive lives. In fact, global life expectancy has increased from about 37 years in the year 1800 to 67 years in the year 2000, for a global average increase of 40 years.

Medicine in the Future

As medicine has developed from rituals, through alchemy and astrology, to modern technology and to the possibility of genetically tailored treatments, one fact has remained and will remain consistent: The human body becomes ill, breaks down, and dying and death continue to occur. Given humankind's desire to prolong life, the human body will always need healing, repair, and medical treatment by specialists.

The future prospect of space travel suggests the possibility of encountering new forms of diseases and ailments as a result of exposure to the outer space environment. This new environment could also put us into contact with alien microorganisms that could cause different types of illnesses. In addition, space travelers may need to be placed in hibernation for long journeys, and muscle strength would need to be maintained in zero gravity conditions. All of these possibilities would open up a new area of space-travel medicine, a new specialty that would deal with these unprecedented medical problems.

John K. Grandy

See also DNA; Dying and Death; Egypt, Ancient; Paracelsus; Time Travel

Further Readings

Ball, P. (2006). *The devil's doctor: Paracelsus and the world of renaissance magic and science.* New York: Farrar, Straus, & Giroux.

Harner, M. (1982). *Way of the shaman* (1st ed.). New York: Bantam.

Porter, R. (2004). *Blood and guts: A short history of medicine.* New York: Norton.

Porter, R. (2006). *The Cambridge history of medicine.* Cambridge, UK: Cambridge University Press.

Sackett D. L., Rosenberg W. M., Gray, J. A., Haynes, R. B., & Richardson W. S. (1996). Evidence-based medicine: What it is and what it isn't. *British Medical Journal, 312*(7023), 71–72.

MELLOR, DAVID HUGH (1938–)

David Hugh Mellor, emeritus professor of philosophy at the University of Cambridge, is known for his important contributions to metaphysics, philosophy of science, and philosophy of the mind, with studies, for example, on probability, time, causation, properties, and decision theory. His work stands in the Cambridge tradition of F. P. Ramsey and Richard Braithwaite, in whose honor he has edited anthologies and the works of Ramsey. Mellor's philosophy of time profits from this broad field of interest and combines them systematically; he is one of the advocates of the "new tenseless theory of time."

Two important theories of the early 20th century have influenced Mellor and, as he claims, the whole modern theory of time: McTaggart's A- and B-series theory of time (1908) and Einstein's special theory of relativity, published in 1905. Agreeing with McTaggart's argument against the reality of the dynamic, tensed view of time, Mellor adopts a theory of time based on the B series, which acknowledges only the static scale of the B series as fundamental for any concept of time. In contrast to McTaggart, Mellor does not conclude that time is unreal. He shows that the concept of spacetime as it is presented in the special theory of relativity does not spatialize time. In fact time differs from space in this concept, and that becomes obvious in the formalization of the theory. Time is therefore a problem in its own right. Mellor's position can be characterized as a B series of time that argues for the reality of time—more precisely: time as the causal dimension of spacetime.

A and B Series of Time Differentiated

The first step on the way to Mellor's theory of time is the differentiation between the A and B series of time established by McTaggart. The A series orders facts in relation to the present moment as past, present, or future. Their relation to each other does not change, but their qualification as past, present, or future changes with the flow of time. The B series orders facts or events only with respect to their successive occurrence, no matter which of them is the present one. Therefore it does not need the concept of a flowing time. In everyday language, facts involving time are usually expressed in A sentences; that is, by using tensed verbs. Mellor argues that all tensed propositions or beliefs have B facts as their truth conditions. The crucial question is what makes tensed sentences, or A sentences like "Peter arrived yesterday," true.

Mellor's answer to this question in general is the following: A sentences have B facts as their truth condition. That means they depend, first, on facts like the time when the sentence is uttered and, second, on whether the event that is mentioned really occurred at the time that the sentence says it did. Both conditions can be expressed in B terms as follows: (1) The sentence was uttered on March 2nd, and Peter arrived on March 1st; (2) Peter really arrived on March 1st. If both conditions obtain, the A sentence is true. If the truthmakers were A facts, they would cause contradictions, because the fact that Peter arrived yesterday would have to obtain in order to make the statement true if uttered on March 2nd and not obtain to make it false when uttered on March 3rd. The central thought in Mellor's B theory is that A sentences need B truthmakers in order not to fall prey to contradictions. He does not want to do away with the way of expressing subjective perspectives of time in A sentences. On the contrary, he recognizes the necessity of A sentences and A beliefs within a concept of agency.

As agents, Mellor explains, we depend on our A beliefs existentially. A true A belief at the right time is needed for an action to succeed. But, first, A beliefs are necessary to cause an agent to act. I need the belief that my train arrives at 2 o'clock, for instance, to cause me to leave the house by 1:30. For the success of this action, my catching the 2 o'clock train, the belief that the train arrives at 2 o'clock must be true. Again, the condition for this truth is not the A fact of it now being 2 o'clock, but the B fact that the train arrives at 2 o'clock. The function of A beliefs is to cause agents to act. Therefore A beliefs are indispensable. Beliefs about what is happening now are needed for any action.

Mellor states that the conviction that what is perceived is also present is not grounded in facts but in pragmatic beliefs. The presence of a perception is usually confused with the alleged presence of the object. This plays no role in everyday life, because the time light needs to travel from most objects to the eye and the time needed to process the information is negligible; in the case of cosmologic events, however, the events we perceive now may have happened millions of years ago. On the one hand, the fact remains that there is no presence in the strict physical sense of the word; on the other hand, as species we would never have survived if we had not taken the prey just seen or perceived the predators following us as present. So it is pragmatic and existentially necessary to have A beliefs.

Mellor explains this necessity in terms of evolution: It was necessary for the survival of humanity and even animals to have such A beliefs. Furthermore he thinks that having an A belief is more basic than having a language. It is not necessary to be able to express an A belief or to have a concept of the self or the present for acting on such a belief. The range of A beliefs is not limited to sentences about time. All subjective perspectives, all beliefs about time, place, and the subjective situation belong to its range. A beliefs are subject-relative, which means they belong to the subject that has them in order to cause it to act: "I" have to believe something at the present moment "now" in order to act. No one would act on someone else's belief or on the belief in true B sentences alone.

Besides granting A beliefs a pragmatic necessity, Mellor also says that A sentences have different meanings from their B analogues and therefore cannot be replaced by B sentences. Although A sentences and their B analogues have the same content according to Mellor, they have a different character, because their relation to their truth conditions differs. While a B sentence, for example, "It is raining at *t* (let *t* be the exact time)" is always true (if and only if it rained at *t*); the A sentence "It is raining now" will only be true if said at the time *t;* if said before or after *t,* it may be false. This is the reason why Mellor says that A sentences "mean the functions from any *B*-time to their *B*-truth-conditions at *t.*" In other words: The truth of an A sentence depends on the position of its token on the B scale of time (when it is uttered) in relation to the B-scale position of its content, the event or fact expressed in the A sentence.

The fact that A sentences constantly change their truth values is crucial for Mellor's version of a B theory of time. Their constant change is caused by the constant change of our A beliefs, which determine our perception or understanding of time. A beliefs about what is now change nearly every moment, and their spatial analogues about what is here change similarly. Through the constant change of A beliefs, Mellor explains how the impression of flowing time arises. Despite the fact that the flow of time does not exist as a property of time itself, it is a psychological truth. The phenomenon of flowing time is a mere construction of our minds, which are constantly concerned with changing A beliefs. It is important to stress the point that the psychological and dynamic character of A beliefs does not invalidate them in their function. According to Mellor, these subjective beliefs, or at least some of them, are also fundamental to a concept of the self, even though they do not deserve such a concept to function as a cause of agency. The concept of the self is a second-order belief. A beliefs belong to the first-order beliefs that make us eat if we perceive food; no concept of self is needed for this belief to cause an action, according to Mellor.

Mellor's tenseless theory of time explains how subjective sentences involving time can be made true by nonsubjective truthmakers. It becomes obvious that he acknowledges the tensed view of time as essential only for time-consciousness and in a pragmatic perspective, but from that he does not infer an ontological relevance of the A series. Only the static B series is of ontological relevance. In order to strengthen the ontological argument, Mellor raises the question of what time is. This question entails, among others, the problems of the difference between time and space and of time as the dimension of change, as well as the question of causation.

Time and Space Contrasted

In fact, if time should be a subject of ontology, then it must not be reducible to something else. In special relativity it could seem as if time had been identified with space. Mellor attempts to show that this is not the case. Even though time resembles space in more than one way, it is not the same thing. The four dimensions of spacetime seem to be treated quite equally, because both space and time are systems of order, and time can be represented as a dimension just like space—the three dimensions of space are combined with the one dimension of time. Nevertheless they are not equal. Mellor explains that a difference between spacelike and timelike separations of events or entities is made in special relativity. The three dimensions of space represent an array of possible ways by which things can be in contact, interact, or fail to do so. The dimension of time has basically the same function, but, in contrast to space, things in time can fail to be in contact although they are in the same place, because they are there at different times, or in different words, it is possible for two things or events to occupy exactly the same place, because they can do it at different times. These kinds of separations are called timelike separations. They are treated differently from spacelike ones. This can be shown in the mathematical formalization of distances in spacetime:

The spacetime separation has a positive sign if it represents a spacelike separation and a negative one if it represents a timelike one. Therefore time differs from space in special relativity.

In the next step, time is described as the dimension of change, which leads to the definition of time as the causal dimension of spacetime. Time defined this way is marked off from space not only in terms of formalization but also by nature. McTaggart characterizes time as the dimension of change. Mellor seeks to defend this view without sharing McTaggart's opinion that the A series, which had just been shown as containing contradictions, is fundamental. Taking the A series as fundamental for change would lead to contradiction in the concept of change.

Mellor wants to give an account of change from the B theory's point of view and in respect to special theory of relativity. The first question that arises here is, What is change? The answer Mellor gives is the following: A thing has undergone a change when it possesses incompatible properties at different times; change is a variation over time in the properties of a thing or an entity. This definition excludes various events that might in general be called changes but that do not meet the conditions of the definition: Mellor claims that spatial variation cannot be called change, because the properties that are subject to change have to be intrinsic, not relational. Changes in relational properties are not changes in the thing; they are a variation in its relation to some other thing. Neither spatial variation nor variation in temporal parts or in relational properties can be defined as change, because there is no change in intrinsic properties of the thing. Mellor also states a difference between things and events: Both terms denote particulars, but events, because they are stretched over a certain period of time, consist of temporal parts, while things do not have temporal parts, they are wholly present at more than one time. This is why only things can undergo change in the strict sense of the word, meaning they can possess incompatible intrinsic properties at different times.

McTaggart's restriction of change to the A series was due to the fact that B facts never change; being true at one time means that a B fact is always true. The sentence "It is raining at *t*" will always be true if it really rained at *t*. Facts that change are A facts: that it is raining now might obtain at the present moment and be false a few minutes later because it has stopped raining in the meantime. McTaggart's reason for change relying on the A series is the possibility of continuity. The continuous change of the A series (of the present state to a past one and so on) constitutes the flow of time, and therefore time can be the dimension of change only if the A series exists. Since the A series contains a contradiction, time as the dimension of change does not exist. Mellor does not share McTaggart's conclusion. Although the B series account of change is being criticized for not being able to explain the continuity of change, Mellor attempts to show that indeed it can. If change is described as the possession of incompatible properties at different B times, which entails that every single B fact at its B time does not change, that does not mean that the succession of different B facts cannot be a continuous change. The facts in themselves need not change for change to occur: they only have to follow each other along a causal chain.

Causality

For Mellor, causality is the basic concept that grants time reality. Time differs from space in spacetime, because only in the temporal order are causes and effects necessarily separated. In space, cause and effect can occupy the same place, but they do not occur at the same time. According to Mellor it is not time that fixes the causal order (that would mean accepting the A series), it is causation that gives time its direction: Time order is synonymous with the causal order. Mellor gives several reasons for adopting a causal theory of time order (see Mellor, 2005). In short, his basic assumptions are that (1) causation links only events separated in time, not in space, because causation is never unmediated, and (2) causes always precede their effects. On the basis of these assumptions, a causal theory can explain the differences between past and future as well as the continuity of change. The causal concept of time explains why, according to Mellor, "We can perceive but not affect the past, and affect but not perceive the future" without the necessity to state modal or ontological differences between past, present, and future.

A profound survey of Mellor's approach to the philosophy of time can be found in *Real Time II;* for a concise and informative entry into his thinking, his article "Time" in the *Oxford Handbook of Contemporary Philosophy* is recommended. A deeper understanding concerning the theoretical fundamentals of the definition of time as the causal dimension of spacetime is given by *The Facts of Causation.*

Yvonne Foerster

See also Causality; Determinism; McTaggart, John M. E.; Metaphysics; Ontology; Time, Relativity of; Space and Time; Spacetime Continuum; Relativity; Special Theory of; Time, Real; Time Travel

Further Readings

Lillehammer, H., & Rodriguez-Pereyra, G. (Eds.). (2003). *Real metaphysics. Essays in honour of D. H. Mellor.* London: Routledge.

Mellor, D. H. (1981). *Real time.* New York: Cambridge University Press.

Mellor, D. H. (1991). *Matters of metaphysics.* New York: Cambridge University Press.

Mellor, D. H. (1993). The unreality of tense. In R. Le Poidevin & M. MacBeath (Eds.), *The philosophy of time* (pp. 47–59). New York: Oxford University Press.

Mellor, D. H. (1995). *The facts of causation.* London: Routledge.

Mellor, D. H. (1998). *Real time II.* London: Routledge.

Mellor, D. H. (2001). The time of our lives. In Anthony O'Hear (Ed.), *Philosophy at the new millennium* (pp. 45–59). New York: Cambridge University Press.

Mellor, D. H. (2005). Time. In F. Jackson & M. Smith (Eds.), *The Oxford handbook of contemporary philosophy* (pp. 615–635). New York: Oxford University Press.

Oaklander, N., & Smith, Q. (Eds.). (1994). *The new theory of time.* New Haven, CT: Yale University Press.

Memory

Memory is the ability to retrieve learned or acquired information. This information can be of previous events, a learned skill, or factual knowledge. Memory is usually distinguished as short-term memory, which is the recollection of recent events, and long-term memory, which is recalling the more distant past. Memory as a biological phenomenon is a record of time.

Ideas and theories about memory have changed in recent decades. Theories regarding memory, much like the theories regarding consciousness, have been profoundly influenced by research in the neurosciences and understanding the functioning of the brain (physiologically and biochemically).

The clinical assessment of memory of the human brain is specified by three categories, which can give insight into the functioning of a person's cognition. First is immediate memory, which functions over a period of seconds. Second is recent memory, which applies over a scale of minutes to days. Third is remote memory, which typically encompasses a period of months to years. These classifications differ only slightly compared with stipulating them as short-term and long-term memory.

Memory can be further classified according to how it is utilized. Working memory is not only classified by the duration of memory retention, but also by the manner in which it is used in daily activities. For example, performing a series of simple calculations would utilize working memory. The actual process of retaining this information (in this case, numbers to be used in a calculation) for short-period use is the working memory, because it is being used at that time. However, working memory is not to be confused with short-term memory, which is memory stored for a short period of time that is not being used functionally.

In addition to classifying memory by the length of time a particular brain is able to retrieve information, it can also be described in terms of implicit (also called procedural) and explicit (also called declarative) memory. Implicit memory is defined as memory that is retrieved automatically, or without conscious involvement. For example, memory for learned skills is claimed to be largely implicit, in that it is automatic. This is in contrast to explicit memory, which requires conscious awareness and intentional recollection to recall. An example of this would be recalling events that took place several years ago, which would require an intentional recollection of that neurological data.

Ivan Pavlov: Early Experiments in Conditioning

Ivan Pavlov (1849–1936) was a psychologist, physiologist, and physician who is well known for his work done in what is known as *classical conditioning.*

Also well known is "Pavlov's dog," a phrase that arose from experiments he performed on dogs. These experiments consisted of producing a stimulus (such as ringing a bell, blowing a whistle, or striking a tuning fork) prior to feeding the dog. This was done repeatedly, and eventually that same stimulus would cause the dogs to salivate even in the absence of food. This process was called *conditioning.*

Early experiments in conditioning were important to the understanding of memory, because the conditioned response relied on the fact that a memory of that stimulus was associated with the presentation of food. Thus, memory is a component of learned behavior.

Human Memory

In establishing an understanding of human memory, four basic elements of memory have been explained: encoding, storage, retrieval, and forgetting. The first element is encoding, the registration of neurological data. This is an active procedure of processing and combining information. For example, while watching a television program, one would see (visual stimulus) and hear (auditory stimulus) information that would be processed by the brain as an event, (e.g., watching the weather report).

The second element is the storage of memory in the brain, which creates a neurological record. Currently, this is understood to take place in three stages: sensory store, short-term store, and long-term store. The sensory store is the perception of the image (e.g., the meteorologist reading the weather forecast), which is thought to last only a split second or just long enough to be perceived by the brain. Short-term store is the storage of this information for only a short period of time, typically only minutes to hours; for example, if someone just entered the room, having missed the weather forecast, and then asked you what it was, your recollection would then be based on short-term store. Long-term store is the storage of that same information hours, days, or years later. The retention of this information for longer periods of time requires rehearsal. In fact, memorization is a method of rehearsal that allows an individual to recall information verbatim.

The third element of memory is the retrieval of memory. This is the recollection of stored information, which is not a random process. In fact, it is an intentional process that is typically in response to a cue, in reaction to a stimulus, or to perform a particular activity. However, it is also thought that memories are reconstructions of the actual event, and these reconstructions can contain errors or inconsistencies in perception when recalled; for example, one might make errors in reporting what the weather forecast was.

The fourth element of memory is forgetting. This is the loss of or the inability to retrieve stored information. There are several theories on forgetting, such as *pseudo-forgetting* (which is held to occur due to ineffective attention in the acquisition phase), and retrieval failure (which is claimed to be an inability to retrieve information at a particular time and, consequently, the inability to be able to recall it at a later time). It is also held that memory loss happens naturally due to decay over time or because of lack of use. Another theory, known as *motivated forgetting,* is an individual's intentional attempt to forget events that are unpleasant or traumatic. This phenomenon was studied extensively by Sigmund Freud (1856–1939); he called these *repressed memories.* He maintained that repressed memories were not lost or forgotten; rather, they were stored in the unconsciousness and are responsible for certain psychiatric conditions that he called *neuroses.*

Long-Term Potentiation

For almost a century, scientists were baffled about how the neurons in the human mammalian brain were able to store memories. In 1973, the first neurological research was published by Timothy Bliss on what he called *long-term potentiation* (LTP), today also known as long-term enhancement. He characterized the phenomena of LTP, which was originally observed by Per Andersen in Oslo, Norway.

While conducting experiments on the hippocampus of rabbits, Timothy Bliss and his colleges discovered that a few seconds of high-frequency electrical stimulation on particular neurons would enhance synaptic transmission in the hippocampus for days and, in some studies, for weeks. This enhanced and prolonged stimulation in the hippocampus was held to be responsible for the formation of short- and long-term memory. Today, researchers of memory concur that the most current evidence supports the role of LTPs in both memory and learning.

Prior to the idea of LTP, the Hebbian theory was the accepted idea of how memory and learning occur. The Hebbian theory was named after neuropsychologist Donald O. Hebb (1904–1985), who stated that the strengthening of the neuronal synapses to one another was primarily responsible for memory and learning. This in part could still be true, and we are learning more about neuroplasticity, which is a process in which the brain changes, or, in this case, strengthens its neuronal connections.

The Neuroanatomy of Memory

Theories about what memories actually are and how memories are actually stored have changed over time. This is mostly due to decades of research in the neurosciences. It is now known that several areas of the brain are required for obtaining, storing, and retrieving memory.

In the human brain, memory is stored and retrieved from what is known as the neural network of the brain. Information from sensory organs travels through specific parts of the brain, is processed, stored, and then able to be recalled at later periods of time. The anatomic regions currently known to be critical to the formation and recollection of memory are the medial temporal lobe, certain diencephalic nuclei, and the basal forebrain.

The medial temporal lobe contains the hippocampus and the amygdala. The hippocampal region is where electrochemical activity converts short-term memory into long-term memory by via LTP. LTP is thought to be a persistent electrochemical increase in synaptic strength following high-frequency stimulation of a chemical synapse.

The amygdala is claimed to rate the emotional importance of a particular experience. For example, a very intense experience, such as pain or pleasure, would create a very strong memory. Conversely, a mild or indifferent stimulus, such as tying a shoelace, may be disregarded altogether and not stored as a lasting memory.

Certain diencephalic nuclei in the dorsal medial nucleus of the thalamus and the mamillary bodies are also involved in memory. This is known because, if these areas are damaged—for example, in thiamine-deficient states or alcohol impairment—then the brain has the inability to recall events. Neurological inactivity in these areas is also noted in Korsakoff's syndrome, a medical condition in which severe impairment is noted in recalling remote memory.

The basal forebrain consists of the basal ganglia and areas called brain-stem nuclei. These structures lie deep inside the brain and consist of the caudate nuclei, lentiform nuclei, portions of the amygdala, and claustrum. Collectively, these areas are involved in voluntary movement and nonmotor learning. It is known that damage to these areas can result in the decline and loss of memory, as well as loss of executive functioning (planning and the ability to pay attention) and loss of ability in set-shifting (the ability to alternate between two or more tasks). This is seen in Parkinson's disease and Huntington's disease.

Neuroplasticity: The Brain Can Change

Neuroplasticity (also known as *cortical plasticity*) is the ability of the brain to form new neuronal connections and to reorganize itself. This can happen in response to certain types of injuries or diseases and in response to new situations and changes in the environment. The concept of neuroplasticity has challenged the previous dogmas that the brain is immutable and that, after a certain age of development, it does not change. Neuroplasticity does allow changes in the brain, and it allows the brain to be incredibly adaptive.

How does neuroplasticity work? In the neuronal network of the brain, each neuron forms several connections with other neurons. Connections that are used infrequently eventually fade away, a process called *synaptic pruning*. Conversely, connections that are used regularly and frequently are strengthened (as proposed by Hebb before there was knowledge of neuroplasticity). In addition to this, neurons can also form new connections to other neurons. It is thought that these new connections are involved in forming long-term memory in response to new information.

It is maintained that the earlier hominid brain was similar to but not as complex as the more modern and evolved *Homo sapiens sapiens* brain. The anatomical and chemical changes in complexity had to have changed over time in order to improve the process of human memory. These changes were in all likelihood induced by natural selection. As our earlier ancestors began to evolve into a hunting and gathering species, an increase in neurological demand was made because of the increased need for communication and the ability to learn and remember more information. In short, an increase in the ability to remember equals an increased chance of survival, and this was provided by neuroplasticity. Neuroplasticity is not exclusive to *Homo sapiens sapiens*. The ability to learn and adapt to new information is apparent in most animals and organisms possessing a nervous system. However, it is not clear how human neuroplasticity differs from the neuroplasticity in other animals.

Virtual Memory

Like the human brain, computers are able to store and retrieve memory called *data*. Early computers used a two-level storage system that consisted of a main memory (RAM), which consisted of magnetic cores, and a secondary (hard disk) memory that was composed of magnetic drums. The problem with this two-level system was that the main memory was very limited, and most programs had to use the much slower hard disk (secondary memory).

In 1959, a one-level storage system known as "virtual memory" was conceived. This new system utilized a special automatic set of hardware and software that kept the majority of the current programs and memory in the faster main memory and, in conjunction with secondary memory, created the illusion of unlimited available memory.

Currently, computers are able to store and retrieve massive amounts of data in seconds, but they need to be programmed to do so. The computer's physical memory is stored using a binary code and can be stored on computer chips, disks, or electromagnetic tapes. This is slightly different from memory storage in the human brain, which is done primarily in the hippocampus using long-term potentiation. The human brain is capable of storing a large amount of memory, but modern computers can store practically unlimited amounts of data with greater accuracy than human memory.

Chimpanzee Memory

Much research has been done on primates, in particular chimpanzees, because of their similarities to humans. The dogma has always been that human executive and cognitive functions are superior to those in the apes. Recently, it was shown in a study done at Kyoto University that young chimpanzees could grasp many numerals at a glance and recall the sequence of these numerals. In most cases, they actually performed at a higher level than mature chimps and humans. This shows that other primates, besides humans, have extraordinary working capabilities for numerical recollection. As with our early human ancestors, this is likely a result of natural selection. Primates with better memory would have an adaptive advantage, and this would increase their chances of survival and reproduction.

Alzheimer's Disease

Several medical conditions that impair memory or cause memory loss have been mentioned already, such as Korsakoff's syndrome, Parkinson's disease, and Huntington's disease. There are several other such conditions including encephalopathy, vascular dementia, stroke, vitamin deficiencies, hypothyroidism, and psychiatric conditions. However, Alzheimer's disease is the most well-known condition that causes memory impairment and loss in humans; this disease is a progressive neurological disorder in which the loss of short-term memory is present in early stages. During the later stages, progressive memory loss will continue, and eventually long-term memory loss takes place.

Much has been learned about the pathology process of Alzheimer's disease. Neurologically, the brain develops extracellular deposits of amyloid-beta protein, intracellular neurofibrillary tangles, and eventually loss of neuron mass. In addition, certain genes have been identified in familial forms of Alzheimer's disease. This suggests that, in the future, perhaps gene therapy may be able to prevent or treat these forms of Alzheimer's disease.

Current treatments can potentially halt the progression of Alzheimer's disease. Medications known as cholinesterase inhibitors have been somewhat effective in treating Alzheimer's patients.

There are also other classes of medications that can help halt the progression of this disease. However, all these medications are very expensive and only slow down the eventual progression of the disease.

Enhancing Memory

Improving or enhancing memory is an interesting topic, because the ability to recall more information accurately and faster would provide an individual with a great advantage. Not having to look up information in a book or journal years after that information has been forgotten would be a tremendous advantage in the work place, pursuing research, completing academic projects, or learning other languages.

No current methods or medications have proven to be 100% effective in improving human memory. Certain didactic methodologies aim at improving the retention of memories (such as facts, words, and diagrams) that can help with learning and scholastics. Herbal medicines like *Gingko biloba* have been shown to improve the circulation in the brain. Proposals have been made that this medicine could, in theory, improve memory, but no conclusive evidence exists as of now. A more complete understanding of the genetics that may be involved in memory could in the future propose the possibility of the genetic enhancement of memory.

It has been well documented that exercise that increases circulation improves memory but does not enhance memory. This is because improved circulation increases oxygenation to the brain. It is also true that a healthy diet gives rise to a healthier brain and thus improved memory. Dietary vitamins, especially B-vitamins and omega-3 fatty acids, are known to maintain healthy memory. Again, maintaining healthy memory does not mean enhancing memory beyond its human capacity.

Our understanding of memory has changed over time, mostly due to neurological discoveries, in particular LTP and neuroplasticity. Likewise, over time, our ability to utilize memory has improved our species' ability to survive. However, human memory is not perfect; neurological information can become distorted, lost, or in some cases repressed. Disease can also degrade the memory process. Continuing biological evolution may gradually result in improvements in human memory. In the shorter term, perhaps our own efforts to understand the neurobiology of memory more completely will lead to future improvements in human memory through new medications or gene therapy.

John K. Grandy

See also Cognition; Consciousness; Information; Intuition; Perception; Sleep; Time, Phenomenology of

Further Readings

Basar, E. (2007). *Memory and brain dynamics: Oscillations integrating attention, perception, learning, and memory.* New York: CRC Press.

Eichenbaum, H. (2002). *The cognitive neurosciences of memory: An introduction.* New York: Oxford University Press.

Grandy, J. (2005). Consciousness. In H. J. Birx (Ed.), *Encyclopedia of anthropology* (Vol. 2, pp. 563–566.). Thousand Oaks, CA: Sage.

Kandel, E. R. (2007). *In search of memory: The emergence of a new science of mind.* New York: Norton.

Pinker, S. (1999). *How the mind works.* New York: Norton.

Shaw, C. (2001). *Toward a theory of neuroplasticity.* Philadelphia, PA: Psychology Press.

Merleau-Ponty, Maurice (1908–1961)

Maurice Merleau-Ponty was a French philosopher in the tradition of phenomenology. He taught at the École Normale Supérieure, later held the chair of child psychology and pedagogy at the Sorbonne, and in 1952 became the successor of Louis Lavelle at the Collège de France. In 1946 he founded the journal *Les Temps Modernes* together with Jean-Paul Sartre, but he withdrew his cooperation in 1955 and subsequently left the editorial board.

Merleau-Ponty developed his thoughts under the influence of various schools of thought, the phenomenology of Husserl and Heidegger being the most important; others were dialectics (Hegel, Marx), existentialism (Sartre) and neocriticism

(Brunschvicg). His theory of time is based on the Husserlian phenomenology and is to be found in two of his major large-scale works: the early work *Phenomenology of Perception* (1945) and the unfinished manuscript *The Visible and the Invisible* (1964), which was posthumously published by Claude Lefort.

In the discussion surrounding his work, it remains contentious whether there is any continuity between the early and the late thought or not. Time as subject is treated quite differently in the two works mentioned above: In *Phenomenology of Perception,* Merleau-Ponty analyzes time in a phenomenological manner, and in *The Visible and the Invisible,* in which Merleau-Ponty seeks for a way of thinking beyond subject-object dualism, time becomes a subject of ontology. But already in his early work, Merleau-Ponty argued against the dualism of subject and object and the theoretical concepts of empiricism and idealism, which emphasize a dualistic way of thinking. Although his analysis of time exhibits a tendency toward a subjective notion, it would be a misinterpretation to speak of a subjective view of time. In fact, Merleau-Ponty's concept of time undergoes continuous development throughout his life's work.

In *Phenomenology of Perception,* Merleau-Ponty argues against the idea that time resembles a flowing river. This image of time is problematic from two perspectives: First, it suggests that time has an existence in itself and, therefore, is something in the world; and, second, that time flows from the past to the future or vice versa. Merleau-Ponty denies that time exists in the world; he says that there is no such thing as succession to be found in the world of things. This position has led to the widespread conviction that he defends a subjective view of time. But time is also not to be found in consciousness, according to Merleau-Ponty. The role of the subject is to unfold time, and it does so from the present, which Merleau-Ponty considers to be the source of time. Therefore, there is no flow of time from the past to the future. Moreover the concept of unfolding or constitution of time is not completely subjective, because contact with the world is necessary. Furthermore, this special kind of constitution does not imply a completion—time is never wholly constituted, it remains *in statu nascendi* as Merleau-Ponty calls it. That means it cannot become an object of complete recognition. Merleau-Ponty understands time in *Phenomenology of Perception* as a dynamic structure that constantly evolves from the primordial interaction of subject and world.

The constitution of time is bound to the present, and the reason for this present-centered view of time lies in the function of the body (the *corps propre*). The concept of the body takes over a transcendental function similar to the transcendental consciousness in Husserl's phenomenology: It is the condition of possibility for perception, because it situates the subject in the world and gives it a perspective. Husserl did not think of the body as basis for perception, because he held that the body is a concept that first has to be constituted in experience before it can function as a means of perception. The difference to Husserl's concept is that the transcendental consciousness is itself transcendent. On the contrary, the body has a transcendental function but is not transcendent itself. The body's significance for time lies in its presence for the subject. It is always present for the subject even if the subject has not attained self-consciousness yet. My own body cannot be compared to an ordinary object in the world, because I am not able to distance myself from it in order to perceive it as a whole thing or from all sides. That means my own body is always present for me, but it can never be wholly presented to me. Because the body is fundamental to perception, it also determines the time-consciousness to evolve from the present moment. But the present moment, just like time as a whole, remains *in statu nascendi;* in other words, it is never complete; the present as the source of time is never fully present, because it constantly evolves.

Merleau-Ponty states that time and the subject are identical, in the sense that their structure is alike. The identification of subject and time is programmatic for Merleau-Ponty: It expresses his aim to describe time from the perspective of the subject without limiting it to a subjective concept. Since time and the subject are not heterogeneous to each other, Merleau-Ponty doesn't require a higher-order subjectivity that synthesizes time to make it available for consciousness.

In *The Visible and the Invisible,* Merleau-Ponty's aim is to give phenomenology an ontological foundation. He stresses the notion of time as dimension of being and even speaks of time as an element in the ancient Greek sense. Time is the element in which structure becomes possible, and structure is necessary for the possibility of being; as a dimension of being, it is fundamental to the structure of subject and object. Merleau-Ponty expands the body-concept to the concept of "flesh." He uses this term to denote the irreducible bond of subject and world on the level of corporeality and perceptual structures. Because he connects the notion of flesh with time (the flesh of time), it becomes obvious that here at the latest he no longer defends a subjective view of time (if he ever had done so). The importance of the present has not lost weight. The flesh as universal structure is not restricted to the perceiving subject but contains both the perceived and the perceiver (perception has its place in between subject and object, it is not only an act of the subject); it is a structure of simultaneity. In the presence of the flesh, past, present, and future are contained simultaneously. The present is itself structured, this richer notion of presence is captured in the term *simultaneity.* In this expanded view of time, Merleau-Ponty also reflects on historicity from various, preferably nondualistic, perspectives. An ontology of time from a Merleau-Pontyan perspective will neither objectify it nor restrict it to subjectivity. Although being cannot become objectified, it is open to philosophical interrogation; therefore, Merleau-Ponty himself calls his ontology an indirect one.

Merleau-Ponty's late thoughts about time remained fragmentary and, as such, open to various interpretations. Nevertheless, they are worth considering for a nondualistic time theory. His philosophy influenced among others the works of Foucault and Derrida (for example, his critique of the metaphysics of presence) and is the subject of philosophical discussions in Europe and to a great extent in North America, where for example the relevance of his theory for the interpretation of recent results in the cognitive sciences is discussed.

Yvonne Förster

See also Bergson, Henri; Derrida, Jacques; Epistemology; Farber, Marvin; Hegel, Georg Wilhelm Friedrich; Idealism; Marx, Karl; Metaphysics; Perception; Ricoeur, Paul; Time, Phenomenology of

Further Readings

Barta-Smith, N. A. (1997). When time is not a river: Landscape, memory, history, and Merleau-Ponty. *International Philosophical Quarterly, 37*(4), 423–440.

Carman, T., & Hansen, M. B. N. (Eds.). (2005). *The Cambridge companion to Merleau-Ponty.* Cambridge, UK: Cambridge University Press.

Dillon, M. C. (1988). *Merleau-Ponty's ontology.* Bloomington: Indiana University Press.

Dreyfus, H. L. (2002). Intelligence without representation—Merleau-Ponty's critique of mental representation: The relevance of phenomenology to scientific explanation. *Phenomenology and the Cognitive Sciences, 1*(4), 367–383.

Kelly, S. D. (2005). The puzzle of temporal experience. In A. Brook & K. Akins (Eds.), *Philosophy and neuroscience* (pp. 208–238). Cambridge, UK: Cambridge University Press.

Merleau-Ponty, M. (1962). *Phenomenology of perception.* London: Routledge.

Merleau-Ponty, M. (1964). *The primacy of perception.* Evanston, IL: Northwestern University Press.

Merleau-Ponty, M. (1968). *The visible and the invisible.* Evanston, IL: Northwestern University Press.

Muldoon, M. S. (2006). *Tricks of time: Bergson, Merleau-Ponty, and Ricoeur in search of time, self and meaning.* Pittsburgh, PA: Duquesne University Press.

Priest, S. (1998). *Merleau-Ponty.* New York: Routledge.

Metamorphosis, Insect

Metamorphosis is a temporal process of development involving the interaction of hormones triggered at particular stages of growth. Metamorphosis of greater or lesser degree is found in most organisms where there is a developmental transition over time in body form between the egg and adult. Insects undergo a particularly noticeable metamorphosis involving distinct stages of development that often occur in different habitats or utilize different food resources. Developmental transitions occur between different juvenile stages and are terminated when the adult stage is reached.

With a relatively inflexible outer integument or exoskeleton, insect growth is only possible through periodic shedding or molting of cuticle between each instar followed by a rapid expansion of a soft, new cuticle until it hardens. This expansion facilitates further growth during each developmental stage or *instar.* At emergence from the egg, most insects are structurally different from their adult counterparts. This difference may be slight or pronounced. Juvenile stages are usually characterized by feeding, growth, and development of external and internal structures such as wings and reproductive organs that are not fully developed until the final molt into the adult. In many insects metamorphosis is confined to a series of instars during a single season or year for those species with an annual life cycle.

Metamorphosis between instars may be confined to a matter of days in species with rapid life cycles (such as insects feeding on ephemeral fungal fruiting bodies) or spread out over many years in long-lived species with an extended juvenile growth period. There can also be considerable variation within a single species. Juvenile development in the wood-boring ghost moth *Aenetus virescens,* for example, may vary from as little as 9 months between egg and adult to as long as 4 years within a single population. The number of instars is also variable between species, and sometimes within species. Many insects, especially those that develop through their life cycle each season, have relatively few instars, with four to five stages being common.

There are several distinct patterns of insect metamorphosis with contrasting developmental patterns that contribute to the structural diversity of a group of organisms that may otherwise have had a more homogenous appearance. Insects that never evolved wings, where juveniles resemble adults and adults also continue to molt, are called *ametabolous* (or *aptergota* = without wings). This development pattern occurs in five primitive insect orders that include springtails and the common silverfish. Most insect orders are *hemimetabolous* (also called *exopterygota* for their externally visible wing development) with a dimorphic life history divided into a series of nymphs that molt through several instars and adults that do not molt. In these insects, wings develop gradually as external wing pads in the older juveniles, and only the adult has fully functional wings (with the unique exception of mayflies, where the final juvenile instar has fully developed wings). Hemimetabolous insects include grasshoppers and their close relatives, such as stoneflies, and true bugs.

Insects with the most distinctive stages of metamorphosis are *holometabolous,* where there are three main stages of development: larva, a pupa, and adult. In this developmental sequence the larva is structurally and behaviorally different from the adults. Compound eyes are usually absent in the larval and pupal stages (where the eyes are otherwise absent limited to several single lenses). The holometabola are also referred to as *endopterygotes,* because the wings and other features develop internally until the pupal stage, when they become everted and visible externally, although they are not fully expanded to the adult structure. Evolution of holometabolous development is widely regarded as a key evolutionary innovation contributing to the comparatively diverse speciation within 11 orders that compose about 75% of all insect species.

Transitions between different stages during metamorphosis involve a sequential web of interacting genetic and biochemical factors and the balancing effects of hormones that effect molting with those that maintain the juvenile stage (juvenile hormone) by preventing the epidermis from depositing adult cuticle in response to the presence of molting hormone. The specific developmental triggers for the secretion of brain hormone are generally not understood. In some cases, metamorphosis is triggered by internal indicators of body growth, such as stretch receptors that indicate a particular level of body expansion has been reached, or the attainment of a critical body weight.

Molting proceeds through three major complementary processes:

1. Old cuticle is separated from the underlying epidermis, and 80% to 90% of this cuticle is reabsorbed through the action of enzymes while a new cuticle is secreted by the epidermis. This process is initiated by release of prothoracicotropic hormone in the corpora cardiaca (or corpora allata in Lepidoptera) of the brain, and this hormone stimulates production of ecdysome in the prothoracic gland. In turn, ecdysome results in the production of the molecule 20-hydroxyecdysome, which regulates the genes that produce new cuticle.

2. Molting is controlled by release of a molting trigger from the epitracheal glands. This hormone stimulates the brain to produce molting hormone

and behavioral changes in the insect as a prelude to molting. The molting hormone also produces a positive feedback loop between the brain and epitracheal glands resulting in a massive release of molting hormone that in turn results in the release of crustacean cardioactive hormone by the ventral ganglia. This regulates the transition from premolting behavior, such as body movements that help separate the overlying cuticle, and molting behavior, which comprises waves of body contractions that continue until the molt is complete.

3. Following the molt, the body undergoes expansion and hardening of the cuticle. Wing expansion in the new adult is facilitated through abdominal contractions forcing blood into the wings, and this activity also stimulates release of bursicon, which further increases the flexibility of wing cuticle and then initiates hardening of the cuticle.

In holometabolous insects, brain and juvenile hormones are both produced during the immature stages until the last immature instar, when juvenile hormone production is either terminated or decreases below a threshold level where metamorphosis into the pupal and adult stage occurs as adult characteristics are no longer inhibited. The adults of most insects do not molt, because the prothoracic glands degenerate either before or after adult emergence, and there is no longer the secretion of molting hormone. In hemimetabolous insects, the transformation between the immature and adult stages is more gradual. Brain hormone includes a number of steroids that act on genes through a receptor-mediated process that determines which genes are activated at a given time and consequently which enzymes and structural proteins are synthesized.

The transition between instars usually takes place over a few minutes as the old cuticle breaks open along the dorsal midline of the thorax and the insect extrudes is body through this opening, with the appendages such as legs and antennae along with the tracheal tubes being the last to pull away. This process is preceded by a separation of epidermal cells from the old cuticle and the secretion of a new cuticle as well as enzymes that digest 80% to 90% of the old cuticle. The insect pumps air into the body, which expands its volume, resulting in the breakage of old cuticle along the dorsal line. The body is extruded out of this break, and the insect pulls itself away from the old cuticle. This is followed by a period of resting as the new cuticle is hardened. It is during this process that the size of the body is expanded so the new instar is larger than the previous instar. The process of metamorphosis begins with hatching from the egg when all insects are small, sexually immature, and lack wings. As the juvenile and adult stages have often diverged evolutionarily in form and function, the juvenile is more efficient at feeding and growth, while the adult is more specialized with respect to dispersal and reproduction.

The size at which molting occurs is not absolute and depends on the size of the insect at the beginning of the instar. In some species and where food is insufficient, the molt may result in a smaller instar or in the retention of a juvenile stage rather than a subsequent instar such as the pupa. Environmental conditions may also modify the amounts or timing of hormone secretion, and where these factors are predictable components of development, they will result in characteristic differences between individuals composing different castes in social insects. Juvenile honey bees will develop into queen bees when fed a diet based on secretions of the nurse bee's mandibular glands, and they will develop into workers when fed higher proportions of hypopharyngeal gland secretions from worker bees. Ants will develop into minor or major workers or soldiers according to the quantity of food that will result in larger juveniles. These size differences affect the quantity of juvenile hormones, which is higher in the final instars of larvae developing into queen bees or major and soldier ants. In aphids, metamorphosis into the final adult form is affected by day length. Under long photoperiods, the largest embryos will develop into parthenogenetic forms, whereas under short photoperiods the embryos will give rise to sexually reproducing forms. The development of parthenogenetic forms is stimulated by the secretion of hormones from cells in the brain that respond to the amount of light passing directly through the cuticle of the head.

John R. Grehan

See also Evolution, Organic; Photosynthesis

Further Readings

Chapman, R. F. (1998). *The insects: Structure and function.* Cambridge, UK: Cambridge University Press.

Heming, B. S. (2003). *Insect development and evolution.* London: Comstock.

Metanarrative

A metanarrative is a theory of history that is said to move in a specific direction and, on the strength of which, confident predictions about the future can be made. Metanarratives have also been called Grand Narratives, or Master Narratives (usually complete with capital letters) and the philosopher Karl Popper spoke of historicism in the same context. The critical factor in a metanarrative, and what distinguishes it from a historical perspective, is the blending of the historical account into an assertion about how the future will unfold.

Several great systems of thought have articulated, or at least assumed, a historical narrative. Marxism and Christianity, for instance, both involve a metanarrative. For example, Christianity speaks of a creator God who made the world and then placed Adam and Eve in it as the most important products of that Creation. Eve's sin meant the expulsion of them and their progeny from paradise and into the world of sin, suffering, and death. People were then offered a way out of this condition when God sent his only son as savior, so those who believe in his salvific efficacy would be saved from death and live in bliss in heaven for eternity. Eventually history will be brought to a close when, at some time in the future, Jesus Christ returns (the Second Coming) to judge the living and the dead and confer punishments and rewards as appropriate. This is a metanarrative in that the theory of history blends seamlessly into a prediction about the future.

Metanarratives have been around for a long time. Ancient and medieval writers frequently spoke in terms of history being a succession of ages. In 725 CE, the Venerable Bede (673–735 CE) wrote of the ages of man in his *De Temporum Ratione* (*On the Reckoning of Time*). Bede followed the most popular route, thinking in terms of the four ages of man. This goes back to the Pythagorean numerology and to the association of the number four with the four seasons, the four cardinal directions, and the four original elements as outlined in Greek philosophy.

In the 12th century, the idea that human history is in fact punctuated by seven ages became more popular. Unlike the four-ages theory, the seven-ages theory was astrological in origin, working on Ptolemy's seven-planet (including the sun and moon) cosmos. It is most memorably recalled for us now in Shakespeare's *As You Like It* (act II, scene 7).

The extraordinary appeal of Marxism in the 19th and 20th centuries lay in the secular treatment it gave to what was fundamentally a religious metanarrative, with its confident belief that socialism would, in the future, be replaced by communism, which will mean that all material contradictions and inequalities will have been resolved.

Metanarratives found their most enthusiastic critics in postmodernist thinkers. Postmodernism was not so much a coherent philosophical movement as a diffuse mood. It remains influential in some humanities' disciplines but, since the second half of the 1990s, has faded from prominence in most areas. The classic definition of postmodernism was given by the French thinker Jean-François Lyotard as "incredulity toward metanarratives." This seemingly reasonable idea was promptly undermined, however, when Lyotard made it clear he did not mean incredulity at all, but outright opposition. It was also apparent that Lyotard was, if unwittingly, assuming a metanarrative of his own. A few sentences after talking of incredulity, Lyotard spoke of "the obsolescence of the metanarrative apparatus" and the time "after metanarratives."

The postmodernist hostility to metanarratives was expressed even more openly by Patricia Waugh. In her introductory essay to an influential anthology of postmodernist ideas, Waugh went further than Lyotard when she added that postmodernism was about the "abandonment of all metanarratives which could legitimate foundations for truth." More than this, Waugh also declared metanarratives were no longer even desirable. Another influential postmodernist, Zygmunt Bauman, spoke of modernity as a "long march to prison," one that was being undone by the "second Copernican revolution" of postmodernist thought, led by Martin Heidegger. And the most radical postmodernists spoke in terms of modernity having been vanquished and the theories that sustained it destroyed, leaving modernity redundant, never to be brought back. At whichever point they were located along the postmodern spectrum of thought, it appeared that, however much they declared their incredulity toward metanarratives, it seemed postmodernists could not free themselves from them.

Much more effective criticism of metanarratives came from Karl Popper's book *The Poverty of Historicism.* Written in 1935, it was not translated into English until 1957 but quickly established itself as influential after that. The key weakness of historicism, Popper argued, was to equate laws of development with absolute trends, which were arrived at by some metaphysical necessity. The historicist went on, Popper claimed, to want to change the course of history by virtue of superior knowledge of the dialectic of history. But this was to put the cart before the horse. While history needs to be written from a preconceived point of view, Popper wrote, this does not mean the historian's preconceived points of view should be taken as historical laws.

Others have wished to retain a place for metanarratives, even if only for their symbolic power. Some feminist thinkers, for instance, have been fiercely critical of the wish to jettison historical accounts of the progressive emancipation of women from patriarchal oppression. By dismissing such an important struggle as simply an arching tale, these critics suggest, the historical reality of those emancipations is jeopardized.

Bill Cooke

See also Bede the Venerable, Saint; Bible and Time; Language; Marx, Karl; Popper, Karl R.; Postmodernism

Further Readings

Bauman, Z. (1993). *Intimations of postmodernity.* London & New York: Routledge & Kegan Paul.

Lyotard, J.-F. (1988). *The postmodern condition: A report on knowledge.* Minneapolis: University of Minnesota Press.

Popper, K. (1957). *The poverty of historicism.* London: Routledge & Kegan Paul.

Waugh, P. (1994). *Postmodernism: A reader.* London: Edward Arnold.

Metaphysics

Metaphysics is one of the oldest, least explicit, and most controversial disciplines in Western philosophy as far as its evaluation is concerned. There has been no widely accepted definition, nor any strict delimitation of its subject and goal so far. In the last 2,500 years of its history, metaphysics was considered to be a basic philosophical discipline covering the question of "what really exists," or dealing with the first principles of being and cognition of all things. On the other hand, it has been questioned and rejected as a useless and nonsense activity. Thus, all of its history can be seen as a process of continual transformations seeded in the tension between its acceptance and the cyclical recurrence of critiques proclaiming its "crisis," "abolition," "termination," or "death." One of the most significant motives of the critique of metaphysics was metaphysics' prevailing tendency not to fully realize the role of time and temporality in the explanation of the world.

Origin of Name

The term *metaphysics,* handed down to the present, arose accidentally in the 1st century BCE when Andronicus of Rhodes, a peripatetic expositor of Aristotle, summarized his unlabeled treatises under the term *ta meta ta physika,* treatises that in his catalogue follow Aristotle's work *Physics.* Aristotle himself most commonly referred to these teachings as "the first philosophy" (*he prote philosophia*) but also as "wisdom" (*sophia*) or "theology" (*theologiké*). Late Aristotelians, especially Alexander of Aphrodisias in the late 2nd and early 3rd centuries CE, understood Andronicus's classificatory meaning of the word *metaphysics* as matter of content—as a teaching dealing with what is *beyond* the physical world, a teaching about the supersensible, transcendent, and intelligible entities (including God). Also in this sense, Simplicius and Boethius used the one-word term *metaphysica* in the 4th and 6th centuries CE, respectively. European philosophical thinking acquired this understanding of the term during the 13th-century period of scholasticism.

History of Metaphysics

Antiquity

In ancient Greece, the development of metaphysics is connected notably with the names of Parmenides, Plato, Aristotle, and Plotinus. Parmenides (540–450 BCE), by arguing that "being is, and nothing is not because nothing cannot be thought of, and so, the thought and the being are the same" and by his follow-up claim that being is one, continuous, changeless, and eternal, founded a qualitatively new variant of philosophical thinking about reality. His "thinking and being" correspondence principle, as well as his distinction between reality being ultimately true and appearing by sense perception, have become determining, and at the same time, limiting considerations for the entire field of classical metaphysics. Plato (428–348 BCE) elaborated on Parmenides' reflections, and in his middle writing period, he argued for the division of the world into two realms. One was *noumenon*—which exists by reason, and therefore is real, being the perfect world of original, eternal, changeless, and intelligible ideas; and the other was *fainomenon*—the imperfect world of appearances and derived, changeable, perceptibles.

Because metaphysics transcends the perceptible and reaches the top of the hierarchically ordered world of ideas—where the Good dwells as an origin of everything, factual thinking about reality is associated with axiological and at last, epistemological aspects. For metaphysics examines the fundamental principles of specific sciences (mathematics) as well. However, for further development of metaphysics, Plato's distinction of a privileged, timeless world of ideas and a disqualified temporal world of particulars became significant; time is understood there as something that metaphysics should overcome.

Aristotle (384–322 BCE), the founder of metaphysics as an independent philosophical discipline, critically elaborated on his predecessors' initiatives. This became obvious in his tendency to rehabilitate the dynamic world of perceptibles. Time as "a number of movement in respect of the before and after" is the fundamental element of entities to which belong motion and becoming in this context. Considering Aristotle's solution of this problem, as well as his delimitation of metaphysics' subject and goals, the factual legacy of his work remained ambiguous. Notably, his understanding of metaphysics, understood at one point as *ontology* and at another one as *theology,* proved to be historically significant.

In the first case, Aristotle distinguished metaphysics as a universal science about "being" (entity) as being science dealing with being as such, from specific sciences always dealing with a specific kind of being (entity). However, his ontology did not declare being to be one and beyond the diversity of things as Parmenides did. It just gives us an account of the most universal kinds of entities in their plurality—that is *a theory of categories.* Existing means to exist in a certain way, that is, in quite a few differentiations—as substance (e.g., a man), quality (e.g., whiteness), quantity, relation, etc.

However, further inquiries led Aristotle to understand metaphysics as a specific science about an ultimate type of entity, that is to say, substance, which testifies what a thing is and has an important position among categories. In reality, however, there are various types of substances, and the most perfect and dignified is the divine one (the self-thinking, eternal, unmoved first mover). Thus, metaphysics becomes a special kind of substantial ontology—*a theology.* Another definition of metaphysics by Aristotle, as a "knowledge of first causes and principles" of entities, led to a similar conclusion. The most crucial of the four causes—material, formal, efficient, and final—is the last one: Everything is carried out for some purpose, due to some good. And because the highest purpose and good is God, metaphysics, as a study of first principles, is (natural) *theology.* Here, the roots of the idea of unity between ontology and theology, according to Heidegger's so-called ontotheology, are to be found. This idea was later elaborated by scholastic philosophy.

It's worth mentioning that Plato as well as Aristotle did not understand metaphysics as a strictly descriptive contemplation on reality. On one hand, there's an axiological aspect—the cognition of reality is always interconnected with the cognition of the good and the values, and on the other hand, there's an epistemological-logical aspect—categories and principles do not only relate to the reality but to our thinking about it as well.

For Plotinus (204–270), metaphysics is mainly so-called *henology*—a teaching concerning the unthinkable and unspeakable, the not-being and the above-being *One* as the basic principle of all things. Out of this *One,* the entire reality emanates following a degenerative descending process. Even time is only a movable, imperfect picture of the eternal spiritual principle. The aim of philosophy is to free man from empirical plurality and temporality and to unify him with the perfect *One.* Plotinus brought strong elements of mysticism into metaphysics.

Middle Ages

There was an important transformation of metaphysics in the Middle Ages, especially with regard to the scholastic reception of Aristotle's work. Creationistic elaborations of its ontotheological traits, mostly performed by Thomas Aquinas (1225–1274), were of determining importance. For Thomas Aquinas, metaphysics, as a constituent of philosophical wisdom (*sapientia*), is a science about being as to what extent it is being (*ens inquantum ens*); hence it is a science covering the entire reality. However, investigating is not restricted to the perceptible world. It gradually proceeds to examinations into the supersensible (soul, angels) and results in rational contemplation on God. For if metaphysics is to understand being as such, first, being's cause must be understood. And the first cause is God, not only in Aristotelian terms as a *final cause of becoming of things,* but as *a cause and principle of being* of all things created. Metaphysics culminates in (rational) theology as in its ultimate ontology. Thomas Aquinas determines time likewise as Aristotle did, as a measure of changes that arise in bodies in respect of the before and after. It belongs to creature, not to God alone.

For such a concept of metaphysics, it was necessary to provide a rational evidence of God's existence; especially an ontological argument was needed. The argument was put forward by Anselm of Canterbury (1033–1109), who deduced an inevitable fact of God's existence out of the concept (essence) of God, understood as something that is the most perfect (and thus timeless and changeless also) and above which nothing greater can be conceived. Otherwise, there would be controversy. This proof, in its various forms, has become a constituent of great metaphysical systems of modern times.

The problem of universals led to serious consequences for metaphysics in the Middle Ages: Do the contents of universal concepts exist in reality or not? The answer to the dispute, with its prehistory in Plato, the Cynics, and Aristotle, resulted in the birth of *realism* (Anselm of Canterbury; William of Champeaux, 1068–1121) and *nominalism* (J. Roscelin, 1050–1120; Duns Scotus, 1264/1270–1308; William of Ockham, 1290–1349). According to the first, universals exist in reality, before particulars exist and apart from them, and only subsequently they appear in particulars, which are derived from them, or after particulars, in human mind. The opposite standpoint, in its most radical variation, supposes that universals are mere words (*nomina*), claiming that only particulars exist in reality. In its moderate variation, nominalism concedes the existence of universals in human mind (P. Abelard's [1079–1142] *conceptualism*). This controversy has endured, in its various forms, up to the present-day metaphysics, philosophy of logic, and mathematics.

Modern Times

In modern times, the classical understanding of metaphysics as a dogmatic teaching covering the problems of soul, world, and God has undergone significant changes in terms of its understanding, which tends to be more critical. René Descartes (1596–1650) was at the beginning of these critical efforts, and Immanuel Kant (1724–1804) concluded them.

Descartes considered the investigation of the highest principles providing certain knowledge of the world to be the goal of his *philosophia prima.* An epistemological motive, which had been of secondary concern for metaphysics until his time, became fundamental for Descartes. By means of methodological skepticism, he ended up asserting two principles. The first one, *cogito ergo sum,* anchors the certainty of every knowledge of the world not in the world itself, as it worked in the old metaphysics, but through intuitive self-evidence of a reasoning subject (*res cogitans*). This anthropological turn provides background for the second principle, *the existence of a perfect God,* as a source for the verity of the external world (*res extensa*). From these principles, like from roots, all efficient knowledge of humankind should grow (physics, mechanics, medicine, morals).

However, part of Descartes's heritage was *res extensa* (material, incapable of spontaneous motion, axiologically neutral) and *res cogitans* (spiritual, autonomous, evaluating) dualism. Primary, attributive determination of the first one is extension. Time (duration) is only secondary, modally characteristic of unattributive motion (alteration of position). It was especially rationalists who tried to offer an answer to varied forms of *mentioned dualism* (the problem of subject-object relations, mind-body problem, absence of values and dynamics in reality, etc.). For Baruch Spinoza (1632–1677), substance is one, that is, *Deus sive natura,* and metaphysics is identical to ethics. N. de Malebranche (1638–1715) presents an occasional correspondence of both substances through God. According to G. W. Leibniz (1646–1716), there are an infinite number of spiritual, that is, spontaneously dynamic and thinking substances.

Another division of metaphysics emerged in 1562 with the Spanish philosopher B. Pereira (1535–1610), who divided it into general (*metaphysica generalis*) and specific (*metaphysica specialis*). This division was systematically completed by Christian Wolff (1679–1754) in 1730. While general metaphysics (ontology) is a basic philosophical science, rational cosmology, theology, and psychology are disciplines of specific metaphysics.

The Modern Ages empiricists, unlike the rationalists, took quite a rejecting stance on the possibility of metaphysics; they demonstrated the empirical groundlessness of its basic concepts based on the thought/being correspondence. Thus, for John Locke (1632–1704), substance is unconceivable apart from the bundle of its attributes. George Berkeley (1685–1753) denies the existence of the external substance (*res extensa*) and David Hume (1711–1776) denies the existence of both external and internal substance (*res cogitans*) and the concept of causality as well, taking it as a result of everyday habit.

Kant rejected all of the previous "dogmatic" metaphysics as a fictitious knowledge, because by exploring the problems of the soul, world, and God, it applied concepts of pure reason in an inadmissible and transcendent way and beyond the borders of possible experience to a thing-in-itself (*noumenon*), which is unconceivable. Thus, metaphysical reason, not being able even to justify the existence of its subjects, had to cope with insolvable antinomies in all of its domains. As becomes obvious, an ontological proof of God's existence is impossible. Existence is not an attribute and cannot be deduced. The *cogito ergo sum* proposition is a paralogism as well.

Thus traditional subjects of metaphysics do not belong to theoretical reason but, like its postulates (the postulate of freedom, immortality of the soul, the existence of God), to a practical one. However, a new critical metaphysics is possible. It is a transcendental theory that does not pursue the objects themselves, but, provided that such cognition is a priori possible, it pursues our forms of their cognition. It is a study of categories, but these are not perceived as the most universal aspects of things (ontologically) but as a priori forms of data being linked in experience, as a "logic" of experience. It includes also contemplations about time, which according to Kant does not belong to things themselves but is one of two a priori principles of pure intuition (the second one is space) that provide humankind with an inner experience and vicariously an experience of external phenomena.

G. W. F. Hegel (1770–1831) took quite a different stance toward metaphysics. Although he blamed metaphysics for not being dialectical, he accepted and absolutized the thought and being correspondence principle. On its basis, he attempted to offer a dynamic, nondualistic view of reality in his absolute idealism. Within the evolutionary conception of reality, he identified ontology (general metaphysics) with (dialectic) logic. Here he describes the self-contradictory process of the self-realization of the freedom of absolute idea, which is developing from the "in-itself and for-itself" stage (*logics* as the overcoming of rational theology), through the "otherbeing in nature" stage (*philosophy of nature* as the overcoming of rational cosmology), to the stage where idea "recurs back from its otherbeing" (*philosophy of the spirit* as overcoming of rational psychology). The essential medium of this process is time ("intuitive becoming"), in which self-realization of the spirit is realized until is rounded off.

Hegel's influence on 19th-century metaphysics was strong but short-lived. In contrast to Hegel's approach, the majority of 19th-century philosophical movements (positivism, neo-Kantianism, etc.), being under the influence of Hume's and Kant's criticism, rejected classical metaphysics as a theory of reality (scientism, gnoseologism). The implication of this was that similar stances toward metaphysics, though not epistemologically motivated, were taken by Søren Kierkegaard (1813–1855), Karl Marx (1818–1883), and Friedrich Nietzsche (1844–1900).

20th Century

In the 20th century, understanding of metaphysics' subject and goals was considerably influenced by a linguistic turn in philosophy. The attitude of analytic philosophy, as a protagonist of this turn, changed. At its inception (until the 1920s), it offered a criticism of certain type of metaphysics (neo-Hegelian) for the sake of another one (G. E. Moore's [1873–1958] realism, or Bertrand Russell's [1872–1970] logical atomism). Later, in the 1950s and 1960s, under the influence of Ludwig Wittgenstein (1889–1951) and the activities of logical positivism, analytic philosophy became radically antimetaphysical. Following Wittgenstein, who claimed that philosophy is not a theory but a criticism of language, logical positivists including Moritz Schlick (1882–1936) and Rudolf Carnap (1891–1970), who understood philosophy as a logical analysis examining the meaningfulness of the language of science, whereas a criterion for meaningfulness resides in the feasibility of the sentences to be empirically verified (verificationism). Metaphysical sentences are not verifiable and thus are meaningless, because they refer to something above or beyond any experience. They do not account for referential but only for expressive function; they reflect their authors' personal sentiments and view of life.

Since the late 1950s, a renaissance of metaphysics within analytic philosophy has taken place, which was mostly associated with initiatives of Willard van Orman Quine (1908–2000) and P. F. Strawson (1919–2006). Quine recognizes metaphysics (or, to be more exact, ontology) as a reflection on what there is. However, it does not fall under the competence of philosophy (which was supposed to turn into so-called naturalistic epistemology) but belongs to (natural) sciences. This metaphysics concerns especially the so-called ontological commitment—a question as to which entities we are committed to adopt for a theory to be true—and Quine offers a specific criterion. Unlike Quine, Strawson allows for a possibility of an autonomous philosophical metaphysics. It resides in the clarification of the nature and relations between the key—the essential, the most universal and irreducible—concepts, which constitute a structural pattern of human thinking (both ordinary and scientific) about the world. Strawson also makes a distinction between descriptive and revisionist metaphysics. The former, which he prefers, settles for the description of the actual structure of our thinking about the world; the latter is engaged in rendering a better one.

The attitude of the 20th-century nonanalytic philosophy toward metaphysics was critical; however, there were efforts to propose a revised conception. Alfred North Whitehead's (1861–1947) nonsubstantial process philosophy (in which there is the fundamental metaphysical unit—event—well-founded temporally) may be an illustrative case, but the formation of ontologies figures significantly as well—for example, the critical ontology of Nicolai Hartmann (1882–1950) or Edmund Husserl's (1859–1938) eidetic science of intentional objects. One of the most important was Martin Heidegger's (1889–1976) fundamental ontology (elaborating on Husserl's phenomenological method), the aim of which is to manifest temporality as the meaning of being. A later important case was Jean-Paul Sartre's (1905–1980) phenomenological ontology.

An original critical response to the peripeties of this school of thought was provided by E. Lévinas. According to Lévinas, the fundamental weakness of previous metaphysics was that in finding the solution to its crucial question of *the same* and *the other* relation, it always turned into ontology. But ontology is egological and solipsistic in nature—in its efforts to master and usurp the other, it reduces it to itself (the same), canceling its peculiar (unique) otherness and exteriority. Such an identification of the other with the same occurs through the incorporation of a middle element between the two, a representative that the same finds in itself. This has been a universal concept since Socrates, including Berkeley's *individual perception* but also Heidegger's *being*, because Heidegger subordinates the relation toward the other (existing) to the relation toward its being. The path to the other is never direct; transcendence toward it is always embraced by immanence. However, metaphysics is not concerned with representations and thus allows the totalization of the other into the same. On the contrary, metaphysics' main effort is to maintain radical exteriority, the totally other. The transcendence is provided by a face-to-face encounter with the other; another human being and the bond established is thus ethical. Metaphysics as ethics, in this sense, precedes ontology as a philosophy of power and constitutes a philosophy of justice.

Totalizing inclinations within philosophy were again rejected in the last third of the 20th century. However, this occurred not by reinterpreting metaphysics as Lévinas did. It was postmodern philosophers, chiefly Michel Foucault (1926–1984), Jacques Derrida (1930–2004), Gilles Deleuze (1925–1995), Jean-François Lyotard (1924–1998), and Richard Rorty (1931–2007) who started proclaiming the death of metaphysics as such. The days of metaphysics as a privileged human (philosophical) activity aspiring to acquire a mirror cognition of *the Reality* in a metanarrative, absolutely valid and ultimate system, are over. Its common topics—truth, reality, humankind, history, mind, language—are no more perceived as being noumenal or transcendental in status. They are temporal—socially, historically, and linguistically determined. Metaphysics does not offer an absolute, genuine picture of reality; it is just one of the numerous narratives of reality, a mere rhetoric. "God's eye view" realism submits to relativism, monism to pluralism, totalization and sameness to the otherness, eternity to time and temporality.

Marián Palenčár

See also Aquinas, Saint Thomas; Aristotle; Becoming and Being; Berkeley, George; Causality; Cosmogony; Derrida, Jacques; Descartes, René; Dialectics; Hegel, Georg Wilhelm Friedrich; Heidegger, Martin; Kant, Immanuel; Kierkegaard, Søren Aabye; Marx, Karl; Nietzsche, Friedrich; Ontology; Parmenides of Elea; Plato; Plotinus; Postmodernism; Spinoza, Baruch de; Theology, Process; Whitehead, Alfred North

Further Readings

Aristotle. (1993). *Metaphysics.* New York: Oxford University Press. (Original work c. 350 BCE)

Carnap, R. (1959). The elimination of metaphysics through the logical analysis of language. In A. J. Ayer (Ed.), *Logical positivism.* Glencoe, IL: The Free Press.

Descartes, R. (1984–1991). Meditationes de prima philosophia (J. Cottingham, R. Stoothoff, & D. Murdoch, Trans.). In *The Philosophical Writings of Descartes* (Vol. 2). Cambridge, UK: Cambridge University Press.

Heidegger, M. (2000). *An introduction to metaphysics.* New Haven, CT: Yale University Press.

Kant, I. (1963). *Immanuel Kant's critique of pure reason* (N. K. Smith, Trans.). London: Macmillan.

Lévinas, E. (1969). *Totality and infinity.* Pittsburgh, PA: Duquesne University Press.

Strawson, P. F. (1964). *Individuals: An essay in descriptive metaphysics.* London: Routledge.

van Invangen, P., & Zimmermann, D. W. (Eds.). (2004). *Metaphysics: The big questions.* Oxford, UK: Blackwell.

Meteors and Meteorites

Both meteorites and meteors are classified more broadly as meteoroids, the simplest definition of which is a small body in space. The term *meteor* is applied to any streak of light in the upper atmosphere that is produced by a small body when it enters from space. They are more commonly referred to as shooting stars. Efforts have been made to predict and record the occurrence of the most impressive phenomena, and there is wide interest in observing them among both scientists and laypeople.

Meteorites are natural bodies that have managed to travel through the atmosphere to land on the surface of the earth. To accomplish this, they must normally be large and dense enough so that they don't crumble into fragments or vaporize completely. They are of great scientific interest because they provide an opportunity to examine extraterrestrial material older than any material found on earth. Whereas many of earth's early rocks have been destroyed by natural geologic processes, meteorites have remained unchanged since the birth of the solar system. It is now generally accepted that research to determine the age of the materials in meteorites will very likely help us to determine the age of our planet.

Knowledge about and attitudes toward meteorites have undergone significant change through the centuries. There is evidence that many ancient people did accept meteorites as heavenly bodies. Both meteors and meteorites inspired a variety of myths and folk beliefs, with the element of superstition emerging prominently during the medieval period. During the 17th and most of the 18th centuries, the possibility that meteorites had originated from beyond the earth was not even considered.

The last decade of the 18th century marked a distinct forward movement in the scientific study of meteorites. Luckily, witnesses were present when several thousand meteorites fell near L'Aigle in northern France. In 1794, Ernst Chladni, a German physicist, published a book asserting that bodies of rock and metal actually do fall from the sky. He is now considered the founder of the study of meteoristics.

From that time on, museums and interested individuals began collecting samples, making further research possible. At present there are around 1,000 samples available for scientific study. The earliest known meteorite still available for study fell near the village of Ensisheim in the province of Alsace in 1492.

Some specific scientific observations still lagged behind. It wasn't until the 1930s and 1940s that scientists concluded that some craters had originally been formed by the impact of meteorites.

Chrondites represent the most primitive kind of meteorite, making up about 80% of the total. They appear to have been formed with the same substances as the earth and other planets. Their composition is also very similar to that of the sun, with the exception that the sun also contains large amounts of hydrogen, helium, and other noble gases.

In the 1950s, Clair Patterson, of the California Institute of Technology, began the process of dating chrondites by a radiometric process that made use of a uranium-lead clock. He found their age to be 4.55 billion years, and the range of 4.5 to 4.7 billion years has been corroborated by five different radiometric dating methods. The basis for all of these methods has been the measurement of the radioactive decay that has occurred. About 70 meteorites have now been accurately dated. Radiometric dating has also been used to determine ages of many other subjects, including the earth, the moon, fossils, early humans, and a variety of geological events.

Comparing chrondites with rocks that appeared to have originated on earth, Patterson found a similar composition of lead isotopes. A logical conclusion is that both may have been formed when silicates condensed from the sun's nebula. It has also been concluded that the universe may be as old as 14 to 17 billion years.

Scientists now generally believe that most meteorites are pieces broken off when asteroids—or minor planets—collided. Asteroids themselves are large chunks of debris that circle the sun in the belt between Mars and Jupiter's orbits. Like the earth, these meteorites are composed mostly of silicates and metals. A few rare meteorite samples may have come from Mars or the moon, probably as the result of a crater being formed by another body impacting those surfaces.

In the 1980s, sediments from around the world were examined and found to contain chemicals of the same age. That suggested that a large body had slammed into the earth approximately 65 million years earlier. It is even possible that our moon might have been formed as the result of a collision between the earth and a body about the size of Mars.

It is estimated that about 30,000 meteorites with a mass larger than 3.5 ounces fall to the earth each year. They are classified by the amount of silicate and nickel-iron that they contain and fall into the three broad categories of irons, stones, and stony irons.

Meteorites are classified as falls or finds, and each is given a name that is usually based on the geographical area in which it fell. To be classified as a fall, the event must be witnessed. Because meteorites usually signal their arrival by a light display and a variety of sounds, it is not surprising that some are observed at the time of impact. Reporting of these events can allow scientific study to begin soon after the event. The largest recorded fall occurred in 1976, in Jilin, Manchuria. The total weight of its pieces was 2 metric tons.

The term *finds* applies to those that are discovered by accident and subsequently identified by the chemical and mineral content or their structure. The largest known meteorite find is the African Hoba meteorite, weighing approximately 66 tons.

The study of known meteorites is expected to reveal even more information about the solar system. The search for meteorites that are yet to be discovered will be ongoing. Two challenges that have been recognized are the need to search in the comparatively neglected desert areas and the need to identify very small meteorites.

Betty A. Gard

See also Comets; Dinosaurs; Extinction and Evolution; Extinctions, Mass; Geology; Nuclear Winter; Paleontology

Further Readings

Bevan, A., & De Laeter, J. (2002). *Meteorites: A journey through time and space.* Washington, DC: Smithsonian Institution Press in association with University of New South Wales Press.

Norton, O. R. (2002). *The Cambridge encyclopedia of meteorites.* New York: Cambridge University Press.

Zandra, B., & Rotaru, M. (2001). *Meteorites: Their impact on science and history.* Cambridge, UK: Cambridge University Press.

Methuselah

The epitome of an extended life span lies in the biblical tale of Methuselah or Metushélach (מְתוּשֶׁלַח / מְתוּשָׁלַח), the longest living Hebrew patriarch. Humankind's quest toward great longevity has been a common mission throughout the generations. From scrutinizing ancient texts to the exploration of advancing modern sciences humankind has endlessly sought a proverbial fountain of youth.

Methuselah is credited with having lived 969 years, making him the oldest person in recorded history, yet little else is stated. He is mentioned in Luke 3:37 (King James Version) when tracing Jesus of Nazareth's lineage and is noted as being the father of Lamech, Noah's father, in Genesis 5:25–27:

> And Methuselah lived an hundred eighty and seven years, and begat Lamech. And Methuselah lived after he begat Lamech seven hundred eighty and two years, and begat sons and daughters: And all the days of Methuselah were nine hundred sixty and nine years: and he died.

In the year of the Great Flood, or the Deluge, he eventually died.

> As long as Methuselah lived, the Flood did not come upon the world. And when Methuselah died, it was withheld for another seven days after his death to fulfill the period of mourning. (Avot d'Rabbi Natan 32:1)

The direct etymology of the word *Methuselah* is "his death shall bring"; this translation comes from the roots *mûth,* meaning "death," and *shelach,* meaning "to send forth." Based on the etymological translation, the year of his death, and his remarkable age, Methuselah has often been cited as an allegory in which God withheld judgment upon humankind for a great period of time.

Some scholars, however, dispute Methuselah's age and theorize that it is in fact merely an error resulting from mistranslations of text and loss of documentation through the ages. Within several early biblical texts, the numbers were written in an archaic, precuneiform Sumerian number system in which the decimal is placed differently than in latter Sumerian number systems. Their theory is that in the year 1700 BCE the scriptures were initially mistranslated into a later Sumerian system and then mistranslated again in 550 BCE when compiling the Hebrew Genesis 5.

Another theory speculates that certain dates are based on a standard using lunar instead of solar cycles. This theory would convert Methuselah's age upon death to 87, and to 15 when he fathered Lamech. However, this lunar theory also has met with skepticism, because it results in other patriarchs, such as Enoch and Mahalel, being merely 5 years old when they fathered their own children.

Other scholars have accepted Methuselah's remarkable age as fact, yet have speculated how such a prolonged age was once feasible. Numerous theologians hypothesize that human life expectancy was shortened due to humankind's falling from God's grace. This credence is marked by the biblical examples of Adam, Methuselah, and Moses. Adam marks the end of possible immortality, Methuselah marks humankind's potential life span preflood, and Moses sets the modern standard of 120 years. A recent theory as to why Methuselah lived to be nearly a millennium old is that the human gene pool experienced a massive bottleneck at the Great Flood. This theory speculates that a drastic loss in population, caused by events such as the Deluge, would result in small gene diversity and a loss of certain inherited gene qualities. This could have resulted in the trait for longevity to be lost or hidden deep within our genetic code. Currently, numerous scientific efforts, such as the Human Genome Project and the Methuselah Mouse Prize, are striving to discover the secrets of increased longevity.

Derik Kane

See also Adam, Creation of; Bible and Time; Genesis, Book of; Longevity; Moses; Noah

Further Readings

Best, R. (1999). *Noah's ark and the Ziusudra epic.* Fort Myers, FL: Enlil Press.

The Holy Bible (King James version). (1999). New York: American Bible Society.

Wieland, C. (1998). Living for 900 years. *Creation Magazine, 20*(4), 10–13.

Michelangelo Buonarroti (1475–1564)

Michelangelo di Lodovico Buonarroti Simoni is universally recognized as one of the greatest artists in history, perhaps the greatest. He excelled as a sculptor, painter, architect, and poet of the Italian High Renaissance. During his long life, Michelangelo created some of the world's most recognizable works of art.

Mastery in a variety of skills has earned Michelangelo the title of *Renaissance man.* He rivaled his fellow Florentine, Leonardo da Vinci, as the embodiment of this role. His most famous sculptures include the *David* in Florence and the *Pietá* in Rome. Although Michelangelo claimed that he did not enjoy painting, he created many remarkable works, most famously the series of panels containing more than 300 figures depicting the Creation, the downfall of man, and the promise of salvation—the entire core of Christian doctrine—on the ceiling of the Sistine Chapel in the Vatican. Michelangelo also designed the dome of Saint Peter's Basilica, which remains the largest dome of any church in the world.

An intensely spiritual man, Michelangelo focused on biblical themes throughout much of his work. He was deeply concerned with the notion of time, depicting such scenes as God's Creation of man and the universe. His fresco *The Last Judgment* portrays the future of humankind, as based on scenes described in Dante's *Divine Comedy.*

Michelangelo Buonarroti was born to a Florentine family in the village of Caprese, Italy, on March 6, 1475. He demonstrated great talent as a painter at an early age while apprenticed to Domenico Ghirlandaio. He studied sculpture under the guidance of Bertoldo di Giovanni and was soon commissioned by the foremost family in Florence, the Medici. Following a brief fall from power of the Medici in 1596, Michelangelo fled to Rome. There at the age of 23 he completed the *Pietá* that now stands in Saint Peter's Basilica in Rome. He returned to Florence between 1501 and 1504, where he met Leonardo da Vinci; the two artists were temperamentally very different and did not become friends. During this time he carved the famous statue *David.*

In 1505 Michelangelo again traveled to Rome on invitation from Pope Julius II, a patron of the arts. Michelangelo reluctantly accepted the commission to paint the ceiling of the Sistine Chapel in the Vatican, a project that would require heroic effort and consume years of his life. This highly renowned work includes masterful depictions of the most significant events from the Old and New Testaments of the Bible. Michelangelo also carved the tomb of Julius II, which includes a powerful statue of Moses. He returned to Florence in 1515 to design the Medici Chapel containing the tombs of Medici princes. These tombs include carved allegories representing day, dawn, dusk, and night.

Michelangelo again returned to Rome in 1534 in a period of political turbulence when Florence ceased to be a republic. Commissioned by Pope Paul III, he painted *The Last Judgment,* in which the souls of humankind either ascend to paradise or are cast into the inferno. In 1546 the pope appointed Michelangelo chief architect of Saint Peter's Basilica in Rome, where he began supervising the construction of its dome. Unfortunately, he did not live to see the dome completed. Michelangelo died on February 18, 1564, at the age of 89, and was buried in Florence.

James P. Bonanno

See also Adam, Creation of; Dali, Salvador; Genesis, Book of; Last Judgment; Moses; Noah

Further Readings

Bull, G. (1995). *Michelangelo: A biography.* London: Viking.

Grimm, H. (1896). *Life of Michelangelo* (F. E. Bunnett, Trans.). Boston: Little, Brown.

Hughes, A. (1997). *Michelangelo.* London: Phaidon Press.

Migrations

Migration can be defined as a total or partial change of location (habitat) and/or movement into new areas for a certain period of time or forever. The term *migration* is widely applied in social and human sciences as well as in biology, geophysics, astronomy, and computer sciences with reference to plants and animals, fish and birds, insects and cells, planets, systems, and data.

In contemporary social and human sciences, migration interpreted as population displacement (translocation) usually is viewed as one of the four basic genres of human activity along with habitation, storage, and creation. The term *migration* is an integral part of the professional terminology of contemporary archaeologists, ethnologists, sociologists, demographers, cultural anthropologists, and geographers. The sphere of its application and meaning seem to be so clear that some reference books consider definition and interpretation of this concept unnecessary.

Nevertheless more careful analysis of the application of the term *migration* indicates that this concept is often applied to processes as they vary along spatial and chronological scales, as well as in their ecological, economic, ethnic, and social consequences.

Migration in Social and Human Sciences

The concept of migration applies to studies of displacements of population (group and individual movements) as well as to dispersion of created artifacts and culture in general. Sometimes anthropological and other data indicate that culture transmission does not accompany its human substrate displacement; in such a case migration of ideas is assumed.

The possibility of culture migration was widely discussed in cultural anthropology at the beginning of the 20th century in the context of several schools and theories of diffusion. The notion of cultural diffusion, understood as spatial transference of cultural phenomena, was put forward by them; human history was interpreted as a series of cultural clashes, adoptions, and transfers. Long-distance contacts, such as international trade and exchange, conquests, and conscious imitation were regarded as basic ways by which certain cultural phenomena and/or artifacts could surmount considerable distances from the point of their primary origin.

The origin and rapid upsurge of genetics during the second half of the 20th century provided the possibility of verifying a hypothesis about the translocation of ideas and artifacts without human displacement; this could be done by comparing the genetic makeups of human populations in certain areas. Such DNA spatial distribution studies, brilliantly developed by Luigi Cavalli-Sforza at the beginning of the 21st century, have resulted in a series of gene-flow theories that have brought studies of human migrations taken in historical perspective to a higher level.

Migrations in Human History

Purposeful changes of habitat were inherent to human beings since the very origins of the species. Human dispersion over the Old World in contemporary prehistoric sciences is interpreted as a result of migrations of early hominids and early human beings. Colonization of the New World and the European North happened at the end of the Pleistocene—at the beginning of the Holocene ias the next important example of large-scale migrations in prehistory. Human migrations often are regarded as the driving forces of agricultural dispersion in the Old World as well as the basic mechanism of Indo-European language origin and dissemination.

The origin of nomads and the formation of the nomadic mode of subsistence, strategy, and cultural morphology, as illustrated by Cimmerians, Scythians, Sarmatians, and other nomadic communities of the early Iron Age, is the next phase of human history connected with migrations. The period between 500 and 700 CE, traditionally called the Migration

Period, was characterized by barbarian invasions all over Europe and was marked by the collapse of ancient world empires and a transition to a new, early feudal phase of human history.

Human migrations of medieval times—the Muslim conquest (the Arab migration across Asia, Europe, and Africa between 632 and 732 CE); the Turkic migration across the Middle East, Europe, and Asia (between the 6th and 11th centuries CE); the long-distance migrations of Mongol and Turkic tribes over eastern and central Europe (between the 12th and 14th centuries CE); the Ostiedlung (movement of German tribes to eastern Europe); and others—in most cases were caused by the necessity of modern ethnic and political structures to find their proper places in the early medieval community. Hence these displacements usually are mentioned among the crucial factors of the reshaping of medieval political structures (early and centralized states and empires), often accompanied by the origin of modern nations and the creation of the contemporary ethnopolitical map, which is well traced over Europe even today. These displacements also are associated with the formation of basic principles of modern geopolitics.

The era of discoveries and the Age of Exploration showed most Europeans the impressive economic potential of newly opened and intensively explored territories (e.g., America, Australia, eastern Asia, Siberia), and so the economic aspect of human migration became its dominant motivation.

The modern era has brought new impact into human migrations, connected, first of all, with displays of religious background of human migration (the great Puritan migration from England to North America in the mid-17th century, the great Serb migration from the territory of the Ottoman Empire to the domain of the Habsburg monarchy from the end of 17th to the middle of the 18th centuries). In addition, the migrations of the age of imperialism and industrialization were motivated by the necessity to improve the quality of life through finding new workplaces and trade opportunities (transatlantic migration of Europeans to North and South America, the mass migration of African Americans out of the southern United States during the first half of the 20th century, exploration of eastern and southern Asia).

World War II and the following cold war demonstrated the importance of the political and ideological background of human migrations based on necessity of free self-identification and self-expression.The contemporary trend in the direction and motivation for human migrations resulted from the the fall of the iron curtain, the end of the cold war, and the economic problems of the developing postsocialist and so-called third world countries of Asia and Africa.

Problem of Causes and Motivation, Diversity of Forms and Cultural Consequences

Migrations have been one of the basic forms of human activity since the origin of humankind, and since that time reasons, backgrounds, and motivation of human displacement have continued to change in accordance with priorities and emphases of human culture and the development of various livelihoods. In contemporary social science, the problem of migration's causes and motivation usually is conceptualized through a system of "push and pull" factors mediated by "barriers and obstacles." Among the most influential push factors, one can distinguish among the restriction of subsistence resources (work shortage, low trade opportunities, low income), threats to human life (environmental disasters, poor medical care, insecurity, armed conflicts), the impossibility of free displays of political, ethnic, and religious self-identification, and restrictions on emotional life, family links, and education. Pull factors, in contrast, provide the possibility to overcome the restrictions imposed by push factors, and to change living conditions for the better.

The concept of human migration currently is applied to a rather wide variety of forms of human displacement that differ by their positions in space and time, by their mode of their personification, and by their motivation—whether migrators travel of their own free will or based on coercion.

Personification of Human Migration: Group and Individual Displacements

Defining the factors underlying instances of migration shows the diversity of its forms and genres; here the opposition of the individual versus the group character of migration is especially acute. History shows examples when translocations of individuals of separate families were the first steps of large group migrations, waves of which involved sometimes the population of the whole administrative unit (immigration to the United States). The opposite has also occurred; in the past one can find situations when a first impressive wave of group colonization was followed only by separate individuals or was not supported at all (primary colonization of Australia).

Detection of the social, ethnic, political, religious, economic, or other background of migrant groups' formation currently is at the center of the fields of demography, cultural geography, and cultural anthropology as well as of other social sciences dealing with the causes that force people to leave their dwelling place and search for another, more suitable one. Environmental and geographic parameters of primary living space were the agencies of principal importance in prehistoric and preindustrial societies, while in the contemporary world these have yielded to sociopolitical agencies.

Infiltration is a peculiar form of small-scale group migration that implies movements (often secret ones) of a restricted number of participants within a restricted time and chronological span. Usually infiltrations take place at the stage when a group is choosing its destination. Also this term could refer to attempts of small groups to cross a state border or guarded space illegally (contemporary Palestinian border infiltration).

Human Migration Distance Scale: Colonizations and Relocations

The geographic and spatial dimension of migrations is another factor giving rise to diverse forms of this process as well as to the variety of its sociocultural implications and consequences. One can distinguish long-distance movements (or colonization) and comparatively rapid movements (or relocation).

Colonization is regarded as durable and in most cases long-distance and repeated (wave) movements of human collectives (never individuals) to previously empty (i.e., unsettled by human beings) places. Their earliest examples in human history are the exodus of *Homo habilis* and *Homo erectus* (the first representatives of the genus *Homo*) from Africa, which happened at the beginning of the Pleistocene. Primary colonizations of America and Australia happened at the end of the glacial period and are regarded as the traditional examples of Upper Palaeolithic (i.e., earliest realized by anatomically modern humans) migrations of such a type. The times of glacial retreat and during the Holocene period had comparable results in the colonization of northern Europe: the Baltic region, Scandinavia, Greenland, and so on. Further colonization processes were directed toward mastering free niches in areas already more or less intensively explored.

The general tendency and intensity of colonization processes, as well as their typical undulating, multistage, and multilevel character, opens the possibility of interpreting them as phenomena with not only historical, cultural, ethnic, and economic consequences but also geographic and ecological consequences. In such a context, colonization is one of the basic features of general human evolution that causes a long-lasting effect on the landscape structure of human natural habitats.

Relocations are regarded as relatively strictly defined in time, isolated changes of habitat realized by a group or by individuals with a particular purpose. This category covers the overwhelming majority of all known historical and contemporary migration cases. Even at first glance one can distinguish at least two sorts of relocation: (1) transmigration, or population displacements that significantly enlarge or totally change the habitat of a particular group, and (2) movements of individuals or human groups of a different genre (social, gender, cultural, ethnic, professional, etc.) within their traditional living space.

Transmigration is more likely to be the means of preserving a group's culture in new territory but in the same ecological situation; in the case of oncoming movements, it is obviously one of the forms of ethnic contacts.

The Timescale of Human Migration

In accordance with the period during which human translocation takes place, it is possible to distinguish daily, episodic, seasonal, cyclic, and pendulous migrations and population movements with long-lasting consequences, or irretrievable migrations.

Daily migration implies traveling between residence places and working places, usually marked by the notion of "daily commuting" or "pendulous migrations." Daily migrations could be long distance (especially in hunter-gatherer societies and, sometimes, in contemporary highly urbanized areas and megalopolises), but in most cases transportation time does not exceed working time. This form of migration is practiced by the majority of the contemporary population and currently is showing a gradual but stable tendency of proportional growth.

The notion of *episodic migration* is used to define business, recreation, shopping, and other trips realized from one center of permanent or semipermanent occupation; such translocations usually are irregular and could engage different routes and paths as well as differ in their spatial scales.

Seasonal and cyclic migrations are practiced mainly by the able-bodied population moving to their temporary working places (often seasonal ones). Such translocations usually cover several months and include the possibility of return to migrants' permanent residential place. Cyclic migration recurrence can be related to employment (e.g., military service, seasonal agricultural jobs) or to peculiarities of economic cycles, which, in turn, usually are geographically and/or environmentally determined (as in the cases of horizontal migrations inherent to nomadic pastoral farming and of vertical migrations in other social groups"). Rural-urban and urban-rural migration also could be regarded as a special case of cyclic migration. In the case of cyclic migration, usually only part of a social group is moving, while the rest of the population (mostly women, children, and the aged) remains at the permanent or semipermanent living place.

Migration with long-lasting consequences, sometimes called irrevocable migration, usually involves long-distance movements and is accompanied by total changes of permanent residence of individuals or a certain group as a whole. In most cases such migration is external and stipulates dislocation not only between administrative units, but also between countries. In the case of colonization, irrevocable migration is the only source of permanent residents in newly explored territories. If displacement is directed to already inhabited and actively explored regions, most widespread forms of migration with long-lasting consequences are emigration, or leaving one's native country to live in another, and immigration, or coming to a foreign country to live. Neither of these last forms of migration could be declared irrevocable as far as durability of their consequences depends on people's free will and the coercive actions of authorities in their former and new countries.

Human Migration as an Act of Free Will Versus a Coercive Action

In the contemporary world, migration flows tend to be regulated by separate countries' authorities and are also subject to international legislation. In the context of the political realities and international relations of the second half of the 20th century and the beginning of the 21st century, a series of human migrations resulting from coercive actions of national governments (forced migration) can be distinguished. These include *deportation,* or coercive expulsion imposed on foreign citizens or individuals without citizenship; *repatriation,* or forced return to the home country of citizens who due to different (mostly unfriendly) circumstances have appeared on the territory of other countries; and *expatriation,* or eviction from the native country usually accompanied by denaturalization.

In contrast to migrations resulting from coercive actions of national governments and international institutions, some forms of human migrations could be regarded as acts of migrants' own free will. Thus, *re-emigration,* akin to repatriation, is realized mostly on the basis of conscious choice and maintained by international legislative acts, such as the Geneva Convention on victims of war. Expatriation also could be the conscious act of protest against certain actions of the native country. All forms of *illegal migration* are derived exclusively from the free will of migrants; nevertheless this free will usually has a fundamentally sociopolitical, economic, or religious background. *Chain migration,* widely applied in the contemporary immigration policy of the United States, allows foreign citizens to naturalize in the country on the basis of the acquired citizenship of their adult relative. The idea of migrants' free will expressed in their displacements is the basis of the concept of *free* or *open migration* with its emphasis on people's freedom to move to whatever country they choose as most preferable for their self-realization. *Brain drain,* or emigration of highly qualified professionals in the field of fundamental and applied sciences and hi-tech, could be regarded as a special case of free migration.

Olena V. Smyntyna

See also Anthropology; Archaeology; Ecology; Economics; Evolution, Cultural; Evolution, Social; Harris, Marvin; Shakespeare's Sonnets; White, Leslie A.

Further Readings

Anthony, D. (1990). Migration in archaeology: The baby and the bathwater. *American Anthropologist, 92,* 895–914.

Appleyard, R. T., & Stahl, C. (Eds.). (1988). *International migration today.* Paris: UNESCO; Nedlands, Western Australia: University of Western Australia, Centre for Migration and Development Studies.

Brettell, C. B., & Hollifield, J. (Eds.). (2000). *Migration theory: Talking across disciplines.* London: Routledge.

Cavalli-Sforza, L. L., Menozzi, P., & Piazza, A. (1994). *The history and geography of human genes.* Princeton, NJ: Princeton University Press.

International migration outlook: Annual report. (2007). Paris: Organisation for Economic Co-operation and Development.

Milton, John (1608–1674)

John Milton was an English poet and political critic, celebrated for his epic poem *Paradise Lost* (1667). In it he deals with the biblical account of Creation and the place of humankind in time. Total blindness late in life forced Milton to write this poem and others through dictation.

Milton was a devout Puritan with a deep interest in the Bible. His writing often deals with religious themes. His masterwork, *Paradise Lost,* is an immense poem in 12 books. It describes God's Creation of Adam and Eve in the Garden of Eden and their subsequent fall and expulsion. The poem describes the biblical story of Creation of the earth and the universe as well as of hell and the emergence of the devil (Lucifer). The devil is given a prominent role in Milton's poem, characterized as an overly ambitious angel banished from heaven. The timeless struggle between God and the devil, or good versus evil, is a major theme in the poem. Eve notoriously yields to the devil's temptations, resulting in the loss of paradise for humanity. Milton also argues for the doctrines of predestination and salvation.

John Milton was born in London on December 9, 1608. He attended Christ's College, Cambridge, from which he graduated with a master of arts degree in 1632. He demonstrated an interest and ability in writing poetry while a student. Following graduation from college, Milton retreated to his family's summer home in Horton and devoted six years to private study and poetry. His works from this period include *Comus* (1634), a masque, and *Lycidas* (1637), written to commemorate the death of a close friend. In 1638 Milton left Horton to take a tour of Europe, meeting the astronomer Galileo Galilei in Florence. He returned to England during the civil war to write political pamphlets defending the Puritan cause. These include *The Tenure of Kings and Magistrates* (1649), which argues that the people have the right to remove a tyrant from power. The Puritans won the civil war, and King Charles I was beheaded.

Milton was married three times. The unhappiness of his first marriage led him to write *The Doctrine of Discipline and Divorce* (1643). He wrote *Areopagitica* the following year in defense of the freedom of speech. The government of Oliver Cromwell appointed Milton to the post of secretary of foreign tongues, where he oversaw the translation of dispatches into Latin. In 1652 Milton suffered a permanent loss of his eyesight. Thereafter his work had to be dictated to an assistant. In spite of this challenge, Milton's final years were highly productive. He published *Samson Agonistes* and *Paradise Regained* in 1671. *Paradise Regained* deals with the future of humankind. These and Milton's other works have had a lasting impact on world literature and have influenced artists and writers for more than 400 years.

James P. Bonanno

See also Alighieri, Dante; Bible and Time; Genesis, Book of; God as Creator; Last Judgment; Poetry; Satan and Time

Further Readings

Levi, P. (1997). *Eden renewed: The public and private life of John Milton.* New York: St. Martin's Press.

Lewalski, B. K. (2000). *The life of John Milton: A critical biography.* Oxford, UK: Blackwell.

Parker, W. R. (1968). *Milton: A biography.* Oxford, UK: Clarendon Press.

Moon, Age of

The moon, a natural satellite of the earth, was formed at approximately the same time as the earth, roughly 4.6 billion years ago. The origin of the moon remains the subject of debate; differing theories have been advanced, each with its merits and defects. Scientists have created a lunar geological timescale and used radiometric dating methods to aid in estimating the age of the moon.

Theories explaining the formation of the moon include the fission theory, capture theory, cocreation, and collision-ejection theory. The fission theory states the moon was a piece of the earth that broke away in an early stage of the planet's development. The circumstances for this event to have occurred are considered somewhat implausible.

The capture theory states that the moon was an asteroid, or another similar spatial entity, that was pulled into orbit by the gravitational field of the earth. This theory is improbable given the size of the moon and the gravitational force needed to trap it.

The cocreation theory states the moon accumulated from the debris orbiting the earth as it, too, was forming. This theory succeeds in explaining the similar ages of the two bodies, but it fails to explain the difference in composition between the Earth and the moon.

The collision-ejection theory is the favored theory; it combines aspects of the different theories into one unified theory. It theorizes that an asteroid, possibly the size of Mars, collided with the earth, ejecting debris from both bodies and merging the fragments into one. The fragments fused over time while captured in an orbit around the earth. This theory is fairly consistent with the known facts about the moon.

The moon has its own geologic timescale consisting of five periods. Unlike the earth's geologic timescale, there are no subdivisions of these periods due to the insufficient amount of information available. Two of the major reasons for this lack of information are the limited accessibility of the moon and the nonexistence of fossils, which helped to detail the geological timescale of the Earth. A lot of what we know about the moon is through distant observations and rock samples returned from numerous lunar missions.

Using relative and radiometric dating methods, scientists have been able to determine that the moon sustained a period of bombardments after the surface solidified. This accounts for the vast number of craters across the moon's highlands. Toward the end of this period, known as the Nectarian period, asteroids 100 kilometers in diameter collided with the moon, leaving enormous craters. These craters were later filled with basaltic lava during the next 500 million years, also known as the Imbrian period. Early earthbound observers believed these lava-filled craters, long since solidified, to be large bodies of water and dubbed them *maria,* or seas.

Scientists tested the rock samples from lunar missions using radiometric dating methods. The age of these rocks varies between samples from the older highlands and the younger maria. The samples range from 3 to 4.5 billion years old. The older rocks confirm that the moon was formed when the earth was in a primordial state.

Mat T. Wilson

See also Earth, Age of; Eclipses; Moon, Phases of; Nebular Hypothesis; Satellites, Artificial and Natural; Universe, Age of

Further Readings

Dalrymple, G. B. (2004). *Ancient earth, ancient sky: The age of earth and its cosmic surroundings.* Palo Alto, CA: Stanford University Press.

Nicolson, I. (1999). *Unfolding our universe.* Cambridge, UK: Cambridge University Press.

Moon, Phases of

The phases of the moon have played an important role in the determination of time. The word *month* is directly derived from the word *moon.*

Traditionally, one month was equivalent to one revolution of the moon around the earth. As the moon orbits the earth, the visible reflective portion (called the lunar disk) seen from the earth varies in predictable patterns. At the beginning of the cycle, the moon is a very thin crescent that increases in size within hours. Because the full cycle of the moon is not exactly 28 days, a day-length based on the rising and setting of the moon results in inconsistent solar positions during the daytime.

Lunar Calendars

A lunar calendar is divided into even periods between phases. In using a lunar calendar, for example, the solstices and equinoxes are never the same. This calendar, utilized by the Greeks, was devised by the philosopher Metones of Athens and bears his name as the Metonic cycle. The Metonic cycle is equivalent to 19 solar years or approximately 235 lunar months. The lunar calendar was eventually abandoned, as it never remained evenly synchronized with the annual solar cycle. The Romans devised their own solar calendar to replace the Metonic calendar.

Lunar Phases

Lunar phasing is the visible change in the daily appearance as the moon orbits the earth. There are four major phases: new, first quarter, full, and third quarter. Additionally, the moon passes through phases as it rotates on its own axis. In modern times, the new moon occurs when the moon is positioned relatively between the sun and the earth. Because the moon orbits the earth at an inclined angle with respect to the earth's revolution around the sun, the moon is not always in a direct line between the sun and the earth (see Figure 1). Occasionally, a solar eclipse occurs when the moon does get into direct position, and the moon's shadow darkens the daytime skies of earthbound observers. At the new moon, the lunar age is zero days. Lunar age is defined as the number of days from the last new moon.

New Moon to First Quarter

As the cycle progresses, the lunar disk continues to gain luminescence (wax), and a crescent shape appears to the right of the lunar face. This stage is known as the waxing crescent and continues for approximately 7 days. To the observer, the moon shifts approximately 12° per day, an effect that causes the moon to rise 48 minutes earlier each day as it lengthens an apparent gap between itself and the sun. Occasionally, when atmospheric conditions are met, the darkened portion of the lunar disk can be seen as a very dull, gray feature in contrast to the illuminated area. This effect, called earthshine, is caused by the reflection of direct sunlight from the earth projected onto the moon (see Figure 2). Earthshine is best viewed when local weather conditions are clear, and the relative humidity is low. Another factor of waxing earthshine magnitude is cloud cover over a location westward of the observer. Clouds reflect more sunlight than land and water reflect.

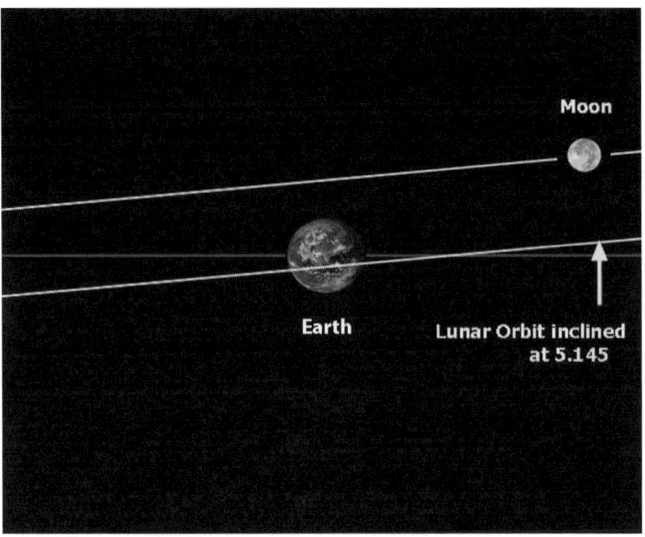

Figure 1 The moon's inclination to the Earth's orbit

Notes: Distances and diameters are not drawn to scale. The visible inclination is drawn to scale.

First Quarter to Full Moon

At the age of 7 days, the moon reaches a relative right angle with the sun and the earth (earth at the apex); the right half of the lunar disk is visible to the terrestrial observer. This phase is the first quarter, and the term quarter refers to two aspects. First, the moon has completed one quarter of its revolutionary period, and second only one quarter of the lunar sphere is visible. At this location, the

Figure 2 Earthshine on the lunar surface

Source: James P. Collins, reproduced with permission.

Note: The lunar cycle in this photo is 2.5 days old.

moon is exactly 90° east of the sun. Once the moment of the first quarter has passed, the moon continues to wax; yet, the crescent shape appears on the left, unlit portion of the lunar disk. This stage is a gibbous stage; therefore, the term *waxing gibbous* is used to describe the luminescent increase from the first quarter. This process continues for 7 days, after which time the moon gets to the halfway point in its revolution.

Full Moon to Third Quarter

The full moon occurs when the entire lunar disk is illuminated at an age of 14 days. Although it is not widely known, the full moon is technically considered second quarter phase. Throughout history, we have applied the term *full moon* to the visible portion of the lunar disk. However, this is a misnomer, as in reality the moon is a sphere, not a disk, so only half the lunar sphere is truly reflecting the sun's light. At this point, given the right circumstances on the inclined lunar orbit, we can see the moon passing into the shadow of the earth, resulting in a lunar eclipse. Lunar eclipses are much more common than solar eclipses due to the earth casting a bigger shadow over the moon than the moon does over the earth.

From the moment of the full moon, lunar disk illumination begins to recede (wane). Marking the second half of its revolutionary motion, the crescent face of the moon is again on the darkened side; however, the darkened portion shifts to the right side of the moon. This stage is termed *waning gibbous.* During the current lunar cycle, 21 days has elapsed.

Third Quarter to New Moon

The moon forms another right angle with the earth and sun and has now reached the third (last) quarter phase. From earth, the left side of the moon is illuminated and is exactly 90° west of the solar position. As lunar disk recession proceeds, the crescent shape returns to the left side of the lunar disk, and the stage is called waning crescent. During this stage, earthshine is prominent again; however, the magnitude of the projection depends on what is eastward of the observer's location. Figure 3 depicts the entire lunar cycle from start to finish. As it completes the nearly 28-day cycle, the moon is visible in the early morning until it returns to its starting position between the earth and sun, at which point none of the lunar disk is visible, as it is lost in solar glare. The new moon, under the right conditions, will provide a solar eclipse on earth.

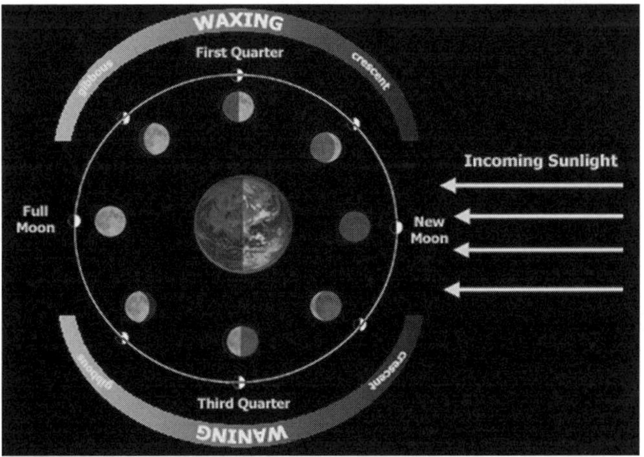

Figure 3 Complete lunar phasing diagram

Notes: As the moon orbits the earth, the sun illuminates half of the lunar sphere. Inside the lunar orbit is a photograph of the moon as seen on the observer's local meridian.

Period Lengths

The lunar sidereal period is one complete lunar cycle. It is equal to 27.321661 days, or the time needed for the moon to revolve around the earth. However, as the moon moves around the earth, the earth is moving around the sun. Observationally, it takes time for the moon to return to the local meridian over its orbital period around the earth. The earth has proceeded in its own orbit, and as a result, the terrestrial observer's moon has not quite reached the same location. When it does, it will have completed one lunar synodic period, which is equivalent to 29.53 days.

During the course of a lunar cycle, the positions of the craters tend to change slightly. Lunar libration is the apparent wobble of the moon as it revolves around the earth caused by the difference in inclination between the earth and moon, the shape of the moon's orbit, and the effect of the earth rotating, creating a parallax shift from the observer's perspective. Although the moon's daily rotation is locked to the terrestrial rotation, these factors result in approximately 10% more surface feature visibility. Time-lapse photography of all the lunar phases results in a slight shift in the daily positioning of lunar features such as craters, mountain ranges, and lunar maria or seas.

Timothy D. Collins

See also Calendar, Gregorian; Calendar, Julian; Earth, Revolution of; Eclipses; Moon, Age of; Seasons, Change of; Time, Measurements of

Further Readings

Cadogan, P. H. (1981). *The moon.* New York: Cambridge University Press.

Guest, J. (1971). *The earth and its satellite.* London: Hart-Davis.

Karttunen, H., Kroeger, P., Oja, H., Pouanen, M., & Donner, K. J. (2003). *Fundamental astronomy* (4th ed.). Berlin: Springer.

Morality

Broadly considered, morality encompasses all beliefs and practices associated with our concerns about humans and human conduct, along with all that we conceive or do to improve ourselves or others, all that we conceive or do to improve the world, and all that we conceive or do under the feeling that we must, should, or ought to, for any reason. It specifically includes all that we conceive and do, or don't do, either for or against what we think of as base, bad, evil, wicked, wrong, immoral, unjust, or cruel, by any name, whether in ourselves or in the world. It also includes our assessments of humans and their effects on the world, including assessments of human actions and human efforts, whether in gardening, governing, writing, dance, or sport. At this broadest level of description, almost all thought either is, or is importantly impacted by, moral thought. Given its emphasis on controlled or directed change, morality is implicated in every aspect, or almost every aspect, of our thinking about time.

Theory and Morality

Because it is otherwise too broad and unwieldy, most theorists restrict their conception of morality so that it encompasses only traditional practices and beliefs that are specifically concerned with human conduct toward persons. But even under this severe limitation, it still names a feature so general and pervasive that one cannot help but think of morality as fundamental to our lives, for nearly every greeting, parting, and discussion exhibits easily identifiable, traditional practices concerned with the well being of persons.

In the most remote reaches of theory, traditional expectations about conduct toward persons are encountered at every level. For example, in the sciences, where we expect to help one another find true statements, we also expect to be able to trust researchers to be honest about their findings. Intentional violations of these expectations about the conduct of colleagues have traditionally been severely dealt with. These dimensions of science as a moral enterprise are complemented by another, in which science is motivated by moral beliefs, such as the belief that science benefits humanity, or that its practice improves one's character. In the austere world of epistemology, morality is found in an ethics of belief formation, which proposes that it is wrong to form beliefs on the basis of inadequate evidence, not because the beliefs formed that way will be unreliable but because forming them that way is unethical.

Morality is also found in questions about the normativity, that is, the compulsory quality, of rules, rule-bound inference, and rational thought. Attempts to reduce epistemology to morality continue to receive many a hearing. Ethics, the branch of philosophy concerned with morality, has often been aided in its studies by metaphysics, the branch concerned with descriptions of reality. Together, they have produced centuries-long discussions about the nature and conceptual status of the self, obligation, value, evaluative judgment, pain, pleasure, envy, sympathy, privacy, freedom, conscience, justice, rights, acts, sin, redemption, and dozens of other morally charged items.

Questions about the spatial and temporal standing of these items are perennial sources of philosophical and theological lucubration. Is obligation extended in space? How, if at all, is it extended in time? Is the will temporally limited? Is sympathy? Is justice? The aging self and responsibility pose many a conundrum. At what point, if any, am I no longer liable for any one of the most minor moral lapses of my youth? And in what sense, if any, could I be responsible already for something I do in the future? If God is omniscient, and thus already knows what I will do tomorrow and every day after that, then am I really free? Doesn't an omniscient God entail a universe without free will and therefore without moral responsibility?

Among the more daunting problems arising for moral theorists who think about time is the problem of moral change. It is challenging merely to say what moral change is a change of. Shall we think in terms of changes in an individual's moral thinking, or that of a society? In both of these directions further questions await. Is it an alternation in conception? Rules? Judgment? Sentiment? Brain structure?

Consider a child who has moved from tolerance for the dismemberment of live grasshoppers to intolerance for it. An answer to the question of what has changed will determine how we explain it. If we think that he has altered his judgment, we might then imagine that he has learned a new rule, so that the set of moral rules he understands has grown by one. We might then hope to explain that acquisition environmentally. Perhaps an adult told him it was wrong or cruel. However, he might have long understood the rule that it is wrong to harm small creatures, but only recently began to conceive of grasshoppers as small rather than large creatures. In that case, his rule set remains the same, and the alteration of his judgment is based on a new way of classifying creatures. Thus, what needs to be explained is not a rule acquisition, but a new application of an old concept.

However, it is possible that his conceptions and rules have remained the same and the change we see is something else, such as a change in sentiment or taste. In that case, we might seek to explain his new attitude in developmental terms. Perhaps he has acquired sympathy for grasshoppers, or grown averse to the sight of struggling animals. This kind of change might then be explained in physiological terms, so that the child appears to acquire new attitudes because his brain is becoming more sophisticated. In this scenario, one will be tempted to conclude that moral changes in the child are not changes of judgment per se, but changes in something prior to and more fundamental than judgments that are in turn reflected in altered judgments.

The development of moral understanding in the individual might be studied by psychologists or biologists, while philosophers and historians are more likely to take an interest in social and cultural levels of description. At this more general level, the difficulties found in the individual case are multiplied and complicated, because a change at this level might involve more factors.

Moral Change

One may tackle the problem of collective moral change by tying the object of study to traditional judgments. At any given point in human history, there is a prevailing or traditional evaluation of typical, widely recognized practices, such as war, slavery, usury, monogamy, homosexuality, and cremation. At any other significantly distant point in human history, the prevailing traditional views of these same practices are likely to be very different as compared to those at the first. When the difference is plainly due to geography, or to a long temporal distance, we can speak of differing moral traditions. We can then restrict ourselves to speaking of moral change only when the difference is found in the same tradition at different times, and we can say that changes in prevailing traditional views are the moral changes we seek to study and explain. Although this is a simple and elegant first move, many difficulties await.

Consider a tradition that formerly disapproved of cremation but now shows a wide tolerance for it. Attempting to explain this change in traditional views will bring our system of thinking face to face with the fact that, as in the earlier case, these differing judgments might reflect a change in something more basic than judgments, such as rules, concepts, or sentiment. In addition, it could be that the alteration in reported judgments reflects a change in social or political conditions, such as an influx of cremation-favoring immigrants or the rise of a cremation-favoring regime. In the first case, demographics can account for the change in prevailing attitudes, while in the second, what we are proposing to treat as moral judgments might be nothing of the sort, because people might privately hold traditional anticremation views while tactfully compromising with the new authorities by publicly expressing only cremation tolerance. This possibility illustrates a difficulty in identifying moral views for study—when does a publicly expressed moral judgment express a moral belief, rather than tact, supererogatory good manners, or a fleeting taste?

Absolutism and Relativism

One means of avoiding all of these problems simply denies that there is any such thing as moral change. Moral absolutism is the thesis that there is only one moral standard that is universal, eternal, and static. Though the standard may remain undiscovered, it is not subject to change. Thus, for example, if slavery is in fact immoral, then it was always immoral, for all persons and at all times, regardless of the fact that it has been widely practiced and tolerated for many centuries and in many lands. For the absolutist, theorists of moral change study only the alterations of fickle human judgment, not morality as it is. That, they believe, never changes, and the problem of moral change is no problem at all, but a mistaken intellectual pursuit.

Ethical relativism, on the other hand, holds that moral standards depend on time and place and thus are subject to change as circumstances change. Morality as the absolutists describe it does not exist. What exists are human contrivances—rules, laws, agreements, conventions, knowledge, and expectations. For the relativist, the existence of morality implies a specific time and place. For the absolutist, it implies no particular human location at all.

Cognitivism and Noncognitivism

If all problems having to do with setting up the study of moral change could be handled, explanations of moral change would encounter further difficulties from the fact that moral beliefs, whether at the individual or cultural level, can be understood in either of two ways that are deeply incompatible.

One way emphasizes rational thought, while the other emphasizes nonrational, historical, or causal processes. The former method implies cognitivism, the view that moral judgments are the kinds of statements that can be either true or false, while the latter goes in the direction of noncognitivism, in which moral judgments, such as "cremation is an acceptable practice," are the kind of statements that cannot be either true or false. If moral beliefs are cognitive, then a decision to break with a traditional view in favor of a nontraditional one can, at least sometimes, be taken because the nontraditional view is or seems true, more plausible, more likely, or more probable on some moral ground. For example, duty might require that cremation be tolerated when duty is conceived as doing what is best for the health of the community, and cremation is discovered to be healthier than its alternatives. Or, where there is oppression of cremation supporters, a sense of justice might demand a more tolerant view of the practice wherever justice is taken to imply equal rights.

Causal and Normative Theories

If they are noncognitive, moral beliefs are held due to causal necessity and not for moral (or any other kind of) reasons. Many theorists account for moral change causally. The Marxist looks for economic forces to explain moral changes, the Freudian looks for psychosexual causes, and the postmodernist seeks underlying hegemonic powers. But all agree that moral beliefs are not held on the basis of moral insight, moral judgment, or morally based rational decisions of any kind.

In contrast to these causal theorists, normative theorists hold that moral changes are, at least in some cases, morally justifiable. If there are justifiable moral changes, it would be because the new attitudes reflect better normative insights or better moral reasoning or cognition. In holding that some moral changes can be morally justified, these views endorse the basic elements of a belief in the possibility of moral progress.

Normative theorists often search for rational grounds for holding to views different from prevailing or traditional moral views. The traditional view is likely to be held by most of the persons native to a given place and time. It is usually beneficial to hold the prevailing view, especially when we have no interests at stake. Nevertheless, rational insights might be capable of creating the conditions for disinterestedly accepting nontraditional moral beliefs. To begin with, one can become aware that norms in one area violate norms accepted in another area, rendering the tradition, which can be thought of as a body of practical knowledge, internally inconsistent. Perhaps the realization that abuse of pets is a felony, while the abuse of livestock is not, creates discomfort and prompts questions.

Other Conditions

In addition to inconsistency, general rules might prompt alternative moral views. For example, 19th-century Quakers strove for moral purity and clear conscience. On this basis they refused slave ownership due to its high likelihood of polluting the soul with unnecessary distresses of conscience. Meanwhile, knowledge of history and alternative traditions offer abundant examples of moral attitudes that contrast with those of our local tradition and can form a basis for criticism of it. A process based in this kind of reasoning appears, for example, when the ancient world's tolerance for homosexuality is employed to criticize less tolerant contemporary attitudes, or when attitudes toward a practice, such as prostitution, in a foreign tradition are appealed to in criticizing local attitudes. In the same way, knowledge of the Quaker attitude toward slavery could supply a moral example for non-Quaker critics of the practice.

Rationally justifiable alterations in traditional moral views might also be caused in part by a change in social conditions. John Langbein has argued that moral attitudes toward judicial torture evolved rapidly toward intolerance only after confessions ceased to be requirements for convictions in the most serious cases. Until this change in the law, moral criticisms of torture could gain little public footing. More recently, Lynn Avery Hunt has argued that the 18th-century appearance of the epistolary novel and the accompanying rise of portraiture allowed large numbers of people to grasp and to empathize with the subjective experiences of people very unlike themselves. On the basis of these new arts, talk of inalienable rights was able to move from the rarefied air of Enlightenment philosophy into the streets and, just as important, into the homes of the powerful.

Bryan Finken

See also Aristotle; Causality; Change; Cognition; Epistemology; Ethics; Evolution, Social; Globalization; God and Time; Humanism; Kant, Immanuel; Marx, Karl; Metaphysics; Postmodernism; Progress; Values and Time; Zeitgeist

Further Readings

Crane, R. S. (1934). Suggestions toward a genealogy of the "man of feeling." *English Literary History, 1,* 205–230.

Hunt, L. A. (2007). *Inventing human rights: A history.* New York: Norton.

Joyce, R. (2006). *The evolution of morality.* Cambridge: MIT Press.

Langbein, J. (1977). *Torture and the law of proof: Europe and England in the ancient regime.* Chicago: University of Chicago Press.

Noonan, J. (2006). *A church that can and cannot change: The development of Catholic moral teaching.* Notre Dame, IN: Notre Dame University Press.

Roberts, R. C., & Wood, W. J. (2007). *Intellectual virtues: An essay in regulative epistemology.* Oxford, UK: Oxford University Press.

Singer, P. (2002). *One world: The ethics of globalization.* New Haven, CT: Yale University Press.

Wallace, J. D. (1996). *Ethical norms, particular cases.* Ithaca, NY: Cornell University Press.

More, Saint Thomas (c. 1477–1535)

Thomas More, in former times also called Thomas Morus, was an English statesman, attorney, and author of humanism. More served as lord chancellor from 1529 to 1532. He was executed on the king's orders for refusing to accept the secession of the English Church under King Henry VIII from the Catholic Church of the papal Holy See. Today More is seen as an ideal of steadfastness in defense of morality and conviction, especially among Roman Catholics. More was declared Saint Thomas More in 1935; Pope John Paul II made him the "heavenly Patron of Statesmen and Politicians" in the year 2000. He is considered founder of the literary school of Utopia, picturing a vision of an ideal society.

More was born on February 7, 1477 or 1478, in London, the eldest son of the eminent and prosperous judge Sir John More. After attending St. Anthony's School and spending some time as a scholar with the lord chancellor and archbishop of Canterbury Cardinal Morton, More entered the University of Oxford in 1492. There he studied history, logic, and the classical languages. In 1494 More left the university to study law in London, where he became a lawyer in 1501 and was elected a member of parliament in 1504.

The following year More married Jane Colt, with whom he had four children. Upon the death of his wife in 1511 he married again.

By 1510 More had become an influential public servant in the city of London. In 1516 he became a diplomat of the crown for a short time; then, in 1521, he was knighted and appointed subtreasurer of the court. In 1523, More became speaker of the House of Commons according to the will of the influential confidant of the king, Cardinal Wolsey. In 1529 More was appointed the first layman in the history of England to become lord chancellor. Being known for refusing any Protestant influence in his new function, he also soon acquired a reputation as a persecutor of so-called heretics.

Starting around 1530 More ran into deep conflict with King Henry VIII, triggered by the latter's attempt to cancel his marriage with Catherine of Aragon and the king's later self-proclamation as the "Supreme Head of the Church," culminating eventually in More's absence on the occasion of coronation of the succeeding queen, Anne Boleyn. In 1532 the king finally accepted More's second request to resign as lord chancellor. Then, in a campaign against him driven amongst others by his successor Thomas Cromwell, More was charged several times and for different arbitrary reasons, imprisoned in the Tower of London, and deprived of any income and land. On July 6, 1535, five days after his sentence in an unjust trial for high treason, More was beheaded.

Literary Work and Influence

Notwithstanding his manifold duties, More left a rich humanistic literary legacy; as an author he was influenced by his correspondence with his friend Erasmus of Rotterdam. More's primary literary achievement is *Utopia* (1516), a thought experiment and a humoristic critique of the zeitgeist in which More flings off the restraints of his own times and describes an imaginary, ideal, tolerant, humanistic society. More thus became the originator of a new literary style and a mode of dealing with time and present circumstances that would exert considerable influence on thinkers in the centuries to come. *Utopia* is said to even have influenced Karl Marx, despite the fact that the communism More pictured was religious.

Furthermore his *History of King Richard III* (1518) attracted attention not only as a dispraise of Richard III but also as a cryptic critique of the royal totalitarianism of the ruling Tudors. This work exerted an influence on William Shakespeare's play *King Richard III*.

Thomas More was beatified in 1886, was canonized in 1935, and received the title "heavenly Patron of Statesmen and Politicians" in the year 2000. Especially his canonization is thought to have been intended by the Vatican to form a strong warning against the inhuman and un-Christian conduct of German National Socialists as well as of Soviet Communism. Thomas More, who preferred to die rather than betray his firm ideals, still serves as a symbol of humanity and a just societal order.

Matthias S. Hauser

See also Christianity; Humanism; Marx, Karl; Machiavelli, Niccolò; Time, Sacred; Utopia and Dystopia

Further Readings

Ackroyd, P. (1999). *The life of Thomas More.* New York: Anchor Books.

Marius, R. C. (1984). *Thomas More: A biography.* New York: Knopf.

Morgan, Lewis Henry (1818–1881)

Although Lewis Henry Morgan (1818–1881) was a successful New York attorney in the mid-1800s, he is remembered as a pioneer in the fledgling science of cultural anthropology. Morgan focused his study on the Iroquois Native American tribe, particularly the Seneca. His objective observations and chronicling of their daily cultural routines became the basis for the study of ethnology as we know it today. He is also credited with applying the idea of evolution to cultural behavior. Accordingly, some have called him the father of American anthropology. Scientists following Morgan's model have recorded the traditions of peoples that have since been acculturated to modern society, and their works have preserved these peoples' customs for future generations to study.

Morgan was born on November 21, 1818, in Aurora, New York. Little was recorded about his personal life. He attended Union College and earned his degree in law in 1840. Lewis married Mary Elizabeth Steele in 1851. His legal practice was very lucrative, and over the span of his career Morgan amassed a small fortune. He was greatly respected among his peers and served in the New York State Assembly and Senate.

Morgan's wealth allowed him to pursue his interests outside of the courtroom. One of his law clients was Ely S. Parker, a prominent Native American engineer and member of the Seneca tribe. Morgan's relationship with Parker inspired him to learn more about the native culture of the region. Setting out to learn firsthand about the customs and traditions of the Iroquois, he spent years immersing himself in the culture of the Seneca people. In 1847 he published his observations in a series in the *American Review* entitled "Letters on the Iroquois." In 1851 he released a longer work, *The League of the Iroquois,* explaining the social and political structures of this tribe. This is considered one of the first ethnographies, or scientific descriptions, of an ethnic group.

These studies led Morgan to think about the social structures of all societies and peoples around the world. He took a broad approach in examining the kinship systems and clan organization of many different cultures. Such a scientific approach to culture and behavior was a new concept in the 19th century. Morgan wasn't interested in individuals or societies in the ancient past but in the interactions of people in his contemporary world. He was also curious about how culture changed through time. As were many of his contemporaries, he was influenced by Darwin's theory of biological evolution and tried to apply it to cultural behaviors.

Morgan theorized that culture progressed unilinearly through time and that there have been three stages in cultural evolution; savagery, barbarism, and civilization. He linked technological progress with social change and published these ideas in *Ancient Society* in 1877. For better or for worse, he is often associated with the movement of social Darwinism, although he seemed to support equality for all ethnicities.

As did members of the legal community, his fellow scientists held Morgan in high regard. In 1879 he was elected president of the American Association for the Advancement of Science. He became an official adopted member of the Iroquois tribe and was given the name *Tayadaowuhkuh,* meaning *bridging the gap.* Morgan died on December 17, 1881.

Jessica M. Masciello

See also Anthropology; Evolution, Cultural; Evolution, Social; Harris, Marvin; Materialism; Time, Prehistoric; Tylor, Edward Burnett; White, Leslie A.

Further Readings

Moore, J. D. (2004). *Visions of culture: An introduction to anthropological theories and theorists* (2nd ed.). Walnut Creek, CA: AltaMira.

Morgan, L. H. (1877). *Ancient society.* London: Macmillan.

Mortality

Mortality is defined as "the quality of being mortal or subject to death." Death is the permanent end or ceasing of the life of a biological organism. A common thread that links all living creatures is the fact that someday they will die. Unlike their fellow creatures, human beings are aware that they are mortal and have devoted much of human history to the pursuit of overcoming this mortality.

One of the favorite themes throughout history in religion, literature, and art has been the theme of immortality or the escape from death and the extension of one's time. The belief in life after death is based on the notion that some part of the human person, usually described as the soul, goes on for an infinite period of time. This is a quest that has produce many stories and is the basis for many systems of belief.

The fascination people have with their own mortality is not a new phenomenon. The literature of the ancient Greeks mentions it frequently. One of the most well-known scenes in ancient Greek philosophy is Socrates' deathbed speech. Most of what we know about Socrates comes from Plato's account of this speech. Stories of the River Styx and Hades illustrate the early Greek ideas of mortality and death.

In the Judeo-Christian tradition, the Hebrew scriptures provide insight into the concept of mortality as understood by the ancient Hebrews, whose focus was very much on this world; little is said about the notion of an afterlife. A similar outlook underlies many of the beliefs and practices found in modern Judaism. With the advent of Christianity, however, ideas of the afterlife took on greater importance. Islam, which shares the Abrahamic tradition with Judaism and Christianity, also finds the basis of some of its beliefs about an afterlife in these early traditions.

In the Asian religious and philosophical traditions, reincarnation is a concept that takes on major importance; to be reincarnated is seen as a form of prolonging life, but in Hinduism and Buddhism, the cycle of endless reincarnation is understood as a form of bondage that the enlightened hope to escape into a state of nonbeing, thus attaining true immortality.

In popular culture, a continuing fascination with mortality and the human desire to escape the finality of death manifests itself in the current interests in ghosts, mediums, and other psychic phenomena. In an age when tangible evidence is often needed for people to believe in something, some are trying to find tangible evidence to prove the existence of life beyond death.

Science, too, is subject to a concern with mortality and the search for ways to overcome it. Studies in genetics have made some previously fatal diseases treatable. Advances in certain medical technologies have made physical life sustainable in cases where it previously would not have been. This has given rise to the debate over exactly what constitutes life.

In the health care and insurance industries, mortality usually refers to the measurement of life expectancy. Mortality rates, which assess the frequency of death occurring annually in a given population, have declined substantially over the last century in many countries because of progress in science, medicine, and sanitation.

Carol Ellen Kowalik

See also Cryonics; Diseases, Degenerative; Dying and Death; Fertility Cycle; Gerontology; Grim Reaper; Life Cycle; Longevity

Further Readings

Lief, J. L. (2001). *Making friends with death: A Buddhist guide to encountering mortality.* Boston: Shambhala.

Tarlow, S. (1999). *Bereavement and commemoration: An archeology of mortality.* Oxford, UK: Blackwell.

Moses

Moses is the most prominent figure in the Hebrew Bible, or Pentateuch. He is noted for receiving the law from God and for leading the Israelites out of bondage in Egypt to the promised land. Being such a pivotal figure, Moses plays an important role in later Jewish, Christian, and Islamic traditions. Through the centuries, Moses has been a prominent subject in art, literature, and, in modern times, film; even, most recently, as a toy action figure based on the young Moses as depicted in an animated movie.

Moses's life as described in the Bible is fairly straightforward. He was a Hebrew born in Egypt, raised by the pharaoh's daughter, and trained in the Egyptian court. Following an incident in which he killed an Egyptian who was abusing a Hebrew slave, Moses fled for his life toward Midian, where he lived 40 years. Moses received a calling to deliver the enslaved Israelites when he encountered God at a burning bush at Mt. Sinai. Upon returning to Egypt, he had several confrontations with an unnamed pharaoh accompanied by a series of

Iconic figure of Moses holding tablets with the Ten Commandments and a short staff.

Source: Library of Congress, Prints & Photographs Division.

divine signs and increasingly severe plagues. Finally, after the deaths of Egypt's first born, the pharaoh allowed Moses and the Israelites to leave Egypt. After a few setbacks, these people reached Mt. Sinai, where Moses met God on the mountain and received the law. Moses spent the rest of his life (about 40 years) wandering in the wilderness with the Israelites, dealing with their stubbornness and sins. After appointing Joshua to succeed him, Moses died on Mt. Horeb at the edge of the Promised Land.

Postbiblical Jewish sources depict Moses as lawgiver, prophet, priest, inventor, philosopher, holy man, and paradigm of a king. Both Philo and Josephus use the exploits of Moses to convince their audience of the viability of the history of the Jews. Artapanus, a 2nd-century BCE Jewish writer, claims Moses led an Egyptian military campaign against Ethiopia, where he won many great battles. Further, he says, Moses introduced the custom of circumcision to Egypt. Other Jewish writers wrote books containing Mosaic legends, such as the *Testament of Moses* and the *Assumption of Moses*. The book of *Jubilees* recounts the events of Genesis 1 to Exodus 12 and adds additional revelations of God to Moses.

Christian sources emphasize Moses's role as lawgiver and as a significant historical figure. He appears at the transfiguration of Jesus along with Elijah (Matt. 17:3ff.) representing the most significant persons in Jewish history. The writer of Jude adds to the Mosaic tradition by telling of a dispute between the Archangel Michael and Satan over the body of Moses after his death (Jude 9). Islamic sources treat Moses in a manner similar to that of the Hebrew Bible; however, it is in his role as a prophet that he is most often mentioned.

The portrayal of Moses in Roman sources is mixed. Generally, he is rendered as the Jewish lawgiver; however, writers like Manetho, Tacitus, and Quintilian blame Moses for teaching practices that go against civilization. Strabo wrote that Moses taught against making any kind of image of God, asserting that Moses left Egypt because the Egyptians made images of their gods.

Moses has been the frequent subject of western art, appearing in many paintings, sculptures, and stained-glass windows. In 1515, Michelangelo completed a marble statue of a seated Moses wearing horns, a pictorial convention now understood to be based on a mistranslation from the Hebrew phrase meaning "radiated light." This statue is the centerpiece for the tomb of Julius II. Moses has been the subject of several modern biographies and scholarly works, most notably Sigmund Freud's 1937 study *Moses and Monotheism*. Interestingly, Freud acknowledges that he was strongly influenced by Michelangelo's statue, which he viewed with fascination during a visit in Rome. In popular culture, Moses has been depicted in several movies, from Cecil B. DeMille's 1923 silent film *The Ten Commandments* to the 1998 animated feature *Prince of Egypt*. In the public imagination, however, the character of Moses is perhaps best known from the actor Charlton Heston's stoic and commanding portrayal in DeMille's 1956 remake of *The Ten Commandments*.

Terry W. Eddinger

See also Bible and Time; Egypt, Ancient; Genesis, Book of; Judaism; Michelangelo Buonarroti; Noah

Further Readings

Britt, B. (2004). *Rethinking Moses: The narrative eclipse of the text.* New York: T & T Clark.

Chavalas, M. W. (2003). Moses. In D. Alexander & D. W. Baker (Eds.), *Dictionary of the Old Testament: Pentateuch.* Downers Grove, IL: InterVarsity.

MULTIVERSES

The idea of multiple universes, or parallel universes that exist next to our own universe, has been discussed by physicists, cosmologists, and philosophers. Many different approaches and theories have been developed to consider the nature of such universes within a physical framework. If such multiverses do exist beyond our own, then one can easily imagine that the element of time could be very different in other universes.

One important theory is based on the picture of cosmic inflation. The idea of cosmic inflation, introduced by physicist Alan Guth at the beginning of the 1980s, describes a phase of rapid expansion of the universe in its very early stages. The concept behind this scenario is that the tension of the vacuum manifests itself in a repulsive gravitational force, and as a result the universe is blown up like a balloon. In more detail, a transition between different vacuum states, starting from a state with repulsive gravity and passing over to its state today, marks the beginning and end of inflation, while the vacuum energy is being released into a hot fireball of particles. After this, the normal cosmological evolution takes place. This so-called vacuum decay is not simultaneous for the overall universe as a result of quantum fluctuations during the decay process. This means that some regions have not yet reached the end of inflation, while others including our own have, resulting in island universes that can also be pictured as bubbles. This process is called eternal inflation and has been studied in detail by Alex Vilenkin; similar approaches also based on inflation have been proposed and discussed by Andrei Linde.

These island universes can be viewed as multiple universes, each following its own laws with its own time. A single universe has its own timeline, completely independent of any neighboring universe. It is possible to define a general "time" that includes all universes, although there are still problems in making any direct predictions. Such island universes can be completely different from our universe or be similar. An infinite number of such universes may exist, and it is possible that elsewhere our universe is repeated, complete with exact copies of ourselves!

Another possibility to describe and interpret multiverses comes from quantum mechanics (QM). Here, the many-worlds view of Hugh Everett and the so-called Copenhagen interpretation are two heavily discussed topics. In QM, any observable values (e.g., a subatomic particle's position and velocity) cannot be measured simultaneously, as stated in the well-known uncertainty principle of physicist Werner Heisenberg. This view is completely at variance with our common experience. In the quantum mechanical world, however, a physical object like an electron can occupy only states that have defined values (eigenvalues) with a certain probability. This probability is described by a wavefunction. Any measurement influences the system, thereby forcing the electron into one of the defined states (wavefunction collapse). In the many-worlds interpretation, the ensemble of possible states represents different universes. This means, if we measure that the electron is located in state A, it will be in state B in another universe. Note that all possible outcomes happen at the same time. Put most simply, one could say that any measurement leads to a number of universes presented by the different possible states.

The Copenhagen interpretation differs from the many-worlds view. Here, a measurement is taken over all possible universes (with different timelines), where we find ourselves in a universe with a high probability.

Many other theories have been proposed, some perhaps more plausible than others, regarding this fascinating subject. For now, all are necessarily consigned to the realm of speculation until such time as we gain the ability to test them by direct measurements.

Veronika Junk

See also Cosmogony; Cosmology, Inflationary; Histories, Alternative; Time and Universes; Universes, Baby; Worlds, Possible

Further Readings

Deutsch, D. (1997). *The fabric of reality.* New York: Penguin.

Guth, A. H. (1997). *The inflationary universe: The quest for a new theory of cosmic origins.* Cambridge, MA: Helix/Perseus.

Vilenkin, A. (2006). *Many worlds in one: The search for other universes.* New York: Hill and Wang.

Mummies

A mummy is the corpse of a human being or animal whose soft tissue has been preserved by either accidental or intentional exposure to airlessness, chemicals, extreme cold, or very low humidity. The presence of any of these conditions halts the growth of bacteria and fungi that would normally cause decay. Mummification freezes a moment in time, giving scientists a unique look at aspects of a culture that existed centuries or millennia earlier. Some mummies are so well preserved that autopsies could be performed on them. Details of diet, dress, hairstyle, tattooing, and more can be witnessed in their original context.

The English word mummy is derived from the Latin *mumia,* which was borrowed from the Arabic *mumiyyah* (bitumen). The Arabic word was also borrowed, from the Persian word *mumiyá,* which also means bitumen. Because unwrapped mummies had blackened skin, it was believed that bitumen, a black, tarry substance that seeps from cracks in the earth naturally in various locations in the Middle East, was used in embalming procedures by ancient Egyptians. Bitumen was widely used for a variety of medical applications in ancient times, so this was not an unreasonable assumption. However, the discoloration was actually caused by resins, used to prepare the body, that blackened over time.

Spontaneous mummification occurs when natural environmental conditions cause preservation without human intervention. It is rare, because very specific conditions are required. Better-known examples include Otzi the Iceman frozen in an Alpine glacier, bog people dumped in the peat bogs of northern Europe where acid and airlessness preserved soft tissue, the Greenland mummies preserved by cold and dry winds, and mummies from deserts in Chile and Egypt where heat and aridity preserved human and animal remains.

In anthropogenic mummification, humans deliberately halt the process of decay for a purpose. Examples of deliberate mummification have been discovered on every inhabited continent on Earth. While ancient Egyptians are the best known for making mummies, they were not the first to do so. The Chinchorros, a sophisticated culture occupying the northern coast of Chile, were embalming and reassembling their dead 7,000 years ago. Further to the north, the Incas used naturally occurring salts and the cold dry air of the region to preserve their mummy bundles. Human sacrifices left for the gods in caves on Andean mountaintops also were mummified in the cold, dry air. The Aleuts off the coast of Alaska processed bodies and dried them in the open air before placing them in caves. Island natives in the South Pacific smoke-cured their dead, covered them with clay, and displayed them in their villages.

Different cultures practiced mummification for different reasons. Some societies believed a person's spirit remains near after death. Mummification would pacify the spirit, hasten it along on its journey, or keep the spirit near for consultation. The preservation of enemies killed in battle gave status to the victor or filled them with the strength of those that perished. Mummification of leaders in some cultures linked the ruler to the gods and made them immortal. Animals would be mummified for use in rituals or to appease the spirit of prey after a successful hunt.

In most cases, mummification was an expensive endeavor, requiring a significant investment of time, effort, and materials. However varied the reasons for intentional mummification among the cultures of the world, the practice illustrates the power of motivation and the desire to endure through time.

Jill M. Church

See also Afterlife; Anthropology; Egypt, Ancient; Immortality, Personal; Museums; Rameses II;

Further Readings

Aufderheide, A. C. (2003). *The scientific study of mummies.* Cambridge, UK: Cambridge University Press.

Chamberlain, A. T., & Pearson, M. P. (2001). *Earthly remains: The history and science of preserved human bodies.* New York: Oxford University Press.

Cockburn, A., Cockburn, E., & Reyman, T. (1998). *Mummies, disease, and ancient cultures.* Cambridge, UK: Cambridge University Press.

Reid, H. (2001). *In search of the immortals: Mummies, death, and the afterlife.* New York: St. Martin's.

MUSEUMS

All museums can be said to have a common origin in certain institutions that began many centuries ago. Whether identifying when a historic event took place, when a species became extinct, or when an art form was popular, museums share a common theme that underlies their particular mission: namely, understanding changes to human, animal, plant, and inorganic artifacts that occurred through time.

The development of museums occurred over millennia, with the first museums originating in Old World nations nearly 2,300 years ago. The earliest forms, such as the Mouseion of Alexandria (Egypt), were sanctuaries that housed a multitude of collections, including gardens with diverse plant species, remains of unique and common animal species, technological innovations, and miscellaneous writings. These institutions also served as centers for discourse between scholars who sought to unravel intellectual puzzles and educate others. Additional museumlike institutions existed in areas of Asia, Africa, and Europe, but many were simply private gatherings of antiquities, far removed from the complex structure of the museums of today and from the view of the general populace.

Private collections remained a major source of collected antiquities until the spoils of war and discoveries of New World voyages, gathered by explorers and military commanders, made their way to Europe to be housed and displayed in museums developed to serve national interests during the 17th, 18th, and 19th centuries. These institutions, developed as part of universities and as individual establishments, included the Louvre in Paris and the British Museum in London and established the practice of exhibiting botanical, zoological, library, art, and other miscellaneous collections. American museums originated during the 18th and 19th centuries, focusing chiefly on natural and local history. By the early 20th century, these institutions were often supported by funds and collections provided by patrons who traveled the world and then donated antiquities they collected in their travels, including religious paraphernalia, paintings, sculptures, coinage, human remains, and ancient weaponry. Some of the unique items brought to the United States during this time included mummies from Egypt, coins of the Roman Empire, Japanese *katanas,* and paintings created by some of the world's most celebrated artists.

In the mid-20th century, museums collectively started to receive new sources of funding, such as grants from government and private endowments, and began to expand their missions to include more systematic exploration of their own existing collections. While a number of museums had devoted efforts to understanding collections throughout history, it was during the 20th century that a widespread reevaluation and modification of museums' research activities took place. These activities led to a deeper commitment to understanding the development of humanity and specific societies through time as well as an analysis of emerging technologies. It is to all these predecessors that contemporary museums owe their existence, public support, and direction.

Each type of museum has different temporal concentrations integrated into their missions and, consequently, their collections. History museums and historical societies are among the prevalent museum types found today and have one of the most uncomplicated focuses on time. Although history museums and historical societies exist that have a world focus in their mission, most museums of this kind concentrate their collection, research, and exhibit practices on their own locality, whether it be a town, city, state, or province. These institutions seek to acquire artifacts and oral history documenting the history of their respective regions, often collecting materials connected to events or people that significantly influenced the course of that society's history. In recent years, history museums and historical societies have placed particular importance on exhibiting the genealogy of families that most affected a region's history and materials documenting the contributions to the society of different ethnic populations through time.

Art museums, meanwhile, concentrate on the collection of works including paintings, sculptures, and print media that document the different historical styles of art. The foci of exhibits in such institutions may be on art techniques (e.g., Cubism or Impressionism), individual artists (e.g., Michelangelo or Picasso), or regions (e.g., Native American art of the Southwest). Whatever the art form highlighted, exhibits usually emphasize time, whether it be the duration of popularity of an art technique or the life of an influential artist.

Science museums and centers place emphasis on innovative technologies while also exploring scientific discoveries, particularly through the fields of archaeology, astronomy, paleontology, and zoology. Science museums and centers place a special focus on the emergence and evolution of individual animal and plant species, animal orders, planetary bodies, and the universe itself. In regard to technological advances, science museums often assess specific technologies and their impact on society, examining what technologies such innovations replaced, and industry progress through time.

Natural history museums are primarily concerned with organisms and their development over time. Such institutions vary in their exhibits, collections, and research goals, which may include understanding and illustrating life forms both in stasis and within an evolutionary context.

Ultimately, all kinds of museums—whether a history museum collecting art created by local artists, a natural history museum displaying watercolor renderings of birds, or a science museum exploring the evolution of dinosaurs through a virtual exhibit—share common collection goals. Museums also show a tendency to amass artifacts or whole collections outside their mandates. That said, all museums place an emphasis on time in their own research, within their exhibit designs, and when educating visitors to their institutions.

Neil Patrick O'Donnell

See also Anthropology; Archaeology; Egypt, Ancient; Fossils and Artifacts; Hammurabi, Codex of; Mummies; Observatories; Planetariums; Rome, Ancient; Rosetta Stone

Further Readings

Alexander, E. P. (1996). *Museums in motion: An introduction to the history and functions of museums.* Walnut Creek, CA: AltaMira.

Burkholder, J. (2006). Museums. In H. J. Birx (Ed.), *The encyclopedia of anthropology* (Vol. 4, pp. 1647–1650). Thousand Oaks, CA: Sage.

Trigger, B. G. (1993). *A history of archaeological thought.* Cambridge, UK: University Press.

Music

Many scholars regard music as the most important of all temporal arts. In ancient Greece music was called *μουσική τέχνη*—the art of the muses, or music drama, as it consisted in a combination of dance, singing, and instrumental music. The word *music* can still refer to musical genres like opera or music drama. There are, however, scholars who employ the notion "music" to refer solely to instrumental, or absolute music. For two reasons, a wider notion of music will be used in this entry, one that includes not only instrumental but also noninstrumental music. First, operas, musicals, and songs are usually referred to as music. Second, the etymology of the word *music* is such that the word has usually been connected with a broad meaning.

If they wish to define what music is in a short and specific phrase, scholars have to face many problems, as there are always musical pieces that are clearly musical but that do not correspond to the definition given. On the other hand, if they put forward a wider definition, it is usually the case that too many examples are included. Are bird songs music? The safest way to define music is to hold that music is everything that experts such as composers, musicologists, or music critics regard as music.

A central idea about music is that it is a temporal art, which implies that music needs to be performed in order for it to exist. All arts that need to be performed for their realization are temporal arts. Drama is a temporal art based upon a text, dance is one based upon corporal movement, and music is one based upon sounds. In temporal arts, an interplay between objective, intersubjective, and subjective time exists. Objective time refers to the time from God's perspective or from the perspective of an eternal realm of ideas in which musical works might exist. The existence of objective time can certainly be doubted.

Intersubjective time, on the other hand, is the time a watch tells us. This kind of time is based upon an interhuman agreement concerning the duration of intervals and how we calculate the duration of time. Human beings agreed to relate time to contingent natural constants like the period of time taken by the earth turning around the sun or the moon turning around the earth. Intersubjective time can be found on musical scores as *tempo,* which is discussed later in this entry.

Another type of time relevant for temporal arts is subjective time. Subjective time depends on our perception of something. If we perceive a musical piece as boring, instants seem to have a longer duration than when we regard the music as entertaining. Another kind of subjective time is related to the stage of life in which the maker or the listener happens to be. Human interests differ significantly between, say, a teenager and an adult in midlife. Concerning the difference between music makers and listeners, some distinctions need to be presented. In the case of music, there are makers of various orders. The first-order maker is often the composer who is responsible for the score. The score needs to be read and interpreted directly by the makers of the second order, whether musicians or a conductor. If the conductor is on the second level, then on the third order would be musicians. Even though the maker of the first order creates the music, the maker of the third order performs it and enables the audience, the receivers of music, to listen to it.

The following section presents some definitions of musical terms concerning the relationship between music and time, such as tempo, meter, and rhythm. Next follows a description of what the philosophers Plato and Schopenhauer put forward concerning the relationship between words and rhythm. Third is a short summary of the history of rhythm in music, and finally comes a discussion of the importance of time for the concept of the musical work.

Musical Notions and Time

Tempo

The *tempo* (plural form *tempi;* Italian for time; from the Latin word *tempus*) determines the basic pulse of the musical piece. Traditionally the tempi were described as follows (from slow to fast): *grave, largo, larghetto, lento, adagio, andante, andantino, moderato, allegretto, allegro, vivace, vivacissimo, presto, prestissimo.* When adjectives are added, the descriptions of the tempi became more precise, for example, *ma non troppo* (but not too much), *con fuoco* (fiery), and *molto* (much). In addition, there are terms specifying the change of tempo. If the tempo is supposed to become faster, the following expressions can be used: *accelerando, stringendo, piu mosso, poco piu.* If the tempo is supposed to become slower, the following expressions can be used: *poco meno, piu lento, calando, allargando, rallentando, ritardando, ritenuto.* Even though most of the time Italian tempo markings have been used, they can also turn up in French, German, or English.

By using words to specify the tempo of a musical piece, the subjective understanding of the respective notions of musicians, conductors, and the composer of a piece become more relevant. In order to fix the tempi, the metronome was invented. By means of the metronome, musical time is related to intersubjective time so that the tempo of a musical piece is specified by clarifying the amount of beats per minute (BPM). As minutes are defined on the basis of a human decision, the tempo of a musical piece that is specified by means of a description of beats per minute includes a relationship between musical and intersubjective time.

Meter and Rhythm

There are two further terms that are important for the temporal structure of a musical piece: meter and rhythm. The musical term *meter* (*μέτρον*; ancient Greek for *measure*) is used to describe the organization of beats within a regular pattern. One single entity of such a pattern is called a *bar.* A *beat,* on the other hand, is the basic temporal unit of a piece—it is a type of pulse that may or may not be heard.

The specific temporal organization of a musical piece is referred to as *rhythm* (*ῥυθμός* ancient Greek for *flow*). The surface structure of a piece, which is called meter, is more constant and regular than the rhythm. However, meter and rhythm are not independent of one another. There are several theories that try to describe their relationship. It can be the case that both meter and rhythm are something similar, that rhythm is subordinate to meter, as rhythm is a meter come alive or that meter is regulated rhythm, which implies that it is a basic structure necessary for rhythm taking its proper shape.

Rhythm and Words

There are several possibilities for how rhythm can come about. In the history of music philosophy, two positions concerning the relationship between rhythm and words have been dominant. On the one hand, there is Plato's position. He defends the superiority of words over sound. His position had an enormous influence on the Florentine Camerata, whose members invented the opera genre. A similar position was also put forward by the composer Richard Wagner in *Opera and Drama*. On the other hand are thinkers like Immanuel Kant and Arthur Schopenhauer, who defend the superiority of sound over words. Concerning other musical aesthetic questions, such as that of the effect of music, those thinkers disagree significantly. Kant stresses the pleasure music provides listeners, whereas Schopenhauer regards the brief liberation from the personal will as the most important effect of great music. A more permanent kind of salvation can only be reached by means of asceticism, according to Schopenhauer.

According to Plato, a song (*melos*) consists of word (*logos*), harmony (*harmonia*) and rhythm (*rythmos*), whereby the word is supposed to be fundamental; that is, rhythm and harmony are supposed to follow the words, because words specify the content of a musical piece. Without words we would not understand the piece and would not know what it is about. It is important that we know what it is about, because music is supposed to convey the idea of the good, and only if words are at the basis of a song can it fulfill this task.

Even though Schopenhauer's philosophy of art includes many Platonic elements, he disagrees with Plato concerning the relationship of words and melody. According to Plato, words come first and determine harmony and rhythm. According to Schopenhauer, the instrumental sounds come first. The genius composer transcends the personal will and gets an understanding of the will in itself, which can directly be represented in the artistic genre of instrumental music. In order for music to have the highest kind of quality, the laws of instrumental music have to be dominant, and the words simply have to follow those laws as in the case of the composition of an opera. Here the aesthetic laws of sounds are supposed to be responsible for rhythm, melody, and harmony.

History of Rhythm

Rhythm has been dealt with theoretically as well as practically since antiquity. This entry provides a summary of the history of rhythm since the baroque era, because the history of music from that time onward includes most of the music we still listen to today. A significant notion in baroque music is the "monody," which then meant music for one voice which is accompanied with chords. The words of the one voice were supposed to bring about the sounds and with it the rhythm. The monody became popular together with the invention of the music drama by the Florentine Camerata at the end of the 16th century. Instead of simply entertaining the audience, music was also supposed to convey values. The members of the Florentine Camerata thought that music can achieve this best by means of the monody. They were inspired by a reading of ancient texts on philosophy of music, especially Plato. It was their intention to revitalize ancient Greek tragedies. They believed that ancient dramas were sung from beginning to end, a concept that scholars now regard as false. As a consequence, the opera genre was invented. Another important aspect of baroque music is its dancelike character. Hence, the general rhythmical stream was more important than each single beat, which had to be embedded in the general, overall structure.

In classical and romantic music, however, the rhythmic aspect within music became more complex. The tension between musical periods and meter became stronger. The meter of a musical piece represented its body, but the rhythm was an expression of the human spirit. Consequently, the rhythm was supposed to represent and deal with the variety of affects that human beings can have. Music was therefore far less schematic than in the baroque era. A particularly impressive representative of that period is Richard Wagner. He developed the concept of the "infinite melody," whereby rhythm also developed into something undetermined and unlimited.

In contemporary music we find an immense variety of rhythmic concepts. According to Adorno, one can distinguish two traditions within 20th-century music. The first is the avant-garde tradition, which is associated with Schoenberg. Various concepts of musical time turn up in this tradition. Here Luigi Nono has to be mentioned, because some of his scores contain no bars. The second is the neoclassical tradition, to which Stravinsky belongs; it consists mostly of linear music in which the rhythmic element is similar to that of 18th-century music, as often the rhythmical figures remain constant throughout a movement. Bartok is another composer who is representative of this tradition.

One might wish to add a third tradition to which one could refer as the postmodern one. It unites elements from various cultures and plays with these elements. The most important postmodern musical movement is called minimal music, and its leading representatives are Philip Glass and Michael Nyman. It is mostly tonal music based on simple harmonies. Of particular importance is the rhythmic element. Very often a type of polyrhythm is used, which means that various rhythms overlap. Polyrhythmic elements are usually associated with jazz and have their origins in African and Indian music. In minimal music, a simple pattern is repeated very often, whereby only simple variations occur. A piece of minimal music is in many cases constituted by uniting various such patterns and variations. Once a pattern is played at a different speed, phase shifting or phasing occurs.

Many non-Western traditions of music are even more challenging concerning the variety of rhythms. Within the Chinese, Japanese, Indian, and African traditions, irrational and polyrhythmic patterns were developed that have recently had an impact on Western minimal music. In India, the *talas* are of particular interest in this respect, as talas are rhythmical patterns that determine a composition of classical Indian music. The tempo of most talas, which are the basis of their classical compositions, can vary.

Musical Works and Time

In musicology one distinguishes between musical works and other musical pieces. A musical work is determined by means of the following qualities: it is autonomous, original, and unchangeable; was created by one composer who is regarded as its origin; is the center of attention when it is being performed; and was created for eternity.

When musical works first came about is a matter of controversy. Some scholars think that the beginning of the musical work is connected to the first formulation of the concept, which was done in the 16th century. Then Nicolaus Listenius used the phrase *opus perfectum et absolutum*. However, many scholars regard the beginning of the musical work tradition as related to its historical representation, and most scholars agree that around 1800 the tradition of the musical work became particularly strong.

There are various reasons for this position: First, before 1800, music was performed on the occasion for which it was written. From then on, older music was performed again; for example, in 1829, more than 100 years after its first performance, Bach's "St. Mathew Passion" was performed again. Second, before 1800, composers were considered not very important, but the occasion for which a piece was written or the person for whom it was composed was of significance. From 1800 onward, for at least 150 years, musical works were the dominant kind of music. After 1950 music became extremely diverse. Musical pieces were composed that were heteronomous (John Taverner, Arvo Pärt) or involved chance elements (John Cage, Iannis Xenakis), which cannot therefore be regarded as proper musical works. Most composers in the Western tradition before 1800 composed heteronomous music. Their music was created for a particular purpose, such as a coronation or a specific religious or royal celebration. These musical pieces were usually performed solely at the event for which they were made. The composition of a musical piece was, therefore, mostly related to a specific contemporary event in the here and now.

Autonomous music, on the other hand, is usually composed for eternity. Composers are sometimes seen as geniuses who manage to put together eternal music, maybe even by having access to an eternal realm themselves. What makes this music autonomous is that the music comes about via a composer who decides for himself which pieces he wishes to realize and which not. Antonio Vivaldi is an exemplary composer of heteronomous music who clearly did not write musical works; he composed about 300 concerts that all sound fairly similar. When he was alive this was not a problem, because then musical pieces were supposed to represent the type to which they belong in an exemplary matter, which is what his pieces did. A musical work, on the other hand, represents a specific solution to a detailed aesthetic problem. In the 20th century the development went so far that many works represent a type of music in itself.

Another aspect of a musical work is that it is original and unchangeable. After it has been created, it remains identical with itself. Iannis Xenakis, however, composed stochastic music, which means that it involves chance elements. It might include the demand that the audience makes a specific sound when the conductor tells them to do so. As the size of the audience is different at each performance, it is clear that his compositions involve chance elements, which is the reason why his works are not musical works. Magical music includes the Indian *raga,* a series of notes upon which a melody of classical Indian music is founded, and the tala, which knows only some vaguely given rules on the basis of which one has to improvise. As improvised music is not eternally fixed, such musical pieces are by definition not musical works. Until the 19th century, German folk songs (*Volksmusik*) were such as to enable people to have a pleasant or cozy evening together, and they were not written by a single composer. Both are reasons for their not being considered musical works.

The musical work represents a means of uniting objective, intersubjective, and subjective time. The composer grasps eternal music (objective time) and puts it together in a score so that it can be performed (intersubjective time), and listened to (subjective time). In any case, the relationship between music and time is a complex one, and it also has to be stressed that this relationship has been neglected by many modern and postmodern musicologists and philosophers of music.

Stefan Lorenz Sorgner

See also Kant, Immanuel; Nietzsche, Friedrich; Plato; Presocratic Age; Pythagoras of Samos; Schopenhauer, Arthur; Wagner, Richard

Further Readings

Begbie, J. S. (2000). *Theology, music and time.* Cambridge, UK: Cambridge University Press.

Hamilton, A. (2007). *Aesthetics and music.* London: Continuum.

Kramer, J. D. (1988). *The time of music: New meaning, new temporalities, new listening strategies.* New York: Schirmer.

Sorgner, S. L., & Fuerbeth, O. (Eds.). (in press). *Music in German philosophy: An introduction.* Chicago: University of Chicago Press.

Mutations

See DNA

Mysticism

Mysticism is a type of religious experience or altered state of consciousness in which a person senses intimacy or union with the source or ground of ultimate reality. Mystical states or experiences are qualitatively different from normal, everyday consciousness. They can be experienced variously as a vision, an ecstatic state, an emptying or silencing of the self, union with God, or absorption into God.

For the monotheistic Western religions, mysticism exists as a movement or school of thought within the religious tradition. For the Eastern world religions, mysticism is the central practice and goal. Mysticism is also a central aspect of primal religions such as shamanism. Some people without an explicit religious attachment also testify to personal mystical experiences; these include Aldous Huxley, Walt Whitman, Carl Jung, and Simone Weil.

Eastern forms of mysticism are directly related to the concept of time. For example, Hindus believe in an endless cycle of death and rebirth called *samsara.* Through the discipline of yoga, one can obtain *Samadhi,* the highest level of spiritual perfection. Through the resulting union of *Atman* (the essential self) with *Brahman* (that which is truly real), one experiences liberation (*moksa*) from *samsara.*

The Western religions of Judaism, Christianity, and Islam do not share the Eastern view of the reincarnation or transmigration of the soul, so they do not seek release from this cycle into a state of existence outside of space and time. However, many thinkers in these religions view God as existing outside of space and time. Therefore, when devotees experience mystical union with God, they often report sensations of the absence of space and time.

The adjective "mystical" (*mustikos*) was used by Christians from the 2nd century onward, but the noun "mysticism" was first used in French (*la mystique*) in the 17th century. Because it is a relatively new term, attempts to define it have varied greatly. The term can be defined broadly as consciousness of the immediate or direct presence of God (which many have claimed to experience). This broad definition would identify mysticism with spirituality or religion in general. It can also be defined narrowly as a union of the self with God or absorption of the self in the Absolute (which few have claimed to experience).

One of the most notable attempts to describe mysticism was William James's list of four characteristics of mystical experiences. First, they are ineffable: Mystics struggle to put their experience into words. Second, they are noetic: Insights gained from the experience inform a person's knowledge and understanding. Third, they are passive: They are experienced as an undeserved gift. Fourth, they are transient: Mystical experiences usually last for a short period of time. The last characteristic has proved to be less convincing to students of mysticism than the first three.

Religious traditions of mysticism have developed practices and disciplines that enable a person to achieve a state of mystical union. They provide systems of initiation and apprenticeship to inculcate mystical values and disciplines. Central to most mystical systems are the practices of meditation and contemplation, which are distinguished from each other in most traditions. In meditation, a person focuses attention and imagination on a religious idea or image. In contemplation, a person suspends the activity of the body and the thought processes of the mind in order to center the spirit on the presence of God. Meditation would be more closely associated with kataphatic mysticism, which utilizes images in order to experience intimacy with God. Contemplation is associated more closely with apophatic mysticism in which the union of the self with God is experienced as negation or absence. The terms *extrovertive* and *introvertive* mysticism are sometimes used to describe similar phenomena.

The mystical traditions in various religions reveal a diversity of practices and goals but also some similarities. Students of mysticism tend to emphasize either the common core found in all mystical traditions (e.g., Walter Stace) or the irreducible differences among them (e.g., R. C. Zaehner). Hindus practice various forms of yoga through which the conscious and the subconscious are mastered by means of moral, physical, respiratory, and mental discipline. In Zen Buddhism, the practitioner contemplates a nonsensical *koan* in order to free the spirit from the domination of the conscious mind. In Kabbalah, adherents mentally manipulate numbers and images in order to annihilate the ego, detach themselves from the physical world, and experience the presence of God directly. Christian mystics have relied upon contemplative prayer, fasting, solitude, and other forms of asceticism in order to focus the mind and spirit on the presence of God within the person. Sufi Muslims practice fasting, sleep deprivation, vigils, dancing, chanting, and contemplation to achieve annihilation of the self.

Mystical elements were present in the teachings of Plato. Centuries later, these were developed more fully by Plotinus, the founder of Neoplatonism, and his later follower Proclus. Plotinus taught that the soul must lose its present identity in order to find a transcendent self in the One or the First Principle. Neoplatonism was a major influence on mystical traditions in Judaism, Christianity, and Islam.

Mystical elements were present from the very beginnings of Christianity in the teachings and practices of both Jesus and Paul. The first Christian theologian to develop fully a theory of mysticism was Origen in the 3rd century, which led to the development of monasticism in the 4th century. Christian mystics have often spoken of three stages on the way to the vision of God: (1) purgation, which is a purification of the flesh and soul brought about through ascetic disciplines such as prayer, fasting, and almsgiving; (2) illumination, which is an enlightenment of the mind by the Holy Spirit; and (3) contemplation or union, which is an unadulterated awareness of the love of God.

Following John of the Cross, some mystical writers have identified a "dark night of the soul" that occurs before the final stage of union is reached. Notable Christian mystics through the centuries include Evagrius, Gregory of Nyssa, Pseudo-Dionysius, Augustine, Gregory the Great, William of Saint Thierry, Bernard of Clairvaux, Bonaventure, Marguerite Porete, Meister Eckhart, Henry Suso, John Tauler, Jan van Ruusbroec, Richard Rolle, Catherine of Siena, Julian of Norwich, Ignatius of Loyola, Teresa of Avila, John of the Cross, Jeanne Guyon, Emanuel Swedenborg, Thérèse de Lisieux, Thomas Merton, Karl Rahner, and John Main.

Some Christian mystics have offered theological and philosophical speculations on the nature of time. In Augustine's view, time was created by God, and it flows from the future into the present and recedes into nonexistence. One may access fleetingly and fragmentarily the experience of timeless eternity by means of a "rare vision" of enlightenment. These experiences result from withdrawal from the sensory world, an interior movement into the depths of the soul, and a movement above the soul to the vision of God. Meister Eckhart taught that the soul must look outside space and time in order to know God, because God exists outside of space and time. Giordano Bruno's mystical reflections on the heliocentrism of Copernicus led him to propose that both space and time were infinite, without beginning or end. Miguel de Unamuno y Jugo posited that the conflict between reason and the desire for human immortality gives rise to the need for faith in God. Pierre Teilhard de Chardin taught that the perfection of humanity through the evolutionary process will culminate in an Omega Point in the future.

Sufi mystics promote the role of single-minded love in the pursuit of God. Two crucial stages on the Sufi path are fana' and baqa'. *Fana'* ("passing away") is the annihilation or nullification of the ego in the presence of the divine, and *baqa'* ("subsisting") is subsisting in the divine reality, which is all that remains. In Iran, the development of the dervish as a method of achieving mystical trance popularized mysticism among all levels of the population. Some of the notable Sufi mystics are Ja'far as-Sadiq, Rabi'ah, Ibn Mansur al-Hallaj, Abu Yazid al-Bistami, Abu al-Qasim al-Junayd, Abu Hamid al-Ghazali, Ibn al-Farid, Ibn al-'Arabi, and Jalal al-Din Rumi.

Through the centuries, Judaism has produced various forms of mysticism. The earliest form of Jewish mysticism is found in Merkabah literature, written between the 2nd and 10th centuries. It promoted meditation on Ezekiel's vision of the heavenly throne room of God so that the seer might ascend through the various levels of heaven (*hekhalot*) until he arrived at the highest heaven where God dwells. Hasidism originated in eastern Europe in the 12th century. Kabbalah mysticism originated in Spain in the 13th century and was spread to the rest of the Jewish world when the Jews were expelled in 1492. Its most notable document was the *Zohar.* A new Hasidism arose in eastern Europe in the 18th century. Important figures in Jewish mysticism are Abraham ibn Ezra, Moses ben Shem Tov de Leon, Isaac of Acre, Abraham Abulafia, Isaac Luria, Dov Ber, Shne'ur Zalman, Aharon Halevi Horowitz, and Nahman of Bratslav.

Psychologists and scientists have tried to explain the phenomenon of mystical experience from the perspective of their disciplines. Sigmund Freud theorized that mystical experiences were illusions caused by a neurotic desire to recapture the infantile bliss of union with the mother. Carl Jung, who described his own mystical experiences in his autobiography, explained them more positively as encounters of the individual unconscious with the archetypes of the collective unconscious. Some psychologists view them as pathological symptoms of schizophrenia, psychosis, epilepsy, or other psychological and brain disorders, but others report that people who experience mystical states possess higher-than-average levels of psychological health.

Recently, neuroscientists have conducted brain-imaging studies of people undergoing mystical experiences. These studies suggest that repetitive, rhythmic rituals deprive the brain's orientation association area of sensory and cognitive input. As a result the brain would not be able to orient the self in its spatial context or identify the boundaries of the body. The mind experiences these sensations as a spaceless and timeless void.

Gregory L. Linton

See also Augustine of Hippo, Saint; Bruno, Giordano; Eckhart, Meister; God and Time; Kabbalah; Maximus the Confessor, Saint; Nicholas of Cusa (Cusanus); Nirvana; Plato; Plotinus; Sufism; Teilhard de Chardin, Pierre; Time, Sacred; Unamuno y Jugo, Miguel de

Further Readings

d'Aquili, E. G., & Newberg, A. B. (1999). *The mystical mind: Probing the biology of religious experience.* Minneapolis, MN: Fortress.

Ellwood, R. S. (1999). *Mysticism and religion* (2nd ed.). New York: Seven Bridges.

Epstein, P. (1988). *Kabbalah: The way of the Jewish mystic.* Boston: Shambhala.

Idel, M., & McGinn, B. (Eds.). (1996). *Mystical union in Judaism, Christianity, and Islam: An ecumenical dialogue* (2nd ed.). New York: Continuum.

James, W. (2008). *Varieties of religious experience: A study in human nature.* London: Routledge.

McGinn, B. (1991–2005). *The presence of God: A history of Western Christian mysticism* (4 vols.). New York: Crossroad.

Mythology

In an effort to comprehend the whole of human experience and the place of humanity in time, ancient civilizations generated myths. The Romans chronicled time as beginning with birth and culminating in death, with great emphasis placed upon the need to be "good" in order to earn eternal life. Myths, then, are largely religious in origin and function and are in fact the earliest records of history and philosophy. Interestingly, myths have been created throughout the course of human existence and have been rendered timeless because they remain an integral part of the culture that framed them. Myths may be classified as traditional stories that deal with time and eternity, nature, ancestors, heroes and heroines, supernatural beings, and the afterlife that serve as primordial types in a primitive view of the world. Myths appeal to the consciousness of a people by embodying their cultural ideals or by giving expression to deep and commonly felt emotions. These accounts relate the origin of humankind, its place in time, and a perception of both the visible and the invisible world.

Why Were Myths Created?

It is not surprising that myths evolved in primitive cultures when people were faced with impersonal, inexplicable, and sometimes awesome or violent natural phenomena and the majesty of natural wonders. In comparison to these wonders, human beings felt dwarfed and diminished. As a result, they bestowed extraordinary human traits of power and personality to those phenomena that most profoundly evoked human emotions. The beginning of time, the miracle of birth, the finality of death, and the fear of the unknown compelled early humans to create deities who presided over the celestial sphere. In time, every aspect of nature, human nature, and human life was believed to have a controlling deity.

Initially, myths of cosmogony illuminated the origin of humankind. Virtually every culture embraces a Creation myth. Myths explain the beginnings of customs, traditions, and beliefs of a given society and reinforce cultural norms and values, thereby depicting what that society regards as good or evil. Myths assist in defining human relationships with a deity or deities. Judeo-Christian-Islamic societies have established a supreme power, a father figure, whereas the Norse tradition restricts the power and purveyance of the gods. Finally, myths help to dispel the fear and uncertainty that is part of the human condition. Fear of the elements may be explained by the activities of the gods. Fear of failure is overcome by reliance on them. Fear of death is often explained as the passage or transition to another dimension or to another domain. Simply stated, myths are a symbolic representation reflecting the society that created them. Although unjustified and unjustifiable, myths take the raw edge off the surface of human existence and help human beings to make sense of a random and threatening universe.

Universal Themes

Myths are seldom simple and never irresponsible. Esoteric meanings abound, and proper study of myths requires a great store of abstruse geographical, historical, and anthropological knowledge. The stories underscore both the variety and the continuity of human nature throughout time. The abiding interest in mythology lies in its connection to human wants, needs, desires, strengths, fears, and frailties. By their nature, myths reveal the interwoven pattern of circumstances that are beyond the control of the mortal and the immortal.

A study of the world of myths imparts greater appreciation for the subtle and dramatic ways that they pervade societies and that particular myths mirror the society from which they were created. Myths are decidedly human in origin. Ironically, it is the human ability to make myths, and the very need to do so, that ultimately sets us apart from other inhabitants of the earth. Only we humans can identify our place on the eternal timeline or calculate our brief appearance in time and space as we search for significance and immortality.

Myths may be drawn from any era and any geographical area. This entry examines components of Greek and Roman mythology, Norse and Teutonic (Germanic) mythology, Asian mythology, and commonalities among these and several other cultures.

Greek Mythology

To people of Western cultures, the most familiar mythology outside of the Judeo-Christian culture is that of Greek and Roman mythology. The mythology of ancient Greece and Rome stemmed from the human desire to explain natural events, the origin of the universe, and the end of time as we know it. Our journey through time commences with birth and continues on as we experience the tribulations and celebrations of life, and it culminates with death. Time encapsulates us. The Greek myths chronicle Zeus and his brothers, Poseidon and Hades, who exacted control of the universe from their father, Cronus, and the Titans, a powerful race of giants. Cronus himself had wrenched control from his own parents, Uranus (heaven) and Gaea (earth). Great epics were recorded of war and peace and proud heroes and courageous heroines who represented the basic cultural values of the Greek people.

Men and women alternately worshipped and feared a ménage of gods and goddesses who traditionally resided on Mt. Olympus. People attributed failure and defeat to the wrath of the gods and success and victory to the grace of the gods. The most powerful of the gods and goddesses were Clotho, Lachesis, and Atropos, the goddesses of destiny. It was they who determined how long a mortal would live and how long the rule of the gods would endure. When a mortal was born, Clotho wove the thread of life. Lachesis measured its length, and Atropos cut the thread at the exact point in time that life would end. Not even Zeus could alter their timeframe.

Roman Mythology

Much of Roman mythology had its roots in Greek mythology, although Jupiter and Mars were part of the Roman tradition long before the Romans interacted with and eventually conquered the Greeks. Subsequent to 725 BCE, the Romans adopted many Greek deities, renaming them, and making them their own. In both Greek and Roman mythology, realms of the universe were delineated. They saw the cosmos in terms of the skies, the earth, the seas, and the lower world. Jupiter (Zeus) ruled the skies from atop Mt. Olympus, where he controlled the movement of the sun, the phases of the moon, and the changes of the seasons. The fairest and the wisest of all immortals, Jupiter, when outraged, hurled lightning and thunderbolts down upon the earth. Neptune (Poseidon), the second-most-powerful god, ruled the seas, and Pluto (Hades), the god of wealth, ruled the lower world.

In Roman mythology, myth and time are closely related. The Romans equate the onset of time with the birth of Rome, which is accredited to Romulus and Remus. While these two were infants, they were set afloat to die in the river by their uncle who feared that they would usurp his power. The children survived their ordeal and eventually founded the city of Rome, named after Romulus, and this founding was considered to be the advent of time. Saturn, the father of Jupiter, Neptune, and Pluto, was believed to be the god of time, birth, and death. The Romans strongly believed that living an honorable life on Earth could earn for them eternal life in heaven and possibly a place among the gods. Evildoers, unless they appeased the gods, would be damned to the underworld, where they would spend eternity. The end of time for man was marked by death and reclaimed for eternity in the afterlife.

As we read these highly entertaining and often spiritually uplifting myths, we learn a great deal about human nature and of our debt to Greek and Roman cultures. Even today, in our struggle to survive, Greek and Roman myths help people to better understand humanity's obedience to a higher power, the relationships of men and women to one another, the power of love and friendships, the horror of war and natural catastrophes, and the passage of time, with death and the afterlife to come.

Norse and Teutonic (Germanic) Mythology

Germanic mythology refers to the myths of people who spoke Germanic dialects prior to their conversion to Christianity. These ancient Germanic people from the continent and England were illiterate. Most of what we do know about the mythology and beliefs of that era comes from literary sources written in Scandinavia and then transcribed into the Old Norse language of Iceland between the 12th and 14th centuries. Two collections of verse, known as the *Eddas,* exist. The earliest, the *Elder Edda* or *Poetic Edda,* contains the earliest Norse mythology; the *Younger Edda* or *Prose Edda* was written by Snorri Sturluson about 1220 CE. In the *Prose Edda,* Sturluson combined a variety of sources with three earlier poetic accounts of the origin of the world in order to create a wholly representative mythology.

In the *Prose Edda* version of the Creation, all that originally existed was a void called Ginnungagap. There was no time, and everything remained still. To the north of the void was the icy region of Niflheim; to the south, the sunny region of Muspelheim. Warm breaths from Muspelheim melted the ice from Niflheim, and a stream of water flowed into the void from which the giant, Ymir, ancestor of the Frost Giants, emerged. Created from drops of the melting ice, Audhumbla, the cow, nourished Ymir and was nourished herself by licking salty frost- and ice-covered stones. The stones were formed into a man, Bori, who was destined to become the father of Odin, Vili, and Ve. The brothers slaughtered Ymir and created the earth from his flesh, the mountains from his bones, the sea from his blood, the clouds from his brains, and the heavens from his skull.

The heavens, according to the *Prose Edda,* were balanced by four dwarfs: Austri, Westri, Nordi, and Sudri, the directions on a compass. Sparks from the fire-land, Muspelheim, became the stars of the sky. This newly created land, named Midgard, was to become the somber home of mortal humans. Even in Asgard, home of the gods, the atmosphere was grave, and the Norsemen believed that the end of time would come in a bleak and horrible way. The only hope was to face disaster and fight the enemy bravely to earn a seat in Odin's castle, Valhalla.

As did so many similar myths, Norse mythology reflected the attitude of the culture that in death there is victory and true courage will not be defeated. The final chapter in Norse mythology is called Ragnabrok, meaning the twilight of the gods. In this period it is prophesied that winter will continue for three years. On the last day, Odin will lead dead heroes in a fierce battle against the Jotuns (trolls) and the power of darkness. Odin, himself, will be devoured by the Fenris Wolf, and the world will become a smoking ruin swallowed by the sea. From this will come new life and a time of peace.

Asian Mythology

The myths of India, China, and Japan are highly complex and sophisticated. They differ from Greek, Roman, and Germanic mythologies in that rather than venerating anthropomorphic deities, the structures of their deities are often polymorphic, intricately combining human and animal forms. The gods and goddesses of the Hindus sometimes take extraordinary human forms, with numerous heads and eyes and arms. Hindus believe that their religion existed before the universe came to be. Their mythology on the afterlife is unique. They believe in karma and reincarnation. Simply stated, karma is the effect of an individual's actions resulting in consequences in present and future lifetimes. At the time of death, to be reincarnated as a human being is thought to be a blessing. Reincarnation as an animal or a plant is reserved for those for whom a spiritual life is not possible.

Deities in Chinese and Japanese myths were also animistic, but these myths were supplanted by mythologies derived from the three great religions: Taoism, Confucianism, and Buddhism (Buddhism having been brought to China from India in 300 BCE). Shinto, the religion indigenous to Japan, borrowed much from Chinese mythology, resulting in a tradition that paralleled that of the Buddhist pantheon. Notably, in Asian cultures, two or more religions may be observed simultaneously, because they are less eschatological but more ethical or philosophical in emphasis.

Generally speaking, before there was heaven or earth, there was chaos, devoid of time, shape, or form. First to materialize was the Plain of High Heaven and the three creating deities. Earth was born, and immortals procreated. It was due to the "divine retirement" of the goddess, Izanami, that the notion of death or life limited by time entered the world. It is interesting to note that there is difficulty in finding a Chinese word equivalent to the English word "time." "Shi" can be associated with time, but the meaning tends to mean "timelessness" or "seasonality." In Asian cultures, time is marked by history, as in the duration of a dynasty. Individually, time is of little concern, because a person's life is viewed as part of an ancestral continuum. Although the tenets of these religions encompass a belief in life after death, the focus is showing people how to live rather than what will happen to them upon their demise. We may deduce, then, that oriental myths not only explain the origin of the universe and the parameters of earthly time, but they also deal with commonly held distinctive aspects and cultural values of each civilization.

In ancient times, throughout history, and even today, the family is considered to be a critical part of oriental society and culture. Honor and obedience to one's parents is related to ancestor worship. During the Han period in China, emperors set up shrines for their ancestors because they believed that spirits could bring blessings to them and to their families. In general, oriental myths connect the actions of deities and other supernatural beings to the everyday actions of men, women, and the natural world around them. Individual gods protected the family, the home, and the country and represented the sun, the moon, and the planets. Myths describe how the islands of Japan were created and deliberately located by the gods in the very center of the world. The two main books of the Shinto religion are the *Koji-ki* and the *Nihon-gi*. *Nihon-gi* explains how all of the emperors of Japan are directly descended from the sun goddess. Today, the rising sun is symbolized in the Japanese national flag.

Commonalities

What the disparate mythologies from all over the world have in common is their heartfelt desire to explain the origin of humankind and to validate its existence. We are searching to satisfy the very human need to explain our relationship with the powerful and mysterious forces that drive the universe. Throughout the world, myths reflect those themes that deal with nature, supernatural beings, ancestors, heroes and heroines, life, death, and time. In Africa, where myths have been preserved mainly through the oral tradition, the natural elements are immortalized through myths. Many versions of Creation stories abound. The Dinka of Sudan believe that the first man and woman were made from clay and put into a tiny, covered pot, where they grew to full height. Australian aboriginal mythology deems that their community and culture were created during dreamtime, "the time before time," when spirited creatures came from the sky, the sea, and the underground to generate mountains, valleys, plants, and animals. We are familiar with the great spirit myth of North American Indians and the time that Native Americans identify for their ancestors in the happy hunting ground. The Aztec people of South America were polytheistic and offered sacrifices to appease their gods. Huitzilopochtli, the great protector of the Aztecs, was portrayed in the form of an eagle, and it was he who deemed where the great pyramid would be built as "the heart of their city and the core of their vision of the universe." From Ireland comes a myth about Cu Chulainn, a hero who could change form to oppose evil forces. A Polynesian myth from the islands of the Kanaka-Maori people centers on Maui, who brought the gift of fire to his people. In ancient Egypt, from pharaohs to peasants, each individual had a god or a goddess corresponding to his or her time and place in society.

As we can perceive, in every era and in every geographical area, myths have evolved as nearly sacred literature, devoid of theology. Each myth is unique to a culture and is in itself a monument to the precariousness of human existence.

As society becomes increasingly more global and less local, people continue to find experiential concepts that are impossible to fathom and beyond human comprehension. The beginning of time, the miracle of birth, the finality of death, and the fear of the unknown compelled early humans to create deities who presided over the celestial sphere. If myths are a symbolic representation reflecting the society that created them, how, then, will the mythology of our times satisfy our collective need to know?

Suzanne E. D'Amato

See also Beowulf; Cronus (Kronos); Rome, Ancient; Tantalus; Wagner, Richard

Further Readings

Baker, A. (2004). *The Viking*. Hoboken, NJ: Wiley.

Bellingham, D., Whittaker, C., & Grant, J. (1992). *Myths and legends: Viking, Oriental, Greek*. London: New Burlington.

Burland, C., Nicholson, I., & Osborne, H. (1970). *Mythology of the Americas*. London: Hamlyn.

Christie, A. (1983). *Chinese mythology*. Feltham, UK: Newess Books.

D'Aulaire, I., & D'Aulaire, E. P. (1992). *D'Aulaire's book of Greek myths*. New York: Doubleday.

DuBois, T. A. (1999). *Nordic religions in the Viking age*. Philadelphia: University of Pennsylvania Press.

Guirand, F. (Ed.). (1959). *Larousse mythologie générale*. London: Batchworth.

Kirk, G. S. (1974). *The nature of Greek myths*. London: Penguin.

Lip, E. (1993). *Out of China: Culture and traditions*. New York: Crabtree.

Poisson, B. (2002). *The Ainu of Japan*. Minneapolis, MN: Lerner.

Wolfson, E. (2002). *Roman mythology*. Berkeley Heights, NJ: Enslow.

Nabokov, Vladimir (1899–1977)

Vladimir Vladimirovich Nabokov, a Russian author and entomologist, was born into a wealthy patrician family in St. Petersburg, Russia, and died in Montreux, Switzerland. Time plays an important role in all of Nabokov's novels, and in all of his works a certain melancholy is noticeable.

Nabokov's father, Vladimir Dimitrijevich Nabokov, was a liberal criminologist, publicist, and politician who was one of the leaders of the Constitutional Democratic Party in Russia before the Russian revolution. His mother, Jelena Ivanova Nabokov, née Rukavishnikov, came from a family of industrialists and land owners. In the Russian October Revolution of 1917, the family lost all its property. The elder Nabokov was one of the leaders in the Duma in the February Revolution, but he had to flee after being captured by the Bolshevists in the October Revolution. The family left Russia for England, where Nabokov attended Cambridge and studied Russian and French literature; later the family moved to Berlin, Germany. In 1922, while shielding a friend from gunfire, his father was assassinated at a political meeting by a reactionary Russian exile.

In 1925, Nabokov married Vera Jevsejevna Slonin, a Jewish Russian. In the same year, his first novels, *Maschenka* and *Korol, Dama, Walet* (*King, Dame, Knave*) were published. In 1937, Nabokov had to flee to France because he was married to a Jew in Germany. In Paris, he wrote his first novel in English, *The Real Life of Sebastian Knight* (1939), was published in 1941. In the same year, his mother died in Prague. In 1940, together with his family, he moved to New York shortly before France was conquered by Germany. From 1941 to 1948, he worked as a lecturer in Russian language and literature and as a research fellow in entomology at Harvard University in Cambridge, Massachusetts. From 1948 to 1959, he was a professor of Russian and European literature at Cornell University, in Ithaca, New York.

Nabokov's first autobiography was published in 1951, and it was republished after many changes as *Speak, Memory* (1966). In 1955, what was to be his most famous work, the novel *Lolita,* was published. In this novel, the cultured Frenchman Humbert Humbert writes his memoirs concerning his love for a 12-year-old American girl. The book was a succès de scandale and a literary sensation; the substantial income from this novel allowed Nabokov to quit his work as a college professor and to move back to Europe, where he and his wife lived in a suite in a hotel at Switzerland from 1961 onwards.

His second famous novel, *Pale Fire,* appeared in 1962 and consists of two parts. The first part is a poem by a fictional author called John Shade. In the second part, the scholar Kinbote, who claims to be the king of a country called Zembla, comments at length on the poem and interprets it as the story of his own life.

In his most famous novel, *Lolita,* there is the contrast between the protagonist and the narrator of the story. Both are one person: Humbert Humbert.

But, the narrator Humbert tries to relive the pleasures he once had by telling the story of his time with Lolita. The pervading melancholia results from the feeling of being small and powerless in comparison with the universal laws of time and the universe. On the other hand, time does not matter for that work of art which can revive memories (even the memories of strangers). Unfortunately, these memories are always deficient.

While Nabokov's concept of time remains implicit in his novels, it is one of the main themes in his autobiography *Speak, Memory: An Autobiography Revisited* (1966). Many of his creative ideas were inspired by his reading of Marcel Proust. Nabokov's main theme in this context is human consciousness. For him, the human life is "a brief crack of light between two eternities of darkness." But in consciousness, a human life (even other human lives) can be preserved in memories. From this viewpoint, literature is a way to preserve memories (although a deficient one) and also a way to prevent others from dying. Moments and persons can be conserved in words, but not without a loss of reality. Consequently, consciousness is superior to literature.

From Nabokov's point of view, literature is a desperate fight against time and darkness, which, in his opinion, follow the consciousness of a human life.

Markus Peuckert

See also Consciousness; Memory; Proust, Marcel

Further Readings

Boyd, B. (1990). *Vladimir Nabokov: The Russian years.* Princeton, NJ: Princeton University Press.

Boyd, B. (1991). *Vladimir Nabokov: The American years.* Princeton, NJ: Princeton University Press.

Connolly, J. W. (Ed.). (2005). *The Cambridge companion to Nabokov.* Cambridge, UK: Cambridge University Press.

Nabokov, V. (1996). *Novels and memoirs 1941–1974.* New York: Library of America.

Nāgārjuna, Acharya (c. 150–c. 250 CE)

Mahayana Buddhist tradition attributes the founding of the Madhyamika School to Acharya Nāgārjuna (c. 150–c. 250 CE), but this attribution is probably incorrect, because there is no designation for such a school until the work of another monk named Candrakirti refers to it in the 7th century CE. Although he is credited with composing other works, Nāgārjuna's seminal text is *Fundamentals on the Middle Way,* in which he advocates a philosophy of the middle way between the extremes of being (eternalism) and nonbeing (nihilism).

This philosophical position means that nothing in the world exists absolutely and nothing perishes totally. Nāgārjuna's middle way is located beyond concepts or speech in the sense that it is transcendental. This philosophical position also means that no specific position is limitless or ultimate. In fact, the ultimate truth is that there is no correct view, final truth, or goal, because all views are flawed. Nāgārjuna's middle way implies rising above clinging to either existence or nonexistence. More precisely, the middle way is the practice of the perfection of wisdom (*prajnaparamita*), or ultimate virtue for a bodhisattva (enlightened being). When one achieves wisdom, one does not arrive at a particular type of knowledge, but one rather reaches a point at which all knowing and theorizing are terminated.

Nāgārjuna makes a distinction between two kinds of truth: conventional and ultimate. The former is valid and useful for practical living, but it is illusory, because it becomes self-contradictory if we push it too far. This is evident with the concept of time. From a conventional perspective, the concept of time represents the past, present, and future that appear as a series of moments; this is associated with human perception and conceptual formulation. The conventional concept of time is formed by the assumption of a self-substantiated reality that binds persons to their own emotional and conceptual habits. In short, for Nāgārjuna this is the realm of ignorance, which involves mistaking things or concepts for what they are not in fact, because ignorance obscures the real nature of things and constructs a false appearance.

In contrast to conventional truth, Nāgārjuna defines ultimate truth as a nondual type of knowledge that involves a contentless intuition. By viewing time from this perspective, we observe it simultaneously much as we view a painting on a wall by seeing the whole of it. This intuitive type of knowledge is beyond ordinary objective knowledge and reason, because it represents dissolution of the conceptual function of the mind, although it does not represent a total rejection of conventional truth. It is the realization that all distinctions, such as the three moments of time, are empty (*sunyata*), which is true of everything in the world.

If time and everything else in the world is empty, there can be no essential distinction between existing things. Nāgārjuna denies the distinction between self-being (*svabhava*) and other-being (*parabhava*). If self-being is the essential nature by which something is what it is and not something else, and other-being owes its existence to something else, Nāgārjuna denies that there is anything that is not dependent or conditioned. The heat of fire, for instance, is never encountered apart from fire, making heat something created and not self-existent. If the true nature of everything is emptiness, there is nothing that is self-existent, because it would have to be necessarily noncontingent and unrelated to anything else, which means that lack of self-existence is the nature of things.

These fundamental philosophical convictions motivate Nāgārjuna to criticize basic categories, such as causation, motion, and time. If all things are conditioned, each phase, for instance, possesses a before (future) and an after (past) relative to it. Using his dialectic that undermines all philosophical positions, Nāgārjuna asserts that each part of his dialectic is a counterpart to the prior step, which results in each part negating and canceling out its predecessor. Nāgārjuna's dialectic moves toward the negation of the final part. By thus disposing of all philosophical views, Nāgārjuna uses his dialectical method as a therapeutic device to cure humankind of its suffering, which is caused in part by its mental and emotional attachment to phenomenal and conceptual entities in preparation for genuine insight into the nature of things.

With respect to the notion of time, the dialectic indicates that neither the present, past, nor future can be seized as absolute, but they have significance only relative to each other. Thus the three moments of time are relational concepts. There is no such thing as the past in itself, the present in itself, or the future in itself. Moreover, time is pertinent merely to this world. If ultimate truth indicates that there is neither past, nor present, nor future, time is a derived notion or a mental construct. There are no individual entities of time.

This position means that time is not an immutable substance that can be grasped and measured. There is also no absolute time that continues to be real apart from successive moments. Time is merely a mode of reference that points to the arising and perishing of events. When a person witnesses these arising and perishing events, he or she names them "time" and draws distinctions among the moments of time in relation to each other.

From Nāgārjuna's perspective, the three moments of time enable one to grasp time as a set of relations. To be located in any particular moment—past, present, or future—means to be dependent upon the location of the other moments. Therefore, the present is, for instance, such only because it is located within the instants of the past and future, which suggests that time is a dependent set of relations among three moments. No single moment of time represents an entity in its own right.

From the perspective of attaining liberation, there is no escape from time, because there is nothing from which to escape. In the final analysis, Nāgārjuna does not deny the commonsense view of time. What he wants to show is that time and other categories are not ultimately real.

Carl Olson

See also Buddhism, Mahayana; Buddhism, Theravada; Buddhism, Zen; Dialectics; Intuition; Time, Nonexistence of

Further Readings

Nagarjuna. (1986). *The philosophy of the middle way: Mulamadhyamakakarika* (D. Kalupahana, Trans.). Albany: State University of New York Press.

Streng, F. J. (1967). *Emptiness: A study in religious meaning*. Nashville, TN: Abingdon Press.

Wood, T. E. (1994.) *Nagarjunian disputations: A philosophical journey through an Indian looking glass* (Monographs of the Society for Asian and Comparative Philosophy, No. 11). Honolulu: University of Hawai'i Press.

NAVAJO

The Navajo, or Diné (meaning "the people"), are the most numerous of the North American Indian tribes, having more than 290,000 people. The Navajo nation ("the big rez"), which is about the size of West Virginia, officially encompasses 25,000 square miles at the juncture of northeastern Arizona, southeastern Utah, and western New Mexico. The reservation itself was created in 1868 by the U.S. government; however, the Navajo live within the four sacred mountains, (Mt. Blanca, Mt. Taylor, San Francisco Peaks, and Mt. Hesperus), a place of great beauty where they feel they belong in accordance with the guidance of their holy ones. This space is actually larger than what is considered the present day reservation.

Life and time are cyclical in the Navajo cosmology, and everything has a place within it. Stories passed down in Navajo culture explain that the people emerged into this location after going on a long and arduous journey during which they passed through four different colored worlds. In the first world (black world), first man and first woman were formed. They passed through the second world (the blue world), the third world (the yellow world), and the fourth world (the glittering world). Quarreling forced them out of each world; they had encountered insect beings, several species of birds, and many mammals, including Coyote, a trickster who is an important figure in Navajo culture. The natural world was put into harmony or balance (*hozjo*) by the Creator or spiritual life force. Finally, a flood brought the Navajo to the fourth world or glittering world of the four sacred mountains. The Navajos are said to have emerged from a hole in the La Plata Mountains.

Early History

Archaeological and linguistic evidence shows that the Navajo migrated from present-day northwest Canada and Alaska to the American Southwest around 1000 CE. Linguistic similarities suggest that the Navajo and Apache tribes were once a cohesive ethnic group, speaking the dialect of the Athabaskan and Apachean language family. Navajos were considered highly adaptive to changing conditions and were able to incorporate things from other cultures. They were famously known to their Pueblo neighbors as traders and raiders. The Pueblo groups exchanged maize and woven textiles for meat and hides of deer, antelope, and elk. The Spanish arrived in the 1500s, bringing with them horses, sheep, and goats of European origin. After Spanish colonization in the 17th century, Navajo life became more sedentary as the people established camps to raise sheep and corn. They learned weaving from the Pueblos and silver crafting from their Mexican neighbors.

Shaping of the Navajo Nation

Anglo Americans and Navajos lived in relative peace during the 1800s until a Navajo leader named Narbona was killed in 1849. In the 1850s the U.S. government began to set up forts in Navajo territory, including Fort Defiance and Fort Wingate. After the Spanish were expelled by the Anglo Americans in the southwestern United States, the Navajos fell under the scrutiny of the U.S. government, which was determined to settle the West. The Treaty of 1868 is a significant event in Navajo history and Navajo/U.S. government relations. Unlike so many other Native American tribes during the 19th century, the Navajo were allowed to return to a portion of their traditional homelands. Attempts were being made to round up tribal groups in an effort to solve the "Indian problem" by creating reservations. In 1863, the dispossession of their lands was a major blow to the Navajo. The U.S. government was trying to prevent raids intertribally and on encroaching settlers. In addition, whites suspected that there were valuable minerals on Diné lands. Colonel Christopher "Kit" Carson was called in with his army to defeat the Navajo. Carson commanded his soldiers to shoot on sight men, women, and children. He also wiped out the Diné food supply, burning crops, killing domesticated animals, and torching houses. In February of 1864, the Navajo began to turn themselves in to army forts in surrender.

The policy of Indian relocation set forth by the U.S. government proved to be a disaster. The Long Walk (to Bosque Redondo) is an especially painful moment in Navajo history. The Diné were forced to walk for 3 weeks for more than 300 miles, and hundreds lost their lives. When the Navajo finally arrived at Bosque Redondo, they faced extreme living conditions. The water was unfit to drink, there was no firewood, and they were rationed poor-quality subsidized food from the U.S. government. Around 9,000 Navajos were relocated and were not allowed to return to their lands until the Treaty of 1868, which many Navajo leaders signed to recover their lands.

Navajo Family and Daily Life

The Navajo tribe is a group of more than 100 separate clans, including the four originals: Towering House, Bitterwater, Big Water, and One Who Walks Around. Families consist of extended kinship networks, with clans being traced through the mother's side ("born to"), and the father's side is acknowledged by saying "born for." Women have an important status in Navajo society. When a man gets married he joins his wife's family, and the wife's brother takes on many roles associated with fatherhood toward her children. Traditionally, marriage within clans is not permitted. Both men and women care for children. The women have land rights, in addition to owning the house, the goats, and the sheep.

When Navajos first meet each other, they state which clan they are from. *Ya'at eeh* is a common phrase, which means "greetings" in the Navajo language. Navajos live in isolation from one another on the landscape. It can be miles between two Navajo sheep camps, the land between can appear to be uninhabited. Physical and personal space is valued. It is estimated that 80% of the Navajo still speak their native language. Language, customs, and lifeways have been preserved and passed down in the Navajo culture despite strong outside Anglo influences to assimilate.

Hogans

Traditional houses are called *hogans*. The houses are eight-sided, domed, and nearly circular in shape. Old timers still use these as a dwelling, while more often in contemporary times these structures are used for ceremony. They are made of wood poles and earth with the doorway open to the east to welcome the morning sun and receive positive energy and blessings. The sun is an important symbol of the divine and Creation; however, the sun itself is not worshiped in Navajo religion. The Navajo find balance with an orientation within the four directions. East represents the dawn and thinking, south signifies planning and what needs to be done, west represents life and how to carry out plans, and north an evaluation and reflection of how to continue of the path of life. ("Before me, behind me, below me, and above me, with balance I pray" is a common morning prayer of centering along with the focus on beauty: "All is beautiful, if everything around you is beautiful, beauty is the way you live.")

Today, many hogans are constructed with modern materials and have windows. Sometimes, small hogans are built for sweat baths. The steam makes the body sweat and a cleansing occurs on both a physical and spiritual level. Additionally, some families build summer hogans and winter hogans near appropriate places where their sheep graze.

According to custom, it is considered rude to greet a person when visiting a hogan without waiting several minutes before entering. It is also considered rude to make eye contact or to shake hands with another with a firm grip. Religion is integrated into all other aspects of daily life for the Navajo. Life itself is considered sacred along with the earth and the idea of maintaining balance or harmony with all things. Life is cyclical, as well as time, and everything has a place and a season in Navajo religion and cosmology. In the Navajo culture there is no such thing as a coincidence; things happen because they were meant to. Religion ceremonies are elaborate and complex, lasting anywhere from 3 to 8 days or more. Religious gatherings include songs, chants, prayers, and sandpaintings. Ceremonies are called *Ways* and the religious leader is a *singer* who sings special songs and makes sandpaintings. Ceremonies were given to the Navajo by the holy ones, who instructed them in how to recite the prayers and songs. Important lessons have been passed down along with history and universal wisdom. Ceremonies are used for healing the sick, for blessing a person, or for celebrating a happy moment, such as the birth of a child. Corn and corn pollen are considered sacred and play an important role in Diné ceremonies. The content of ceremonies includes much sacred knowledge with great power, and parts of ceremonies remain secret and are not discussed outside the ceremony.

Sandpaintings

Sandpaintings are "a place where the gods come and go" in the Navajo language and are used in healing ceremonies and to connect with nature. These elaborate creations are filled with symbolism and often contain representations of the first man and first woman from Navajo cosmology. After the ceremony, the pictures are swept up and the sand is taken away.

Ways

There are several Way ceremonies that serve a certain purpose in Diné culture. The Night Way is a 9-day healing ceremony where friends and relatives gather around the sick person and songs, and prayers, are offered and sandpaintings are created. The enemy Way is used when a Navajo returns from a non-Navajo society to cleanse that person of foreign influences. The Blessing Way is a ceremony that is used to be sung over someone. It is unique, because it is not used for healing; rather, it is used to foster good luck, good health, and blessings relevant to a person's life.

Navajo Code Talkers

During World War II, 3,600 Diné men and 12 women entered military service. The Japanese had been able to decipher all U.S. military codes until the Navajo marines created a code in the Navajo language that was an impenetrable means of secret communication. This made the contributions of these Diné soldiers indispensable to the U.S. military. In 1982, President Ronald Reagan declared August 14th Navajo Code Talkers Day, and in July 2001, 29 code talkers were given Congressional Gold Medals.

The Navajo Today

The Navajo Nation today is a mixture of traditional and modern. There are houses and hogans, sheep farmers and engineers, silversmiths and nurses. However, much is preserved due to the organization of tribal government and Navajo schools with curriculums relevant to Navajo culture and staffed by Navajos, including Diné College. The Diné nation continues to grow and flourish in the 21st century despite some major dark points in its history. The Navajos have proven to be highly adaptable and have overcome adversities. Their determination and philosophy of life allows them to continue and to "walk in beauty."

Luci Maire Latina Fernandes

See also Pueblo; Sandpainting; Time, Cyclical

Further Readings

Iverson, P. (2002). *Diné: A history of the Navajo.* Albuquerque: University of New Mexico Press.

Kluckhohn, C., & Leighton, D. (1962). *The Navahos.* Cambridge, MA: Harvard University Press.

Underhill, R. M. (1956). *The Navahos.* Norman: University of Oklahoma Press.

Nebular Hypothesis

At the beginning of the 17th century, Johannes Kepler was able to illustrate that the planets tended to move elliptically. Subsequently, the publication of Isaac Newton's laws of motion and gravitation in 1687 marked the first systematic scientific approach to examining the origin of the solar system. Then, in 1755, the German philosopher Immanuel Kant proposed the theory that the solar system had its beginnings as a cloud of dispersed particles of both dust and gas and that it is a product of centrifugal and centripetal forces. In 1796, the French mathematician Marquis Pierre-Simon de Laplace refined the theory further. He described the original state of the solar system as a hot, rotating nebula. As the mass cooled and contracted, the nebula assumed a flattened shape. The sun was formed at the center, with rings of gaseous material surrounding it. Planets then condensed from the rings. By the same process, moons formed around planets. This theory seems to explain why planets generally move in the same plane and direction.

It was at this point in time that the theory first became known as the "nebular hypothesis." It has since sometimes been referred to as the Kant/Laplace nebular theory, because Kant apparently arrived independently at the modified version of the hypothesis about the same time that Laplace did so. Laplace thought that the theory supported the predictability of the universe, while Kant believed that it indicated the universe was likely to change through time.

Though the nebular hypothesis has been examined and modified through the subsequent years by the scientific community, Kant's original brilliant concept can still be said to serve as an important component of current theories on planet formation.

Early in the 20th century, several British and American scientists pointed out definite deficiencies in the nebular hypothesis and proposed that planets were formed by a rare encounter of a star and the sun. In the mid-20th century, these star encounters were shown to be impossible, as the gaseous material involved would naturally dissipate rather than condense as planets. Therefore it was generally concluded that the formation of planets and stars must take place during the same process. Scientists have indeed noted that planets tend to form around newborn stars, and they now refer to the disks of dust and gas they have observed around these stars as "protoplanetary disks."

Naturally, scientists have always been interested in determining how long it took the universe and specific planets to form. A common theory has been that there are two stages in the formation. It is during the first stage of accretion that small, rocky planets such as Earth form. Solids collide and stick together, with gases forming atmospheres around these smaller planets.

A smaller planet must have the time to grow large enough for the second stage to begin and for larger, gaseous planets to form. There is more limited opportunity for these larger planets to form, because the gas itself might disappear in a few million years. On the other hand, the formation of the smaller planets can continue more slowly for up to 10 or hundreds of million years.

It has been speculated that the disks around smaller planets disappear after 3 to 5 million years. Yet that time period may well be too limited to permit the formation of larger planets, such as Jupiter. A number of scientists now accept the theory that other solar systems with similar sized planets must be fairly common (and planets larger than Jupiter have actually been noted), but more research is needed to attempt to explain exactly how Jupiter—and all the other planets—did form.

Two theories of how the gas giants formed have been proposed. "Core accretion" would result in these giants forming relatively slowly, because a large, solid core would be necessary to attract a large quantity of gas. "Disk instability" proposes that a cold disk could break up on its own if it is cold and dense enough, resulting in gravitational abilities. The latter process could produce protoplanets in only hundreds or thousands of years.

The original concept that the nebular hypothesis involved a disk forming around a condensed center is still held regarding what is now commonly referred to as the Solar Nebula. As gas and dust collapsed toward the center, kinetic energy was formed and the temperature rose to the point of producing a nuclear reaction and the subsequent birth of the sun.

Theories have also been developed about the differences between the inner and outer planets. Their size appears to involve how much water is available and whether planets are far enough away from the sun for ice to form. Planets located at a far enough distance from the sun can acquire more solid mass and attract large amounts of many other elements, including the abundant elements hydrogen and helium. These larger planets formed beyond what is referred to as the "snow line." As is the case with many credible theories, there have been questions about whether or not this process is inevitable. Discoveries and observations of other solar systems, such as 51 Pegasi, indicate enough definite differences from our solar system to raise further questions about exactly how planets are formed.

Current efforts to verify or modify the body of knowledge about how planets are formed are employing many avenues for research. Some, such as Richard H. Durisen's attempt to update Laplace's theory, make use of computer simulations while still identifying dense gas rings as the mechanism of planet formation. Other research focuses on areas beyond our own solar system, such as the Hubble Space Telescope's production of images of protoplanetary disks around stars in the Orion Nebula—about 1,600 light-years away. Continuing observation by astronomers is likely to yield greater understanding of the process of planet formation.

Betty A. Gard

See also Kant, Immanuel; Laplace, Marquis Pierre-Simon de; Planets; Stars, Evolution of; Telescopes

Further Readings

Durisen, R. H. (2005). Rings of creation. *Mercury, 34*(3), 12–19.

Schilling, G. (1999). From a swirl of dust, a planet is born. *Science, 286*(5437), 66–68.

Weintraub, D. A. (2000). How do planets form? *Mercury, 29*(6), 10.

NEOGENE

The Neogene, a term introduced by Moritz Hörnes in the mid-19th century, is a period in the geochronological scale and a system in the chronostratigraphic scale. This dual procedure (time interval and corresponding rock record) is used by earth scientists to subdivide geologic time in deciphering the history of the earth. The Neogene is the last period/system of the Cenozoic era/erathem and, accordingly, the most recent one of the earth's history. It began 23 million years ago, at the Paleogene-Neogene boundary, and it ends at the present. Under the current proposal of the International Commission on Stratigraphy, it includes four epochs/series: Miocene, Pliocene, Pleistocene, and Holocene. The classification and interpretation of the last two epochs/series, as well as a Quaternary sub-era, have been, and still are, a matter of debate. An important consequence is that the scientific community inherited different notions for the Neogene. Thus, the end of the Neogene has been variously interpreted to be at the Pleistocene–Holocene, Pliocene–Pleistocene, and Tertiary–Quaternary boundaries.

Marine microfossils are the backbone of the subdivision of the Neogene into its constituent ages/stages. Complex mammal evolution under the influence of major continental separations and climatic change and orbital forcing cyclicity in sediments and oxygen isotopes records in the Atlantic and Mediterranean (supported by the Australian–Antarctic marine magnetic polarity scale) provide a precise and highly accurate Neogene timescale.

The Neogene Earth looked much like our own. However, the relatively similar distribution of landmass between then and now masks some dramatic changes. Some continental motion took place during the Neogene, the most significant event being the counterclockwise rotation of the Arabian Plate, connecting Africa and Eurasia and cutting off the remnants of the old Tethys. India collided with Asia, giving rise to the Himalayan Mountains and the connection between North and South America.

In the south, a continuous circumpolar current circled Antarctica, isolated from other landmasses. Thus both poles were thermally isolated from warm equatorial waters, and (perhaps for the first time since the Ordovician) both poles accumulated heavy coverings of ice. At the same time, the virtual closing of east-west circulation through the Mediterranean Sea and between the Americas changed the hot, circulating currents. Thus, during the Neogene the world became much drier and cooler, culminating in the Pleistocene ice ages and the harsh conditions of the present day.

The world dried out. Huge deserts developed in North Africa and Central Asia. Grasslands expanded and quickly replaced the thinning forests. During the Neogene, birds and mammals evolved considerably, and the dawn of the genus *Homo* occurred. However, most other animals not needing grasses were relatively unchanged. Grasses are poor fodder: tough, low in nutrients, high in tooth-destroying silicates. Consequently, herbivorous species were smashed or utterly changed. Some grazer species emerged and evolved high-crown teeth. Ruminants diversified, and cranial appendages appeared. Nevertheless, their predators followed them into extinction or transformation. The later Neogene saw the creation of an entirely new type of hunter, the pursuit predator. The pursued developed their own responses: herd behaviors, seasonal migrations, and big bodies adapted for speed and endurance in open country.

Another line of adaptation led to small-bodied generalists (rodents, raccoons, rabbits, and opossums) and their predators, the foxes, cats, dogs, and snakes. These generalists were mainly unspecialized herbivores or omnivores with partially fossorial habits, strong territoriality, and high reproductive rates. Theirs was the ability to exploit many resources within small, locally, or temporarily favorable conditions.

The Miocene epoch (23 to 5.3 million years ago [mya]) or "less recent" is so called because it contains fewer modern animals than the following, Pliocene, epoch. The Miocene is the longest epoch of the Neogene. During the late Miocene the island continent of India slammed into Asia, pushing up the Himalayas. Elsewhere, the western American cordilleras, the Alps, and the Caucasus rose as well. One of the best known events in the marine realm is the Messinian salinity crisis at the end of the Miocene. The rise of mountains in the western Mediterranean combined with the global fall of sea level due to formation of the Antarctic ice cap sealed the western end of the Mediterranean for about 600,000 years. During this time, the Mediterranean Sea virtually dried up, forming enormous evaporite deposits. When the present Strait of Gibraltar was ultimately opened, the Atlantic would have poured a vast volume of water into the Mediterranean drying basin, resulting in a giant waterfall, much higher than 1,000 meters and far more powerful than Niagara Falls. On the other side of the African continent, three major rifts opened in roughly an east to west sequence. These events were probably related to the counterclockwise rotation of the Arabian plate. The Miocene was a time of warmer global climates, but during the mid-Miocene (14 mya), a marked drop in temperatures occurred and further led to the buildup of the East Antarctic ice cap.

The Miocene was a time of huge transition, the end of the ancient world, and the birth of the more recent sort of world. Two major ecosystems first appeared during the Miocene: kelp forests and grasslands. It was also the high point of the age of mammals. Also, this period saw animals that had evolved on different continents during the Eocene and Oligocene spread via land bridges.

The Pliocene epoch (5.3 to 1.8 mya), compared to previous epochs, was a relatively brief period of only 3.5 million years. The name *Pliocene* means "more recent." During this time, the earth approached its current form, with ice caps, relatively modern geography, modern mammals, and the evolution of hominids. Continents had taken up their present positions. Both North and South America were drifting northward. However, South America was moving somewhat faster, related to a shift in the Caribbean tectonic plate. Thus, a permanent land bridge between the Americas developed in the mid-Pliocene, allowing mammals to migrate across. The closing of the Isthmus of Panama isolated the waters of the Gulf of Mexico and separated the marine biota of the east and west coasts. This tectonic episode had major consequences for global temperatures, because warm equatorial ocean currents were cut off and the climate became cooler and drier. At the same time, the Himalayan uplift accelerated the unfolding cooling process.

The Pleistocene epoch (1.8 to .011 mya) is known as the ice age, because this short epoch witnessed a dramatic, continued cooling, culminating in a series of advances and retreats of the ice as the climate fluctuated between cold (glacial) and warm (interglacial) periods at periodicities fitting Earth's orbit cycles (Milankovitch cycles). The sea level rose during the melting of the glaciers; then land bridges, created during cooler periods when glaciers sequestered more water, enabled the migration of animals and humans across continents. The term *Pleistocene* ("most recent") was coined for strata with 90% to 100% present day species. Animals and plants were basically modern species, although distributions were unusual. The great mammalian megafauna flourished. Many giant mammals evolved and lived on all continents. During the Pleistocene, the hominid tendency to increase brain size and hence intelligence continued, and finally modern man (*Homo sapiens*) emerged.

The Holocene epoch covers the last 11,500 years of the Neogene period. The term *Holocene* means "completely recent" and refers to the present geological epoch. The Holocene represents a marked climatic warming phase corresponding to the present interstadial (warm period between glaciations) phase. All other ages, epochs, and eras are represented by natural evolutionary and geological phenomena. The Holocene in contrast is distinguished by being the epoch during which there has been an exponential growth in human population and knowledge. Human activities have had a marked, and for the most part extremely detrimental, effect on the rest of the biosphere.

Beatriz Azanza

See also Chronostratigraphy; Earth, Age of; Evolution, Organic; Fossil Record; Geologic Timescale; Geology; Glaciers; Ice Ages; Paleogene; Paleontology; Plate Tectonics; Stratigraphy

Further Readings

Gould, S. J. (Ed.). (1993). *The book of life: An illustrated history of the evolution of life on Earth.* New York: Norton.

Gradstein, F. M., Ogg, J. G., & Smith, A. G. (Eds.). (2004). *A geologic time scale.* Cambridge, UK: Cambridge University Press.

Stanley, S. M. (2004). *Earth system history* (2nd ed.). New York: Freeman.

Wicander, R., & Monroe, J. S. (2003). *Historical geology: Evolution of Earth and life through time* (4th ed.). London: Brooks/Cole.

Nero, Emperor of Rome (37–68 CE)

Nero Claudius Caesar was the fifth of the five Julio-Claudian emperors and one of the most notorious. Two millennia after his death, his name continues to conjure images of a cruel, self-indulgent tyrant who "fiddled while Rome burned." He is also legendary for being one of the first rulers to order the persecution of a small religious sect known as the Christians. Despite the overwhelmingly negative views of Nero and his rule, he represents an important period in the history of Rome. Nero was the last emperor with a hereditary link to Julius Caesar. He reigned at the end of a century of peace. After his death, civil war broke out. Because Nero had ordered the death of any relative who might inherit the throne after him, when he died the throne was open to anyone with the power to claim and keep it.

Early Years

Nero was born in Antium in December, 37 CE, and was named Lucius Domitius Ahenobarbus. His father, Cnaeus Domitius Ahenobarbus, was a member of a distinguished noble family of the republic. His mother, Agrippina the younger, was the daughter of Germanicus. When Lucius was 2 years old, his father died. The reigning emperor, Gaius (Caligula), brother to Agrippina, seized his inheritance and banished mother and son to the Pontian Islands, where they lived in near poverty. Caligula and his wife and infant daughter were killed in 41 CE. His uncle Claudius, a far milder ruler, ascended to the throne and recalled his niece and her son from exile. Agrippina, a very ambitious woman, promptly arranged a proper education for her son.

In 48 CE, Claudius had his wife Messalina executed for adultery. The following year he married his niece Agrippina, and she furthered Lucius's prospects by having him betrothed to his stepsister Octavia (whom he married 4 years later). Lucius completed his education under the tutelage of the eminent Stoic philosopher Lucius Annaeus Seneca. In 50 CE, Agrippina persuaded Claudius to formally adopt her son, securing his place as heir to the throne. Lucius's name was officially changed to Nero Claudius Drusus Germanicus.

Emperor Nero

Claudius died in 54 CE—probably poisoned by his wife. Nero claimed the throne with the support of the praetorian prefect Sextus Afranius Burrus. Agrippina acted as regent to the 16-year-old emperor. Nero's first few years as ruler were stable, led by the sound guidance of Burrus and Seneca. Nero announced that he would model his rule after that of Augustus, a very prestigious and respected ancestor. Nero applied himself to his judicial duties, granting more freedom to the senate, forbidding the killing of gladiators and criminals, lessening taxes and the extortion of money by provincial governors, and making reforms to legislation.

Difficult decisions and administrative pressures eventually caused Nero to withdraw. He devoted himself to pleasures: chariot racing, singing, poetry, acting, dancing, and sexual activity. Seneca and Burrus attempted to keep his performances private and the government running smoothly. Agrippina was furious about (some say jealous of) Nero's conquests. She also deplored her son's interest in Greek art. Nero grew hostile toward his mother as news of her virulent gossip came back to him.

Nero's life reached a turning point when he took a new mistress named Poppaea Sabina. Agrippina supported his wife Octavia, who was naturally opposed to this latest affair. Nero responded by making various attempts on his mother's life. After three unsuccessful attempts to poison her, the ceiling over her bed was rigged to collapse while she slept. That also failed, so a boat was constructed that would break apart and sink in the Bay of Naples. However, Agrippina managed to swim ashore as the boat sank. An exasperated Nero finally sent an assassin who clubbed and stabbed her to death in 59 CE. Nero reported to the senate that his mother was plotting his death and he had no choice but to retaliate. The senators had never approved of Agrippina and did not question her removal.

Nero celebrated his freedom with even more contests, festivals, and orgies. He sang, acted, and played instruments in public. Performers were considered unsavory, so it was an outrage to have an emperor on stage. In 62 CE, Burrus died from an illness. He was succeeded as praetorian prefect by two senators who were corrupt and encouraged Nero in his excesses. Seneca found the situation uncontrollable and resigned. Nero's life became a series of scandals. He divorced Octavia and had her killed later that year. He then married Poppaea, who was also killed by Nero a short time later.

Despite Nero's behavior in Rome, the empire as a whole was relatively peaceful and prosperous until one of the greatest disasters in Rome's long history occurred. In July of 64 CE, the great fire of Rome ravaged the city for 6 days. When it was finally contained, 10 of the 14 districts of the city had been reduced to rubble and ashes. After the fire, Nero claimed a vast area and began construction of his "Golden Palace." Given the vast size of this complex, it could never have been built before the fire. The Roman people began to have suspicions about the source of the blaze. Nero, always desperate to be popular, looked for a scapegoat to blame. He found a new, obscure religious sect called the Christians. Many were arrested, thrown to the wild animals in the circus, burned to death, or crucified.

Beginning of the End

A large-scale conspiracy planned by a number of senators to remove Nero from the throne was discovered in 65 CE. The 27-year-old had been emperor for more than a decade, and the only positive result of his reign was the strict building codes put into place after the great fire, making Rome a safer and more attractive city. Nero's response to the conspiracy was to take an extended tour of Greece while his prefect decimated the senatorial ranks by execution or ordered suicide. In the year 68 CE, another revolt began in the provinces. Galba, an elderly provincial governor, claimed the throne and marched his troops toward Rome. Nero was abandoned by the few supporters he had remaining. The senate ordered Nero's death. Nero heard and chose to commit suicide instead, with the assistance of his secretary. He died on June 9 in the year 68 AD. Even at the very end of his life, Nero remained more concerned with his artistic pursuits than affairs of state. His last words were "*Qualis artifex pereo!*" or, "What an artist the world loses in me!"

Nero ruled when there had been over 100 years of relative peace. Despite his antics in Rome, the people in the rest of the empire lived well enough to prevent unrest. Before Nero's death, he had every relative who might ascend to the throne killed, to assure that a child with his blood would rise to power. After he died childless, civil war broke out, because the throne was open to anyone who claimed it. The last of the Caesars was succeeded by a series of provincial strongmen with personal armies.

Jill M. Church

See also Caesar, Gaius Julius; Christianity; Rome, Ancient

Further Readings

Champin, E. (2003). *Nero.* Cambridge, MA: Harvard University Press.

Grant, M. (1985). *The Roman emperors: A biographical guide to the rulers of imperial Rome 31 BC–AD 476.* New York: Charles Scribner's Sons.

Suetonius. (2003). *The twelve Caesars.* London: Penguin Classics.

Nevsky, Saint Alexander (1220–1263)

During his short life, Saint Alexander Nevsky, born Aleksandr Yaroslavovich, made his mark in history as one of Russia's best-known Christian military commanders; he protected Russia against European invasion during the Middle Ages. He was heralded as savior of the Russian Orthodox Church.

A military strategist of the time, Nevsky's triumphant defense against the invasion of Russia by the Swedes, the Teutonic Knights, and the Lithuanians saved Russian culture. Many historians believe that, through his collaboration with the Mongols, Nevsky was able to save Russian lives and land and also saved Russia from Roman Catholic control, thus preserving the Orthodox faith. He also did much to advance the centralization of the Russian government.

Nevsky's role as defender of Russian land from the German and Swedish feudal lords began when he was at the tender age of 16, when he became the duty-bound prince of Novgorod. His legendary victory over the Swedes, who in 1240 attempted to block Russia's access to the Baltic at the Battle of Neva, earned him both the name of Nevsky and a place in history by raising him to legendary status. The military tactic of a surprise lightning attack in this battle ensured Nevsky's victory. This was an especially significant event because it prevented an all-out invasion of Russia by the Swedes. This victory further strengthened the young prince's political aspirations.

When the invasion of Russia was again at hand, Alexander Nevsky again went to war. His victorious battle with the German Knights in the Battle of the Ice in 1242 was a significant historical event of the Middle Ages, for it was in this victory that foot soldiers first defeated mounted knights, a military tactic that was to become a timeless battle strategy. This victory and the victory over the Lithuanians in 1245 established Nevsky as a one of the greatest military leaders of that time and ensured the survival of Russia.

In 1246, Nevsky faced a problem that would require his skills not on the battlefields but in diplomacy. The problem concerned a loss of Russian independence. At this time, with the Mongols' conquest of eastern Russia, Nevsky wisely cooperated with them. He was made Grand Prince of Vladimir (1252–1263), and his cooperation with the Mongols allowed him to protect the Orthodox Church from aggression, to spare Russians from further hardship, and to achieve some stability in northern Russia.

Nevsky died in 1263. The end of his leadership was a loss deeply felt by the country, but his heirs ruled Russia until 1917. With his successors and descendants, princes came to be monarchs in Moscow. In death, he continued to be an influence in Russian history. In the 14th century (1381), Nevksy was elevated to the status of a local saint, and in 1547 he was canonized by the Russian Orthodox Church, because his collaboration and intercession with the Mongols helped to maintain Russia's way of life and religious freedom and prevented much bloodshed. This endeared him to the Russian people.

The Alexander Nevsky Monastery, founded in 1710, now includes some of the oldest buildings in the city of St. Petersburg and burial places for some of the giants of Russian culture. In 1725, one of the highest Russian military decorations, the Order of Alexander Nevsky, was established to revive the memory of Alexander's struggle with the Germans, and in 1836 a triumphal arch was erected in his memory and a principal street was named for him in St. Petersburg. In 1937, Sergei Eisenstein made a classic film to promote Russian nationalism using Nevsky as the subject. The Soviet Order of Alexander Nevsky was introduced during the Great Patriotic War of 1942.

Several Russian naval vessels were named for him, including the 19th-century frigate *Alexander Neuski.* Without Nevsky's leadership the history of Eastern Europe and thus of the world might have been very different.

Joyce K. Thornton

See also Attila the Hun; Genghis Khan

Further Readings

Christiansen, E. (1998). *The northern crusades* (2nd ed.). New York: Penguin.

Commire, A. (Ed.). (1994). *Historic world leaders* (Vol. 3). Washington, DC: Gale.

Mitchell, R., & Forbes, N. (Trans.). (1970). *The chronicle of Novgorod, 1016–1471.* New York: AMS Press.

Presniakov, A. E. (1970). *The formation of the great Russian state* (A. E. Moorehouse, Trans.). Chicago: Quadrangle Books.

Newton, Isaac (1642–1727)

Sir Isaac Newton, English natural philosopher, mathematician, and physical scientist, revolutionized the theoretical concepts regarding the physical laws that govern the universe. The impact of his contributions to the understanding of the mathematical perspective of motion, space, and time provided the means to further challenge both the Aristotelian foundations of science and ecclesiastical authority over science. Additionally, Newton's metaphysical distinctions based on his theoretical principles drew criticism from many theologians and philosophers of the Enlightenment traditions, including rationalist and empiricist.

Praised as a scientist and ridiculed as a philosopher, Newton attempted to construct a metaphysical bridge in order to reconcile the mechanics of the universe with God, but his attempt was vastly underscored or dismissed. For Newton, there had been a conflict between traditional or orthodox religion and science that he now attempted to overcome in presenting his own unorthodox metaphysical approach to God and God's significance for the universe and humankind. Newton's major works include *Philosophiae Naturalis Principia Mathematica* (1687), *Opticks* (1704), and *Arithmetica Universalis* (1707).

Portrait of Sir Isaac Newton (1642–1727) by Sir James Thornhill. A mathematician, physicist, and English astronomer, Newton proposed that gravity was a universal force and that the sun's gravity was what held planets in their orbits.

Newton's scientific, albeit philosophical, explorations in the laws that govern the physical universe are steeped in cursory observations, rational speculations, and mathematical computations. Regarding the temporal nature of the universe and its relation to physical matter, the abstract notions of motion and space are irrevocably united within the conceptual framework of a theoretical and sometimes theological framework. Beyond the eternal, infinite, and geocentric concepts set forth by Aristotle (384–322 BCE), Ptolemy (c. 90–168 CE), and, theologically, Saint Thomas Aquinas (1225–1274 CE), the scientific advancements of Nicolaus Copernicus (1473–1543), Galileo Galilei (1564–1642), and Johannes Kepler (1571–1630) changed and influenced previous perceptions of space, motion, and more important, time. Newton influenced and integrated advancements in both science and philosophy within his perspective; he furthered scientific advancements by creating a mathematically systematic approach to explain the natural phenomena of the universe. These secular contributions to science and mathematics, along with a theistic cosmological perspective, had secured the separation of physics from theology. Interesting and yet obscure, the theistic tendencies within Newton's cosmology possess a confounding blend of ontology and teleology.

Newton provided a scientific explanation for the symbiotic relationship of gravitation and motion. In the explanation of the physical phenomena observed on this planet and in the universe, he calculated that gravitation had an effect on physical bodies, from earthly objects to celestial bodies. In resolving the terms of motion, Newton had stated three laws:

1. Motion is constant until an external force is applied.
2. Force can be calculated by the relationship between mass (m) and acceleration (a); thus F = ma.
3. Every action has an equal and opposite reaction.

These explanations in terms of motion have unique implications for the concept of time. Time and space, interrelated and reflected in Newton's established laws, are not in constant and fluid motion. Objects exist in absolute time, and absolute space is perceived in relation to motion. This would question the exact nature of space—space being a void or some reactionary force—and the conceptualization of time in terms of the interactions between the mass and velocity of bodies. Furthermore, a lack of knowledge regarding physical composition, especially of celestial bodies, and its relationship to space and motion, would have greater implications. Although Newton provided mathematical computations to support his rational speculations, his metaphysical implications drew criticisms from both philosophers and scientists with differing theological and philosophical perspectives. However, as a precursor to the work of Albert Einstein (1879–1955), these speculations would lead to monumental contributions to theoretical physics.

The concept of time and its relationship to our own species' developmental ontology and teleology are easily tracked within the historical accounting of philosophical ideals. Reflected in these ideals, self-awareness and temporal perception are uniform; some are supported in part by mathematics. Time becomes an integral part of human existence far beyond the physical mechanics of the universe. For Newton, this reality in the concept of time is often underscored. Within his theistic cosmology, Newton acknowledged that a divine power existed that was both infinite and eternal. Space, similar to this divine power, was infinite yet containing finite objects. These celestial bodies, placed within the confines of space, are not only maintained by a divine force but also explained in terms of mechanical operations of gravity within space. The necessity for a stable system, both theologically and cosmologically, became paramount. The infinity of the universe, as opposed to sidereal boundaries of space, precludes any philosophical distinction and definition of what exists beyond the boundaries of time and space. The distribution of stars and planets, albeit complete solar systems, are finite within the infinite. For Newton, the temporal nature of the material universe, involving creations, recreations, and dynamic (yet stable) changes, was beyond the depiction of time necessitated by religious scripture. The age of the solar system was steeped within the immensity of time. Regardless of the duration of time, the material universe and the mechanical operations of gravity and motion remained to be the sole factor in Newton's argumentt for the existence of God.

The cosmological ideas of Newton concerning the sensorium of God, though heretical, depicted an infinite universe governed by the mechanical operations of an intelligent and divine power. Contrary to the views of his day, the previous Aristotelian order and operation of the universe was replaced by a mathematically ordered universe that reflected this divine power. Motion was considered not as a result of the actions of an unmoved mover but of natural laws of gravity, acceleration (velocity), and space. Although Newton never attributed this design to the Greek notion of *logos,* the natural order and laws that govern the solar system and stars was reduced to a theistic "proof" of God's existence, although Newton's perspective on the basic structures of Christianity remains elusive.

Regardless of Newton's theistic beliefs, the universe, stars, solar systems (perhaps multiple systems), and planets were much older than depicted in the Bible. With the errors within scripture being pervasive, it would be interesting to speculate on Newton's own temporal view of our species. The concept of time, regardless of scientific advancements, becomes a personal and universal reality by means of which human beings are aware. Motion and its related mechanical principles illuminate this concept of time and our relation to the external world. Newton's contributions to understanding of gravity, motion, and abstract notions of time and space (finite and infinite) were critical for the foundations for modern physics.

David Alexander Lukaszek

See also Einstein, Albert; Einstein and Newton; Galilei, Galileo; God, Sensorium of; Newton and Leibniz; Space, Absolute; Time, Absolute;

Further Readings

Cohen, B. (Ed.). (1958). *Isaac Newton's papers and letters on natural philosophy and related documents*. Cambridge, MA: Harvard University Press.

Janiak, A. (Ed.). (2004). *Philosophical writings/Isaac Newton*. New York: Cambridge University Press.

McGuire, J. E. (1983). *Certain philosophical questions: Newton's Trinity notebook*. New York: Cambridge University Press.

Newton, I. (1999). *The principia: Mathematical principles of natural philosophy*. Berkeley: University of California Press. (Original work published 1686)

Newton and Leibniz

The controversy between Isaac Newton (1642–1727) and Gottfried Wilhelm von Leibniz (1646–1716) was primarily over their views of space and time. There had been some claims among Newton's followers that Leibniz had plagiarized from Newton, particularly regarding the calculus. It was later proved that there was no plagiarism, but that these two geniuses, standing on the shoulders of those (like Johannes Kepler, for example) who had preceded them, had each made the conceptual leap independently to the calculus. However, Newton (somewhat paranoid) may have retained some feelings of resentment toward Leibniz, and thus does not respond to him directly about their differing views; rather, the differences are aired primarily in the correspondence between Leibniz and Samuel Clarke, a follower of Newton.

Newton had expressed his views on space and time in his *Principia Mathematica* (1686), in the scholium following the section on definitions. He explains why he did not include in the definitions time, space, place, and motion, because they were well known to everyone. However, people commonly held prejudices regarding these concepts, so he expounds his technical definitions. "Absolute, true, and mathematical time, of itself, and from its own nature, flows equably without relation to anything external." It is also called duration. Relative, or common, time is a sensible measure of duration (by means of motion). "Absolute space, in its own nature, without relation to anything external, remains always similar and immovable." Relative space is somehow a measure of this absolute space. It is important to look at both concepts together, because they share characteristics that may help in understanding what Newton means. Space (absolute space) is a kind of permanent container in which things may come and go, appear and disappear. Imagine a huge matrix on which one might move pieces (a kind of cosmological chess board); the pieces may move and change, but the board always remains. By analogy, absolute time is always flowing, and does not depend for its existence on the motion or change of bodies. Change implies a before and an after in time, which therefore presupposes time, so it must be more fundamental, an objective reality, according to Newton. He says that absolute time would exist even if there were no motion, or a lapse between motions.

Leibniz denies that space and time are absolute; he argues that they are relative, that they are relations. Space is the order of perceptions of monads (a monad is Leibniz's basic metaphysical, indivisible substance, a concentration of energy, a kind of mind) that coexist; time is the ordering of a monad's different perceptions. Because space and time depend on monadic perceptions (of the world), they are ideal (phenomenal), not real. Just as other relations (such as "smarter than") do not have an independent existence, but are dependent on the entities compared, neither do space and time have an independent existence. Time is more fundamental; for a monad, the present is represented clearly, the past and the future more obscurely. Space is then the ordering of coexisting monads (or aggregates of monads) at the same time. Leibniz argues that there is a continuum of monads in this best of all possible worlds. (God would have created nothing less; existence is maximized in the best possible world.) If there had not been a best possible world, God would not have had a sufficient reason to create anything.

A controversy between Leibniz and the Newtonians began in 1705 over these serious disagreements. It culminated in 1715–1716 with an exchange of letters between Leibniz and Samuel Clarke, a follower of Newton. It began with a criticism of Newton's position sent by Leibniz to Princess Caroline of Wales; Clarke responded on behalf of Newton, Leibniz wrote back in defense, and so on. There were five letters written by Leibniz, five replies by Clarke, and it ended without resolution, at the death of Leibniz.

Leibniz begins his first letter by attacking Newton's materialism (of atoms and void) and its implications for religion. He argues that Newton's system implies that God is a kind of unskillful watchmaker who has to fix and adjust his creation periodically and has need of space as his sense organ. Clarke responds that God needs no medium of perception, because he is omnipresent. Clarke argues that a God who does not attend to his creation would be like a king who ignores his kingdom and lets it run on its own. Leibniz responds that God, in his wisdom, foresaw everything and has made the best possible world machine, which consequently does not need his intervention; his action, as supreme ruler, is that of continually preserving his creation.

Leibniz's fundamental principle is that of sufficient reason (nothing happens without a sufficient reason why it is so and not otherwise), a rational self-evident truth. If space were a real absolute being, as Newton says, all points in space would be indistinguishable, and there would not be a sufficient reason why God placed bodies in one arrangement rather than its opposite (east instead of west, for example). Similarly with absolute time, its parts would be uniform, so there would not have been a sufficient reason why God created everything *when* he did, and not sooner or later. Clarke replies that God's will is the sufficient reason for creating how and when he did. He attacks the idea of space as relational by saying that if God reversed the position of the stars with that of the moon and Earth, on a relational account, it would be the same, which is contradictory. Or if God moved the material world in a straight line at any speed, on the relational account it would continue in the same place and time (which is absurd).

Leibniz calls Clarke's examples impossible fictions, and says that Clarke did not understand the nature of relational space, by supposing that a world would be moved against an absolute background space and time, which is what is denied. Two indiscernible states are the same state, says Leibniz, and so "'tis a change without a change." Another fiction is that God might have created the world sooner, because God does nothing without a sufficient reason. Leibniz says that the order of bodies (aggregates of monads) makes space/situation possible, just as the succession of bodily states makes time/duration possible. If there were no creatures, there would be no space and time. Clarke reiterates that there can be identical parts of space that are yet distinct, and two points in time that are identical; quantity of time can be greater or less, yet the order of temporal events could be the same. Leibniz's principle of continuity denies this; between any two states there must be another state, so greater elapsed time implies more successive, distinct states.

Clarke argues that Leibniz's universe destroys freedom; everything would be determined. Leibniz replies that he has argued at length, in his *Theodicy,* that God's foreknowledge is compatible with free choice. God, in his wisdom, simply created (thought) the one possible world full of free creatures that was the best "actualizing their free natures." What God did not create, in thinking the best world, is nevertheless possible.

Leibniz argues that absolute space and time would be infinite and eternal and so independent of God, thus possibly greater than God (which is contradictory). If being *in* space and time is necessary to God, he would then be in need of them and thus limited (but God is without limits). Motion depends on a change that can be observed. Time without things, Leibniz says, is only an ideal possibility; to say the created world might have been created sooner is not intelligible (if it *could have* been created sooner, God would have done so, because that would mean more existence). Space and matter are different, but inseparable; time and motion are distinct, but inseparable, says Leibniz.

Stacey L. Edgar

See also God and Time; Leibniz, Gottfried Wilhelm von; Newton, Isaac; Space, Absolute; Time, Absolute

Further Readings

Ariew, R. (Ed.). (2000). *Leibniz and Clarke: Correspondence*. Indianapolis, IN: Hackett.

Ariew, R., & Garber, D. (Eds. & Trans.). (1989). Monadology: Principles of nature and grace. In *Philosophical essays/G. W. Leibniz*. Indianapolis, IN: Hackett. (Original work published 1714)

Newton, I. (1934/1966). *Sir Isaac Newton's mathematical principles of natural philosophy and his system of the world* (A. Motte, Trans., 1729; F. Cajori, Rev., 1966). Berkeley: University of California Press. (Original work published 1687)

Nicholas of Cusa (Cusanus) (1401–1464)

Nicholas of Cusa (Kues), a German cardinal of the Roman Catholic Church, philosopher, jurist, mathematician, and astronomer, is widely considered as one of the greatest geniuses of the 15th century. He received a doctorate in canon law from the University of Padua in 1423. His ideas influenced philosophical, political, and scientific thought and anticipated the work of astronomers Nicolaus Copernicus and Johannes Kepler.

Generally speaking, Cusa's theory of time exemplifies a Christian Platonism whose elements already had a long tradition (in, e.g., the work of Saint Augustine of Hippo, William of Conches, Albert the Great, and Saint Thomas Aquinas). Cusa's elaborations of this material, however, are sometimes quite original. Space permits inclusion here of only a few examples from among his many discussions of the topic.

In *De venatione sapientiae* (1463), Cusa begins his exposition of the subject by dealing with eternity, the image of which is time (*imago aeternitatis*). Eternity itself (*aeternum*) has to be distinguished from "perpetuity" (*aevum et perpetuum*): Whereas eternity is a mode of duration that has absolutely no beginning and no end, such that it is the peculiar mode of being of God who can himself be called eternity (*aeternitas*), perpetuity designates a duration that has no beginning and no end within the realm of time. On the other hand, the subjects of perpetuity, like the heavenly entities (*caelestia*) and the objects of pure thought (*intelligibilia*), transcend the realm of coming-to-be and perishing. Those passages, where Cusa distinguishes two senses of eternity—the pure eternity of God and the derivative eternity of the perpetual elements of the cosmos—have to be understood according to the same passages; that is, they also reflect the distinction between eternity and perpetuity. Cusa describes eternity as the realm of things that have a being as possible things (*posse fieri*). The existence of such *possibilia* is a necessary condition for God's being able to create the sensible world. The eternal objects themselves, however, are not created, but initiated (*initiata*). In this respect, Cusa echoes the Scotistic model of eternal ideas within God, which are the elements of the world in possibility while they are necessary in their own form of existence as *possibilia*. Time, on the other hand, has the connotation of change, of coming-to-be and perishing—in other words, of the sensible world. Only in this respect, Cusa sometimes also uses the Aristotelian definition of time as a measure of motion. The creation of the sensible, or timely, world is described in Neoplatonic terms as the unfolding (*explicatio*) of that which exists not unfolded (*complicite*) in the intelligible realm of perpetuity. The intelligible world is already structured in a way that enables God to create our world in its beauty. It is probably because of this connection of both the intelligible and the sensible world that Cusa sometimes calls the world itself eternal (*aeternus*) without qualification: the world in its entirety, that is, including the intelligible world, has no beginning in time; rather, it is the realm within which time begins and ends. In the eternity of God himself, on the other hand, the difference between the two worlds of being able to coming-to-be (*posse fieri*) and existing actually (*esse actu*) is suspended, because he alone is what he can be (*possest*). It is obvious from these remarks that Cusa's theology stays fairly close to the classical Neoplatonic system from late antiquity, though he reformulates it in terms that show some traces of the earlier Christian discussions. In ancient Neoplatonism, however, these ideas about time and eternity were closely connected with the idea of an eternal cosmos, which contained only a limited number of souls that migrated from one body to another one. Thus, it is no surprise that Cusa has difficulties explaining in which way the eschatological elements of Christian thought—that is, the resurrection of the dead—can be explained as related to time. In his early work *De docta ignorantia,* he explains that at the time of resurrection we will arrive, because of the end of all motion, at a place beyond time (*supra tempus*). But this process of transcending that way of being into which we have been born cannot be explained by philosophical reasoning, and consequently Cusa in his subsequent works does not give a new treatment to this question.

In his sermon *De aequalitate* (1459), Cusa approached the problem of time from another point of view. Here, he discusses the relation between soul and world in a way that is clearly inspired by Augustine's treatment of the topic in *Confessiones XI:* The soul sees that it is "timeless time" (*anima videt se esse intemporale tempus*). This is because it apprehends time to exist only in the world of change. While the soul sees itself as an active element of this world, and consequently as an element of time, it understands too that it is in its essence separate from all temporality, because it can find in itself both the past and the future moments of time, which do not exist in the world of change. Insofar as all three stages of time are present to the soul, each of them can be called "perfect time" (*tempus perfectum*). According to Cusa, the past is the memory (*memoria*) of the soul, the present its intellect, and the future its will. All of them depend, insofar as they are in the soul, on memory as their reason of being (*quia est*), but what they are (*quid est*), they are as objects of the intellect; in the will they are like an intended goal (*in intento fine*). Though the soul can liberate through these activities the instants of time from its necessary connection with elements of the world, its own existence is still timely, because the soul itself is directed toward the moments of the successive time. In this regard, it recognizes itself as different from eternity: It is in a timely way free from corruption, but not absolutely free from it as God is. It is an analogy to eternity (*similitudo aeternitatis*), but not itself eternity.

In principle, the more perfect intelligences (which must probably be understood as the angels, but the concept clearly echoes the Neoplatonic doctrine of *nous*) are in the same situation of existing timeless in time. But while they always have an active awareness of their natural state, the human soul has to strive for such an awareness by reducing its dependence upon the timely world. The relation between time and the human soul is treated again in the second book of *De ludo globi* (1463). In this work, the timelessness of the soul is explained by its intention to reach its highest good, the science of God. In its fundamental "first intention," the soul is free of any change, but one can speak about change insofar as the soul's "second intentions"; the soul's actual strivings are subject to many changes that, however, do not infect its timeless and changeless essence. This solution reminds the reader of the theories of Albert the Great (who is the interlocutor of the cardinal at this place) and of Aquinas's theories of a natural striving for the good that cannot be lost. From Cusa's explanation of *De aequalitate,* the new account differs insofar as it presupposes a difference between timely change and substantial timelessness in the soul itself.

Matthias Perkams

See also Aquinas, Saint Thomas; Aristotle; Bruno, Giordano; Eternity; Plotinus; Teilhard de Chardin, Pierre

Further Readings

Bellitto, C. M., Izbicki, T. M., & Christianson, G. (2004). *Introducing Nicholas of Cusa: A guide to a Renaissance man.* Mahwah, NJ: Paulist Press.

Jaspers, K. (1964). *Anselm and Nicholas of Cusa* (H. Arendt, Ed.). New York: Harcourt Brace Jovanovich. (Original work published in German, 1957)

Nietzsche, Friedrich (1844–1900)

Friedrich Nietzsche has emerged as perhaps the most influential thinker of the recent past. To a significant degree, this is due to the fact that he took time seriously in terms of both cosmology and ethics. Nietzsche offered a dynamic worldview that rejected the entrenched Aristotelian philosophy and Thomistic theology of Western civilization. His provocative writings contained scathing criticisms of modern European culture, particularly its religious beliefs and social morals (all decadent values, as he saw them).

Nietzsche spent his formative years in Röcken and Naumburg, Germany, where he developed a lasting interest in music and literature. At universities, his academic concerns shifted from classical philology to ancient philosophy. He became fascinated with the early culture of Greece, especially the fundamental idea of Heraclitus, which maintained the cyclical flux of all reality. Furthermore, Nietzsche stressed the necessary value of feelings and emotions (over the use of reason) for human creativity and fulfillment. His own emerging ideas were greatly influenced by the seminal writings of the philosophers Ludwig Feuerbach and Arthur Schopenhauer.

After his studies at Bonn and Leipzig, Nietzsche became a professor at the University of Basel, Switzerland. Due to chronic illnesses, he left the university after 10 years and became a solitary wanderer in the mountains of southern Europe. The following years gave the philosopher free time to rigorously reflect on the place of our species within both Earth history and sociocultural development. He wrote a series of ingenious books, his masterpiece being the four-part poetic work *Thus Spake Zarathustra* (1883–1885).

Over time, Nietzsche wrote blistering criticisms of Christianity that dismissed all the basic beliefs of traditional theism. He boldly proclaimed that "God is dead!"—guaranteeing himself a permanent place in Western thought. He remained an unabashed atheist his entire life.

It was Charles Darwin the scientist who awoke Friedrich Nietzsche the philosopher from his dogmatic slumber. Although he benefited from reading Darwin's writings on evolution, Nietzsche's own interpretation of organic evolution offered startling philosophical speculations that were far beyond the views of his scientific naturalist contemporaries. The Darwinian struggle for survival (existence) became the Nietzschean struggle for power (creativity). The iconoclastic Nietzsche also called for a rigorous reevaluation of all values, because he saw religion, democracy, communism, and utilitarianism promoting values that were reducing human beings to a collective mediocrity. Consequently, he stressed the value of those superior individuals who are unencumbered by the vacuous ideas and false beliefs of the inferior masses.

One of Nietzsche's most important ideas, that of eternal return, is the theory that there is infinite time and a finite number of events, and eventually all events will recur again and again.

For Nietzsche, dynamic reality is essentially the will to power. As such, all the objects of this evolving universe are composed of vital energy as units of force. This is a strictly naturalistic stance that gives no credence to philosophical idealism or theological spiritualism. Throughout time, this will to power continuously creates all those objects that fill this evolving universe. If a steadfast observer with a high-powered telescope had witnessed, over billions of years, organic evolution on Earth from the surface of our moon (the process rapidly accelerated like a time-lapse film), then he or she would have experienced life forms exploding into an astonishing diversity of plant and animal species: One-celled organisms are followed by multicellular life forms, invertebrates precede vertebrates, and fossil apes give rise to human beings. Briefly, the creating universe includes creative evolution.

In terms of organic evolution, Nietzsche held that the human animal is a temporary link between the fossil apes of the remote past and the overbeings who will emerge from our species in the distant future. Moreover, in the creative sweep of biological history, he claimed that the overbeing to come will be as advanced beyond our species as the human animal of today is advanced beyond the worm! Quintessentially, this incredible progress will be made in terms of intellectual development (rather than merely temporal changes in physical characteristics). As a result, Nietzsche concluded that the overbeings to come are the meaning, purpose, aim, and goal of organic evolution on earth. These forthcoming noble beings will devote their superior intellect to creating artistic works and new values.

In early August 1881, while walking along the lake of Silvaplana near Sils-Maria in Switzerland ("6000 feet beyond man and time"), Nietzsche came upon a huge pyramid-shaped boulder. In a flash of intuition, he grasped a provocative perspective. This incident caused him to reflect upon time and, consequently, to develop his colossal idea of the eternal recurrence of this same universe. The resultant awesome temporal perspective that now occurred to him justified (so he thought) his philosophy of overcoming; the struggle for existence is worthwhile if one's personal life somehow continues to exist throughout time in this material reality. Nietzsche was delirious with joy over his immortality-granting idea. He even contemplated studying the natural sciences in order to empirically demonstrate the scientific truth of eternal recurrence.

Nietzsche argued that cosmic time is eternal, while cosmic space is finite. Moreover, he held that there is only a finite amount of matter and energy in this universe. Therefore, only a finite sequence of objects and events may take place in a cosmic cycle. But if time is eternal, then this finite cosmic cycle will repeat itself again and again. In fact, Nietzsche argued that this identical cosmic cycle has repeated itself an infinite number of times in the past, and it will repeat itself an infinite number of times in the future. Thus, material reality is the eternal recurrence of this identical universe, with no progressive evolution from cycle to cycle.

If true, then the ramifications of this cosmology are staggering for human existence. In general, it meant that everything that has ever existed will appear again in the same finite sequence throughout eternal time. In particular, it means that Nietzsche will never pass out of existence forever, since he will eternally return as the same individual living the exact life in every detail in each finite cycle. It also embraced the evolution of morality from the premorality of fossil apes to the metamorality of the future overbeings yet to emerge. In short, for our species today, eternal recurrence challenges one to seriously consider both the choices one makes and the values one holds.

Nietzsche's pervasive "Yes!" to dynamic reality, and therefore to cosmic time, is an affirmation of life with far-reaching implications for ethics, morals, and values beyond good and evil. It commands that every choice a person makes is a decision for eternity. In short, once is forever! Because an individual has no knowledge of his or her previous life, it is as if one is free to make choices, but this freedom is, in fact, an illusion. Moreover, each moment of existence has eternal value in this universal wheel of time.

Friedrich Nietzsche claimed that he wrote in blood and philosophized with a hammer. His this-worldly stance repudiated the human, all too human ideas and moralities of common life. In his dynamic viewpoint, neither species nor values are fixed in nature. Nietzsche offered an astounding interpretation of time and human existence. His optimistic philosophy is ultimately grounded in a cosmic perspective that teaches the eternal recurrence of this same universe. He held that time is an endless series of finite cycles, with each cycle being absolutely identical to all the other cycles. His extraordinary vision of cosmic time remains both a challenging and an essential frame of reference for scientists, philosophers, and theologians.

H. James Birx

See also Cosmology, Cyclic; Eternal Recurrence; Ethics; Evolution, Organic; Feuerbach, Ludwig; Heraclitus; Nietzsche and Heraclitus; Presocratic Age; Schopenhauer, Arthur; Time, Cyclical; Values and Time; Wagner, Richard

Further Readings

Klossowski, P. (1997). *Nietzsche and the vicious circle.* Chicago: University of Chicago Press.

Köhler, J. (2002). *Zarathustra's secret: The interior life of Friedrich Nietzsche.* New Haven, CT: Yale University Press.

Nietzsche, F. (1993). *Thus spake Zarathustra* (H. J. Birx, Ed., T. Common, Trans.). Amherst, NY: Prometheus. (Original work published 1883–1885)

Pearson, K. A. (1997). *Viroid life: Perspectives on Nietzsche and the transhuman condition.* London: Routledge.

Safranski, R. (2001). *Nietzsche: A philosophical biography.* New York: Norton.

Sorgner, S. L. (2007). *Metaphysics without truth: On the importance of consistency within Nietzsche's philosophy.* Milwaukee, WI: Marquette University Press.

Vaihinger, H. (1924). *The philosophy of "as if."* New York: Harcourt Brace.

Nietzsche and Heraclitus

The philosopher Friedrich Nietzsche (1844–1900), although extremely critical of many thinkers, completely accepts Heraclitus's concept of time and develops it further. When Nietzsche read Heraclitus he felt at home, because concerning the form and content of being, Heraclitus's philosophy is similar to that of Nietzsche. Concerning the content of being, both defend a natural philosophy of permanent change that leads to a never-ending process of overcoming and of creation and destruction of forms. Nietzsche specifies the content of being further by developing the metaphysics of the will-to-power. (*Metaphysics* here is understood as a description of the nature of the world, which in Nietzsche's case is monistic and this-worldly.)

Both thinkers agree, however, not only concerning the content of being but also concerning its form, which is cyclical. In Nietzsche's case this form is referred to as the *eternal recurrence,* whereas Heraclitus calls it the *great year,* which is supposed to have a (metaphorical) duration of 10,800 years. Nietzsche stresses the similarity of his concepts to Heraclitus's in various published and unpublished writings. In *Ecce Homo* he stresses that traces of his concept of the eternal recurrence can be found in Heraclitus, and in *Thus Spake Zarathustra* he himself employs the notion of the great year in order to explain his understanding of the form of being.

Eternal Recurrence, Time, and Salvation

Many Nietzsche scholars today (such as Volker Gerhardt) stress the ethical relevance of Nietzsche's concept of the eternal recurrence. However, the fact that he clearly compares his concept to Heraclitus's metaphysical one in his published writings shows that it was meant as a metaphysical one. This reading of the eternal recurrence gets further support from his plans to study physics in Paris for 10 years in order to prove his concept scientifically, as his friend Lou Andreas-Salome pointed out, and from passages in his notebooks that he did not publish himself in which he puts together arguments with which he tries to prove the eternal recurrence philosophically. All of these arguments fail or are insufficient. At the least they do not establish what they are meant to establish. From the premises he mentions, he cannot infer the eternal recurrence by necessity. This, however, does not mean that his arguments cannot be improved so that the premises actually imply the eternal recurrence of everything. In addition, current scholars who stress the ethical relevance of the eternal recurrence are surely correct in doing so. The eternal recurrence is of immense ethical importance, as it is Nietzsche's theory of salvation, and it can give meaning to people who do not believe in a Christian afterlife but rather in a this-worldly concept of existence.

Eternal recurrence implies that whatever you have done and will do will recur identically. You lived the very same life before you were born, and you will lead it again in the very same manner. This implies that your life will not be over when you die but that you will return again and again, and you will meet the very same people you know now, and you will have to go through all the pains and pleasures you have experienced and will experience. You will not remember that you have lived the very same life before, but if you hold this concept, then you can feel comfort in realizing that you will experience all the wonderful events you have experienced again and again. These events are not over and done with. On the other hand, a life might have been so terrible that the concept of eternal recurrence can be unbearable.

What is important concerning salvation on the basis of this concept is that you experience one moment that you can affirm completely. Once you have had such a moment, then all other moments before and after this one are justified by means of this one moment, because all the other moments have been and are necessary in order for that moment to occur, according to Nietzsche. By saying yes to one moment, you also affirm all the pain and suffering you have to go through to reach it. However, if you experience a moment that you can affirm completely, you are saved on the basis of this concept. Therefore, the eternal recurrence or the great year increases the importance of any single moment immensely. Scholars who focus mainly on the ethical aspect of this concept stress that it is like an existential imperative: One ought to live as if one's life were to recur forever without this actually having to be the case. This does not correspond to Nietzsche's way of thinking. He was a classical philologist who was trained to think metaphysically and ontologically and did so himself because he understood that if you wish to have the ethical side of the concept, it needs an ontological foundation. Hence, he considered proving it scientifically and attempted to argue for it philosophically.

Circular Time and Identity

Nietzsche's eternal recurrence and Heraclitus's great year, understood philosophically, have implications concerning ontology and the philosophy of time. Concerning ontology, the concept can be seen as the form of being, the circle in which all other worldly events occur. Concerning the philosophy of time, the concept implies the circularity of time. If we take the notion of the circularity of time seriously, then it does not imply that we have the *x* plus *n*th (whereby *n* has to be a natural number) circulation but only one. Hence, we cannot talk about infinite time but rather circular time, time that is in itself circular. According to this concept, everything that occurs, occurs infinitely often. However, it would be more appropriate not to say that anything occurs infinitely often but rather that the very same thing occurs again and again. In the end, it is not the same but rather the identical that happens again. The circularity of time therefore implies that there is the relation of identity in its strongest Leibnizian sense between the various circulations of the world. The qualities that are valid in one circulation are also valid for all others. Here we also have to include spatiotemporal qualities.

Nietzsche tried to prove this concept philosophically. He argued, for example, thus: The law of the conservation of energy demands eternal recurrence. However, this is clearly false. The law of the conservation of energy alone does not necessarily imply this concept. If we assume the conservation of energy but also that the amount of energy is infinite, then we have an infinite number of possible combinations. However, an infinite number of possible combinations excludes the possibility of the eternal recurrence. If we assume that energy is finite but space exists separately from energy and is infinite, then again we have an infinite amount of combinatory options. Even if it were the case that energy is finite and time and space were qualities of energy, the eternal recurrence does not follow by necessity, as it could be the case that the law that is responsible for the change of the energy is itself subject to change, which excludes the possibility of the occurrence of the eternal recurrence.

Even if we further assume that energy is finite, time and space are aspects of energy, and the law underlying the change is determined and unchanging, the eternal recurrence does not follow by necessity, as there is the option that energy can turn up in an infinite amount of sizes, which also precludes that eternal recurrence takes place. Only when we assume, in addition to the premises just mentioned, that energy can turn up only in a finite amount of sizes, eternal recurrence follows, as the law of the conservation of energy (which is already included in our premises) also entails the reversibility of all states of the world—another necessary premise. By adding the premises just listed, however, we significantly expanded the argument suggested by Nietzsche.

It has to be noted that the notions "energy" and "force" that Nietzsche employs can be substituted by the notion "substance." There is a fairly wide range of words that can be exchanged for one another. However, the meaning of the words depends on the context in which they appear. Hence, it is not usually the case that one word has only one meaning in Nietzsche's writing, which is why one must be extremely cautious when interpreting him. Nietzsche's worldview is based on one substance, to which he refers using a great variety of words. This one substance is not unified but consists of various will-to-power quanta, to which he also refers as "force" or "energy." When various will-to-power quanta are unified, will-to-power constellations come into existence. In his notebooks he tried to develop a will-to-power metaphysics. By doing so Nietzsche was the only thinker who further developed Leibniz's monadology. However, Nietzsche's was based upon only one, nonunified substance, as "will can only act upon will." Because one type of substance is all there is, time and space have to be aspects of it and cannot exist separately from it. In addition, this substance is finite, as Nietzsche tries to banish all aspects of infinity from his philosophy.

In practice, this undertaking has the following implications: An infinite force cannot exist; therefore the existing force has to be finite. Infinite divisibility cannot exist; therefore the divisibility of force has to be finite. As a further premise we thus get the following: Force is not divisible infinitely often but turns up only in certain quantities. The logic or law of change demands that it be based upon a certain, specific law, which Nietzsche described in detail in his will-to-power metaphysics. It is decisive for eternal recurrence that such a law exists and that it is unchanging. Furthermore, the reversibility of all states of the world at an instant has to be given. Nietzsche solves this problem by reference to the law of conservation of energy. All the reflections just mentioned were not made by either Nietzsche or Heraclitus. However, they correspond to Nietzsche's and Heraclitus's way of thinking. We can summarize the foregoing as follows:

1. There is only one substance.
2. Time and space are aspects of one substance but do not exist independent of it.
3. The complete amount of substance is finite.
4. The divisibility of the substance is also finite, that is, there is a minimal amount in which the substance exists that is not divisible any further, whereby the size is different from zero.
5. There is a certain law that is the basis for change of the one substance.
6. The reversibility of all states of the world at an instant has to be assumed.

Conclusion: In such a system, substance can have only a finite number of states.

Once all states are present, then, one state has to recur, as the possibility of recurrence is given. If one state of the world at an instant recurs, and if the law governing the change does not change either, all other states of the world have to recur in an identical manner. As infinity is excluded, we cannot talk about infinite time either. As time is an aspect of the one substance, it also has to be identical with the previous time. (The word "previous" here has to be understood solely as a metaphor.) Such an understanding of time can be applied only if time is circular and finite. On that basis, the objection that this cycle is different does not apply either, as it is not different. This objection goes as follows: This cycle is different from all the other cycles, for we are in this cycle and we are conscious of being within this cycle, which does not apply to the cycles *n* plus 1 or 2. Hence this cycle cannot be identified with the other cycles. However, individuals who argue in this manner do not take the notion of circularity seriously, because if they did, they would know that there is only this one cycle.

Space

The concept of eternal recurrence also has interesting implications for the concept of space. According to both Nietzsche and Heraclitus, there is only one substance, which changes on the basis of one determined law that represents one aspect of the substance. All things and qualities are aspects of the one existing substance; that is, space and time are such aspects, too. Because the amount of substance is finite, it is also limited at its end. Limitation can only come about by something other, so that substance would have to be limited by another something. If such a something limited the substance, then it would exist outside of the substance. However, there can be only one substance.

In addition, substance can interact only with itself; two absolutely different substances cannot interact. Hence, the one substance cannot be limited by another substance. It cannot be limited by a vacuum either, as this would imply that there is an empty space. However, space exists only as an aspect of the substance, so empty space cannot exist. As substance can be limited neither by another substance nor by empty space, it has to be limited by itself. If the substance is limited, and if it can be limited only by itself—the amount of substance being finite and space being one aspect of the substance, then space can only be regarded as curved. Even though the concept of curved space was developed only about 150 years ago by first Georg Friedrich Bernhard Riemann and then Einstein, it has to be implicitly included in the philosophy of all those thinkers like Heraclitus and Nietzsche who defended the great year or the eternal recurrence of everything.

Science

In addition, we can argue that even on the basis of contemporary physics, the premises necessary for the occurrence of the eternal recurrence to occur are such that corresponding premises can be found there:

1. Time and space have aspects of ones subtance: Since Einstein formed his general theory of relativity, this has been the dominant scientific paradigm.
2. Energy or substance is limited: Scientifically, this is a very probable premise. If the density of matter is sufficiently high, the Friedmann equations, which can be deduced from Einstein's equations of the general theory of relativity, demand a closed universe. If the mass of neutrinos were zero, then when they are not moving, the density of matter of the universe would be great enough for a closed universe.
3. Force is limited, which implies that it is being limited in amount and in its divisibility: Again, a comparable premise can be found in contemporary physics. Energy can turn up only as an integer multiple of the Planck constant.
4. There is a determined law responsible for change: Physicists attempt to find a grand unified theory (GUT) in which they hope to unite the four basic forces: the weak force, the strong force, the electromagnetic force, and gravitation. The GUT can be seen as the determined law demanded by eternal recurrence.
5. The reversibility of all states has to be given: The law of the conservation of energy alone might be sufficient to imply this premise. However, one can also argue against the validity of this premise by referring to the second law of thermodynamics, which implies the permanent striving for entropy. However, there are natural scientists, such as Benjamin Gal-Or, who hold that the second law of thermodynamics might just be connected to the expansion of the universe. We might have antientropy during the collapse of the universe.

Given the foregoing comparison between the premises of the eternal recurrence and of modern science, we can see that even from a scientific standpoint, eternal recurrence bears some plausibility. However, this does not mean that eternal recurrence can simply be identified with the physical theory of the big bang and the expansion and contraction of the universe, because here problems concerning the understanding of singularity come in. Stephen Hawking's and Roland Penrose's understanding of singularity is dominant today, and it causes problems for this kind of understanding of the eternal recurrence. This again does not mean that all cosmological interpretations of the eternal recurrence have to fail. With the eternal recurrence or the great year, Nietzsche and Heraclitus developed concepts of time that are very different from our everyday perception of time, but once grasped, these concepts inspire a lifelong fascination.

Stefan Lorenz Sorgner

See also Cosmology, Cyclic; Eternal Recurrence; Heraclitus; Nietzsche, Friedrich; Presocratic Age; Time, Cyclical

Further Readings

Danto, A. (1965). *Nietzsche as philosopher.* New York: Columbia University Press.

Sorgner, S. L. (2001). Heraclitus and curved space. *Proceedings of the Metaphysics for the Third Millennium conference,* Rome, September 5–8, 2000 (pp. 165–170).

Sorgner, S. L. (2007). *Metaphysics without truth—On the importance of consistency within Nietzsche's philosophy* (Rev. ed.). Milwaukee, WI: Marquette University Press.

Nirvana

Nirvana is from a Sanskrit word that means "extinguishing." It is also spelled *nibbana.* The idea of nirvana is found most prominently in Buddhism, although Hinduism alludes to the idea. Buddhist doctrine includes the idea of *karma* (one's actions), *dukkha* (suffering), and *samsara* (reincarnation). Buddhists live a life filled with karma, acquired through positive and negative actions. When a Buddhist dies, the life is continued through rebirth; the condition of the rebirth is determined by the karma acquired in previous lives. In addition, Buddhism teaches that life is filled with dukkha. The acknowledgment of dukkha ties a Buddhist to the stream of samsara. By realizing that the essence of a Buddhist is not the accumulated lives lived but instead is separate from the experiences of the past and present lives, a Buddhist experiences enlightenment, which leads to nirvana. Through a reduction in karmic actions plus an awareness that a state of being exists beyond the physical one of samsara, the Buddhist can attain nirvana.

Nirvana is neither a physical nor a spiritual place; therefore it is beyond questions of where and when. No description of nirvana has been given, although descriptions of its attributes exist. In the *Dharmapada Sutra,* the Buddha described nirvana as the highest happiness (or bliss). Yet this happiness and bliss is not to be confused with that state experienced while in the process of transmigration, but with a state of calmness that transcends the earthly experience. A Buddhist who is "in" nirvana will not "leave" nirvana; death does not again occur. Siddhartha Gautama (or Gotama; 5th century BCE) realized how to reach enlightenment without aid and so became the Buddha. Upon the death of his last life (or his "cessation") Siddhartha entered into a state of nirvana. He is currently "in" nirvana and will no longer die or return to a physical form.

Entrances into nirvana, and the definition of nirvana itself, differ between the two larger groups within Buddhism: Theravada and Mahayana. Theravada ("doctrine of the elders") Buddhism teaches that each Buddhist must attain nirvana individually. The Buddha left teachings that can guide Buddhists toward reducing suffering and negating karma, but these must be attained alone. In addition, only a monk can attain enlightenment. Upon being enlightened, that is, upon realizing that illusion of rebirths and of suffering, the Theravada Buddhist becomes an *arhat,* or one who will not be reborn and enters into nirvana. Consequently, Theravada Buddhism necessitates a long period of rebirths in which to reach nirvana. For this reason, it is also known as Hinayana, or "lesser raft" Buddhism, because of the smaller number of Buddhists who reach the state of nirvana during a set time period.

Mahayana Buddhism teaches that each Buddhist has help from those who have been enlightened and decide to remain in the physical world of rebirths in order to aid other Buddhists. These previously enlightened individuals are called *bodhisattvas* (as opposed to arhats in Theravada Buddhism.) Tibetan Buddhism, a branch of Mahayana Buddhism, gives these bodhisattvas the name of lama, with the chief lama being known as the Dalai Lama. The highest goal in Mahayana Buddhism is to reach enlightenment yet decline to enter into nirvana, instead choosing to return in future rebirths. These bodhisattvas experience death and rebirth, but it is their choice to delay entrance into nirvana in order to aid others.

Within the Mahayana Buddhism tradition exists a writing known as the *Nirvana Sutra* (or *Mahaparinirvana Sutra* or *Maha-nirvana Sutra*). Mahayana Buddhists believe this is the last teaching of the Buddha. This writing teaches a different understanding of nirvana. The *Mahaparinirvana Sutra* also draws a distinction between nirvana and mahaparinirvana. Nirvana is the realization that suffering is an illusive aspect of one's physical lives. Mahaparinirvana is the realization of one's Buddha-nature. This idea of the Buddha-nature is unique to Mahayana Buddhism. The Buddha-nature exists in everyone and is the capacity of understanding reality. Therefore, the state of Mahaparinirvana consists of realizing the true nature of suffering plus an awareness of one's Buddha-nature.

The word *nirvana* is found in Hinduism. Hinduism also teaches the doctrines of karma and rebirths. The idea of a break in the cycle of rebirths differs from Buddhism, though, for in Hinduism one is connected with Brahman, the impersonal spirit. The word *nirvana* as a Hindu idea occurs primarily in the *Mahabharata,* an epic poem within Hinduism, and specifically a writing within the *Mahabharata* called the *Bhagavad Gita.* Within the *Bhagavad Gita,* the idea is linked with the Hindu idea of moksa or "release" from the rebirths: "Only that [one] whose joy is inward, inward his peace, and his vision inward shall come to Brahman and know nirvana. All-consumed are their imperfections, doubts are dispelled, their senses mastered, their every action is wed to the welfare of fellow-creatures: Such are the [ones] who enter Brahman and know nirvana." Thus, the Hindu ideal is for liberation from the passions and acknowledgments of a physical life; when this occurs, one experiences moksha or nirvana, and is united with Brahman. Consequently, the idea of nirvana as found in Hinduism differs from the nirvana as taught in Buddhism, although both are similar in that nirvana is the experience of and result from being released or enlightened.

Buddhist doctrine also includes the concept of parinirvana. *Parinirvana* is the term given to the death of one who has been enlightened yet chooses to remain in a physical state a while longer. In Mahayana Buddhism, the day that celebrates the Buddha's final death is known and celebrated as Parinirvana (or Nirvana) Day. The Buddha died in present-day Kushinigar, India, where today exists a parinirvana temple commemorating the site where the Buddha left his last physical life (died). Inside the temple lies a statue of the Buddha at his moment of parinirvana, or moment when he ceased his earthly lives and entered into nirvana.

Mark Nickens

See also Buddhism, Mahayana; Buddhism, Theravada; Buddhism, Zen; Hinduism, Mimamsa-Vedanta; Hinduism, Nyaya-Vaisesika; Hinduism, Samkhya-Yoga; Reincarnation

Further Readings

Harvey, P. (1990). *An introduction to Buddhism: Teachings, history and practices.* New York: Cambridge University Press.

Rahula, W. (1974). *What the Buddha taught.* New York: Grove Press.

Noah

Noah is the main character of the cataclysmic flood story told in the biblical Book of Genesis (chapters 6–9). Noah represented the last generation of the era of the antediluvian Creation and became the father of the current world order, thus a link between the ancient past and the present.

Noah's Ark, *by the Französischer Meister ("The French Master"), Magyar Szépművészeti Múzeum, Budapest. c. 1675. The ark story told in Genesis has parallels in the Sumerian myth of Ziusudra, which tells how Ziusudra was warned by the gods to build a vessel in which to escape a flood that would destroy humankind. The deluge story is one of the most common folk stories throughout the world.*

According to the Book of Genesis, Noah lived nine generations past the first created humans—Adam and Eve. Adam and Eve introduced sin into the world by disobeying God when they ate from the tree of knowledge of good and evil. This sin escalated throughout Creation to the point where all humanity was wicked by Noah's time. Then God regretted he had made humanity and decided to destroy the world with a flood. However, God chose to spare Noah, the only righteous man, and his family. God ordered him to build a massive ark upon which Noah, his family, and a pair of animals representative of each species would be spared from a watery annihilation. Noah obeyed, built the ark, and survived the deluge. After the flood, the ark landed on Mt. Ararat, a mountain located somewhere in the area that is now eastern Turkey, where Noah and his family disembarked.

Noah represents the best and worst of humans. He represents the best because before the flood God deemed him alone as righteous and blameless of all the people in the world (Gen 6:9). He is the only one who deserved to be spared from destruction. However, he also represents the worst of a corrupt world. After the flood, Noah became drunk and passed out. Ham, one of Noah's sons, sinned when he "saw the nakedness of his father" (Gen 9:22). Therefore, through Noah sin carried over from the initial Creation into the postdiluvial world.

Noah is one of many heroes of ancient Near Eastern flood stories. Ziusudra, hero of the Sumerian flood story, overheard the gods plotting to wipe out humanity. He was saved from a 7-day and 7-night flood by making a boat. Afterward, he offered a large sacrifice to the gods. Being pleased, the gods gave him eternal life. Atrahasis, hero of the Akkadian flood story, is warned by the god Enki that the gods intended to flood the earth and destroy humanity. Atrahasis built a boat and escaped the flood. Utnapishtim, hero of the flood story in the Gilgamesh epic, has an experience similar to that of Atrahasis, except after the flood Utnapishtim offered a sacrifice to the gods, who were so thankful that they gave Utnapishtim and his wife eternal life and placed them in a distant location separate from the rest of humanity.

The story of Noah has two timeless elements. First, the story has a theological connection, that is, the seriousness of sin and the grace of God. This theme is the basis of the flood story and is prominent throughout the Hebrew Bible. Second, the story of Noah has inspired some people to search for the remains of the ark, in a sense to find a modern physical connection with the ancient past. Some people believe that the ark landed on the Agri Dagh peak in the Ararat mountain range. However, to date no evidence of the ark has been found.

Terry W. Eddinger

See also Adam, Creation of; Bible and Time; Genesis, Book of; Moses; Sin, Original

Further Readings

Hunt, J. H. (2003). Noah. In T. D. Alexander & D. W. Baker (Eds.), *Dictionary of the Old Testament: Pentateuch*. Downers Grove, IL: InterVarsity.

Westermann, C. (1987). *Genesis 1–11: A commentary* (J. J. Scullion, Trans.). Minneapolis, MN: Augsburg Fortress.

Nostradamus (1503–1566)

Michel de Nostredame, more commonly known as Nostradamus, is criticized by some and acclaimed by many as a seer whose 942 quatrains (poems with four lines each) transcend time and predict the future. Born on December 14, 1503, in Saint Remy de Provence in the south of France, Nostradamus was the first son and one of nine children born to Reyniere de St. Remy and grain dealer and notary Jaume de Nostredame. He is best known for his book *Les Prophecies* (1555) and *The Almanac,* a set of annual predictions published until his death on July 2, 1566.

Historically, the Nostredame and St. Remy families had strong academic ties. Several family members were doctors and scholars. Although originally of Jewish descent, the family converted to Christianity in 1502 as a consequence of the ascension of Louis XII. Starting at the age of 15, Nostradamus followed family tradition and attended the University of Avignon for more than a year. An outbreak of the black plague forced the university to close its doors, leading Nostradamus to pursue a career as an apothecary. In 1529 he entered the University of Montpelier to study for a doctorate in medicine, but upon the discovery of Nostradamus's occupation as an apothecary, he was expelled. Despite his lack of university credentials, many of his colleagues and publishers still addressed him as Doctor.

Following his expulsion, Nostradamus married and had two children, but his wife and children succumbed to the plague in 1534. Thirteen years later, in 1547, Nostradamus married a wealthy widow named Anne Ponsarde Gemelle, and the two had six children: three sons and three daughters.

In 1550, *The Almanac* was first published, and Michel de Nostredame became widely known under the name Nostradamus. *The Almanac* was so successful that he decided to publish his predictions annually.

Today, many admirers of Nostradamus believe that he accurately predicted such events as the great fire of London, the rise of Adolf Hitler, the crash of the American space shuttle *Challenger,* and the September 11, 2001, attack on the World Trade Center and the Pentagon. Overall, Nostradamus made 6,338 predictions, many of which, however, never came to fruition. In any event, skeptic and follower alike cannot contest that Nostradamus left his mark; his prophesies remain an ongoing source of speculation and debate. In the wake of the events of September 11, 2001, an Internet hoax cited a Nostradamus quatrain that, it was claimed, predicted the fiery attack. Later, to the delight of skeptics, the quatrain was proven false. As both skeptics and neutral observers have noted, Nostradamus's prophecies are ambiguous because they fail to provide specific dates; therefore these predictions are open to numerous and contradictory interpretations.

Jennifer R. Fields

See also Futurology; Prophecy; Toffler, Alvin

Further Readings

Hogue, J. (1987). *Nostradamus & the millennium: Predictions of the future.* New York: Doubleday.

Lemesurier, P. (1997). *The Nostradamus encyclopedia: The definitive reference guide to the work and world of Nostradamus.* New York: St. Martin's.

Nothingness

The term *nothingness* denotes the result of specific negation of a reality (metaphysical or ontological meaning) and/or of a value and validity of something (axiological meaning). Its content is close to the terms "nothing," "nonbeing," "emptiness," and "vanity."

As opposed to the common meaning of the word "nothing" ("nonbeing"), which relatively negates the attributes, state, or existence of a particular object (or a category of objects), *nothingness,* in its ontological meaning, is concerned with absolute negation of the whole of entities, thus of any particular object (or category of objects) at all. For example, in the sentence "Yesterday, there was a glass on the table, but today there is nothing there," by "nothing" we mean only the so-called relative nothing—the absence of the glass in the given place and not the fact that there is no other particular object in the place of the expected glass. In contrast, in the sentence "The world was created from nothing," by "nothing" we mean exactly the absence of any particular object (category of objects), absolute nothing, nothingness. In the first case, the absence (nothing) of something is the presence (existence, entity) of something else, other than the negated thing. In the second case, this does not apply: Absence (nothing) is the absolute opposite of any presence (existence, entity). Because of this negation of all that exists, the concept of nothingness is often considered the result of illegitimate abstraction and therefore unacceptable for rational thinking. To favor it during the explanation of reality is considered to be (ontological) nihilism. This applies even more so to its stronger alternative, which negates not only all reality but also possibility (so-called negative, or absolute nothingness).

Nothingness in its axiological meaning is in human experience and its reflections thematized in several ways. Depending on which aspect of a value it negates (positive, negative, or both), the basic positions of value nihilism vary from the extreme pessimistic, for example when we sigh along with the preacher, "Vanity of vanities! all is vanity"; through the more moderate one, which answers the question of what humanity is with "A Nothing in comparison with the Infinite, an All in comparison with Nothing" (Pascal); all the way to the mystic ideal of a mind emptied of values (neutral), for example in the state of nirvana. Depending on the kind of the negated value, we differentiate between ethical nihilism (negating the obligatory validity of moral values), religious nihilism (denying the existence of the sacred or gods), political nihilism (negating the obligatory validity of social order), logical nihilism (denying the existence of truth), and gnoseological nihilism (rejecting the possibility of knowledge).

The concept of nothing, or nothingness, as an ultimate explanation principle of reality and human life, played a more important role historically in Asian religious and philosophical thinking than it did in the thinking of Western civilization. It was also to a great extent particular to mystically inclined thinking rather than to rationalistic systems. For example, in the ancient Chinese doctrine of Laozi (5th century–4th century BCE), the basic principle of reality, *tao,* is empty—nameless and formless; later in the writings of Zuang-zhou (4th century BCE) it is the emptiness itself—*xu.* In ancient India, it was mainly the teaching about *nirvana* (Hinduism, Buddhism), the content of which is the emptying of the mind of an individual of partial contents and its transition into the state of "nonbeing." But most of all the doctrine of emptiness (*sunyata*), understood in an ontological as well as axiological way, developed in Mahayana Buddhism. The concept of nonbeing (*abhava*), also in absolute meaning, can be found as one of the categories even in the logically tuned system Vaisesika. Later it was Zen Buddhism that attempted a synthesis of the ways of Buddhism and Taoism, also in the area of the understanding of emptiness as a central concept. In Jewish mysticism, a similar function of holy nothingness was carried by the concept of *ein sof.*

Unlike oriental doctrines, ancient Greek philosophy was more hostile toward the concept of nothing and nonbeing (nothingness). According to Parmenides of Elea (540–450 BCE), being (*on*) is and nothing (nonbeing—*me on*) is not, because it is unthinkable, as thinking and being are the same thing. Nothing (nothingness) as the absolute opposite of being should have been excluded from the rational discourse. But already Democritus (450–370 BCE) was forced in his philosophy, in an effort to explain plurality and motion, to postulate alongside being (plenitude, atoms) also the existence of nonbeing—of empty space. Even Plato (428–348 BCE) admits in a certain way to the existence of nonbeing, in order to explain, for example, the possibility of an error in human knowledge, because an erroneous statement is a statement about the nonexistent. Nonentity (nothing), however, is not the opposite of being, as it was with Parmenides, but it's a part of it; it is not absolute nonbeing, but merely relative—it is other being, the being of other, because each determined entity, as identical with its own self (*tauton,* is also at the same time different from another determined entity (*heteron*). For example, motion *is* motion (itself), but it *is not* stability (something other). In another context, in an attempt to clarify the changeable nature of the perceptible world, Plato again admits to the existence of nonbeing (matter), in which this world participates, as being similar to being itself (ideas). Nonbeing not only *is not* any determined other, but also it is not itself; it is presented here as a generalized principle of an undetermined otherness. Later, for Aristotle, with his differentiation of matter (possibility) and form (reality) as the necessary aspects of the existence of every object, nonbeing is a not yet realized entity, *ens potentialis,* which lacks some or all attributes of a future thing; it is rid of its positive determination (lack of being, *privatio*).

Greek philosophy, despite its plurality in the understanding of nonbeing, in principle did not overstep its rationalist rule according to which "out of nothing, nothing comes" (*ex nihilo nihil fit*). On the contrary, medieval Christian thinking acknowledged the concept of the absolute nothing, nothingness as a legitimate element of discourse about reality and man, by the fact that on the basis of its credo, it explained the origin of the world as "creation from nothing" (*creatio ex nihilo*). It was mainly Saint Augustine of Hippo who, in an attempt to demonstrate the unlimited nature of God's power, theoretically justified this change, whereas he also endowed nothingness with a morally sacral dimension. Nothingness was worthless, connected to evil and sin. As a result of this, all creatures carried its dual seal—ontological as well as axiological.

A new aspect was added to meontological problematics by thinkers influenced by Neoplatonic mysticism, most of all Pseudo-Dionysios (5th century CE), Meister Eckhart (1260–1327), and Nicholas of Cusa (1401–1464), who on the grounds of negative theology applied the concept of nonbeing (nothing) through various ways even to God himself. However, in their case, it was not nothingness in the sense of absence or lack of being (or its determination), but the opposite, nothingness "resulting" from surplus of being, because the infinite and absolute God, according to them, cannot be identical with being that is specified as an entity by finite determinations. God is beyond it and above it; he is incomparably (even in terms of worth) more than being itself. If, despite all differences, this thinking was closer to

Plato's ancient message, Aristotle's reflections on lacking (*ens potentialis, ens rationis*) were developed mostly through scholastic tradition (Saint Thomas Aquinas, 1225–1274; Francisco Suarez, 1548–1617; and others). Later it was followed by the classics of modern philosophy—Baruch de Spinoza (1632–1677), G. W. Leibniz (1646–1716), and Immanuel Kant (1724–1804).

Significant for further development of these problematics were Georg Wilhelm Friedrich Hegel's reflections on nothing (and negativity as such). Hegel (1770–1831) in a quasi-rationalistic reception of mystical views, gets almost as far as negation of the logical law of contradiction through a provocative statement that "pure being and pure nothing are therefore the same," whereby their truth is becoming. Here, the negativity of nothingness becomes the driving force of dialectics. These reflections, together with his master-slave dialectic, significantly inspired existentialists (Martin Heidegger, 1889–1976; Jean-Paul Sartre, 1905–1980; Albert Camus, 1913–1960; and others), as the main successors of the meontological problematics in the 20th century. According to Heidegger, who in his ontological difference distinguishes between entity and being, nothing is the absolute negation of the whole of entities. However, one does not encounter it in thinking, but in anxiety. Nothing, nonbeing (which paradoxically is the veil of being) enables human being (*being-there, dasein*), because for one to be means to be held out into this nothing, to transcend one's self. Later, Sartre uses nothingness, as opposed to being in-itself, as being for-itself, that is, empty, contentless, and undetermined consciousness that allows human freedom, because paradoxically "it is what it is not and it is not what it is." This approach received great criticism from analytically oriented philosophy and particularly from logical empiricism. According to Rudolf Carnap (1891–1971), violation of logical rules of language lies behind Heideggerian reflections, when for example the word *nothing* is used as a name of an object, whereby it is a form of negation, used to create negative existential statements. All thoughts of a similar kind are therefore in their ultimate effects meaningless.

Marián Palenčár

See also Aquinas, Saint Thomas, Aristotle; Augustine of Hippo, Saint; Becoming and Being; Eckhart, Meister; Hegel, Georg Wilhelm Friedrich; Kant, Immanuel; Leibniz, Gottfried Wilhelm von; Metaphysics; Nicholas of Cusa; Ontology; Parmenides of Elea; Plato; Spinoza, Baruch de; Time, Nonexistence of

Further Readings

Carlson, E., & Olsson, E. J. (2001). The presumption of nothingness. *Ratio, 14,* 203–221.

Gale, R. M. (1976). *Negation and non-being.* Oxford, UK: Blackwell.

Heidegger, M. (2000). *Introduction to metaphysics* (G. Fried & R. Polt, Trans.). New Haven, CT: Yale University Press. (Original work published 1953)

Sartre, J.-P. (2003). *Being and nothingness: A phenomenological essay on ontology.* New York: Routledge. (Original work published 1943)

Novels, Historical

The historical novel is a genre of literature whose story is set during a period that predates the author's own time, often by a significant number of years. A historical novel generally involves substantial research by the author concerning details of the period. The genre became widely popular during the 19th century Romantic period, advanced by great novelists such as Sir Walter Scott.

The purpose of the historical novel extends beyond that of entertainment, though many excel at this in their own right. Authors have often intended to deliver a message, advance a cause or ideology, or popularize history and present a time period to the public; none of these intentions is necessarily exclusive of the others. Historical novels are commonly set during eventful periods in human history, depicting a conflict or a transitional moment in time. Some historical novels span a lengthy duration and may include many accurate details about the past.

A prominent example of a historical novel that deals with the notion of time is Mika Waltari's *The Egyptian,* published in 1945. First written in Finnish, *The Egyptian* is set during the reign of the pharaoh Akhenaton, more than 3,000 years ago. The novel is centered on a fictional character, Sinuhe, the personal physician to the pharaoh, who recounts the tale of the pharaoh's decline and fall. The tale also parts from Egypt and describes Sinuhe's extensive travels throughout the ancient world. Published immediately following World War II, Waltari's novel was intended to explore the violence and brutality of the human condition and to imply that this has changed little from ancient to modern times.

The legendary Greek conqueror Alexander the Great has been depicted in numerous works of literature. Nikos Kazantzakis's *Alexander the Great,* written in the 1940s, is one modern example intended primarily for a younger audience. Kazantzakis was already a prominent author and philosopher by the time he penned the novel. *Alexander the Great* is a flattering depiction of the hero from a Greek author with a sense of pride in Greek history. However, Kazantzakis does not entirely succumb to glorification of the hero, presenting his faults and human qualities as well. The modern reader can thus relate to an ancient heroic figure.

The intention of delivering a powerful message through a classic work is exemplified by Henryk Sienkiewicz's historical novel *Quo Vadis: A Narrative of the Time of Nero.* Published in 1895, *Quo Vadis* is the Polish author's most famous work. The story is centered on a romance between a Roman patrician (Marcus Vinicius) and a young Christian woman (Lygia) at a time when Christians faced violent persecution by the Roman authorities. The novel conveys a strongly Christian message, implicit in the title as well as in the vivid depictions of Christian martyrdom. Sienkiewicz received the Nobel Prize for Literature in 1905. *Quo Vadis* has withstood the test of time, having been adapted to several film interpretations.

Popular interest in the time of imperial Rome and the birth of Christianity is evident by the reception of *Ben-Hur: A Tale of the Christ* in 1880. The author, Lew Wallace, was an American Civil War general, politician, and novelist. The book has been acclaimed for its accurate descriptions of the Holy Land of 2,000 years ago, though Wallace had never set foot there. Set during the reign of Tiberius, *Ben-Hur* is a tale of a Jewish aristocratic named Judea Ben-Hur who is falsely accused of murder by a Roman officer, Messala. Like *Quo Vadis,* the story deals with a pivotal moment in time as Christianity emerges within the Roman Empire. *Ben-Hur's* exceptional popularity helped to make the historical novel a popular literary and cinematic genre in the United States.

The author of *Ivanhoe,* Sir Walter Scott, is often credited as the father of the historical novel. Scott's 27 historical novels established the standard structure of the genre and greatly influenced later writers. His interest in the European Middle Ages is reflected in *Ivanhoe,* published in 1819. The story is set in 12th-century England during the time of King John. *Ivanhoe* is not only a tale of chivalry, combining fictional characters and actual events, but also a critique of the persecution of Jews in England. *Ivanhoe* helped to rekindle popular interest in the Middle Ages during the 19th century.

Leo Tolstoy deviated from the conventional novel with *War and Peace,* published as a series between 1865 and 1869. This ambitious story is set during the Napoleonic period and specifically during the Russian campaign. The Russian author spurned the "great man" paradigm of history in favor of capturing the daily human struggles during warfare. As such a human, Napoleon does not fare well in Tolstoy's depiction, whereas the personal interactions within Russian society form the narrative. This notion of "history from below" is central to Tolstoy's understanding of time and the movement of events.

Perhaps the most significant antiwar novel written is Erich Maria Remarque's 1929 classic *All Quiet on the Western Front.* Remarque's novel was timely; it was published in German during the interwar period. The novel is set during the First World War and is narrated by a young German soldier, Paul Baumer, who is engaged in the infamous trench warfare emblematic of the conflict. Baumer experiences the horrors of war and comes to recognize the deception of blind nationalism. The novel breaks sharply from the traditional portrayal of warfare over time, in which it has generally been glamorized. So powerful is the message in *All Quiet on the Western Front* that the novel was banned as subversive by the Nazi party in 1933.

A host of other distinguished novels could be added to this list. All of the aforementioned historical novels have film adaptations, which have also contributed to the popularity of the genre. Historical novels continue to connect readers to the past and to the passage of time.

James P. Bonanno

See also Dostoevsky, Fyodor M.; Flaubert, Gustave; Novels, Time in; Tolstoy, Leo Nikolaevich

Further Readings

Buckley, J. H. (1967). *The triumph of time: A study of the Victorian concepts of time, history, progress and decadence.* Cambridge, MA: Belknap Press of Harvard University Press.

Lukacs, G. (1978). *The historical novel* (H. Mitchell & S. Mitchell, Trans.). London: Humanities Press.

McGarry, D., & White, S. H. (1963). *Historical fiction guide: Annotated chronological, geographical, and topical list of five thousand selected historical novels.* New York: Scarecrow Press.

Saintsbury, G. E. B. (1971). *The historical novel.* Folcroft, PA: Folcroft Library Editions.

Novels, Time in

Time in novels is based on a fundamental duality. Günther Müller, a German literature theorist, was the first who reflected thoroughly on this duality. In 1948 he introduced the opposition between *erzählte Zeit* (story time) and *Erzählzeit* (narrative time), a literary terminology that has gained international acceptance. *Story time* designates the chronology of the events told—that is, the time of the story—whereas *narrative time* means the time of the narrative presentation of the story; in other words, the time of the plot. The temporal complexities that result from the relation between story time and narrative time were studied in detail by Gérard Genette, a French literary theorist who is primarily associated with the structuralist movement. In his canonical work, *Narrative Discourse: An Essay in Method* (1972), Genette proposes that time in novels may be classified in terms of order, duration, and frequency. We shall follow Genette's terminology, because it fills the need for a systematic theory of time in narrative texts.

The first main category discussed by Genette is the category of *order.* It designates the discrepancies between, on the one hand, the temporal order of a series of real or fictitious events connected by a certain chronology and, on the other hand, the temporal arrangement of these events in the narrative discourse. Genette pays attention to the fact that events occur in one order but are narrated in another. This discordance between the temporal succession of events in the story and their arrangement in the plot is called *anachrony.* Anachronical relations between story and plot are realized mostly by means of analepses and prolepses. *Analepses* are narrative episodes that take place earlier than the temporal point of departure of the narrative into which they are inserted. They are narrative retrospections or backflashes. The framing narrative is the first narrative, containing the analeptic sequence as the second narrative. *Prolepses* are temporal anticipations, narrative episodes that take place later than the temporal point of departure of the narrative, into which they are inserted. The framing narrative that contains the proleptic sequence is the first narrative, whereas the framed episode is the second narrative. Yet a careful analysis of the temporal order of narrative texts is not finished with the mere identification of analepses and prolepses, which must be further specified in respect to reach and extent. *Reach* designates the temporal distance of events told in the analeptic (proleptic) sequence from the moment in the story when the narrative was interrupted to make room for the anachrony. To determine the reach of analepses (prolepses) the reader must ask, "How long is the temporal distance between first and second narrative?" *Extent* is used to describe the duration of the story that is covered by the analeptic (proleptic) sequence. In this case the reader's question is, "How long is the period of time told in the second narrative?" In addition, analepses (prolepses) may contain further anachronies. These narrative sequences already framed by an anachronic episode are analepses (prolepses) of second degree.

Duration is the second main category discussed in Genette's *Narrative Discourse.* The duality of story time and narrative time allows novelists to control precisely the speed of their narratives. Duration is defined by the relation between the length of the story (measured in seconds, minutes, days, etc.) and the length of the text used to describe it (measured in lines, pages, chapters, etc.). The discrepancies between the length of the story and the length of the text are *anisochronies* (variations in speed). Genette determines four types of speed variations: summary, scene, pause, and ellipses.

- *Summary* means that the period of time told in a story is much longer than the narrative sequence used to describe it. Summaric episodes may cover 50 years in three or four sentences. Usually they are used for the transition between two scenes; they form the background against which scenic presentations stand out.
- *Scene* designates the correspondence between the period of time told in the story and the length of the narrative that describes it. Most often scenic episodes are characterized by dialogues, they present important conversations between the protagonists as well as dramatic events in their real length.
- *Pause* means that the speed of a narrative text is remarkably slowed down. Descriptive pauses are realized by epic ecphrasis or digression. The length of the narrative discourse of a pause may cover eight or nine pages, though it tells nothing about the further development of the story. Descriptive pauses stop the temporal progression of the story while the narrative discourse moves on constantly.
- *Ellipses* are gaps in the temporal continuity of the story. The period of time left out may cover hours, days, years, and more. For a precise knowledge of ellipses, readers have to consider the elided story time. Thus, their first question is to know whether the temporal elision is explicitly announced by quick summaries of the type "some time later" (explicit ellipses) or left out without any comment (implicit ellipses). Their second question is to know whether the duration of these time gaps is indicated (definite ellipses) or not (indefinite ellipses).

Frequency is the last main category discussed by Genette to describe the complex relations between story time and narrative time. It designates the relation of repetition between the events told in the story and their presentation in the narrative discourse. The frequency of a narrative text determines its temporal rhythm. Genette works out four types of frequency: narrating once what happened once, narrating *n* times what happened *n* times, narrating *n* times what happened once, and narrating once what happened *n* times. The first two types are singular narratives. In singular narratives the number of events in the story corresponds exactly to the number of statements about these events in the narrative text: If something happens once in the story, it is mentioned once in the narrative discourse; if something happens twice in the story, it is mentioned twice in the narrative discourse, and so on. The third type of frequency is the repeating narrative. Repeating narratives are characterized by a discrepancy between the number of events told in the story and the number of statements about these events in the narrative. For example, something happened only once, yet it is mentioned two times or more in the narrative discourse. The last type of frequency is the iterative narrative. In iterative narratives regular events are reduced to one single utterance of the type "Every Monday he caught the train at 11 p.m."

Genette's *Narrative Discourse* exclusively analyzes the literary presentation of time with regard to the form of narratives. Yet time can also be the thematic center of a novel. When we study time as the theme of a narrative, we no longer refer to its form but to its content. In this respect a fundamental distinction of time is the one between objective and subjective time. *Objective time* designates the regular succession of minutes, hours, and days that is measurable with a watch or a calendar. It is a consistent kind of time without deviations. *Subjective time* is based on the perception of characters in a novel. It depends on their individual experience, capricious imagination, or rambling memory. Subjective time changes constantly; hours are stretched into months, years compressed into days.

A careful textual analysis of the duality between story time and narrative time has to consider the specific characteristics of both kinds of time, especially in their relation to form and content. In the following section (Time as Form) we shall focus on some of the most famous novels of world literature in which the literary presentation of time decisively determines the narrative form. The final section, Time as Content, discusses renowned novels in which time is a thematic center. Although the presentation of time in novels has to be considered separately in respect to form and content, the two aspects are not to be isolated from each other. It is exactly their correspondence, or their significant contrast, that constitutes the temporal complexity of a narrative piece of writing.

Time as Form

Thomas Mann's *The Magic Mountain* (1924) is one of the most prominent novels in which the presentation of time in the story is mirrored in the form of the narrative. The protagonist's subjective experience of, and his reflection on, time is formally expressed by variations in speed. Hans Castorp, an ordinary young man, takes a short break between his final examinations in engineering and his first job. He plans a 3-week visit to his cousin Joachim, who stays at a sanatorium in the Swiss mountains because he suffers from pulmonary tuberculosis. Though Hans's visit begins as a holiday trip, it turns into a stay of 7 years that is abruptly ended by the outbreak of World War I.

The spatial distance between the mountains and the so-called flatland turns the sanatorium into a hermetic place. The rhythm of time on the mountain is completely different from the rhythm of time in the plain where the young protagonist comes from. At first, Hans views life on the mountain from the perspective of the flatland. He accuses Joachim of wasting his time in the sanatorium and does not understand why for him and the other patients the smallest temporal unit is a month. Yet life in the sanatorium rapidly benumbs Hans's sense of time. Soon he has experienced all the details of clinical routine and knows by heart its daily, weekly, and even monthly rhythms. The protagonist begins to live in a perpetual present, lacking even the temporal clues of nature, because in the mountains all kinds of weather appear in disorder throughout the year. Paradoxically, the extratemporal present on the mountain inspires Hans Castorp to reflect about the essence of time and to question temporal categories he has previously taken for granted. The new patient learns by experience that time in the flatland (objective time) is completely different from the psychological experience of time on the mountain (subjective time). The protagonist's complete absorption in atemporality ends with the outbreak of the First World War: He has to do military service and dies, presumably in a battle.

Hans's subjective experience of time and his isolation in an eternal present find their formal expression in a significant acceleration of speed. The length of the text that describes the first 3 weeks on the mountain is about seven times longer than the summary of the following 3 weeks. Apart from Hans's discussion with the Italian humanist Lodovico Settembrini and his opponent Leo Naphta, the remaining time of the protagonist's stay in the sanatorium (altogether 7 years) is summarized even more briefly. Another outstanding event in respect to time is Hans Castorp's dreamlike vision of life and death in a blizzard during a solitary skiing expedition. For the presentation of this event, the narrative speed slows down again, just to accelerate even faster afterward. These variations of speed illustrate clearly that Mann tried to find a formal equivalent for the temporal experience of the protagonist.

James Joyce pursued a similar idea when he wrote *Ulysses* (1922), whose composition of time depends primarily on the author's new approach to characterization. Instead of describing the characters from outside, Joyce puts himself in their place and depicts them from inside. This radical internalization is called "stream of consciousness." It tries to present the thoughts and feelings of literary figures exactly as they pass through the characters' minds. Joyce was the first writer to use a stream-of-consciousness presentation continuously in his narratives. This new approach to characterization significantly influences the presentation of time in *Ulysses*. In respect to the story, internal characterization leads to an emphasis on subjective time. Objective time provides merely an external framework for the temporal structure of Joyce's novel. The major action takes place in the temporal experience of the main characters (Leopold Bloom, his wife Molly, and Stephen Daedalus). The intertextual references to Homer's *Odyssey* enrich the presentation of time in *Ulysses* on a symbolic level. Each episode of Joyce's novel has its parallel in the *Odyssey*. Yet the Irish author summarizes Odysseus's wanderings, which lasted around 10 years, in just 1 day: The action of *Ulysses* takes place exclusively on June 16th, 1904.

Because the story of Homer's epic poem is implicitly integrated into *Ulysses*, time in Joyce's narrative assumes a mythological quality. In respect to the form of *Ulysses*, the consequent use of the stream-of-consciousness technique provokes a deceleration of speed in the narrative discourse: Whereas the story time covers just one day, the narrative time is significantly longer. It is impossible to read Joyce's novel in precisely 24 hours, even if readers want to. The literary presentation of the character's thoughts and feelings consumes much more time than their actual succession in reality. Thus it is precisely the technique of the stream of consciousness that shapes the conception of time in *Ulysses* in respect to both form and content.

In contrast to the 1-day summary of *Ulysses,* the events in Homer's *Odyssey* (c. 700 BCE) cover a temporal period of approximately 10 years. Yet the passage of time is not reported continuously. The *Odyssey* is full of anachronic sequences that switch back to the protagonist's past or implicitly hint at his future. The anachronic presentation of time allows Homer to portray the wanderings of Odysseus exclusively in their final and decisive phase, which lasts about 40 days. All events that precede this starting point of the first narrative (from the fall of Troy until the protagonist's shipwrecking on Ogygia) are told in analeptic sequences by Odysseus himself or in the songs of a minstrel. A similar temporal order applies also to Homer's earlier work, the *Iliad* (c. 750 BCE), that centers around a quarrel between Achilles and Agamemnon. Having evoked the conflict between the characters that the narrator proclaims to be the starting point of his story, he switches back about 10 days to reveal the cause of the quarrel. The first analeptic sequence is thus inserted in the very beginning of Homer's epic poem. It reveals the cause of the conflict in about 140 retrospective lines. A closer look at Homer's epics makes clear, therefore, that the temporal structures of the *Iliad* and the *Odyssey* are based fundamentally on anachronies.

Returning to 20th-century literature, the most complex literary presentation of time can be found in Marcel Proust's *In Search of Lost Time* (also known as *Remembrance of Things Past,* 1913–1927). The reflection on time in the *Search* is exceptional in respect to form and content. As in Mann's *Magic Mountain* and Joyce's *Ulysses,* subjective time determines the story of Proust's narrative. In the *Search,* time assumes the form of involuntary memory (*mémoire involontaire*). The narrator-protagonist, Marcel, tries to remember his life. He learns that only the involuntary memory is able to bring back the complete image of the past immediately. The voluntary memory (*mémoire volontaire*) is under rational control and thus remains superficial. A canonical example of the workings of the involuntary memory is the so-called madeleine passage in the opening of the *Search.* A discrete similarity between present and past (in this case the unexpected flavor of a piece of madeleine cake that Marcel dips into a cup of tea) transports Marcel immediately back to his childhood at Combray. The involuntary memory evokes an overpowering recollection and grants Marcel a glimpse into the essence of the past.

Proust's literary presentation of the involuntary memory was significantly inspired by the philosophy of Henri Bergson. His philosophical study *Matter and Memory* (1896) appeared when Proust was 25. In the second chapter of his work, Bergson distinguishes two kinds of memory: the memory of habit and the pure or spontaneous memory. He argues that the latter is independent of our will. Proust's poetics can be called "Bergsonian" insofar as he implicitly adopts the philosopher's distinction between an intentional and an unintentional form of remembrance.

The involuntary memory also determines the narrative form of the *Search.* Although remembrance is a temporal process, the basic structure of the novel seems to be static and lifts the action out of time. The impression of atemporality is primarily evoked by the iterative character of the narrative form of the *Search:* Marcel tells us once what happened every Sunday in Aunt Léonie's household; he tells us once about the daily walks with his family. Regular repetition has turned these ordinary actions into a ritual. It is the iterative narration of ritualized events that evokes the impression of immovability and makes Marcel's remembrance of the past eternal.

The last important literary technique for the presentation of time we shall discuss is digression. The complexity of time in Laurence Sterne's *Tristram Shandy* (1759–1767) results primarily from the exuberant use of rhetorical digression. Digressions are excurses from the primary line of a narrative. They slow down the narrative speed or even cause it to stagnate. Rhetorical digressions thus traditionally assume the function of pauses: In a digressive sequence, the time of the story is stopped while the narrative discourse moves on. In *Tristram Shandy,* however, the use of digression is driven to the extreme and undermines its classical function. Digression causes an inversion of the ordinary flow of time: Having set himself the task to omit nothing of his life that is relevant, the narrator-protagonist,

Tristram, has to admit its impossibility. Up to the middle of the fourth volume of the novel, he has gotten no farther than the first day of his life. The more Tristram tells about his life, the more he will have to tell. He dwells upon long excurses while his own life is constantly proceeding but remains undocumented. Sterne's exuberant use of digression turns it into progression: His novel is digressive and progressive at the same time. Sentences are interrupted abruptly and not finished until 30 pages farther on. Yet the time in between is stuffed with oddities and strange encounters that enrich the characters by adding new details of their past lives. By driving digression to the extreme, the author creates temporal paradoxes that completely suspend the linear flow of time.

Rhetorical digression also shapes the presentation of time in Herman Melville's *Moby Dick* (1851). Although *Moby Dick* is a work of fiction, Melville included chapters that are largely concerned with an almost scientific discussion of whales. These chapters are known as episodes of cetology (from the Greek nouns *cetus* "whale" and *logos* "knowledge"). They significantly slow down the narrative speed of the story that focuses on Captain Ahab's furious obsession to kill Moby Dick, the white whale, who swallowed his leg. The cetology chapters form a story of second degree that parallels Captain Ahab's obsession on a more abstract level: They present the narrator's untiring effort to bring the mythic power of Moby Dick under rational control by scientific reflection. Yet, as Ahab drowns in a fight with Moby Dick, the intention of Ishmael, the narrator, remains unfulfilled, too. All the scientific approaches fail to give him an adequate account of the white whale's ineffable mystery and strength. The conception of time in *Moby Dick* is thus determined to a significant degree by rhetorical digression. In contrast to *Tristram Shandy,* however, digression keeps its traditional function in *Moby Dick:* It does not generate the whole novel but remains a deviation from the primary line of the narrative.

Time as Content

This section discusses in chronological order certain novels in which time is a central theme of the story and thus primarily relevant in respect to the narrative's content.

The content of Ovid's *Metamorphoses* (completed in 8 CE) unites three kinds of time: primeval time, mythological time, and historical time. Ovid starts his epic poem with the story of Creation. He proceeds with a literary description of the mythological history of the world and ends his narration in his own time under the rule of Augustus. The major theme of the *Metamorphoses* is, as the title already suggests, transformation or change. Divine and human beings are physically transformed into animals or plants. The story of each single transformation is intended to give a reason for the existence of things. Metamorphosis thematically unites the loosely connected episodes. It is a process in time that demonstrates that continuous change is the fundamental principle of the mythological as well as the terrestrial world. Containing versions of many of the most famous myths of Greece and Rome, Ovid's epic poem has become one of the cornerstones of Western culture.

Dante's *Divine Comedy* (c. 1300) opposes two time schemes: the historical time of the terrestrial world on the one hand, and the time of salvation history on the other. In the middle of his life, Dante (the author, narrator, and protagonist) has lost the straightforward pathway to God. To be purified again, he has to cross the three empires of the beyond: hell, purgatory, and paradise. Dante enters the eternal sphere as an individual person, though his figure has a representative meaning, too: Dante's voyage beyond the grave is also the voyage of humankind. By assuming both an individual and a representative meaning, the figure of the narrator–protagonist unites historical time and eternity. This double meaning applies also to the voyage that is not a mere movement through time and space but a spiritual journey. We can measure Dante's travel in historical time: It lasts about 7 days. Yet, what is more important than the measurement of time in days is the traveler's process of purification during his transition from hell to paradise. This process is spiritual and thus situated outside of time, in eternity. However, the narrative discourse shows no traces of reflecting this extratemporality. The narrative form of Dante's epic poem can be described as a regular alternation between scene and summary. The *Divine Comedy* thus illustrates a strict division between the representation of time in respect to its content and the temporal conception of the narrative discourse.

Temporal paradoxes are thematically discussed in Lewis Carroll's book *Alice's Adventures in Wonderland* (1865). Following a white rabbit down a rabbit hole, Alice finds herself in a dreamlike world where the logic of reality is completely abandoned. In Wonderland even time runs differently so that Alice joins, for instance, a never-ending tea party in which it is always six o'clock. The temporal absurdities the young girl is confronted with are discussed in the story. They find, however, no corresponding expression in the narrative discourse that shows the classical alternation between scene and summary.

Gustave Flaubert's *The Temptation of Saint Anthony* (1874) is about Anthony the Great, a recluse who lives isolated on a mountain top in the Egyptian desert. The novel describes one night in the life of Anthony during which he is besieged by carnal temptation and philosophical doubt. Asceticism and meditation completely suspend the protagonist's sense of time and space. In this hallucinatory mood Anthony is confronted with the vision of a primordial earth. He witnesses the first signs of inanimate and animate nature as well as the birth of humankind. In addition, the saint is haunted by several allegorical figures (among them Lust and Death) who want to change his belief that isolation is the truest form of worship. In *The Temptation of Saint Anthony* Flaubert tried to summarize the history of the world in the visions of one night.

The discussion of time as memory is an important theme in Fyodor M. Dostoevsky's *The Brothers Karamazov* (1880). Whereas the primary line of narrative centers around the story of a parricide, the topic of memory is discussed primarily in the most important side story of Dostoevsky's novel. It focuses on a group of schoolboys throwing rocks at one of their peers, Ilyusha. In the course of the novel, however, Ilyusha dies. His funeral is discussed at length in the epilogue of *The Brothers Karamazov.* Alyosha, the youngest of the Karamazov brothers, makes a speech near the stone where Ilyusha's parents wanted to bury their son. Alyosha asks the boys to keep Ilyusha as well as the day of his funeral in their memories forever. He proclaims that especially the memory of one's childhood is the best kind of education. Memory shall unite Ilyusha, his friends, and Alyosha forever. With Alyosha's memorial speech Dostoevsky emphasizes the importance of living memory, that is, an active kind of commemoration that constantly tries to evoke the dead as companions of the living.

Time as history is the thematic center of Henryk Sienkiewicz's *Quo Vadis. A Narrative of the Time of Nero* (1895) whose story takes place in antiquity. *Quo Vadis* is the most famous historical novel of the late 19th century. It tells of a love between Lygia, a young Christian woman, and Marcus Vicinius, a Roman patrician. The action is set around 64 CE in the city of Rome under the rule of Nero. Before writing *Quo Vadis,* Sienkiewicz thoroughly studied the history of the Roman empire. He filled his novel with historical conflicts and characters to evoke the time of Nero as authentically as possible. The tradition of the historical novel experienced a great revival with Umberto Eco's *The Name of the Rose* (1980), perhaps the most famous historical novel of the 20th century.

In contrast to historical novels that deal with a distant past, science fiction novels discuss imaginary technological or scientific advances that may determine our life in the future. With *The Time Machine* (1895), H. G. Wells set the cornerstone for this literary genre. His narrative is concerned with the concept of time travel using a vehicle that brings the traveler into the past or the future. The physical constitution of time is discussed at length in the framework story of *The Time Machine:* The time traveler, a nameless amateur inventor living in London, explains to his evening guests that time is nothing but the fourth dimension of space. He is convinced of the fact that there must exist a suitable apparatus that can move back and forth in time.

Eventually the inventor constructs such a machine that takes him to the year 802701 CE, where he is faced with a society that has diverged into two branches: There are the peaceful, childlike, but unintelligent Eloi on the one hand, and the intelligent but bestial Morlocks on the other. Both species are of subhuman intelligence. Another adventure brings the time traveler to a future that is 30 million years from his own time. He sees the last traces of life on a dying Earth, where the only sign of life is a black creature with tentacles. From a last voyage in time, the inventor never returns.

The Time Machine also set the ground for the literary tradition of dystopia in the 20th century. Dystopian narratives present the picture of an imaginary society that is worse than our own that the majority of us would fear to live in. Dystopian images are for the most part visions of a future society. Wells's narrative is both a science fiction novel and a dystopian novel that presents the author's vision of a troubled future.

The basic conception of time in Arthur Conan Doyle's *The Lost World* (1912) resembles to some degree the conception of time in Mann's *The Magic Mountain*. It tells the story of an expedition to a hidden plateau in South America where dinosaurs and other extinct animals are still alive. Again it seems to be spatial isolation that provokes a significant change of time and takes the explorers back into a prehistoric past. Conan Doyle's narrative is among the first science fiction novels of the 20th century.

James Hilton's utopian novel *Lost Horizon* (1933) is a further narrative that centers on the specific conception of time in a geographically isolated place. Hugh Conway, a member of the British diplomatic service, is among four kidnap victims who are brought to Shangri-La, a lamasery in the mountains of Tibet. They receive a very friendly welcome by the monks, who firmly believe in an ethics of moderation. Like the sanatorium in *The Magic Mountain*, the valley of Shangri-La is a peaceful but isolated place that ignores the actions of the outer world. Conway and the other victims soon realize that the temporal rhythm in Shangri-La differs to a large degree from the rhythm of time outside the valley. This dichotomy is a further aspect that resembles the conception of time in *The Magic Mountain*, though a closer look reveals that time in the Tibetan lamasery also differs significantly from time in the Swiss sanatorium: Whereas Hans Castorp experiences an extreme acceleration of time, the temporal rhythm of life in Shangri-La is remarkably slowed down. This deceleration of time affects primarily the physical processes of the human body. By a combination of drugs, meditation, and a special diet, the metabolism itself is slowed. Thus the inhabitants of Shangri-La may arrive at an age of several hundred years. In contrast to the artistic masterpiece of Mann that reflects the protagonist's subjective experience of time in the form of the novel, the deceleration of time finds no corresponding expression in the narrative form of the *Lost Horizon*.

Verena Kammandel

See also Alighieri, Dante; Bradbury, Ray; Carroll, Lewis; Chronotopes; Clarke, Arthur C.; Dostoevsky, Fyodor M.; Flaubert, Gustave; Homer; Joyce, James; Mann, Thomas; Ovid; Proust, Marcel; Shangri-La, Myth of; Sterne, Laurence; Tolstoy, Leo Nikolaevich; Verne, Jules; Wells, H. G.; Woolf, Virginia

Further Readings

Bergin, T. G. (Ed.). (1967). *From time to eternity. Essays on Dante's* Divine Comedy. New Haven, CT: Yale University Press.

Fluchère, H. (1965). The mind and the clock. In H. Fluchère, *Laurence Sterne: From Tristram to Yorick* (pp. 90–129). London: Oxford University Press.

Genette, G. (1990). *Narrative discourse. An essay in method*. Ithaca, NY: Cornell University Press.

Kristeva, J. (1996). *Time and sense: Proust and the experience of literature*. New York: Columbia University Press.

Kumar, U. (1991). *The Joycean labyrinth: Repetition, time, and tradition in* Ulysses. Oxford, UK: Clarendon Press.

Nakin, P. J. (2001). *Time machines: Time travel in physics, metaphysics, and science fiction*. New York: AIP Press.

Westfahl, G. (Ed.). (2002). *Worlds enough and time: Explorations of time in science fiction and fantasy*. Westport, CT: Greenwood Press.

Now, Eternal

The eternal now is a notion often linked with the nature of God, according to Western theology and according to various mystical, Asian, and so-called New Age traditions.

Anicius Boethius (480–c. 524 CE) and Saint Thomas Aquinas (c.1225–1274 CE) conceive of God as existing outside of time. They argue that the divine essence involves a perpetual present or now, without succession of moments. God exists in a kind of unchanging specious present. Such a divine essence could not thus be described as existing "before" any event in time. It could exist, instead, as intimately present at every moment of our time, while being itself outside of time. Even Jesus, the alleged physical embodiment of God, is said to have asserted that "Before Abraham was, I am" (instead of "I was"). Paul Tillich (1886–1965) in his book *The Eternal Now* claims that a timeless eternal now is what makes possible our sense of the temporal present now. For, looked at objectively, time as a succession of moments does not exhibit any present or any now. Yet we experience a now. The experience of a present now, Tillich claims, is made possible only by the breaking through of a timeless eternal present (of God, and of our essential divine self) in the time sequence.

A much-repeated metaphor meant to shed light on the paradoxical relation between a divine eternal now and our successive time is the image of God as an immovable point at the center of a circle, a center point equally distant from every point on the circumference. The circumference represents the moving, successive, spread-out nature of time. Such a circumference would have to be imagined as infinitely long and as not necessarily returning to close upon itself. Also, in this image the distance between God, as center point, and each point on the circumference would have to be imagined as nonexistent—to establish the immediacy of God's presence in every moment of time.

Philosophical Challenges

One of the challenges for this notion of a timeless divine specious present, or eternal now, is to preclude it from being static or frozen, thus lifeless and nonconscious. Can there be nondurational, nonsuccessive consciousness? Does consciousness not require some dynamism or activity? Those who answer in the positive, like the philosopher Josiah Royce, tend to insist that our durational present should remain the model for understanding the divine consciousness and its eternal now. These critics understand the divine state as still somehow temporal, although this temporal present would be much wider in scope, perhaps infinitely wider, than our human specious present. Would such a divine durational now be a temporal flow that is parallel to our time? Or would it involve some sort of temporality, some hypertime, unknown to us? Could this divine now be, instead, some sort of nontemporal, nonsequential, dynamism? Does the latter notion make logical sense? Are we to take refuge in paradox here?

It is worth noting that 20th-century physics has also addressed the notion of time and of the present now. The theory of relativity proposed a union of space with time. According to a widespread interpretation of this union, though it is still debated, time becomes assimilated to a fourth dimension of the static continuum "spacetime." Some philosophers and scientists have objected to this static-like interpretation of time. Even Albert Einstein argued that in relativistic spacetime, the time dimension is not equivalent to the spatial dimensions. He preferred viewing space as being dynamized by this union with time, instead of time being spatialized by it. For instance, simultaneity becomes relativized in Einstein's theory. This means that the notion of "instantaneous space," that is, the notion of all events occurring at the same time across all space, becomes problematic. Such a set of "simultaneous" events cannot be extracted unambiguously from the four-dimensional world process. This is because a point-event occurring before, at the same time, or after, another point-event depends in part on the standpoint from which the sequential observation is made. (The exceptions might be cases of causally related events.) Thus in relativistic physics, unlike Newtonian physics, past and future are not separated by a durationless three-dimensional "now" instantaneously spread across the universe. Past and future are instead better viewed as separated by a four-dimensional region of "elsewhere."

Recently there has been a resurgence of nonphilosophical and nontheological experience-based claims regarding some form of timeless eternity. These accounts derive partly from the New Age movement since the 1950s—a dispersed movement characterized by eclectic nontraditional mysticism-based spirituality. Similar claims derive also from the now widespread phenomena of near-death experiences.

The former movement includes a number of alleged spirit channelers reporting on pantheistic claims to the effect that God is in everything and that by turning inward we can access God's and our own (and more real) timeless nature. This viewpoint advises that beneath our ordinary temporal realm there is another and more basic one, often characterized as a divine eternal now. The metaphysics underlying this claim seems to be that the ultimately timeless divine reality opts to manifest itself as the temporal and spatial multitude of things we call the universe (and perhaps as many other universes as well). The "Seth" books by Jane

Roberts are among the most popular examples of this New Age account of reality.

There is, in addition, an extensive recent literature regarding the *experience* of some form of timelessness during alleged "near death experiences" (experiences one has during certain traumas, and sometimes in hospitals through moments while one is mistakenly declared clinically dead). During parts of such experiences—while reviewing one's whole life "instantaneously," or while engaging in some form of "instantaneous" thought-travel—many people claim that their sense of time slows down drastically, or stops altogether, while their awareness of countless events continues. Some claim that this represents the experience of an eternal now, and they often attribute this eternal now to a feature of the postlife divine environment.

Carlo Filice

See also Aquinas, Saint Thomas; Becoming and Being; Boethius, Anicius; Einstein, Albert; Eternal Recurrence; Eternity; God and Time; Tillich, Paul; Time, Perspectives of

Further Readings

Boethius, A. (1969). *The consolation of philosophy* (V. E. Watts, Trans.). London: Penguin. (Original work c. 524 CE)

Craig, W. (2001). *Time and eternity: Exploring God's relationship to time.* Wheaton, IL: Crossway Books.

Leftow, B. (1991). *Time and eternity.* Ithaca, NY: Cornell University Press.

Roberts, J., & Butts, R. F. (1994). *Seth speaks: The eternal validity of the soul.* San Rafael, CA: Amber-Allen.

Nuclear Winter

Nuclear winter is a term used to describe the potential environmental and climate effects resulting from a large-scale nuclear war. On December 23, 1983, five scientists, Richard P. Turco, Owen B. Toon, Thomas P. Ackerman, James B. Pollack, and Carl Sagan, published a paper in the journal *Science* that has come to be known as the TTAPS Study. The paper raised concern over the short-term and long-term consequences of dust, smoke, radioactivity, and toxic vapors that would be generated by a nuclear war. Although with the end of the Cold War in the 1980s, the threat of nuclear war has subsided, it should never be dismissed completely.

In the article, the five scientists concluded that exploding just one half of the combined nuclear weapons of the United States and the former Soviet Union would throw billions of metric tons of dust, soot, smoke, and ash into the atmosphere. In each explosion, most of this dust would be carried up by the nuclear fireball itself, and some of it would be sucked up the stem of the mushroom cloud. Even a more modest explosion on or above cities would produce massive fires like those in Hiroshima and Nagasaki at the end of World War II. These fires would consume wood, natural gas, and a wide variety of combustibles. The resulting smoke would be far more dangerous to the earth's climate than the dust; the smoke, the scientists argued, could produce a blanket of air pollution so thick that it would have the potential to block more than 80% of the sunlight that would otherwise reach the northern hemisphere.

As a result, they claimed, severe worldwide climate changes could occur, including prolonged periods of darkness and below-freezing temperatures, making the average land cool to 10°C to 20°C; continental interiors could cool by up to 20°C to 40°C, with subzero temperatures possible even in summer. There would also be the potential for violent windstorms. The combination of cold temperatures, dryness, and lack of sunlight would also cripple agricultural production and destroy ecosystems, putting most of the world's population at risk of starvation, according to the 1985 report by the International Council of Scientific Unions. Other studies suggest that even a small nuclear war would devastate the earth.

Severe climate change may have been a factor in the demise of the dinosaurs toward the end of the Mesozoic era. There is evidence that the end of the Mesozoic era saw changes in climate resulting in a pronounced drop in temperatures, similar to a nuclear winter. Major volcanic eruptions may have produced enormous quantities of smoke and ash that blocked the sunlight over major portions of the earth's land mass; alternatively, as some geological evidence suggests, an extraterrestrial object, most likely a meteor or asteroid, may have struck the earth, throwing up huge quantities of debris into the atmosphere and blocking out sunlight for a time. Such an event would have created a winter condition that killed plants and larger animals. This scenario is similar to that of a nuclear winter that could follow a major nuclear war.

The nuclear winter theory has been the subject of some controversy. Efforts were made by government and military scientists to play down the possible consequences. They argued that the effects would not be nearly so severe and began talking of a "nuclear autumn." In 1984, the U.S. National Research Council publicly stated that it agreed with the ideas advanced in the *Science* article; however, in 1985, the U.S. Department of Defense issued a report saying that while the nuclear winter theory might be valid, it would not change defense policies.

Today, although the threat of nuclear war has receded somewhat, the continued existence of nuclear weapons is a reminder that the possibility of a nuclear winter cannot be entirely dismissed.

Patricia Sedor

See also Dinosaurs; Ecology; Extinction, Mass; Extinction and Evolution; Sagan, Carl

Further Readings

Fisher, D. E. (1990). *Fire and ice: The greenhouse effect, ozone depletion and nuclear winter.* New York: HarperCollins.

Grinspoon, L. (1986). *The long darkness: Psychological and moral perspectives on nuclear winter.* New Haven, CT: Yale University Press.

Rowan-Robinson, M. (1985). *Fire and ice: The nuclear winter.* New York: Longman.

Sagan, C., & Turco, R. (1990). *A path where no man thought.* New York: Random House.

Observatories

Throughout history, humans have watched the stars in the night sky. Observatories were built to aid in the study of celestial objects as a means of measuring time. Many calendars are based upon the information gathered through these observations, including the Julian and Gregorian calendars. Different cultures and countries built observatories of different kinds to study the skies and track their calendars.

Prehistoric and Early Observatories

Arguably the most well-known prehistoric observatory is Stonehenge in Wiltshire, England, a circle of massive stones arranged to align with certain celestial events. Stonehenge was used as a way of tracking the moon, stars, and eclipses. There are a number of other Neolithic structures throughout Europe similar to Stonehenge in design and serving a similar purpose.

Native Americans built rock pattern structures called medicine wheels, or spiritual healing sites, that are strongly connected to the night skies. The three most notable medicine wheels are located at Big Horn, Wyoming, in the United States and at Moose Mountain in central Saskatchewan and east of Calgary in Majorville, Alberta, Canada. Most medicine wheels are created of stones located on top of hills or mountains, and they consist of a central stone from which spokes radiate toward a rim.

The Mayans have written records of astronomical observations, and several of the Mayan buildings were architecturally designed around the heavens or for the purpose of observing the skies. One such building is the Caracol Tower in Chichen Itza, which has three windows used by Mayan astronomers for observations. The Mayan people used the information gathered through these observations as a source for creating the Mayan calendar.

Muslim Observatories

In medieval Islam, there existed a desire to retain and elaborate upon the knowledge created by the ancient Greeks. Constructed in Baghdad, in what is now Iraq, during the Abbasid era of Islam, the House of Wisdom, a research center with observatory, enabled astronomers to create updated charts of planetary motion based on Ptolemy's research.

Many observatories were destroyed or abandoned due to superstition or political conflict. One such research facility, also known as the House of Wisdom, was constructed in Cairo, Egypt early in the 11th century CE. The observatory was destroyed 100 years later due to the superstition held by the populace toward planetary observation and political strife. The Istanbul Observatory, built 500 years after the Egyptian House of Wisdom, met a similar fate for similar reasons.

Not all Islamic observatories met such ill-fated demise. The two most successful observatories in the Islamic world were located at Maragha, in modern-day Iran; and at Samarkand, in modern-day

El Caracól observatory at Chichen Itza, dedicated to the study of astronomy, consists of a tower erected on two rectangular platforms.
Source: Kristine Kisky/Morguefile.

Uzbekistan. The Maragha observatory was originally constructed with astrological, rather than astronomical, intent. The major accomplishment associated with the Maragha observatory is the Ilkhanic tables charting the motion of the planets, the moon, and the sun. Fifteenth century astronomer Ulugh Beg used the ruins of the Maragha observatory as a model for the observatory at Samarkand built in the 15th century CE, where a sextant with a 120-foot radius was constructed. This instrument allowed for a resolution of arc seconds in its measurements, which was not to be surpassed until the invention of the telescope 200 years later. Astronomers at the Samarkand observatory recorded a list of nearly 1,000 stars visible from their location. Today, a crater on the moon is named after Ulugh Beg.

Asian Observatories

The Beijing observatory, built in the 15th century CE, was occupied by astronomers charged with watching the night skies. Over the course of 3,000 years, these astronomers had recorded data on hundreds of astronomical activities, including lunar and solar eclipses, meteor showers, and comets. The Chinese, in the 1st century CE, were the first to build automated instruments to measure celestial objects. The Chinese used a water clock as the timer for these devices, creating the first known clock drive. In the 1670s, Jesuits convinced Chinese astronomers to add Western instruments to the Beijing observatory. Observational activities were limited due to a conflict of interests. The Chinese astronomers' interests in observations were only in revising the current calendar. The church's stance against any activities associated with heliocentric ideas, and the Jesuits' strict adherence to this position, also prevented further contribution.

India's most notable observatories were constructed in the 18th century under the direction of the Hindu prince Jai Singh. He constructed the largest instruments he could to increase accuracy. One tool he commissioned to be built was an 88-foot-tall sundial. Updating Ulugh Beg's star data, Singh added 4 degrees, 8 arc minutes to the ecliptic longitudinal measurements to accommodate the earth's precession for the past 300 years.

European Observatories

Bernard Walther built the first notable European observatory in Nuremberg, Germany, in the mid-1470s CE. Observational data collected at this observatory noted a slight discrepancy in the Alfonso Tables, the most popular star tables in Europe from 1300–1500.

Galileo is often credited with inventing the telescope, which was in fact created in 1608 by the Danish spectacle maker Hans Lippershey. Galileo, however, was the first, in 1609, to turn the device to the skies, using the invention as an astronomical observational device and making a number of improvements to Lippershey's model. Galileo's observations through the telescope changed the study of astronomy. Galileo's most notable observations include the surface features of Earth's moon and four of the moons of Jupiter. Galileo noted the surface of Earth's moon was abundant with mountains and dark spots believed to be large bodies of water, which he named *maria,* or seas. Galileo also observed the phases of Earth's moon and those of Venus.

Tycho Brahe's 16th century observatory, built on the Danish island of Hven, was named Uraniborg and housed a number of instruments used in observations. Brahe is credited with recording the most accurate observational data of his time, prior to the invention of the telescope. Johannes Kepler assisted Brahe, and after Brahe's death, Kepler used the data gathered by Brahe to create his laws of planetary motion.

The Paris Observatory, built in 1667, became a model for national observatories. National observatories, such as the Paris and Greenwich observatories, were dedicated to gathering observational data for national interests, including improving navigation and calendars. By creating accurate tables of star positions, and with the invention of the chronometer, seafaring vessels were capable of determining their geographical location at sea.

Private observatories were also constructed, usually by wealthy individuals or organizations. In 1781, William Herschel's discovery of Uranus won favor with King George III, and Herschel was soon appointed as the king's astronomer. Herschel built his own telescopes, creating them with a high-enough resolution to open the field to galactic astronomy. Inspired by Herschel, Johann Schröter built the largest observatory of its time in Europe in the small town of Lilienthal, Germany. Schröter sketched the surfaces of the moon and of Mars, creating an interest in the studies of planetary astronomy.

Lowell Observatory, Clark Dome, in Flagstaff, Arizona. The observatory's stated mission is to pursue the study of our solar system and its evolution; to conduct pure research in astronomical phenomena; and to maintain public education and outreach programs to bring the results of astronomical research to the public.

Source: Library of Congress, Prints & Photograph Division.

American Observatories

Following in the footsteps of early national observatories, younger nations also built their own modern observatories, some shaped by the field of astrophysics. One such nation, the United States, erected the United States Naval Observatory in 1839.

While attending the Massachusetts Institute of Technology, George Hale invented a device known as a spectroheliograph, which is capable of photographing the sun. Hale built a solar observatory in his parents' backyard, located in a Chicago suburb. Hale accepted a position at the University of Chicago, and the construction of the university's Yerkes Observatory in Wisconsin was completed in 1897. The Yerkes Observatory was one of the first observatories constructed for the purpose of studying astrophysics rather than average celestial observations. Hale also planned the construction of the Mt. Wilson Observatory in California in 1904.

Conclusion

After the success of the Mt. Wilson Observatory and the Palomar Observatory in California, it was recognized that mountaintop locations near the ocean were ideal for observatories. Most observatories are now constructed upon higher elevations in an attempt to observe above, rather than through, the atmospheric interference. Today, observatories use not only optical telescopes but also radio telescopes and X-ray and gamma-ray telescopes. These telescopes allow us to "see" wavelegths outside of visible light. While radio telescopes can be used on Earth, most X-ray, gamma-ray, infrared, and ultraviolet *wavelengths* are blocked by the atmosphere and are best observed from space, through space-based observatories such as the Hubble Space

Telescope. Optical telescopes also capture their best images from space, due to the lack of interference from the atmosphere.

Mat T. Wilson

See also Calendar, Gregorian; Calendar, Julian; Galilei, Galileo; Light, Speed of; Planetariums; Sundials; Telescopes

Further Readings

Aveni, A. F. (1977). *Native American astronomy.* Austin: University of Texas Press.
Heilbron, J. L. (1999). *The sun in the church.* Cambridge, MA: Harvard University Press.
North, J., & Porter, R. (1995). *Norton history of astronomy and cosmology.* New York: Norton.

Old Faithful

Nathaniel P. Langford and Gustavus C. Doane, of the 1870 Washburn Expedition, coined the name Old Faithful to testify to this geyser's punctual regularity. Old Faithful is one of thousands of thermal features in Yellowstone National Park that result from a subterranean magma spike—where magma rises to within 40 miles of the earth's surface, compared to an average distance of 90 miles. This unique feature accounts for the park's Upper Geyser Basin, which contains nearly one quarter of Earth's geysers.

Old Faithful Geyser, Yellowstone National Park, erupting.
Source: Photo by Ansel Adams, 1941, courtesy of the National Park Service.

Predicting Eruption Time

Yellowstone boasts six grand geysers, each ejecting water spouts exceeding 100 feet. One of these giants, Old Faithful, ranges in height from 105 to 185 feet. It is a popular attraction; crowds flock to see it year round. For visitors to better understand the geyser, general rules have been devised to help predict eruption times.

Joseph Le Conte's claim that Old Faithful erupted hourly became replaced by a generalization saying eruptions lasting under 4 minutes are followed by another in 40 to 60 minutes, and eruptions over 4 minutes are followed in 75 to 100 minutes. Over time, more detailed studies yielded more specific rules. In general, an eruption of 1.5 minutes will be followed in 45 minutes. Approximately 65 minutes of rest follow a 3-minute eruption, and it takes 86 minutes to rebuild after a 5-minute eruption. Old Faithful's eruptions, on average, last 4 minutes.

Many factors upset the balance of geysers over time. People can cause immense damage. Even within two years, from the 1870 expedition to a formal survey in 1872, noticeable damage had been done to Old Faithful's cone by specimen collectors. More damage has ensued as the geyser sees 3 million visitors yearly. A 1980s cleanup project removed debris from the geyser itself, as visitors sometimes throw coins and other objects in. These activities all affect a geyser's stability.

Earthquakes serve as an additional factor that could contribute to lessening the reliability of Old Faithful. Seismographs record up to 215 tremors on the Yellowstone Plateau each year. Large earthquakes noticeably affect the balance of geysers. Before a 1959 quake, eruptions occurred at an average 65-minute interval. This quake and another in 1975 lengthened intervals, and a 1983 quake increased the average to 78 minutes.

Although geysers come and go, one should not prematurely conclude Old Faithful has yet become less predictable. Correlations and frequent observation continue to yield reliable estimates even as rest intervals increase. Though Old Faithful now erupts less often, it still maintains an uncannily consistent schedule.

Jared Nathaniel Peer

See also Geology; Plate Tectonics; Wegener, Alfred

Further Readings.

Bryan, T. S. (1995). *The geysers of Yellowstone.* Niwot: University Press of Colorado.

Bryan, T. S. (2005). *Geysers: What they are and how they work.* Missoula, MT: Mountain Press.

Schreier, C. (1992). *A field guide to Yellowstone's geysers, hot springs and fumaroles.* Moose, WY: Homestead.

Olduvai Gorge

Olduvai Gorge remains one of the most recognized archaeological sites in the world. It has provided, and continues to provide, vital information to researchers seeking answers about the origins of humanity and its evolution through time.

Wilhelm Kattwinkel, a German entomologist, stumbled across Olduvai Gorge in 1911 in northern Tanzania. The location is a canyon approximately 40 kilometers long with walls standing nearly 100 meters high that showcase nearly 2 million years of history. Extensive investigations at Olduvai Gorge began shortly afterward, yielding an array of lithic tools and fossilized animal remains amongst which were the remains of early hominids, including those of *Australopithecine* (*boisei*) and *Homo habilis* specimens. The variety of nonhominid remains discovered at Olduvai Gorge, found both separate from and amongst those of hominids, includes giraffe, antelope, and elephant. Collectively, the remains and associated tools continue to provide researchers with crucial data regarding the activities of early hominids and the overall development of humanity through time. For the latter issue, Olduvai Gorge helped put Africa in the forefront of human origins research, leading to other significant finds within the region. This includes Mary D. Leakey's 1979 discovery of *Australopithecus afarensis* footprints in Laetoli, Tanzania, which provided evidence of bipedalism approximately 3.6 million years before the present (BP).

A host of researchers investigated Olduvai Gorge's assemblages after Kattwinkel's discovery in 1911. However, it is the Leakey family of researchers in particular who are most noted for their excavation of and reporting on Olduvai Gorge. The Leakeys, particularly Louis, Mary, and Richard, are a multigenerational family of scholars who continue to spearhead human origin investigations as they did throughout the 20th century. Although their fieldwork has been curtailed in recent years, the Leakey family's collective impact on our understanding of the development of the human species through time is still strong. As for their collection and analysis of fossil and lithic material recovered from Olduvai Gorge, it was the parents, Louis and Mary, who initiated their activities throughout the early and middle portions of the 20th century. With the help of their children, the Leakeys uncovered evidence of early hominids evolving within Africa, including *Australopithecines* and *Homo habilis,* and of changes in lithic technology over time. Ultimately, the Leakey family's impact on the study of human origins and time itself is unparalleled. Through their examination of Olduvai Gorge and its various sites, the Leakeys helped determine multiple stages of hominid evolution and expand our contemplation of human evolution to accept the long duration of time through which humanity changed from early hominids to humankind's present form.

The finding of Olduvai Gorge was in and of itself an important discovery for archaeologists, paleontologists, and other researchers interested in the evolution of species and the changes of technologies developed by hominids. Specifically,

Olduvai Gorge is an archaeological site located in the eastern Serengeti Plains, northern Tanzania. Some of the oldest remains of early hominids have been found in this ancient gorge.

Source: Chris Crafter/iStockphoto.

however, there are a few discoveries in particular that set Olduvai Gorge apart from other archaeological sites pertaining to prehistoric populations. The first such find was uncovered in 1959.

After years of excavating in Olduvai Gorge, Mary Leakey uncovered the remains of an early hominid, which was later designated as *Zinjanthropus boisei.* These thick-boned and almost complete skeletal remains, discovered in 1959, were later dated to 1.8 millions years BP and ultimately classified as *Australopithecus boisei.* As stated earlier, this discovery not only provided a broader understanding of the development of the human species from earlier primate forms, it also provided an impetus for investigating Africa for evidence of human origins and evolution. That fact alone is grounds for securing Olduvai Gorge's place among the most important archaeological sites throughout the world, although it also provided other landmark discoveries.

Decades of excavation, analysis, and interpretation on the part of the Leakey family helped direct the attention of researchers toward Africa in the quest to unravel the mysteries of human evolution. The Leakeys' reputation, which was enhanced by the uncovering of *Zinjanthropus boisei,* was further cemented by the early 1970s discovery of human remains that were eventually classified *Homo habilis.* It was the first set of *Homo habilis* remains to be discovered and further hinted at Africa's being the location where humanity developed.

A final Olduvai Gorge discovery that requires mentioning is the Oldowan Tradition. With all the early hominid remains and ancient tools recovered from the gorge, it ultimately became possible for researchers to determine stages of technological development on the part of hominids along with the stages of biological development that humanity endured. The Oldowan Tradition, a term coined by the Leakey family, included lithic tool forms resembling rocks with flakes removed to create a cutting or puncturing edge; this tool kit included cores and flakes, both possibly used as tools and weaponry. While rudimentary in design, the Oldowan Tradition showed a propensity toward technology on the part of early hominids unmatched by other organisms. As for Olduvai Gorge's temporal significance as related to this technology, the gorge provided researchers with an idea of when the technology surfaced and for how long it survived.

Additional evidence of early hominids has been discovered elsewhere, including Laetoli (Tanzania), Ethiopia, and South Africa. Yet Olduvai Gorge, which continues to yield data regarding human evolution, retains lasting significance. By helping to guide researchers toward Africa in their efforts to ascertain how modern human beings evolved from earlier hominids, the discoveries at Olduvai Gorge were crucial in helping to provide a foundation for the study of humanity's development through time.

Neil Patrick O'Donnell

See also Anthropology; Archaeology; Fossils and Artifacts; Geology; Paleontology

Further Readings

Fagan, B. M., & DeCourse, C. R. (2005). *In the beginning: An introduction to archaeology.* Upper Saddle River, NJ: Pearson Prentice Hall.

Leaky, L. S. B. (1951). *Olduvai Gorge: A report on the evolution of the hand-axe culture in beds I–IV.* Cambridge, UK: Cambridge University Press.

McCarthy, P. (2006). Olduvai Gorge. In H. James Birx (Ed.), *Encyclopedia of anthropology* (pp. 1451–1452). Thousand Oaks, CA: Sage.

Price, D. T., & Feinman, G. M. (1993). *Images of the past.* Mountain View, CA: Mayfield.

Omega Point

See Teilhard de Chardin, Pierre

Omens

An omen, also known as a portent, is a sign that is believed to foretell a future event, which may or may not be supernatural in nature. From earliest times, omens have been given credence in the world's cultures and folklore. Although usually classified according to the generic terms "good" and "bad," an omen is more likely referred to in the foreboding sense, to indicate something sinister that has yet to occur.

The first recorded omens are those of the ancient Babylonians and Assyrians. Both of these cultures believed that the future could be foretold and controlled. Because religion was integral in these ancients' lives, and the omens were thought to be directly from the gods, appeasements could be made in an attempt to stave off the impending calamity. Priests skilled in the arts of omen reading and divination, known as *baratu,* would interpret the portents. These portents could be found in the sky, in animal entrails (known as extispicy), and in the weather, among other sources. The omen could be as simple as a lightning bolt hitting a tree or as complex as a pregnant snake circling a statue, laying her eggs, and dying right after. Each of these meant something different and required the baratu to interpret them, though the meaning may have been explicitly clear.

Ancient Greece and Rome also were filled with omens. In ancient Rome, before official state business was conducted, omens or the auspices (special omens observed in birds, either involving their flight in the sky or observations of the bird in general) were taken. One such auspice involved the observation of a sacred chicken's choice of whether or not it would eat food placed in front of it by an augur (a priest specially trained in auspicy). The chicken would even accompany armies to battle in a cage and the auspices would be taken before battle. A famous omen from ancient Rome involves the consul Publius Claudius Pulcher before his attack against the Carthaginians. The chicken refused to eat the grain laid before it, which was interpreted as a bad omen, and consequently as it being an inopportune time to attack the Carthaginians. Knowing his crew would find this an unfavorable omen, Claudius threw the chicken overboard. Subsequently, the Romans suffered a terrible defeat, with almost all the ships under Pulcher's command sunk.

A more popular category of omens are those concerning the weather. One such modern omen of this type is observed every year in the United States and Canada on February 2nd. It is colloquially known as "Groundhog Day" and involves the observation of a groundhog's shadow. If the groundhog fails to see his shadow because it isn't a bright day, winter will end very soon. If in fact, he does see his shadow, due to the sun being out in that particular moment, then the groundhog will be frightened, run back into his hole, and winter will continue for at least 6 more weeks. Although more of a tradition now than an actual example of a prophecy, it remains classified as a bad omen if the groundhog sees his shadow and a good one if he does not.

Supernatural and paranormal omens exist as well. These may be in the form of dreams, visions, or apparitions. One particularly frightening omen, found in the folklore of Ireland, involves the spectral banshee, or "otherworld woman," that appears before certain Irish families, then weeps and wails to portend the impending death of one of the family members. Another popular omen in this category is visions seen in the sky. One such example of this is Constantine I's legendary vision, in which he observed the Christian cross along with the words "by this sign you will be the victor." Whether or not this truly happened is subject to debate, though it is interesting to note Constantine's devotion to the Christian religion after his victory.

Another interesting omen category involves the appearance of astronomical occurrences that include, but are not limited to, comets, eclipses, and shooting stars. These particular omens sometimes signify notable births, deaths, and other significant events. A shooting star after a funeral may be confirmation that the deceased will be warmly accepted into an afterlife. A famous example, often referred to by astrologers, is the astrological chart of Princess Diana. She was married to Prince Charles on a solar eclipse date, and the day before her death was another day when a solar eclipse occurred.

Among the most frightening types of omens are those that are believed to signify the apocalypse, or the end of the world. Every culture seems to have some notion of this, and it usually has religious connotations. For example, in Christian literature, one is expected to see the sun go dark and the moon to not give off light. Another popular omen considered to foretell the end of the world, successfully adapted into the aptly named film *The Omen,* is the appearance of the Antichrist. The Y2K bug was the source of some fear for a while also, some believing it would result in a technological catastrophe that would ultimately lead to the end of the world.

It is important to note that omens can be culture dependent and not universal. For example, in the United States, it is considered that one will have bad luck if a black cat crosses his path, but in the United Kingdom, the effect is good luck. Likewise, there are certain omens that do seem to be universal. Omens that fall into the supernatural and paranormal category are almost always considered bad omens. Appearances of spectral warnings seem to be viewed with much anxiety and are considered to be signs of impending disaster.

Omens continue to be read and misread by people all over the world, in accordance with local traditions. The continuing fascination with omens may lie in humankind's uneasiness with what is yet to come. Whereas skeptics will attribute to chance any bad events that may follow the appearance of an omen, others continue to view them with fear and dread.

Dustin B. Hummel

See also Apocalypse; Nostradamus; Prophecy

Further Readings

Buckland, R. (2003). *Signs, symbols and omens: An illustrated guide to magical and spiritual symbolism.* St. Paul, MN: Llewellyn.

Waring, P. (1998). *A dictionary of omens and superstitions* (New ed.). London: Souvenir Press.

Ontology

The term *ontology* (from Greek *to on, ontos*—being, entity; *logos*—concept, science) usually denotes: (a) a philosophical discipline that studies being (entity) as being (entity), that is, being in general; (b) the ontology of a theory: the kind of entities that should exist if the given theory is true. One of the fundamental problems of ontology (particularly in its first meaning) is the question about the relation between being and becoming and thus the question about the place and role of time in the explanation of reality.

As a philosophical discipline, ontology has existed at least since the time of Aristotle (384–322 BCE), who in his *Metaphysics* claims that one of its tasks is to investigate "being as being and the attributes that belong to this in virtue of its own nature." While the investigation itself is old, the name is relatively new. It was not created until the 17th century as a result of the efforts of ontology to emancipate itself as an element of metaphysics in relation to its other disciplines, most of all however in relation to (rational) theology. The first to have used it was the German scholastic Rudolf Goclenius (1547–1628) in his work *Lexicon Philosophicum* dated 1613, but it was Christian Wolff (1679–1754) who definitively introduced it into philosophical terminology and thus the period's intellectual awareness. In his work *Philosophia Prima Sive Ontologia* from 1730, he identified ontology as a fundamental philosophical discipline within general metaphysics. While the latter describes being (entity) in general, the disciplines of specific metaphysics are concerned with its partial domains, such as God, soul (humanity), and nature.

Historically, there are three basic and interconnected areas of problems that differentiate themselves within ontology, and these may be briefly delimited by these three questions: (1) *What is being?* (2) *What really exists?* (3) *What exists?*

What Is Being, or, What Does It Mean to Be, to Exist?

Although it may seem that this question would be central to ontology, most philosophers gave up on any serious research in this direction as something problematic, perhaps even impossible. Already Aristotle, in his polemics with Parmenides (540–450 BCE), considered it disputable to think of being as such—that is, as the most general concept in a single meaning, and he emphasized its polysemic nature. Blaise Pascal (1623–1662), for example, pointed out the danger of circular definition (*circulus in definiendo*) in such generally understood being (as it can only be determined with the help of the word "is"). Immanuel Kant (1724–1804) argued that being, or existence, is not an attribute (predicate). There is no empirical attribute within the concept of an existing object by which it would differ from the concept of a similar, but nonexisting, object. With similar intentions, modern logic solves the problem of statements about existence; for example, it transforms the statement *Man is* into a formally correct form of the statement with an existential quantifier, that is, *There exists a thing that is a man.*

Despite difficulties in thinking about being (existence), Martin Heidegger (1889–1976) attempted early in the 20th century to build a fundamental ontology, the main aim of which is exactly to seek the meaning of being. As opposed to the previous philosophy (ontotheology), which forgot about being, replaced it with entity, and would explain one entity with the help of another (even divine) entity, it is necessary to clearly distinguish an entity from the being of this entity (*ontological difference*). The key to the understanding of the meaning of being is the analytics of a particular type of entity—human being, being-there (*dasein*), through the medium of objectless forms of thinking—existentials. The project of fundamental ontology remained unfinished, but it inspired phenomenological, existentialist, and hermeneutical thinking in the given field (Jean-Paul Sartre, 1905–1980; Maurice Merleau-Ponty, 1980–1961; and others).

What Really Exists, or, Which Things Do Really Exist?

Although the question "What does really, therefore truly, ultimately, exist?" is logically a version of the question "What exists?" it is historically older and it has been perhaps the most typical of ontological questions in traditional metaphysics since its beginnings. It presumes a split of reality into two ontologically as well as axiologically unequal fields: (a) a privileged, true reality (*noumenon,* substance) which exists in-itself, as a autonomous, changeless, and independent of all else, and (b) a secondary, ontologically less valuable phenomenon (*fainomenon, akcidencia*), which, as nonautonomous and changeable, is derived from the first. It is mainly the ontology of middle-period Plato (428–348 BCE) that is paradigmatically known in this sense, with his differentiation of a perfect world of intelligible ideas and an imperfect world of empirical particulars, which have reality only to that extent to which they have a part in the former.

It is the same with time as a movable and imperfect picture of eternity. In his understanding of the *first philosophy* as a teaching about a divine substance, Aristotle supported such a model of an axiologically saturated ontology when he placed against each other the perfect entity of an unmoved first mover and the hierarchically lower, by their form dependent, and derived entities. Unlike the first one, their characteristic is time as a number of movement with regard to before and after.

The dispute between idealism and materialism also became classic in this sense. Idealism considered matter as merely a manifestation (other-being) of an absolute spirit, and time as intuitive becoming was the essential but imperfect and temporary manifestation of this spirit (G. W. F. Hegel 1770–1830), or a bundle of perceptions (ideas) of the human mind (G. Berkeley, 1685–1753). Materialism considered consciousness, mind, and spirit as merely a product, an attribute or a function of matter. This traditional, substantialist ontology model dominated until the times of René Descartes (1596–1650), or Kant, after which it became the subject of perennial criticism. In the 20th century, this criticism came mainly from analytical philosophy, but also from Heidegger, Alfred North Whitehead's (1861–1947) process philosophy, Nicolai Hartmann's (1882–1950) critical ontology, and others.

What Exists, or, Which Kinds of Things Do Exist?

At the beginning of this investigation, more neutral and unreductionist—compared with the previous one—was Aristotle, with his understanding of ontology as teaching about categories, that is, about the most general kinds (predicates) of things. Categorical analysis has been a component of philosophy until the present day, but historically, both the understanding of the nature of categories and their selection have changed within it. While Aristotle understands them first and foremost realistically (they are the attributes of things themselves), and he lists 10 of these—substance, quality, quantity, etc.,—Kant understands them transcendentally, as fundamental forms of our cognition of things, so far as is possible a priori, and there are 12 of these—unity, plurality, totality, etc. Time too is a priori transcendental and does not belong to things in themselves but is one of two principles of pure intuition (the second one being space) that allow humans' inner experience and vicariously an experience of external phenomena. Similar thoughts are those of Edmund Husserl (1859–1938) in his ontology, as an eidetic science about intentional objects in general with time as "world horizon" that allows a contact between humans and the whole of existence, whereas Bertrand Russell (1872–1970) as an analytic philosopher presumes that the structure of the world (atomistic facts) exposes itself to us through the structure and categories of language (atomistic propositions).

A stimulating conception of so-called critical ontology—realistically understood categorical analysis that investigates the forms of being gained by observation of reality and the relations between them—was created by Hartmann in the 20th century. Hartmann understands reality in its gradual unfolding from an inorganic level through organic and psychic levels to a spiritual level, whereby he presumes that these levels are categorically heterogeneous. Every one has specific categories that cannot be reduced to categories of other levels, neither downwards (criticism of materialism) nor upwards (criticism of idealism). All beings are understood to be dynamic as becoming; time is a more fundamental determination of reality than space.

The development of linguistically and mainly analytically orientated philosophy in the 20th century has led to specific reasonings about the ontological implications of language, which in the second half of the century resulted in discussions about so-called ontological commitments. According to the protagonist of this discussion and the author of the term, Willard van Orman Quine (1908–2000), the answer to the ancient ontological question "*What exists?*" does not come from philosophy but from science. Because acceptance of a scientific theory always presumes some ontology, it also therefore postulates some kind of entities that should exist if the theory is true. Quine also formulates a concrete criterion for specification of thus-formed ontological commitments of theories. If the theories are formulated in a standard language of first-order predicate logic, then for them "to be means to be the value of a bound variable." However, as he points out, this criterion is not absolute—the given theory can be satisfied by several different ontologies, and it is not possible to definitely decide which one of them is valid (the principle of ontological relativity). Quine holds a perdurantist understanding of persistence of things in time, according to which a concrete particular is an aggregation of its temporal parts. Such an understanding is anchored in an eternalist conception of time that (unlike the presentist one) supposes that all time dimensions, not only the present, are ontologically real.

Marian Palenčár

See also Aristotle; Bergson, Henri; Descartes, René; Heidegger, Martin; Hegel, Georg Wilhelm Friedrich; Husserl, Edmund; Idealism; Kant, Immanuel; Materialism; Merleau-Ponty, Maurice; Metaphysics; Nietzsche, Friedrich; Parmenides; Plato; Plotinus; Presocratic Age; Russell, Bertrand; Schopenhauer, Arthur; Spinoza, Baruch de; Whitehead, Alfred North

Further Readings

Aristotle. (1993). *Metaphysics.* New York: Oxford University Press. (Original work 1st century CE)

Hartmann, N. (1953). *New ways of ontology.* Chicago: Regnery.

Heidegger, M. (1996). *Being and time.* Albany: State University of New York Press.

Inwagen, P. van. (2001). *Ontology, identity, and modality: Essays in metaphysics.* Cambridge, UK: Cambridge University Press.

Quine, W. V. O. (1964). On what there is. In W. V. O. Quine, *From a logical point of view.* Cambridge, MA: Harvard University Press.

Oparin, A. I. (1894–1980)

Aleksandr Ivanovich Oparin, a Russian biochemist, was noted for his contributions to the explanation for the origin of life. Profoundly influenced by Charles Darwin (1809–1882), Oparin presented a theoretical foundation that stressed both a materialistic and mechanistic explanation for both planetary formation and the evolution of life on this planet. His understanding of astronomy, chemistry, geology, biology, and philosophy had allowed for a comprehensive view of a temporally evolving world and humankind's place within it. Oparin's underlying principles encompassed not only the chemical processes that constituted the precursors to and the emergence of life but also the immense evolutionary time that was necessary for its formation. Oparin is best known for his major works *The Origin of Life* (1938) and *Genesis and Evolutionary Development of Life* (1968).

The biological history of this planet could be found within its remote geological history. Against prevailing philosophies and theologies, Oparin viewed the emergence of life as a result of chemical synthesis and external influences of our planet's developing environment. Acknowledging and encompassing the universe in its totality, this chemical synthesis and development of cellular organisms are unique to this planet within the cosmos. In terms of theoretical or cosmological origin paradigms, Oparin rejected the concepts of autogeneration, cosmozoa, spontaneous generation, panspermia, and vitalism as possible explanations. For Oparin, only inorganic matter existed in the beginning of Earth's development. This evolving matter would later emerge as organic. The implications are certain: life emerged from inorganic matter. The saga of symbiotic relationships between and among organic and environmental (inorganic) matter is as complex as life itself. Nevertheless, the origin of life evolved from simple beginnings.

According to Oparin, the evolution of carbon compounds, especially hydrocarbons, was necessary for a biogenic synthesis of organic substances. The formation of proteins and the development of amphoteric electrolytes allowed for multiple reactions with water. The subsequent complexity of protein and protein molecules allowed for greater organization and even greater complexity. This billon-year process resulted in the primordial "soup" by which greater complexity would slowly evolve. Although it would be over 2 billion years before the earth would attain single-celled life, the protobionts stage encompassed the emergence of coacervates, coenzymes, enzymes (including genetic information), and anaerobes. Additionally, changes in atmospheric conditions allowed for further evolution and emergence of aerobes. Over this span of nearly a billion years developed greater complexity and symbiotic relationships that resulted in multicellular life.

Oparin's speculations offered a unique perspective on the relationships within the inorganic matter from which life itself emerged. Stressing chemical action and reaction, the chance "environment" in which greater organization and complexity took place makes life unique within a highly improbable universe containing life. This point has two implications. First, the idea that life emerged from inorganic matter over billions of years implies a rejection of anthropocentric philosophies and theologies. Second, it answers the question, "Are we alone in the universe?" with a degree of implausibility. Today, design theories and the "God gene" are alternative explanations for the philosophical questions of human existence. There is very little doubt that Oparin would reject any version of these ideas and their manipulation of science. The principles set forth by Darwin, when applied in a comprehensive manner, would exclude these assertions. Oparin, a Darwinian evolutionist, understood both the temporal nature of organic and inorganic processes and their implication for humankind. From inorganic to organic, life, especially human life, is distinctive but not necessarily unique within this dynamic material universe.

David Alexander Lukaszek

See also Darwin, Charles; DNA; Evolution, Chemical; Evolution, Organic; Life, Origin of; Materialism

Further Readings

Oparin, A. I. (1953). *Origin of life.* New York: Dover Press.

Oparin, A. I. (1968). *Genesis and evolutionary development of life.* New York: Academic Press.

Orwell, George (1903–1950)

George Orwell was the pen name of Eric Arthur Blair, an English novelist and journalist. He is best known for his political satires *Animal Farm* (1945) and *Nineteen Eighty-Four* (1949). The two novels criticize totalitarianism and social injustice.

Nineteen Eighty-Four was one of the most popular and widely quoted novels of the 20th century. It describes a bleak picture of a future world ruled by totalitarian regimes. Individual freedom and privacy are suppressed in the tightly controlled society described by Orwell. "Big brother is watching" is one of the many famous lines that have been frequently quoted from the book. The Stalinist regime of the Soviet Union was the real-life archetype for Orwell's idea.

George Orwell was born in Motihari, Bengal, India, on June 25, 1903, into a lower-middle-class family. He attended the prestigious Eton School in London. In 1922 he returned to India and joined the British imperial police. In India and in Burma Orwell developed a strong antipathy toward imperialism and class division. He resigned from the imperial police and returned to Europe, where he moved between England and France, living a life of poverty. This experience gave him the material to write the memoir *Down and Out in Paris and London* (1933). His other works of the period include *A Clergyman's Daughter* (1935), and *The Road to Wigan Pier* (1937), in which Orwell describes his strong socialist sympathies. In 1936 he traveled to Spain and fought alongside the Republicans in the Spanish Civil War. He also became highly critical of communism during this period, expressing these sentiments in *Homage to Catalonia* (1938).

Orwell returned to England in 1937 and became a productive journalist, writing numerous articles and essays before and during World War II. He denounced Nazism but opposed war with Germany. In 1944 he completed *Animal Farm,* a brilliant satire in which barnyard animals overthrow their oppressive human master, only to divide into a new social order thereafter. The pigs assume power over the other animals and establish privileges for themselves. Orwell wrote *Animal Farm* to parody the Russian Revolution and the dishonesty of leaders such as Joseph Stalin. Orwell's book was especially relevant during a time in human history marked by ideological struggles and totalitarian regimes. He believed that fascism and communism were not inevitable outcomes in time, but were dangerous ideas that should be resisted. Many lines from the book continue to be quoted in popular culture, including "All animals are equal, but some are more equal than others." Government's actions hostile to a free society are often referred to as "Orwellian."

Orwell was married twice and had one adopted son. He suffered from tuberculosis during the later years of his life. He died in a London hospital on January 21, 1950.

James P. Bonanno

See also Bradbury, Ray; Clarke, Arthur C.; Futurology; Novels, Time in; Toffler, Alvin; Verne, Jules; Wells, H. G.

Further Readings

Bowker, G. (2003). *George Orwell.* London: Little, Brown.

Davidson, P. (1996). *George Orwell: A literary life.* New York: St. Martin's.

Shelden, M. (1991). *Orwell: The authorized biography.* New York: HarperCollins.

Ovid (43 BCE–17 CE)

The poet Ovid lived in Rome under the reign of the Emperor Augustus in the 1st century BCE. He is best known for the *Metamorphoses,* an epic poem chronicling the history of the cosmos from creation to his own era. He incorporated many ancient myths and legends into this work, many of which had never been recorded. In doing this he preserved centuries of oral history for future generations to enjoy.

Ovid was born in 43 BCE in central Italy near the Abruzzi Mountains. At age 16 he left for Rome to study rhetoric. By 18 he had become a judge but was unsatisfied with a career in law. While in Rome he developed a passion for poetry and decided to make his living as a writer. He also found happiness with Fabia, his third wife, with whom he had one daughter.

Ovid began his career by writing love poems, and his first public work was titled *Amore*. His works were very popular, because they referred to the daily activities of "modern" young people. This was very unusual for poets of his time. As his career progressed, Ovid began writing more erotic poems. He considered himself an instructor for young lovers. This sometimes explicit poetry got him into trouble with the law. To counteract the amoral trend of Roman society, Augustus created new morality laws to help reinstate family structures. On hearing of these laws, Ovid decided to write a poem based on mythological stories he had enjoyed as a child.

The *Metamorphoses* was completed in 8 CE and consists of 15 books, 12,000 verses, and 246 legends. The poem's title comes from the focus on transformation stories in which gods and spirits are changed into plants and animals. The first transformation story is the changing of Earth into Man, and the work ends with the story of Julius Caesar becoming a star. Ovid wrote it in dactylic hexameter, the meter closely associated with epics and the poets Virgil and Homer. However, the content is not that of a traditional epic. Instead of chronicling the story of a single protagonist, it spans all time, from Earth's creation from chaos right up to the reign of Augustus. The theme is the same as in all of Ovid's poems, love. Each story seems to glorify love—the emotion or the god Amor (Cupid).

Ovid's attempt to evade punishment for his early risqué poetry was in vain. He was exiled to Romania, where he is still considered a national hero. Because of his immense popularity with the Roman public, he was allowed to keep his fortune and continued to write and publish until he died in Tomis, now Constanţa, in 17 CE.

Ovid's greatest work influenced many of the writers that would follow him. Dante mentions him twice in his great *Divine Comedy.* Many scholars believe that Shakespeare's famous *Romeo and Juliet* was based on *Pyramus and Thisbe,* which also features in *A Midsummer Night's Dream.*

Jessica M. Masciello

See also Caesar, Gaius Julius; Lucretius; Poetry; Rome, Ancient

Further Readings

Ovid. (2005). *The metamorphoses* (F. J. Miller, Trans.). New York: Barnes and Noble Classics. (Original work c. 8 CE)

Rădulescu, A. (2002). *Ovid in exile.* Iaşi, Romania: Center for Romanian Studies.

Paleogene

The Paleogene is a geochronological and chronostratigraphic unit of the Cenozoic era/erathem. It is a period in the geochronological scale and a system in the chronostratigraphic scale, and it is placed between the Cretaceous and Neogene periods. It began 65 million years ago, at the Cretaceous-Paleogene (K-Pg or K-T) boundary, and ended 23 million years ago, at the Oligocene-Miocene (O-M) boundary; thus, the Paleogene lasted 42 million years. It consists of three epochs and/or series: Paleocene, Eocene, and Oligocene.

The Paleogene followed the Cretaceous period and began with the Cretaceous-Paleogene mass extinction event. There is paleontological evidence of abrupt changes in flora and fauna in this event (most often referred to as the K-T boundary mass extinction), including the total extinction of dinosaurs, ammonites, belemnites, cephalopods, and rudist molluscs; and the catastrophic mass extinction of planktic foraminifers, calcareous nannofossils, corals, bivalves, brachiopods, fishes, mammals, and other reptile groups. The Paleogene is most notable as being the period in which mammals and birds were diversified, exploiting ecological niches untouched by the previously extinct dinosaurs. Both groups evolved and came to dominate the land. Mammals evolved considerably into large forms in terrestrial and marine environments; birds evolved into roughly modern forms in an airborne environment.

During the Paleogene, global tectonic processes continued that had begun during the Mesozoic era, with the continents drifting toward their present positions. Although these were gradual processes, the drifting of the continents caused significant paleoclimatic and paleoceanographic turnovers during the Paleogene. The former components of the old supercontinent Gondwana continued to split apart, with South America, Africa, and Antarctica-Australia pulling away from each other. Africa moved north toward Europe, slowly closing the occidental Tethys Ocean until it disappeared during the Eocene, and uplifting the Alps during the Oligocene. Similarly, India initiated its rapid migration toward the north, until it collided with Asia, narrowing the oriental Tethys Ocean, folding the Himalayas, and forming the Indian Ocean. The Tethys Ocean vanished during the Paleogene, becoming today's Mediterranean Sea, the remnant of that old ocean. The northern supercontinent Laurasia began to break up during the Eocene, with Europe, Greenland, and North America drifting apart. The tectonic splitting of the Greenland and Norwegian seas increased the submarine volcanic and hydrothermal activity (North Atlantic flood basalts) during the Paleocene-Eocene transition. Antarctica and Australia began to split in the late Eocene, and South America and Antarctica in the Oligocene, which allowed the formation of the circumantarctic current.

The climate of the earliest Paleogene was slightly cooler than that of the preceding Cretaceous. Nevertheless, the temperature rose again in the late Paleocene, reaching its highest point at the Paleocene-Eocene (P-E) boundary. A sudden and extreme global warming event occurred in the P-E boundary, 55.8 million years ago, called the Paleocene-Eocene Thermal Maximum (PETM). It was an episode that lasted less than 100,000 years, very rapid in geologic terms, and it caused an intense warming of the high latitudes (up to 7° C) and a mass extinction in the benthonic fauna of the bathyal and abyssal oceanic environments (mainly benthic foraminifera). It is hypothesized that PETM was caused by runaway greenhouse effect due to a sudden release of methane from oceanic hydrates. This methane flux and its oxidation product carbon dioxide could be of a magnitude similar to that from present-day anthropogenic sources, creating the sudden increase of greenhouse warming. The main cause of this short-term change may be related to the reorganization of tectonic plates that produced an increase of volcanic activity (mainly in the North Atlantic) as well as significant paleogeographic and paleoceanographic turnovers. Among these last, the most important was the closing of the Tethys Ocean with the formation of vast areas of shallow epicontinental seas. This may have been responsible for the shift in the locus of ongoing deep-water formation from cold and nutrient-depleted deep waters produced in the polar (Artic and Antarctic) regions to warm, saline, and oxygen-deficient deep waters formed in Tethyan evaporative basins. The stability of these methane hydrates depends on temperature, and, therefore, it is possible that the abrupt deep sea warming induced a shift in sediment geotherms.

The climate continued to be warm and humid worldwide during the early and middle Eocene, with tropical-subtropical deciduous forest covering nearly the entire globe (even in Greenland and Patagonia) and ice-free polar regions covered with coniferous and deciduous trees in a temperate environment. The equatorial areas, including the Tethys Ocean region, were characterized by a tropical, hot, and arid climate. The Eocene global climate was the warmest and most homogeneous of the Cenozoic, but the climatic conditions began to change in the late Eocene. A global cooling, initiated toward the end of the Eocene, occurred during the Oligocene. This cooling caused gradual extinctions along the Eocene-Oligocene (E-O) transition, between 39 and 33 million years ago, that drastically affected land mammals and vegetation. Tropical areas, such as jungles and rainforests, were replaced by more temperate savannahs and grasslands. The E-O cooling episode was the most recent transition from a greenhouse (Cretaceous to Eocene) to an icehouse (Oligocene to present-day) climate mode. The cause of this climatic cooling was the establishment of the circumantarctic current that isolated the Antarctic. Feedback mechanisms, such as the formation of the Antarctic icecap, a substantial drop in sea level, and an increase in the earth's albedo, drove rapid climate change, which eventually led to the Pleistocene glaciations.

During the first epoch of the Paleogene, the Paleocene, the flora are marked by the development of modern plant species, including the appearance of cacti and palms. The flowering plants or angiosperms, which first appeared in the beginning of the Cretaceous, continued their development and proliferation. Along with them evolved the insects that fed on these plants and pollinated them. During the PETM and Eocene, the high temperatures and warm oceans created a tropical, humid environment, with forests spreading throughout the globe from pole to pole. By the time of the climatic cooling of the late Eocene and Oligocene, deciduous forests covered large parts of the northern continents (North America and Eurasia, including the Arctic areas), and tropical rainforests held on only in equatorial South America, Africa, India, and Australia. The tundra stretched out over vast areas of Antarctica, and open plains and deserts became more common.

Because of the dinosaur extinction at the K-T boundary, the reptiles were reduced to palaeophid snakes, soft-shelled turtles, varanid lizards, and crocodilia. With the extinction of marine plesiosaurs and ichthyosaurs, sharks became the chief ocean predators. Birds began to diversify during the Paleocene, but the most modern bird groups appeared during the Eocene and Oligocene, including hawks, owls, pelicans, loons, and pigeons, among others. The fossil mammal evidence from the Paleocene is scarce, but it is characterized by an evolutionary radiation of small mammals, mainly insectivorous species, including monotremes, marsupials, multituberculates, and primitive placentals, such as the mesonychid. The most important radiation of mammals occurred during the climatic optimum of the Eocene, when there appeared new and modern groups such as artidactyls and perissodactyls. Early forms of many other mammalian orders also appeared, including primates, bats, proboscidians, rodents, and cetaceans. Several mammal groups were extinguished during the cooling episode of the E-O transition, including mesonychids and creodonts.

Ignacio Arenillas

See also Cretaceous; Fossil Record; Geologic Timescale; Geology; K-T Boundary; Neogene; Paleontology

Further Readings

De Graciansky, P. C., Hardenbol, J., Jacquin, T., & Vail, P. R. (Eds.). (1998). *Mesozoic and Cenozoic sequence stratigraphy of European basins.* Tulsa, OK: Society for Sedimentary Geology (SEPM).

Gradstein, F. M., Ogg, J. G., & Smith, A. G. (Eds.). (2004). *A geologic time scale 2004.* Cambridge, UK: Cambridge University Press.

Paleontology

Paleontology (from Greek: *palaeo,* "old, ancient"; *on,* "being"; and *logos,* "speech, thought") is the study of ancient life. Life appeared on Earth about 3,550 million years ago in the oceans, subsequently evolved from simple bacteria-like cells to complex multicellular forms, and colonized the land. Countless adaptations resulted in a great diversity of biological forms and in addition changed the planet itself. We have learned about extinct organisms through the examination of fossils, the visible evidences left behind by them and preserved in rocks and sediments. Fossils include mineralized, carbonized, mummified, and frozen remains of bodies after death, or of cast-off parts, normally of the skeleton or portions, such as teeth, that became partially mineralized during life; the preservation of soft tissues, however, is extremely rare. Many other fossils consist of casts or impressions, tracks, burrows, fossilized feces (coprolites), as well as chemical residues.

People have collected fossils ever since recorded history began, and probably before that, but the nature of fossils and their relationship to life in the past became better understood during the modern era as a part of the changes in natural philosophy that occurred during the 17th and 18th centuries. The emergence of paleontology, in association with comparative anatomy, as a scientific discipline occurred at the end of the 18th century, when Georges Cuvier clearly demonstrated that fossils were left behind by species that had become extinct. Paleontology therefore is the study of fossils throughout geological time. The totality of fossils, both discovered and undiscovered, and their placement in sedimentary layers or strata is known as the *fossil record.* The fossil record ranges in age from the Holocene, the most recent geological epoch that began 12,000 years ago and continues until present, to the Archean eon, which extends from about 3.8 to 2.5 billion years ago.

Fossils vary in size from the microscopic (microfossils), such as fossilized shells of unicellular organisms, to those of gigantic proportions, such as the fossil bones of dinosaurs. Micropaleontology studies microscopic fossils, including organic-walled microfossils, the study of which is called *palynology.* The study of microfossils requires a variety of physical and chemical laboratory techniques to extract them from rocks and the use of light or electron microscopy to observe them. Macrofossils are usually studied with the naked eye or under low-power magnification, but the observation of fine skeletal details often needs high-powered magnification.

The study of macrofossils is undertaken by several specialties. Invertebrate paleontology deals with fossils of animals with no vertebral column, while vertebrate paleontology deals with those of animals with a vertebral column, including fossil hominids (paleoanthropology). Paleobotany undertakes the study of macrofossils of plants. There are many developing specialties, such as paleoichnology (the study of trace fossils), molecular paleontology (the study of chemical fossils or biomarkers), and isotope paleontology (the study of the isotopic composition of fossils).

Two of the most important portions of knowledge that paleontologists obtain from fossils include first how they were formed (taphonomy), that is, the process of fossilization through which some material or information was incorporated from the biosphere to the lithosphere; and second, what the organisms were that produced them (paleobiology), as well as how and where they lived and what their evolutionary history was. The source information for this purpose is the biology and ecology of present-day organisms, applying the uniformitarian principle. Fossils usually contain morphological information that allows us to recognize most of them as living organisms and then to identify and classify them according to the Linnaean taxonomy, as well as to study their relationships to other taxa.

Fossils are generally found in sedimentary rock with differentiated strata representing a succession of deposited material. To place the fossils in context in terms of the time, setting, and surroundings in which the organisms lived, paleontologists require knowledge of the precise geological location where the fossils were found and details of their source rock strata. Paleontology has provided important tools for both geologists and biologists.

In geology, fossils are important in the analysis of the order and relative position of strata and their relationship to the geologic timescale, and also in correlating successions of rock strata from the same time interval across the globe. The deep time of Earth's past has been organized into a timescale composed of various units that are usually delimited by major geological or paleontological events, such as mass extinctions. In a pioneering application of stratigraphy, at the end of the 18th century, William Smith in England and Georges Cuvier and Alexandre Brongniart in France made extensive use of fossils to help correlate rock strata in different locations. They observed that sedimentary rock strata contain particular assemblages of fossilized flora and fauna and that these assemblages succeed each other vertically in a specific, reliable order that can be recognized even in widely separated geologic formations (principle of faunal succession). Later, Darwin's theory of evolution closely described the causal mechanism of the observed faunal and floral succession preserved in rocks.

This principle is of great importance in determining the relative age of rocks and strata by using the fossils contained within them (biostratigraphy). As the distribution and diversity of living organisms are limited by environmental factors, and as vestiges of biochemistry of the original organism and isotopic signatures of ancient environments are preserved on fossilized skeletal remains, fossils provide an insight into the environment once inhabited by living organisms (paleoecology) and help in the interpretation of the nature of ancient sedimentary environments and the diagenetic processes undergone by the rocks that contained them. The primary economic importance of paleontology lies in both applications to geology. The study of the fossils, especially microfossils, contained in a rock remains one of the fastest, cheapest, and most accurate means to determine the age and nature of the rocks that contain them or the layers above or below. This information is vital to the mining industry and especially the petroleum industry.

In biology, fossils are the most direct evidence of the evolution of life on Earth; they have helped to establish evolutionary relationships and to date the divergences between taxa (phylogeny). An expanding knowledge of the fossil record encouraged the formulation of early evolutionary theories. In fact, Darwin himself collected and studied South American fossils during his trip on the H.M.S. *Beagle*. After Darwin's evolutionary theory was published in 1859, much of the focus of paleontology shifted to understanding lineage evolution, including human evolution. George Gaylord Simpson and, later, Stephen J. Gould played a crucial role in incorporating ideas from paleontology to evolutionary theory.

Fossils indicate long-term patterns of biodiversity in the geological past. The story of the development of life on Earth, of the biosphere, forms the subject of paleontology. Modern paleontology sets ancient life in its context by studying how, over this vast time span, life has adapted to a changing world; this change is barely discernible during a single human lifetime. Long-term physical changes of global geography (continents and oceans pushed by plate tectonics, mountains formed and eroded) and long-term fluctuations between hot and cold climates (ice ages driven by orbital factors, warm periods in response to rapid increases in atmospheric carbon dioxide) triggered changes in living things: Populations, species, and whole lineages disappeared, and new ones emerged. Ecosystems have responded to these changes and have adapted to the planetary environment in turn. These processes continue, and today's biodiversity is affected by these mutual responses.

Very few species, known as living fossils, survive virtually unchanged for tens or hundreds of millions of years. Most species today appeared very recently in geological terms. It has been estimated that more than 95% of species that ever lived have become extinct. Paleontology has shown that extinction is a natural process that generally happens at a continuously low rate. Throughout geologic history, very few mass-extinction events have occurred in which many species have disappeared in a relatively short period of geological time. Thus, paleontology evidences the fragility of the world. Humans appeared only about 2.5 million years ago, and several human species have become extinct. Modern man appeared very recently in geological terms, no more than 200,000 years ago. At the end of the last glacial period, around 12,000 years ago, many mammals weighing more than 40 kilograms (megafauna) disappeared. There is a debate as to the extent to which this extinction event can be attributed to environmental and ecological factors, to the onset of warmer climates, or to human activities, directly by overkilling megafauna or indirectly. Megafaunal extinctions continue to the present day, and this deteriorating situation is being referred to as "the sixth mass-extinction."

The present biodiversity of an area is the consequence of the natural evolution of the species in its dating back to more than 3 million years ago. This evolution has been conditioned, in many ways, by geological history and certain other natural phenomena. But for the first time, a single species—ours—appears to be almost wholly responsible for an extinction crisis. Natural environments are now so degraded that we must go back in time to true known natural environments to understand natural processes. Paleontology furnishes an extensive database that, when integrated with neontological data, allows us to define models that explain better the past and present biodiversity and that would be useful in prospective studies.

As a consequence, the strategies aimed at the protection of biodiversity should also take into account the preservation of paleobiodiversity (paleontological heritage) and of geological materials (geological heritage and geodiversity), which constitute proof of past natural processes. Although fossils are also preserved in museums and private collections, the paleontological heritage exists in the natural environment as fossil sites. These sites compose an irreplaceable and finite resource for science, education, and recreation. Paleotourism, as part of adventure tourism, is poised for dramatic growth in the decades ahead, a fact directly related to the demography of wealthy nations. As this industry grows in years to come, it will be important for scientists and government officials to work together with the local inhabitants of the fossiliferous regions to create effective partnerships to educate the public and to protect and develop our paleontological heritage. This element of natural and cultural heritage is vulnerable to abuse and damage and therefore needs safeguarding and management to ensure its survival for future generations.

Beatriz Azanza

See also *Archaeopteryx;* Dating Techniques; Dinosaurs; Fossil Record; Fossils, Interpretations of; Geology; Stromatolites; Trilobites

Further Readings

Briggs, D., & Crowther, P. R. (2001). *Palaeobiology II.* Osney Meads, Oxford, UK: Blackwell Science.

Foote, M., & Miller, A. I. (2007). *Principles of paleontology* (3rd ed.). New York: Freeman.

Gould, S. J. (Ed.). (1993). *The book of life: An illustrated history of the evolution of life on earth.* New York: Norton.

Prothero, D. R. (2004). *Bringing fossils to life: An introduction to paleobiology.* Boston: McGraw-Hill.

Prothero, D. R. (2007). *Evolution: What the fossils say and why it matters.* New York: Columbia University Press.

Paley, William (1743–1805)

William Paley, English churchman, theologian, moral philosopher, and apologist, is best known for his "watchmaker analogy," a classic argument for the existence of God, the Creator. From its publication in 1802, Archdeacon Paley's famous book, *Natural Theology,* influenced the Creation/evolution debate, which became especially lively from Darwin's era until the present. Few issues related to the study of time hold more significance.

Paley was born in Peterborough in 1743, the son of a vicar and schoolmaster. To prepare himself for the ministry, Paley enrolled in Christ's

College, Cambridge, in 1758, from which he graduated and where he later became a fellow and tutor. (Nearly 70 years later, Charles Darwin also enrolled in Christ's College, lived in rooms formerly occupied by Paley, and studied—and admired—the latter's writings.) Ordained in 1767 and married in 1776, William Paley advanced through clerical ranks and held various appointments. In 1782, he became archdeacon in Carlisle. There he began the process of expanding his Cambridge lectures on apologetics and ethics for publication.

Though he published several other books, Paley's fame rests on four works, all of which exerted considerable influence in his lifetime—and beyond. His first study, *The Principles of Moral and Political Philosophy* (1785) became a standard textbook at Cambridge. He taught a form of ethical utilitarianism and opposed the slave trade. In 1790, Paley published his second book, *Horae Paulinae, or the Truth of the Scripture History of St. Paul,* a defense of the Bible's historical nature. The third book, *A View of the Evidences of Christianity* (1794), another study on Christian apologetics, sorted and updated important material from earlier authors and achieved wide acclaim. In contrast to Hume, he supported the historicity of biblical miracles.

He published a fourth volume in 1802 under the title *Natural Theology, or Evidences of the Existence and Attributes of the Deity Collected from the Appearances of Nature.* Paley regarded this last book as the most important, a logical predecessor to the rest. Through an orderly arrangement and readable style, Paley—like his predecessors who advanced the cause of natural theology (e.g., John Ray, William Derham)—compiled a series of case studies to support the teleological argument for the existence of God, also known as the argument from design. In his *Natural Theology,* still in print after more than two centuries, Paley pointed primarily to complex parts and systems of human anatomy as arguments in favor of "intelligent design." As a watch's functioning components imply the existence of a watchmaker, he suggested, the biological realm reflects the work of a purposeful designer.

Critics point to various flaws in this famous analogy (e.g., it begs the question by assuming that the watchmaker must be the God of the Bible) and claim that Darwin's conclusions have obviated the need for an external, divine artificer. Others still find Paley's simple premise compelling. As the church reacted to intellectual challenges of that era (e.g., deism; Enlightenment writers such as Hume and Kant), Paley offered hope to readers who believed in a Creator with a personal interest in the universe.

Gerald L. Mattingly

See also Bible and Time; Creationism; Darwin, Charles; God and Time; Gosse, Philip Henry; Scopes "Monkey Trial" of 1925; Teleology; Watchmaker, God as

Further Readings

Brooke, J. H. (1991). *Science and religion: Some historical perspectives.* New York: Cambridge University Press.

Dillenberger, J. (1960). *Protestant thought and natural science: A historical interpretation.* Nashville, TN: Abingdon Press.

Eddy, M. D. (2004). Science and rhetoric of Paley's *Natural Theology. Literature and Theology, 18,* 1–22.

Fyfe, A. (2002). Publishing and the classics: Paley's *Natural Theology* and the nineteenth-century scientific canon. *Studies in History and Philosophy of Science, 33,* 729–751.

LeMahieu, D. L. (1976). *The mind of William Paley: A philosopher and his age.* Lincoln: University of Nebraska Press.

McGrath, A. E. (1999). *Science & religion: An introduction.* Malden, MA: Blackwell Science.

Nuovo, V. (1992). Rethinking Paley. *Synthese, 91,* 29–51.

Paley, W. (2006). *Natural theology.* New York: Oxford University Press. (Original work published 1802)

PANBIOGEOGRAPHY

In 1964 the biogeographer and evolutionist Leon Croizat published a book titled *Space, Time, Form: The Evolutionary Synthesis.* The title of the book presented a renewed emphasis on the role and significance of space and time in the evolutionary process and in understanding evolutionary history. Many representations of the theory of evolution from the time of Darwin's *On the Origin of Species* (1859) assumed that space and time together constituted a separate environmental container through which organisms moved and evolved. Croizat pointed out that this perspective resulted in an erroneous understanding of the evolutionary process.

For most people, time is perhaps the most compelling element of evolution that directly links the present with the past, principally through the geological fossil record. Fossilized organisms are identifiably related to those of the living world, either at a general level of organization or at more specialized levels such as those of genera and species. These fossils have contributed to the idea that living species have a history of ancestral species, some of which are preserved in the fossil record. However, this record alone was not necessarily enough to convince everyone of evolution by descent with modification—including Darwin himself. But it was during his world voyage on H.M.S. *Beagle* that Darwin found that some fossils in Argentina were more closely related to organisms currently living in that region than to those of other areas. This geographic juxtaposition of temporal records and biological relationships provided Darwin with a critical insight that helped lead him from a creationist to an evolutionary perspective. So it is no surprise that the very first sentence in Darwin's 1859 book began with the observation that the distribution of organisms and the geological relationships of the present to the past inhabitants of South America "seemed to throw some light on the origin of species."

Darwin's discovery pointed to a key aspect of evolution, that time and space are causally interrelated. But this interrelationship was taken largely for granted in much of evolutionary theory until nearly a century later, when Leon Croizat developed his unique approach to evolution called panbiogeography. In this approach Croizat did what no one else had ever done before. He tested Darwin's theory of evolution through the comparative study of animal and plant distributions, whether living or fossil. Animal and plant distributions provide a direct representation of time and space in evolution. The spatial component is represented by their location, while the temporal component is represented by their differentiation or divergence as well as the spatial correlation of distributions with tectonic features associated with earth history.

In recognizing the integral relationship of time with geographic location and the evolution of biological form, Croizat proposed the representation of evolution as the summation of their individual and combined effects by the following equation: Evolution = space + time + form. This formulation showed that the study of evolution was effectively the study of how all three elements are interrelated and affect each other, rather than just the study of a purely physical (biological) process. Time now becomes part of the evolutionary process rather than just a temporal record of evolutionary events. Because of this integral relationship, Croizat regarded the process of biological evolution (which he referred to as "form-making") in space over time as fundamental for the whole of biology in both its theoretical and practical aspects.

Croizat's approach to time as an aspect of space has its historical background in new ways of thinking about time and space that were developing at the transition between the 19th and 20th centuries, particular in Italy where Croizat was born and spent his formative years concurrently with the emergence of the Italian futurist movement. Futurists challenged the conventional assumptions of space and time as absolute categories by developing space and time as relational concepts with physical form. This was further developed in challenging presence and absence as independent and localized concepts. Croizat's panbiogeography showed that the full meaning of an organism at any one place and time is always permeated by a phylogenetic, morphological, ecological, or biogeographic counterpart or complement that is located somewhere else in space and time. The trace of this vicariant counterpart is represented graphically as a line or track that shows the connection of organisms in space and time. In this context, organism and environments are not absolute entities, but biogeographic and ecological relationships where spacing and temporalization are the ways in which replication of the past in the present influences the future.

By comparing the geographic distributions of animal and plant species, Croizat concluded that evolutionary differentiation of biological form (e.g., speciation) results in related taxa (species, genera, families, etc.) occupying different geographic sectors without any of these taxa having individually moved to those locations. This process was made possible by their common ancestor already having a distribution range that encompassed all the descendant locations. Each descendant came to occupy different areas through their biological divergence over different parts of the ancestral range. Croizat called this process *vicariant form-making*. These various taxa may be assigned any one of a number of different taxonomic ranks (species, genus, family, etc.), but Croizat was adamant that time in evolution is not tantamount to age as expressed in any particular taxonomic group. He argued, for example, that even though one might assume that a genus comes before a species because the genus is made up of species, in reality a genus could hardly exist independently from at least some of its species, so that the two ranks are effectively contemporary in origin. In this way Croizat attempted to distinguish between absolute time as an overall process inherent to the spatial evolution and differentiation of taxa, and relative time as the relationship between a specific taxonomic rank and its place in evolutionary rather than absolute time, so a species in one group may be as old as a genus or family in another group.

Key questions of time in evolution include the estimation of evolutionary rates of differentiation and the provision of a temporal scale for divergence between lineages. To address these questions, evolutionists often refer to the fossil record. The fossil record provides a general geological timescale for the first appearance of various lineages in the fossil record. Fossils can represent only the minimal age of fossilization for a recognized group of organisms. Fossils contain no information on whether or how long a group existed before the appearance of their earliest fossil. For example, the earliest known bird fossil, dated at about 150 million years old, may show that birds had evolved by this time, but the fossil contains no information as to how much earlier birds originated or the age of the common ancestor of birds and their nearest dinosaurian relatives. The fossil record is replete with examples of organisms for which the fossil record extends the organism's history tens of millions of years further than previously existing records did.

Method

Panbiogeography provides a spatial method for estimating temporal divergence and understanding evolutionary rates by correlating animal and plant distributions with tectonic features involved with geological history. Tectonic formations such as spreading ridges, plate boundaries, transform faults, and zones of uplift or subsidence are all indicators of geological process that underlie the geological topography now occupied by plant and animal distributions. These features can be geologically dated through radioactive decay rates to provide a geological timescale for their formation. A temporal correlation is often made for fossils that are imbedded in dated geological strata, but this approach may also be used for living taxa. This geological correlation technique may have significant implications for dating the origin of taxa where the fossil record is sparse or lacking altogether, and it may even lead to controversial challenges to accepted ages of origin based on the fossil record.

The possible temporal implications for evolutionary rates and age of origin may be illustrated by the spatial correlation between the 200-million-year-old Triassic fossil mollusk *Monotis* and the modern flowering plant genus *Coriaria*. The fossil distribution of *Monotis* overlaps with extensive circum-Pacific geological terranes (geological strata that have a different origin than other strata immediately adjacent) as well as a series of terranes extending through central Asia to Europe. This fossil range is spatially comparable with the modern distribution of *Coriaria* with the exception of western North America, where the plant is absent and no fossil representatives are known. The overall spatial correlation with circum-Pacific terranes may suggest that the modern distribution of Coriaria is as old as the Triassic, or it may have a more recent origin that was still affected by the subsequent geological history of those terrains that previously influenced the Triassic distribution of *Monotis*. The controversial aspect of this spatial correlation between geology and biology is that it suggests the genus *Coriaria* may be older than the earliest known fossil flowering plants recorded from the early Cretaceous (125–130 million years ago). Although some interpretations of the fossil record allow for an earlier origin for the evolution of flowering plants, the idea that some modern genera may also be this old would be widely viewed as problematic if not impossible.

Croizat's Mesozoic Theory

Spatial correlations between modern distributions and the earth's tectonic features (including spreading ridges, faults, and ocean basins) led Croizat to conclude in the 1950s that the origin of most groups of plants and animals distributed between continents originated in the Mesozoic. He proposed that their ancestors were already widely distributed before the Mesozoic, and so they are now isolated between continents, because these land areas have since become isolated. Croizat argued that some modern plant groups may have originated within the Jurassic, while others originated later in the Cretaceous. Even within continents he suggested that many groups originated in the Tertiary and survived the Pleistocene glaciations, an evolutionary model that was later to gain greater support from researchers but at the time was rarely considered.

The Mesozoic theory was strongly opposed by the influential theorists of Croizat's time, such as George Gaylord Simpson and Ernst Mayr, who looked to a much more recent origin of modern life and therefore had to appeal to theoretical migrations, whereby a vast range of animals and plants had to embark on a globetrotting series of migrations in different directions all over the globe to establish themselves on the different continents. Over the last 3 decades, Croizat's model has become widely accepted, although it remains controversial for many groups where other biogeographers believe that the plants or animals in question are of recent origin. A Mesozoic origin for modern life also has critical implications for understanding the mass extinction of dinosaurs and other groups at the end of the Cretaceous that has been attributed to a comet impact. The correlation of modern plant and animal distributions with Mesozoic tectonic structures may suggest that the ancestors of these groups survived the extinction event.

The expanded biogeographic timescale for plant and animal evolution was even proposed for the origin of animals and plants on oceanic islands such as Hawai'i and the Galapagos. Even though these islands were only a few million years old, Croizat argued that they inherited life that occupied earlier islands or island groups that no longer existed in the immediate vicinity. This model was later corroborated by the discovery that the Galapagos and Hawai'i, for example, are the latest formations in a series of volcanoes generated at a stable hotspot, and that both hot spots have a history of volcanic eruption extending back at least 90 to 100 million years. If these hot spots came into contact with mobile island arcs or microcontinents, some of their inhabitants may have colonized the volcanoes and continued to persist at the hot spots by sequentially migrating onto new volcanoes as they appeared, while the older islands were moved away by plate transport and as they eroded finally submerged beneath the sea. Because life is able to colonize new landscapes, it is possible for a young geological surface to support a very ancient biota, whether in reference to recent volcanic islands, volcanoes within a continent, or newly emergent land covered by recent oceanic sediments (e.g., mudstone, limestone).

The Mesozoic model and extended timescale for evolution has recently come into conflict with the popular application of molecular clocks. Molecular clock methods rely on the establishment of a temporal rate of molecular difference between related organisms as a function of time. In order to link a divergence rate to a particular temporal difference, it is necessary for the molecular divergence to be calibrated against a known geological age. This is most often accomplished by using a fossil representative and then extrapolating the relative age for those species for which there is no fossil representative. This method has resulted in the divergence of many groups being calculated as later than a particular geological event (such as the separation of continents occupied by the group), with the conclusion that the origin of the divergence postdated the earlier geological connection, so the current disjunction (such as between different continents) must be the result of recent dispersal or migration. This line of reasoning has been shown to be erroneous, because the molecular divergence date is calibrated by a fossil that can only provide a minimal divergence date. Any molecular dates applied by extrapolation to other taxa must, therefore, also represent minimal, not maximal, divergence estimates. Molecular divergence estimates that postdate a geological event do not, therefore, falsify the possibility that the geological event was actually involved with a divergence that was underestimated by the molecular clock. Molecular clock divergence dates may, however, provide potential falsification of a later geological event. The widespread reference to molecular clock divergence estimates as a falsification of Croizat's biogeographic model is, therefore, unfounded.

Panbiogeography and Theories of Divergence

As a final example of the panbiogeographic perspective on time and molecular approaches to evolution, one may contrast the chimpanzee and orangutan theories of divergence between humans and their nearest living great ape relatives. According to molecular similarity, chimpanzees are our nearest living relatives. Based on the molecular clock theory calibrated by the fossil orangutan relative *Sivapithecus* at about 13 million years, or by other primate fossils, the divergence between humans and chimpanzees has been estimated from as little as 4 million years to as much as 10, with 6 to 8 million years being often favored. In the absence of a fossil record for chimpanzees or gorillas, there is no other corroboration of this divergence estimate. The orangutan theory of relationship would establish an entirely different timescale, with divergence between orangutans and humans occurring at least 13 million years ago. This temporal alternative is also supported spatially with fossil orangutan relatives being distributed around parts of the Mediterranean, central Asia, and eastern Asia. These distributions, along with that of the fossil hominids of East Africa, are largely vicariant, and this would suggest that the temporal differentiation of hominids, the fossil apes, and the orangutans all occurred in their respective areas from a common ancestor that was already widely distributed over all localities. Subsequent extinction of many of these lineages by about 9 million years ago (some apparently the result of climate change and loss of forested environments) resulted in the apparent discrepancy between the origin of hominids in Africa and the origin of modern orangutans in southeastern Asia. It is only through the triple consideration of biological affinity (the evolutionary relationships), the temporal history (in the fossil record), and the spatial distribution of living and fossil taxa that the modern geographic disconnection can be understood as a spatial artifact resulting from the extinction of geographically intermediate forms.

John Grehan

See also Darwin, Charles; Evolution, Organic; Extinction and Evolution; Fossil Record; Fossils, Interpretations of; Phylogeny

Further Readings

Craw, R. C., Grehan, J. R., & Heads, M. J. (1999). *Panbiogeography: Tracking the history of life.* New York: Oxford University Press.

Croizat, L. (1958). *Panbiogeography.* Caracas, Venezuela: Author.

Croizat, L. (1964). *Space, time, form: The biological synthesis.* Caracas, Venezuela: Author.

Matthews, C. (1990). Panbiogeography [Special issue]. *New Zealand Journal of Zoology, 16.*

Sermonti, G. (1988). Panbiogeography [Special issue]. *Rivista di Biologia Biology Forum, 81.*

Pangea

Pangea (Greek for "all earth") was a supercontinent that gradually formed during the early Paleozoic era (beginning approximately 450 million years ago) and broke apart beginning in the Middle Jurassic era (approximately 180 million years ago). At its most complete state in the Early Jurassic, Pangea consisted of nearly all of the world's continental crust sutured together into a single giant landmass. This landmass was C-shaped, centered on the equator, and surrounded by a single giant ocean, termed Panthalassa. Pangea enclosed an eastward-facing body of water called the Tethys Sea, of which the modern Mediterranean is a much-reduced remnant, and was home to vast subtropical deserts. The existence of a single giant continent had profound effects on life, and many important evolutionary events occurred on Pangea, including the radiation of the dinosaurs and the early evolution of birds and mammals.

During the early history of geology it was thought that Earth was a static, stable planet whose surface remained largely unchanged through time. However, beginning in the early 1900s,

German meteorologist Alfred Wegener accumulated substantial evidence showing that the continents had not always occupied their present positions. Wegener discovered that South America, Africa, India, Australia, and Antarctica shared a suite of unique Mesozoic fossils, including a tropical plant flora characterized by the fern *Glossopteris* and a reptile fauna that included the tusked, pig-like *Lystrosaurus*. Because modern animals do not range across all continents, owing to the existence of oceans, mountain ranges, and other barriers, Wegener hypothesized that these landmasses must have been linked during the Mesozoic era (225–65 million years ago), and have since moved to their present, widely divergent positions. Additionally, Wegener chronicled closely matching rock units shared by Africa and South America, as well as evidence of former equatorial climate belts and glaciations shared between now divergent continents. Taken together, these facts suggested to Wegener that all of the continents had once been joined together into a supercontinent, which he named Pangea. Initially Wegener's ideas were controversial, as he could provide no plausible mechanism for continental motion. However, further research in the years after World War II roundly supported Wegener's observations and firmly established the existence of a Paleozoic-Mesozoic supercontinent. Today, many geologists believe that Earth is characterized by a supercontinent cycle, in which these giant landmasses form and disintegrate roughly every 500 million years.

The formation of Pangea was gradual and can be traced to the breakup of a previous supercontinent, Pannotia, approximately 750 million years ago. Pannotia split into three large landmasses, including a large southern platform called Gondwana and the more northern Proto-Laurasia, which subsequently split into several smaller landmasses. Two of these smaller landmasses, termed Laurentia and Baltica, collided in the Late Ordovician (about 450 million years ago) and were joined several million years later by Avalonia, a slice of crust comprising present-day New England, Nova Scotia, and Great Britain. Meanwhile, Gondwana was fragmenting into many small landmasses that periodically collided with the Laurentia-Baltica-Avalonia landmass, known as Euramerica. These smaller collisions occurred throughout the Devonian, Mississippian, Pennsylvanian, Permian, and Triassic periods (about 420–200 million years ago), until nearly all of the world's continental crust was completely sutured together during the Early Jurassic.

The breakup of Pangea began during the Middle Jurassic, approximately 180 million years ago. During this time North America moved to the northwest away from Africa and South America, opening the central Atlantic Ocean. North America remained connected to Europe and Asia, and together the land that is now these continents composed the landmass of Laurasia. Laurasia rotated clockwise and completely split from the southern expanse of Pangea, now termed Gondwana and composed of present-day South America, Africa, India, Australia, Antarctica, and Madagascar. Further disintegration of these two large landmasses occurred during the Cretaceous period (about 145–65 million years ago). South America split from Africa to form the south Atlantic Ocean, with the two continents gradually separating from south to north. There is some controversy about the timing of this separation, as most geophysical models place the breakup at about 100 million years ago, but recently published reconstructions suggest it may have occurred much earlier, perhaps up to 140 million years ago. Around the same time as the South America–Africa split, India and Madagascar rifted away from Antarctica and Australia, forming the eastern Indian Ocean. India separated from Madagascar about 95 million years ago and moved rapidly northeastward before colliding with Asia near the end of the Cretaceous. In the Late Cretaceous, North America and Europe rifted apart, forming the North Atlantic Ocean, and Australia and Antarctica separated. Further motion during the Cenozoic brought the continents to their modern positions.

The existence of a single large landmass had profound effects on geography, climate, and life. The numerous continental collisions that built Pangea produced many mountain ranges, including the Appalachians, which were formed by the Silurian collision between Laurentia and Avalonia and subsequently raised by the Mississippian collision between northwest Africa and Euramerica. Alteration in wind and precipitation patterns led to vast inland subtropical deserts. These deserts, along with an extensive central Pangean mountain range, formed substantial barriers for floral and faunal migration. On the whole, however, the presence of a single landmass allowed for easy dispersal of organisms, and many plant and animal species (including Wegener's *Glossopteris* and *Lystrosaurus*) had a cosmopolitan distribution during the Mesozoic.

Many important evolutionary events also happened on Pangea. Dinosaurs first evolved as small, swift predators during the Late Triassic (about 225 million years ago), and the key to their subsequent rise to dominance may have been their ability to thrive in the dry interior of Pangea. The first mammals and birds also arose during the Late Triassic and Early Jurassic, and the early evolutionary history of both groups played out on Pangea. Furthermore, two major extinctions occurred during the existence of Pangea: the Permo-Triassic mass extinction, history's largest mass die-off in which perhaps 95% of all marine species were killed; and the end of Triassic extinction, which decimated both marine and terrestrial communities. Although several different hypotheses have attempted to explain these events (including extraterrestrial impact for the Permo-Triassic), it is possible that the altered climate regimes and ease of migration on Pangea were contributing factors, at least for terrestrial extinctions.

Stephen L. Brusatte

See also Catastrophism; Extinctions, Mass; Geology; Paleontology; Plate Tectonics; Uniformitarianism; Wegener, Alfred

Further Readings

Scotese, C. R. (2004). Cenozoic and Mesozoic paleogeography: Changing terrestrial biogeographic pathways. In M. V. Lomolino & L. R. Heaney (Eds.), *Frontiers of biogeography.* Sunderland, MA: Sinauer.

Smith, A. G., Smith, D. G., & Funnell, B. M. (1994). *Atlas of Mesozoic and Cenozoic coastlines.* Cambridge, UK: Cambridge University Press.

Wegener, A. L. (1924). *The origin of continents and oceans.* New York: Dutton.

Ziegler, A. M., Scotese, C. R., & Barrett, S. F. (1983). Mesozoic and Cenozoic paleogeographic maps. In P. Broche & J. Sundermann (Eds.), *Tidal friction and the earth's rotation II.* Berlin: Springer-Verlag.

Pantheism

See Bruno, Giordano

Paracelsus (1493–1541)

Philippus Theophrastus Aureolus Bombastus von Hohenheim, otherwise less exhaustively known as Paracelsus, was a medieval alchemist, astrologer, occultist, and physician who was well ahead of his time. The title Paracelsus, literally meaning "greater than or equal to Celsus," pays homage to the Roman encyclopedist Aulus Cornelis Celsus (25 BCE–50 CE), whose only extant work, *De Medicina,* deals largely with archaic medical practices. Stories of Paracelsus's ego and arrogance would lead one to believe that Paracelsus accepted this title readily. His interest in time is evident in his numerous prophecies.

Born in Switzerland some time in late 1493 (conflicting accounts exist as to whether the month was November or December), Paracelsus was the son of a German physician and chemist and a Swiss mother. In fact, his earliest medical training was likely with his father, until more formal arrangements were made later on at the University of Vienna, where he graduated with a bachelor's degree in 1510. A doctorate was later procured, but the university he received it from is unknown, although some believe it to be the University of Ferrara.

After completing his research, Paracelsus adopted a wanderer's life for some 10-odd years. His journeys took him across Europe and landed him in Russia, where he was taken prisoner by the Tartars. Gaining the favor of the ruler of the Tartars, Paracelsus eventually was enlisted to escort the ruler's son to Constantinople.

It was in Constantinople that Paracelsus's latent quest for hermetic knowledge was nourished by Arabian adepts, although Paracelsus was a devout Catholic. Perhaps this was his first encounter with the mentioning of *alkahest,* the universal element and solvent that Paracelsus believed to be the philosopher's stone, the pursuit of all alchemists. It was the belief of many alchemists that the philosopher's stone, in addition to being able to transmute other metals into gold, was also the key to immortality.

After returning to Europe, Paracelsus began his practice of medicine as an army surgeon. His contempt for contemporary medical tradition would become apparent after his witnessing of many amputations. Paracelsus's approach to medicine was all natural, similar to some homeopathic practices that exist today. His knowledge of alchemy allowed him to clean wounds with chemicals rather than allowing them to become gangrenous and require amputation. In fact, he would become the first to name the element zinc in 1526, after the German word for pointed: *zinke*. This combination of alchemical knowledge with medicinal practice would become the precursor to the modern discipline of pharmacy.

Also, like many other physicians and alchemists of the time, Paracelsus consulted astrology to help treat various ailments that he encountered in his practice. He believed in the idea that humankind was governed by the movement of celestial bodies. A medical astrologist, Paracelsus would base diagnoses on natal charts and Zodiac signs, with the understanding that each of the 12 signs of the Zodiac corresponded to various regions of the body.

Paracelsus's background in alchemy and astrology would allow him to become acquainted with much mystical knowledge. This is confirmed by an account of Paracelsus mentioning that he had created a homunculus, the so-called little man. This artificially created human was purportedly one foot tall and carried out menial tasks for its creator. As unbelievable as this may sound, humankind's current preoccupations with creating robots and other artificial intelligence show similar pursuits.

Perhaps the most striking of all of Paracelsus's mystical cornerstones are his prophecies. Like the more popular prophecies of Nostradamus, Paracelsus's prophecies seem to be heavily abstracted. A majority of them seem to hint at the course of the Reformation and other ecclesiastical affairs. One particular prophecy seems to predict Napoleon Bonaparte's imprisonment on Saint Helena.

Scholarly dispute exists as to whether Paracelsus's prophecies are prognostications of future historical events or whether they are merely a collection of allegories pertaining to the evolution and development of the human soul. The human soul and its relation to the cosmos was certainly an essential part of Paracelsus's teachings. Most likely both of these arguments have some foundation, as many seers of the time cloaked their messages in layers of symbolism to escape religious persecution.

Paracelsus, however, was one who was constantly plagued with persecution, and even his vast mystical and medical knowledge could not cure this malady. After being threatened with imprisonment for fining a clergyman with a medical fee, Paracelsus abandoned his post at the University of Basel and returned to his former nomadic lifestyle. His dedication to his medical practices and pursuit of hermetic knowledge continued. Years later, in 1536, his *Die Grosse Wundartzney* ("The Great Surgery Book") would be published, thereby enhancing his reputation.

On September 24, 1541, Paracelsus died in Salzburg. His wish to be buried at St. Sebastian Church in Salzburg was honored. After his death, his work continued to be studied for its remarkable value in medicine and other areas. Almost 4 centuries later, the psychiatrist Carl Jung praised Paracelsus for his bombastic nature. Without Paracelsus's stubbornness, he said, it is unlikely that the practice of medicine would have progressed as fast as it did.

Dustin B. Hummel

See also Healing; Medicine, History of; Philosopher's Stone; Prophecy; Zodiac

Further Readings

Bale, P. (2006). *The devil's doctor: Paracelsus and the world of Renaissance magic and science.* New York: Farrar, Straus, & Giroux.

Goodrich-Clarke, N. (Ed.). (1999). *Paracelsus: Essential readings.* Berkeley, CA: North Atlantic Books.

Haarman, F. (1993). *Paracelsus: Life and prophecies.* Whitefish, MT: Kessinger.

Paradigm Shifts

See Darwin and Aristotle

Parmenides of Elea (c. 500 BCE)

Parmenides of Elea was one of the most influential of the Presocratic philosophers. He was born before 500 BCE in Elea, a Hellenic city on the southern coast of Italy where he founded the School of Elea. Unfortunately, his only known work, the didactic poem later titled "On Nature," written in hexameter verse, is extant only in fragments totaling approximately 150 of what were originally around 3,000 lines. It deals with the topic of time indirectly by separating real being from the influences of time.

The poem was divided into an introduction and two main sections about "the way of truth" (*aletheia*) and "the way of opinion" (*doxa*). In the proem the narrator describes his ascent to the home of an unnamed goddess from night to day, unusual for human beings. With the help of the sun maidens, he passes the gates of night and day and meets the goddess, who reveals to him the natures of truth and opinion, that is that which "is" and that which "is not." According to her instruction, real inquiry via pure reason (*logos*) is only possible concerning that which "is," because it is a true, real, unchanging, ungenerated, imperishable, indestructible, continuous whole that keeps remaining in being. But one cannot know or name that which "is not," because it is the opposite of that which "is." Ordinary human beings believe that which "is not," namely, the physical cosmos including sun, moon, earth, and the stars, to be real. But they are mistaken, due to their using deceptive sense perception. They trust the illusion that perceived or imagined things actually "are," and even give them names.

Although Parmenides does not explicitly lay down a philosophy of time, the characterization of that which "is" and which "is not" indicates his point of view. Scholars have different opinions on this subject, but they agree that according to Parmenides, unchanging being is never affected by time, because it has neither a "before" nor an "after," both of which are indissolubly linked with change and process. Because the "is" has no duration, but is "now," the question arises if being should be adequately described as atemporal or even atemporal eternity. Surely one is not allowed to identify the "now" of being with the "now" in the realm of opinion that is embedded in the process of coming-to-be and passing away. Whereas the latter "now" is changing all the time, the "now" of being remains always the same without lasting in time. Because being is free from doxical time, it seems correct to understand it as atemporal. It can also be addressed as eternal, insofar as being eternal is usually characteristic of the revered realm that has neither beginning nor end, and insofar as eternity is not meant as everlastingness here but as a continuous present beyond ordinary time.

The Parmenidean duality of appearance and reality as well as the connotations for a philosophy of time considerably influenced Plato and Neoplatonic thinkers, as is apparent in Plato's *Sophist* and *Parmenides,* as well as in Proclus's commentary on the latter dialogue. Both philosophers treated Parmenides' work with extreme respect and saw themselves as adherents of his ontology, but they introduced further differentiations, such as the ontological comparative and the explicit distinction between real eternity, everlastingness, and time.

Anja Heilmann

See also Anaximander; Anaximines; Becoming and Being; Empedocles; Heraclitus; Presocratic Age; Pythagoras of Samos; Thales; Xenophanes

Further Readings

Meijer, P. A. (1997). *Parmenides beyond the gates: The divine revelation on being, thinking and the doxa.* Amsterdam: J. C. Gieben.

Mourelatos, A. P. D. (1970). *The route of Parmenides: A study of word, image, and argument in the fragments.* New Haven, CT: Yale University Press.

Tarán, L. (1971). *Parmenides: A text with translation, commentary, and critical essays.* Princeton, NJ: Princeton University Press.

Parousia

The word *parousia* is borrowed from the Greek; it means "presence" or "arrival." It is used in Christianity to refer to the coming of Jesus; the word is also commonly used in the Christian Bible.

In I Corinthians 16:17, Paul writes of the parousia of several Christians to Corinth. In II Corinthians 7:6, Paul writes of the parousia of Titus to Corinth. These two passages refer to people who are coming to Corinth. But the word is also used to denote a time in the future when Jesus will come back to earth, as in Matthew 24:27, 37, and 39 and I Thessalonians 2:19, for example.

The timing of Jesus's parousia has caused debate. Christians who lived soon after Jesus believed that his parousia was imminent. I John, which many scholars believe was written in the 90s CE, states, "Dear children, this is the last hour." Revelation 1:3, also written in the 90s, states: "Blessed are those who hear it and take to heart what is written in it, because the time is near." As time passed and Jesus did not return, Christians nevertheless continued to anticipate his arrival. Among the most famous of the predictors was William Miller, who predicted that Jesus would return in 1844. Many Christians in the 19th century believed him. After Jesus failed to return, some formed the Christian denomination the Seventh-Day Adventists. Currently, some Christians believe that the parousia of Jesus will occur in their lifetimes.

Over time, Christians have debated the nature of Jesus's coming: Some link it with the "rapture," or Second Coming of Jesus. These events are not synonymous. The rapture is best described in I Thessalonians 4:16, 17: "For the Lord himself will come down from heaven . . . and the dead in Christ will rise first. After that, we who are still alive and are left will be caught up together with them in the clouds to meet the Lord in the air. And so we will be with the Lord forever." At this parousia, Jesus will not come to earth but will call Christians to him while suspended in the air.

The Second Coming of Jesus is recorded in Matthew 24 and Revelation 19:11ff. This event forecasts Jesus's coming to earth with a resultant battle against the forces of evil. The result of this battle will be a victory for Jesus, and Christians will rule in peace with Jesus. The amount of time that Christians will spend in peace with Jesus is debated, with some believing that this will last forever and others believing it will last for 1,000 years (Revelation 20:4).

Not all Christians believe that the parousia of Jesus will be a physical one, or even one in some future time. Some Christians believe that the events spoken of in Matthew 24 and the Book of Revelation have already occurred, that they were referring to persecutions meted out by the Romans. Others believe that these events are not meant to be taken literally but refer instead to spiritual phenomena and situations.

Mark Nickens

See also Apocalypse; Ecclesiastes, Book of; End-Time, Beliefs in; God and Time; Last Judgment; Religions and Time; Teilhard de Chardin, Pierre; Time, Sacred

Further Readings

Frykholm, A. J. (2007). *Rapture culture: Left behind in evangelical America*. New York: Oxford University Press.

McGinn, B. (1998). *Visions of the end*. New York: Columbia University Press.

Peloponnesian War

The Peloponnesian War (431–404 BCE) was more than a classic conflict between two Greek city-states and their allies; it was also a clash of ideologies—a democracy (Athens) and an oligarchy (Sparta)—and of opposing views on the nature of the world. This war is an example of conflict between two militaries with very different and incompatible styles of fighting; thus, it serves as an example of adaptation (or the lack of it). A large volume of information, including contemporary accounts of the war, has survived, providing greater detail than is known about any other ancient war. Because of the opponents' political and military differences and the amount of information available, modern war colleges study the Peloponnesian War in terms of strategies, consequences of actions, and politics.

History of the War

The cause of the Peloponnesian War dates back to the time of the Greek wars with the Persians in the early 5th century BCE. Sparta, with its totally land-based military, had taken command of the Hellenic League. The seafaring powers resented

Sparta's leadership, especially over naval forces. Athens, a sea power, accepted leadership of a new league of sea-power states called the Delian League in 479–478 BCE. Sparta remained the leader of the land-power states called the Peloponnesian League. Tension mounted between Athens and Sparta until war broke out in 460 BCE. The causes of the conflict, sometimes called the First Peloponnesian War, include Spartan suspicion of the growing Athenian empire, the defection of the city of Megara to Athens, and the Athenian construction of a long wall to the sea. The war continued until the both sides ratified the Thirty Years' Peace in 445 BCE.

The conflict restarted in 431 BCE, beginning the Peloponnesian War. Historians divide the war into three phases: the Archidamian war (431–421), the Sicilian campaigns (420–413), and the Decelean war (413–404). In the Archidamian war, Sparta launched brief land invasions into Attica in hopes of scaring Athens into capitulation. This strategy did not work, for two reasons. First, Sparta was far from home, which made sustaining operations difficult. Second, Sparta's slave class, the Helots, was always at risk of revolting when the army was away. Therefore, the Spartan army tended to keep excursions brief in order to prevent a Helot uprising. Athens, following the advice of their army general, Pericles, chose to fight a passive war, to remain behind her strong city walls, and to avoid a direct land battle. Athens did raid the Peloponnesian coast using her superior navy. Their aim was to harass Sparta and her allies in the hope that the Spartans would withdraw to Peloponnesus and sue for peace. Neither strategy worked nor could one side exploit the other side's weakness in order to win a decisive battle. This period concluded with the Peace of Nicias in 421 BCE.

The peace reflected poorly on Sparta in that its allies thought Sparta had sold them out; however, it gave the Athenians time to rest, rearm, and reevaluate their strategy. Athens, with a new general named Alcibiades, a new fleet, a replenished treasury, and a pacified empire, decided to take the offensive. Athens convinced Argos, Sparta's greatest rival, to join the alliance. Athens sent troops to Argos for a battle with Sparta, but they were too few in number. Sparta won the battle and regained the momentum; Athens had lost an opportunity to win the war.

Next, Athens chose to attack Sparta's supply lines from the west. This move began the Sicilian campaigns. Syracuse, an ally of Sparta, posed a threat to Athenian allies on Sicily. Athens sent a fleet to destroy Syracuse; however, poor decisions on strategy and slow movement to action gave Syracuse time to prepare its defenses and cost the Athenians their opportunity to achieve their goal. This campaign, far from home, ended disastrously for Athens. Athens tried to fight a land battle instead of relying on her strength—the fleet. This led to the destruction of the Athenian fleet, the capture of almost the entire army, and the defection of her enthusiastic general, Alcibiades, to the Spartans.

The Decelean phase of the war was a time of desperate strategy for both sides. Sparta made an unlikely alliance with the Persians, while the Athenians reverted to a defensive strategy. Furthermore, Sparta took to the seas, an unusual strategy for a land-based army, to cut off the Athens supply lines at the Hellespont. The Athenian fleet, once again under the command of Alcibiades, was able to put together a string of victories, but none of them was decisive. Finally, in 405 BCE, the Spartan general Lysander caught the Athenian fleet on shore. He knew not to face the Athenian fleet head on but chose to fight Athens on Spartan terms, that is, to battle a moored Athenian fleet. Lysander won a decisive victory, destroyed 168 ships of the Athenian fleet, and captured over 3,000 sailors. This, the Battle of Aegospotami, probably is the best example in the war of perfectly using one's strength against another's strength. This battle effectively ended the war.

The Peloponnesian War devastated Greece, from which she would never recover. Not only did the war bring an end to her Golden Age, a time when she was the cultural and intellectual center of the world, but it also left the countryside ravaged and the general populace struggling for survival. The Greeks lost their place as a regional power and, in 371 BCE, lost their independence to the Macedon king Philip II.

The War's Lessons

The Peloponnesian War has timeless elements that are relevant to the modern world. First, the war represents a battle between vastly differing political ideologies—a democracy versus an oligarchy. This political difference bred distrust, which led to alliance building and then to war and destruction. Because these same political differences exist in modern times, the Peloponnesian War serves as a paradigm and an example of what can go wrong.

Second, the mismatch between the two sides' military capabilities—with Sparta strong on land and Athens strong at sea—resulted in two phenomena that still have relevance in the 21st century. First, because neither alliance could effectively attack the other's center of gravity, each initially opted for a war of attrition that sapped resources without essentially harming the opponent in any serious way. Second, the opponents clearly saw the need for joint or coordinated operations, that is, operations on land and sea simultaneously. The Spartans ultimately succeeded in joint operations, while the Athenians failed. This concept of joint operations has become fully ingrained in the Western military system, especially since the rise of the European navies in the 16th century. In addition to the areas of land and sea, modern warfare adds air, space, and the cybernetic world to the arena of joint operations.

Terry W. Eddinger

See also Alexander the Great; Plutarch; Thucydides; Weapons

Further Readings

Bagnall, N. (2006). *The Peloponnesian War: Athens, Sparta, and the struggle for Greece.* New York: St. Martin's Press.

Strassler, R. (Ed.). (1998). *The landmark Thucydides: A comprehensive guide to the Peloponnesian War* (R. Crawley, Trans.). New York: Touchstone.

PENDULUMS

A pendulum is a suspended body that swings back and forth or oscillates about a fixed point. The simple pendulum consists of a heavy object called a bob suspended from a string, rope, wire, or light cord that, when given an initial push, will swing back and forth under the influence of gravity over its central or lowest point. Its swing or oscillation is regular and periodic. It is believed that the famous scientist Galileo Galilei (1564–1642) observed the regularity and periodicity of the pendulum's motion as early as 1583 by comparing the movement of a swinging lamp in a cathedral in Pisa, Italy, with his pulse rate.

Pendulums have made a valuable contribution to our understanding and appreciation of the concept of time since Galileo's observations and subsequent analysis of their motion led to the utilization of pendulums in clockwork, providing the world's first accurate measure of time. Prior to the advent of the pendulum, time was measured by the use of natural periodic phenomena, which appeared to be steadily rhythmic but varied widely from system to system and place to place. For example, use was made of the sun (sundials), flowing water (water clocks or clepsydra), flowing sand (sand clock or hourglass), burning candles and incense, and then early mechanical clocks in which weights would slowly drop, turning a gear

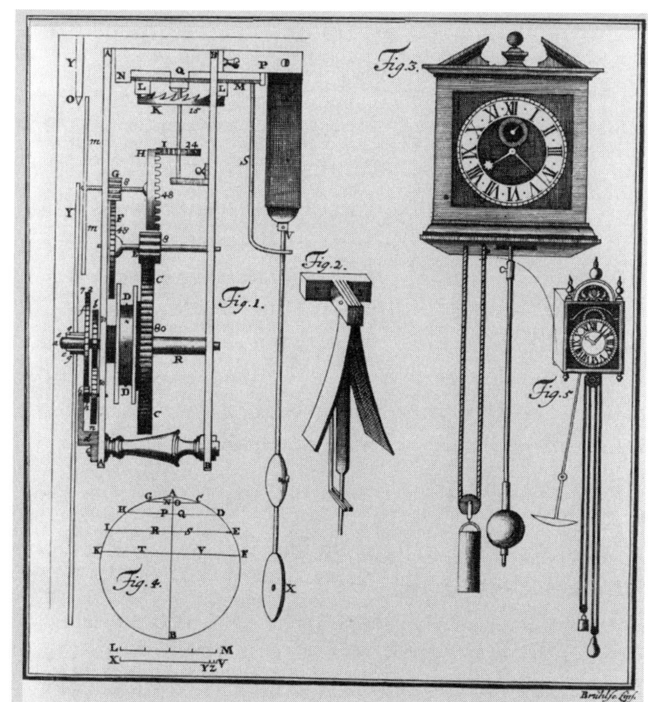

The mechanical system of a clock driven by weights and pendulum (1725). The Dutch mathematician and astronomer Christiaan Huygens was the first to successfully produce the first pendulum clock in 1656.

Source: Library of Congress, Prints & Photographs Division.

to move the clock's hand and display the time. Before the pendulum, these early mechanical clocks were probably the most accurate, and yet it was not unusual for these clocks to be off by as much as 2 hours per day! It is easy to imagine the colossal negative effects of this on society as a whole, particularly on work practices and on scientific investigations.

Within 30 years of the first use of pendulums in clocks, the average error of measuring time went from about 15 minutes a day to less than 10 seconds. By the early 1900s, time could be measured to within a hundredth of a second a day because of the pendulum. The implications of this ability to accurately measure time for the first time in history were significant, resulting in considerable technical, social, cultural, and scientific advances. Indeed, some believe that it is this (which was facilitated by the pendulum) which ushered in the scientific revolution, so creating the foundations of modern science and society. Rapid advances in our knowledge about ourselves, nature, the earth, and the entire universe would have been impossible without accurate time measurement (and therefore the pendulum).

The time required for a pendulum to swing back and forth (one complete oscillation) is called the *period* of the pendulum. This time is constant at a fixed location on earth. It does not depend on the weight of the bob or on the size of the swing (for small swings). It is dependent only on the length of the pendulum. It is this invariant time for one oscillation that makes pendulums useful for regulating the action of several devices, especially clocks. For example, one well-known use of the pendulum is as a metronome to aid musical students to keep time.

The motion of the idealized so-called simple pendulum with a light string and frictionless system is referred to as simple harmonic motion, because it is continuous and repetitive, oscillating in such a way that it swings out an equal distance on either side of a central point, with its acceleration toward the central point proportional to the distance of the pendulum from it.

The term *pendulum* is often used outside of timekeeping and science to describe systems or situations where anything undergoes regular shifts or reversals in value, direction, attitude, opinion, or state. For example, one may hear about the "pendulum of public opinion" spinning from one extreme to the next. It is also used to describe systems that use any kind of "to and fro" process. For example, *pendulum arbitration* is a method of arbitration where one side makes a proposal, and then the other suggests their proposal, with the arbitrator finally choosing one of the proposals, which then becomes binding on both sides. A technique for delivering supplies or for rescue using a helicopter, where a rope or hoist cable is set swinging by rocking the helicopter or pushing and pulling the rope, is called a *pendulum maneuver.*

The Period of a Simple Pendulum

The simple pendulum is an idealized or mathematical model of a practical pendulum for which it is assumed that the

- ring, cord, rope, wire, or rod from which the bob is suspended is so light that it is considered massless.
- string, cord, rope, wire, or rod from which the bob is suspended is inextensible; that is, of unvarying length; in simple terms, it must not stretch.
- bob is a point mass (i.e., its mass must be concentrated at a point).
- amplitude of the swing, that is, the angular displacement, is small (less than about 10 degrees) and thus the motion is simple harmonic (see θ in Figure 1).
- pendulum swings only in the vertical plane.
- effects of the air in which it moves are negligible.
- effects of the pivot or point of suspension are negligible.

In the simple idealized pendulum, when it is displaced from its rest position, only two forces act on the pendulum as shown in Figure 1. The first force is the tension in the cord acting upward along the cord, and the second force is the weight of the bob acting directly downward. The weight is given by mg, where m is the mass of the bob and g is the acceleration due to gravity. The weight consists of the two components, F_1 which balances the tension in the cord and F_2 which acts along the path of the pendulum. The force F_2, acting to restore the pendulum to its original position is equal to $-mgx/L$, where x is the displacement and

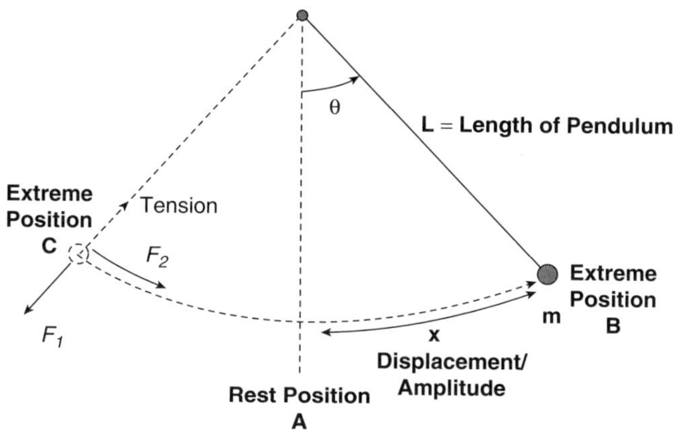

Figure 1 Principle of the simple pendulum

L is the length of the cord. The value of F_2 is always negative since it acts in the opposite direction to the displacement.

The law governing the motion of the idealized simple pendulum was discovered in the late 1500s by Galileo and states mathematically that the period of oscillation of a pendulum varies directly with the square root of the length of the pendulum and inversely with the force of gravity (acceleration due to gravity). The time, T, required for a complete swing (the period) is described as follows:

$$T = 2\pi\sqrt{\frac{L}{g}}$$

where L is the length of the pendulum and g is the acceleration due to gravity. Because π is a constant and g is a constant at a fixed location, the period of the pendulum varies only with its length. In other words for a pendulum of constant length, at a fixed location, the period is constant. Much more sophisticated mathematics is required to describe the motion of an actual physical or complex compound pendulum.

History of Pendulums

By some reports, the pendulum was used as early as 132 CE in a seismoscope or earthquake recording instrument invented by a Chinese philosopher named Chang Heng. Thus, almost certainly, people have been using pendulums for thousands of years in devices and possibly looking at swinging pendulums. Indeed, the notebooks of Leonardo da Vinci (1452–1519) include a theoretical design of a silent pendulum mechanism that could be used to regulate the motion of a clock.

Yet, the first detailed studies of the motion of the pendulum were reported only in the 17th century, by some leading scientists of the time. These included Galileo Galilei, Christiaan Huygens (1629–1695), Isaac Newton (1643–1727), and Robert Hooke (1635–1703). Although the extent of the contributions to pendulum theory and clock making of the various scientists, particularly Galileo and Huygens, is sometimes the subject of debate, it is generally accepted that it was Galileo who, at different times, made four entirely new claims about pendulum motion, as follows:

1. The period varies with the length of the pendulum (and later with the square root of length—the *law of length*).
2. The period is independent of the size of the swing or amplitude (the *law of amplitude independence*). (Huygens and Mersenne later showed this to be true only for pendulums with small amplitudes.)
3. The period is independent of the weight of the bob (the *law of weight independence*).
4. For a given pendulum length, all periods are the same (*Law of Isochrony*).

Christiaan Huygens's Contribution

It is thought that the ideas of Galileo inspired the Dutch scientist Christiaan Huygens to patent a mechanical clock in 1656 that employed a pendulum to regulate its movement. Others believe that it was the theologian and mathematician Marin Mersenne (1588–1648) who gave Huygens the idea to use the pendulum as a timing device. Whatever his inspiration, Huygens is credited with making the first pendulum clock, and after this invention, he continued to work on increasing the precision and stability of these clocks. He also devoted much time to the construction of pendulum clocks for use on ships in the open sea. The construction and use of pendulum clocks quickly spread. Indeed, in 1658, only 2 years after Huygens obtained the patent for the first pendulum clock, a clock maker from Utrecht named Samuel Coster built a church pendulum clock with a guaranteed accuracy. Indeed, by then most major towns in Holland had pendulum tower clocks.

In Huygens's memoirs, published in 1673 and entitled *Horologium Oscillatorium,* he gave a detailed description of pendulum clocks and their theory. There, he confirmed the earlier observation by Marin Mersenne that the period of a pendulum does vary with the size of the swing, and that Galileo's observation of amplitude independence was accurate only for small swings or amplitudes.

Much later, around 1675, Christiaan Huygens (with Robert Hooke) invented the coiled balance spring to make clocks more mobile. The modern version is now sometimes called the hairspring, and it is still found in some of today's wristwatches. Although this mechanism depends on the elastic properties of a spring and not a pendulum, the underlying physics of the device is the same as that of the pendulum.

Huygens has also been credited with being the first scientist who observed and described the "synchronization phenomenon." In 1665, he noticed, while lying in bed sick observing two clocks hanging on a wall in his room, that the pair of pendulum clocks that hung from a common support had synchronized: Their oscillations coincided perfectly, and they moved always in opposite directions. He described this synchronization as the "sympathy of two clocks."

This discovery was probably just as important as his invention of the first pendulum clock, because synchronization phenomena are frequently found in engineering, nature, science, and everyday life. Systems as varied as recurrent viral epidemics (e.g., childhood infections), orbits of celestial bodies, singing crickets, flashing fireflies, and applauding audiences display synchronization. Synchronization has today become a popular research topic.

Other Scientists' Contributions

Apart from Huygens and Galileo, several other scientists of the time used pendulums to increase our knowledge of the earth and to construct useful devices. The English scientist Robert Hooke was responsible for suggesting (as early as 1666) that the pendulum could be used to measure the force of gravity. He was also the inventor of the conical pendulum, consisting of a weight attached to a light rod, which was made to execute circular motion. He applied his observations of this pendulum to his study of orbital planetary motion. His observations of the conical pendulum and its application to planetary motion were also thought to have influenced the development of Newton's laws of motion.

In 1671, during an expedition to Cayenne, French Guyana, Jean Richer demonstrated the variation in the force of gravity with location on the earth by observing that the period of a pendulum was slower at Cayenne than it was at Paris.

For pendulums that used metal wire or rods to suspend the bob, it was observed that the length of the pendulum changed with temperature due to expansion of the metal with increased temperature. This, of course, caused the period to vary and was to be avoided in making an accurate clock. George Graham in 1726 designed a type of "compensated" pendulum having a glass hollow bob filled with mercury suspended by a steel pendulum rod. In his design, the thermal expansion of the mercury balanced or compensated for the thermal expansion of the pendulum rod. This came to be known as Graham's pendulum.

John Harrison, a British clockmaker, expanded on this idea of compensation for changes in the length of pendulums due to temperature changes by inventing a pendulum with up to nine alternating steel and brass rods that was known as the gridiron pendulum.

An experiment with an unusual pendulum has been credited with contributing to the popularization of pendulums and science in the early 19th and 20th centuries. This experiment was also used to demonstrate, for the first time in a simple way, that the earth indeed rotated on its axis. In 1851, Jean Bernard Léon Foucault suspended a pendulum (later named the Foucault pendulum) from the dome of the Panthéon in Paris. The bob of the pendulum was a 62-pound iron ball suspended from an iron wire 220 feet long. Foucault set it in motion, causing it to rock slowly back and forth. To record the swing of the pendulum, he attached a pointer to the ball and placed a ring of damp sand on the floor. From the patterns in the sand, the pendulum appeared to rotate, leaving a slightly different trace with each swing, thus illustrating that the earth rotated on its axis.

There were many more developments in pendulum clocks over the next century, and these led in 1889 to Siegmund Riefler's invention of a clock made with a nearly free pendulum, which attained an accuracy of a hundredth of a second a day, becoming the standard in many astronomical observatories. A more accurate free pendulum clock was later invented by W. H. Shortt in 1921. Interestingly, the Shortt clock contained two pendulums, one called the master and the other the slave. The National Institute of Standards and Technology (NIST) based in the United States had used the Riefler clock from 1904 until 1929 as the national time standard. It was briefly replaced by the Shortt double pendulum clock before the NIST finally switched to an electronic timekeeping system.

Types of Pendulums

Following Galileo's analysis and Huygens's work, several different types of pendulums were designed and constructed for use in timekeeping and in other applications. Indeed, Newton's pendulum (sometimes called Newton's cradle) survives today as a very popular desk and table ornament. It consists of five metal spheres as bobs suspended, so they all touch one another along the same line. One or more of the metal bobs are drawn back and released, causing the released spheres to swing and collide with the remaining spheres. Other types of pendulums include these:

- *Astatic pendulum:* a pendulum that is not restricted to swinging in a plane.
- *Ballistic pendulum:* a device consisting of a heavy block suspended by rods that measures the velocity of a projectile, such as a bullet, by calculating the angle of the pendulum's swing.
- *Barton's pendulum:* in its simplest construction, a device that uses approximately 10 different pendulums hung from one common string.
- *Compensated pendulum:* a clock pendulum that has mechanisms that are designed to maintain constant pendulum length in spite of temperature changes. One example of this type of pendulum using zinc and steel was used in the Big Ben clock of London, England.
- *Cycloidal pendulum:* a device that consists of a heavy weight suspended by a cord that hangs between two cycloid-shaped metal constraints.
- *Double pendulum:* a pendulum that has been attached to the end of another pendulum to make what is often called a chaotic pendulum.
- *Galitzin pendulum:* a very large horizontal pendulum that is used to measure changes in the direction of the force of gravity with time, serving as the basis of a seismograph.
- *Horizontal pendulum:* a pendulum that moves in a horizontal plane, such as a compass needle turning on its pivot.
- *Kater's (reversible) pendulum:* a pendulum that can be supported from either of two movable knife edges that is used to accurately measure the acceleration of gravity.
- *Long period pendulum:* a pendulum used for measurements of gravity, in seismometers, and for vibration isolation. In 1901, one such pendulum, 4,440 feet long and with a period of 70 seconds, was made by the faculty of the Michigan Technological University.
- *Magnetic pendulum:* consists of a permanent magnet that is suspended and free to turn in a horizontal plane in a magnetic field with a horizontal component.
- *Torsional/torsion pendulum:* This consists of a horizontal disk suspended by a wire attached to its center. When the disk is twisted and released, it returns to its original position. The disk continues to spin first one way and then the other.

Other Uses of Pendulums

Child Development/Cognitive Development

The pendulum was used by the Swiss psychologist Jean Piaget in investigations of the cognitive development of children. This involved testing the children's knowledge of the physical world by observing their understanding of the laws governing the motion of a simple pendulum. The pendulum test has been used in several investigations to identify the onset of formal operational thinking in children.

Divination

Pendulums are often used as a so-called divination device or tool in unscientific methods of searching for subsurface water, minerals, or other hidden materials. In a process sometimes known as dowsing, a forked twig or pendulum is often used as the "dowser" or "divining rod." Reportedly, this was first practiced in Europe during the Middle Ages. Although it is not widely recognized by science, dowsing has been used commercially and in archaeology. In particular, dowsing with a pendulum is sometimes called *radiesthesia* and is used in many fields, including alternative medical diagnosis, and in the location of missing persons or objects.

Hypnosis

Pendulums are sometimes used as tools to induce the so-called altered state of consciousness characteristic of the hypnotized subject. The use of a swinging pendulum to bring about a trancelike state is well known in popular culture.

Impact of Pendulums

The pendulum has played an important role in the development of science, culture, and society. Indeed, the developments in timekeeping and watchmaking that were made possible by the pendulum were so significant that the century from 1660 to 1760 is referred to as the British horological revolution. The pendulum provided the world with the first accurate measure of time. The accuracy of mechanical clocks went from within half an hour a day to a few seconds per day in just a few years. This dramatic increase in the accuracy of measuring time enabled new types of accurate measurements in fields as diverse as astronomy, navigation, and mechanics. Certainly, accurate measurement of time led to the determination of longitude. Because latitude was already known, the accurate mapping of the earth was made possible. Some believe that the British horological revolution, which began with Huygens's invention of the pendulum clock, played a major role in Britain's supremacy at sea and may have led to the industrial revolution in Britain.

The pendulum also played a key role in establishing many laws of motion in physics, including the collision laws. It was used to determine the value of the acceleration due to gravity *(g)*. By using the pendulum to determine the variation in *g* at different parts of the earth (specifically at the poles and equator), it was possible to deduce the earth's shape. This led to the creation of the science of geodesy.

Foucault's pendulum provided evidence for the rotation of the earth on its axis and also played a role in the popularization of science in the 19th and early 20th centuries. The many significant scientific and social developments that occurred because of the accurate measurement of time that became possible with the advent of pendulum clocks are the subject of several published works.

Jennifer Papin-Ramcharan

See also Clocks, Mechanical; Divination; Earth, Rotation of; Experiments, Thought; Galilei, Galileo; Harrison, John; Longitude; Newton, Isaac; Time, Cyclical; Time, Measurements of; Watches

Further Readings

Aczel, A. D. (2003). *Pendulum: Léon Foucault and the triumph of science.* New York: Atria Books.

Baker, G. L., & Blackburn, J. A. (2005). *The pendulum: A case study in physics.* Oxford, UK: Oxford University Press.

Barnett, J. E. (1998). *Time's pendulum: The quest to capture time—from sundials to atomic clocks.* New York: Plenum.

International Society for the Study of Time, Fraser, J. T., Lawrence, N. M., & Haber, F. C. (1986). *Time, science, and society in China and the West* [The Study of Time V]. Amherst: University of Massachusetts Press.

Macey, S. L. (1980). *Clocks and the cosmos: Time in Western life and thought.* Hamden, CT: Archon.

Matthews, M. R., Gauld, C. F., & Stinner, A. (2005). *The pendulum: Scientific, historical, philosophical and educational perspectives.* Dordrecht, The Netherlands: Springer.

Matthys, R. J. (2004). *Accurate clock pendulums.* Oxford, UK: Oxford University Press.

Newton, R. G. (2004). *Galileo's pendulum: From the rhythm of time to the making of matter.* Cambridge, MA: Harvard University Press.

Pikovsky, A., Rosenblum, M., & Kurths, J. (2001). *Synchronization: A universal concept in nonlinear sciences* [Cambridge Nonlinear Science Series, No. 12]. Cambridge, UK: Cambridge University Press.

Roberts, D. (2003). *Precision pendulum clocks: The quest for accurate timekeeping.* Atglen, PA: Schiffer.

Perception

When considering the concept of time perception, we must first ask the following question: Is it right to talk about perception when dealing with time? We must, in this context, remember James J. Gibson's remark to the effect that time, as such, is not a distinct stimulus and note that no one has so far discovered a sense or sensory organs that function directly to process time-related stimuli, nor do we know much about the information we use in the process that eventually generates a sense of duration. We know that there is a relationship between certain hormones, such as melatonin and dopamine, and the activities of some biological pacemakers, but this does not by a long stretch amount to a definition of *time perception*. Strange as this hiatus in our knowledge may seem given the crucial role of time in the adaptation of any living organism, we cannot as a result talk about time as we do about visual, auditory, or tactile stimuli, for instance.

Let us remember that the term *perception* refers to the process whereby a stimulus in some form of physical energy reaches dedicated sensory organs, in which, then, a process takes place that translates the energy into neural signals. These signals, in turn, reach a certain brain area via the nervous system in which they are decoded in order to create a sensory experience. Even though brain research has been making progress in the identification of brain areas associated with time-related processes, it is still wholly impossible to describe a perceptual track for the creation of time-related experience. So far, the research for a clear, stable, and meaningful connection between the pace of activity of any type of biological pacemakers or of any cyclical biological activity such as, for example, daily temperature fluctuations, on the one hand, and the properties of duration estimations of events, on the other, has been in vain. Thus for instance, research conducted in "time-free" environments, where individuals are located in deep caves or isolated bunkers without any connection to the world and with no way of obtaining external information so that, in estimating duration, they have to rely exclusively on their internal senses, did not succeed to find a significant relationship between, say, the circadian rhythm and duration estimations. Given this situation, many researchers tend to assume that the sense of time is a product of information processing or a cognitive process that utilizes information from various sources, at least while speaking of humans. If this, indeed, is the case, then it would be more adequate to talk in terms of a process of time estimation or judgment rather than of "time perception." We will, nevertheless, continue using the latter notion, but taking account of its above explained meaning.

We should also remember that the experience of time has a number of dimensions, the most prominent of which is that of duration, but which also include the order of events in time, simultaneity of events as opposed to successiveness, perception of rhythm, and more. Most of the research is concerned with duration, and hence we shall focus on it below and more particularly on prospective time judgment of short—seconds or minutes long—durations in humans. The reason for this focus is that prospective duration judgments of short intervals resemble the "perception" notion much more than duration judgments of long intervals or perception of other temporal dimensions. Temporal order perception, for example, is not determined solely by the discriminability of two events' duration but, to a large extent, by the meaning of the stimuli. Long durations, unlike short ones, are related to long series of events that are not simply an integral property of a single event, and the processes involved in experiencing and remembering the duration of a series of events are probably consciously controlled. This is not the case regarding brief durations, where duration information may be assumed to be an integral part of the experience of a single event

Psychophysical Properties and Sensitivity of Time Perception

Due to the high dependency of time perception on contextual factors and on stimuli's parameters, it is not easy to portray a clear-cut and simple picture of the psychophysical properties or of the sensitivity of time perception. Hence, only a general description is outlined in this section.

If an event lasts for less than a few milliseconds, it is perceived as instantaneous and without duration, but if, for example, two auditory stimuli are presented dichotically with an interval greater than several milliseconds between onset times, successiveness will be experienced. This is the case regarding the auditory modality, which is the most sensitive of all modalities to time. The sensitivity to time is lower in the visual modality. For example, if two binocularly presented stimuli occur with an interval of about 44 milliseconds or less between the onset times of the two, they are perceived as a single stimulus. However, if two 100-millisecond visual stimuli strike spatially adjacent receptors on the retina, an apparent movement from one position to the other will be experienced when the interval between the onsets of the two stimuli is as short as 3 to 10 milliseconds. A different experience of a flicker will be reported if the two stimuli strike the same retinal receptor. Successiveness will be experienced, in dependence on stimuli's parameters, only when the interval between the onsets of two 100-millisecond stimuli is longer than 120 milliseconds. This indicates that two successive stimuli can be discriminated reliably as separate events only if they are at least 20 to 25 milliseconds apart. Stroud (1995) suggested that psychological time is not a continuous dimension but rather consists of discrete chunks, called "perceptual moments," that are the psychological units of time. He estimated that the duration of a psychological moment is about 100 milliseconds. Other researchers estimate this duration to be 30, or 60 to 70, milliseconds in length. In any case, the psychological moment notion suggests that its duration should be the shortest possible perceived duration, an idea supported by the visual masking phenomenon. Most studies report that only beyond the range of 100 to 150 milliseconds will people be able to discriminate time intervals as being of different durations.

Judgments of time periods in the range from about 500 milliseconds to a few minutes tend to be monotonously and linearly related to objective durations with a slope of about 1.0, or in other words, duration estimates are a power function of actual duration with an exponent slightly less than 1.0.

Judgments of duration usually vary approximately up to 10 percent around the mean. However, it was found that accuracy can be significantly improved by extended practice.

The Impact of Sensory Modality

Psychophysical studies demonstrate modality differences in time perception. In both absolute judgments and discrimination tasks, people typically estimate auditory stimuli as longer than visual ones. Evidence from electrophysiological studies revealed greater sensitivity to duration in the auditory than in the visual modality.

The Impact of Contextual Factors

Duration estimations involve processes that are highly sensitive to contextual factors. A major contextual factor is the duration estimation paradigm. Estimating short durations is called *prospective* if a person knows before the beginning of a target interval that its duration should be estimated, and *retrospective* if a person is not aware of the need to estimate duration until the target interval has ended. Block differentiated between *experienced* and *remembered* time in relation to the two paradigms, respectively.

It was J. J. James who asserted that duration in passing lengthens when one pays attention to the passage of time itself, but duration in retrospect lengthens as a function of the complexity of the memories that the time affords.

Indeed, the ability to pay attention to the passage of time under prospective conditions is a major distinction between the processes that underlie timing in the two paradigms. A 1997 meta-analytic study revealed that there are some fundamental differences between duration estimations under prospective and retrospective conditions. For example, prospective judgments are longer than respective retrospective judgments in about 16% of all cases and tend to be more accurate. In both cases the judgments tend to be underestimates of objective durations. A major difference between the two paradigms is revealed when the nontemporal information processing load is manipulated during a target interval. Whereas retrospective estimates lengthen as the amount of nontemporal information increases, prospective estimates shorten as the nontemporal information processing load increases. Such fundamental differences imply that somewhat different mechanisms subserve duration judgments under prospective and retrospective conditions.

Models of Time Perception

Models of time perception vary according to the type of timing mechanism suggested.

Whereas time perception of very brief durations is grounded primarily in models of biopsychological and sensory-perceptual processes, the experience of time regarding periods in the range of seconds and minutes seems to be a manifestation of temporal information processing. This realization occurred only after it was recognized that the experience of time cannot be explained solely in terms of biological pacemakers.

Types of Duration Judgment Models

Models of duration judgment can be classified as models with and without an internal clock mechanism. Models that do not incorporate the notion of an internal clock are based on some sort of a neural or cognitive mechanism that is able to represent the passage of time. Neural-network models are one example. These models do not assume any dedicated circuit or area for encoding duration. Rather they rely on de facto neural activation responding to stimuli in order to represent its temporal features.

Some of these models assume that the activity of some neural nets in the brain can represent a certain time interval by the state of the neural net itself. This time interval can later on be reproduced by reactivating the net until it reaches the former state. Other models propose that timing is based on the detection of simultaneous activity across multiple neural inputs or the detection of patterns of neural activity associated with a specific time interval.

Despite the appealing nature of neural-net and brain models, many researchers believe that models that do not include a timer of any sort are unable, at the present time, to account for many of the characteristics of temporal behavior, like its scalar properties.

Clock models assume the existence of a biological clock, a pacemaker, or a cognitive mechanism with the characteristics of a physical clock such as the regular and constant emittance of a pulse of signals in a constant pace.

Scalar Timing Theory (SET)

Scalar timing or scalar expectancy theory (SET) is usually considered to be the most completely developed model of timing. It was originally developed by behavioral psychologists like Gibbon and Church in order to account for animals' timing behavior, which is characterized by the ability to exhibit temporal judgments of a wide range of durations, as in the case of time-based schedules of reinforcement.

SET postulates an internal clock, a memory store, and a decision mechanism. The clock itself consists of a pacemaker, a switch, and an accumulator. The pacemaker produces pulses at regular intervals. The switch acts as a gate to the pulses and either passes them or blocks them from arriving at the accumulator, in which the pulses are counted. The operation of the switch can start and stop the timing of an event. Because the pacemaker produces pulses at a regular pace, the animal's internal clock measures time along a scale that is linear with objective time, and perceived duration is a monotonic function of the total number of pulses that were counted in the accumulator.

The two major properties of scalar timing are

1. mean accuracy, which implies that mean measures of time behavior vary linearly with respective objective time intervals, and
2. scalar property of variance, which implies that timing sensitivity remains constant as duration timed vary. This property can be considered to be a form of Weber's law.

Indeed, it was found that animal timing behavior reflects these properties.

Recently, some researchers, like Wearden, found evidence for both scalar properties in human adults and in children. The internal clock notion can also provide an explanation for the auditory modality dominance over the visual modality in temporal processing. It was assumed that the internal clock is running faster for auditory than for visual stimuli, and as a result more pulses are accumulated for similar objective durations in the auditory accumulator, and for this reason subjective time is perceived as longer in the auditory modality.

Cognitive Models of Retrospective Time Judgments

Most models of retrospective duration judgments are cognitive ones and do not incorporate a notion of an internal clock. These models are confined to memory processes. The basic idea behind such models is that retrospective duration judgment relies on retrieval of information, believed to represent the target interval, from memory. The models differ in the definition of the nature of that information, but all of them assume that retrospective duration judgments are a function of the amount of relevant information that was successfully retrieved. While Ornstein's storage size model does not specify the type of information, the contextual change model (Block) claims that estimates of durations are based on the number of changes observed during an interval, or in Fraisse's words, "Psychological duration is composed of psychological changes." The model assumes that alternating cognitive operations increases the sense of subjectively experienced change and in consequence increases the subjective experience of duration. An elaboration of the contextual change model is the segmentation model introduced by Poynter. The assumption here is that contextual changes are coded as significant markers that segment experience and provide subjective referents for the estimation of the passage of time. As a result, the more an interval is segmented by meaningful stimuli, the more its retrospective duration estimates increase.

The weakness of cognitive models of retrospective duration judgment without an internal clock is rooted in its inability to provide an explanation for empirical findings, which show that under constant conditions, retrospective duration estimates are linearly correlated with objective durations. Note that even a completely "empty" interval will not be retrospectively estimated as a "duration-less" interval. In order to account for this property of retrospective duration estimates, Zakay suggested that retrospective duration experiences reflect some sort of an integration between the output of the memory retrieval process with the output of some internal clock mechanism.

Cognitive Models of Prospective Duration Judgment

Under prospective conditions, a person may intentionally encode temporal information as an integral part of the experience of the time period. Due to this characteristic, most models are relying on some sort of an attentional process in order to account for prospective duration judgments.

Models that do not assume the existence of an internal clock posit the existence of a cognitive counter that may be viewed as a processor of temporal information. The counter is incremented as a function of the amount of temporal information being processed, which depends on the amount of attentional resources allocated to it. The more attentional resources allocated for timing, the more units of temporal information are being processed and the higher the experience of prospective duration will be. Because timing competes constantly with current nontemporal tasks for attentional resources, the end result is that when one has to prospectively judge the duration of an interval during which an easy nontemporal task has to be performed, the prospective duration estimate will be higher as compared with a similar interval during which the performance of a difficult nontemporal task is required (see Time, Illusion of). The weakness of attentional models without an internal clock is that the definition of temporal information and the meaning of allocating attentional resources for temporal information processing are vague and unclear. These models are also unable to provide strong support for scalar properties of prospective duration judgments. In order to cure this situation, Zakay and Block introduced the attentional gate model (AGM), which is an elaboration of SET.

What is added to the components of SET is an attentional gate, which is a unit that enables more pulses from the pacemaker to pass through to the counter when more attentional resources are allocated for timing and that reduces the number of passing pulses when the amount of attentional resources is reduced. Thus, according to the AGM, allocating attention for temporal information processing means opening the attentional gate wider, and temporal information processing consists of accumulating and counting the number of pulses emitted by the pacemaker.

Thus, the AGM provides an explanation for scalar properties of prospective duration judgments while enabling us to take an evolutionary perspective on the development of timing mechanisms in humans.

Conclusions

It can be concluded that time perception in humans is a complex cognitive activity rather than a real perceptual process, except for very brief durations of less than 1 second. Subjective time in the context of intervals longer than 1 second can be understood only by taking into account complex interactions among all the relevant contextual factors in both the internal and in the external environments within which duration is being judged.

At the moment it seems that models that incorporate the notion of an internal clock poses better explanatory power regarding the variety of timing phenomena, in both animals and humans, as compared with models without an internal clock.

It is plausible, however, that models based on brain mechanisms that do not operate in a clock-like manner will be able, in the future, to surpass the former type of models and will have higher explanatory power.

Dan Zakay

See also Cognition; Consciousness; Memory; Psychology and Time; Time, Phenomenology of; Time, Subjective Flow of

Further Readings

Block, R. A. (Ed.). (1990). *Cognitive models of psychological time.* Hillsdale, NJ: Lawrence Erlbaum.

Grondin, S. (2001). From physical time to the first and second moments of psychological time. *Psychological Bulletin, 127*(1), 22–44.

Wearden, J. H. (1991). Do humans possess an internal clock with scalar timing properties? *Learning and Motivation, 22,* 59–83.

Zakay, D. (2005). Attention and duration judgment. *Psychologie Bulletin Francaise, 50,* 65–79.

Zakay, D., & Block, R. A. (1997). Temporal cognition. *Current Direction in Psychological Science, 6,* 12–16.

Permian Extinction

After more than 100 million years of relative stability during the Carboniferous and the Permian eras, this last period ended 251 million years ago with the largest extinction event in the Earth's history. This biotic crisis is known in paleontology as the Permian-Triassic (P-T) boundary extinction event, sometimes popularly called the Great Dying. It was more devastating than the much more famous K-T boundary extinction event, when the dinosaurs were extinguished. It has been estimated that as many as 52% of families and 90% to 95% of species were lost, far more than were lost in the K-T extinction, in which 11% of families and 75% to 80% of species were extinguished. Some authors have considered that perhaps 99.5% of individuals died as a result of the event. The primary marine and terrestrial victims included the fusulinid foraminifera, trilobites, rugose and tabulate corals, blastoid echinoderms, acanthodians, placoderms, and pelycosaurs, which did not survive beyond the P-T boundary. Other groups that were substantially reduced include the bryozoans, brachiopods, nautiloids, ammonites, sharks, bony fish, crinoids, eurypterid arthropods, ostracodes, and echinoderms. Terrestrial fauna affected included insects, amphibians, reptiles, as well as the dominant terrestrial group, the therapsids (mammal-like reptiles).

During the Carboniferous and Permian, life flourished with crinoids, nautiloids, and ammonites. Corals and fishes dominated the oceans, and amphibians and reptiles progressively invaded the terrestrial environment. This scenario changed in the end of the Permian due to causes that still are under debate. Many causes have been proposed, including meteorite impacts, volcanic activity, glaciations, and fluctuations in sea level. It is known that the formation of the supercontinent Pangea occurred in the Permian, collecting all the earth's major landmasses; that it extended from the North to South poles, and that it caused an effect on ocean currents. These large continental landmasses created climates with extreme variations of heat and cold (continental climates), and the deserts were widespread on Pangea. One of the first scenarios proposed for explaining the Late Permian extinctions was the reduction of shallow continental shelves as a result of the formation of this supercontinent. Such reduction would cause an ecological competition for space, acting as an agent for extinction. In fact, it is suggested that the marine environment was more affected than the terrestrial one, estimating that more than 95% of marine species and only 70% of land species became extinguished. Although this is a viable hypothesis, it is known as the formation of Pangea and the putative destruction of the continental shelves occurred in the Early and Middle Permian—that is, unrelated to the Late Permian mass extinction.

A second possible mechanism for the Permian extinction was severe climatic fluctuations produced by concurrent glaciation events on the North and South poles, and subsequent sea level changes. There is sedimentological evidence of significant cooling and drying in temperate latitudes, such as thick sequences of dune sands and evaporites, and prominent glaciation in the polar latitudes, such as glacial tillites.

The hypothesis for the Permian extinction most broadly accepted by paleontologists posits an increase in volcanic activity. There is abundant evidence that massive flood basalts from magma output contributed to rapid climatic turnovers and environmental stress. A massive eruptive event spanning the Permian-Triassic transition, about 252 to 250 million years ago, formed the famous Siberian Traps, a large igneous province in Siberia that covered over 200,000 square kilometers. Extensive pyroclastic deposits suggest that numerous large explosive eruptions occurred during or before the eruptions of basaltic lavas. The combination of a worldwide ash cloud and sulphates in the atmosphere might have initiated sudden climatic changes. Dust clouds and acid rain might have disrupted photosynthesis and caused the collapse of the food chains, triggering the P-T mass extinction. Later, the carbon dioxide emitted by the Siberian Traps eruptions caused the possibly cyclical climatic warming (greenhouse effect).

The main challenge to this hypothesis is the need to prove that the emissions of dust and aerosols were enough to explain the extinction event. Estimations suggest that the Siberian Traps eruptions occurred within a period of no less than 200,000 years, doubling the atmosphere's carbon dioxide content and raising global temperatures by 1.5° C to 4.5° C. This seems to be insufficient to explain the P-T extinction, but the greenhouse effect might have exponentially increased due to the release of methane from oceanic methane hydrates trapped in deep-sea sediments. Carbon isotopic analyses suggest giant methane hydrate gasification across the P-T boundary, and oxygen isotopic analyses indicate that global temperatures increased by about 6 °C near the equator and therefore by more at higher latitudes. Methane is a greenhouse gas about 62 times as powerful as carbon dioxide. Because the stability of the methane hydrates depends on temperature, it is possible that a light deep-sea warming caused a giant transfer of methane from oceanic hydrates to the atmosphere, increasing the greenhouse effect and triggering the P-T extinction event.

There are questions as to the velocity of the extinctions during the Late Permian event, because some specialists consider that those extinctions were compatible with a more gradual mass extinction model and with the gradual volcanic theories. However, other studies indicate that the extinctions occurred very abruptly, consistent with a catastrophic, possibly extraterrestrial, cause. These data suggest that an impact event (asteroidal or cometary) accompanied the P-T extinction event, as was the case for the K-T boundary event. P-T boundary impact evidence has been reported, including chondritic fragments in Antarctica, shocked quartz, fullerenes trapping extraterrestrial noble gases, and several potential impact craters: the 120-kilometer-diameter Woodleigh crater (western Australia), the 250-kilometer-diameter Bedout crater (northeastern coast of Australia), and even the giant 500-kilometer-diameter Wilkes Land crater (Wilkes Land, Antarctica). It has been suggested that the large-scale volcanism in the Siberian Traps was triggered by some of these impacts. Nevertheless, for all proposed craters, the size, impact origin, and precise age of the impacting object have not been yet completely demonstrated.

Therefore, the real initial cause for the Permian mass extinction event is under debate. At present, the most broadly accepted hypothesis considers a multicausal scenario including climatic cycles between glacial and greenhouse conditions, volcanic eruptions, dissociations of methane hydrate, and probable meteoritic impacts.

José Antonio Arz

See also Armageddon; Catastrophism; Extinction; Extinction and Evolution; Extinctions, Mass; Fossil Record; Geology; Glaciers; Global Warming; K-T Boundary; Meteors and Meteorites; Paleontology; Pangea; Uniformitarianism

Further Readings

Becker, L., Poreda, R. J., Basu, A. R., Pope, K. O., Harrison, T. M., Nicholson, C., et al. (2004). Bedout: A possible end-Permian impact crater offshore of northwestern Australia. *Science, 304,* 1469–1476.

Erwin, D. H. (1993). *The great Paleozoic crisis: Life and death in the Permian.* New York: Columbia University Press.

Ward, P. D., Botha, J., Buick, R., De Kock, M. O., Erwin, D. H., Garrison, G. H., et al. (2005). Abrupt and gradual extinction among Late Permian land vertebrates in the Karoo Basin, South Africa." *Science, 307,* 709–713.

Petrarch, Francesco (1304–1374)

Francesco Petrarch (Petracco) was an Italian poet and scholar widely recognized as the "father of humanism." He revived European interest in the knowledge of antiquity through his writings and collection of ancient texts.

Petrarch marks a transition in time from the Middle Ages to the Renaissance. Deeply critical of the neglect that ancient texts received in Europe during the Middle Ages, he sought to personally collect and translate ancient writings, visiting monasteries and churches around Europe. Were it not for his work, the Latin writing of ancient historians such as Cicero might be largely unknown today. Petrarch thus preserved the historical record for future generations. He referred to the centuries after the fall of Rome as the Dark Ages and personally laid the foundation for the rebirth of learning now known as the Renaissance.

Petrarch was born in Arezzo, Italy, on July 20, 1304. His family moved to Avignon in 1309 in support of Pope Clement V and the Avignon papacy. Petrarch studied at Montpellier and Bologna where he developed an interest in Latin literature. Following his studies he held various clerical posts and devoted much time to writing. His first major work, *Africa,* is a Latin epic about the Roman general Scipio Africanus. Petrarch considered this his greatest work, revising it throughout his life. His other Latin works include *De Contemptu Mundi,* an imaginary dialogue between the author and Saint Augustine, and *Itinerarium* (1358), a guide to the Holy Land. Petrarch saw no contradiction between humanism and faith, expressing his own religious faith in *De Vita Solitaria,* in praise of monastic life.

Petrarch wrote hundreds of poems in Italian. Most of these are compiled in his famous *Canzioniere,* a book of songs. In these poems, which consist primarily of sonnets and odes, Petrarch expresses his devotion to a beloved woman named Laura. He pursued Laura until her death and afterward wrote of her as a guide to God. Petrarch achieved great fame with his *Canzioniere.* In 1341 he was crowned as poet laureate in Rome. This tribute is often cited as marking the transition to the beginning of the Renaissance period, as Petrarch was the first poet in centuries to have received the honor. He traveled widely throughout Europe on diplomatic missions, befriending both Dante Alighieri, author of the *Divine Comedy,* and Giovanni Boccaccio, author of the *Decameron.* Together with the writings of these two prominent authors, Petrarch's writing helped to render the Tuscan dialect as the standard Italian language. He also maintained a wide correspondence, exchanging numerous letters with scholars around Europe. By his role in initiating the Italian Renaissance, Petrarch set in motion a series of revolutions over time that changed the course of world history. He has often been referred to as the first modern man.

Petrarch died on July 19, 1374, a day short of his 70th birthday. He is buried in Padua.

James P. Bonanno

See also Alighieri, Dante; Augustine of Hippo, Saint; Christianity; Humanism; Poetry

Further Readings

Bergin, T. G. (1970). *Petrarch.* New York: Twayne.

Bishop, M. (1963). *Petrarch and his world.* Bloomington: Indiana University Press.

Philo Judeaus (c. 20 BCE–c. 50 CE)

Philo Judeaus, also known as Philo of Alexandria, was a Jewish philosopher born in Hellenic Alexandria, Egypt. His writings are a unique blend of Greek philosophy and Jewish teachings and are the only surviving manuscripts from the culture of Hellenistic Judaism. Philo's work reflects a deep concern with time and the meaning of life.

Philo's writing gives much insight into the experience of the Jews within the Roman world. He was a contemporary of Jesus of Nazareth but makes no mention of Jesus in his work. Philo does discuss Pontius Pilate, the Roman governor of Judea, as well as the Roman emperors Augustus, Tiberius, Gaius Caligula, and Claudius. Around 40 CE Philo was chosen by the Jews of Alexandria to lead a delegation to Rome to protest injustices committed against them. Philo was highly regarded within his community because of the wisdom and knowledge, as expressed in his extensive philosophical writings.

Little is known about the private life of Philo. Scholars believe he was born around 20 BCE to a wealthy and influential family in Alexandria. He later wrote negatively about affluence. Philo most likely received a traditional Jewish education as well as schooling in the writings of the Greeks. His writing shows the influences of Plato, Aristotle, and the philosophies of stoicism and cynicism. Philo's work can be divided into three general categories: (1) discussions of Jewish law, (2) popular works, and (3) philosophical essays. The discussions of Jewish law include Philo's *Allegorical Commentary on Genesis,* an interpretation of the Ten Commandments and the lives of the prophets. Popular works include the *Life of Moses,* intended for a wider audience. Philosophical essays consist of treatises on a variety of issues, such as *On Providence* and *On Animals.* Philo believed that the Torah contains both literal and allegorical meaning.

The Temple of Jerusalem still stood during Philo's lifetime and had great meaning to him. He made at least one pilgrimage there. He believed that Gentiles could not be excluded from Judaism and that Jewish teachings had universal application. Philo never departed from his strongly held Jewish beliefs, and he often serves as a bridge between the ideas of the Jews and the Greeks. Clearly the teachings of both the Jews and the Greeks have had a profound impact on the development of knowledge over time. Both schools of thought deal with the nature of humankind and its place in the universe. Philo's work stands as the most important link between the two.

The Jewish historian Josephus wrote an account of the delegation to Rome led by Philo. The emperor Caligula had ordered statues of himself as a god erected in the Jewish temples of Alexandria. The Jews naturally viewed this as a deliberate provocation. Philo met with Caligula directly, but the two did not find common ground. Scholars place the date of Philo's death around 50 CE.

James P. Bonanno

See also Christianity; Egypt, Ancient; Judaism; Law; Rome, Ancient

Further Readings

Bentwich, N. (1910). *Philo-Judaeus of Alexandria.* Philadelphia: Jewish Publication Society of America.

Sandmel, S. (1979). *Philo of Alexandria: An introduction.* New York: Oxford University Press.

Williamson, R. (1989). *Jews in the Hellenistic world: Philo.* Cambridge, UK: Cambridge University Press.

Philoponus and Simplicius

The beginning or the eternity of the world and infinity or finitude of time is a central topic in the philosophy of late antiquity, especially in the debate between Christians and pagans. This quarrel started for the first time in the Neoplatonic school of Alexandria in the 6th century CE between John Philoponus and Simplicius, who were the philosophically most talented pupils of Ammonius Hermeiou. Simplicius preserved the orthodox Neoplatonic doctrine (Ammonius and his master Proclus always held to the eternity of the world), whereas the Christian Philoponus opposed this view. It is astonishing that the grammarian Philoponus (he called himself John the Grammarian and edited most of Ammonius's lectures on Aristotle's writings) argued without Christian presuppositions and personal disparagement; Simplicius, however, usually a very modest and well-educated philosopher, very rudely called Philoponus's arguments "rubbish" and accused him of "bragging and contentiousness." Obviously, they had never met personally (most probably, Simplicius had been working in Athens long before 529 CE, when the academy was closed by Justinian; Philoponus apparently never left Alexandria). Philoponus argued against the eternity of the world in his commentaries on Aristotle's *Physics* (probably written in 517 CE) and *Meteorology*, then in *On the Eternity of the World, Against Proclus* (*De aeternitate mundi contra Proclum*, written in 529 CE; this treatise refutes 18 arguments from a lost treatise written by Proclus about the eternity of the world). The writing *Against Aristotle* (*Contra Aristotelem*), which can be dated between approximately 530 CE and 534 CE, is preserved only in fragments. The first five books contained Philoponus's criticism of Aristotle's theory of the fifth element, the sixth book his criticism of Aristotle's theory of eternal movement, and at least two further books contained reflections about a Christian theory of divine creation. Simplicius's answer to Philoponus can be found mainly in his commentary on Aristotle's *De caelo I* and *Physics VIII*.

Philoponus attacks the eternity of the world by demonstrating inner contradictions in Aristotle's theory of time and eternity and by refuting Aristotle through Aristotle himself. One argument goes as follows: The eternity of the world is incompatible with Aristotle's definition of movement, because movement is the act of what is movable in potency, that is, the movable in potency exists prior to the movement. This implies that the eternal movements (e.g., the heavens' circular movements) have some movable in potency prior to them (e.g., the heavens), if the movable in potency is always anterior to the movement. Philoponus concludes that the Aristotelian definition of movement is not universal. Simplicius defends the universality of Aristotle's definition of movement by making a difference between infinite and finite movement: In the case of finite movement, the movable is still there, if the movement has finished; in the case of infinite, eternal movement, only one state of movement is prior to another state. For instance, if the sun is in Aries, then it is the movable, which is potentially in Taurus.

Further, if any first movement is excluded, Philoponus argues that all present movements become unintelligible, because every movement presupposes an infinite number of previous movements; we could not avoid a *regressus in infinitum*. Moreover, all present movements are added to those of the past; that leads to the evidently absurd notion of an infinite constantly increasing. The same problem arises concerning the future: If time and movement infinitely continue in the future, there would be an infinite body with infinite power. But that is not possible, so the world could not exist indefinitely in the future. The core of this argument is Philoponus's attack on Aristotle's notion of infinity: Aristotle contends that infinity is merely potential and never actual. For if you divide a line or a duration, you can actually mark off only a finite number of divisions, either physically or mentally. There is only a potential infinity of divisions, inasmuch as infinity exists through a process of dividing one point (or one now) after another; it is the same with the infinity of numbers.

Philoponus attacks this notion of infinity by several arguments. First, the universe must have had a beginning, or it would by now have traversed an actual infinity of years. The second argument is this: If you suppose an actual infinite number of years up to this year, next year will be an infinity plus one year. So the infinity is increasing. Simplicius says Aristotle had already anticipated Philoponus's objections, for he had pointed out that the past years have finished, so you do not get an actual infinity of them existing. That implies that time and movement are not an actually infinite quantity, but their infinity means there is a possibility of transcending every given limitation. The most fundamental difference between Philoponus and Simplicius is this: Whereas, Simplicius's infinite time is a circular indefinite repetition of finite times, Philoponus's notion of time is linear. However, the rejection of Philoponus's argument appears difficult in the context of Aristotle's philosophy of nature if you want to preserve the singularity of the individual parts of time, for instance, days or hours.

Philoponus's arguments against the eternity of the world were repeated by Bonaventure in the 13th century, after the arguments had been elaborated by Islamic philosophers. Finally, the dispute between Philoponus and Simplicius has an equivalent in Kant's doctrine of the "antinomy of pure reason" in his *Critique of Pure Reason,* especially the "first conflict of transcendental ideas." One branch of the antinomy is equivalent to Philoponus's argument (in Kant's words, "The world has a beginning in time"); this and the opposite argument ("the world has no beginning in time") is equivalent to Simplicius. Probably, there is not any "immediate effective historical connection" between the Alexandrian school quarrel and Kant's cosmological antinomy, but it shows that in this quarrel, "Greek thinking comes to the limits of its own presuppositions."

Michael Schramm

See also Aristotle; Christianity; Eternity; Infinity; Kant, Immanuel; Time, Cyclical; Time, Linear

Further Readings

Sorabji, R. (1987). Infinity and the creation. In R. Sorabji (Ed.), *Philoponus and the rejection of Aristotelian science* (pp. 164–178). London: Duckworth.

Verbeke, G. (1982). Some later Neoplatonic views on divine creation and the eternity of the world. In D. O'Meara (Ed.), *Neoplatonism and Christian thought* (pp. 45–53). Norfolk, VA: International Society for Neoplatonic Studies.

Philosopher's Stone

The philosopher's stone is a substance, tincture, or item created through alchemy that is supposed to change base metals into more precious ones and has the ability to extend one's life, cure sickness, and even grant immortality. Western traditions of alchemy are more closely associated with the concept of the philosopher's stone; the Eastern traditions of alchemy were more closely associated with the concept of the elixir of life.

Europeans probably discovered Islamic alchemy following the influence brought about by the Crusades. Islamic alchemy has its roots in Alexandria, which combined various traditions of both Greece and Egypt. The Western traditions were much more involved with the idea of changing base metals into higher or noble metals (gold or silver from lead) with longevity as a secondary effect. That is not to say that there was no interest in longevity, but that wealth was the primary motivator for many alchemists in Western Europe.

There were some European alchemists, however, who pursued the aspects of longevity as well as those of healing. Paracelsus (1493–1541) was both a physician and alchemist, using alchemy to cure the sick and pursuing the art in a different manner than others before him. He was one of the first to separate the pursuit of transmuting metals (*alchemia transmutatoria*) from that of healing (*alchemia medica*). Paracelsus is also considered the founder of toxicology. He died at the age of 48, possibly of cancer, somewhat young for one devoted to pursuing longevity and healing.

Another well-known alchemist associated with the philosopher's stone is Nicolas Flamel (1330–1418). He began a career as a scrivener but found a small book on alchemy that he wanted to understand and began his career as an alchemist. Several books are attributed to him and describe his pursuit of the philosopher's stone, but many scholars now believe that some of these works were written later by others to lend legitimacy to the field of alchemy and the pursuit of longevity. There are myths surrounding the death of Flamel and his wife; notably, the legend that he did not die but faked his death and is still living today on his discovery of the philosopher's stone. Again, these legends are attributed to those who want to add to the legitimacy of the field of alchemy.

As with the Eastern tradition of alchemy, the Western tradition became more and more philosophical or transcendental. The idea that the transmutation of a human was less a physical process and more of a spiritual exercise, and not achievable through any physical process, became important as the role of science became more prevalent. Scientists continued to make discoveries regarding the immutability of metals and, by extension, humans also could not physically transmute. Therefore, they had to do so internally, through mysticism, philosophy, or some other transcendental means.

In one respect, Nicolas Flamel and others have achieved a kind of immortality. As seen in some of today's popular media (e.g., the *Harry Potter* movies, and the *Da Vinci Code* novel by Dan Brown), Flamel has lived on, although not in the way he intended.

Timothy Binga

See also Elixir of Life; Longevity; Paracelsus; Shangri-La, Myth of; Youth, Fountain of

Further Readings

Franklyn, J. (1935). *A survey of the occult.* London: Arthur Barker.

Holmyard, E. J. (1957). *Alchemy.* Baltimore, MD: Penguin Books.

Thorndike, L. (1923–1958). *A history of magic and experimental science.* New York: Columbia University Press.

Philosophy, Process

See Whitehead, Alfred North

Phi Phenomenon

The phi phenomenon is a type of apparent movement or an illusion of movement. It is also known as *stroboscopic movement.* In fact the phi phenomenon consists of three types of apparent movement: beta, gamma, and delta movements. For reasons of space, we shall elaborate only on the first, the beta movement. As in the case of the tau and kappa illusions, the phi phenomenon illustrates the complex nature of the interrelations between the perceptions of movement, distance, and time.

The apparent movement occurs when two physically distinct stimuli—say, two light points—are alternately displayed at a frequency that exceeds a certain threshold value. Under such conditions, a continuous movement from one stimulus to the other is perceived. The apparent movement's characteristics depend on a number of factors: the physical spatial distance between the two stimuli; the physical intensity of the stimuli, which determines the amount of energy reaching the relevant sensory apparatus in the perceiver; and the duration or time span between appearance of one stimulus and the next. If the distance and intensity factors are kept constant, then when the time span between the first stimulus and the second is approximately 20 milliseconds, this will result in a perceptual experience of simultaneity. When the time span is increased to approximately 60 milliseconds, this will result in a perception of continuous movement between the two stimuli. At approximately 200 milliseconds, the resulting perception will be of one stimulus followed by the next, without any movement between them.

By 1915, the psychologist A. Korte had defined the conditions for apparent movement as follows:

- If the intensity of stimuli is kept constant, then the time span needed for optimal apparent movement changes in direct proportion to the changes in distance between the stimuli.
- If the time span between stimuli is kept constant, then the required distance between them in order to generate optimal apparent movement changes in direct proportion to stimuli intensity.
- If the distance between stimuli is kept constant, then the stimulus intensity required for optimal apparent movement is inversely proportionate to the time span between the first stimulus and the second one.

Some researchers, however, have questioned the relevance of light intensity—in the visual case—for apparent movement.

Beta movement is a continuous apparent movement between two, closely adjacent static light points that alternate lighting up. Continuous movement is perceived when the frequency at which the lights go on and off exceeds a certain threshold value, which depends on the distance between the two light points. Some researchers believe that light intensity is a significant factor here as well.

Beta movement is used, for instance, on billboards that are actually made up out of tiny static light points: Each couple of adjacent light points switches on and off at an appropriate frequency and thus the flowing effect we are all familiar with is created.

Cinema is the ultimate application of the beta phenomenon. The audience is presented with projections of static images at a frequency that ranges between 12 and 24 images per second, and this is what creates the experience of continuous movement.

Another manifestation of beta movement is its tactile variant: This occurs when an apparent sense of movement is experienced between two adjacent locations on the skin when they receive alternating stimulations. Here, too, the felt quality of the resulting apparent movement will depend on factors of distance between the two locations, the time span between one stimulus and the next, and—according to some researchers—on stimulus intensity. The existence of the tactile phenomenon indicates that the phi phenomenon is a fundamental physical principle that is not specific to one sensory modality.

The first to describe the phi phenomenon was Max Wertheimer (1912), a Gestalt psychologist. The phenomenon served as early empirical evidence for Gestaltist psychologists' claim that perceptual experience cannot be explained by means of a one-to-one relation between proximal stimulus and sensory process. There is no fit, in the case of the phi phenomenon, between the perceived movement and the external physical stimuli, which are actually static. In the visual case, even though there is no motion of a retinal image, there is a perceptual experience of movement. And so, the basic argument of the Gestaltists is that a perceptual experience does not constitute a simple mapping of the external stimulus. Perceptual processes, in fact, reflect rules of perceptual organization that the perceptual system imposes on the external stimulus.

The phi phenomenon has not been fully and exactly explained to date. Explanations relating to the visual variant refer to the persistence of vision and to the existence of a degree of temporal overlap in the activity of adjacent receptors on the retina.

Persistence of vision occurs as a retinal image continues to exist for one sixteenth of a second after the disappearance of an external stimulus. Hence, if a second stimulus occurs, creating a retinal image on an adjacent receptor within that time span (i.e., within less than one sixteenth of a second), what emerges is a temporal overlap between the activities of the two receptors. This then leads to the emergence of the perception of apparent movement.

Dan Zakay

See also Film and Photography; Memory; Perception; Psychology and Time; Time, Illusion of

Further Readings

Sekuler, T. R. (1996). Motion perception: A modern view of Wertheimer's 1912 monograph. *Perception, 25,* 1243–1258.

Steinman, R. M., Zygmunt, P., & Pizlo, F. J. (2000). Phi is not beta, and why Wertheimer's discovery launched the gestalt revolution. *Vision Research, 40,* 2257–2264.

Photography, Time-Lapse

The purpose of using time-lapse photography is to speed up events that normally take considerable time. A simple example is the opening of a flower. In a sense, it is just the opposite of the process used in creating slow-motion pictures. Both processes involve a technique for intentionally altering a normal duration of time in order to learn more about the subject by closer examination. But, whereas slow-motion focuses only on the specific subject being filmed at the time, time-lapse photography can record processes that in real time take or months or years to be completed. With both techniques, humans are able to effectively manipulate time for their own practical or artistic purposes.

In time-lapse photography, the photographer takes a sequence of pictures at a slower rate than the standard 24 frames per second used by the movie industry. The special skills needed to perform time-lapse photography include being able to discern how much time should be allowed to lapse between each photograph being taken, so as to record discernable changes that are occurring. Time intervals may need to be adjusted depending on the particular changes the subject is undergoing, but the process will usually involve taking individual pictures over 24-hour periods for as long as is needed to record the entire process. The effects of both temperature and light must also always be considered.

John Ott is generally considered a pioneer in time-lapse photography. What began as a hobby in his high school days in the late 1920s developed into a career in which his skills became influential. Because this type of photography was relatively unexplored, he had to use whatever equipment was available and to devise improved equipment himself until he was satisfied that he was photographing subjects in their most realistic setting. Starting with a Brownie camera and a timer made from kitchen clock works, he eventually built an automatic plastic greenhouse. He developed the ability to take microscopic pictures, as well as a process known as "total spectrum lighting."

Ott witnessed growing interest in uses of time-lapse photography beyond entertainment and advertising, such as applications in horticulture and medicine. He received numerous honors, including an honorary degree from Loyola University, and eventually became a faculty member in the Department of Horticulture at Michigan State University. He worked as a researcher for various companies, including Quaker Oats and General Electric.

In working with Walt Disney on the film *Secrets of Life,* which was to include a segment on the growth of an apple, Ott discovered that ordinary glass would not transmit all the ultraviolet and shorter wavelengths needed to accurately record the normal process. He was finally able to complete the assignment by substituting special plastic materials.

In his film *Our Changing World,* he wanted to depict the orderly progressions of earth and the creation of life and its development. The power of the single cell had always impressed him. He noted that humankind has been on earth a much shorter time than plants and that plants and animals tend to respond similarly to light.

Though Ott had said that his work was often slow and discouraging, his pioneering efforts were very impressive. His advances encouraged others to devise and use more sophisticated techniques and equipment. Those working with the process continued to emphasize the importance of correct lighting and determining the most appropriate time intervals between pictures.

Time-lapse photography has been used in many scientific studies, such as research on glacier motion in Glacier National Park, studies of sleep patterns, studies of slow-acting geologic processes in natural settings, studies of movement in plants, and studies of weather phenomena. In addition to exploring the natural world, there are also many other potential applications of time-lapse photography with respect to human activities, such as monitoring business projects and procedures, and studying the urban environment and urban renewal. An example of a successful project related to the urban landscape can be found in the work of Camilio Jose Vergara, a photographer, sociologist, and ethnographer. For 30 years he recorded the changes in inner-city neighborhoods in New York, Newark, Chicago, Detroit, and Los Angeles.

Perhaps one of the most dramatic uses has occurred at Ground Zero, the site of the World Trade Center disaster. Documentary filmmaker Jim Whitaker initiated Project Rebirth in the spring of 2002, with the goal of recording what was happening in this area. As the project continued, cameras were installed at all four corners and at ground level. Some results of the ongoing process are available on the Project Rebirth Web site. The process involves taking one frame every 5 minutes for 7 days a week; the final result will be viewable within a span of 20 minutes.

Betty A. Gard

See also Film and Photography; Perception; Time, Illusion of

Further Readings

Abeid, J., & Arditi, D. (2002). Time-lapse digital photography applied to project management. *Journal of Construction Engineering and Management, 128*(6), 530–535.

Kinsman, E. M. (2006). *The time-lapse photography FAQ: An introduction to time-lapse photography.* Retrieved June 7, 2007, from http://www.sciencephotography.com/how2d02.shtml

Ott, J. (1958). *My ivory cellar: The story of time-lapse photography* (3rd ed.). Winnetka, IL: Twentieth Century Press.

Photosynthesis

Photosynthesis is the process by which organisms convert light energy into chemical energy in the form of carbohydrates. The inputs of the chemical reaction are light energy, carbon dioxide, and water; the outputs are carbohydrates and oxygen.

The overall reaction, which has many intermediate steps, is written as follows:

$$\text{Light energy} + CO_2 + H_2O => (CH_2O) + O_2$$

The sun is the main source of light for the process. Photosynthetic organisms break down the bonds in the resulting carbohydrates to obtain the necessary energy for life-sustaining functions. Plants, algae, and some bacteria are the known organisms capable of photosynthetic activity. They all produce pigments, specialized proteins that capture energy when exposed to light. Numerous photosynthetic organisms have developed adaptations to regulate the timing of photosynthesis. By lengthening the time spent in photosynthesis per day or changing the time of day when photosynthesis occurs, organisms improve the efficiency of photosynthesis and their ability to survive.

Locations and Functions of Pigments

The location of pigments in photosynthetic organisms depends on whether the organism is prokaryotic (does not have a cell nucleus or organelles) or eukaryotic (has cell nucleus and organelles). The prokaryote *Halobacterium halobium* and other photosynthetic bacteria have pigments embedded in their cell membranes. Prokaryotic blue-green alga has pigment proteins inserted in a more complicated system of stacked membranes interior to the cell wall. Higher plants, such as needle-leaved plants and flowering plants, have a specialized organelle for photosynthesis within the plant cell, the chloroplast. The double-membraned organelle contains photosynthetic membranes that are embedded most commonly with the pigments, chlorophyll-a and chlorophyll-b.

Pigments are essential to photosynthesis, because they can absorb energy from photons, the units of light energy. Each pigment absorbs a characteristic wavelength, which is a stream of photons. For example, chlorophyll-a absorbs wavelengths in the range between 550 and 700 nanometers (nm, 1×10^{-9} meter), and bacteriochlorophyll-a in bacteria absorbs wavelengths between 470 and 750 nanometers.

Pigments efficiently absorb energy because they contain chemical bonds that accommodate fluctuating levels of energy. The characteristic carbon rings in pigments include many double bonds. Carbon atoms joined by double bonds share their electrons; thus the electrons are not strongly attracted to a particular carbon nucleus and move in a loose cloud around the entire molecule. When photons strike a pigment, their energy is accepted by the pigment's electrons, which can easily move from a lower energy level to a higher one in the cloud of electrons. Chlorophyll-a has five carbon rings with a total of 10 double bonds, making it an excellent acceptor of energy from light. The pigment can either donate the energized electrons to other molecules or release the energy from the electrons as longer wavelengths than those the pigment absorbed.

Structures in Photosynthesis

Organisms have structures in their photosynthetic membranes called *reaction centers* and *antennae*, respectively, both of which are necessary for photosynthesis to occur. The reaction center is composed of the unique pigments capable of initiating the chemical reactions of photosynthesis by donating electrons to molecules within cells; the pigments are bacteriochlorophyll-*a* in bacteria and chlorophyll-*a* in algae and plants. Scientists have identified special forms of these chlorophylls that are responsible for the actual work of changing light energy into chemical energy in the reaction centers. The chlorophylls are P870 in bacteria and P700 and P680 in algae and plants, where P stands for pigment and the number refers to an absorption wavelength. However, the specialized chlorophylls cannot absorb enough light energy on their own to drive photosynthesis; they are fed energy by the antennae.

The antenna structure in membranes is the locus of light energy absorption and concentration. It is composed of accessory pigments that generally can absorb shorter wavelengths than P680, P700, and P870 can. Examples of accessory pigments are bacteriochlorophyll-*b* (absorbs 400 nm–1020 nm wavelengths) in purple bacteria and chlorophyll-*b* (absorbs 454 nm–670 nm wavelengths) in higher plants. Accessory pigments capture light energy and then release it to other accessory pigments or chlorophylls in the antennae as longer wavelengths, but these accessory pigments are not capable of donating electrons to other molecules. The accessory pigments pass along longer wavelengths to each other until the waves reach a length that can be absorbed by the specialized chlorophylls in the reaction center.

A substantial number of accessory pigment proteins are needed to feed a reaction center with enough light energy to drive photosynthesis. Over 300 molecules of chlorophyll-*b* are needed to funnel enough light energy to activate one molecule of chlorophyll-*a* in the reaction center of a typical higher plant.

Adaptations in Photosynthesis

Photosynthetic organisms are capable of making photosynthesis more efficient by regulating the time spent in photosynthesis per day or changing the time of day when photosynthesis occurs. Some higher plants can change their leaf position over the course of a day to track the sun's movement. This adaptation allows the plants to increase the number of hours per day spent in direct sunlight and maximum light absorption. Experiments have confirmed that this behavior, called *diaheliotropism,* increases the efficiency of photosynthesis.

Plants that live in hot, dry climates, such as cacti, have developed an adaptation of photosynthesis that allows parts of the process to occur at a different time of day than in the majority of plants. Generally, all steps of photosynthesis occur during daylight, including the intake of carbon dioxide through stomata, which are openings in the leaves of plants. The majority of plants take in carbon dioxide and initially fix the carbon into a compound called 3-phosphoglycerate. However, plants in hot, arid regions lose water at a high rate when the stomata are open, so many have developed crassulacean acid metabolism (CAM) to avoid dehydration. CAM plants open their stomata only at night, initially fix carbon dioxide into malic acid, and then store the acid. During the day, CAM plants close their stomata, break down the malic acid to release the carbon dioxide, and then proceed with photosynthesis. The CAM adaptation makes it possible for plants to withstand long periods of drought.

Erin M. O'Toole

See also Chemical Reactions; Chemistry; Ecology; Global Warming; Seasons, Change of; Trees

Further Readings

Das, V. S. R. (2004). *Photosynthesis: Regulation under varying light regimes.* Enfield, NH: Science Publishers.

Heldt, H.-W. (2005). *Plant biochemistry* (3rd ed.). Burlington, MA: Elsevier Academic Press.

Lawlor, D. W. (2004). *Photosynthesis* (3rd ed.). Oxford, UK: Taylor and Francis.

Phylogeny

A phylogeny is an evolutionary history of an organism or group of organisms; it may be interpreted as a genealogical tree, an ancestor and descendant lineage, or as systematic relationships of form within a classification scheme. Phylogenies are studied principally in the fields of phylogenetics and systematics.

History of Phylogenetics

Phylogeny was discussed in detail by the 19th-century German morphologist Ernst Haeckel, who proposed a biogenetic law (or the law of recapitulation). The biogenetic law states that phylogeny, or the evolutionary history of an organism, is recapitulated through its ontogeny, or the development of an individual organism in embryo. The subsequent rejection of Haeckel's law was a significant move away from using mechanical explanations or causes, such as embryonic development, to explain the relationship between organisms. Haeckel's most significant contribution was that of the phylogenetic tree (*Phylogenetisches Stambaum*), the now universally accepted way to depict genealogical relationships.

A phylogenetic tree may depict hypothetical ancestor-descendant relationships, sometimes called a *transformation series,* between groups of organisms (species, genera, and families) or their characteristics, through time. Such phylogenetic trees have been popular tools of paleontologists who use them to establish so-called ghost lineages between similar-looking fossils throughout the stratigraphic record. Phylogenetic trees were challenged in the early 20th century by the German-speaking systematic morphologists, led by Adolf Naef. The evolutionary relationships that phylogenetic trees were claimed to depict were based on linking similar-looking organisms that overlapped through time, rather than considering relationships of form.

The systematic morphologists considered *homologues* (different manifestations of the same morphological structure) to be a sounder basis for the discovery of relationship than the assembly of ghost lineages. If organisms are related, their characters are homologous, that is, the same; as opposed to analogous, that is, similar but not the same. Naef's trees related organisms only at the terminal branches, rather than depicting hypothetical lineages, with organisms (hypothetical or real) at both the nodes and tips. Homologous organisms belonged to "natural groups or classifications" that share a greater relationship among themselves than they do to any other group.

The rejection of phylogenetic trees and the concomitant support for natural groups was criticized by Anglo American phylogeneticists such as George Gaylord Simpson and Ernst Mayr, who defended the depiction of lineages in phylogenetic trees rather than the discovery of natural groups, which challenged some traditional taxonomic groups. Anglo American phylogenetics, however, changed considerably in the latter half of the 20th century when the work of Willi Hennig, a German entomologist, was translated into English.

Phylogenetic Systematics

Hennig's *Phylogenetic Systematics* attempted to resurrect Haeckel's systematic phylogenetics by reintroducing the causal mechanisms that had been rejected by Adolf Naef. Hennig's phylogenetic systematics combined Haeckel's transformational viewpoint—but at the level of character rather than taxon—with Naef's trees of relationships to form ancestor-descendant schemes of relationship with organisms only at the tips, and character transformations leading from the nodes to the tips. The resulting trees attempted to group homologous organisms into "natural" or monophyletic classifications based on a causal mechanism, thus combining Haeckel's phylogenetic tree with Naef's systematic morphology.

Phylogenetic systematics developed into a numerical method by incorporating the principal notion of *phenetics,* that is, similarity concepts, with a causal mechanism to find optimal trees.

Phylogenetic systematics, later referred to as *cladistics,* underwent a revolution in the work of Gareth Nelson by returning to systematic morphology. Pattern cladistics rejected causal homologies and ancestor-descendant relationships as uninformative and misleading, because they introduced bias into phylogenetics. The pattern cladists, led by Ronald Brady and Gareth Nelson, considered monophyly to indicate "natural groups," which can be used to test existing taxonomies rather than to identify causal relationships (a common ancestor). The resulting diagrams, called cladograms, could represent numerous lineages but only a single classification. Hennig's elimination of paraphyly and its connection made with ancestry by cladists such as Colin Patterson helped to define phylogenetics as a science of classification based on the relationships of form.

Molecular Phylogenetics

Molecular phylogenetics is the study of amino acid or DNA sequences and how they may be related among different organisms. The field has grown exponentially and amassed a significant volume of data. Unlike phylogenetic systematics, molecular phylogenies tend to consist of individual character trees (relationships between organisms based on a single character) and are used to hypothesize recent genealogies in populations as well as ancestor-descendant relationships in species. Despite its popularity, very little theoretical work has been done on the relevance of homology of DNA sequences. Molecular phylogenetics, however, has progressed methodologically and technologically in such issues as alignment of sequences and in mapping the similarity distances in phenetic methods.

Phylogenetic Classification

Phylogenies may be interpreted as explicit evolutionary pathways, natural groups (classifications), or a combination of both. The latter has caused the most controversy in its claim for phylogenetic classifications. Recent debate has focused on defending lineages rather than classifications in taxonomy. A *nonmonophyletic* group (also known as a *paraphyletic* or *polyphyletic* group) is an artificial or incongruous set that shares greater relationship to other groups than to its own. A proposed lineage may be paraphyletic and therefore contradict any given natural classification. Reptiles are an example of a paraphyletic group that exists in name only, not within a natural classification. The defense of paraphyletic groups in classification reflects the battle between the Anglo American paleontologists and systematic morphologists in the early 20th century, during which classification and hypothetical lineages were confused.

Phylogenetic Biogeography

Phylogenies have been used in biogeography (the study of biotic distributions) during three periods: in the late 19th century, with the advent of natural selection as a viable mechanism for species evolution (e.g., Haeckel); in the 1960s, with the onset of Hennig's phylogenetic systematics; and in the late 20th century, with the use of molecular phylogenies. The same method has been used in each of these periods, namely that of proposing a center of origin and drawing the direction of dispersal and/or vicariance events (allopatry) on a phylogenetic tree.

Since the late 19th century, fossils were used to date such events within any given phylogenetic tree. The method is still widely practiced today (i.e., using a molecular clock). The only difference between these periods is the data used. Nineteenth-century phylogeneticists relied on fossils, mid-20th-century phylogeneticists on the morphology of extant taxa, and 21st-century molecular systematists on molecular data.

Malte C. Ebach

See also Cybertaxonomy; Darwin, Charles; DNA; Evolution, Organic; Haeckel, Ernst; Huxley, Thomas Henry

Further Readings

Kitching, I., Forey, P. L., Humphries, C. J., & Williams, D. M. (1998). *Cladistics: The theory and practice of parsimony analysis* (Systematics Association Publications No. 11). Oxford, UK: Oxford University Press.

Nelson, G., & Platnick, N. I. (1981). *Systematics and biogeography: Cladistics and vicariance*. New York: Columbia University Press.

Williams, D. M., & Ebach, M. C. (2008). *Foundations of systematics and biogeography*. New York: Springer.

Piaget, Jean (1896–1980)

Jean Piaget was a Swiss philosopher and psychologist whose principal research interests were in epistemology and developmental psychology. He believed that to understand knowledge, one must look at its psychological origins and how it evolves as children become adults. His research led him to the epistemological stance he deemed *constructivism*—the position that knowledge is constructed from experience over time.

During work in Alfred Binet's lab at the Sorbonne, Piaget noticed that children of the same age consistently made the same mistakes on intelligence tests. Later, after very careful observations of children during which he would ask questions or assign tasks to elicit behaviors that would give him insight, Piaget noted that children were organizing and reorganizing the world as they gained more experience. He concluded that children were not simply imitating or regurgitating what they were told or observed; they were creatively interpreting the world based on their past experiences at a fairly constant rate, which could be organized into stages.

According to Piaget, children proceed with this process of interpretation by constructing and revising gestalt-type schema, which allow for future recognition of patterns that have been experienced. With this view in mind, Piaget conducted experiments to determine specifically how the knowledge of time is constructed in children. His results are outlined primarily in *The Construction of Reality in the Child* (1937) and *The Child's Conception of Time* (1946).

These experiments led Piaget to conclude that children begin with an egocentric conception of time where duration (number of minutes, years, etc.) depends on speed of action—the faster one goes the less time one spends on an action.

Eventually children develop a distinction between duration and succession (past, present, future; before and after), which allows them to see that the flow of time is constant. For Piaget, the fact that children seem to move from special relations constituting the concept of time to the more complex notion of independent time flow showed that space was a more basic concept from which the concept of time is constructed.

Many have criticized Piaget's epistemology and conception of time. J. T. Fraser, a prominent author on time who debated Piaget, held that there must be some intuition of time in children. Later, other prominent authors critiqued Piaget for unwarranted generalizations and other experimental problems. However, his importance to the progression toward understanding time in psychology and philosophy is evident in his influence on such thinkers as Jürgen Habermas, Thomas S. Kuhn, and contemporary constructivist epistemologists, as well as reactions from such prominent thinkers as Noam Chomsky and Hilary Putnam.

Kyle Walker

See also Cognition; Consciousness; Education and Time; Epistemology; Kuhn, Thomas S.; Language; Memory; Psychology and Time; Time, Teaching

Further Readings

Cohen, D. (1983). *Piaget: Critique and reassessment.* New York: St. Martin's Press.

Singer, D. G., & Revenson, T. A. (1997). *A Piaget primer: How a child thinks* (Rev. ed.). Madison, CT: International Universities Press.

Piltdown Man Hoax

In 1912, fragments of a skull and jawbone were found in a gravel pit at Piltdown, a village in East Sussex, England. When assembled, scientists believed the specimen to be the "missing link" between ape and human, providing solid proof of the theory of evolution. Forty years after its discovery, Piltdown man became exposed as the Piltdown hoax, one of the most notorious frauds in the history of science.

The Find

A laborer working in the Piltdown gravel pit in the early 1900s claimed to have found a piece of a skull. He passed it on to Charles Dawson, a local solicitor and well-known amateur archaeologist. Dawson found additional fragments at the site in 1911, and presented them to Sir Arthur Smith Woodward, keeper of geology at the British Museum. Interested in the finds, Woodward returned to the site with Dawson, where they recovered additional skull fragments and half of a lower jawbone. The same pit also produced a few fossil animal bones and a tooth. In December 1912, they presented their reconstructed skull to the Geological Society of London as a new type of early human, *Eoanthropus dawsoni,* or "Dawson's Dawn Man." The bone of the skull was unusually thick and stained with age, implying primitiveness, while having the shape and larger size of a modern braincase. However, the mandible associated with it was far more simian than human. This apparent ape-man found with the bones of extinct mammals in a Pleistocene gravel bed was exciting news for English paleontologists. Until that point, all fossil human remains had been found in various locations on the Continent, especially Germany and France. England could now claim a place on the evolutionary tree even earlier than these other hominids.

Additional finds were made at Piltdown through 1915, including additional animal bones, stone tools, and a second skull (also found by Dawson) 2 miles away from the original site. Interestingly, no more finds were made after Dawson's death in 1916.

The reconstruction of Piltdown man was challenged from its introduction. The hinge joining the jaw to the skull was conveniently missing, causing some experts to doubt that the skull and jaw were from the same individual. Others developed a completely different model from the pieces. In the 1920s, Franz Weidenreich, an anatomist, examined the specimen and reported that it was a modern human cranium and an orangutan jaw with filed teeth. He was correct, but it took 30 years for paleontologists to admit it. As more and more hominid finds were made around the world in the following years, including *Homo erectus* and *Australopithecus, Eoanthropus* was pushed aside.

Not only did it not fit into the increasingly clear evolutionary tree, but also no other specimen was ever found resembling it.

Exposure of the Hoax

Joseph Weiner, an anthropology professor at Oxford University, has been given credit for exposing the hoax. In the early 1950s, he attended a paleontology congress in London. Piltdown man was hardly mentioned once again, for not "fitting in." The possibility of fraud dawned on him. After meticulously gathering evidence, conducting interviews, and using recently developed tests on the bones themselves, he exposed the forgery in 1953. A new dating technique, the fluorine absorption test, was developed to determine age and had been applied to the Piltdown fossils in 1949. The results established that the remains were, in fact, relatively modern, but they were still assumed to be genuine. In 1953, with the fluorine test more advanced, the Piltdown remains were retested. It was determined that the cranium was from the Upper Pleistocene and approximately 50,000 years old, while the mandible and tooth were modern. Another new test devised by American scientists analyzed nitrogen content to determine age and corroborated these results. In 1959, however, the recently discovered carbon-14 dating technique was applied to the bones. The skull was shown to be between 520 and 720 years old, and the jawbone a bit younger.

Eventually it was proven that the Piltdown site had been salted with bones and artifacts from a variety of sources. The hoax had succeeded for 40 years due to a variety of circumstances. The find gave experts evidence supporting a theory they believed to be true at the time. The scientists closest to Piltdown were experts, but not in hominid evolution. In the early 20th century, there were no chemical tests or dating techniques. Analytical tools were primitive by today's standards. At the time of the find, there were no hominid fossils other than the earliest Neanderthal remains for comparison. Finally, the Piltdown bones were kept locked away in the British Museum as valuables, so most scientists wishing to study them had to rely on pictures, sketches, or poor-quality X-rays.

Identity of the Forger

The identity of the perpetrator has never been proven. The reason for the entire hoax is also debated. This hoax was deliberately and systematically carried out over a number of years with no obvious motive. About two dozen suspects have been suggested over the intervening years, and there is plenty of circumstantial evidence implicating several of them.

The most obvious suspect is Dawson himself. He had no formal training but a lot of luck at finding unique artifacts. He was present at Piltdown when all the major finds were made. Once Piltdown was exposed, scientists reexamined his other finds, and 46 objects credited to him turned out to have questionable backgrounds or were outright forgeries. He appears to have appropriated the finds of others as his own, and many of his writings were plagiarized. Most people agree he was involved, but question the presence of an accomplice. Woodward is also suspect. He was Piltdown's strongest supporter and refused to allow some of the simplest scientific tests to be given to the specimens, which would have exposed the forgery immediately. Sir Arthur Conan Doyle, author of the Sherlock Holmes stories, is often presented as a suspect. He was a neighbor of Dawson, an amateur fossil hunter, and a participant in the digs at Piltdown. It has been argued that his novel, *The Lost World,* has clues referring to the hoax. Pierre Teilhard de Chardin has also been accused. He was a Jesuit, a theologian, and an anthropologist. As Dawson's friend, he participated in the work and was present for many of the key discoveries. He had traveled to locations in Africa where some of the frauds originated, and his later recollections of the events at Piltdown were very vague.

The most recent accusations, however, have been against Martin Hinton, zoology curator at the Natural History Museum in London. He worked under Woodward at the time of the hoax and had a public conflict with him over salary. There was also professional rivalry between the two, and Hinton may have perpetrated the hoax to embarrass his colleague. He was known for creating elaborate practical jokes. In the mid-1970s a trunk was found in a loft of the museum with Hinton's initials on the lid. Among other things, there were a variety of bones and teeth stained with chemicals identical to those used on the Piltdown finds. The chemical recipe was apparently created by Hinton and presented in an 1899 scientific paper.

The identity of the culprit and the reasons for creating such an elaborate scheme may never be proven, but there is an important lesson to be learned. The hoax was successful for decades, owing to inconsistent examination and analysis. Once an expert had established the importance of the find, it was accepted uncritically. Scientists embraced the Piltdown find because it supported the prevailing beliefs of the time. Scientists are still human, and ambition, pride, and rivalry can all come into play.

Jill M. Church

See also Anthropology; Dating Techniques; Evidence of Human Evolution, Interpreting; Hominid-Pongid Split; Teilhard de Chardin, Pierre

Further Readings

Millar, R. (1972). *The Piltdown man.* New York: St. Martin's Press.

Russell, M. (2003). *Piltdown man: The secret life of Charles Dawson and the world's greatest archaeological hoax.* Stroud, Gloucestershire, UK: Tempus.

Walsh, J. E. (1996). *Unraveling Piltdown: The science fraud of the century and its solution.* New York: Random House.

Werner, J., & Stringer, C. (2003). *The Piltdown forgery* (50th anniversary ed.). Oxford, UK: Oxford University Press.

Planck Time

The *Planck time* has a value of 5.39121×10^{-44} second (current uncertainty 40×10^{-44}). It is the smallest time that can be operationally defined, that is, measured even in principle. The *Planck length* has a value of 1.61624×10^{-35} (current uncertainty 12×10^{-35}) meter and represents the smallest length that can be operationally defined.

By international agreement, the distance or length, L, between two points in space is defined as the time, t, it takes for light to travel between the points in a vacuum, multiplied by a constant, c

$$L = ct, \quad (1)$$

where c = 299,792,458 meters per second is the speed of light in a vacuum. (This number is exact by definition.) In order to measure t, and thus L, we need a clock with an uncertainty Δt no larger than t. The time-energy uncertainty principle says that the product of Δt and the uncertainty in a measurement of energy in that time interval, ΔE, can be no less than $\hbar/2$, where $\hbar = h/2\pi$ and $h = 6.6260693 \times 10^{-34}$ joule-second (current uncertainty 11×10^{-34}) is Planck's constant. That is,

$$\Delta Et \geq \Delta E \Delta t \geq \frac{\mathrm{h}}{2} \,. \quad (2)$$

Thus,

$$\Delta E \geq \frac{\mathrm{h}}{2t} \geq \frac{\mathrm{h}c}{2L} \,. \quad (3)$$

This energy is equivalent to the rest energy of a body of mass m,

$$\Delta E = mc^2. \quad (4)$$

Equation 5 implies that within a spherical region of space of radius L, we cannot determine, by any measurement, that it contains a mass less than

$$m = \frac{\mathrm{h}}{2cL} \,. \quad (5)$$

Now, a spherical body of mass M will be a black hole if its radius R is less than or equal to

$$R = \frac{2GM}{c^2}, \quad (6)$$

where $G = 6.6742 \times 10^{-11}$ cubic meters per kilogram per square second (current uncertainty 10×10^{-11}) is Newton's gravitational constant. This is called the *Schwarzschild radius.*

Consider a body of mass m given in Equation 5. Its Schwarzschild radius will be

$$L_{PL} = \left(\frac{\mathrm{h}G}{c^3}\right)^{\frac{1}{2}}, \quad (7)$$

which is called the Planck length. Notice it is simply the length formed from the three basic constants in physics, $\hbar$, c, and G. It represents the smallest length that can be operationally defined, that is, defined in terms of measurements that can be made by any instrument. If we tried to measure a smaller distance, the time interval would be smaller, the uncertainty in rest energy larger, the uncertainty in mass larger, and the region of space would be experimentally indistinguishable from a black hole. Because nothing inside a black hole can climb outside its gravitational field, we cannot see inside and thus cannot make a smaller measurement of distance.

Similarly, we can make no smaller measurement of time than the Planck time,

$$t_{PL} = \frac{L_{PL}}{c} = \left(\frac{\mathrm{h}G}{c^5}\right)^{\frac{1}{2}}. \quad (8)$$

The Planck time and Planck length are the most basic units of time and space. Although distance and time are assumed continuous variables, they are fundamentally discrete. However, because physics experiments have not yet even come close to probing space and time on the Planck scale, treating them as continuous remains a good approximation.

General relativity, the theory of gravity introduced by Albert Einstein in 1915, has so far passed every empirical test to high precision. However, not being a quantum theory, it can be expected to break down at the Planck scale, where it will have to be replaced by a quantum theory of gravity, still not developed.

The Planck time represents the earliest time that can be operationally defined for our universe on the positive side of the time axis. However, this does not mean that "time began" at that moment. Nothing forbids, and time symmetry implies, another universe at earlier times, on the negative side of our time axis.

Victor J. Stenger

See also Attosecond and Nanosecond; Einstein, Albert; Light, Speed of; Newton, Isaac; Quantum Mechanics; Time, Measurements of; Time, Operational Definition of; Time, Symmetry of; Time, Units of

Further Readings

Stenger, V. J. (2006). *The comprehensible cosmos: Where do the laws of physics come from?* Amherst, NY: Prometheus.

Planetariums

A planetarium is a device for artificially depicting the night sky, showing the relative positions and motions of the sun, moon, and planets. Modern planetariums are theaters, usually dome shaped, that employ elaborate equipment, including projector systems and lasers, for educating and entertaining the public about astronomy and for training nautical and military personnel in celestial navigation.

The lineage of the planetarium can be traced back to ancient Greece. An ancient mechanical calculator used to accurately determine astronomical positions was found in 1900 by sponge divers in what is now referred to as the "Antikythera wreck," off the Greek island of the same name (located between Kythera and Crete) and has been dated to about 150–100 BCE. This technology was extraordinarily complex for the time, and it has no known precursor and no successor or equivalent until the 18th century CE. This movement, or one similar, with its high level of sophistication for that and most other eras, is widely believed to have been used in Archimedes' construction of a primitive equivalent to the modern planetarium. Rather than providing public entertainment or instruction to navigators in training, the creation of Archimedes was used to predict the movements of the sun, moon, and planets as known at that time and also to approximate their relation to each other at various phases and points in time.

Giovanni Campano, more readily identified as Johannes Campanus, was an Italian astrologer, mathematician, and astronomer of the 13th century. In his *Theorica Planetarum,* Campano describes, and, more importantly, provides direction on how to assemble, a planetarium incorporating the astronomical knowledge at that time. Given the instructions left on how to build this piece, it can be stated that there is a very high correlation between the "planetarium" of Campano and the orrery of today. (The orrery is so named for the Earl of Orrery; Orrery is a location within Ireland, and an 18th-century Earl of Orrery had one constructed.) An orrery is a three-dimensional mechanical device that depicts the relative positions and motions of the planets and moons in the solar system, as well as their relation to each other,

An orrery or planetarium designed by George Adams showing the relative positions of the planets in relation to the sun (1799).

Source: Library of Congress, Prints & Photographs Division.

as based upon the presumption of Copernican heliocentrism. These pieces usually owe their movement to a large clockwork mechanism with a sphere (representing the sun) at its core and with a distinct and specific representation of a particular planet at the end of each of its arms. Given the small size of orreries as constructed by Campano and his predecessors, it appears that these devices were mostly used for personal curiosity, knowledge, and recreation, as they were not large enough to be of any true service to a crowd. One obvious limitation of this form of representation is its complete inability to replicate or depict the backdrop of stars and constellations.

Early 19th-century England provided the backdrop for Adam Walker and his Eidouranion, the name given his very large orrery, approximately 20 to 25 feet in height, with a proportionate width. Walker's Eidouranion provides the first documented usage of an orrery for either educational or entertainment purposes, with his lecture incorporating both facets into a simulated presentation on the heavens. While not extremely precise in its representation, this show provided the audience an opportunity to encounter elements of time beyond an immediate number, day, and date. In relaying the parallels of heavenly occurrences such as a lunar phase, or planetary alignment, with definite cycles and events (e.g., seasonal change resulting from the earth's positioning and alignment in relation to the sun), Walker can be seen as one of many to have helped establish the depth and permanence of events of this world by incorporating the heavens as support. As Walker's popularity, and presumably wealth, began to grow, others such as William Kitchener and his Ouranologia began to take their rather inaccurate orreries on the road, forgoing scientific display for sensationalism and the awe of large crowds.

The preeminent German optics firm of Carl Zeiss found itself in a unique situation at the turn of the 20th century. Working within the firm's compound in Jena, Germany, were astronomer Max Wolf, former director of Heidelberg's Baden Observatory, and Franz Meyer, chief engineer of optical works for Zeiss. Both men, in conjunction with Oskar von Miller of Munich's Deutsches Museum, looked to create a representation of the night sky free of movement created by overt force (e.g., the mechanically driven arms of an orrery). The result of their ingenuity and labor was a projector that produced the movements of planets and stars, without aid from bulky and visually obtrusive rails, supports, and the like. Instead, once centrally mounted, their optical projector was capable of projecting images upon the surface of a hemispherical ceiling, and in 1923 the first modern-day planetarium projected its representation of the heavens upon the inside of a dome erected on the roof of the Zeiss building. Given the complexities associated with production of a planetarium (named for the device used to project images, and not necessarily the hemispherical room or building in which the projector is housed; commonly the entire unit—projector[s]— and dome, are referred to as a planetarium) and the sterling reputation of Carl Zeiss, most every planetarium produced or in use upon the globe prior to World War II could be directly traced to the Zeiss factory of Jena.

Projectors such as those introduced by Zeiss use a hollow sphere with a light contained within as their primary means of projecting the appearance of the heavens. This "ball of stars" contains a tiny hole for each star being represented, with the location and relation of "star holes" being computed to near perfection when compared with their authentic form in the true night sky. To simulate the planets and their movements, another projector is commonly used to superimpose these images upon the starry backdrop.

Currently, digital projectors are beginning to appear in more and more planetarium settings. The cost of upkeep is considerably less than that of the traditional "star ball" models, and synchronization of various projectors (e.g., a star projector and a separate planet projector) is not required, as all solar/celestial data are stored and represented by one computer and its corresponding digital projector. Much like digital projectors in a lecture hall or elsewhere, images of the night sky are displayed as pixels, with higher pixilation resulting in a better viewing accuracy.

Daniel J. Michalek

See also Galilei, Galileo; Observatories; Telescopes

Further Readings

King, H. C., & Millburn, J. R. (1978). *Geared to the stars: The evolution of planetariums, orreries and astronomical clocks.* Toronto, ON, Canada: University of Toronto Press.

Marche, J. D. (2005). *Theaters of time and space: American planetaria, 1930–1970.* New Brunswick, NJ: Rutgers University Press.

Planets

Astronomy, one of the oldest sciences of humankind, always provided orientation in space and time: Cardinal directions (east, north, west, south) are defined and obtained by basic astronomical measurements. Time and calendar issues are also definable and measurable by astronomical observations: One "year" is the period the earth needs for one full revolution around the sun (originally, before the Copernican revolution, it was seen the other way around), and one "month" is roughly the time our moon needs to orbit the earth. The currently most widely used calendar system, the Christian calendar in use in Europe, North and South America, and many other parts of the earth, is based mainly on the motion of the earth with respect to the sun. Other cultures have developed slightly different calendars based either on the moon (e.g., the Moslem calendar) or a combination of sun and moon (e.g., the Jewish calendar).

We count seven days per week because, a long time ago, people considered "seven" objects as "planets" or "planet-like objects," namely the real planets, which could be observed by the naked eye before the invention of the astronomical telescopes: Mercury, Venus, Mars, Jupiter, and Saturn, as well as the other two large visible bodies in the solar system, the sun and our moon, together seven objects, hence also the names of the seven days of the week:

Sunday, as the original first day of the week, is the day of the sun, the brightest object in the sky, often even worshipped as a god in several ancient cultures.

The word *Monday* obviously refers to the moon.

Tuesday is named after Mars, the god of War (notice in French, Italian, and Spanish, the words for Tuesday are still close to that for the Roman God Martius (for Mars), namely *Mardi, Martedi,* and *Martes,* respectively) and it originally comes from *Tiwes dag* or *Tyr dag,* from the old Teutonic word *Tyr* for Mars.

Wednesday is named after the Roman God Mercury (in Romanian, the day is still known as *Miercuri),* and the word Wednesday itself comes from *Wodan dag* for the Teutonic god Wodan.

In Roman times, the fifth day of the week (*Thursday*) was known as *dies Jovis,* after their god of thunder and chief of the gods, Jupiter, where Thursday itself comes from *Thunor dag,* the day of the Teutonic God Thor.

The Romans named another day after their goddess of beauty, Venus, and called it *dies Veneris* (still similar in French). When Germanic tribes invaded England more than 500 years ago, they imposed their goddess upon that day and called it *Frigedaeg,* now Friday.

And finally, *Saturday* is obviously called after Saturn.

Nowadays, both time and the unit *second* are defined by the speed of light. Previously, a second was defined by the atomic clocks, and also earlier as one certain small part of a day, that is, one small part of a revolution of the earth. Still, astronomical observations are important for fixing "time": Due to tidal interaction among the sun, earth, and

An upright shot of the planets at Hayden Planetarium in New York City. Since February 2000, the planetarium has been one of the two main attractions within the Rose Center for Earth and Space.

Source: Daria Peleg/iStockphoto.

moon, the rotation period of the earth is very gradually slowing down.

Historical Background

The definition of a planet has changed over the centuries, always following new astronomical observations and new understanding. The word *planet* comes from the Greek word for wanderer, meaning a wandering or fast-moving starlike object (e.g., the old Arabic name for the Egyptian capital Cairo is "Al Qahira" for "the backwards wandering," meaning Mars). As mentioned above, a few hundred to 3,000 years ago, people could see, by the naked eye, seven objects apparently moving fast in the sky (compared to the "fixed stars"), incorrectly thought to orbit around the earth in the center, namely the sun, Mercury, Venus, the moon, Mars, Jupiter, and Saturn. The next planet known today behind Saturn, called Uranus, is also visible to the naked eye during clear and dark nights, and may be visible when Uranus is close to the sun and the earth is roughly in between the sun and Uranus. This is called *opposition:* when an outer planet like Uranus is brightest as seen from the inner planet like Earth, but no such reports are known so far, possibly because Uranus moves only slowly and is quite faint.

During the Renaissance period in general and the so-called Copernican revolution in particular, it became clear, through a number of new observations, that the old theory placing the earth in the center of the universe was not perfect. Those observations became possible with the invention of the astronomical telescope. In 1609, Galileo Galilei observed the phases of Venus and craters on our moon and discovered moons around Jupiter (first called "Medici planets" or "Medici stars" after his supporters of the Italian Medici family, and now known as "Galileian moons"). All this together favored an alternative explanation, putting the sun in the center of our solar system and having the planets orbiting around the sun. At this moment, it also became clear that the earth is orbiting the sun and, hence, was now seen as a planet. Later, two more planets were discovered beyond Saturn, namely Uranus and Neptune.

While it was always possible to estimate the orbital periods of planets around the sun by their periodic appearance and disappearance in the sky, it was originally difficult to measure distances between the planets, or from Earth to either its moon or the sun. The distance between Earth and the sun is now called the *astronomical unit,* which is about 150 million kilometers. The first good estimates of such distances were obtained a few centuries ago by observing eclipses of the sun by the inner planets Venus and Mercury, which happen only very rarely (usually only once or a few times per century): One has to measure exactly either the angular distance between the apparent path of the planet across the solar disk, as seen from two different locations on Earth, or the exact times of ingress and egress of the planet moving in front of the sun. These four so-called contacts must be observed and measured from different locations on Earth with as large as possible a distance in between them, for example from South Africa and Europe. A few centuries ago, it was still difficult to coordinate such efforts and also to run precise clocks. A first observation was done in 1639, a Venus transit. After several attempts, the first good values for the distance between Earth and the sun were obtained in 1761 and 1769—these values also giving evidence about the size of the sun. Together with the laws of gravity just determined by Isaac Newton and their application to the solar system by Johannes Kepler, these values immediately yielded all distances between each of the planets and the sun.

Toward the end of the 18th century, the so-called Titius-Bode law was found and discussed: According to this law, the distance from planet to planet roughly doubles with each planet reached as one moves further away from the sun; for example, Saturn is roughly twice as distant from the sun as Jupiter, Uranus is roughly twice as distant as Saturn, and so on.

However, from Mars to Jupiter, the distance increases roughly by a factor of four, so that there would be space for one more planet. Even the famous philosopher G. W. F. Hegel wrote his dissertation about this problem at the University of Jena in Germany. Many astronomers were already hunting for this new planet. Then, in January 1801, an object was found at the expected distance from the sun, called Ceres, and celebrated as a new planet. However, soon afterward, more similar objects were found, all at a similar distance; a few decades later, the solar system had more than 20 "planets." It was also found that these new objects were smaller than all other previous planets, so it was decided to call them "minor planets" (a new class of objects). Hence, objects celebrated and counted as planets were removed from the list of planets by a new definition.

Early in the 20th century, another new object was discovered and celebrated as a new, ninth planet, called Pluto, located most of the time beyond Neptune, but sometimes crossing its orbit.

Planet Redefined

In August 2006, the general assembly of the International Astronomical Union (recognized by the UN as the international body to define and name celestial objects) discussed, among many other issues, the definition of "planet" again. The definition was prompted by two new discoveries in the 1990s: (1) Minor bodies like Pluto were discovered near and beyond Pluto; and (2) planets around other stars, so-called extrasolar planets or exoplanets, were discovered, and they were apparently different in many respects from the solar system planets, so that a new definition seemed necessary.

After lengthy and heated debates, a new definition was confirmed by a majority vote. This definition reads as follows:

> The IAU resolves that planets and other bodies (except moons) in our solar system be defined into three distinct categories. A "planet" (see note 1, below) is a celestial body that (a) is in orbit around the Sun, (b) has sufficient mass for its self-gravity to overcome rigid body forces, so that it assumes a hydrostatic equilibrium (nearly round) shape (see note 2, below), and (c) has cleared the neighborhood around its orbit.
>
> A "dwarf planet" is a celestial body that (a) is in orbit around the Sun, (b) has sufficient mass for its self-gravity to overcome rigid body forces, so that it assumes a hydrostatic equilibrium (nearly round) shape (see note 2, below), (c) has not cleared the neighborhood around its orbit, and (d) is not a moon. All other objects (see note 3, below)—except moons—orbiting the Sun shall be referred to collectively as "Small Solar System Bodies."
>
> Note 1: The eight planets are Mercury, Venus, Earth, Mars, Jupiter, Saturn, Uranus, and Neptune.
>
> Note 2: An IAU process will be established to assign borderline objects into either dwarf planets or other categories.
>
> Note 3: These objects include asteroids, trans-Neptunian objects, comets, and other small bodies.

This says mainly that objects that (a) are in orbit around the sun (i.e., not being moons of planets), (b) have at least a certain mass (to be round by gravitational effects), and (c) have maintained their orbits by their own gravitational forces are the "planets" of our solar system. According to this definition, Pluto is not a planet (anymore). However, this was not the first time that an object was deleted from the list of planets (see above regarding Ceres and the other "minor planets"). This effect, however, is highly controversial, and it is very well possible that the definition will be changed again soon.

The Solar System

There are now eight planets in the solar system.

Mercury, the innermost known planet, is also the smallest known planet in our solar system with a diameter of less than 5,000 kilometers. (Pluto is smaller, but it is not a planet anymore according to the new definition.) It has a rotation period of 59 days, which is about two thirds of its orbital period around the sun (88 days); hence one "Mercury-day" is equal to two "Mercury-years." Mercury does not have an atmosphere that is comparable to that of Earth, and its surface is similar to that of the moon. Two thirds of its material and mass is made of iron. According to Einstein's general theory of relativity, the orbit of Mercury should change slowly: The location of the perihelion (the point in the planet's orbit at which it is the smallest distance from the sun) moves by a small angle of 43 seconds or arc per century, which has been confirmed observationally.

Venus needs 225 days for orbiting the sun (compared to 365 days for one Earth orbit around the sun). The rotation of Venus around its own axis is retrograde, that is, in the rotational direction opposite to the direction in which it revolves around the sun, and one such "Venus-day" lasts 243 days; that is, it is longer than one "Venus-year." Venus has a dense atmosphere consisting mostly of carbon dioxide and nitrogen, and it has strong pressure on the surface, from where one would never be able to see the stars in the night sky through the dense clouds.

Mercury and Venus, as planets inside the earth's orbit, orbit the sun faster than the earth does and are often close to the sun, as seen from Earth. Hence, they are observable either in the evening sky just after sunset or in the morning sky just before sunrise; that is, Venus is also called the "morning star" or the "evening star." The Greeks called Mercury "Hermes" when it appeared as the evening star and "Apollo" when it appeared as the morning star; Venus was similarly called either "Hesperus" or "Phosphorus," respectively.

Earth is the third planet from the sun; it needs 365 days for a complete orbit around the central star and 24 hours for one rotation. Its atmosphere consists mainly of oxygen and nitrogen. This planet is the only one known so far to harbor living beings like plants, animals, and intelligent life.

The fourth planet is called Mars. It has an orbital period of 687 days and a rotation period of 24.6 hours, so that a "Mars-day" is only slightly longer than a day on Earth. Its thin atmosphere consists mostly of carbon dioxide and nitrogen, but this atmosphere is not identical to that of Earth. There is frozen carbon dioxide and water ice at the poles, but no fluid water has yet been detected. However, some surface structures look like dry river beds and may indicate that fluid water was present some billions of years ago. It is not impossible that life has formed on Mars, too, but no clear evidence for life on Mars has been found yet. Mars is orbited by two small moons, called Phobos and Deimos, with 8-hour and 30-hour orbital periods, respectively. Like the moon of Earth, their rotation is bound: Their orbital period equals their rotational period; they are rotating around themselves only by orbiting their planet and always show the same side to their planet.

The innermost four planets are also called terrestrial planets, as they are all made mostly of solid material like Earth (*terra*). Between Mars, the fourth planet, and Jupiter, the fifth planet, there is a large gap where many small bodies are orbiting the sun. These are called minor planets or sometimes *planetoids,* because they are physically like terrestrial planets; that is, they are rocky objects. They are also called *asteroids,* because in the sky they look like the stars looked when they were discovered, namely pointlike (as opposed to the planets of our solar system, which appeared to be extended on the sky even in naked-eye observations, because of their larger size and smaller distance from observers).

The four outermost known planets (Jupiter, Saturn, Uranus, and Neptune) all are larger in size than the terrestrial planets, mostly because of their large atmospheres and only small solid or fluid cores. (In the case of Jupiter, there may not even be a core at all.) Hence, they are called the "gaseous giant planets."

Jupiter is the largest planet in our solar system; it has a diameter of 143,000 kilometers and a mass of 318 times the mass of Earth. It needs almost 12 years for one orbit around the sun, but only 10 hours for a rotation around itself, as can be observed with even a small telescope because of the moving large red spot in its outer atmosphere. Given its diameter, mass, and composition (mostly molecular hydrogen), it is not absolutely clear whether it has a solid or fluid core or possibly even no core, that is, no solid surface. If it has a core, the core could have a mass of a few or maybe 10 Earth masses. Due to contraction, Jupiter is still radiating more energy to outer space than it is receiving from the sun. This giant planet also has a small ring system and a large number of moons, probably a few dozen; new small moons are still being discovered. The four largest moons were originally discovered by Galileo, when he observed Jupiter for the first time with a telescope. These four moons (Io, Europa, Ganymede, and Callisto) are called the Galileian satellites.

Saturn is twice as far from the sun as Jupiter is. Saturn is known mostly for its large ring system. It also has a large number of moons. Saturn needs 29.5 years to orbit the sun and has a rotation period of 10 hours and 40 minutes. It has a solid core of a few Earth masses and a large atmosphere made mostly of molecular hydrogen gas. Saturn is the second largest planet (120,000 kilometers in diameter) and the second most massive (95 times the mass of Earth) in our solar system.

All planets from Mercury to Saturn (including Earth) have been known for several thousand years to most cultures on Earth, because they can be observed by the naked eye. The outermost planets, Uranus and Neptune (as well as Pluto), were discovered after the invention of the telescope. While Jupiter and Saturn are called "gas giants," Uranus and Neptune are also gaseous planets that can be seen as "ice giants."

Uranus was discovered (and recognized as a planet) in 1781 by William Herschel. Others had observed it before but did not recognize that it as a planet. Uranus is also a gaseous planet with a central solid core, but in total it is only 15 times as massive as Earth. Uranus needs 84 years to circle around the sun, and one "day" on Uranus lasts around 17 hours. Uranus's atmosphere consists of 83% hydrogen, 15% helium, and 2% methane. So far, 21 moons have been discovered (and astronomers are still counting). Uranus also has a small ring system as discovered by the Voyager satellites.

Neptune is the outermost known (and accepted) planet. It was observed by Galileo in 1612, but he did not recognize it as a planet. Because of apparent deviations in the orbit of Uranus, both John Couch Adams and Urbain Le Verrier predicted the existence of another planet theoretically and tried to forecast its rough location in the sky. Later, in 1846, the observer Johann Gottfried Galle in Berlin, Germany, searched that area of the sky for a small moving object and discovered Neptune within a few hours. Neptune needs 165 years for one full circle around the sun. One "day" on Neptune lasts 16 hours. Neptune has a small solid core, a large gaseous atmosphere composed mostly of molecular hydrogen, and a total mass of 17 times the mass of Earth. Neptune, like all gaseous planets in our solar system, has moons and rings.

The object Pluto was discovered in 1930 and celebrated as a new planet, but it was deleted from the list of planets in the 2006 definition of *planet* by the International Astronomical Union.

The new definition of *planet* is formulated for the solar system, but it can and should be applied analogously to other planetary systems around other stars. However, there is as yet no consensus or definition for the upper mass limit of planets. Such an upper mass limit, however, would be very important for extrasolar planets, to be able to decide whether they are planets or so-called brown dwarfs.

In history, the two definitions for a planet worked for about 200 years: The first definition worked from the Copernican revolution to the discovery of Ceres and other minor planets (which now form the asteroid belt between Mars and Jupiter); the second definition, excluding the minor planets, was in effect again for about 200 years until 2006.

Both the problem regarding Pluto and the missing upper mass limit for planets may very well lead to a new definition at one of the next meetings of the International Astronomical Society, which holds a general assembly every three years.

Ralph Neuhäuser

See also Copernicus, Nicolaus; Earth, Revolution of; Earth, Rotation of; Laplace, Marquis Pierre-Simon de; Nebular Hypothesis; Planets, Extrasolar; Planets, Motion of; Relativity, General Theory of; Time, Relativity of

Further Readings

Corfield, R. (2007). *Lives of the planets: A natural history of the solar system.* Cambridge, MA: Basic Books.

Encrenaz T., Bibring, K.-P., Blanc, M., & Barucci, M.-A. (2002). *The solar system* (3rd ed., S. Donlop, Trans.). Berlin: Springer-Verlag.

Karttunen, H., Kröger, P., Oja, H., Poutanen, M., & Donner, K. J. (Eds.). (2007). *Fundamental astronomy.* Berlin: Springer-Verlag.

Maunder, M., & Moore, P. (1999). *Transit: When planets cross the sun.* (Patrick Moore's practical astronomy series). London: Springer-Verlag.

Planets, Extrasolar

The term *extrasolar planets* or *exoplanets* stands for planets outside our solar system—that is, planets that orbit other stars, not our sun. Planets in our solar system are defined as objects with enough mass to be spherical and round by their own gravity and to be alone on their orbit around the sun; in other words, to be the dominant object in a particular orbit and not to be a moon or asteroid

Most exoplanets are discovered by observing the stellar motion around the common center of mass of the combined star-and-planet system, that is, by observing somehow the motion of the objects in orbit around each other. This is typically done by measuring precisely the periodic variation of certain values, such as radial velocity or brightness, with time. For example, the first extrasolar planets were found with this timing technique around a pulsating neutron star.

The recent definition of "planets of our solar system" by the International Astronomical Union deals mainly with the question of the minimum mass for an object to qualify as planet and excludes Pluto. This matter was raised by the fact that more and more objects similar to Pluto were discovered by larger and larger telescopes. The questions of maximum mass and formation of planets were left out in this new definition, possibly partly because there is not yet a consensus in the international community. For a discussion of extrasolar planets, however, the maximum mass is very important in order to classify an object as planet or nonplanet and to distinguish between planets and brown dwarfs.

Both planets and brown dwarfs are substellar objects in the sense that they are less massive than stars so that they cannot fuse normal hydrogen (as stars do to produce energy and to shine for a long time). Brown dwarfs, while they cannot fuse normal hydrogen (which has an atomic nucleus of just one proton), can burn deuterium (heavy hydrogen, which has an atomic nucleus of a proton plus a neutron) so that they are self-luminous for a few millions of years until the original deuterium content is depleted.

The upper mass limit of planets can be defined either through the lower mass limit for deuterium fusion, which is around 13 times the mass of Jupiter (depending slightly on the chemical composition) or by the mass range of the so-called brown dwarf desert (as discussed in the following paragraphs).

We will next discuss the different exoplanet discovery techniques by chronological order of success and thereby also discuss the properties of objects found so far.

Radial Velocity

For a few thousand years, speculations have existed as to whether other stars can have their own planets. Both Giordano Bruno and Nicholas of Cusa answered this question positively a few hundred years ago. However, not until 1989 was the first object discovered that could really be an extrasolar planet and that is today still regarded as planet candidate. This first extrasolar planet candidate was discovered serendipitously by the so-called radial velocity technique: The velocity of the motion of an object directly toward us or away from us (in one dimension) is called *radial velocity* and can be measured by the so-called Doppler shift of spectral lines. Atoms in the atmosphere of stars can absorb light coming from the interior of the star at a certain energy, frequency, or wavelength for each kind of atom or ion, producing absorption lines in the spectrum of the star. If the star moves away from us, such lines are said to be *red-shifted;* if the star is approaching us, they are called *blue-shifted;* in either case, their wavelength is different from the normal wavelength (larger for red-shifted).

When a second object, like a planet, orbits around a star, actually both objects orbit around their common center of mass. Hence, also, the star wobbles: It sometimes approaches us, sometimes flies away from us. This can be observed as periodically changing radial velocity. The period of the variation gives the orbital period, and the amplitude of the change in radial velocity yields the mass of the companion. However, because the inclination between the orbital plane and our line of sight is normally not known, only a lower mass limit is known. Therefore, such low-mass companions detected by the radial velocity technique are to be seen as planet candidates; they could have a mass above the maximum mass for planets, making them perhaps brown dwarfs or even low-mass stars. The first such case was published in 1989 by Latham, Stefanik, Mazeh, Mayor, and Burki, namely a companion with minimum mass of 11 Jupiter masses around the star HD 114762. This planet is called HD 114762 b, following the convention that the first planet found around a star is called by the name of the star plus a lowercase *b* behind the star name (thus, a lowercase *c* for the next planet, etc.). This planet may very well have a true mass above 13 Jupiter masses, in which case it could be regarded as brown dwarf.

The first object discovered around a sunlike normal star, which is almost certainly a planet, is called 51 Peg b and is a planet with about half the mass of Jupiter as minimum mass found by Mayor and Queloz around the star 51 Peg in 1995.

The radial velocity can nowadays be measured with an accuracy of about 1 meter per second, so that planets with minimum mass of a few Earth masses can be detected by the wobble they produce on a low-mass star. In the time of one Jupiter orbit, that is, 12 years, since the important discovery of 51 Peg b, about 250 planet candidates have been discovered. In some cases, several planet candidates are orbiting a single star, in some other cases, individual planet candidates are found in binary stars. Because the high precision of the radial velocity technique has been available only since about the early 1990s, planets with more than 20 years of orbital period have not yet been discovered; one needs to observe at least one orbital period. Many planet candidates found so far orbit their stars within only a few days, which is quite different from our solar system, where the innermost planet Mercury needs several months to orbit the sun. The high number of planet candidates with short orbital period, however, can be seen as observational bias, because such planets also introduce a larger wobble on their stars due to Johannes Kepler's and Isaac Newton's laws of gravity.

The mass range of all these planet candidates shows a strong peak at about one Jupiter mass and a strong dip at around 20 to 30 Jupiter masses, even though this method would be biased toward more massive companions, because they have a stronger effect on their central star. Across the range of planet candidates, there are about 250 planet candidates all with masses below about 20 Jupiter masses, then almost no objects with minimum mass between about 20 and 70 Jupiter masses (that is, there are only a few brown dwarfs), and then again a large number of stellar companions with minimum mass above 70 Jupiter masses. This paucity of brown dwarfs identified with the radial velocity technique is called the *brown dwarf desert,* and the dip in the mass spectrum is deepest at around 20 to 30 Jupiter masses. Either the lower mass range of the brown dwarf desert or the minimum mass for deuterium burning can be used as an upper mass limit for planets, if one would define the upper mass limit for planets.

Pulsar Timing

At the end of the lifetime of a massive star, after most of the material is burned by fusion, the star collapses due to its own gravity, then forms a very dense and compact object made up mostly by neutrons, called a *neutron star,* while the rest explodes due to a rebound as supernova. A neutron star typically has about 1.4 times the mass of our sun but a diameter of only 20 to 30 kilometers. Such neutron stars rotate very fast, sometimes even about 100 times per second, sometimes once in few seconds. Most known neutron stars emit strong radio emission along their rotation axes (beams), which appear pulsed due to the fast rotation. Such objects are called *pulsars.* We should keep in mind that so-called pulsars are not pulsating, but rotating fast. One can measure the rotation period with both high precision and high accuracy. In the case of the pulsar called PSR1257, Wolszczan and Frail discovered sinusoidal variations of the millisecond pulses in 1992 and interpreted these variations to be caused by low-mass objects in orbit around the neutron star, each with a mass equivalent only to about that of Earth. This discovery of pulsar planets (by pulsar timing) came as a big surprise, because planets were not expected around neutron stars; it is still dubious as to whether planets can survive the supernova explosion, and it is unknown whether the objects found around PSR1257 are remnants of the explosion or were formed afterwards.

In the case of planets or planet candidates discovered by the radial velocity technique, the inclination of the orbit of the planet around the star in not known. One way to determine this inclination would be to use a transit. A transit occurs when the planet orbits around the star into our line of sight; the planet moves in front of the star once per orbit (and behind the star also once per orbit). When the planet is in front of the star (that is, in front of the spatially unresolved stellar disk), a small part of the stellar light is blocked by the planet. This event is called *transit* or *eclipse*. Such events also happen in our solar system; for example, as seen from Earth, the inner planets Mercury and Venus can follow a path directly in front of the sun, which can even be observed as spatially resolved. The transit light curve enables observers on Earth to measure the inclination of the orbit and also the radii of stars and planets. Then, one can determine not only the true mass of the companion (planet or brown dwarf), but also, from mass and radius, its density.

The first case for which this was successfully observed was HD 209458 in the year 2000, the first radial-velocity planet candidate confirmed to be a true planet (and found to be a gas giant planet with low density like Jupiter). About one Jupiter orbit after the discovery of 51 Peg b, about 33 transiting planets are known (as of November 2007), most of which have also been discovered first by the transit, then confirmed as planets by radial velocity. In a few cases, also the secondary transit is detected; this is a small decrease in the total combined brightness of star plus planet (one should keep in mind that in all such cases, the planet is not seen directly) when the planet is behind the star. From the difference in brightness between the time of secondary transit and the time immediately before and/or after the transit, one can indirectly determine the brightness of the planet.

Astrometry

Whereas the radial velocity technique measures the wobble of the star due to the orbiting planet in just one dimension (radial), one can measure the wobble in the two other dimensions by *astrometry*, very accurate and/or precise determination of the position of a star on the sky. Our sun as seen from about 30 light years' distance also moves slightly in the sky due to Jupiter orbiting it, but this is a very small effect, less than .001 of an arc second (the moon has a diameter of 1,800 arc seconds). The star GJ 876 was the first for which this wobble was detected, using the fine guidance sensor of the Hubble Space Telescope, confirming the radial velocity planet candidate GJ 876 b to be a real planet with just about two Jupiter masses. In the meantime, a few more planet candidates were confirmed by astrometry, and also one radial velocity planet candidate was found to be a low-mass star. Other observing programs have started, using both ground-based and space-borne telescopes, wherein one tries to discover such a wobble in stars where no planets or candidates have been found by other techniques.

Direct Detection

All previous techniques—radial velocity, astrometry, and transits—cannot determine which photons are coming from the stars and which photons from the planet; that is, they cannot directly detect (or see) the planet. While stars are bright and self-luminous due to fusion, planets are very faint, mostly shining only due to reflected light, and they are also very close to their respective stars so that they cannot be detected or seen next to the much brighter star. Young planets, which are still contracting and/or accreting matter, are self-luminous and, hence, not that faint, so that it could be less difficult to directly detect a young planet next to a young star. Several observational campaigns were started around the turn of the millennium in 2000 with the Hubble Space Telescope and ground-based 8- to 10-meter telescopes.

In the case of the ground-based observations, the earth's atmosphere is another problem, leading to the twinkling of stars, so that we obtain images lacking the best possible image quality. With the new technique of so-called adaptive optics, one can de-twinkle the stars: flatten the disturbed wave front with a deformable mirror in the telescope. With such a technique used at the 8-meter Very Large Telescope of the European Southern Observatory in Chile, a few companions to young stars have been found since 2004 that could really be young planets detected directly. The first such case was the star GQ Lupi with its companion GQ Lupi b detected by Neuhäuser, Guenther, Wuchterl,

Mugrauer, Bedalov, and Hauschildt. In such cases, it is more difficult to be sure about the exact mass of the companion, because the orbital period is several hundreds of years, so that these few objects could also be low-mass brown dwarfs.

Microlensing

According to Albert Einstein's theory of general relativity, mass or matter deforms space, so that a light ray moving close to matter would be diverted. One would see a ring of light (Einstein ring) around the object. If such an Einstein ring is not resolved spatially due to small mass and/or small angular resolution, one would still see the background object being brightened by the foreground object, the gravitational lens. Such an event is called *microlensing*. If a binary lens (a star plus a planet) were to move—as seen from Earth—directly in front of a background single star, one would observe a double-peaked light curve, one brightening due to the star and one brightening due to the planet. This way, one also can detect planets at great distances—thousands of light years away. There are a few cases where such a double-peaked light curve has been observed that could possibly be due to a planet. However, in all such cases, due to the large distance and hence small brightness, the nature, mass, and distance of neither the lens (or the primary object in the double lens, the star) nor of the lensed background object are known, so that the mass of the secondary object in the lens (possibly a planet) cannot be determined without great uncertainty. The mass of the companion is determined from the mass of the primary (unknown) and the brightness ratio of the two peaks. Practically, such events can never be observed again nor confirmed.

Timing of Pulsating Stars

Some stars toward the end of their normal lifespan (i.e., after most of the light material is burned) are pulsating: They periodically increase and decrease their volume and, hence, brightness. Such a pulsation is observable as periodic brightness change. With precise observations, one can detect a small periodic variation in the pulsations, which can be explained by a wobble of the star due to an orbiting low-mass object. This is very similar to the pulsar timing and radial velocity technique. In the case of the pulsating star V391 Peg, such a variation was detected recently that can best be explained by a planet candidate with three Jupiter masses as minimum mass. This star has burned all its hydrogen already, has expanded to the red giant phase, has lost large amounts of its material, and is now again contracted to become a so-called white dwarf. This is the first time that a planet candidate has been detected in a star after the red giant phase (except for the pulsar planets). This case shows that planets can survive the red giant phase. Our sun will undergo this red giant phase in a few billion years, when it will then expand enough to swallow Mercury and Venus. It is not yet clear what effects this will have on the earth, but it is likely that all life will be extinguished.

All these different techniques to discover planets have resulted in several hundred planets and planet candidates, including some planetary systems, where several planets orbit the same star. (Updates on planet discoveries can be found on www.exoplanets.org.) Planetary systems consist not only of the planets and their host star but also of minor bodies like asteroids, comets, and moons, and often if not always also of a circumstellar disk with dust remaining from the formation phase. This is also the case in our solar system, where dust in the so-called zodiacal disk can be observed on dark moonless nights due to reflection of sunlight on dust particles; such dust debris disks can also observed around other stars, with or without planets.

All the planets discovered so far have masses of at least several Earth masses and are much different from Earth. It is not yet possible to detect earthlike planets. Such discoveries may be possible in the future, either by the use of larger telescopes or by ground- or space-based interferometry, using a combination of several telesopes.

Another eminent question is the habitability of exoplanets. So far, no signs of life have been found on other planets, neither in our solar system nor elsewhere. It is difficult not only to define life, but also to detect earthlike planets—to say nothing of earthlike or even nonearthlike life on distant planets. However, it may well be that life could form either on some already detected planets or on their moons, if these exist.

Ralph Neuhäuser

See also Bruno, Giordano; Laplace, Marques Pierre-Simon de; Nebular Hypothesis; Nicholas of Cusa (Cusanus); Planets, Extrasolar; Planets, Motion of; Pulsars and Quasars; Telescopes; Time, Planetary

Further Readings

Boss, A. (1998). *Looking for earths: The race to find new solar systems.* New York: Wiley.

Clark, S. (1998). *Extrasolar planets.* New York: Wiley.

Cole, G. H. A. (2006). *Wandering stars: About planets and exo-planets: An introductory notebook.* London: Imperial College Press.

Dvorak, R. (2007). *Extrasolar planets: Formation, detection, and dynamics.* New York: Wiley.

Latham, D. W., Stefanik, R. P., Mazeh, T., Mayor, M., & Burki, G. (1989). The unseen companion of HD114762—A probable brown dwarf. *Nature, 339,* L38.

Mason, J. (2007). *Exoplanets: Detection, formation, properties, habitability.* New York: Springer.

Mayor, M., & Queloz, D. (1995). A Jupiter-mass companion to a solar-type star. *Nature, 378,* 355.

Neuhäuser, R., Guenther, E. W., Wuchterl, G., Mugrauer, M., Bedalov, A., & Hauschildt, P. H. (2005). Evidence for a co-moving sub-stellar companion of GQ Lup. *Astronomy and Astrophysics, 435,* L13.

Wolszczan, A., & Frail, D. (1992). A planetary system around the millisecond pulsar PSR1257+12. *Nature, 355,* L145.

Planets, Motion of

As viewed from Earth, the other planets in our solar system exhibit some apparent motional changes as each revolves around the sun. These oddities long puzzled a succession of civilizations as observers contemplated the structure of the universe. For thousands of years, it was common belief that the earth was the center of the universe, around which all other objects revolved. Many times, planets were seen as omens, not necessarily benevolent ones; they could also represent destructive prophecies. Most times, this led to self-fulfilling prophecies, but through time humankind has finally come to better understand the position of celestial objects.

The inferior planets (Mercury and Venus) obey a different set of observational parameters than do superior planets (those outside the earth's orbit). Inferior planets pass from inferior conjunction counterclockwise to greatest western elongation (GWE), superior conjunction and greatest eastern elongation (GEE). Superior planets, on the other hand, have no inferior conjunction, but rather a point of opposition. Additionally, they have observational points called eastern and western quadratures. In the sky of the observer, planets move regularly from west to east against the background of stars; this is called *prograde motion.* As part of orbital mechanics, this direct motion reverses as an inner body catches up to, and surpasses, the outer one. This is called *retrograde motion* (see Figure 1).

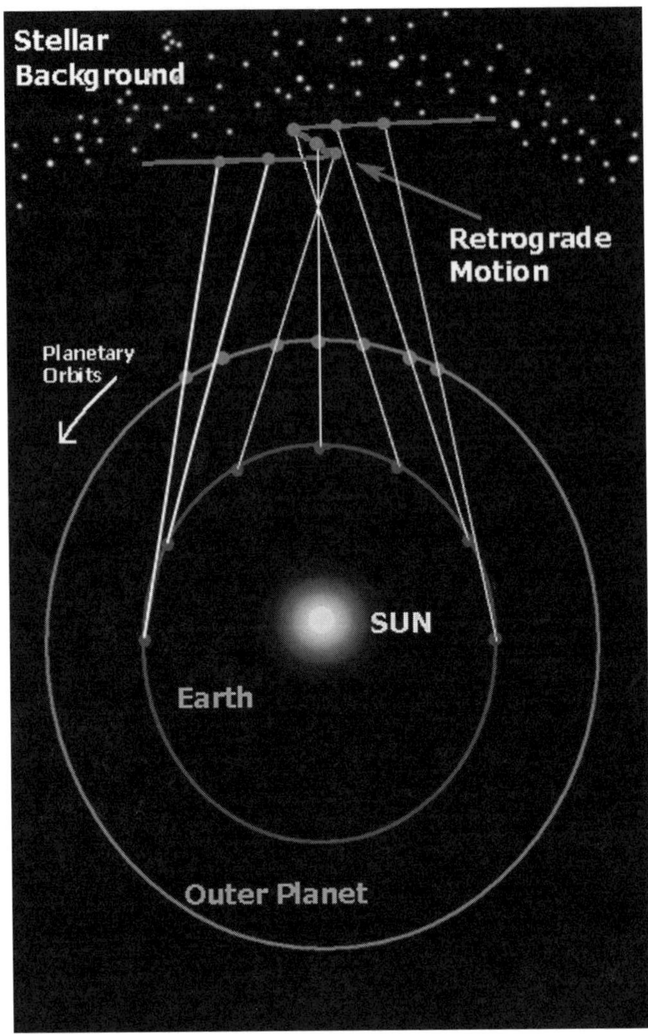

Figure 1 Retrograde motion of a superior planet

Note: Prograde motion stops midcycle, reverses, then resumes.

Inferior Planets

Every planet exhibits a consistent orbital behavior. A planetary conjunction occurs when two or more planets are in the same or opposite orbital location with respect to each other and the sun (see Figures 2a and 2b). Assuming a level plane of orbit, the sun and conjunctive planets would be in a straight line.

Inferior conjunction occurs when the inferior planet passes between Earth and the sun. In Figure 2a, the inferior planet enters into inferior conjunction with Earth. As it surpasses Earth, it will be lost in the glare of the sun. On rare occasions, the inferior planet will appear from Earth to transit the sun. At GEE, the motion of the inner planet will appear to slow, and then it will begin retrograde motion and pass behind the sun to superior conjunction. Once beyond GWE, the inferior planet will return to prograde motion. If this trajectory were seen from a location in space near Earth (which would remove the effects of Earth's daily motion), the observer would view the entire motion of the inferior planet around the sun. Note that a conjunction can occur when the two planets are at any point in their orbits; it is not necessary for them to be at perihelion or aphelion.

The inferior planets are visible between their conjunctions, but at the conjunctions they are lost in the solar glare of daylight. Each of the elongation points is related to the terrestrial cardinal direction that the planet is located in relative to the sun. As a result, observing an inner planet as a morning star prior to sunrise means that the inferior planet is in proximity to GEE. As it emerges from behind the sun, it becomes an evening star after sunset, moving toward greatest western elongation (GWE).

It is noteworthy to remember that none of these objects is fixed, including Earth. Another terrestrial observational feature is that inferior planets go through phasing similar to that of the moon. The full phase is superior conjunction, and the new phase is inferior conjunction, at which neither point can the planet be seen unaided.

Superior Planets

The outer planets, however, have a slightly different set of observational characteristics (see Figure 2b). At conjunction, the outer planet appears to pass behind the sun, while at opposition it will rise at sunset, similar to a full moon. These points are called quadratures, and they are the points where the superior planet forms a right angle with the earth and sun. At eastern quadrature, prograde motion continues but in decreasing angular measurements. Terrestrially, the superior planet sets 6 hours after the sun. As Earth approaches eastern

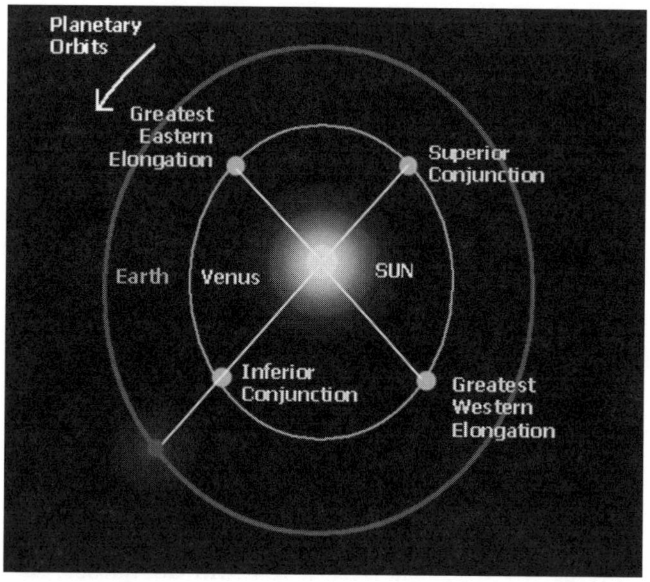

Figure 2a Inferior planetary motion

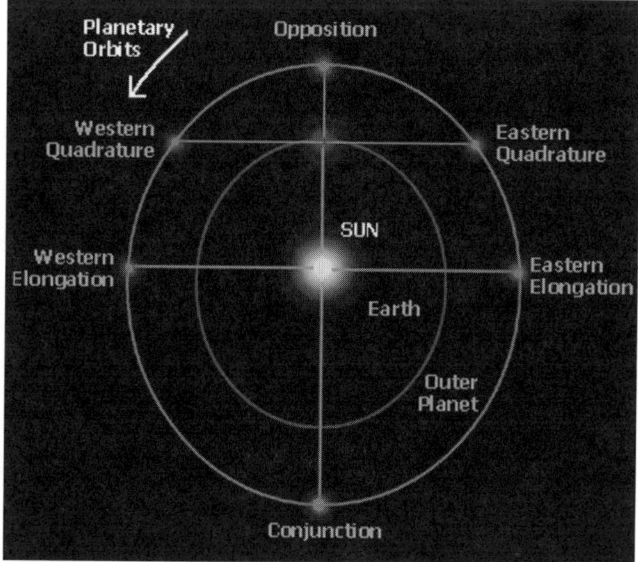

Figure 2b Superior planetary motion

quadrature, the magnitudes of prograde measurements decrease until the daily progression of the outer planet motion apparently stops and reverses into an east-to-west direction (see Figure 1). Retrograde motion continues through opposition but returns to prograde motion shortly after Earth has passed the superior planet, and the prograde distances appear to increase. Earth then continues to western quadrature, where the planet appears due south in the dawn sky. Finally, Earth returns to conjunction with the superior planet and the sun.

Predictability

A planetary alignment occurs when planets visibly cluster in the terrestrial sky. Although it is common for people to call these planetary groupings conjunctions, this is a slight misnomer. There are many planetary motions that occur, and are all part of nominal orbital mechanics. Because all planetary orbits are consistent, it is possible to predict alignments accurately. Computer-driven applet models present a three-dimensional look at when a conjunction or opposition will occur, as well as various alignments, and make it possible to predict similar occurrences in our solar system. These models are located on several Web sites on the Internet and located by utilizing a search engine.

Throughout history, planetary alignments have been blamed for the start of many wars and have been the cause for mass hysteria. In May 2000, Mercury, Venus, Mars, Earth, the moon, Jupiter, and Saturn aligned together in the western sky. News agencies reported several concerns regarding the coming tectonic, magnetic, radioactivity, and tidal catastrophes that were going to occur as a result. Over time, this has always been the case when such alignments occur. In the modern technological era, panic and hysteria is even more apt to spread with the usage of the Internet, rather than word of mouth, to carry the message. Furthermore, the enhanced ability of predicting such events gives hoaxers time to prepare their allegations. Since 2003, the Mars hoax occurs every August as Earth nears opposition with Mars; spreading mostly by e-mail to millions of people, the message claims that Mars will appear in the sky as large as the moon.

In 1974, John Gribbin misused the term *planetary alignment*, in the book *The Jupiter Effect*. The book was about the rare grouping of all nine planets being on the same side of the sun. (This actually occurs once about every 200 years.) The result would be that their combined gravitational effect would cause chaos with the sun and create massive earthquakes and floods on Earth. After reading the book, many people believed the hypothesis, and a great panic ensued. Gribbin tried to publicly quell the panic and reinforce the fictional nature of the book.

Despite the attempt to attribute terrestrial destructive forces to predictable solar system events, they simply occur as a result of the physics of motion around the sun. As each planet orbits the sun, the optical illusions of prograde and retrograde motion will be seen relative to the planetary location of the observer. Future alignments of planets will occur because they are mathematically connected, not as a precursor to chaos. Each planet takes a specific period of time to complete a solar revolution that will remain relatively constant throughout the life cycle of the solar system.

Timothy D. Collins

See also Copernicus, Nicolaus; Galilei, Galileo; Planetariums; Planets; Planets, Extrasolar; Telescopes; Time, Planetary

Further Readings

Gribbin, J., & Plagemann, S. H. (1974). *The Jupiter effect: The planets as triggers of devastating earthquakes*. New York: Macmillan.

Guest, J. (1971). *The earth and its satellite*. London: Hart-Davis.

Hetherington, N. S. (2006). *Planetary motions: A historical perspective* Westport, CT: Greenwood Press.

Karttunen, H., Kroeger, P., Oja, H., Pouanen, M., & Donner, K. J. (Eds.). (2003). *Fundamental astronomy* (4th ed.). Berlin: Springer.

Plate Tectonics

From the beginnings of science, it was thought that Earth was a static, stable planet whose surface remained largely unchanged through time. This view radically changed during the 1960s, as an array of improved analytical techniques and an influx of new observations revealed that Earth's surface is in a state of constant change. This new approach to understanding the earth is known as plate tectonics, and it holds that the outer skin of our planet is divided into several plates whose motion results in mountain building, earthquakes, volcanism, and other geological events. Understanding the processes of plate tectonics has allowed scientists to systematically explain the history and structure of the earth and to study both past and modern geological events in a unified and rigorous fashion. The philosophical shift from viewing Earth as static to viewing it as a dynamic planet governed by plate tectonics is also regarded as a prime example of a paradigm shift in science.

Early Speculation

The first recorded suggestions of a dynamic and changing Earth were offered by 16th-century philosophers and geographers, who noted the congruence between the Atlantic coastlines of Africa and South America. In 1596 the geographer Abraham Ortelius argued that the Americas were once conjoined with Europe and Asia, but later "torn away" by earthquakes and other catastrophes. In recent years, historians of science have discovered nascent hints of plate tectonics in the writings of Francis Bacon, Scottish philosopher Thomas Dick, noted French scientist Comte de Buffon, German explorer Alexander von Humboldt, and Benjamin Franklin. However, it wasn't until the early 20th century that scientists began to assemble strong evidence that the surface of the earth has changed over time, as well as a coherent hypothesis to explain how. Much of this data set was articulated by Alfred Lothar Wegener, a German meteorologist who assembled widely divergent lines of evidence into an understandable theory of continental motion.

Wegener and Continental Drift

Like the early geographers before him, Wegener was intrigued by the closely matching Atlantic coasts of South America and Africa. After reading a paper describing similar Paleozoic fossils from these two continents, Wegener launched a massive literature search in the hopes of finding additional data to support the concept of continental drift. The data he uncovered were varied and wide ranging. Not only did the coastlines of South America and Africa match, but so too did the coasts of Newfoundland, England, parts of Greenland, and Scandinavia, especially when the outlines of the continental shelf were taken into account. Additionally, Wegener discovered that South America, Africa, India, Australia, and Antarctica shared a suite of unique Mesozoic fossils, including a tropical plant flora characterized by the fern *Glossopteris* and a reptile fauna that included the tusked, piglike *Lystrosaurus*. Modern animals do not range across all continents, because it is often impossible to disperse across oceans and other barriers. This suggested to Wegener that these continents were linked during the Mesozoic era (225–65 million years ago), and have since moved to their present, widely divergent positions. Furthermore, the presence of tropical fern fossils in Antarctica makes no sense if the continent has always occupied a polar position.

Wegener supported these observations with several additional lines of evidence. This included closely matching rock units shared by Africa and South America, the distributions of former equatorial climate belts (as shown by coals and fossil reefs shared by the five aforementioned lands), and the locations of past Paleozoic glaciations. Taken together, these facts suggested to Wegener that all of the continents had once been joined into a supercontinent (which he named Pangea) that later split into two larger fragments. The first fragment, termed Laurasia, included North America, Europe, and Asia, while the second landmass, called Gondwana, consisted of Africa, South America, India, Madagascar, Australia, and Antarctica. Over time these two fragments further split into the individual continents recognized today.

Wegener presented his hypothesis of continental motion in a series of lectures and journal articles in 1912. Three years later he outlined his ideas in a short, 94-page book, *Die Entstehung der Kontinente und Ozeane,* which was subsequently revised three times and translated into English as *The Origin of Continents and Oceans.* The notion of continental drift, which overturned much of the conventional geological wisdom of the day, was initially dismissed by critics as untenable, largely because Wegener could provide no plausible mechanism for continental motion. In his book, Wegener suggested that the centrifugal force resulting from the Earth's rotation, or possibly the pull of gravity from the moon, drove the lighter, granite-rich continents through the denser, basalt-rich oceanic crust like a ship plowing through water. However, eminent Cambridge geologist Harold Jeffreys, one of the most respected scientists of his time, did the calculations and found these forces insufficient to move something as large as a continent. In response to Wegener's evidence, many in the scientific establishment suggested that now-sunken land bridges allowed for floral and faunal interchange between separate continents, and shifts in climate regimes explained the locations of past glaciations and similar rock units between South America and Africa.

At the time of Wegener's death during a 1930 expedition to Greenland, his hypothesis was openly ridiculed and his scientific credibility scorned. However, although Wegener would never know it, his hypothesis was later verified as a new age of science dawned in the shadow of World War II. Over the course of the 1960s a handful of earth scientists from across the globe instituted a scientific "revolution" that molded Wegener's observations into the theory of plate tectonics. Today, plate tectonics is regarded as the grand unifying theory of geology, and it helps to explain everything from animal dispersal and mountain building to volcanism and earthquakes.

Supporting Data

Important data supporting Wegener's evidence for continental motion came from studies of paleomagnetism. As lava cools into solid rock, tiny crystals of the magnetic mineral magnetite are "locked" into position, thereby recording the direction of Earth's magnetic field at the time of the rock's formation. Using trigonometric equations, geologists can take these data and determine the latitude at which a certain rock formed. This procedure was applied to igneous rocks across the globe, and it was discovered that the latitudes at which the rocks formed were usually different from the latitudes they occupy today. At first it was thought that the earth's magnetic pole simply wandered over time, thus causing the discrepancy in paleolatitude measurements. To test this hypothesis, scientists began compiling "polar wandering curves," charts of a continent's latitudinal position over time (enabled by recent advances in radioactive dating of rocks).

If the continents were fixed and the magnetic pole was actually wandering, the curves of different continents should match throughout geological history. However, extensive data sets showed the opposite to be the case: Each continent had its own unique polar wandering curve, which indicated that the magnetic pole is essentially fixed and the continents have drifted relative to each other. Interestingly, the polar wandering curves of North America and Europe were found to align during the early Mesozoic, indicating that these continents were moving together. Similar matches for other continents during this time support the existence of a supercontinent, as originally suggested by Wegener.

Paleomagnetic data strongly supported the reality of continental motion, but several questions about the form and cause of such motion still remained. Many of these questions were capriciously answered during the era of undersea exploration immediately after World War II. Leading this research was Princeton geologist Harry Hess, who discreetly took measurements with a fathometer while captaining a Navy transport ship in the Pacific during the war. Hess's surveys detailed the topography of the seafloor, and revealed that a series of long mountain ranges, deep trenches, and extinct volcanoes littered the deep abyss. Later evidence suggested that a large percentage of the world's earthquakes were occurring in these submerged mountains, hinting that the ocean bottom was a dynamic place.

Over time Hess became convinced that new seafloor was currently being formed in the mid-ocean mountain ridges that occurred in the centers of the Atlantic and Pacific, as well as other oceans and seas. In a landmark 1962 paper, Hess suggested that new basaltic seafloor was produced at the ridges and spread away symmetrically in both directions, a process called "seafloor spreading" by geologist Robert Dietz. This process was soon supported by a slew of additional observations. Magnetometers deployed by the Allies during the war had revealed strange magnetic patterns on the ocean floor, namely, the strength and direction of the magnetic field followed zebra-striped patterns that not only paralleled the midocean ridges, but also were symmetrically the same on both sides of the ridges. This pattern puzzled scientists until Cambridge geologists Drummond Matthews and Fred Vine directly dated much of the seafloor, and found that the absolute dates were also symmetrical about the ridges, with rocks becoming older with increased distance. Additionally, the amount of sedimentary cover and degree of erosion of extinct volcanoes increased systematically and symmetrically away from the ridge, as would be expected if the seafloor gradually increased in age. These and other observations only confirmed what Hess, Matthews, and Vine suspected: that new seafloor was created in midocean ridges and gradually spread outward in both directions.

The Theory Outlined

The coherent theory of plate tectonics, a unification of Wegener's continental drift and Hess's seafloor spreading, was largely pieced together during an 8-year period in the 1960s. Today, geologists view the earth as composed of two principal layers: the brittle, outer lithosphere (composed of the crust and upper mantle) and the denser, warmer asthenosphere (composed of the lower mantle and core). The lithosphere, akin to the fragile shell of an egg, is broken into some 20 distinct plates, which are rigid but deformable at their edges. These plates can contain both dense oceanic crust, which is rich in magnesium and iron, and lighter continental crust, which is enriched in silicon and aluminum. The interaction of plates produces many characteristic geological phenomena and is the subject of intense study by geologists and geophysicists.

When two plates meet, one of three general interactions occurs: They can move away from each other (divergent), move toward each other (convergent), or slide past each other (transform). When plates move away from each other at divergent boundaries, seafloor spreading takes place, and new oceanic crust is formed. Midocean ridges are an example of a divergent boundary, as are the rift zones of continental interiors (such as the present East African Rift). If a rift is successful and a piece of continuous continental crust is split in two, seafloor spreading commences and further pushes apart the continents. This process explains the mechanism of continental breakup, such as the splitting of the Mesozoic supercontinent Pangea into the seven main continents of today.

When plates come together at a convergent boundary, the result is more complicated. If denser oceanic crust meets lighter continental crust, the oceanic crust is subducted beneath the continent, often giving rise to volcanism. This process explains the so-called Ring of Fire, which follows many prominent subduction zones along the Pacific Rim. The subduction of oceanic crust is often associated with mountain building, most prominently the formation of the Andes of South America, which have resulted from the subduction of the Nazca Plate under the South American Plate. Additionally, subduction zones are frequently regions of deep and powerful earthquakes and are the source of the recurrent large tremors that rock Alaska and the northern Pacific. On the other hand, since continental crust is too buoyant to be subducted, the collision of two continents results in intense crustal wrinkling and thickening, producing mountains. Continental collision formed the Appalachians during the Paleozoic and today is responsible for the continuing uplift of the Tibetan Plateau, which formed via the collision of India and Asia during the Late Cretaceous and Early Tertiary.

Finally, when two plates slide past each other, earthquakes frequently occur, as manifested by the San Andreas Fault of California and the Anatolian Fault of Turkey. Such boundaries are among the most seismically active regions on Earth.

Despite this understanding of plate interactions, the exact driving force of plate tectonics is still poorly understood. Wegener originally proposed several possible mechanisms for continental motion, most of which were dismissed by critics as untenable. Years after Wegener's death, noted British geologist and physicist Arthur Holmes resurrected Wegener's suggestion that convective currents in the waxlike asthenosphere drive the motion of the overriding lithosphere, a hypothesis widely supported today. More recent research has added further details, and suggests that gravitational and frictional forces on subducting plates, as well as "push" from new extruded material at midocean ridges, also contribute to plate motion.

The ultimate driver of many of these forces is radioactive decay in the earth's core.

The "revolution" of plate tectonics, occurring between the first publication of Wegener's heretic views in 1912 and the accumulation of convincing paleomagnetic and oceanic data in the 1960s, is a prime example of a paradigm shift in science. When Wegener first proposed his theory of continental drift, nearly the entire geological community regarded Earth as static and unchanging, with current landmasses occupying the same position throughout the entirety of geological history. As would be expected, Wegener's views caused quite a stir in scientific circles, inspiring several symposia and publications that roundly criticized and dismissed continental drift. Respected geologist R. T. Chamberlin ridiculed the theory as being "of the foot-loose type," while Edward Berry bluntly labeled Wegener's method as "unscientific." Some of this criticism was pure obstinacy, but the majority of geologists chastised Wegener's failure to provide a plausible mechanism for continental motion. It was only several decades later, after new instruments and techniques revealed irrefutable evidence from paleomagnetism and oceanic surveys, that the scientific community accepted the reality of mobile continents. By this time, an increased understanding of radioactivity allowed for a plausible driving force: mantle convection. With the acceptance of plate tectonics came a new and open opportunity for understanding the processes and history of the earth, and it finally allowed for reasonable explanations of mountain building, volcanism, and earthquakes—geological phenomena that had puzzled geologists of the "static Earth" camp.

Conclusion

The theory of plate tectonics is the grand unifying theory of geology, a complete, understandable, working theory of the earth. Unlike most traditional sciences, geology deals with large-scale patterns and processes that operate over unthinkable lengths of time. Early geologists found it necessary to describe the geological history and processes of Earth based on events that occur in the present. Although intuitively rational, this "uniformitarianist" method of thinking prevented many geologists from recognizing complex processes that are difficult to observe, such as mountain building and earthquakes. An understanding of plate tectonics allows modern geologists to place volcanism, earthquakes, rifting, and other phenomena into a rigorous theoretical framework. The surface of the earth is divided into several plates that constantly move as new crust is generated at midocean ridges, driven by convective currents in the mantle. This basic set of processes has defined Earth for millions of years, differentiates our planet from other bodies in the solar system, and has governed both the physical and biological evolution of our world throughout geological history.

Stephen L. Brusatte

See also Catastrophism; Geology; Pangea; Stratigraphy; Uniformitarianism; Wegener, Alfred

Further Readings

Hallam, A. (1973). *A revolution in the earth sciences.* Oxford, UK: Oxford University Press.

Hess, H. H. (1962). History of ocean basins. In A. E. J. Engel, H. L. James, & B. F. Leonard (Eds.), *Petrologic studies: A volume to honor A. F. Buddington* (pp. 599–620). New York: Geological Society of America.

Oreskes, N. (2003). *Plate tectonics: An insider's history of the modern theory of the earth.* New York: Westview.

Sullivan. W. (1974). *Continents in motion: The new earth debate.* New York: McGraw-Hill.

Wegener, A. L. (1924). *The origin of continents and oceans.* New York: Dutton.

Plato (c. 427–c. 347 BCE)

Along with his teacher Socrates and his pupil Aristotle, Plato is recognized as one of the most influential thinkers in ancient philosophy. Plato deals with the notion of time in different contexts. He describes time as "a moving image of eternity" in his philosophy of nature (in his *Timaeus,* 37c–47c): the temporal structure of the cosmic universe aims at imitating the unchangeable eternal realm of the ideas, of which the real world is a reflection. Plato also develops (in the dialogue the

Statesman / the *Politikos,* 268d–274d) the idea of a cyclic dimension of the historical time, which consists of two "world ages" that alternate for eternity. The two ages have different characteristics and are seen as a result of a change of the overall inner motion of the cosmos. Finally, Plato aims at a deeper understanding of the paradoxes of movement that are linked to temporal becoming (in the dialogue *Parmenides,* 151e–157b). The main background for Plato's understanding of time is his doctrine of ideas, according to which the temporal world is in constant change; but "behind" this change lies the realm of unchangeable ideas, which exists outside of space and time.

Born in Athens, Greece, into an old aristocratic family, Plato was hindered in his pursuit of a political career by the turbulent politics of his times. In his youth he was a pupil of Socrates for 8 years, until Socrates' trial and execution in 399 BCE. After the death of Socrates, Plato's extended travels led him to Cyrene and Egypt and thus to Euclid of Megara, then to lower Italy, and to the Pythagoreans. In 385 Plato founded his famous school, the academy in Athens (which was only to be closed in 529 CE). He traveled several times to Sicily, where he attempted to put his political ideas into practice. He discarded this, however, after he fell into disgrace, became enslaved, and had to be redeemed by his pupils.

The platonic works that are preserved—besides the letters and the *Apology*—are all written in the form of dialogues, most of which depict Socrates debating with his interlocutors to seek the truth. Socrates' quest for a deeper understanding of the virtues and the nature of the good is presented as an antidote to the relativism and the empty rhetoric of the Sophists. In *Epistemology,* Plato aims thus at a distinction between true knowledge (*episteme*) and mere belief (*doxa*). In his political works he depicts a perfect state that is described as a corporate state, in which the people with the highest wisdom—the philosophers—should rule. Most influential for western metaphysics was Plato's doctrine of ideas, which also serves as a background for his cosmological notion of time.

Plato's Notion of Time

Plato develops his cosmological ideas in the dialogue *Timaeus* in the form of a mythical explanation of the creation of the universe. To understand this myth, one needs to look at Plato's doctrine of ideas. The core of this doctrine is the distinction between the realm of individual entities, which can be perceived by the senses and are in a constant state of change, and the eternal realm of unchanging ideas, which can only be perceived by reason. The constant change and motion of the elements of the empirical world make it impossible to secure general knowledge about them. In contrast, the eternal ideas form the realm of proper being, about which knowledge of mathematical certainty is possible. Ideas and individual entities are connected through a relation of "participation": An individual thing is only cognizable and has "being" to the degree in which it participates in a timeless idea. Something can for example be considered beautiful only if it participates at the idea of the beauty, if it is an instantiation of the general concept of beauty.

This distinction between ideas and the realm of real objects signifies on the one hand a relation of hierarchy: The archetype is more real and perfect than the image. On the other hand: the image resembles the archetype: Individual things can thus be regarded to be concrete instantiations of abstract ideas in space and time. It becomes clear that the distinction between unchanging ideas and the realm of the empirical world by Plato is also often regarded as a distinction between the timeless realm of eternal truth (like mathematical truth), in which no movement or change in time occurs, and the empirical world, in which no individual object remains the same, and a change over a period of time affects each of them.

The creation of time itself as a part of the visible universe is depicted in a myth told by Timaeus about the creation of the world. According to this myth, the godlike "demiurge" (craftsman) formed the cosmos out of preexisting matter guided by the unchangeable eternal ideas. He attempts thus to create an image of these ideas that is as nearly perfect as possible. The cosmos as a whole is seen as one perfect living being containing an invisible spirit and the visible parts of the universe (its body). The cosmos is created in the image of the perfect living creature (*zoon*) out of previously unstructured elements in a way that nothing is left outside of it. Because there is thus no exchange of matter or forces between an "outside" and an "inside," modes of spatial or temporal change cannot be attributed to the cosmos as a whole. The universe is thus eternal and does not grow in shape or size, and in this sense it also does not grow older (*Parmenides,* 33b). In the same line of argument also, the soul of the cosmos is declared to have an endless intelligent life for all times (ibid., 36e).

In order to make this image as perfect as possible, the demiurge creates "time" as a moving picture of eternity. The main idea here is that the cosmos cannot (as being created) be "eternal" in the same sense that the realm of the ideas is eternal (i.e., not affected by time), but it can mirror the idea of eternity through a cyclical periodical movement according to mathematical proportions. This movement is what we call time. One can illustrate this idea by comparing it to a thought we find in the *Symposium:* Individual living creatures are mortal and cannot be eternal, but they imitate the eternal idea through reproduction as a periodic act of the recreation of their species. Even though the individual is mortal and affected by time, the species—as the more general concept—is eternal and unchangeable.

Time is furthermore for Plato linked very closely to the realm of astronomy. It is created together with the heavens, and it finds its perfect expression in the periodic movements of the planets. The parts of time are therefore defined through astronomical relations: Days, months, and years are the natural elements of time, which are themselves determined by the movement of the earth, the moon, and the sun. In addition to this, Plato points out that there are numerous other parts of time, related to the movements of the other planets, that are—though often unknown to men—of equal importance and beauty. The most perfect cosmological "year" is completed when all planets have the same position again and the movement starts anew.

We can thus see that time aims at depicting eternity in the form of a periodic and cyclic movement, in the same sense as the circle is considered to be the perfect shape and circular movement as the most perfect type of movement. This cyclical understanding of time is characteristic of ancient Greek thought. Only in later periods (e.g., in early Christianity) did the idea of an overall direction of history culminating in salvation become important, eventually replacing the idea of an eternal return of everything by the notion of a clearly defined "progress."

Besides the parts of time (the days, months, years, etc.), Plato also discusses the modes of time, the past (the "it was"), the present (the "it is") and the future (the "it will be"). He insists that the "it was" and the "it will be" can only be applied to things within the world and not to the eternal realm of the idea, of which only the "it is" is an appropriate formulation. Linked to this idea is Plato's struggle to understand the processes of "becoming" and "changing" in general that all lead to logical paradoxes, discussed already in the Presocratic philosophy. In this context the Eleatic philosophy is very relevant, ascribing real being only to the things that "are" and remain always the same and thus denying the possibility of change. Change is thus seen as an illusion. We can see how Plato on the one hand replicates elements of this idea in his dualistic doctrine of ideas, but how he tries, on the other hand, to reconcile abstract dualism by depicting time itself as a "moving image of eternity" that itself tries to some extent to bridge the gap between the atemporal world of ideas and the temporal world of finite objects.

Cyclical Nature of History

In the *Politikos,* the question of the ideal statesman is discussed. In this context a myth of two world ages is reported by the "stranger," who takes the lead in this dialogue. Because the cosmos has a visible bodily part, it is submitted to forms of movement. Because there cannot be a movement to another place—as there is no "place" outside the cosmos—this movement takes the form of an inner rotation of the whole cosmos. This rotation is set by the gods in one direction at the beginning of the universe. In this age, determined by the overall direction of the rotation, men were born from earth and lived under the guidance of gods a pleasurable life dedicated to philosophy. This period is called the age of *Kronos* (time). (Kronos is in the Greek mythology the father of Zeus.)

After a long period of time, this age comes to its natural end when the cosmos has finished its movement in one direction. Now the gods release the cosmos to start spinning in the opposite direction: The planets and the sun move "backwards." In this second world age, the age of Zeus, nature is left to its own guidance. Men are born from men and will be buried in earth when they die. The further away this period moves from the first age, the more knowledge about the Good gets lost. Eventually this period will again come to its end, and the direction of rotation will change once more. This cyclic alteration keeps continuing for eternity. According to this myth, it is important to know in which of the two periods one is living, as the definition of the good statesman differs for both ages.

Again it can be seen that a cyclical understanding of time is prominent in Greek tradition, although Plato combines this thought with a specific direction for each age. The age we humans currently live in—the age of Zeus—is depicted as an age of decay that on the one hand moves further and further away from the golden age of the past. On the other hand—due to the eternal alternation—this age, like others, will come to an end and will be replaced by a new period "under the guidance of gods."

Logical Paradoxes of Motion and Time

In the *Parmenides,* Plato discusses central aspects of his doctrine of ideas. The main character is the aged Eleatic philosopher Parmenides, who explains to the very young Socrates his doctrine of the one being. In the first part of the dialogue, the doctrine of ideas is discussed; in the second part, several dialectical exercises are carried out by Parmenides that contain logical analyses and problems of the "one."

In this dialectical exercise, opposing theses are discussed and developed out of each other. Different interpretations have been suggested for these difficult passages; the main purpose is very likely to reject the radical dualism of the Eleatic opposition of the "unchangeable, atemporal one" and the moving unstable empirical world of changes of the other side: Zeno (the second Eleatic philosopher in this dialog, who represents this strict opposition) is becoming more and more a side-figure of the dialogue, as the main attempt seems to be to try to bridge those two realms and come to a more concrete understanding of the ontological principles of being (the ideas) and the changes happening in reality.

While the "one of being" is shown to be atemporal, every individual thing is said to have some relation to time. In this context, "existence" is identified as "being present" at a certain time: Everything that exists in time is, however, also capable of "becoming" or—more generally—of changing, especially, of course, of growing older. "Being" is thus deeply linked to time, as "is" means that something exists in the now, while "has been" means something existed in the past; "will be" finally refers to some existence in the future. Plato discusses the apparent paradox that things can thus grow older than they were while at the same time being younger than what they will be in the future.

Plato's notion of time was very influential in ancient philosophy. Its cosmological explanation (in the *Timaeus*) was adopted by Plotinus in later Neoplatonism, though in his explanation of time the relation to astronomy and mathematics gets more and more lost. Christian thought reinterpreted the myth of the *Timaeus* as a myth of creation, understanding the "timeless ideas" as a part of God's eternal mind. Aristotle's theory of "dynamis," and his distinction between the "actual" and the "potential," respectively, between the "substance" and his "attributes," can be seen as an attempt to solve some of the paradoxes linked to the question of how things can change in time and still remain the same (so that Socrates can grow older and still be "Socrates").

Andreas Spahn

See also Aristotle; Aristotle and Plato; Becoming and Being; Cosmogony; Demiurge; Idealism; Paremenides of Elea; Plotinus; Presocratic Age; Zeno of Elea

Further Readings

Brumbaugh, R. S. (1990). *Plato on the one.* New Haven, CT: Yale University Press.

Cooper, J. M., & Hutchinson, D. S. (Eds.). (1997). *Plato: Complete works.* Indianapolis, IN: Hackett.

Cornford, F. M. (1997). *Plato's cosmology: The* Timaeus *of Plato.* Indianapolis, IN: Hackett.

Kraut, R. (1992). *The Cambridge companion to Plato.* Cambridge, UK: Cambridge University Press.

Wright, M. R. (Ed.). (2001). *Reason and necessity: Essays on Plato's* Timaeus. Swansea, UK: Classical Press of Wales.

Plotinus (c. 205–270 ce)

Plotinus was born in the Nile Delta region of Egypt; he studied philosophy in Alexandria and later, Rome. He did not see himself as an original philosopher in the modern sense, but rather as interpreter of the truth that was first elucidated by Plato. Thus he became the founder of Greek Neoplatonism. Central issues of Plotinus's thinking are these: The first principle is the one; that is identical with the good; from the one, the intellect proceeds, which can be identified with being, and from the intellect proceeds the soul. These are the three principles or hypostases of reality.

In his treatise on eternity and time (*Enneade* III 7), Plato's theory of time developed in the *Timaeus* is the framework for Plotinus's own inquiry, but Aristotle's views intensively stimulate his notion of time in this world. Plato's definition of time as the "moving image of eternity" indicates that time can be described only in the context of eternity. Aristotle's definition of time as the "number of motion" shows that time is something closely linked to movement and to the counting soul. Plotinus's main thesis is this: Eternity is the life of the intellect, and time is the life of the soul.

His treatise on eternity and time starts with the presupposition that eternity (*aion*) is linked to the intelligible world of eternal being and time (*chronos*) to our sensible world of becoming. Eternity and intellect both include the same, because both are "most venerable" (*semnotaton*). While the intelligible includes everything just as a whole includes its parts, eternity includes the whole all at once (*homou*), that is, simultaneously and not as parts. Although rest corresponds to eternity as motion does to time, eternity is not identical with rest; otherwise the intelligible world would be limited to only one of the five concepts in Plato's *Sophist;* the other four (substance, motion, the other, the same) would be excluded. The first definition of eternity is this: "Eternity is the life, which belongs to that which exists and is in being, all together and full, completely without extension or interval" (all translations by Armstrong). This definition is founded on Plato's "living being" in the *Timaeus,* which is contemplated by the demiurge. This is described as "eternal" and as "always existing in the same state."

For Plotinus, eternity is not to be identified with the intellect or the intelligible world but it is related to the totality of the intelligible life. This life that is eternity is not identical with the intelligible, but is a manifestation of it: "Eternity is not the substrate but something which, as it were, shines out from the substrate itself." Eternity does not come to the intelligible from outside but it is from it and with it; that is, "The nature of eternity is contemplated in the intelligible nature existing in it as originated from it." So eternity is an aspect of the intelligible as much as beauty or truth and is very close to being like a Plotinian intelligible form. It is a true whole in such a way that it is deficient in nothing, with neither past nor future.

Whereas the universe has a future and hastens by its everlasting circular movement to everlasting existence by means of what is going to be, the complete and whole substance of reality is something that is always existing—from "always existing" (*aeion*) is derived eternity (*aion*). Eternity is described as "state [*diathesis*] and nature [*physis*]" of complete reality, so Plotinus held the thin line between giving eternity a precise ontological status and seeing it as a quality. The difference between "everlastingness" and "eternity" is as follows: Eternity is the substrate from which everlastingness manifests itself, but eternity is the substrate with the corresponding condition manifested. So the final definition of eternity is this: "And if someone were in this way to speak of eternity as a life which is here and now endless because it is total and expends nothing of itself, since it has no past or future . . . he would be near to defining it."

Plotinus's examination of time is based on other philosophers' theories of time, namely these: (a) Time is *movement* (e.g., Plato interpreted by Eudemus, Theophrastus, and Alexander as identifying time with the movement of the heavens); (b) time is *what is moved* (the heavens themselves); and (c) time is *something belonging to movement* (e.g., the Stoic view of time as extension of motion, the Epicurean definition of time as accompaniment of motion, or the Aristotelian definition of time as number of motion). For Plotinus, time proceeds from the transcendence of the intellect. As eternity is the life of the intellect (*nous*), time is the life of the soul, which is to be identified with discursive reason (*dianoia*) and which descends from the hypostasis of the intellect. "Before" the descent of the soul, time was "at rest with eternity." By "a restless active nature which wanted to control itself and be on its own," the soul "transferred what it saw to something else" and "moved on to the 'next' and the 'after.'"

Just as soul constitutes itself as an image of the intellect and then produces the physical world as an image of itself, therefore also soul constitutes time, which is its own life as an image of eternity, and then creates as an image of itself the sensible world in time. Time then exists on two levels: First, "time is the life of soul in a movement of passage from one way of life to another." Second, it is the measured time in the physical world. The life of soul is discursive reason; that is, the movement from one idea to another. Therefore, the notions of "before" and "after" are present even in the intellect, so that "before" and "after" signify the order of causality, not the temporal sequence. Therefore, Plotinus's theory of time connects Plato's definition of time as "moving image of eternity" with Aristotle's definition of time as "number or measure of motion."

According to Plotinus, Aristotle failed to explain the nature and origin of the preexistent measuring number. Plotinus was far from suggesting an exhaustive rejection of Aristotle; instead he only complained that Aristotle's writings lacked sufficient clarity, because they were addressed to an internal school audience. With regard to Plato, Plotinus only wanted to correct misinterpretations: Instead of equating the heavens with time, Plato means that the sphere and the planets "manifest" time. Thus, time as a distinct interval measured by the movement of the heavens could be used as a measure, but time itself is not a measure; it should rather be called what is measured. By integrating Plato's and Aristotle's theories of time in his own system of hypostases, Plotinus succeeded in explaining the substantial nature of time and in differentiating between the metaphysical notion of time itself and the physical measurement of time.

Michael Schramm

See also Aristotle; Becoming and Being; Bruno, Giordano; Cosmogony; Eternity; Lucretius; Nicholas of Cusa (Cusanus); Plato; Rome, Ancient; Teilhard de Chardin, Pierre

Further Readings

Smith, A. (1996). Eternity and time. In L. P. Gerson (Ed.), *The Cambridge companion to Plotinus* (pp. 196–216). Cambridge, UK: Cambridge University Press.

Strange, S. K. (1994). Plotinus on the nature of eternity and time. In L. P. Schrenk (Ed.), *Aristotle in late antiquity* (pp. 22–53). Washington, DC: Catholic University of America Press.

Plutarch (c. 46–c. 120)

Plutarch, a classical Greek writer, lived during the period when the Roman Empire ruled the Mediterranean region, including Greece. Born into an aristocratic and influential family, Plutarch spent much of his life in Chaeronea, his birthplace in Boeotia. In his travels he visited Athens, Egypt, and Italy, often teaching and lecturing in Rome. Later in life, he would found an academy before joining the priesthood of the Oracle at Delphi, as he was a devout believer in the ancient pieties as well as an astute student of antiquity. Plutarch received his education in Athens, where he composed many essays and dialogues and became a writer and thinker versed not only in philosophy but also in science and literature. His works have greatly influenced Western understanding of classical culture, especially in comparing and contrasting exceptional individuals who lived in ancient Greece and ancient Rome.

Plutarch is most renowned for his historical works, which focus on the heroic lives of those who shaped both the classical and Hellenistic ages in Greece. His *The Rise and Fall of Athens: Nine Greek Lives,* and *The Age of Alexander* and *Makers of Rome: Nine Lives* remain widely read as staples of a classical education, although they are sometimes viewed as flawed in their historical methodology. Plutarch often includes what some historians regard as an excess of critical commentary in his accounts; moreover, Plutarch was not a contemporary of many of his subjects and thus relied on secondary sources who did not necessarily witness the events they recorded, so his historical accuracy is sometimes questionable. His style of weaving into his stories a great deal of legend or myth can confound the modern reader seeking detailed historical explanations. Consequently, Plutarch's chief value may lie not in his fidelity to historical detail but in his embodiment of the spirit of the classical Greek and Roman ages.

A philosopher by trade, Plutarch habitually made moral judgments about the character of his subjects on the basis of their deeds; thus, his historical accounts were perhaps secondary to the moral example they provide the reader. Of special interest to Plutarch was the comparison between the great lives of classical Greece and those of his contemporary Romans. His *Parallel Lives* illustrates similarities in excellence of character, or lack thereof, for the purposes of providing social commentary and in a way acting as a moral compass to the leaders of his time.

For his efforts as both biographer and philosopher, Plutarch was honored by Emperor Hadrian with a government appointment in Greece. Until his death, Plutarch continued to travel between Greece, Rome, and Egypt.

Garrick Loveria

See also Alexander the Great; Aristotle; Caesar, Gaius Julius; Peloponnesian War; Rome, Ancient

Further Readings

Lamberton, R. (2001). *Plutarch.* New Haven, CT: Yale University Press.

North, T., & Mossman, J. (Trans.). (1999). *Plutarch: Selected lives.* Ware, UK: Wordsworth.

Poetry

The inexorable passage of time and corollary facets of change, mortality, memory, and nostalgia are rich themes in literature and, particularly, in poetry. Poetry, in the Aristotelian sense, seeks to recreate human experience and capture both its sensory and abstract essence. Unlike prose, poetry relies on a separate set of conventions emphasizing rhythm and sound patterns as well as figurative language and symbolism to achieve a multilayered effect. Common symbols evoking associations with time include references to the phases of the moon and tides, the rising and setting of the sun, the cycle of seasons, and shifting sand in its natural state or as a calculation of time through an hourglass. Such concrete references intensify awareness of the inevitable passage of time, which, although an abstract concept, is measured concretely through human constructs.

A common theme in Western poetry is *carpe diem,* or "seize the day." This metaphor is derived from the Latin *carpere,* "to pluck or grab," and *die,* meaning "day." The phrase is generally attributed to the Roman lyric poet Horace (65 BCE–8 BCE) also known as Quintus Horatius Flaccus. Horace used metaphors from nature to advise against procrastination and urged his readers to enjoy the present, "Be wise, strain the wine; and since life is brief, prune back far-reaching hopes! Even while we speak, envious time has passed: Pluck the day, putting as little trust as possible in tomorrow" ("*carpe diem, quam minimum credula postero*"). This view focuses on the present and encourages making the most of each day and stage of life. The adage has often been extended to embody a hedonistic sense of enjoying pleasure without concern for the future, because time, and ultimately death, are destroyers of life's joys.

The *carpe diem* theme was popularized during the European Renaissance (14th to 17th centuries), which saw a rebirth in learning and emphasis on humanism. This intellectual revolution was marked with great achievement in art, music, and literature. It also followed the devastating Black Plague of the 14th century, which wiped out at least one third of the population of Europe from Italy to Norway. The suffering and fragility of human life, as well as its potential for great accomplishment, made *carpe diem* an anthem to life and living it fully.

English poet and playwright William Shakespeare (1564–1616), long considered one of the greatest writers in the English language, often reflected on mortality and the overarching power of time in his sonnets and plays. His 38 plays were written predominantly in iambic pentameter, or blank verse, as were his 154 sonnets. In Sonnet 18, Shakespeare uses his pen to defy time and bring immortality to his love: "So long as men can breathe or eyes can see,/So long lives this, and this gives life to thee." In Sonnet 30, Shakespeare expresses remorse over wasted time: "When to the sessions of sweet silent thought/I summon up remembrance of things past,/I sigh the lack of many a thing I sought,/And with old woes new wail my dear time's waste." But knowledge of mortality can even intensify love, as in Sonnet 73: "This thou perceivest, which makes thy love more strong,/To love that well which thou must leave ere long." And Sonnet 45 expresses human resistance, as well as powerlessness, against the ravages of time: "And nothing 'gainst Time's scythe can make defence/Save breed, to brave him when he takes thee hence."

Shakespeare also made numerous references to time in his plays, both comic and tragic, as illustrated by the clown's refrain in *Twelfth Night:* "What is love? 'Tis not hereafter;/Present mirth has present laughter, What's to come is still unsure:/ In delay there lies no plenty" (act 3, scene 2). A darker reference is made in *The Tragedy of King Richard II* when Richard laments: "I wasted time, and now doth time waste me:/For now hath time made me his numbering clock" (act 5, scene 5). Whether expressing hope, defiance, regret, or stoicism, Shakespeare grappled frequently with the relentless nature of time and its effect on the human condition.

Other 16th- and 17th-century English poets also addressed the *carpe diem* theme, particularly the Cavalier poet Robert Herrick (1591–1674) and the metaphysical poet Andrew Marvell (1621–1678). In "To the Virgins, Make Much of Time" Herrick urges, "Gather ye rosebuds while ye may,/Old time is still a-flying,/And this same flower that smiles today,/To-morrow will be dying." In "To His Coy Mistress," Marvell pleads, "Now let us sport us while we may,/And now like amorous birds of prey, Rather at once our time devour/Than languish in his slow-chapped power." These works impart a sense of immediacy and become an argument for satisfying desire in the moment. This sentiment was frequently expressed by 20th-century poet Edna St. Vincent Millay (1892–1950) who recognized the short lifespan of passionate love, as in sonnet "XI": "I shall forget you presently, my dear,/So make the most of this, your little day,/Your little month, your little half a year."

Regret over wasted time recurs frequently in the works of modern and contemporary poets looking back on misspent youth or oblivious to the passage of time until it is too late. British writer Christina Rossetti (1830–1894) in "A Daughter of Eve" laments: "A fool I was to sleep at noon,/And wake when night is chilly . . . /Oh it was summer when I slept,/It's winter now I waken." A profound sense of loss is conveyed by 20th-century Welsh writer Dylan Thomas (1914–1953) when he writes powerfully of his youth and disregard for the power of time. In "Fern Hill," Thomas recreates the idealized rural paradise of his early life, which he describes as "green and golden" while "Time let me play and be/Golden in the mercy of his means." But as an adult he realizes, "Time held me green and dying/Though I sang in my chains like the sea." Another of Thomas's poems, "Do Not Go Gentle Into That Good Night," pleads to his dying father, "Do not go gentle into that good night./ Rage, rage against the dying of the light." Thomas's tone conveys not an acceptance of the inevitability of time and death but, rather, resistance even against an insurmountable power.

American writers Archibald MacLeish (1892–1982) and Emily Dickinson (1830–1886), although separated by almost a century, both pondered the unfathomable concept of eternity. Dickinson, who led a most private and solitary life, often grappled with time as an abstract and her own place in the continuum, as in "Behind Me Dips Eternity": "Behind Me—dips Eternity—/ Before Me—immortality—/Myself—the Term between." Her use of punctuation to create pauses reminds the reader of the subtle transition by which past, present, and future flow into one. Archibald MacLeish, a more public figure who was artistically and politically active, echoed a similar voice in "An Eternity": "There is no dusk to be,/There is no dawn that was,/Only there's now, and now,/ And the wind in the grass." Both poems reinforce the significance of being fully alive in the moment and the realization that humans can comprehend the universe only through sensory experience in the present.

Walt Whitman (1819–1892), influential and iconoclastic American writer, applied his theory of an "Oversoul" to create a vision of the interconnectedness of human life and the natural world, which flow as one through time. In his masterwork, "Crossing Brooklyn Ferry," Whitman uses the movement of the river, clouds, and sunlight descriptively and metaphorically, as the ferry and its passengers flow together, not only through physical space, but also through time: "Others will enter the gates of the ferry, and cross from shore to shore . . . /A hundred years hence, or ever so many hundred years hence, others will see them,/Will enjoy the sunset the pouring in of the flood-tide, the rolling back to the sea of the ebb-tide." In Whitman's interpretation, the oneness of humanity is manifest through shared experience over time; therefore, time and the cycles of nature become unifying, rather than separating, elements.

The 19th-century American transcendental movement protested reliance on the empirical world and promoted intuition and spiritualiy over doctrine and pure reason. A leading writer and thinker in this movement, Ralph Waldo Emerson (1803–1882) personifies time in his poem "Days": "Daughters of Time, the hypocritic Days, . . . /Bring diadems and fagots in their hands/To each they offer gifts, after his will,—" This verse ponders how time brings new days that can be wasted or fulfilled, either through chance or the exercise of free will. Emerson juxtaposed the lasting quality and value of a jeweled crown with the ephemeral bundle of sticks that offer a brief light before turning to ashes. Transcendentalists encouraged close observation of nature, as expressed in Emerson's essay: "To the attentive eye, each moment of the year has its own beauty, and in the same field, it beholds, every hour, a picture which was never seen before, and which shall never be seen again." Time, then, creates an ever-changing, but impermanent, beauty.

American transcendentalism is strikingly similar to Eastern views of nature as expressed in traditional Japanese haiku, a 300-year-old classical form that embodies the awareness of time by capturing the moment, particularly the transition of seasons. This traditional form, whose structure is based on elements of sound in Japanese, has been adapted in English to 17 syllables arranged in three lines of 5, 7, and 5 syllables each, respectively. It is one of the shortest but most effective forms of verse and focuses on a fleeting moment or, otherwise, unnoticed event. One of the revered writers of haiku was Basho, born Matsu Kinsaku (1644–1694), a poet, teacher, and Zen philosopher. His poetry compresses time, observing the human life span as reflected in microcosms of nature and the significance of events that mark the transition from one season to another: "The first soft snow!/Enough to bend the leaves/Of the jonquil low." The reader may extend the metaphor, complete the chain of events, and transfer the meaning to human experience.

The concise, but powerful, form of haiku has been adapted to other languages and cultures as a means of escaping time-bound expression and characterizing time as a continuum, or what theologian Paul Tillich (1886–1965) has referred to as "The Eternal Now." Poetry serves as a bridge, transcending time and space and reaching back to the past and forward to the future. As renowned Japanese writer Natsume Soseki (1867–1916) expresses in his haiku: "On New Year's Day/I long to meet my parents/as they were before my birth." Imaginative time travel is also suggested by British poet T. E. Hulme (1883–1917) in "Image": "Old houses were scaffolding once/and workmen whistling." And British poet James Elroy Flecker (1884–1915) projects his greeting forward in his six-stanza quatrain "To a Poet a Thousand Years Hence": "Since I can never see your face,/And never shake you by the hand,/I send my soul through time and space/To greet you. You will understand."

As a literary form, poetry exposes the human mind and heart grappling with the concept of time by recreating the past, capturing the present, and imagining the future. American-born writer T. S. Eliot (1888–1965) considers the nature of time in *Four Quartets,* which poses the possibility of alternate realities, or "What might have been and what has been/Point to one end, which is always present." Each section of the Quartets is related to one of the four basic elements of air, earth, water, and fire, and in "Burnt Norton" (No. 1 of *Four Quartets*), Eliot frames the conundrum powerfully: "Time present and time past/Are both perhaps present in time future,/And time future contained in time past. If all is eternally present/All time is unredeemable."

Linda Mohr Iwamoto

See also Alighieri, Dante; Chaucer, Geoffrey; Donne, John; Eliot, T. S.; Lucretius; Shakespeare's Sonnets

Further Readings

Bevington, D. (2003). *The complete works of Shakespeare* (5th ed.). New York: Longman.

Ferguson, M., Salter, M. J., & Stallworthy, J. (Eds.). (1996). *The Norton anthology of poetry* (4th ed.). New York: Norton.

Poincaré, Henri (1854–1912)

Henri Poincaré belongs to a small group of brilliant scientists who, living at the turn of the 20th century, made fundamental contributions to mathematics, physics, and philosophy. It has been said that Poincaré's mathematical knowledge comprised the whole mathematics of his time, and his lectures on theoretical physics show that he also had an encyclopedic overview of this field. Poincaré began his career as an engineer, but he quickly became famous for his great discoveries in mathematics, particularly in geometry and topology. Poincaré's results had a great impact on the history of mathematics and theoretical physics. For example, his consideration of the mechanical stability of systems composed of three bodies that attract each other by gravity led to the development of chaos theory. Poincaré's views of philosophical aspects of science, which he never presented in a systematic fashion, cannot be easily categorized as belonging to one of the main philosophical schools. This is also true of his idea of the relativity of time, which must be reconstructed from some of his papers collected in *Science and Hypothesis* (1902), *The Value of Science* (1906), *Science and Method* (1908), and the posthumous *Last Essays* (1913).

Jules Henri Poincaré was born on April 29, 1854, in Nancy, France. He was an outstanding pupil and made a very successful career in the French system of elite universities. His first academic degree was in mining engineering, and he worked for a short time as a coal-mining inspector. Yet after receiving his doctorate in mathematical sciences from the University of Paris in 1879, Poincaré began swiftly to publish important mathematical results, which brought him renowned chairs, particularly in mathematical physics, at different universities in Paris beginning in the mid-1880s. Besides being admitted to countless French and international scientific academies and societies, Poincaré was also an important member of the Bureau des Longitudes in Paris that organized the global coordination of clocks necessary for the production of precise maps. When he died prematurely in Paris on July 17, 1912, Poincaré was hailed as one of the most renowned researchers of his time. His fame as one of the last scientists who could make most important discoveries in different fields of research and reflect on his research philosophically continues to the present day.

Poincaré's consideration of time starts, like that of Einstein, with a seemingly innocent question: How can we judge objectively that two events are happening simultaneously? Any answer presupposes that it is possible to objectify the flow of time that our consciousness experiences qualitatively. Poincaré denies that human beings, scientists included, have access to some kind of absolute time beyond the scientific means of quantifying time by measurement instruments. If every process in nature were to be equally slowed down, we could not detect this deceleration, because the processes in the instruments for measuring time would also be slowed down.

Poincaré's criticism of the idea of absolute time exemplifies his philosophical stance on fundamental problems of science. He is convinced that the conceptual system by which we describe, analyze, and explain natural phenomena is a convention in the following sense: It is always possible to use, instead of the given conceptual system, other ones that can fulfill the same task. What convention we choose depends on pragmatic criteria, particularly on how simple it is to apply a conceptual system to the phenomena we want to understand. So conventions are neither true nor false; they are disguised definitions that prove to be more or less convenient for certain purposes.

To measure time, a periodic process in nature, such as the rotation of the earth on its axis, is needed whose cycles we suppose to be of the same duration. Yet this is only approximately true: The rotation of the earth, for example, is very gradually slowing down due to the friction caused by the tides. To explain this slowing down, and to detect it empirically, science presupposes the validity of well-known physical theories, namely thermodynamics and Newtonian mechanics. Their validity must not be affected however we define our measure of time, because we rely on these theories in the very act of defining. When we conventionally select one among the possible measures of time, we do this in such a way that the fundamental laws of nature can be mathematically formulated as simply as possible. As scientists, we ought not to say that one clock is right and the other one wrong; instead, we should say that it is more convenient to use this clock and not that one.

What events we suppose to be synchronous is also a matter of convenience. We choose such criteria of synchronicity as will allow the simplest formulation of the laws of nature we are using when we explain the physical processes by which we make local time measurements and by which we communicate the results of these measurements. Exploring the physics of time means, according to Poincaré, to investigate measurement processes in order to establish conventions that we can use both for an understanding of time and for an explanation of how our clocks would behave if they were perfect measuring instruments.

So far we have discussed only the philosophical sense of the relativity of time in Poincaré's thought. The question of whether Poincaré should be regarded, together with Einstein, as the discoverer of the special theory of the relativity of time and space has been intensely debated. The historical evidence seems to weigh against Poincaré: He relied too much on physical concepts (like the ether) that stood in the way of understanding Einstein's new conception of time and space. For Poincaré, the physical principle of the relativity of time is also a matter of convention, because it makes, for the sake of convenience, the following assumption that turns out to be only approximatively true: Two bodies that are very far away from each other can be described in different frames of reference, because they do not influence each other.

Stefan Artmann

See also Einstein, Albert; Einstein and Newton; Newton, Isaac; Time, Measurements of; Time, Relativity of

Further Readings

Galison, P. (2003). *Einstein's clocks, Poincaré's maps: Empires of time.* London: Hodder and Stoughton.

Greffe, J.-L., Heinzmann, G., & Lorenz, K. (Eds.). (1996). *Henri Poincaré: Science and philosophy.* Berlin and Paris: Akademie Verlag and Albert Blanchard.

Zahar, E. (2001). *Poincaré's philosophy: From conventionalism to phenomenology.* Chicago and La Salle, IL: Open Court.

Polo, Marco (1254–1324)

Marco Polo was a Venetian traveler and merchant who traveled to China at a time when Europeans knew little about Asia. Polo served the Mongol Emperor Kublai Khan for 24 years before returning to Venice. He attained great fame with the publication of his account of the journey, *Il Milione.* This book helped to inspire Europe's Age of Exploration.

Marco Polo was born in Venice, Italy, into a family of merchants. Venice was an important center of trade in the medieval world. Venetian merchants maintained trading posts as far away as the Black Sea, where goods arrived along the ancient trade route known as the Silk Road. Marco's father, Nicolo Polo, and uncle, Maffeo Polo, had traveled across Asia on a trade expedition around the time of Marco's birth. They traveled deep into China, which Europeans called *Cathay,* and they met the Mongol ruler Kublai Khan. The Khan received them very courteously and invited them to return again with Western scholars who could teach Christianity to his people. Nicolo and Maffeo Polo returned to Venice in 1269 and prepared for a second expedition that would include Marco.

In 1271, the Polos began the long journey to the Far East along with two Christian missionaries. The two missionaries became frightened while passing through Armenia and abandoned the mission. Seventeen-year-old Marco Polo, along with his father and uncle, continued the trip alone. It took more than 3 years of travel before the Polos again met Kublai Khan, after traveling much of the way on camels. Marco Polo, with his knowledge of four languages, became a valuable diplomat in the service of the Khan. He traveled on missions throughout Asia, and for 3 years served as a government official in the Chinese city of Yanh-Chow (Janguy). After 24 years away from Venice, the Polos finally decided to return home. The Khan was very reluctant to let them go.

The Polos arrived in Venice in 1295 and were scarcely recognized. They returned to a Venice that was at war with the rival city of Genoa. Marco Polo commanded a Venetian galley during the conflict but was taken prisoner by Genoa following a naval defeat. In prison, Polo wrote an account of his travels across Asia and his relationship with the great Khan. With the help of Rustichello of Pisa, Polo's story was translated into French and completed in 1298. Polo called his book *The Description of the World,* but the name *Il Milione* was more commonly applied. The book became very popular and widely distributed in a time before the advent of the printing press. The Polos were not the first Europeans to travel across Asia, but they became the most well known. By the time the Polos returned to Venice, the advance of the Turks had already cut off trade routes to China. Europeans would in time sail west to reach China and India. Following Marco Polo's release from prison in Genoa, he returned to Venice and became a wealthy and respected merchant. He never left Venice again and died at nearly 70.

Marco Polo's experience had a lasting impact on the contact between Europe and the Far East over time. Historians believe that his book influenced later explorers such as Christopher Columbus.

James P. Bonanno

See also Columbus, Christopher

Further Readings

Forman, W., & Burland, C. A. (1970). *The travels of Marco Polo.* New York: McGraw-Hill.

Hart, H. H. (1967). *Marco Polo: Venetian adventurer.* Norman: University of Oklahoma Press.

Larner, J. (1999). *Marco Polo and the discovery of the world.* New Haven, CT: Yale University Press.

Pompeii

Pompeii was an ancient city of the Roman Empire, located near the Bay of Naples in the Campania region of Italy, a few miles southeast of Mt. Vesuvius. Situated at the mouth of the Sarnus (modern Sarno) River, Pompeii was a popular resort town for wealthy Romans and a busy trade center. In the year 79 CE, Mt. Vesuvius erupted, destroying Pompeii, Herculaneum, Stabiae, and many smaller communities. Today Pompeii is one of the most important archaeological sites in the world. Most cities have many layers of occupation and stages of development, but the burial of this entire city under 30 feet of ash and debris has perfectly preserved for all time an ordinary day in the lives of ordinary people of the Roman Empire.

History of the City

The site of Pompeii was likely first settled by Oscan-speaking descendants of Neolithic people in the 8th century BCE. The strategic river location eventually came under the influence of Greeks who had settled across the bay. Pompeii and Herculaneum remained the center of Greek occupation until the 5th century BCE, when Samnites descended from the north and seized control. Rome drove the Samnites from the region in the 4th century BCE and claimed Pompeii. The city was permitted to keep its own language and culture, but its inhabitants were not granted citizenship or given any privileges. This situation was maintained for centuries, until the citizens of Pompeii had a chance to join with other rebels in an attempt to win freedom from Roman oppression. Sulla, a brilliant Roman general, eventually defeated the Campanians and took Pompeii and Herculaneum in 89 BCE. The rebels were granted Roman citizenship despite their defeat, but many liberties were taken away. Rome strategically housed army veterans in the area to maintain order.

The rich natural resources available in the area and easy access to the sea enabled Pompeii to flourish. As Rome became increasingly prosperous, the standard of living in Pompeii increased proportionally. The entire region along the coast of the Bay of Naples became a popular vacation destination for wealthy Roman citizens.

Mt. Vesuvius remained dormant throughout these many centuries of occupation. There was no way for Pompeians to know that they were living at the base of a volcano that had buried a Bronze Age settlement under 20 feet of debris directly beneath them in 1780 BCE. In the year 62 CE, a severe earthquake resulted in catastrophic damage to both Pompeii and Herculaneum. Repairs were still being made 17 years later when the city was destroyed.

The Eruption of Vesuvius

Roman science in the 1st century was based more on mythology than geology. Mild earth tremors were common and were not alarming to the population. Fumaroles—vents of volcanic gas escaping through the crust to relieve pressure—were seen from a distance and reported to be giants roaming the land. The gods would defeat the giants and trap them under the mountain, which would cause the earth to shake. There was no correlation made between seismic activity and volcanic activity.

Major seismic activity began on August 20, 79 CE. A series of quakes increased in frequency over the following 4 days. Animals were restless, and springs near the mountain ran dry. In the middle of the day on August 24, Vesuvius suddenly erupted. A huge column of superheated gas, rock, and ash blasted straight up from the top of the mountain. As the cloud cooled, it spread and drifted with the wind. A vivid eyewitness description of the initial explosion exists in letters written by Pliny the Younger, who witnessed the phenomenon with his uncle from 15 miles away. As the cloud began to collapse, the sky darkened and pieces of pumice rained down on Pompeii. This "Plinean" phase of the eruption lasted about 18 hours. The initial fall of ash, rock, and pumice deposited more than 9 feet of debris on the town. A few unfortunate individuals were struck down by falling stones or collapsing roofs, but there were few deaths on the first day. It has been estimated that perhaps 80% of the population escaped the region by leaving immediately. The 2,000 or so people who decided to take shelter and wait for the event to pass were the unlucky ones.

The true lethality of the eruption occurred in the final few hours of the second day. Surges of pyroclastic material and superheated gas roared down the mountain, reaching the city of Pompeii in minutes. There was no escape. All remaining residents were asphyxiated in seconds. A series of surges deposited an additional nine or more feet of ash and rock, effectively entombing the town. The complete and instantaneous burial of the town protected it from looting and natural weathering. When dawn broke on August 26, all was quiet and still. The entire region of Campania was buried under a thick layer of ash. The eruption changed the course of the river and raised the sea beach, masking the original location of the city. Over ensuing years, the memory of Pompeii and its neighboring cities faded into obscurity.

The Rediscovery of Pompeii

By the 4th century CE, Pompeii no longer appeared on maps. The area was called Civitas. The ash that had buried the region over 200 years earlier had become fertile soil. Farmers moved back into the area and planted their grapevines and olive trees. In 1709, some well diggers unearthed marble. The prince in charge of the region was building a new villa nearby, so the workers continued to strip the marble until the villa was completed. The marble was from the façade of the theater in Herculaneum. In 1738, King Charles III took control of the region and sent workers to the same location in search of additional treasure. Systematic studies and surveys of the area were carried out. Work began in the vicinity of Pompeii in 1748. An inscription identifying the site as Pompeii was uncovered in 1763. The methodical, deliberate work at these sites in the mid-18th century can be considered the origin of the field of archaeology.

Much of the earliest digging at both cities was often haphazard and careless. Workers were unskilled or were treasure hunters. Unauthorized digging was halted in 1860 when Giuseppe Fiorelli, an Italian archaeologist, became director of the excavations at Pompeii and Herculaneum. Pompeii was divided into regions, which were carefully cleared and documented. Fiorelli realized that there were cavities in the ash with bones in them. He took a revolutionary step in analyzing human remains at Pompeii. Rather than removing the skeletons, he had workers carefully fill the cavities with plaster. After the plaster dried, the volcanic debris was chipped away, leaving a cast of the person's dying moments. By studying the individuals, their dress, and the possessions they carried, researchers could reconstruct their last few minutes of life.

A museum was eventually established at Pompeii to house artifacts and dozens of the casts that were created. Pompeii became a mandatory stop on the grand tour of Italy that was so common among the European aristocracy. The richly colored murals that were exposed triggered a revival of neoclassical art. Well-appointed British homes often included an Etruscan salon, with décor copied from Pompeii.

While the public buildings of Pompeii are impressive, it is the private homes that set this site apart. The fine ash that smothered everything proved to be a remarkable preservative. Archaeologists have uncovered jars that still have food inside. Eighty-one loaves of bread were still in the oven of a bakery. Graffiti is intact on the walls of buildings. Shops and restaurants still have everyday tools in context. City planning and actual land use can be evaluated. Pompeii provides a unique and important source of information on many aspects of daily life during the Roman Empire. Much is written about the wealthy senators and citizens of Rome, but the vast majority of the Empire actually consisted of smaller towns and villages. Pompeii provides a snapshot of social, economic, religious, and political life of ordinary citizens of the ancient world frozen at a single moment in time.

Jill M. Church

See also Anthropology; Archaeology; Dating Techniques; Fossils and Artifacts; Museums; Rome, Ancient

Further Readings

Coarelli, F. (Ed.). (2002). *Pompeii.* New York: Riverside.

Deem, J. M. (2005). *Bodies from the ash: Life and death in ancient Pompeii.* New York: Houghton Mifflin.

Grant, M. (1987). *Cities of Vesuvius: Pompeii and Herculaneum.* New York: Macmillan.

Panetta, M. R. (2004). *Pompeii: The history, life, and art of the buried city.* Vercelli, Italy: White Star.

Popper, Karl R. (1902–1994)

Karl R. Popper was a British-Austrian philosopher and theorist of science. He had significant influence upon 20th-century philosophy with his contributions to epistemology, philosophy of science, and social theories. His central works include *The Logic of Science* (1934), *The Open Society and Its Enemies* (1945), and *The Poverty of Historicism* (1957). This last work contains an explication and criticism of historical scholarship as it was being practiced. Popper argued that such scholarship possessed a metaphysical determinism that assumed that time had a definable shape, such as "progress." He rejected sociology's historical determinism that portrayed time itself as a vector of a teleological conception of history.

As the founder of critical rationalism, Popper challenged the methods of empirical science and the inductive approach of logical positivists such as the "Vienna Circle" philosophers devoted to Moritz Schlick. Because observation in itself is insufficient for establishing scientific validity, the critical rationalists proposed the "principle of falsification." In contrast to the classical empiricism that strove for perfect validity, critical rationalists argued that science is doing its best work when it puts forward theses as statements that are capable of being proven wrong. Despite any number of arguments, assertions can never be entirely verified. The conclusion is that scientific knowledge does not improve by positive evidence. Instead, according to the principle of falsification, the point is to prove all the false hypotheses wrong by finding compelling counterexamples. "Verification" is a mistaken notion. Theories that withstand scientific scrutiny are "more valid," according to Popper, and falsifiability is a much better criterion of demarcation between scientific and nonscientific theories. In the area of social philosophy, he likewise demonstrated the invalidity of "truth claims" for ideologies, be they sociohistorical or political theories. Preferring a motif of indeterminism, his political ideas merged into a liberalistic pleading for a pluralistic, liberal, and democratic "open society."

Popper's scholarly work focused mainly upon what scientists are doing when they are doing "science," and his reflections on methodology mark the central starting point of his thinking. Critical rationalists argue that any scientific theory is abstract and conjectural by nature, making it difficult to identify secure knowledge. Therefore, genuine scientific theories require a specific logical structure:

openness to testability and revisability. By being structurally open to a trial-and-error process, that is, by being falsifiable, new conceptual theories are far more effective than dogmatic assertions of truth by an establishment. Falsifiability provides both a new approach to the challenge of ultimate justification and a method of distinguishing metaphysical statements from theories of scientific validity and reliability. In this way, we see how Popper's critique of historicists is a methodological one.

In *The Poverty of Historicism,* Popper's critique goes against approaches that make historical predictions their main objective and thereby attempt to grasp universally valid patterns of history. Because historicism argues that historical events underlie inexorable laws of development and advance toward an ultimate and discernable end, historicists strive to empirically establish an authoritative chronology of events. Given this position, sociology becomes a mere science of theoretical history in which time assumes an orientating function. At the same time, historicists hold that the method of universalization is not transferable to the social sciences. As a result, social laws must be structured differently from common generalizations. Thus, to be valid for all human history, such laws have to transcend the succession of time. These epoch-spanning laws of historical development now provide the basis for large-scale forecasts that can be identified as prophecies and distinguished from technological predictions. While the former refer to unchangeable events and characteristically serve forewarning purposes, technological predictions, which Popper prefers, are of constructive value, entailing useful instructions or guidelines for dealing with situations most likely to come.

According to Popper, a holistic conception of society that follows unique evolutionistic laws of succession is scientifically unsupportable. Holists mistake tendencies and singular theorems for universal laws. But, because each scientific law must be verifiable empirically, and because the laws-of-development hypothesis and evolutionism are both confined to observing unique events, these theorems cannot get universalized in terms of laws and scientific predictions.

In order to specify his objections to historicist methods, Popper criticized the interpretation of time as a criterion for the elevation of "static laws" to "connective laws" akin to Auguste Comte's and John Stuart Mill's distinction between laws of coexistence (time-independent) and laws of succession (time as a constitutive coefficient). Indeed, laws can contain statements about dynamic processes, but they can never point beyond individual cases or subsume singular, basically distinct causal connections under one rule. Assertions about periodicity, such as the circle of the year, are bound to incorporate/consider a number of single laws that cannot be abstracted, such as those of the earth's rotation, thermal inertia, etc. For this reason, the perception of dynamic succession or development can be due only to metaphorical illustrations of certain tendencies, not to scientific laws.

Over and above this, Popper points out that historicist doctrines provide a theoretical breeding ground for authoritarianism and totalitarianism. First, he refers to Plato's ideal state as resulting in socially immanent and decadent tendencies by maintaining a strictly hierarchical form of government, stifling all change for the sake of political stability. Second, Popper refers to Hegel's dialectic progression of history as being directed toward a perfection that, in practice, results in the glorification of the Prussian monarchy. Third, he argues against the Marxist economic oversimplification of social spheres.

Rejecting such forms of collectivism, Popper proposes the autonomy of the social sciences and a method of analyzing historical events by reference to the behaviors of individuals. Unlike holistic approaches, a "technological" social science recognizes the incalculability of human characteristics. It allows both for the logic of particular circumstances and the unintended side effects. Learning from this methodological individualism, political institutions can be designed in a way to prevent incompetent leaders from wreaking havoc. Similarly, Popper advocates what he calls "piecemeal social engineering." Instead of large-scale utopianism, social and political planning has to proceed stepwise in consideration of human fallibility and the irreversibility of time. By emphasizing the value of democracy as the nonviolent generation and deposition of political leadership, Popper was a contributor to democratic theory from a functionalistic angle. Because of these ideas, Popper strongly defends an open society that is capable of monitoring social phenomena and minimizing misery to enable its members to take part in a broad landscape of critical discourse.

Christiane Burmeister

See also Critical Reflection and Time; Epistemology; Law; Marx, Karl; Plato; Russell, Bertrand

Further Readings

Catton, P., & Macdonald, G. (Eds.). (2004). *Karl Popper: A critical appraisal.* London: Routledge.

Magee, B. (1968). *Karl Popper.* Tübingen, Germany: Mohr.

Miller, D. (1994). *Critical rationalism: A restatement and defence.* Chicago: Open Court.

Popper, K. (1956). *The open society and its enemies.* Princeton, NJ: Princeton University Press.

Popper, K. (1960). *The logic of scientific discovery:* London: Hutchinson.

Popper, K. (1999). *The poverty of historicism.* London: Routledge.

Posthumanism

See Transhumanism

Postmodernism

In few areas is it more problematic to arrive at a clean definition than it is for postmodernism. Not the least of the problems is the resistance of postmodernists to being defined, or, in many cases, to admit the authority or even the possibility of definition as an activity. Before attempting a definition, a series of useful distinctions can be drawn. However, postmodernists are very critical of scientific theories and empirical evidences. Consequently, they do not take seriously conceptions of time in terms of evolution and relativity (among other temporal frameworks).

Postmodernity and Postmodernism

The first important distinction to be made is that between postmodernity and postmodernism. Postmodernity is a name given to a period of history, and postmodernism is the body of theory that has developed to explain that period. Opinions differ about when postmodernity is supposed to have begun, with dates ranging between 1968 and 1973. All agree that conditions in the world have changed since then. In international politics, the sites of authority have fractured from the relatively straightforward conflict of the Cold War to the multipolar, less predictable, and more confusing international situation of today. Other fundamental changes have been made to our styles of work, with entirely new industries, work hours, and arrangements than were the case before. The assumption underlying the notion of postmodernity is that the world has seen not simply changes in style but a fundamental shift in the way the world operates, a shift that has been to the disadvantage of predictability, order, and rationality.

Those who argue for 1968 as a convenient date for the onset of postmodernity point to the student riots of that year, when the post–World War II baby boomer generation, the best-fed, best-educated generation in world history, spurned the cultural conventions they grew up with and demanded change. These revolts did not demand a specific set of political, social, and economic changes; they demanded *change,* as a general rejection of the old. The impact of the baby boomer generation is very significant, but even it was swept along by the broader transformation taking place after the events of 1973. This was the year when the international economy changed forever as a result of the oil shocks brought on by a newly radicalized Organization of Petroleum Exporting Countries (OPEC). The oil shocks were provoked by the oil-producing countries, most of which were from the Muslim world, wanting to express anger at the continued support by the West of Israel, which had been decisive in the recent Yom Kippur war. This marked a fundamental change in the balance of economic and political power and a dramatic shifting in priorities in global politics and economics. The widespread social changes that have occurred in the West are largely a product of these developments. Postmodernity can perhaps best be dated, therefore, from 1973.

It is important to note that "postmodernity" as discussed here is nothing more than a title given to a period by historians, like "Renaissance" or "Dark Ages." Others have called this period Late Capitalism. Still others have rejected the idea that the events of 1973 mean we can no longer speak of "modernity." This debate becomes a question about the periodizing of history and the usefulness of such a procedure.

Postmodernism as Reactions to Postmodernity

We have seen, then, that postmodernity and postmodernism are not related. Postmodernism is, as the suffix implies, an "ism," a series of reactions that have arisen to explain the times known to some as postmodernity. It is quite consistent to see value in postmodernity as a historical term while also rejecting the postmodernism that has arisen as a series of reactions to explain postmodernity. Most introductory essays to postmodernism begin with the insistence that any attempt at a comprehensive delineation and understanding of postmodernism is inherently contradictory and bound to fail. To overcome this problem, most characterizations of postmodernism portray it not as a coherent body of doctrine but as a general mood, or, as described in the previous paragraph, a series of reactions. What is more, the series of reactions are often expressed in negative terms.

Postmodernism is almost entirely confined to universities and literary circles. And within universities, some faculties are more likely to be favorable to postmodernism than others. Most branches of the sciences are hostile, whereas departments of English or French literature, architecture, visual arts, and sometimes anthropology and sociology are more likely to react favorably; opposition to postmodernism in philosophy departments, particularly in English-speaking countries, has grown since the mid-1990s.

In addition to these divisions between faculties, there is also a division based on geography. We have already noted this division among philosophy departments at universities. There are two general strands to the postmodernist reactions to postmodernity. One strand of postmodernism is content to *acknowledge* the changes that have taken place since 1973 and to try to understand them. The other, more radical strand is more willing to *celebrate* those changes. Broadly speaking, this fissure divides postmodernists in the Anglo American academic community from those from the Continental (mainly French) academic community. This is made more complicated when we note that this division is by no means exclusively a geographical one. The two styles of postmodernism, while having a geographical dimension, are principally divided by a series of contrasting emphases with respect to some important questions. The first of these dividing points centers around their attitudes toward philosophical analysis. Anglo American postmodernists may be cautious about analysis as a tool but still acknowledge its value. Postmodernists of the Continental variety, by contrast, are more inclined to view analysis as part of the problem they are looking to overcome than as a useful tool to gather new and better understandings.

A second major area of divergence can be seen in the attitudes toward metanarratives. Most postmodernists, whether Anglo American or Continental, see postmodernity as a time when traditional metanarratives broke down. A metanarrative is a general intellectual framework through which we view history. For example, the communist metanarrative is of a progressive, and inevitable, succession from feudalism, through capitalism to socialism, and, in the future, on to communism. A Christian metanarrative, by contrast, proceeds from a primitive innocence in Eden, followed by the Fall and the condition of sin and death among humans, relieved by the redemption of Jesus Christ, and which will be brought to an end when Christ returns to reward the faithful and judge the damned. The most oft-quoted definition of postmodernism comes from the French thinker Jean-François Lyotard, who defined it in terms of incredulity toward metanarratives. The differences lie in their contrasting attitudes to the end of metanarratives. As noted above, most postmodernists from the Continental tradition are likely to celebrate the end of metanarratives. They claim that not only are metanarratives no longer needed, they are no longer welcome. For with the end of metanarratives comes the end of the tyranny of history, and, more particularly, the end of the presumption of scholars to be chroniclers of historical fact. Instead, postmodernists argue, we have a mass of assertions and claims that can all compete for our attention as best they may, with little difference in the outcome of one being chosen over another.

Underlying this hostility to metanarratives is the postmodernist hostility to what they see as the dominating and bullying nature of science and reason. The central claim of postmodernists of the Anglo American strand is that rationality is a historically conditioned faculty, which means that different historical epochs will produce different notions of rationality. This is not a point many opponents of rationality would dispute. Indeed, many of the people derided by postmodernists, such as the philosopher Bertrand Russell, made these points long before postmodernism was an intellectual force. Others, usually of the Continental strand of postmodernism, are more extreme and seek to condemn what some call "legislative reason" as being inherently domineering and oppressive. In contrast to this, they say, postmodernism is a liberation from rationality, because it is a liberation from all socially constructed bounds and norms. As one prominent postmodernist put it in 1992, postmodernism is about splitting the truth, the standards, and the ideal into what has been deconstructed and what is about to be deconstructed and denying in advance the right of any new doctrine, theory, or revelation to take the place of the discarded rules of the past. David Harvey, in an influential postmodernist work published in 1991, enthused that postmodernism wallows in the fragmentary and chaotic currents of change, as if that is all there is. In a similar vein, Gregory Bruce Smith identified postmodernism as fundamentally a sign of disintegration, transition, and waning faith in what he called the Enlightenment project.

In defining postmodernism, then, we need to take account of these two strands of interpretation. The more moderate version of postmodernism is the assertion that we cannot *know* anything; we can only interpret, and any interpretation we make can only express our partial and narrow perspectives. The more radical version of postmodernism is the assertion that not only can we not know anything, but also our claims to knowledge are bound to be hegemonic and impatient of dissent, which means that our ability only to interpret from our partial and narrow perspectives constitutes a liberation from that tendency toward hegemony, which is particularly evident in science and rationality.

Behind the differences of emphasis and geography, there is a common political heritage for much of postmodernism. This is because postmodernism is a phenomenon of the political left. The reason postmodernists are anxious to write the obituary of metanarratives is that many of them were, earlier in their lives, proponents of variations of socialism and communism, which, as we have seen, have strong metanarrative components. This can be traced back to the disillusionment of German socialists after the failure of the Spartacist revolt in Berlin in 1919. Not only did socialist revolution fail, but also the parties of the left were then seen to fail as well, as they showed themselves incapable of resisting the rise of Nazism. By their slavish following of the dictates of Moscow, the parties of the left had betrayed their purpose. The school of Marxist-inspired criticism, known as the Frankfurt School, traced this progressive failure of each institution that had once been a source of hope.

After the Second World War, the Soviet Union joined the ranks of gods that failed. Between the Soviet invasions of Hungary in 1956 and Czechoslovakia in 1968, the promise of communism as a metanarrative of progressive liberation from alienation became ever more difficult to sustain. The choices of the left after 1968 were stark: acknowledge that capitalism and the metanarratives that sustain it had triumphed, or create a system of criticism that denied the legitimacy of any metanarratives at all. Postmodernism adopted the latter approach. The differences of emphasis between the Continental and Anglo American styles of postmodernism were most pronounced on the question of the degree to which Western culture could be salvaged. Critics like Theodor Adorno (1903–1969) and Max Horkheimer (1895–1973) were among the more pronounced cultural pessimists. In *Dialectics of Enlightenment* (1947), Adorno and Horkheimer saw modernity as inherently self-destructive, carrying the germs of its own dissolution and decay. There was a straight line, they argued, from the Enlightenment to Nazism and Stalinism. Later heirs of this high level of pessimism included the Polish-born sociologist Zygmunt Bauman (b. 1925), who argued that modernity is a long road to prison, and Michel Foucault (1926–1984), who saw all society as engaged in a war against people to impose on them a sterile understanding of what it is to be human.

The more moderate postmodernists in the Anglo American schools, however, were not prepared to go this far. Their general argument was to see value in the humanist tradition of the Enlightenment, merely wanting to excise from it the faults and excesses they perceived it brought with it. Advocates of this approach include the English philosopher Michael Luntley and the New Zealand sociologist Barry Smart.

Postmodernists agree on their suspicion of metanarratives, diverging only on the degree of opposition they are prepared to express. Another area of general agreement is in their attitudes toward science. Anglo American and Continental postmodernists agree that science involves a metanarrative, which they question. The metanarrative of science speaks of cumulative knowledge over time, leading to ever clearer understandings of the universe. And, just as important, postmodernists question the claim of science to being grounded in objective truth. But the attitudes toward science reveal the greatest splits between the Anglo American and the Continental styles of postmodernism. The Anglo American attitudes are best articulated in the works of Richard Rorty (1931–2007), whose critique has been directed against foundationalism, the practice of claiming that one's theories rest on objective foundations of truth. The Continental branch of postmodernism, once again, has wandered much further afield in its attitudes toward science. Indeed, it is this area that has provoked the strongest opposition. Working from an epistemic relativism, many Continental postmodernists have pressed the claims that science is a metanarrative like any other, with no special claims to authority, and indeed, meriting extra levels of criticism precisely because of those claims of authority they perceive scientists making.

The Decline of Postmodernism

Postmodernism began to attract some significant critical attention after about 1994. Much of the criticism was led by scientists and by philosophers who take science seriously. Leading the way was a devastating critique by Norman Levitt and Paul Gross, *Higher Superstition: The Academic Left and Its Quarrels With Science* (1994). Other significant milestones included *Derrida and Wittgenstein* (1994) by Newton Garver and Seung-Chong Lee, *The Poverty of Postmodernism* (1995) by John O'Neill, and *The Flight From Science and Reason* (1996) edited by Paul Gross, Norman Levitt, and Martin W. Lewis. But what publicized the decline of postmodernism was what has become known as the Sokal hoax. In 1996 Alan Sokal, a professor of physics at New York University, submitted a paper to the postmodernist journal *Social Text*. The paper, titled "Transgressing the Boundaries: Toward a Transformative Hermeneutics of Quantum Gravity" supposedly demonstrated that the laws of science are nothing more than social constructs—one of the central points of Continental postmodernism. The *Social Text* editors made the mistake of not submitting the article to the usual process of peer review, and on the day the article appeared, Sokal announced in the journal *Lingua Franca* that the article was a hoax.

The hoax and the controversy that followed was a major embarrassment for the editors of *Social Text* and for the wider claims of postmodernism to academic credibility. Sokal followed up his hoax with a scorching attack, coauthored with the Belgian physicist Jean Bricmont, on French postmodernism. The work was published in the United States under the title *Fashionable Nonsense: Postmodern Intellectuals' Abuse of Science* (1998).

Alongside this counterattack, some of the more moderate Anglo American thinkers who at one time had expressed support for elements of postmodernism have since stepped back. In Britain, Christopher Norris, a prominent defender of the works of Jacques Derrida, has more recently launched important critiques against the relativism inherent in postmodernist thinking. And in the United States, the late Richard Rorty, without having changed his position, expressed his regret over having used the term "postmodernist." Most recently, a newer movement called critical realism has arisen as the successor to postmodernism. The intellectual initiative has clearly now passed to the critics of postmodernism.

Bill Cooke

See also Derrida, Jacques; Enlightenment, Age of; Epistemology; Humanism; Metanarrative

Further Readings

Bauman, Z. (1993). *Intimations of postmodernity.* New York: Routledge.

Docherty, T. (Ed.). (1993). *Postmodernism: A reader.* New York: Harvester Wheatsheaf.

Gross, P., Levitt, N., & Lewis, M. (Eds.). (1997). *The flight from science and reason.* New York: New York Academy of Sciences.

Jencks, C. (1996). *What is postmodernism?* London: Academy Editions.

Lopéz, J., & Potter, G. (Eds.). (2001). *After postmodernism: An introduction to critical realism.* New York: The Athlone Press.

Lyotard, J.-F. (1988). *The postmodern condition: A report on knowledge.* Minneapolis: University of Minnesota Press.

O'Neill, J. (1995). *The poverty of postmodernism.* New York: Routledge.

Smart, B. (1992). *Postmodernity.* New York: Routledge.

Sokal, A., & Bricmont, J. (1998). *Fashionable nonsense: Postmodern intellectuals' abuse of science.* New York: Picador.

Waugh, P. (Ed.). (1994). *Postmodernism: A reader.* London: Edward Arnold.

PREDESTINATION

Predestination is the idea, found within Islam, Hinduism, and Christianity, that human actions are predetermined by a divine being or superhuman force. This is not to be confused with foreknowledge, the idea that God knows outcomes before they occur. Foreknowledge leaves the decision to the individual, God merely knows the result beforehand. In predestination, God makes the decision as to the outcome of a situation; the individual has no choice.

Islamic thought contains the doctrine of *qadar.* In this doctrine, God (Allah) is understood to be all-powerful and to have control over all that happens. Allah must allow or even cause events to happen; no action occurs apart from the will of Allah. In this sense, predestination occurs when Allah decides to act, with resultant effects on individuals. Hinduism, with the idea of one's *karma* (actions) resulting in the determination of future lives, places responsibility on the individual; nevertheless, gods can act with resultant effects on humans, which is understood as an effect being predetermined for an individual. Thus predestination occurs when a god acts preemptively upon an individual.

In Christian theology, the most prominent example of predestination is found within Calvinism. This doctrine is named after John Calvin, a Protestant Reformation leader of the city of Geneva in the mid-1500s. Calvin was not the first Christian, however, to promote the idea of predestination. In the 300s the Christian theologian Augustine promoted the doctrine that God must make the initial step toward a relationship with a person. This was attributed to original sin (introduced when Adam and Eve disobeyed God in the Garden of Eden) having removed the option of individuals making a move toward God. After the fall of Adam, God's initiating grace was necessary for an individual to move toward God. In addition, God does not choose to give this grace to all, but only to the "elect," those individuals he has chosen to receive his grace. Pelagius, a contemporary of Augustine, developed the counterdoctrine that each person has free will to accept or decline Christ. These two thinkers wrote against each other's doctrines, and Augustine's view eventually prevailed. Pelagius was subsequently denounced by councils and popes. Yet Augustine's idea did not gain a strong following in Augustine's lifetime nor in the Middle Ages. It did not receive much attention until the Protestant Reformation in the 16th century.

Calvinism became closely associated with predestination in part through Calvin's writing about it in his most famous work, *The Institutes,* and partly through the Academy of Geneva, which drew Christians from throughout Europe who were interested in learning about Calvin's ideas.

Although Calvin originated a variety of Christian doctrines, his name is most closely associated with predestination. In addition, Christian predestination is associated with the acronym TULIP, which reduces the doctrine of predestination and Calvinism to five points: Total depravity of humanity, Unconditional election, Limited atonement, Irresistible grace, and Perseverance of the saints. Total depravity of humanity holds that all people are unable to please God through their own volition. Unconditional election states that God chose who would become a Christian and who would reject Christianity; the individuals had no role in the decision. Limited atonement is the idea that Christ's atonement (sacrifice to bring people to God) only included those chosen to become Christians; Christ's death on the cross did not include those God did not choose to become

Christians. Irresistible grace carries the idea that God's decision is irrefutable; those God choose cannot choose not to become a Christian. Perseverance of the saints is the idea that those God will ensure the perseverance of those chosen to be Christians.

While the idea of TULIP is associated with John Calvin, the acronym was not developed until decades after Calvin's death. The situation that spurred the development of TULIP was a debate in Holland over the predestination ideas of Calvin and the free will (the idea that each person chooses to accept or not accept Christ) idea of Jacobus Arminius. After Arminius's death, his followers reduced his free will ideas to five points. Later, at the Synod of Dort in 1618, the adherents of Calvin's idea of predestination countered by reducing Calvin's predestination ideas to five points: The acronym TULIP was the result.

Among Christian denominations today, the Presbyterians, reformed churches, and some Baptist groups accept the idea of predestination. Those who hold to predestination have differing ideas of when God made the choice of individuals. The idea that God made the choice before the fall of Adam and Eve is known as *supralapsarianism.* The idea that God made the choice after the fall of Adam and Eve is known as *sublapsarianism.* The Synod of Dort upheld the sublapsarianism doctrine. In comparing the doctrines of predestination and free will, they are known more specifically as double predestination and single predestination. Double predestination denotes the idea that God chose each person to either be a Christian or a non-Christian (double); single predestination denotes the idea that God predestined the way of Christ, and each person can either choose or reject Christ.

Mark Nickens

See also Calvin, John; Causality; Christianity; Determinism; Islam; Predeterminism; Teleology

Further Readings

Articles of the Synod of Dort. (2005). Michigan Historical Reprint Series. Ann Arbor: University of Michigan Library. (Original work published 1618)

George, T. (1988). *Theology of the reformers.* Nashville, TN: Broadman Press.

PREDETERMINISM

Predeterminism is, literally, the idea that events are determined in advance. All future actions and events are already decided or known (by God, fate, or some other force). This concept is closely related to determinism, and sometimes the terms are used interchangeably. More often, however, predeterminism is categorized as a specific type of determinism. In this case, determinism is defined more generally, as the idea that every event is caused by a chain of past events, without a specific being or force behind it.

Ancient Views

The idea of future events being already decided can be found in many cultures and religions throughout history. One early representation of this view is demonstrated by the Fates (also called the Moirae)—three goddesses in Greek mythology who controlled everyone's life span. As written by the early Greek poet Hesiod, each goddess's name indicated her purpose. Clotho was the spinner, who spun the thread of a person's life. Lachesis was the apportioner, who decided the length of the thread (thus determining the length of the life). Atropos was the inevitable (literally, the "unturning"), who cut the thread at the point where the person was to die. The Moirae acted independently of the other gods, and in some stories even Zeus was powerless against them.

A predeterministic view of the world was espoused by the Essenes (a separatist sect of Judaism) who settled in the Qumran region (near the shore of the Dead Sea) between 110 BCE and 70 CE. According to their beliefs, God determined everything that would happen in the universe before he created it. In accordance with God's plan, there were two ways: that of light (good) and that of darkness (evil). These two ways, which exist throughout the universe, are in constant conflict, which is held in control by God.

Christian Theology

The idea of God being all-knowing has created some questions in Christian thinking, for God's foreknowledge would seem to deny humans their free will. Although humans have knowledge of the world, their knowledge is limited and fallible. According to most monotheistic religions, God is infallible and knows everything (even the future) with certainty. This would seem to create a paradox if one were to believe in both God's foreknowledge and the idea of human free will.

Fifth-century philosopher and theologian St. Augustine of Hippo wrote that because God is omniscient, he therefore knows every future action that is going to happen. We cannot behave in any way other than the way God knows we will. However, Augustine felt that these ideas were not a threat to free will. In essence, God knows about future events because they are going to occur; they do not occur because God knows about them. Our future actions do not become involuntary because of God's foreknowledge, just as our remembering a past event does not make it involuntary. God is eternal, so he sees all of time in a way similar to the way we view the present. The terms *before* and *after* cannot be applied to him.

Other theologians concurred with St. Augustine's view that God's knowledge of future events does not take away from man's free will. Seventeenth-century American theologian Jonathan Edwards wrote that God's foreknowledge does not cause events, but it does make them inevitable. Saint Thomas Aquinas stated that God's knowledge is not in itself a cause of anything in the same way that a sign points to an object without actually causing the object to appear. Regarding the idea of God causing specific events to happen, he wrote that God does not intervene in the world in the sense that he enters into an occurrence of which he was not originally part. God is not absent from anything in the world to begin with, and therefore everything that happens does so because God made it that way.

Despite his agreement with Saint Augustine, Aquinas also wrote "On Evil IV," which includes 24 arguments against the view that human beings have free will. Yet he says that God working in all things is not incompatible with freedom. The human will can act, but the action is initiated by God. Although the action does originate outside of the person, the action is not forced upon him or her.

Martin Luther, who began the Protestant Reformation, felt that God didn't merely know about future events, he controlled them. He wrote that for humans to have the freedom to performing their own actions would be incompatible with God's omnipotence, and to believe such a thing was blasphemy. As human beings, all of our actions are evil unless we have the help of God's grace. It is not within our power to do good, and we are controlled by either God or Satan, who determine our actions.

One way to reconcile the seemingly opposing viewpoints of God's omniscience and human free will is to view future events simply as possibilities. God knows all the possibilities, and humans, by their choices, determine which possibility will occur.

Another possibility is that God can change his mind with regard to future events. At certain points in the Bible (such as Jonah 3:1–10 and Jeremiah 26:17–19), God first stated that he would destroy a city or a group of people but later relented due to human actions.

Jaclyn McKewan

See also Augustine of Hippo, Saint; Determinism; Hesiod; Luther, Martin; Predestination; Teleology

Further Readings

Boyd, G. A. (2001). *Divine foreknowledge: Four views.* Downers Grove, IL: InterVarsity Press.

Hasker, W. (1989). *God, time, and knowledge.* Ithaca, NY: Cornell University Press.

Presocratic Age

The conventional term *Presocratic* refers to philosophical and natural scientific thought in ancient Greece prior to the philosophical career of Socrates; hence the Presocratic period denotes the earliest stages of Greek philosophy. Years of this period began with the oldest recorded fragment of philosophical speculation in Greece; Thales reached his acme, or prime of life, in 585 BCE. This is the most natural beginning point for this period of early philosophy, which was still inseparable from myth. Nevertheless, one major focus of Presocratic thought was the concept of time. Socrates began his philosophical mission in around 434 BCE, in his 35th year when his friend Chaerephon returned from Delphi, which would be one sensible endpoint of the Presocratic age. While Socrates philosophized, there were still around him followers of Parmenides, Heraclitus, Pythagoras, and Democritus, among others.

Time in Presocratic Greece

In the Presocratic world of the Greeks, as across ancient civilizations, the sky was the universal timepiece. The time length from sunrise to sunrise was the natural basic unit of recording time. Each day was divided into 12 parts, as was the night, which would change in length by the season. Time within the daylight hours would have been told by sun dial; time across day and night would have been measured by a terra cotta water clock (*clepsydra*).

The ancient Greeks used a lunar/solar calendar, using both sun and moon, of 12 months. Each month was a lunar cycle. In addition, there were alternate months with 29 or 30 days and alternate years of 354 or 384 days. Extras days or even months were inserted in the calendar to prevent accumulated error. Nautical Greeks navigated and estimated time during the night by the movement of the constellations, planets, moon, and stars. Each city-state kept its own calendar, maintained by magistrates, which began at the new moon. Athenians began the year in summer, and it ran from July to June. Years were marked off into 4-year cycles, called *olympiads*. The first year of Olympiad 1 was 776 BCE. The cycles of the Milky Way would have been the final, limiting hand on the cosmic clock (by repeating its four-fold pattern in its own time). There were also planetary conjunctions and eclipses recorded. Astronomers used such cycles to correct accumulated error in the calendar.

Early Greeks had a circular notion of time; at the beginning of time was a primordial event that began the subsequent cycles of time. Human beings celebrated the passing and coming-to-be of cycles with festivals that harkened back to the first days (*in illo tempore, ab origine,* as Mircea Eliade put it). And indeed many months on the Greek calendar were named after festivals. Time was thought of as a cycle of festivals rather than as an abstraction. The early Greeks adopted much from Egypt, including calendar making. Greek civilization had gone through the Bronze Age during the Homeric period, but their folk religions bore witness to a distant Neolithic past, as well. Olympian sky gods had displaced mother earth goddesses and agricultural gods along with the Titans.

Greek Myth of Time (Kronos)

First-born in the long genesis of the world was Chaos, according to Hesiod, then "wide-bosomed Earth" as a dwelling for the gods. Tartarus, lower hemisphere of the future cosmos, was born next, and after it, Love, "most beautiful by far." From Chaos was born "dark-robed Night," which gave birth to Upper Air and Day. Born first to Earth was Heaven, "eternal dwelling place" for the blessed gods. Earth and Heaven were of one form initially, but separated later in the act of birthing. Heaven and Earth created children, who were sequestered away by Heaven, who had grown to become a wicked father. Earth conceived a plan and made a scythe of flint for one child, Kronos, to use in ambushing and castrating Heaven. Kronos threw his father's male member into space, but its scattered blood created on Earth the first generation of gods. Thus the child Kronos took his father's place in the cosmos. Acting on a prophecy of his own doom, Kronos ate each of his children born of Rhea, lest they overthrow him. But Zeus was hidden from Kronos by a stratagem, and he grew to overthrow his father. Kronos was forced to disgorge his children, who then battled the Titans, defeating them and ostracizing them to Tartarus. This introduced the age of the Olympians.

Herodotus and History

No survey of the Presocratic period, especially with regard to conceptions of time, would be complete without mention of Herodotus. With the *Histories* by Herodotus, the Greeks attained a historical sense of time. Events in the story of mankind gave a linear shape to time; in his *Histories*, there is a succession of more or less remarkable events, which makes up a more or less cogent narrative, and which undid the circularity of time so prevalent in the mythological mind. Herodotus was born in 484 BCE and died between 430 and 420 BCE.

Presocratic Philosophy

Eminent scholar Kathleen Freeman has listed nearly 100 Presocratic thinkers, of whom mostly mere names survive. The best-known and best-documented philosophers of this period include Thales, Anaximander, Anaximines, Pythagoras and members of the Pythagorean school, Heraclitus, Parmenides, Xenophanes, Zeno, Empedocles, Democritus and Leucippus.

Thales

Thales was an obscure figure about whom almost nothing is known. His acme was around 585 BCE, and that marks the beginnings of the Presocratic period of philosophy. He apparently wrote a text on nautical astronomy, but only the title survives. Attributed to him, as well, were "On the Solstice" and "On the Equinox." If he indeed wrote a work on nautical astronomy, he would have been able to read the night's sky as a navigational and timekeeping device perhaps expertly. And since Thales is said to have predicted an eclipse, he was observing patterns in time that few then understood.

Concerning the famous saying attributed to Thales—all is water—we might speculate that this was an early natural scientific reduction of all phenomena to water, either figurative or literal. This accords with the notion of a Milesian school of natural explanation. But Thales is also supposed to have said all things are filled with gods, which, in either a figurative or literal sense, is not at all natural scientific.

If applied to the topic of time, the alleged sayings of Thales would imply, first, that time is water; second, that time is filled with gods. If Thales believed in Kronos, he could say that time is one of the gods, but we have no evidence for this conjecture, either way. More likely is Aristotle's reading that soul is intermingled with all things, because it is the principle of motion. This would entail that soul is the principle of time as well.

Thales might have suggested that time is water, because time was measured by a terra cotta water clock, at least for some ancient Greeks, or by the tides. But this is a speculative reading, as well. Hesiod reported that Thales held Earth to float upon water. Now if Thales suggested that all things begin in water, that the universe *came from* water, as a first principle, which is the likely meaning according to Aristotle, time would be measured by water in a truly cosmic sense. Whatever foundational ideas Thales might have had about time are lost to us forever, however.

Anaximander

Anaximander came from Miletus 25 years after Thales, but not even mere titles of his works, if he wrote any at all, survive. Only a few fragments of his ideas remain. Fragments 1, 2, and 3 deal explicitly with the nature of time. The indefinite is the "original material" of existing things, according to Fragment 1.

In addition, "the source from which existing things derive their existence is also that to which they return at their destruction, according to necessity." The reason for this is that "they give justice and make reparation to one another for their injustice, according to the arrangement of time." Fragment 2 tells us that the indefinite is ageless and everlasting, while Fragment 3 calls the indefinite immortal and indestructible.

And so the image of the cosmos that we infer here is that at the origin of all things was an indefinite out of which came all the existents of the cosmos. While in existence, these existents right a wrong, the nature of which is hotly contested and is determined by the exact rendition of the passage one uses (some saying that Anaximander means existents repay a debt to each other, some saying he means that existents repay a debt to the indefinite). We are told by the fragment that time and necessity arrange this retribution. Then presumably measured time ceases, all things return to the source, and eternity continues on unabated.

Now if this interpretation is correct, then Anaximander has drawn the distinction between *eternity* and *measured time*. It would further seem to indicate that time is both endless (indestructible) and measured. Time and necessity sort out the eternal from the original. The rest is measured time. Further, measured time seems finite (as an accidental feature) and *for the sake of justice* (as its essence). If this is correct, then the entire drama of existence seems to be a morality play; we, as existents, have chosen to affirm the I and negate the Thou. Measured time begins only with our self-separation from the indefinite, and it exists only for the sake of restoring order and establishing justice, at which point all measured time ceases and eternity reigns supreme again. Perhaps the earth remains for eternity, because it is like a stone column, for Anaximander, but he may have been referring only to its structure or its appearance to the gods. In the absence of authoritative works, one may only speculate about possible interpretations; this is the risk run by saying anything substantial about the Presocratic thinkers.

From Fragment 1 we may legitimately infer that Anaximander subscribed to what we might now call a temporal realism, meaning that he treated time as a real thing and as eternal and indestructible. What is far less clear is how he further conceived of time: Is time merely reified, or did Anaximander also deify and worship it? If so, did he anthropomorphize it? Is Kronos meant here, in the naive mythical sense, along with Dike and Ananke? It seems incorrect, however, to read Anaximander in terms of alien religions; we may only say that he believed in necessity, time, and justice, as well as the indefinite.

Anaximines

Anaximines is known to have been from Miletus, hence he was presumably another Milesian natural observer, yet next to nothing, a single sentence, survives of his one writing. He was in his prime of life around 546 BCE. His observations that the soul is air, and that air surrounds the whole universe, give us no solid insight into his notion of time. If we may attribute the ideas of compression and dilation to Anaximines, however, then time would be measured most objectively by those physical natural processes.

The problems of change and motion were the major forum by which the Presocratics dealt with time. Change and motion were problems to the Greek mind. They sought a mechanism of change whether it be water, air, fire, or other. How do generation and degeneration come about? Coming-to-be and passing-away? Some thinkers (Anaxagoras, Parmenides, Zeno, the Pythagoreans) believed that mind, not a process of nature, accounted for change and motion. But they all seemed to strive to explain the phenomenon of change.

For the Presocratics, the question of time came down to developing a concept of *origins*. They still believed in the first times of mythical consciousness (*in illo tempore, ab origine*), as recorded in Hesiod and Homer, but reworked into an early natural scientific idiom. Later scientific notions, therefore, should not be anachronistically attributed to the Presocratics.

Pythagoreans

The figure of Pythagoras is largely apocryphal, but he was a religious reformer and cult leader. He was born on Samos but moved to Croton, perhaps in Olympiad 62. His place in the sequence of Presocratic thinkers is especially difficult, and the later Pythagoreans may have come considerably after the earliest members of the community. A conservative range for the acme of Pythagoras would be between Olympiads 62 and 69 (between 528 BCE and 500 BCE). Thus his prime of life was very close to being directly after the acme of Parmenides and Heraclitus.

Pythagoras refused the friendship of a local tyrant, Cylon of Croton, and the religious leader and his followers were cast out from Croton. They resettled in Metapontum, but the Cylonians followed and apparently massacred the majority of Pythagoreans. The survivors further resettled in Rhegium and Greece. The last Pythagoreans were active around 366 BCE. The ancients seemed confused about the relation between Zeno and Empedocles, on one hand, and the Pythagoreans, on the other. Some sources claimed that Empedocles was a follower of Pythagoras; others said he followed Parmenides. These sources also maintained that Zeno followed first Parmenides, then Pythagoras.

Pythagoras was apparently the first Greek to deify number, that is, he held number to be sacred and worthy of study as a universal ethical force. Number is self-moving and thus the principle of the psyche. Pythagoras said that happiness comes from knowledge of the "perfection of the numbers of the soul," and his followers took the study of number as their single concern. Their study encompassed arithmetic, geometry, music (harmonics), and astronomy. As Pythagorean scholar David Fideler has analyzed their curriculum, arithmetic was the study of number in itself; geometry was number in space; music was number in time; and astronomy was number in space and time. As a consequence, time was seen as number in harmonic motion, or rhythm. Number was considered immanent, not transcendent, to time.

The Pythagorean account of time would probably have involved a number of separate and perhaps conflicting elements. Time was number in harmony, of course. But Diogenes Laertius reported that Pythagoras divided the human life span into segments of 24 (childhood, youth, middle age, and old age), which corresponded to the proportions of the four seasons. And all accounts of the life of Pythagoras include a notion of reincarnation, that is, a sequence of lifetimes, which may have formed a circle (the *monad*) in its full development.

In the Pythagorean table of opposites, "one" and "plurality" are opposites, as are "at rest" and "moving." The one is time at rest, or eternity; its opposite is time as it moves, or fluid time.

Among Pythagoreans, the Tetraktys, a triangular figure made from arranging dots as the first four integers, was considered to be holy ("heaven") and to be the universal order (Kosmos and Pan). Pythagoras held number to be the source and first principle of all things. The monad (unity) was seen as the basis of all number and not a number itself. The dyad contains movement, growth, birth, and other aspects suggesting time in motion. Indeed, the dyad is also Rhea, wife of Kronos. If the monad is the producer of time (just as it is called the "space-provider"), then the dyad is time in motion. Both types of time, monad and dyad, together produce the triad, which is "the all" and "everything."

"Time, and wind, and the void" define where each thing exists, said Pythagoras. But time was also said by Pythagoras to be the sphere surrounding the world. Perhaps we may think of local and global time as an analogy; there is macrocosmic and microcosmic time. The tetrad, the square circumscribed by a circle, contains the nature of change. The integer 10 was called the *decad*. One of its aspects is "eternity (*aeon*)." The decad returns to the monad and shares its qualities. "All things accord in number," said Pythagoras, and his followers elaborated arithmetic and geometric evidence for his insight.

As time is an elemental part of the universe, and as time is number, it follows that time is divine. "Turn round when you worship," was a Pythagorean maxim meaning that we should "adore the immensity of God, who fills the universe." Hence, God transcends time but is also immanent to it.

Since the Pythagoreans ascribed all discoveries to Pythagoras himself, we cannot, in principle, discern the founder from his followers in terms of teachings. Some later Pythagoreans may have believed in circular time. According to Stobaeus, Philolaus, one of the last Pythagoreans, said that "the single world is continuous, and endowed with a natural respiration, moving eternally in a circle, having the principle of motion and change; one of its parts is immovable, the other is changing. . . . The one is entirely the domain of mind and soul, the other of generation and change." This echoes the fragment of Anaximander, of course. Much later, though, Friedrich Nietzsche ascribed a version of eternal recurrence of the same to the Pythagoreans.

Timaeus

One supposed Pythagorean astronomer was Timaeus of Locri, though the figure may well be legendary. Looking back into time immemorial, Plato's Timaeus said, "The sight of day and night and the months and the revolutions of the years have created number and have given us a conception of time, and the power of inquiring about the nature of the universe. And from this source we have derived philosophy, than which no greater good ever was or will be given by the gods to mortal man." Timaeus further said, "The night and the day were created, being the period of the one most intelligent revolution. And the month is accomplished when the moon has completed her orbit and overtaken the sun, and the year when the sun has completed his own orbit." Timaeus told his audience that God created the sun and moon, as well as five planets, "in order to distinguish and preserve the numbers of time." As for the stars,

"Their wanderings, being of vast number and admirable for their variety, make up time." But this cosmology is more likely Platonic than Pythagorean.

Religious and Ethical Implications

We should never forget the religious and ethical consciousness of the Pythagoreans; they were united as a community by a set of rules of behavior, however esoteric. Every idea from Pythagoras had an ethical implication. There is an ethical meaning to time. Because mind becomes more aware of number, there is a gradual enlightenment of the individual soul and the society. The universe is the setting for a spiritual transformation, which must be seen as immanent to time, not a mere set of accidents. There is a Logos immanent to time, and a sort of wisdom.

Time is also a beautiful harmony; it is the tempo of the universe. Twenty-three hundred years later, Karl Ernst Ritter von Baer (see entry) suggested that if the rate of human perception were greatly increased, we would hear the orbits of the planets as music. There is a beautiful, higher, nobler understanding of time for the Pythagoreans, because time itself is number, and hence divine. To study number is the same as to study time and divinity.

Heraclitus

Life

With the figure of Heraclitus, we have the earliest Greek thinker with a rather developed philosophy of time. He reached his acme around 500 BCE. Heraclitus almost certainly wrote a book, but his fragments come down to us only through testimonies and traditions. Those fragments are themselves curious puzzles of a hermit's outlook on the ancient world. Though only 126 fragmented passages remain, most of them confront the problems of time, change, motion, and coming-to-be. Fragments from Heraclitus evidence his astronomical knowledge, which entailed an ability to tell time from the stars.

Heraclitus believed in necessity, justice, and time, as did Anaximander. The process of nature contained within itself a deeper Logos, for Heraclitus, and there was a distinctive flavor of morality in his view of the world. Character is destiny, for man. To the extent that we systematize and rationalize Heraclitus, we lose him. His worldview was deeply personal and emotional. His darkest sayings are probably impenetrable by the modern mind. But his metaphysical and religious beliefs do make some sense within the larger Greek cultural milieu.

Heraclitus taught a metaphysic of absolute nonpersistence in change, which he took, figuratively and/or literally, as Fire. Heraclitus may have meant that all things are fire (a "first principle," like Thales's water), or that a ring of fire encircles the remainder of the universe (a most inclusive thing, like Anaximines's air). Or Heraclitus may have seen fire not as the genesis but as the final goal of the universe. Everything must dry up and burn to ash due to the cosmic fire.

As a theory of time, then, Heraclitus's thought denied the ultimate reality of being. The universe burns away like a great bonfire over the long night. What exists solely is becoming; the energetic spark of the instant in time when the power of the universe actualizes itself, and just as soon is gone. There is no smallest uncutable particle in space and likewise no ultimate atom of time. So far down, or up, as one may go in looking, there is only change, not being. There is no ultimate individual or collective. This is as true of cosmic time as it is true of the moment. Heraclitus grasped, for himself, the great truth that all things must pass, and thus discovered the conflagration that is time.

Perhaps the deepest, darkest truth that we get from Heraclitus is that "passing away" comes to all, great and small, and the universe, ultimately, becomes a pile of dust. "This ordered universe, which is the same for all, was not created by any one of the gods or of mankind, but it was ever and is and shall be ever-living Fire, kindled in measure and quenched in measure." And "the fairest universe is but a dust heap piled up at random." Hence Fire is the beginning and final fate of the universe, a great Fire that slowly burns itself out. There is, for Heraclitus, a light that guides the universe and makes meaningful the events of the past, his Logos. It is lightning-like in its enlightening power. It illuminates truth and makes knowledge possible. The deeper question must be, what is the meaning of this cosmic Fire, if its end is only to burn out?

There is a bolt of meaning blazing in the universe, a deeper Logos evident for all to see, but incredibly, no one is willing to see it, according to Heraclitus. Souls are best preserved in a dried body, but man finds pleasure in fluids. Some souls even become soggy and moldy. They cannot see the truth, because they do not look for it. To find meaning, one must look for it. Yet, paradoxically, when we look for meaning, we find something else. "If one does not hope, one will not find the unhoped-for, because there will be no trail leading to it and no path."

Fools cannot see the reason that guides the cosmos, but some reason (Logos) gives meaning to the random pile of dust, according to Heraclitus. Hence his darkest moments give way to a dawn of insight and meaning. These dark sayings are balanced with Heraclitus's conviction that a Logos permeates the universe, and it is open to our access, if only we look for it. Logos (meaning) is everywhere in abundance.

Heraclitus was a hermit, but he was also a warrior in spirit, who saw *ares* (war) as the first principle of existence. War is the father of all things, perhaps in that the cosmos began in strife. Fire, war, Logos, misanthropy; these came together in Heraclitus.

Absolute Nonpersistence

Absolute nonpersistence in change, perhaps the core notion of Heraclitus, is evidenced by the famous saying, variously interpreted and translated as "Those who step into the same river have different waters flowing ever upon them," "It is not possible to step twice into the same river," or "In the same river, we both step and do not step, we are and we are not." This image of a river suggests time as a flow, and in a single direction; hence, the notion of an arrow of time. But the river may well be thought of as the flow of motion through space; there are no eternally existent elemental parts; the entire process of the universe in change absolutely does not persist; each smallest particle, as well, is itself consumed by change and ceases to be.

"The sun is new each day" suggests that each day is a whole relative unto itself, and that there is no highest order of time period. And so a human epoch would be an instant to a god; the size of the sun is the breadth of a man's foot. No two moments are alike, nor does time seem to repeat itself, if this saying be taken literally. Day and night are one; there is no essential difference between days. Hence the quality of time is uniformity, perhaps an indefinite nature, too.

Time Is Fire

Time is a transformation in things from one element to another, continually (re)mixed. Yet fire somehow rests from change in man. What marks off day from night is the sun. Equinoxes and solstices of the sun mark off the seasons, "the hours that bring all things." And the sun stays within the solar ecliptic, "otherwise the Furies, ministers of justice, will find him out," according to Heraclitus. The sun is no wider than a man's foot, perhaps, it might be added, from a man's earthbound vantage point. But surely the sun, seen much closer, would be enormous. Hence (apparent) size is relative to vantage point. Or perhaps he meant to suggest a macrocosm-microcosm relationship—that there may be universes containing our own at indefinite orders of magnitude, in which our own sun would be as wide as a man's foot. . . . But what is the sun, if not fire?

Time is not merely *like* a universe-sized conflagration slowly burning out, it *is* exactly that. Time is the fiery consumption of the universe even until man and earth perish, as measured by the rate of conflagration. All manner of corruption, degeneration, passing away, are manifestations of fire, however minute. Time is itself an elemental force like water, air, and earth or metal, but destructive, corrosive, annihilating of the others. Hence fire ultimately turns all to ash. It dries up the water of Thales. It uses up all the air of Anaximines, which purportedly surrounded the universe. It burns up all the earth and wood and melts all metal until only ash remains. All the elements are exchanged for fire, and then fire yields to the various other elements. Perhaps at this time the entire selfsame affair begins again. Or perhaps he meant that there is a cycle of fire similar to a nitrogen cycle. These fragments may have required initiation into a mystery or esoteric teaching to become comprehensible.

Reciprocal Time

Sleep and waking, life and death, youth and old age, are measures or periods of time that Heraclitus used to show the reciprocal nature of time. Opposites gradually develop into each other. Sleep gives way to waking, which eventually gives way to the former. There are different temporal phenomena in reciprocal states; distortion of time in dreaming, the quickened pace of time in old age, and so on. Each might be said to run according to its own unique clock. Hence, we may reasonably read Heraclitus to mean that time is epicyclical change.

The basis of Heraclitus's ontology is the notion that existents "scatter and again combine" and "approach and separate." These processes mark time or even constitute it. But they are coals in the wind, slowly burning out, like the universe at large. There is a relativity in sequence between the points of a circle in space. Surely it occurred to Heraclitus that points in circular time would be equally relative. Every moment would be the beginning and end of time, consequently. And hence, every day is the same, and astrology, consequently, is fundamentally faulty.

Fragment 52

"Time is a child playing a game of draughts; the kingship is in the hands of a child." This simile is Heraclitus's most explicit word on time, as distinct from "change" or "universe." Time is a child playing because it plays at random, not by strategy and tactics. Hence a child playing at draughts would be moving the pieces without rules and only at whim. Time is the kingship at stake. Why is time a kingship? It is the very being of life, its fire.

Heraclitus is said to have spurned the company of adults and played games with children. But he did not romanticize youth; rather he only believed adulthood to be worse. In another fragment, he shed some small light onto how he envisioned children in charge of a kingship. "The Ephesians would do well to hang themselves, every adult man, and bequeath their city-state to adolescents. . . ." So time has been handed to the least capable out of default by miserable, failed beings, or so it would seem, if the parallel is sound. How is it consistent, though, to say that time is random or worse, and yet the cosmos (which, after all, is only the spatial-material complement to that same time) is guided by Logos? This is evidently another of Heraclitus's antinomies, which will end by saying that they are both random and not random, both according with Logos and not according with it.

Parmenides

Parmenides was born around Olympiad 53. Parmenides was at his acme in his early 60s, at the same time that his antipode, Heraclitus, was also in his prime (as was Pythagoras). Parmenides was one of the most influential of the Presocratic thinkers. At roughly 20 years of age, he was instructed by Anaximander, who at that time was at his acme, or prime of life. The metaphysics of Parmenides took from Anaximander the distinction between being and nonbeing, the distinction between eternity and measured time, and the distinction between the unlimited and the limited. Anaximander apparently considered all six terms as real. Parmenides, however, identified being and truth with eternity and the unlimited, and identified nonbeing and illusion with measured time and the limited. Few known thinkers in the Presocratic world had doubted the testimony of the senses until the figure of Parmenides. And Anaximander's cosmology of three distinct spheres nested in each other was amenable to Parmenides' vision of a spherical "one."

When he was 30, Parmenides met the itinerant Xenophanes of Colophon in Elea. They shared a religious mysticism and a love of poetry. Xenophanes criticized polytheism and anthropomorphism, and Parmenides was a metaphysical monist and a mystic. Close in time to these events, Melissus of Samos became a student of Parmenides. Though the particular dialectic employed by Melissus differed from that of Parmenides, the conclusions were the same: Being is one, eternal and unmovable. Melissus denied time, using a dialectic of negation, like the master, but evidently without achieving a novel contribution, as did Zeno, the better known student of Parmenides. And thus the school of Parmenides at Elea was founded.

Parmenides and his school at Elea evidenced three influences: Anaximander, Xenophanes, and the Pythagoreans. The students of Parmenides may have been influenced by a Pythagorean named Ameinias. However tempting it may be to see in Pythagoras's monad an equivalent of Parmenides' "one," this seems unlikely. Rather than distinct cosmological or ontological doctrines, the similarity between the schools of Elea and Croton is one of lifestyle. The Parmenidean way of life probably resembled the Pythagorean way in that both were lives of religious mysticism and contemplation.

The "Prologue" by Parmenides, which has survived apparently complete, is an exquisite example of inspired philosophy, putting the audience at a spiritual crossroad as spellbound spectator. And the subsequent metaphysical teaching is crystal clear: Being can only be a timeless unchanging thing, perfect and eternal. "For there is not, nor will there be, anything other than what is, because indeed destiny has fettered it to remain whole and immovable."

Because "thought and being are the same," and because "it is necessary to speak and to think what is; for being is, but nothing is not," Parmenides concluded that only reason can bring us to the truth. This reason is the immortal higher path on which Parmenides found himself at the beginning of his "Prologue," and to which he was brought by the "daughters of the sun."

Nonbeing cannot be thought, and therefore is unreal. Nor can we know change; coming-to-be and passing-away are delusions of the senses. What is changeable is temporary and unknowable. Change is unreal, and hence also any notion of measured time. The apparent omnipresence of change is an illusion created by the senses, according to the teachings of Parmenides. Many humans believe the testimony of the senses, but that path is one of darkness and mere opinion. And so Parmenides adopted the Pythagorean opposites of unlimited/limited; one/many; rest/motion; and light/dark. But he identified the unlimited, one, rest, and light with being, while he identified the limited, many, motion, and dark with nonbeing and opinion. In this manner Parmenides collapsed the Anaximandrian and Pythagorean tables of opposites; they are mere names, for him, rather than categories of being.

Anaxagoras

Anaxagoras of Clazomenae wrote one book, and his acme was around 460 BCE. He was born in 500 BCE and came to Athens at 20 years of age. He philosophized in Athens after fleeing the Persians and was active at the same time as Empedocles, with whom he shared many ideas. Anaxagoras set about his mission early in life and pursued it in Athens, perhaps for 50 years, at which time he was forced into exile on charges of "impiety." His life mission was to create a natural history of the heavens. He adopted Anaximines' notion of stars as burning objects fixed on a wheel, but advanced it by suggesting that they are masses of burning stone or metal, which can detach and fall to Earth (as an explanation of the comet at Hellespont). He suggested that there is an ether surrounding the universe and that by the force of the rotation of this ether, vast stones are pulled into orbit, set afire, and recognized as heavenly bodies. Without a clear concept of momentum, Anaxagoras conceived of this force as speed. He hypothesized that Earth is closer to the moon than the sun. He compiled facts on eclipses and suggested that the light of the moon came from the sun and not the moon itself. He hypothesized that the sun is larger than the Peloponnesus, and that the moon has ravines and plains and is made of earth. He even suggested that the Milky Way comprised the light of many stars, which are shaded from the sun by the earth.

Anaxagoras had a clear notion of elemental entities, called *homeomeries,* which are originally mixed, but then separate off in a winnowing motion, like unto like. This primal mixture is a chaos of countless elements, because the elements are actual qualities rather than abstract categories of qualities (an infinite number of seeds in no way resembling one another). In contrast, Empedocles waxed poetic about this primal union. They shared the proposition that elements always have been and always will be. This implied an infinity of the past during which all elements still existed, and in precisely the same number. And it is logically consistent to say that these elements existed in a primal confusion only if there is a directionality of time, and a sense of a *point of origin.* Infinite time is consistent with directionality, but infinite time seems paradoxical with a sense of primal origin. Time, for them both, was an awkward mixture of cyclical and linear notions.

Anaxagoras was evidently very concerned about drawing the metaphysical implications of his vision; a set of propositions of great/small, mixing/separating, and infinite/finite compose several of the fragments, but so far as we may judge, there is no driving dialectical discipline behind it. Indeed, the account of separation told by Anaxagoras is a tissue of contradictions that would drive a careful logician to distraction. Anaxagoras, we might say, saved Zeno the work of reducing his position to contradiction. This must suggest that he was only partially concerned with systematic consistency and far more concerned with communicating a vision of the world and how it operates at its farthest and highest reaches, presumably, the enlightened natural scientist's view.

By emphasizing mind, Anaxagoras was still something of an echo of Anaximander. Empedocles used mind as an explanation but sought to account for the motions of the heavens more than the motives of love and strife. He studied the heavens as Socrates studied Logos, or as Empedocles studied earthly nature: methodically, minutely, patiently, and with an open mind. Yet, by including mind, Anaxagoras surely implied that there is a purpose or design to the universe. Otherwise, why would mind be necessary, if we are explaining the orbits of masses of molten metal? The essence of mind, in contrast to the elements, is purpose. The universal mind had a purpose, presumably, behind the mixture and separation, the formation of orbits, and the rise of life on heavenly bodies, though the fragments spare few details.

As for Earth itself, it is flat and hovers on air it has compressed underneath itself. Thus Anaxagoras required the notion that air was constrained (as in a clepsydra) and tremendously pressurized under the weight of Earth. Unfortunately we lack a fragment explaining *what* constrains the air under Earth, and so on. Here Anaxagoras may have relied on the analogy, drawn by Parmenides, between Earth and a pillar (but once again, what holds up the pillar?). Such analogies did little in the face of Zeno. Most notably, Anaxagoras could not seem to decide whether the whole is one or many.

Zeno

Life

Melissus of Samos was an important follower to Parmenides, but beyond question, the most famous pupil was Zeno of Elea. Conflicting stories have been told of Zeno having been a resistor against tyranny and a victim of torture, but facts are scarce. If Plato's account in his dialogue *Parmenides* is not fictionalized, then a young Socrates heard a middle-aged Zeno and an elderly Parmenides hold forth. (But since so much scholarship has rested on that passage, if Plato's account in *Parmenides* is just good fiction, as one alternative careful chronological calculation suggests, then a much different sequence to the Presocratics must be conceived.) One source suggested that Zeno wrote his book in his youth, but even this is uncertain. The traditional date of his acme is 450 BCE. Almost nothing trustworthy is known of Zeno's life, but his enormous contribution to Western thought, especially logic, could hardly have been greater, for Zeno should be recognized as the discoverer of dialectical thinking.

Unfortunately, only fragments from Zeno's literary production survive. It is part of an ongoing debate about Zeno as to whether he wrote one or more books. And there exists no reliable way to reconstruct his arguments into a single logical system, though it seems likely that there was such a scheme. Testimony about Zeno of Elea comes from Plato's *Parmenides* and Aristotle's *Physics,* yet a complete image of Zeno the logician must always escape us. Reportedly, Zeno had some 40 dialectical syllogisms (*epicheirêma*) to indirectly support various Parmenidean teachings. He began each dialectical syllogism with a hypothesis such as "things are many," and then he reduced the assumption to absurdities. Because the initial hypothesis resulted in a contradiction, we are justified in concluding its opposite proposition. In short, he assumed the proposition of his opponent and showed that it yielded a contradiction, and then he concluded his own proposition, a technique called "indirect proof" or *reductio ad absurdum* by later logicians.

A Sample Epicheirêma

In Fragment 3, Zeno argued that if things are many, they must be both (a) finite and (b) infinite in number:

(a) If there are many things, it is necessary that they are just as many as they are, and neither more nor less than that. But if they are as many as they are, they will be limited.

(b) If there are many things, the things that are, are unlimited; for there are always others between the things that are, and again others between those. And thus, the things that are, are unlimited.

The conclusion to Zeno's dialectical syllogism is that if we assume that there are many things, the same thing will be both limited and unlimited. We may feel justified in assuming that Zeno would then have drawn the consequence that such a contradiction is absurd. And so the supposition which inspired the dilemma—that things are many—must be rejected.

Zeno's Paradoxes of Motion

Aristotle seemed to suggest that Zeno posited four principle paradoxes in the conclusion that motion does not exist. But how these paradoxes fit together with other parts of his logic is unknown.

1. *The Stadium.* Zeno argued that before a runner can traverse the stadium, he must reach the halfway point, and so on (presumably, that the runner must reach half way to the half way mark, and so on). But this would mean that the runner would have to cover an infinite number of points to traverse the stadium. Yet it is absurd to think he would cover an infinite number of points in a finite amount of time. Therefore the runner can never really traverse the stadium. Zeno's paradox of the stadium means to reach the conclusion that *motion does not exist.* It seems entirely likely that this notion would have led, by a dialectic of being and nonbeing, to the Parmenidean proposition that all things are one. Aristotle, the sole source, commented that Zeno ignored the difference between points *infinite in their extremes* and points *infinite in their divisibility.* Though a runner cannot traverse the first sort of infinity, he can traverse the second such infinity (and hence indeed traverses the stadium), Aristotle concluded, at least preliminarily. The runner presumably touches each point he passes over (i.e., the track), Aristotle worried, so if the track is infinitely divisible, it still poses a problem that he would cover an infinite number of points in a finite amount of time. Aristotle would finish his remarks in the *Physics* by saying, "In a way it is but in a way it is not" possible to touch an infinite number of points in a finite time.

2. *Achilles and the Tortoise.* Apparently Zeno proposed the paradox that Achilles would never be able to overtake a tortoise, even though he were moving at a much greater speed. Zeno arrived at the paradox by the same line of reasoning that he used with the stadium argument. That is, Achilles would have to reach the midway point between himself and the tortoise before he could overtake the tortoise. But by the same logic, Achilles would also have to reach the midpoint between himself and the midway mark, and so on. Achilles would never be able to touch an infinite number of points to get to the tortoise. And so, paradoxically, Achilles will never be able to overtake a tortoise.

3. *The Arrow.* Relying on Aristotle's account of the paradox of the arrow, Zeno started with the premise that anything occupying a place exactly equal to its own size is at rest. Zeno then added only one additional premise, that at any present moment, what is moving occupies a space precisely equal to its own size. Zeno concluded from these two premises that at any present moment, what is moving is at rest. To this conclusion Zeno added a new premise, that what is moving always moves at the present moment. Zeno drew his conclusion that what is moving is always at rest. Another source, Diogenes Laertius, told his readers, "Zeno abolishes motion, saying, 'What is in motion moves neither in the place it is in nor in one in which it is not.'" Thus, the arrow shot from a bow and seeming to travel at high speed is at rest!

Aristotle's evaluation of Zeno's arrow paradox proved highly valuable to subsequent philosophy of time. For Zeno's paradox of the arrow relied on the assumption that time is a succession of discrete *nows*. But Zeno was wrong, according to Aristotle, precisely because "time is not composed of indivisible *nows*, no more than any other magnitude." His idea was that a continuum, a line, for example, is not composed of indivisible points that somehow touch and make the line continuous. In short, there are no indivisibles in time, or no time atoms, because *time is not composed of points in a line.* The implications of this rejection of atoms of any sort would require centuries, and even millennia, to play out, though. For, if time does not consist of a succession of discrete *nows*, of what *does* it consist?

4. *The Moving Rows.* Zeno's fourth and final paradox requires the following diagram and legend:

AAAA

Δ BBBBE→

←ΓΓΓΓ

A	stationary bodies
B	bodies moving from E toward Δ
Γ	bodies moving from Δ toward E
Δ	starting place
E	goal

By Aristotle's rendition, Zeno presumed two equally spaced columns of men (B and Γ) marching in opposite directions in a stadium. One group marches from Δ to E, while the other group marches from E to Δ. They march at equal speeds past a set of stationary columns of men (A). Apparently Zeno then argued that the marching columns would pass each other twice as fast as they would pass an equal number of men in a stationary position. From this Zeno inferred that "the half is equal to the whole," meaning that the marching men were, paradoxically, moving both half as fast and fully as fast, at the same time. But this is absurd. Therefore, the marching men were not really in motion.

It seems likely that Zeno presumed that each marching row would be across each stationary column for an equal period of time, as Aristotle suggested. But Aristotle then went on to claim that Zeno's fallacy lies in ignoring that the amount of time required to pass half of the A's was equal to half of the time required to pass all the Γ's. But things are not so clear-cut, as Kirk, Raven, and Schofield brilliantly demonstrated: "Zeno needs only to get us to accept the plausible idea that if a body moves past *n* bodies of size *m,* it moves a distance of *mn* units; simple arithmetic will then show that moving *mn* units will take half the time of moving 2*mn* at the same speed." And so Zeno would have been able to show, after all, that the men are marching at full speed and half speed, at the same time.

Or Zeno might have answered Aristotle's criticism by assuming, instead of columns of marching men, indivisible bodies moving at a speed such that each would be across from a stationary body (another indivisible body) for but an instant. This would escape Aristotle's comments (for what would it mean that each B must pass each Γ in half of an indivisible time?) while raising interesting questions about space, time, and atomic bodies.

Evaluation of Zeno

Even as fragmented isolated arguments of a larger system of logic, what did remain of Zeno's work(s) has caused thinkers headaches for centuries, however, and Zeno's paradoxes of motion remain awe inspiring in a timeless way.

From the Parmenidean point of view, Zeno's project was highly successful. It showed the absurdities implicit in the hypothesis of motion, as our five senses testify to it. To combat Zeno's challenge, Aristotle attacked the rather commonsense notion of time as a continuum composed of discrete present moments. But this already gives the victory to Zeno, because he has proved his point of logic: Motion, as it is given to us by the senses, is not what it seems. Now the modern mind hastens to translate Zeno's paradox of marching rows into relativity theory. But once again Zeno has already won, for Einsteinian relativity defies common sense and the testimony of the senses, and so motion is not what it seems. And an additional devastating Zenonian point of logic came from Kirk, Raven, and Schofield: "If the distance a body moves is simply a function of its position relative to other bodies, is there any absolute basis for ascribing movement to *it* at all?"

Or to use the Parmenidean idiom, mortals fooled by the illusion of motion "wander knowing nothing, two-headed . . . carried along, deaf and blind at once, dazed, undiscriminating hordes, who believe that to be and not to be are the same and not the same."

Empedocles

Life and Influences

Empedocles of Agrigentum (Sicily) lived from 484 to 424 BCE, according to the traditional chronology, and was born into nobility. He was evidently an orator, scientific thinker, poet, and democratic figure in the face of tyranny. Regal in dress and imposing in manner, Empedocles built a legend around himself, claiming to have a golden thigh and to be divine. He wrote two poems, "On Nature" and "Purifications," both of which survive in fragments. Empedocles was in his acme around 450 BCE. His exact relations to the other Presocratics are unknown.

In the surviving fragments, he claimed to have been inspired by "much-wooed white-armed maiden muse." And he suggested that he had an elixir of life against old age; it could also bring the dead back to life.

As a general observation, Empedocles wrote, we come to be familiar only with what we immediately experience, and each of us takes our own sum of experience to be the totality of knowledge. But we learn little to nothing of the whole, and when we die, our knowledge dies with us. Only by the muse's inspiration, presumably, was Empedocles able to shake humankind out of this ignorance. And so Empedocles created his own version of the ageless philosophical distinction between ignorance and true knowledge and indeed gave voice to an ancient skepticism about human knowledge by painting humankind in a fallen state of knowledge.

What we can infer from the fragments of Empedocles, it seems, is a rather naive (uncritical) realism toward change and motion, which, like his natural science, is most reminiscent of Anaximander. A chronological question then raises its ugly head: Which, if either, was the teacher of the other? There is no philological way to settle this issue.

Yet there is no evidence of a clear distinction of eternity and measured time, as would be found in Anaximander's ideas, unless Empedocles' obscure "realm of piety" counts as eternity. Indeed, in Empedocles there is a world of change at the expense of a world of eternity. If so, this would have been an irreconcilable difference from Anaximander.

Some ancient authors reported that Empedocles was a Pythagorean, and there is significant evidence for the claim. While the surviving fragments from "On Nature" seem inspired by the natural science of Anaximander, those from "Purifications" suggest a strong Pythagorean influence on the ethics and religion of Empedocles. Allegedly, Empedocles publicized the secret doctrines of Pythagoras and was expelled, as was Plato. Or perhaps Empedocles was an itinerant member of both Pythagorean and Parmenidean communities for however long a period.

Many scholars see a parody of Parmenides' poem in the verses of Empedocles. (If Empedocles was reacting to Parmenides' challenge against motion, he should be placed contemporary with, or after, perhaps immediately after, Parmenides.) But other ancient authors claimed that Empedocles was lampooning Xenophanes. One story held that he jumped to his death into Mt. Etna.

Account of Nature and Change

Empedocles lived in his own world of the elements (earth, air, fire, water) but also the Olympians (Zeus, Hera, Hephaestus, Aphrodite, "sharp-shooting Sun and gracious Moon"), "Titan Aether," and the realm of piety, as well as the cosmic forces, love and hate. His world was further populated by an array of nymphs ("Sunshine, Speed, Delay, Beautiful, Ugly"), goddesses (Iris, Chthoniê), and the like ("lovely Infallibility" and "dark-eyed Uncertainty"), all anthropomorphized and personified in a rather naive way. Evidently, Empedocles identified Zeus with fire, Hera with air, Aïdoneus with earth, and Nestis with water.

These primal elements and forces work through the world by a "double process"; scattering (moving from one to many) and gathering (moving from many to one). This beginningless and endless scattering and gathering is the objective natural course of the universe. Thus Empedocles' fragments give evidence of a rather detailed observational natural science. His remarkable speculations about the origins of the species have resulted in him being considered as something of an ancient Darwin. But his natural science was always interwoven with the fantastic. So far as his geography went, Earth is most comparable to a "roofed cave," this term having, doubtless, an ethical meaning, as well. As to physiology, the blood around the heart, according to him, is "thought in mankind," put there by fortune. The body decomposes at death into the elements, he said, but Empedocles evidently also believed in metempsychosis, as did Pythagoras, and claimed to already have been a girl, a bird, and a "dumb sea-fish." The destiny of a soul can thus be said to be divided into a long sequence of lives.

Empedocles held that the elements never came into being and can never pass away, suggesting that *time is infinite both in the past and in the future.* More explicitly, in fragment 16 he referred to "infinite time," which will never be emptied of love and hate. In this sense he has broken with the ancient mythical notion of cyclical time in favor of a rather linear conception of time.

Time is an objective force of nature, not a figment of the human mind. If there were no "infinite time," all of nature would be at a standstill, forever. But the elements are real and pervasive, for Empedocles. These elements appear most closely related to what modern metaphysics calls "concrete universals." Thus Empedocles adopted a sort of naive realism about *time and motion as they appear to the five senses.* He did not doubt the testimony of his own eyes, gifts from Aphrodite, who, having built "tireless eyes" from the elements, "fastened them together with clamps of affection." His own unique prototype of qualitative pluralism implied that *time can be measured by the double processes of scattering and gathering*. Perhaps the transformations of individual elements would provide us, in theory, with elemental clocks, too, but that is speculative.

Unfortunately, Empedocles gives a rather contradictory account of *how time operates*. Some fragments suggest that time operates by strict necessity; there is an "oracle of necessity." Yet his account of the origin of species implied that time operates by chance, "for things can be mingled at random." Motion occurs without a mechanism (i.e., from the cosmic-force love) yet leads to mechanical results (events and things of nature). This paradox is typical of the worldview of Empedocles.

Significance

Now if Empedocles really was responding to the Eleatic school, he did not enter into a dialectical battle with Zeno, his contemporary; he did not, at least in the surviving fragments, directly confront Parmenides' theory that motion does not exist. He offered no defense of the senses vis-à-vis the arguments of Parmenides. Rather, he developed his own account of time and change—qualitative pluralism—in isolation; if he had serious rivals, he did not name them or hint at their theories.

Indeed, it may be correctly noted that Empedocles took the way of opinion, as described in the poem by Parmenides, but in an uncritical way, leaving him without having refuted or moved beyond the problem of motion.

The only way to interpret any remotely Parmenidean influence in the writings of Empedocles would be to take his elements as όντα (self-existing, eternal, incorruptible "thing")—a stretch, because Parmenides took there to exist only one όντα. Rather, it seems that Empedocles was most engaged in the thoughts of Anaxagoras. Anaxagoras took countless qualities to be elemental homeomeries; Empedocles allowed only four. But both believed that macroscopic objects comprise masses of microscopic particles.

In his atomic speculations, Empedocles went well beyond Anaxagoras. Now the absolute number of particles of each element in the universe, and hence the total amount of particles, remained finite and constant; now time operated by a conservation of matter and a constant motion. Similar notions of mixing and blending linked Anaxagoras and Empedocles, of course. With the introduction of infinite time in the past, there is no sense of mind as the author of an original motion in the sense of Anaxagoras, however (unless love and strife are theoretical remnants of mind; but why, then, are love and strife necessary, if mind suffices?). Empedocles' theory of effluences (microscopic openings in matter) allowed him at least the possibility of a remarkable insight into time and change at an atomic level. If time is matter in motion, then microscopic motion would require a microscopic scale of time. And his acceptance of a void in the effluences certainly second-guessed Anaxagoras and tested the waters for Democritus and Leucippus.

Indeed, Empedocles certainly set the stage for the subsequent atomists. Yet his insistence that love and strife are authors of motion is a hard doctrine to resolve with an atomistic understanding of motion, unless the atoms themselves are energized by these cosmic forces (an equally speculative claim).

Leucippus and Democritus

Many thinkers laid groundwork for the atomist school of thought in Athens. Leucippus and Democritus, both of Abdêra, were preceded by the so-called Milesian school of natural philosophy (Thales, Anaximines, Anaximander); the Pythagorean cosmology of number; the elemental ontology of Empedocles and Anaxagoras; and the dialectic of Zeno and Parmenides, along with their notion of όντα. Each source provided an idea or principle for constructing an elegant theory of atoms.

Leucippus reached his acme around 430 BCE, and Democritus did so around 10 years later. Leucippus may have coined the first terms or concepts of atomism: *atoms* (literally, indivisible), *great void, contact* (arrangement), *aspect* (position), and *rhythm* (form), among others. Leucippus wrote one book, *The Great World Order,* and possibly a brief work called *On Mind,* from which this important principle comes to us as fragment 2: "Nothing happens at random; everything happens out of reason and by necessity." The fact that an atomist would write anything "on mind" suggests he believed in a mind behind the cosmos of atoms and void, but we know no details.

Democritus may have plagiarized from Leucippus; either way, we should consider Leucippus as the earliest known atomist. Democritus produced an evidently enormous literary estate. His works were arranged into the categories of ethics, natural science, mathematics, music, technical works, and a long series of monographs on causes, all of which are known only by title in various fragments. According to fragment 5, in *Small World Order,* Democritus apparently attacked Anaxagoras as an original thinker and as a student of nature. Another work carried as its title *Pythagoras,* but nothing is known of it. It is certain that Democritus invented an astronomical calendar, and evident that he saw in nature a periodic cycle of seasons, migratory birds, weather patterns, (mis)fortune, and more. The fragments of his calendar that remain show a highly scientific mind working with a folk wisdom of unknown antiquity. We are told in fragment 12 that the "great year" of Democritus (and Philolaus the Pythagorean) was considered to be 82 years with 28 intercalary months. Democritus was, then, familiar with the problems of calendar making and timekeeping in general.

Fragments on Atoms

Perhaps the most famous declaration of Democritus is, "Atoms and void [alone] exist in reality." This first historical formulation of atomism remains readily understandable into the 21st century, even if the current idioms are quarks and space. Democritus had discovered, for the Greeks, the notions of elemental particles of matter and of empty space. He envisaged atoms as indivisible chunks of matter existing in space. These smallest particles, in great numbers, constitute the object of sensory experience, that is, macroscopic objects. Though they were microscopic, Democritus imagined his atoms to have many of the same characteristics as the macro objects; extension in three spatial directions; arrangement; position; and duration in time, motion, speed, and form. Thus the microscopic or microcosmic particles are the indivisible, smallest parts of a whole physical body. It was not until Sir Isaac Newton and his contemporary Roger Joseph Boscovich later came to question this extended particle ("corpuscular") assumption in the theory of atomism (though Newton retained his commitment to the existence of some extended matter in the universe) that such reasoning would become suspected of a fallacy of division.

Democritus's affirmation of a void goes far more against common sense than does his affirmation of atoms. Indeed, some sort of atomism makes common sense, but the notion of a void was roughly that of vacuum; common sense believes nature abhors one, and thus a void cannot exist. Earlier thinkers from Anaximines onward gave a metaphysical stature to air or vapor. And where space between bodies was considered by the Presocratics, it was considered to be filled by the ether. Even the effluences of Empedocles sealed up and left no void. And so the genius of Democritus becomes evident even in the paltry number of his fragments.

Atomism poses its own epistemological problem: If atoms truly are beyond our senses, how do we know their nature, characteristics, or even their actual existence? Democritus faced the epistemological challenge head on by allowing his theory to remain just that, a theory, rather than involve himself in contradictions. And so he wrote, "We know nothing accurately in reality, but [only] as it changes according to the bodily condition, and the constitution of those things that flow upon [the body] and impinge upon it." Applied to his own atoms and void, this meant that we cannot know atoms directly but only through their physical effects at a macroscopic level. Until the instrumental period of science, thinkers would be able to do little more than theorize about atoms.

Yet this, for Democritus, is ultimately true knowledge. For there are two types of knowledge; sensory knowledge, which he called the "bastard," and "one genuine" sort of knowledge, which is mind comprehending ideas. The encyclopedic knowledge Democritus evidenced in his literary corpus suffices to show his regard for learning and abstraction. But he denigrated the testimony of the senses, a tradition among such diverse men as Heraclitus, Parmenides, and Pythagoras. Our senses alone will only mislead us. And so it must have been something of a puzzle to him as to how to reconcile his blanket rejection of the senses with his own attribution of properties in macroscopic (sensory) bodies to microscopic bodies. An equally vexing problem could be posed; if Democritus is certain of the existence of his own mind in the process of discerning physical affects of atoms, then something more than just atoms and the void exist.

What seems certain is that Democritus envisaged *time as the motion of matter through space*. This is far more elegant than the vision of any predecessor. He discovered the dependence of knowledge on a specific level of being; "humanity is a microcosm." Certainly, the universe as a whole is the final macrocosm; but humanity finds itself to be a miniature universe unto itself. So much more so is the atomic world; it is a microcosm at a second remove, and the final one, for Democritus. Applied to time, then, we would expect an entirely different frame of reference for time at micro- and macrocosmic levels.

Thus there must have been an uncertainty principle built into his theorizing about atoms, as about everything else: "We know nothing in reality; for truth lies in an abyss." Nor can we know causality, even in a world of physical bodies; "I would rather discover one cause than gain the kingdom of Persia," said Democritus. These epistemological concerns about natural science made Democritus a timeless philosophic and scientific figure.

Atomic Time and Zeno's Paradox

Democritus discovered the notion of the smallest, indivisible particle. He envisaged the atom as extended, and thus likely also considered a line to be composed of points with no space between them. If this is so, then he likely also believed time to be composed of a succession of discrete present moments. This would all mean that Democritus would have been susceptible to Zeno's paradoxes in both space and time. In short, Democritus was dependent on Euclid's conception of space and time, but therein lay the paradox. Almost certainly Democritus referred to Parmenides and Zeno when he wrote, "It has often been demonstrated that we do not grasp how each thing is or is not." But no fragment offers a hint of reply.

Again, the problem Democritus faced was due to his notion of corpuscular matter, and likely a notion of discrete present moments. The solution here, as with Zeno's paradox of the moving rows, would have been to envision the atoms as dimensionless points, not extended in space but emitting forces or powers. Unless Democritus took that route, though, he would have been saddled with Zeno's paradox, for his atoms would not be able to move, and the void could never be crossed. In fact, he would have been unable to show how time ever began in the first instant. To leave corpuscular atomic theory, though, would have meant crossing over to point-particle theory, a move we have no evidence he made.

Zeno's paradox is ultimately mathematical in nature, and the discovery of the calculus of infinitesimals finally silenced him, but only many centuries later. And so calculus was of no help to Democritus. What Democritus did achieve was to create the scientific type—entirely liberated from mythology and metaphor—within his theory and his own person. After his appearance, science would continue a relentless, if halting, progress. In the meantime, Socrates and his followers, including the young Plato, were already taking their places on the stage of philosophy, and thus the so-called Presocratic period of Greek thought came to an end.

Greg Whitlock

See also Anaximander; Anaximines; Cosmogony; Cronus (Kronos); Eliade, Mircea; Empedocles; Heraclitus; Herodotus; Mythology; Parmenides of Elea; Pythagoras of Samos; Thales; Xenophanes; Zeno of Elea

Further Readings

Fideler, D. (Ed.). (1987). *The Pythagorean sourcebook and library* (K. S. Guthrie, Trans.). Grand Rapids, MI: Phanes Press.

Freeman, K. (1948). *Ancilla to the Presocratic philosophers.* Cambridge, MA: Harvard University Press.

Kirk, G. S., Raven, J. E., & Schofield, M. (1983). *The Presocratic philosophers* (2nd ed.). Cambridge, UK: Cambridge University Press.

Nietzsche, F. (2001). *The Pre-Platonic philosophers.* Urbana: University of Illinois Press.

Robinson, J. M. (1968). (1968). *An introduction to early Greek philosophy.* Boston: Houghton Mifflin.

Wheelwright, P. (Ed.). (1966). *The Presocratics.* Indianapolis, IN: Odyssey Press.

Zeller, E. (1980). *Outlines of the history of Greek philosophy* (13th ed.). New York: Dover Publications.

Prigogine, Ilya (1917–2003)

It is unusual that a winner of the Nobel Prize in chemistry succeeds in arousing a continuing interest of nonscientists in his research. Ilya Prigogine, who won that most prestigious award in 1977, is the exception that proves the rule. His study of the thermodynamics of irreversible processes not only is important in the history of science but also may have far-reaching implications for the philosophical understanding of time, which Prigogine discussed in a series of popular books read by a wide audience.

Ilya Romanovich Prigogine was born on January 25, 1917, in Moscow. Because his family had trouble with the Soviet regime that had been established after the October Revolution, the Prigogines emigrated in 1921 to Germany and in 1929 to Belgium. Prigogine studied chemistry at the Free University in Brussels, where he also became professor after the Second World War. In 1959, he was appointed director of the International Solvay Institutes in Brussels and professor at the University of Austin in Texas. There he also founded, in 1967, the Center for Statistical Mechanics and Thermodynamics. For his scientific work, Prigogine was honored not only by the Nobel Prize but also by more than 50 honorary degrees as well as by a Belgian title of nobility, viscount. On May 28, 2003, Prigogine died in Brussels.

Prigogine's principal field of research was thermodynamics, the science of energy and its transformations. Classical thermodynamics studies closed systems—systems that are thermically isolated from their environment and whose entropy can, according to the second law of thermodynamics, only be constant or increase. On a macroscopic level this means that closed systems become more and more disordered. Yet everyone can observe physical systems that, at least temporally, show an increase in order, for example, organisms. These are open systems; that is, systems that, by constantly exchanging energy with their surroundings, increase their internal order. To develop and maintain the complexity of their bodies, organisms must eat, breathe, and excrete. It is, however, not the case that every open system is able to increase its own complexity; only open systems that are far from thermodynamic equilibrium self-organize (*equilibrium* means a state in which the forces acting on the system are in balance). Prigogine called nonequilibrium systems *dissipative structures*, and he tried to apply their thermodynamic analysis not only to organisms but also to social systems.

A characteristic feature of dissipative structures is that they are historical systems: The processes by which they originate, develop, preserve themselves, and dissolve are irreversible. Prigogine's theory of nonequilibrium systems can thus be regarded as a physics of history. He considered his model of dissipative structures to be more realistic than the classical models of closed systems and near-equilibrium open systems. Consequently, regarding the basic ontological categories of natural science, Prigogine argued passionately for a revolution whose maxim is succinctly expressed by the title of one of his books on philosophy of nature, *From Being to Becoming* (1980). In this work, two concepts of time are distinguished. The concept of external and universal time refers to time as a parameter of motion; it is used when we want to communicate about processes and use clocks as instruments to quantify the succession of events by counting conventionally fixed units. The concept of internal and individual time refers to time as a property of the development of physical systems; it is used when we want to determine the intrinsic age of a system by counting the number of irreversible transformations of a certain type that have been generating the present state of the system.

In the tradition of philosophers like Peirce and Whitehead, Prigogine criticized the standard worldview of physics for being founded on the conceptual separation of the arbitrary initial conditions of a system from the dynamical laws its behavior follows: This way of objectifying nature cannot capture the essentially historical nature of all that is existing. Because dissipative structures are continuously exchanging energy with their environment, Prigogine thought that we humans, after having understood that we are such structures, will be able to develop a new understanding of ourselves as insolubly being a part of nature. The anthropological implications of his new ontology of nature Prigogine explained in his most famous philosophical book, which he wrote with the philosopher Isabelle Stengers, *Order out of Chaos: Man's New Dialogue With Nature* (1979). Whereas his philosophical convictions are, to say the least, quite controversial, Prigogine's far-from-equilibrium thermodynamics has become one of the cornerstones of research into complex systems and their natural historicity.

Stefan Artmann

See also Becoming and Being; Chemistry; Entropy; Time, Units of; Whitehead, Alfred North

Further Readings

Balescu, R. (2003). Obituary: Ilya Prigogine (1917–2003). *Nature, 424,* 30.

Nicolis, G., & Prigogine, I. (1977). *Self-organization in non-equilibrium systems.* New York: Wiley.

Prigogine, I. (1997). *The end of certainty.* New York: The Free Press.

Prime Meridian

The prime meridian is an imaginary half-circle reference line extending north and south from one pole to the other and passing through Greenwich, England. It marks 0° longitude and divides the western and eastern hemispheres of the earth. The line that meets it at both poles and is exactly 180 degrees away from the prime meridian is the international date line. The prime meridian is the standard basis for determining time throughout the world. It is the starting point for all the world's time zones. This meridian is the location of UTC (universal time coordinated) as well as Greenwich mean time (GMT). Noon GMT is defined as the time at which the sun crosses the Greenwich (prime) meridian.

An international agreement in 1884 established the prime meridian in its current location. This line runs through the transit circle telescope, built in 1850 by Sir George Biddell Airy, the seventh astronomer royal, at the Royal Observatory Meridian Building in Greenwich, England. The crosshairs in the eyepiece of the transit circle telescope precisely define 0° longitude for the world. It is located at 51° 28' 38" north latitude. As the earth's crust is moving very slightly all the time, this exact position moves very slightly.

Longitude is a measure of both time and location on the earth. As the earth spins on its axis, a specific location and time can be determined relative to the prime meridian. This north-south line marks noon of the day that begins at the international date line and is the origin from which east or west longitude is measured. One degree of longitude equals 4 minutes of time the world over, but the distance of 1° longitude varies depending on the latitude of the location. Every 15° east measured from the prime meridian marks another hour later; every 15° to the west marks another hour earlier.

History

Lines of latitude and longitude appeared on maps at least 3 centuries before the Christian era. Hipparchus was the first astronomer to determine the difference in longitude and chose Rhodes as the location for his prime meridian, that is, his 0° east or west. By 150 CE, Ptolemy plotted grid lines on 27 maps of his first world atlas. The equator was set from previous astronomical observations, and he chose a line running through the Canary Islands as the prime meridian for his maps.

Later mapmakers moved the prime meridian to the Azores, Cape Verde Islands, Rome, Copenhagen, Jerusalem, St. Petersburg, Pisa, Lisbon, Rio, Tokyo, Paris, and Philadelphia. This placement of the line that marks 0° east or west is purely political, as it is not based on any natural phenomenon.

Beginning in 1667, and continuing for over 200 years, French cartographers used the longitude line running through the Paris Observatory to be the prime meridian. Other meridian lines were given names such as the Rose Line but were never used universally. The English king Charles II founded the Royal Observatory at Greenwich in 1675 for the purpose of improving navigation, and its longitude became another prime meridian.

The development of the accurate chronometer by Englishman John Harrison made it possible for English ships to determine their location based on comparison of their time and the time at a known longitude. Harrison was a clockmaker and not an astronomer, but he was able to solve the problem of determining longitude at sea. This was recognized in 1773.

Ocean charts drove the need for one internationally agreed-upon prime meridian. In 1871, the first International Geographic Congress (IGC) met in Antwerp, Belgium. The general view was that the navigation passage charts for all nations should use the Greenwich meridian as zero, so when ships exchanged longitude at sea, locations and times would match. This did not apply to land maps and coastal charts. In 1875, the second IGC met in Rome. France was willing to accept the Greenwich prime meridian if the rest of the world would accept the metric measurement system. In October 1884, at the invitation of U.S. President Chester Arthur, the International Meridian Conference was held in Washington, D.C., and 41 delegates from 25 nations voted 22 to 1 for Greenwich as the location of the prime meridian. Santo Domingo was against it, and France and Brazil abstained. The United States had already chosen Greenwich as the basis for its national time zones. Seventy-two percent of the world's commerce already used sea charts that used Greenwich as the prime meridian.

International commerce and communication required one standard. Since 1884, the world has had only one prime meridian.

Ann L. Chenhall

See also Harrison, John; Latitude; Longitude; Time, Measurements of; Time, Teaching; Time Zones

Further Readings

Barnett, J. E. (1998). *Time's pendulum: The quest.* New York and London: Plenum Trade.

Raymo, C. (2006). *Walking zero: Discovering cosmic space and time along the prime meridian.* New York: Walker.

Sobel, D. (1995). *Longitude: The true story of a lone genius who solved the greatest scientific problem of his time.* New York: Penguin Books.

Progress

Progress is most easily defined as forward movement toward a destination. The theory of progress, however, is commonly defined as the forward, irreversible pattern of change throughout human history, the destination being the future. In its simplest form, progress is known to be change for the better. Although it may seem like an easy concept, it is widely debated and easily refuted based on its reliance on subjective matter—a lack of empirical proof about what is indeed better.

A critical aspect of progress is time. History is critical in evaluating progress, and history would not have been written without change. Without looking back in time on the condition of human life, science, and morality, it is impossible to know if advances have been made; thus, all progress must be measured relative to the past. Progress throughout history is not continual but rather takes place over time through setbacks and dissonance. In addition, goals, hopes, and dreams change over time, making what could have been measured as progress a century prior, now obsolete.

Progress is a relatively recent idea as applied to society as a whole. In ancient times, the idea was foreign, and in fact, most civilizations viewed the state of being as declining rather than improving. The idea was that life had been perfect in a former golden age and slipped into the silver, then bronze ages, identifying a downgrading of humanity. French historian Charles Perrault said in 1687 that there appeared to be little left to learn and that there was no reason to envy future generations.

The idea of progress took hold in the 18th century in the work of philosophers, scholars, and theorists such as Immanuel Kant, Marquis de Condorcet, and Georg Wilhelm Friedrich Hegel, among others. Prior to this, there was little hope that the world was in fact on its way to becoming a better place. This was the shift from the idea that man had been created perfect, and there was a push to live up to that standard to the idea that intelligence equaled progress, and the more knowledge each generation could obtain, the more forward movement, or progress, would be seen. Most of these thinkers agreed that progress had been made throughout history and was continuing in their respective times. They did not, however, share opinions on the ultimate goals of society and what was to be the ultimate goal.

Hegel saw history as progress toward freedom and the chaos of history as having great purpose toward the future. Kant wrote about a human race looking into the future and advancing to a "universal civil society," with maximum individual freedom afforded to all. Condorcet used mathematical methods in his approach to the social and moral sciences and believed in progress and perfectibility. His Enlightenment theory specifically addresses the likelihood and reality of perfectibility being attained within human history.

The theory suggests that the movement of change is irreversible in direction and that although it is not considered to be a continuous stream of forward change, due to occasional regression or stalled movement, the ultimate result is positive change. The pace of progress is not easily defined, as it varies widely with such fluctuations in direction and cannot be adequately measured in terms of time.

The opposition to this idea of eventual forward movement is quickly dismissed by the cliché "history repeats itself." There are those who subscribe to the thought that history is in fact cyclical, and events, or basic forms of events, reoccur over time. For this reason there is a school of thought that does not accept the idea that history is an upward climb toward that which is better; rather, this school holds that history will return in a similar form in the future. And then there are those still who agree that there is a definitive pattern of history that also moves in an irreversible direction but who argue that change is for the worse.

Among its supporters, many wonder whether progress is eternal, or if at some point it will level off to a point from which no further advancement can be made. Progress is also proven to be poorly defined in that there is no solid agreement as to the breadth of progress. Some argue that it is limited to human inventions, productions, and intellect. Others would say that it also should be inclusive of the improvement of the very nature of man himself.

Many who disagree with the theory on the whole have varied reasons for its dismissal.

The idea of progress is sometimes debated as immeasurable, and therefore its legitimacy is questioned. The argument does not center on the facts of history, which are knowable, but that the determination that an improvement has been made over earlier times is subjective. Many factors can complicate the ability to identify progress, including but not limited to the subjective nature of any determination of what is better. Some may see the automobile as evidence of human progress, while environmentalists may argue that its role in pollution and its drain on the natural resources of the earth actually make things worse.

In science, at least, it is commonly understood that there is progress. Seldom is it disputed that advances have been made, and it would be difficult to argue that curing disease and prolonging life through scientific research and newer medicines are not to be viewed as positive. Although their value is largely undisputed, even scientific advances cannot be seen to represent unequivocal progress; they are simply the closest thing to measurable progress that exists.

Amy L. Strauss

See also Enlightenment, Age of; Hegel, Georg Wilhelm Friedrich; Kant, Immanuel; Marx, Karl; Medicine, History of; Spencer, Herbert; Teleology; Teilhard de Chardin, Pierre

Further Readings

Easterbrook, G. (2003). *The progress paradox: How life gets better while people feel worse.* New York: Random House.

Lasch, C. (1991). *The true and only heaven: Progress and its critics.* New York: Norton.

Pirie, M. (1978). *Trial & error and the idea of progress.* La Salle, IL: Open Court.

Van Doren, C. (1967). *The idea of progress.* New York: Praeger.

Prophecy

Prophecy has played a significant role in world religions, laying the groundwork for fulfillment of both religious and secular expectations of the future. The term *prophecy* derives from the Greek *pro-*, meaning "before" and *phanai,* "to speak," implying a foretelling. The term has become closely associated with supernatural abilities, and the word *prophet,* derived from the Greek *prophete,* has come to mean "one who speaks for another," usually understood to be a divine entity. Prophets channel revelations that include predictions of the future as well as admonitions to return to the laws of a belief system in order to avoid disastrous consequences. Often referred to as "seers," prophets receive their messages through various means, including trancelike states, dreams, visions, or agitated conditions induced by music or dance. Others receive special knowledge through the interpretation of physical signs, including astrological signs. Prophets have also been viewed as teachers and religious leaders. Three of the world's major religions describe their most venerated leaders as prophets. Judaism and Christianity revere the prophets of the Hebrew Bible, and Islam regards Abraham, Noah, Moses, and Jesus as respected prophets but Muhammad (571–624 CE) as the last and greatest prophet.

In the ancient world, the Persians, Assyrians, Chinese, Celts, Indians, Egyptians, and others sought out prophets who were often associated with temples and royal courts. During the later Hellenistic period, prophets, called oracles, were attached to shrines, one of the most important being the Oracle of Apollo at Delphi. The word *oracle,* derived from the Latin *orare* meaning "to speak," particularly in a public forum, refers both to the seer and to the shrine, and the term can be used interchangeably. The temple at Delphi was dedicated to Apollo, the Greek god of medicine, healing, poetry, music, and prophecy. Unlike most other Greek gods, Apollo had no direct Roman counterpart; however, later Roman poets often referred to him as Phoebus. The archaeological remains of this site are located near Mount Parnassus in the valley of Phocis, which the Greeks considered the center of the earth and the universe. The name Delphi is believed to be derived from the Greek *delphus,* meaning "womb," reinforcing its sacred origins.

The priestess of the oracle at Delphi was known as the Pythia and was consulted before all major undertakings, particularly battles. However, the oracle did not directly predict the future but rather gave advice and counsel, usually couched in ambiguous language that might be interpreted in various ways. Historic records dating to the 9th century BCE recount visits to the oracle by important Greek leaders, such as Lycurgus, Solon, Croesus, Lysander, and Philip II of Macedon. By 191 BCE the shrine at Delphi had come under Roman power, and the oracle continued to be consulted by Roman leaders, including Nero and Hadrian. It has been estimated that the pronouncements of the oracles at Delphi and other shrines had an important effect on the shaping of the Greco-Roman hegemonies.

Prophetic teachings play a particularly essential role in Judaism, Christianity, and Islam, which emphasize a personal relationship with God and a sense of individual moral responsibility resulting in future judgment. Judaism is based on the Hebrew Bible (*Tanach*), which is traditionally divided into three parts: the Law (the first five books, called the *Torah*); the Prophets; and the Writings or Wisdom Books. Particularly important are the later prophets, men of position such as Isaiah and Jeremiah, who spoke against religious hypocrisy as well as secular practices of their times, setting them in opposition to established religious tradition. Writing between the 8th and 3rd centuries BCE, they pointed to threats by Assyria, Egypt, and later Babylonia as God's sign that Israel needed a return to more ethical conduct to avoid disaster. The so-called minor prophets, men of humble origins, such as Amos, Hosea, and Micah, also spoke out against social inequality and served as the moral conscience of the community. Other important prophets of the Hebrew Bible (in Christianity, the Old Testament) include Daniel, Samuel, and Elijah.

The Old Testament book of Isaiah is particularly significant in Christian theology and contains the foretelling of a young woman who shall conceive a son and call him Immanuel, Hebrew for "God is with us" (Isa 7:14). Christians consider this a reference to Jesus Christ, the Messiah; however Judaism contends the prophecy has not yet been fulfilled and that believers await the first coming of a savior. The Hebrew prophets solidified the concepts of a messiah and a millennial period that is incorporated into Christianity and, to some extent, Islam. The only purely prophetical book of the Christian New Testament is The Book of Revelation, also referred to as the Apocalypse of St. John, in which John describes the events culminating in the Second Coming of Christ, the final judgment, and the establishment of a new world of peace and justice. Islamic tradition holds that prophets were sent by God to each nation, and all received revelations from God. The prophets who received *Sharia* (a divine code for life) that was ultimately written down and collected into holy books are also referred to as messengers. More than 25 prophets are mentioned in the Qur'an.

Prophecy is an important component in spiritual traditions throughout the world. Prophets pronounce visionary accounts of the future, which may or may not be predictions, but are rather interpretations of consequences based on present action. This particularly applies to beliefs regarding the ultimate future of humanity, and most of the world's major religions contain Messianic prophecies. Judaism, Christianity, Islam, Hinduism, Buddhism, and Native American religions all contain spiritual traditions that foretell the coming of a savior who will return, or send another in his place, to restore justice to the world.

In some beliefs, such as Taoism and ancient Mayan belief, time is cyclical rather than linear, and prophecy is based on predictable recurring events. In the contemporary secular world, researchers continue to investigate claims of some persons' apparent ability to access secret information and predict future events through means not scientifically explainable. Parapsychology is an academic field of study that seeks to determine the nature of those abilities by which some individuals are evidently able too see, hear, or predict events that are not apparent to others; these phenomena include clairvoyance, telepathy, extrasensory perception, precognition, and regression in time. Whether such powers are divinely inspired, supernatural, or scientifically explainable, or whether they turn out to be merely hoaxes, the desire to ascertain the future remains a powerful human motive.

Some modern scholars and theologians look for prophets in the contemporary context. In the tradition of Old Testament prophets and teachers, modern-day visionaries are often social activists, such as Martin Luther King, Jr. and others worldwide who call upon society to reform and promote peace, justice, and tolerance. In this view, then, the future is not to be foretold to passive listeners, but rather created by concerted action.

Linda Mohr Iwamoto

See also Apocalypse; Bible and Time; Christianity; Ecclesiastes, Book of; Futurology; Islam; Judaism; Nostradamus; Omens; Paracelsus; Qur'an; Revelation, Book of; Toffler, Alvin; Verne, Jules; Wells, H. G.

Further Readings

Fontenrose, J. (1981). *The Delphic oracle, its responses and operations, with a catalogue of responses.* Berkeley: University of California Press.

Ramsey, W. M. (1986). *Four modern prophets.* Louisville, KY: Westminster John Knox Press.

Proust, Marcel (1871–1922)

The French novelist and essayist Valentin Louis Georges Eugène Marcel Proust wrote essays and reviews as well as two novels before he became famous for his novel *À la recherche du temps perdu* (translated variously as *Remembrance of Things Past* or *In Search of Lost Time;* in this entry, *Recherche*).

Proust was the son of Adrien Proust, a prominent physician of that time, and his wife Jeanne née Weil, daughter of a wealthy Jewish family. In 1881 Marcel suffered his first attack of asthma, a disease that was to accompany him from that time onward and that finally led to his death at the age of 51. After baccalauréat and military service, he began to study law and political sciences in Paris. He finished these studies in 1892 and 1893, respectively, but against his father's will decided not to start a career in the foreign ministry and instead to begin a new study of "lettres," which could be described as a general study of humanities. Proust chose philosophy as his main subject and received his licentiate in 1895. In 1896 *Les plaisirs et les jours,* a collection of prose writings, was published but not widely recognized. In the same year, he began to write his first novel, *Jean Santeuil.* He stopped working on it in 1904; the fragments were not published until 1952. Many themes of the *Recherche* can already be found in these fragments.

Between 1896 and 1908 Proust published some translations of the British art historian John Ruskin. In 1907 he began the essay "Contre Sainte Beuve" (published posthumously in 1954), which today is considered to be his preliminary stage of the *Recherche.* "Contre Sainte Beuve" gradually changed its nature to finally become *À la recherche du temps perdu* from 1909 onward. The novel consists of 15 volumes, which are divided into 7 parts. The first, *Du côté de chez Swann* (*Swann's Way*) was published 1913, to be followed by *A l'ombre des jeunes filles en fleurs* (*In the Shadow of Young Girls in Flower*) in 1918, *Le côté de Guermantes* (*The Guermantes' Way*) in 1920, and *Sodome et Gomorrhe* (*Sodom and Gomorrah*) in May 1922. Proust died in November 1922, leaving enough material behind to publish three more novels posthumously. The fifth part of the novel, *La prisonnière* (*The Prisoner*) was published in 1923; the sixth, *Albertine disparue* (*The Fugitive*) in 1925, and the last, *Le temps retrouvé* (*Finding Time Again*) in 1927.

The main subject of the *Recherche* is time. The novel describes the life of the protagonist Marcel from his own perspective. The protagonist is not identical with Proust, but there are many similarities to Proust's own life.

The novel wants to give an answer to the question, "What happened to the time lost?" Proust is influenced by the model of time described by the French philosopher Henri-Louis Bergson (1859–1941). Bergson divides memory into two parts, which are named *memoire* (memory) and *souvenir.* Memoire is the cold intellectual side of the memory, while souvenir is an involuntary evolving memory. This *memoire involontaire* (involuntary memory) is often caused by sensations such as a smell or a stone touched by the foot. The perhaps most famous example for souvenir in the *Recherche* is the moment in which Marcel eats a madeleine (small French cake) dipped into tea. This trivial experience triggers a revival of the protagonist's whole life in Combray. Time therefore is not simply a linear experience; it is possible that "lost time" may be brought back to life again by an involuntary remembrance. The lost time comes back as it really was for the one who experienced it.

The *Recherche* itself is an *édifice immense de souvenir,* in which the reader follows Marcel on his way through time described from his perspective. Time is preserved in the memory of the people, and the time they are preserving is lost for the world once they have died. In the last part of the novel, *Le temps retrouvé,* Marcel decides to write down his memories, and with them the world as he knew it, in a book. The novel thus contains a kind of circle, in which the reader finds out that he was following Marcel's search for a way to preserve time, which is found in the end. The solution found is to write a novel, which Marcel decides to do. The work to be written after the story of the novel ends is the same novel that was read by the reader.

For Proust, then, the only way to preserve lost time and all of its souvenirs is through the making of art. The *Recherche* is such a piece of art and the answer to the protagonist's question. It lets the reader relive Marcel's memories, or at least grants access to a version of them, thereby retrieving the time as experienced by the protagonist.

Markus Peuckert

See also Bergson, Henri; Consciousness; Dostoevsky, Fyodor M.; Goethe, Johann Wolfgang von; Joyce, James; Mann, Thomas; Memory; Novels, Time in; Time, Phenomenology of; Tolstoy, Leo Nikolaevich; Woolf, Virginia

Further Readings

Albert, C. (2003). *Monsieur Proust* (B. Bray, Trans.). New York: NYRB Classics.

Bernard, A.-M. (2004). *The world of Proust, as seen by Paul Nadar.* Boston: MIT Press.

Carter, W. C. (2000). *Marcel Proust: A life.* New Haven, CT: Yale University Press.

Psychology and Time

The dimension of time is essential for human cognitive functioning. Because events are perceived over time and actions evolve over time, the ability to process temporal information is essential to monitor the environment and control our behavior. Although we speak of a "sense of time," there exists no sensory organ for the perception of the passage of time or duration comparable to our sensory perception of colors or shapes, sounds or melodies. Time is not a concrete entity in the world we experience. Our brain must actively construct temporal relationships from the sensory information we perceive.

The Experience of Time

In his *Confessions,* Saint Augustine thus concludes his analysis of the experience of time: "It is in thee, my mind, that I measure times." He saw the perception of time as constructed by our self; that is, the experience of time represents the mental status of the beholder, reflecting one's cognitive state and emotional well-being. The accuracy and precision of subjective time estimation is linked to overall cognitive functioning, that is, to attention and memory processes. Our sense of duration depends on the degree of attention we pay to the passage of time and memories of comparable time intervals. Our subjective well-being also strongly influences how time is experienced. Time seems to fly during pleasant activities, but to drag during periods of mental distress. Our sense of time depends on an intricate interplay between specific cognitive functions and influences of our momentary mood states.

The perception of time can be classified in numerous ways, but two concepts form the building blocks of our temporal experiences—*succession* and *duration*. The perception of succession refers to the sequential characteristics of events; in other words, their temporal order. The perception of duration refers to the time interval subjectively experienced between two events or to the persistence of an event over time. The taxonomy of elementary temporal experiences derives from these two basic concepts and comprises the perceptual phenomena of simultaneity, successiveness, temporal order, the subjective present, and duration.

The first three of these experiences relate to the temporal properties of two events that are perceived as happening simultaneously or nonsimultaneously. For instance, our sensory systems have different temporal resolutions for the detection of nonsimultaneity. The highest temporal resolution (the lowest threshold of detection) is observed in the auditory system, where acoustic events 2 to 3 milliseconds apart are detected as nonsimultaneous. The somatosensory system has a slightly higher threshold, whereas the visual system has the lowest temporal resolution, with a threshold approximating 20 milliseconds. Interestingly, the detection of nonsimultaneity of stimuli is not perceptually sufficient to indicate their temporal order. Although we may be aware that two events did not occur simultaneously, we are still unable to tell which one of the two events occurred first.

The temporal-order threshold is more comparable across senses. The onset of two events, independent of sensory modalities, must be at least 20 to 40 milliseconds apart before an observer can reliably indicate their temporal order. The temporal-order threshold thus represents a fundamental perceptual limit. The temporal succession of events over time (the order in which they occur) can only be perceived if they occur approximately 20 to 30 milliseconds apart.

Regarding the perception of duration, a perceptual mechanism that integrates separate successive events into a unit or perceptual gestalt has repeatedly been suggested. We do not just perceive individual events in isolation, but automatically integrate them into perceptual units (or *gestalts*). While listening to a metronome at a moderate speed, we do not hear a train of individual beats but automatically form perceptual units, such as 1–2–3, 1–2–3, etc. These are mental constructs—physically speaking, they do not exist. The duration of this temporal integration mechanism, referred to as the subjective or specious present, seems to be limited to 2 to 3 seconds. Experimental investigations have demonstrated that temporal intervals can be accurately reproduced when their duration does not exceed 3 seconds. Intervals longer than 3 seconds are usually reproduced shorter than the presented interval.

In sensorimotor synchronization (a task in which a sensory signal has to be precisely synchronized with a motor act), if the interstimulus interval is longer than approximately 2 seconds, performance breaks down. Anticipation of the signal is no longer possible. Other examples come from bistable perception (e.g., the Necker cube), where a single stimulus leads to two mutually exclusive perceptual interpretations that alternate in time. The spontaneous rate of perceptual alternation approximates 3 seconds. Based on these findings, researchers have established the categorical distinction between *perception of duration* (for intervals up to 3 seconds) and *estimation of duration* (for longer intervals). Events lasting only a few seconds are processed on the present perception as a whole, whereas longer intervals must be estimated from memory.

A different classification of time phenomena, which still encompasses the above-mentioned elementary time experiences, includes three interrelated dimensions: (1) time estimation, (2) time awareness, and (3) time perspective.

1. Abilities to estimate time are measured by the precision in estimating clock time, that is, in experimental settings in which subjects are requested to tell which one of two presented stimuli was longer or, in real life, when we have to judge whether water has boiled for 3 minutes.
2. Time awareness is the subjective impression of time as moving quickly or slowly, that is, the

feeling that time passes too fast when we are having fun or too slowly when we are desperately waiting for something specific to happen.

3. Time perspective refers to the conception of a past, a present, and a future.

Because we experience time in these three-dimensional aspects, time perspective is a fundamental dimension in the construction of subjective time, which emerges from the cognitive partitioning of human experience into past, present, and future. Human endeavor bridges the past, present, and future, as the memory of past events can be used to plan and act in the present to reach future goals. In this respect, present time is experienced through attention, past time is defined by memory, and future time exists as expectation and prediction.

Psychological and Neural Models of Time Perception

From a systemic perspective, organisms must interact with the environment on different timescales. Experience and behavior are interactions that span time ranges in microseconds (localization of sound in space), tens of milliseconds (detection of succession, phoneme recognition), hundreds of milliseconds (motor control, speech segmentation), seconds-to-minutes (conscious perception of duration), and circadian rhythms (the daily sleep-wake cycle). Additionally, annual cycles influence psychological states and determine an individual's physiological functions. The anticipation of future decades of human life is relevant for long-term decisions, for example, when a young person saves money in a retirement plan, thereby opting for a momentary loss of money to gain future benefits. This classification of relevant timescales between the range of milliseconds and decades encompasses two aspects of time that have been described since ancient times. Time appears to be either cyclic (it repeats itself periodically) or linear and irreversible (time flows in one direction only and one moment does not repeat itself). The cyclic aspect of time is apparent in the recurrence of days and the repetition of the seasons. Nature has adapted to these cyclic properties by establishing endogenous biological clocks in organisms. The circadian (about one day) clock as well as circannual clock regulate fundamental aspects of physiology and behavior and are entrained (synchronized) by the periodicity of light.

Over the past few years, a body of evidence has shown that different time perception mechanisms are associated with different timescales. For example, on a variety of sensorimotor tasks, temporal integration windows of around 250 milliseconds and of around 1 second have been postulated to mark the transitions between different timing systems. It has been suggested that an automatic timing system is responsible for shorter intervals (milliseconds). Such a system can measure time without attentional modulation, whereas a cognitively controlled timing system for intervals longer than a second requires more strongly cognitive circuits in the brain. Furthermore, in relation to the concept of the specious present, intervals up to 2 to 3 seconds have been shown to be processed differently than intervals exceeding this time range. Even for longer intervals, it is conceivable to assume distinct processes for different durations. For example, the estimation of a one-hour interval, but not the estimation of a minute, is related to an individual's circadian rhythm.

Cognitive models of time perception in the seconds-to-minutes range distinguish two fundamental perspectives in time estimation. In prospective time estimation, an observer judges the duration of an interval she or he is presently experiencing. The observer's attention is directed to time while a particular duration is being estimated. In retrospective time estimation, the observer a posteriori estimates a time span that has already elapsed, and, therefore, duration has to be reconstructed from memory. The most prominent cognitive models of prospective time estimation assume an "internal clock" with a pacemaker producing subjective time units (analogous to the ticks of a clock). The time units produced are registered only when attention is actually concentrated on time. As a result, an attentional gate opens, and time units are fed into a counter. When attention is distracted from time—for example, when one gets absorbed in another task—the attentional gate closes, and time units are not recorded. The number of units that have been recorded during a physical time period is then compared with a memory store. The unit counts are compared with the stored representations of time periods, which can be verbalized as seconds or minutes. If more attention is directed to time during the time interval to be estimated, duration is experienced as being longer.

In retrospective time estimation, the duration of a time interval that has already passed has to be judged. In contrast to prospective duration estimation, the observer estimates duration from the amount of processed *and* stored memory contents. In general, the more different events that occur during a time span, the longer the experienced duration. According to this model, the subjective impression of a long time interval depends on the activity of a person with diverse experiences. These two components of time estimation help explain the time paradox familiar to everyone.

The same physical time period can be judged differently, depending on a person's prospective or retrospective perspective. For example, the 10 minutes spent in a dentist's waiting room may seem eternal. We may overestimate the duration of that period simply because, under those circumstances, we pay a lot of attention to the passage of time (prospective estimation). However, when looking back at that waiting period, nothing really remarkable occurred and, therefore, this time span shrinks to a negligible period of time (retrospective estimation). In a different example, when we spend 10 minutes talking to an attractive person, we are not paying attention to time, as we are absorbed in the conversation. Time even seems to pass too quickly. However, in retrospect, we have memorized so many stimulating and emotional moments that we experience this time period as having lasted considerably longer.

The neural mechanisms underlying the perception of time have yet to be identified. Nevertheless, investigations of the neural basis of time perception in the seconds-to-minutes range support the hypothesis of the involvement of brain circuits embedded in the frontal lobes of the cortex and the basal ganglia. These circuits are modulated by the dopamine transmitter system and are thought to be critical for temporal processing.

Evidence for this anatomical hypothesis comes from studies of patients with brain lesions and from investigations using brain imaging techniques. Patients who suffered an infarction in their frontal lobes or traumatic brain injury predominantly affecting frontal brain areas display impaired estimation of time intervals. In healthy volunteers, brain activation studies using a functional magnetic resonance imaging (fMRI) technique show that the processing of duration is associated with activation in the right prefrontal cortex and the striatum (as part of the basal ganglia).

Regarding the neurotransmitter systems, animal and human studies indicate that both dopaminergic agonists (drugs that stimulate the system, like methamphetamine) and antagonists (drugs that inhibit the system, like haloperidol) influence the temporal processing by increasing and decreasing clock speed, respectively. Additional evidence for the involvement of the dopamine system in time perception comes from patients with Parkinson's disease. These patients have decreased dopaminergic function in the basal ganglia and show deficits in the discrimination of temporal intervals. Hence, overall, studies show that the integrity of the dopamine neurotransmission within the fronto-striatal circuitry of the brain is an essential component for the processing of time in the seconds-to-minutes range.

The experience of time is also influenced by hallucinogenic drugs such as LSD and psilocybin, which have similar pharmacological properties acting on the serotonergic transmitter system. People report the feeling of a slowing down of the passage of time and often overestimate time intervals, as minutes can appear to be hours, or time seems to stand still. Marijuana also affects the subjective sense of time, as the user's time estimates can become inaccurate. Individual reports are complemented by experimentally controlled investigations, which explain their findings within the framework of the dopamine hypothesis of time perception. The serotonergic hallucinogens psilocybin and LSD probably act indirectly on dopaminergic transmission. Moreover, marijuana can also influence dopamine activity via cannabinoid receptors on nerve cells of the dopamine system in the basal ganglia.

Other brain regions play a decisive role in the range of milliseconds. For example, the important role of the cerebellum in temporal processing has repeatedly been emphasized. Patients with damage to cerebellar structures exhibit severe deficits in time perception and motor timing of short intervals lasting only hundreds of milliseconds.

Psychophysical and neuroimaging studies in humans, as well as neurophysiological recordings in animals, indicate that a network of cortical areas contributes to the processing of time intervals in this short time range. In a widely distributed network spanning distinct regions of the brain, neural populations within each region may encode duration as a result of specific biological time-dependent changes. Time-dependent changes result from intrinsic neural properties, such as short-term synaptic plasticity. This view postulates that a centralized pacemaker-accumulator system (the "internal clock") is not involved in the processing of time when intervals reach up to hundreds of milliseconds.

In contrast, a stronger localization of functions (i.e., the involvement of a focal area in the brain) has been postulated for the perception of temporal order. Many studies have shown that patients with lesions in the left hemisphere and aphasia (as opposed to patients with lesions in the right hemisphere) have difficulties perceiving the temporal order of two events. These patients require longer intervals between two stimuli before they can distinguish their temporal order.

Children with specific language deficits have difficulties detecting the sequence of events in the milliseconds range. These observations suggest a causal relationship between the detection of temporal order and language processing on the phonological level. According to the phonological processing hypothesis, rapidly occurring elements in the speech signal have to be adequately processed for the identification of stop consonants (e.g., *d* in *dry* or *t* in *try*). In language-impaired individuals, a basic auditory deficit in temporal-order detection leads to impairments in phoneme identification, which is anatomically traceable to temporo-parietal areas in the left hemisphere. Such results demonstrate how intimately related temporal processing mechanisms are with other cognitive functions. In this specific case, a deficit in perception of temporal order has a detrimental effect on language processing. As will be discussed below, disturbances in cognitive functioning and emotional processing can also have profound effects on the experience of time. Any change in a person's mental status has an effect on the processing of time, as minute as the change may be.

Cognitive and Emotional Views of Time

The perception of the passage of time with changing pace is part of our daily experience. We do not perceive time as a steady-paced flow; the subjective passage of time varies considerably. The prospective model of time estimation can help explain several everyday time phenomena in relation to attention. An individual who is concentrating on the passage of time in an uneventful or unpleasant situation, such as nervously waiting for something specific to happen, experiences a slower passage of time and overestimates duration. When, however, the same person is suddenly distracted from concentrating on time, as when he or she is having fun, time seems to pass more quickly, and duration is more likely to be underestimated. According to the cognitive models, the way attention is concentrated on time during a given time span leads to the collection of more or fewer subjective time units. When we devote more attention to time, more time units are collected in a hypothetical counter. When we concentrate more on distracting events during a given time interval, fewer time units are collected, leading to the impression of a shorter duration. According to this theory, the allocation of attention determines the perception of time.

However, this reasoning does not clarify why individuals in certain circumstances actually focus their attention more strongly on the present time. It can be assumed that concentrating less on the passage of time (and perceiving it as passing more quickly) is associated with general psychological well-being. This state is associated with the ability to have meaningful thoughts and engage in self-rewarding activities.

Boredom-prone persons, for example, more often perceive time as passing slowly, even when they are busy performing a task. Institutionalized individuals, like those who live in residential homes for the elderly and whose days are highly regulated and monotonous, experience time as passing very slowly. There is ample evidence that depressed patients perceive a slowing of the pace of time and tend to overestimate durations in time-judgment tasks. Cancer patients who experience high levels of anxiety overestimate duration and report that time passes too slowly. In more general terms, these studies suggest that the overestimation of time intervals is a sign of emotional distress that draws attention away from meaningful thoughts and actions and directs it to the passage of time. Individuals who are institutionalized and/or suffering from emotional distress are not able to engage in their usual role function and have to spend much of their time in situations where they are not emotionally or cognitively engaged, like a series of waiting periods. Both the feeling of meaninglessness and the absence of a stimulating environment can manifest themselves in a state of distress or an existential vacuum that leads to the complaint of the slow passing of time.

Related to the time perspective (past–present–future), a similar entanglement of cognitive and emotional processes manifests itself in decision making. Everyday we have to make choices that have immediate, as well as delayed, consequences. A student may have to decide whether to go to a party tonight or to stay home and study for an exam scheduled tomorrow (with the possible later higher reward of passing the exam). We frequently have to make decisions based on predictions of rewards on different timescales. The process of deciding whether to opt for an immediate reward or for a delayed, but higher, reward is strongly related to scholarly and professional success. To function effectively, we must voluntarily postpone impulsive urges for immediate gratification and persist in goal-directed behavior to achieve positive outcomes in the future. Time perspective is considered a personality trait affecting cognition, emotion, and goal-related actions of an individual.

Ideally, we are able to flexibly switch temporal perspectives according to the situational demands; for example, we resist the temptation of the social gathering tonight because we can anticipate the future consequences. Some individuals, however, have too strong a focus on the present (related to stronger impulsivity and the inability to curb momentary urges) and are inadequately able to cope with everyday demands. For example, a stronger focus on the present and less on the future is highly related to risky behavior, such as gambling, careless driving, unprotected sex, or drug abuse. However, too much emphasis on the past and on the future are also dysfunctional, as they deprive one of important hedonistic features in life that can only be experienced at the present moment. A balance of temporal perspectives is healthy, both psychologically and physically.

A time phenomenon that haunts nearly every adult can be explained by the retrospective models of time estimation. A common complaint is, "Time passes too quickly!" During childhood and adolescence time seems to have a much slower tempo. People even complain that, as they age, time continuously speeds up. At least two factors, but probably more, influence this subjective perception. One lies in the time perspective of an adult, which, due to the individual's social orientation, can be characterized as continuously changing with advancing age. For the elderly, the length of the future time perspective generally decreases when measured by the number of years for which plans are made or personal events anticipated. A shorter future temporal perspective, however, can also influence the awareness of time. Studies show that elderly subjects who report fear of death feel more pressured by the passage of time. Men experiencing a midlife crisis are confronted with the shortening of their time perspective and also experience more time pressure.

A second factor that causes time to fly as we grow older lies in factors influencing retrospective time estimation. Prospective time estimation models cover the experience of the momentary passage of time. An association between age and the speed of time or the subjective duration of a time span, however, typically refers to longer intervals, like weeks, years, or even decades. A common disquieting discovery of an adult is "Have I already been working here for 10 years?" Looking at the past necessarily invokes memories and leads to a personal account of one's life. As predicted by the model of retrospective time estimation, we construct time in reference to the amount of changing information processed and stored. The more the contextual changes stored in memory during a given time span, the longer the experienced durations. Thus, time periods in one's life filled with a lot of changes are perceived as being subjectively longer than periods of time when few changes have occurred.

This provides the second explanation for why a lot of adults complain that time speeds up as they get older. During the course of a lifetime, many people get into the rut of everyday habits. The number of new experiences and openness to lifelong learning often decreases with age. If experiences do not differ significantly from day to day, they are not specially stored in memory. The routine of getting up in the morning, going to work, having lunch in the canteen, going home, watching television, and going to bed will not be stored as memorable events. Thus, in retrospect we have lost all these days, which results in a contraction of subjective lifespan.

Childhood, adolescence, and early adulthood, in contrast, can be characterized as periods of life during which new experiences occur on a daily basis, and skills and knowledge are acquired at a high rate. In retrospect these time periods are filled with many events, so the time period seems comparatively long. This explanation also offers a remedy available in a person's propensity to counteract the tendencies of repetition and custom. Although the general tendency toward routine cannot be fully avoided with increasing age, willingness to change habits and redefine one's goals throughout life will not only stimulate an individual in the here and now but will also lead to a subjective slowing down of the passage of time when we look back on a period of our life.

Applied Research in the Psychology of Time

After initial research in the late 19th century and the early 20th century, the topic of time perception was neglected in the field of psychology for many decades. However, over the last 20 to 30 years, a considerable body of knowledge has been collected regarding the processes underlying the experience of time. Diagnostic tools have been developed to assess disturbed temporal processing mechanisms in various neurological patient populations. Patients with acquired focal brain lesions following a stroke or severe traumatic brain injury, patients with Parkinson's disease or with temporal-lobe epilepsy, and patients with schizophrenia have all demonstrated specific impairments in time estimation and motor timing. These findings are discussed in relation to alterations in several cognitive domains related to time processing—specifically, a pacemaker-accumulator type internal clock, memory, and attention.

A considerable body of applied research concerning possible training procedures exists, albeit in one domain only and controversially discussed—neuropsychological training procedures for improving temporal processing abilities in subjects with language disorders. Based on the diagnostic findings that patients with aphasia and children with language-learning impairments have difficulties detecting the temporal order of stimuli, training methods in temporal-order detection have been developed to improve not only the ability to perceive temporal order but also to improve basic language competence. In these computer-based feedback training procedures, the temporal order of two successively presented acoustic stimuli has to be identified at decreasing interstimulus intervals (time between two stimuli). The interstimulus interval is lowered when the sequence of stimuli can be reliably identified. In this way, thresholds for the detection of temporal order can be sufficiently decreased. It has been shown that successful training can also improve phoneme discrimination ability, which may well depend on a high acuity in temporal resolution. It is likely that similar training procedures will be developed in the future to improve temporal processing functions in patients with other kinds of neurological syndromes. Due to the interdependence of time perception and cognition, novel time perception deficits—related to the taxonomy of temporal experience—in various neurological and psychiatric patient populations are likely to be detected.

The experience of time could become a topic in psychological assessments by providing an additional approach to diagnose the mood states of various patient groups, such as individuals with depression or cancer patients with anxiety or depressive symptoms. As outlined above, the feeling that time passes too slowly is an indicator of psychological distress resulting from an inability to focus on meaningful thoughts and to engage in purposeful actions. In contrast, individuals who are able to distract themselves during a stressful period not only experience a faster passage of time but also experience less mental distress. The reported findings in the psychology of time emphasize the need to address the search for meaning as a coping factor in psychotherapy. Time perception could, for instance, be addressed in psychotherapeutic approaches as a means to positively influence psychological well-being and related subjective experiences.

As the experience of time provides insight into a person's state of mind, the investigation of time perception could also provide crucial insight in the understanding of the mechanisms of self-control and impulsiveness, as in children with attention deficit hyperactivity disorder or in individuals abusing drugs. For example, drug abusing persons exhibit a stronger present time perspective and are more impulsive than abstainers. Existing studies show that alcoholics have a weaker future perspective than social drinkers, and heroin addicts have shortened time horizons, as they are less likely to set goals for the future. Even substance use (alcohol, tobacco, marijuana) in college students is related to a shorter future perspective and a stronger present orientation. At the same time, addicted individuals, such as smokers craving a cigarette or patients addicted to stimulants like methamphetamine or cocaine who are being treated in a drug-treatment program, perceive an indicated time interval to pass more slowly and overestimate it.

The model of cognitive time perception links these separate characteristics of interest. As individuals addicted to a drug show a stronger emphasis on the present time (as opposed to the future), they consequently pay more attention to the momentary passing of time. Allocating attentional resources to time leads to a subjective slowing of time and to an overestimation of duration. An altered sense of time could be one reason why impulsive individuals do not delay gratification and act on impulse. For such a person, the benefit of resisting temptation might lie subjectively too far in the future. Based on drug users' general difficulties with decisions that involve delayed rewards (these difficulties extend into many aspects of life), it has been suggested that treatment programs should develop intervention strategies that cognitively restructure the perception of these choices to shape more adaptive and health-promoting behavior. Moreover, in the therapeutic process, the setting of positive rewards and goals should not be extended too far into the future and should adapt to the individual's shortened time horizon.

The study of psychological time is only slowly entering mainstream contemporary psychology. The experience of time is dependent on numerous factors. Some of these are the individual characteristics of the person, the physical properties of events that mark the time intervals, mental contents that are processed during the time interval, and, last but not least, whether subjects were instructed to estimate duration before or after a target interval is presented (prospective vs. retrospective time perception, respectively). As it is hard to control for all these factors, results from simple time perception tasks are often difficult to explain. Nevertheless, the measurement of time perception will become part of the standard methodological inventory in psychology. First, and on a very personal level, the experience of time has an existential dimension for all of us, because we have only a limited time to live. Second, knowledge of the psychology of time can also pay off on a very pragmatic level.

The following is a story reported not long ago in the media. The management of a hotel had to decide whether to replace an old elevator, an action that would have been very costly. Hotel guests had repeatedly complained that the elevator moved too slowly. A little trick based on knowledge of the experience of time stopped these complaints. Mirrors were hung up in the elevator. Although the elevator still moved at the same pace, nobody complained any more. The hotel guests, now distracted by their own images, no longer focused on the passage of time and did not consider the time spent in the elevator too long at all.

Marc Wittmann

See also Amnesia; Augustine of Hippo, Saint; Cognition; Consciousness; Memory; Perception; Time, Phenomenology of

Further Readings

Draaisma, D. (2004). *Why life speeds up as you get older: How memory shapes our past.* Cambridge, UK: Cambridge University Press.

Levine, R. (1997). *A geography of time: The temporal misadventures of a social psychologist, or how every culture keeps time just a little bit differently.* New York: Basic Books.

Pöppel, E. (1997). A hierarchical model of temporal perception. *Trends in Cognitive Science, 1,* 56–61.

Strathman, A., & Joireman, J. (Eds.). (2005). *Understanding behavior in the context of time: Theory, research, and application.* Mahwah, NJ: Lawrence Erlbaum.

Wittmann, M., & Lehnhoff, S. (2005). Age effects in the perception of time. *Psychological Reports, 97,* 921–935.

Zakay, D., & Block, R. (1996). The role of attention in time estimation processes. In M. Pastor & J. Artieda (Eds.), *Time, internal clocks and movement* (pp. 143–164). Amsterdam: Elsevier.

Pueblo

The early Spanish explorers of North America, presumably impressed with the affinity of the southwest indigenous peoples to their great houses of multilevel apartment-like villages and cliff dwellings, gave the name Pueblo to these tribes, meaning both village and a people or nation. The Pueblo people compose 25 tribes, including the Hopi, Zuñí, Acoma, and Tao, with languages stemming from three branches: the Uto-Aztecan, the Tanoan, and the Keresan. They live in one of the oldest continuously settled regions of North America, in the asperities of the arid plateaus of Chaco Canyon and other parts of the Four Corners area (so named because it comprises the corners of four states: New Mexico, Arizona, Utah, and Colorado) and the surrounding vicinity, also known as the House Made of Dawn. They are presumed to be descended from the first group of migrants to enter the Americas. The ancient Pueblo have many nomenclatures: Desert Archaics, Hisatsinom (Hopi: "people of long ago"), and Anasazi (Navajo: "ancient enemy"). These ancestors, like many ancient astronomers, considered the celestial movements sacred and attempted to harmonize with the cosmos and the six sacred cardinal directions (North, South, East, West, Zenith, and Nadir). The result manifested into building plans and architectural locations that contained apparent stellar alignments in addition to colors and symbols with assigned directional and cosmic representations.

Desert archaic calendar. Rock etchings can be found along the Gunnison River, mainly of the Ute tribe. Also represented are pictographs of the Pre-Anasazi: the Desert Archaics.
Source: Barbara Jetley.

The Desert Archaics to Pre-Hisatsinom

From 9000 to 8000 BCE, a nomadic tribe referred to as the Desert Archaics occupied the mesas throughout the hunter-gatherer period. Their knowledge of celestial cycles appears to have been well established: Petroglyphs representing constellations line many rock walls throughout and surrounding the House Made of Dawn. Scholars posit these storyboard images to be recordings of astrological events, histories, and calendars. Upon those high plateaus is a unique view of celestial movements. It can be presumed that the various stellar patterns were observed and calculated from ancient days, for around approximately 700 CE the Hisatsinom culture, whose people resided in pit houses, begin construction on the near five-story complex archaeoastronomy dwellings, or great houses, with major structures designed with internal solar and/or lunar alignments.

Hisatsinom Culture

The great houses in the House Made of Dawn were once thought to have been haphazardly constructed, but based upon closer examination by scholars such as Anna Sofaer, it is thought that the structures in the Chaco Canyon area were planned and executed with the purpose of recording solar, lunar, and other important stellar movements. Scholars posit that these structures were low in population save for ceremonial times, when pilgrimages were made from the surrounding 150 communities (and perhaps further), which were connected via a precisely constructed set of roads. The structures of the great house are oriented north–south and east–west to form a three-community house that shares in a lunar minor alignment, representing the moon's rise, its meridian, and other occasional observances. In addition, there is a solar and lunar observatory at Fajada Butte that was used to mark solstices, equinoxes, and the 18.6-year cycle of the lunar northern and southern extremes. The Chaco Canyon phenomenon thus embodies the ancient Puebloans' worldview.

The *kiva,* a round, partly underground structure that forms the nucleus of each great house, remained part of the Pueblo religion and culture. The kiva represents the world below or sacred womb, while the kiva's *sipapû,* a hole or indentation in the floor, symbolizes the "earth navel" believed to be the entrance for the people and kachinas or spirits. Their legends of creation begin with a speculation of endless space, elucidating how Spider Woman created a web-matrix for the foundations of the earth and sky. Hopi legends tell of four cycles the humans evolved through as Spider Woman led the journey, the last journey guiding them from the world below through the sipapû.

Pre-Puebloan Culture

A mass exodus from the great houses occurred around 900 to 1300 CE. It brought about new or altered religions; for the Hopi it was the advent of the kachina celebrations, with the New-Fire Celebration in November to start the New Year. The kachina season officially commences in the halcyon days of winter solstice, with rituals for each month of the nearly 7-month season, which concludes in July. The year's remainder operated through the snake or flute season.

In the Hopi tradition, all things animate and inanimate have a kachina. The masked Puebloan dancers symbolically represent these spirits wearing celestial regalia representing the morning star, the sun, and the earth. The symbols within paraphernalia, pictographs, and sand paintings have many cosmic representations: eagle—sun or sky god; fire—life's creation; spider—earth; bear—west; mountain lion—north; wildcat—south; wolf—east, and so forth. Colors can moreover represent the cardinal directions: yellow—north, white—east, green/blue/black—west, and red—south. Like the sand designs of the Tibetans, the Puebloan sand paintings are formed exclusively for ceremonial purposes and swept away at the observance's conclusion.

There were annual, biannual, and quadrennial celebrations. The biannual festivals alternated between the snake and flute societies; the snake during odd years, the flute during even years. Quadrennial celebrations accompanied priestly initiations, and for the duration of that year each of the festivals was extended; for example, a 1-day ceremony lasted 5 days, or a 5-day ceremony lasted 9 days. The extended ceremonies possibly served as training periods for the new priests, who were educated in astrological movements and accompanying traditions.

Pueblo Culture

The Spaniards attempted to eradicate the Puebloan belief system, and so struck at the core by destroying numerous kivas, including filling one with sand and positioning their church atop it, in effect preserving the "kiva beneath the altar." Many atrocities were suffered by the Puebloans at the hands of the Spaniards during early colonization, although the Pueblo tribes held fast on several occasions, including the revolt of 1680. Despite such efforts to supplant indigenous belief, the Pueblo have continued their ceremonial observances. In the late 19th century, at Eldon Mesa, anthropologist Alexander M. Stephen witnessed a winter solstice observance of the sun's arrival at the great house called House of the West.

Contemporary Puebloans

The destruction of the sacred Spider Woman hill in 1984 during a highway construction project enraged the Puebloan communities; in response, after the Puebloans received the assistance of LaVan Martineau, assurances were given that the remaining rock images within the Clear Creek Canyon would be recorded and preserved. Similar to other indigenous peoples, the Puebloans are participating in mainstream society while holding fast to their traditional customs and actively researching their ancient culture.

Pamela Rae Huteson

See also Anthropology; Chaco Canyon; Navajo; Sandpainting; Solstice; Totem Poles

Further Readings

Eggan, F. (1990). *Social organization of the western pueblos.* Chicago: University of Chicago Press. (Original work published 1973)

Fewkes, J. W. (1985). *Hopi katcinas.* Mineola, NY: Dover. (Original work published 1903)

Patterson-Rudolph, C. (1997). *On the trail of Spiderwoman.* Santa Fe, NM: Ancient City Press.

Smith, W. (1990). *When is a kiva? And other questions about southwestern archaeology* (R. H. Thompson, Ed.). Tucson: University of Arizona Press.

Sofaer, A., Zinser, V., & Sinclair, R. M. (1979). A unique solar marking construct. *Science, 206*(4416), 283–291.

Stephen, A. M., & Parsons, E. W. C. (1936). *Hopi journal of Alexander M. Stephen.* New York: Columbia University Press.

Watson, D. (1955). *Indians of the Mesa Verde.* Mesa Verde National Park, CO: Mesa Verde Museum Association.

Pulsars and Quasars

Pulsars and quasars are celestial objects that often emit radio waves but can also emit energy in the visible light, X-ray, and gamma-ray ranges of wavelengths. Unlike quasars, pulsars are rapidly spinning neutron stars that often have a rotational period so regular that, as timepieces, they are more accurate than atomic clocks. Although quasars are neither stars nor timepieces, they are among the most distant observable objects in space, so distant that observing them is akin to looking at the past.

First discovered in 1967 by English radio astronomer Antony Hewish and his Irish graduate student Jocelyn Bell, pulsars are the result of a supernova explosion of a midsized star. The ensuing gravitational collapse of such a star forms a neutron star. Pulsars get their name from the regular pulses of radio waves that emit from their magnetic poles. They behave like lighthouses, but with radio waves in place of flashes of light. Each pulsar has a period, the time it takes for it to rotate once; this period can range from a millisecond to 8.5 seconds. As time passes, and the pulsar ages, the period will grow longer, meaning that the pulsar's spinning slows as its magnetic force increases. One type of pulsar is the millisecond pulsar, which has the shortest period. Its weak magnetic field allows for greater rotation speed. The periods of millisecond pulsars are stable and predictable enough to trump the timekeeping of atomic clocks. Another type of pulsar is the binary pulsar, which is one star in a binary star system in which two stars revolve around each other. Because of their atomic makeup—neutron-saturated nuclei—pulsars may have the mass of the sun but are extremely dense, sometimes less than 20 miles in diameter. For this reason, identification of their periodic pulses is often the only way these objects are discovered and observed.

Another source of radio emissions, quasars are pragmatically named, the term originating from the acronym QSRS (quasi-stellar radio source). Some can be radio-quiet, instead emitting X-rays or

This artist's concept depicts the pulsar planet system discovered by Aleksander Wolszczan in 1992. Pulsars are rapidly rotating neutron stars that are the collapsed cores of exploded massive stars. They spin and pulse with radiation, much like a lighthouse beacon.

Source: NASA Jet Propulsion Laboratory.

The Hubble Space Telescope achieved its 100,000th exposure June 22, 2002, with a snapshot of a quasar that is about 9 billion light-years from Earth. The quasar looks as bright as the star, because it produces a tremendous amount of light from a compact source.

Source: NASA Goddard Space Flight Center.

gamma rays. Quasars appear starlike in that they can be observed optically as a point, or star, of light. However, their emissions are redshifted, which refers to their spectra's longer wavelengths. Redshift is a quality often attributed to objects that are accelerating away from the observer, and cosmological redshift refers to the infinite expansion of space-time and matter since the big bang. Because quasars are the most distant objects observed, their optical light appears faint. However, calculations including distance and speed led to figures of great magnitude, and quasars are sometimes posited to emit a trillion times the radiation of the sun. The most popular theory of the source behind this energy is that quasars are spinning disks of dust that surround a massive black hole. Given the great distance and theorized relationship to the big bang, quasar emissions are arriving from the cosmological past. Unlike the timepiece pulsar, the quasar is a preserver of time, displaying a picture of long ago.

Karlen Chase

See also Big Bang Theory; Black Holes; Clocks, Atomic; Light, Speed of; Observatories; Telescopes; Time, Galactic; Time, Measurements of; Time, Sidereal; Universe, Contracting or Expanding; Universe, Evolving; White Holes; Wormholes

Further Readings

Audouze, J., & Israel, G. (Eds.). (1988). *The Cambridge atlas of astronomy.* New York: Cambridge University Press.

Dickinson, T. (2004). *The university and beyond* (4th ed.). Buffalo, NY: Firefly Books.

Punctuality

Punctuality is defined as the act of arriving at or completing a task, event, or engagement at or before a previously designated time. Perhaps most easily thought of as being "on time" or "on task," punctuality, or the lack of, is in most cases an individual trait or tied to an individual event or task. It is, however, a more widespread problem across some cultures.

By far, punctuality is a much higher priority in countries that are largely industrialized, such as the United States, China, Germany, and the United Kingdom. For countries such as Spain, India, and those in Latin America and the Middle East, however, punctuality is a much lesser priority. When a culture relies heavily on clocks, time dictates the start of events, meetings, and the like. It is in these cultures that watches, clocks, and timepieces are generally kept in unison. In event time, the focus is less on time and more on the event itself, meaning that people living by event time will see an event through before carrying on to the next.

In fact, Latin American governments have recognized the general issues with being on time as a widespread problem that affects their ability to compete globally. Ecuador, in particular, estimates that as much as 10 percent of its GDP is lost to lateness. Ecuadorians live largely based on event time, better known locally as Ecuadorian time. In 2003, Ecuadorians were encouraged to synchronize their watches in an effort to eliminate tardiness. Similarly in Peru, a national campaign called *la hora sin demora,* or "time without delay," was launched to encourage its citizens to be more timely.

International business and travel offer further insight into the punctuality differences across the world. Detailed statistics are kept for all types of commercial travel, including airplanes and trains. Schedules and connections are dependent on their punctuality, and reputations can be damaged by excessive lateness. To that end, Japan, a country that is hypersensitive to the clock and the need to be punctual, has developed such a reliance on the adherence to schedules that its transportation workers fear reprimand for delays. Commuter trains connect with only minutes to spare before their riders must make their next connection. Such a focus on punctuality has been blamed for accidents, such as the one that occurred in 2005 when nearly 100 people were killed because the commuter train's engineer was trying desperately to make up 90 seconds.

Punctuality is to an extent a relatively modern concept. Clocks and standardized time have only been in use for little more than a century. Prior to that, there were wide variations in the keeping of time, thereby making punctuality nearly impossible. Clocks were set relative to the sun's position, making them inaccurate. It was not until a global standardized time plan went into effect that there was a gauge of timeliness.

Time is of the essence in many modernized countries, where sales of time-saving devices and gadgets have skyrocketed over the past 20 years. The pace of the industrial world has quickened, although its preoccupation with speed and attention to the clock do not necessarily mean increased punctuality.

Amy L. Strauss

See also Comets; Eclipses; Globalization; Old Faithful; Solstice; Sunspots, Cycle of; Time, Measurements of; Timepieces; Watches

Further Readings

Brislin, R. W., & Kim, E. S. (2003). Cultural diversity in people's understanding and uses of time. *Applied Psychology: An International Review, 52*(3) 363–382.

Morrison, T., & Conaway, W. A. (2006). *Kiss, bow or shake hands: The bestselling guide to doing business in more than 60 countries.* Avon, MA: Adams Media.

Pythagoras of Samos (c. 569–c. 496 BCE)

Pythagoras was a classical Greek philosopher, mathematician, scientist, and mystic from the Aegean island of Samos. Known as the "father of numbers," he is best known for the Pythagorean theorem, which carries his name. Over the centuries, mathematicians, architects, artists, and musicians have built upon Pythagoras's theories, making those theories the very foundation of the modern understanding of music, astronomy, and mathematics.

Pythagoras was the son of Mnesarchus, a merchant from Tyre. As a child, he traveled widely with his father and gained a broad education from his experiences. He studied with three philosophers in particular—Pherecydes of Syros, Thales, and Anaximander. These men introduced Pythagoras to mathematics, cosmology, astronomy, and geometry. In 535 BCE, Pythagoras moved to Egypt and studied with Egyptian priests at a temple in Diospolis. The rites, rituals, and secrecy he learned in Egypt became part of the religious school he set up later in Italy. The Persians captured Pythagoras when they invaded Egypt in 525 BCE and took him to Babylon. He studied with the magi in Babylon where he learned about Babylonian mysticism, arithmetic, and music.

After his release, Pythagoras returned to Samos, where he set up and ran a school for a few years. Then he moved to southern Italy in 518 BCE, to a village named Croton where he established a secret religious society, emphasizing virtue and strict rules of conduct. The members of this society, known as Pythagoreans, were vegetarians, took vows of silence, and participated in purification rites. They believed in transmigration of the soul, meaning souls are continuously reincarnated as persons, animals, or plants. The Pythagoreans' way of life sought to release the soul from this cycle. Pythagoras lived in Croton until a village noble plotted against him and forced Pythagoras to flee. He went to Metapontum, where he died.

Pythagoras was deeply interested in mathematic principles, especially the abstract philosophy of numbers. He believed that everything, even the order of the universe, could be defined numerically, that is, in terms of harmonic ratios. He applied this theory to mathematics, music, and astronomy. He discovered that music has proportional intervals based on the numbers one through four. Therefore, he held that the universe is based upon the sum of these numbers—10. Pythagoras, or one of his followers, discovered that when the side of a square is squared, the resulting figure equals the area of the square. From here came the concept of the square root.

Although its true origin is uncertain, Pythagoras generally gets credit for developing the so-called Pythagorean theorem, that is, the hypotenuse squared of a right-angled triangle is equal to the sum of the squares of the sides ($a^2 + b^2 = c^2$). Unfortunately, none of his writings has survived to corroborate this association. Egyptian and Babylonian documents show that both of these societies knew about the theorem well before Pythagoras was born; however, some ancient sources believe Pythagoras was the first person to prove the theorem. Pythagoras believed that numbers connected music and math through ratios. Legend tells that Pythagoras discovered a relationship between tonal sound and simple ratios when he passed a blacksmith shop one day and heard the sound of the hammers beating on various size anvils. This principle of tonal harmony and size ratio is called Pythagorean tuning.

Another discovery usually accredited to Pythagoras, or his followers, is the golden ratio. This ratio is a line divided into two parts such that the longer section (a) has the same ratio to the shorter section (b) as the entire length (a + b) has to the longer section (a). In the 20th century, this ratio began to be called *phi*, named after the Greek letter in the name of a Greek sculptor Pheidias who employed the ratio. Numerically, the ratio equals about 1.618034. Since the Renaissance, many artists and architects have used the golden ratio, or its corollary the golden rectangle, in their work, believing that this measure is aesthetically pleasing.

In astronomy, Pythagoras influenced the modern understanding of the movement of heavenly bodies. He held that the planets rotated on an axis and revolved around a central point. He also held that mathematics could explain these movements. Furthermore, he was one of the first people to understand that the moon orbited the earth, that the earth was round, and that Venus was both the evening and morning star.

Pythagoras's work in mathematics, music, and astronomy was revolutionary in his day and forms the basis of much of our understanding of the universe. His discovery of timeless principles truly makes him the "father of numbers."

Terry W. Eddinger

See also Anaximander; Anaximines; Empedocles; Heraclitus; Parmenides of Elea; Presocratic Age; Pythagoras of Samos; Thales; Xenophanes; Zeno of Elea

Further Readings

Kahn, C. H. (2001). *Pythagoras and the Pythagoreans: A brief history.* Indianapolis, IN: Hackett.

O'Connor, J. J., & Robertson, E. F. (1999.) Pythagoras of Samos. Retrieved November 3, 2007, from http://www-groups.dcs.st-and.ac.uk/~history/Printonly/Pythagoras.html

Riedweg, C. (2005). *Pythagoras: His life, teaching, and influence* (S. Rendall, Trans.). Ithaca, NY: Cornell University Press.